Data Mining
Using SAS® Enterprise Miner™

Data Mining Using SAS® Enterprise Miner™

Randall Matignon

Amgen, Inc.
South San Francisco, CA

WILEY-INTERSCIENCE
A John Wiley & Sons, Inc., Publication

Published by John Wiley & Sons, Inc., Hoboken, New Jersey.
Published simultaneously in Canada.

For general information on our other products and services or for technical support, please contact our Customer Care Department within the United States at (800) 762-2974, outside the United States at (317) 572-3993 or fax (317) 572-4002.

Wiley also publishes its books in a variety of electronic formats. Some content that appears in print may not be available in electronic format. For information about Wiley products, visit our web site at www.wiley.com.

Library of Congress Cataloging-in-Publication Data is available.

ISBN 978-0-470-14901-0

Printed in the United States of America.

10 9 8 7 6 5 4 3 2 1

Table of Contents

Acknowledgements

I am dedicating this book to my parents Leon and Carol, and my brother Emil and his wife Aling. I would also like to express my extreme gratitude to many of the same people acknowledged in my first book, *Neural Network Modeling using SAS Enterprise Miner*. I would like to acknowledge Dr. Heebok Park, Dr. Eric Suess, Dr. Bruce Trumbo, Dr. Julie Norton, Dr. Elliott Nebenzahl, Dr. Dean Fearn, Dr. Ward Rodriquez, Dr. Michael Orkin, Dr. Richard Kakigi, and Dr. Richard Drogin from the California State University East Bay (formerly California State University Hayward) statistics faculty, who have influenced me greatly in the field of statistics, for their helpful comments in the process of writing this book. I would like to acknowledge the Clinical Data Management team at Amgen and the incredibly talented and exceptional SAS programmers – Amit Khare, Steve Hull, Ram Govindaraju, Mickey Cheng, Steven Verschoor, Priya Saravanan, Michael Damiata, and Mehul Shukla, and the CDM management team of Mark Stetz, Andrew Illidge, Banvir Chaudhary, Aza Teh, Lori Baca, and Kelly Valliant. I would also like to thank the SAS Institute, Inc. for granting me permission to present several tables in my book and introducing the SAMPSIO.HMEQ data set that is a part of the SAS software. Also, thanks to the SAS Institute technical support team that was instrumental in providing documentation and answering questions that I had with Enterprise Miner and some of the statistical procedures and option settings. Finally, I am extremely grateful to John Wiley & Sons, Inc. for their patience while I completed my book and for turning it into a reality.

The various SAS programs that are referenced within my book can be viewed from my website at www.sasenterpriseminer.com. My website will list various articles that were extremely helpful to me when completing both of my books. In addition, my website will provide a complete listing of the training and scoring code that is listed within the book. The purpose of presenting the training and scoring code is to make it easier to interpret the internal calculations that are performed within the nodes in generating the results and understanding the various concepts of the results that are generated from the various nodes in Enterprise Miner.

Randall Matignon

Introduction

Data Mining Using SAS Enterprise Miner introduces the readers to data mining using SAS Enterprise Miner v4. This book will reveal the power and ease of use of the powerful new module in SAS that will introduce the readers to the various configuration settings and subsequent results that are generated from the various nodes in Enterprise Miner that are designed to perform data mining analysis. This book consists of step-by-step instructions along with an assortment of illustrations for the reader to get acquainted with the various nodes and the corresponding working environment in SAS Enterprise Miner. The book provides an in-depth guide in the field of data mining that is clear enough for novice statisticians or the die-hard expert.

The process of extracting information from large data sets is known as data mining. The objective in data mining is making discoveries from the data. That is, discovering unknown patterns and relationships by summarizing or compressing the data in a concise and efficient way that is both understandable and useful to the subsequent analysis. Extracting information from the original data that will result in an accurate representation of the population of interest, summarizing the data in order to make statistical inferences or statements about the population from which the data was drawn and observing patterns that seem most interesting, which might lead you to discover abnormal departures from the general distribution or trend in the data; for example, discovering patterns between two separate variables with an usually strong linear relationship, a combination of variables that have an extremely high correlation on a certain variable, or grouping the data to identify certain characteristics in the variables between each group, and so on. In predictive modeling, it is important to identify variables to determine certain distributional relationships in the data set in order to generate future observations or even discover unusual patterns and identifying unusual observations in the data that are well beyond the general trend of the rest of the other data points.

The basic difference between data mining and the more traditional statistical applications is the difference in size of the data set. In traditional statistical designs, a hundred observations might constitute an extremely large data set. Conversely, the size of the data mining data set in the analysis might consist of several million or even billions of records. The basic strategy that is usually applied in reducing the size of the file in data mining is sampling the data set into a smaller, more manageable subset that is an accurate representation of the population of interest. The other strategy is summarizing the variables in the data by their corresponding mean, median or sum-of-squares. Also, reducing the number of variables in the data set is extremely important in the various modeling designs. This is especially true in nonlinear modeling where an iterative grid search procedure must be performed in finding the smallest error from the multidimensional error surface.

The potential of data mining used in statistical analysis is unlimited. However, it would be a mistake to depend on data mining in providing you with the final solution to all the questions that need to be answered. In addition, the accuracy of the final results in data mining analysis depends on the accuracy of the data. The most common term that is often used is called "garbage in – garbage out". This is particularly true in data mining analysis where the chance of this phenomenon of happening will occur more often than not due to the enormous amount of data at your disposal. In some instances, discovering patterns or relationships in the data might be due to the measurement inaccuracies, distorted samples, or some unsuspected difference between the erroneous data set and the actual representation of the population of interest.

Often, the choice of the statistical method to apply depends on the main objective of the analysis. For example, in marketing research with the disposal of customer and transactions data, it is usually important to interpret the buying behaviors of the various customers in order to focus our attention on the more effective promotions that will result in an increase of sales that can be accomplished by performing association analysis. It might be important in identifying and profiling the various buying habits of these same customers who might be placed into separate homogenous groups based on the total sales of the various items purchased, which can be performed by applying cluster analysis. Maybe the goal to the analysis might be predicting or forecasting future sales, which can be performed by applying regression modeling. Nonlinear modeling such as neural network modeling might be considered, which does not require a functional form between the predictor or input variables and the response, outcome or target variable to the model, with the added flexibility of handling an enormous number of input variables in the model. Two-stage modeling might be performed by classifying customers in purchasing certain items, then predicting these same customers based on their corresponding estimated probabilities, which are then applied in the subsequent modeling fit to predict the amount of total sales of these same items. Maybe it might be important to the business in developing a credit scoring system in order to establish whether or not to extend credit to customers based on certain numerical scores that are above or below some threshold value, which can performed by applying interactive grouping analysis.

SAS summarizes data mining with the acronym SEMMA, which stands for sampling, exploring, modifying, modeling, and assessing data as follows:

Sample the data from the input data set that is so large that a small proportion of the data can be analyzed at any one time, yet large enough that it contains a significant amount of information to perform the analysis.

Explore the data to statistically, and visually discover expected relationships and unexpected trends while at the same time discovering abnormalities in the data to gain a better understanding of the data.

Modify the data in creating, selecting, and transforming the variables or even the entire incoming data set to perform various modeling selections or certain statistical modeling techniques in preparation for the subsequent data mining analysis.

Model the data by applying various modeling techniques in seeking a certain combination of variables that reliability predicts the outcome response.

Assess the data by evaluating the usefulness and reliability of the results from the data mining process.

Data mining is an iterative process. That is, there is a sequential order to the listed categories in the SEMMA data mining analysis. Typically, you would first **sample** the data to target our population or determine the population of interest to the analysis. Second, you might **explore** the data to visualize the distribution of each variable in the analysis to validate the statistical assumptions such as normality in the distribution of the selected variable or determine patterns and relationships between the variables. Third, you might **modify** the variables to prepare the data for analysis by transforming the variables to satisfy the various assumptions that are critical to many of the statistical methods so that these same variables may be used in the analysis. The next step is that you might **model** the data by fitting a statistical model to generate predictions and forecasts. And, finally, you might **assess** the accuracy, interpreting the results and comparing the predictability of the various modeling designs in selecting the best predictive or classification model. Once the best model is selected, then you might want to generate prediction or forecasting estimates through the use of a scoring function that can be applied to a new set of values that might not necessarily consist of the target variable in the data. It is important to realize that many of the previously mentioned data mining steps might not be applied at all or some of these steps might be applied any number of times before the goal of the data mining analysis is finally achieved.

The SEMMA design and data mining analysis is constructed within the process flow diagram. Enterprise Miner is designed so that very little SAS programming experience is needed in constructing a well-built SAS reporting system. The reason is that the process of constructing the process flow diagram within Enterprise Miner is performed by simply dragging icons on to a GUI interface desktop window, then connecting these same icons to one another, all within the same graphical diagram workspace. And yet a data mining expert can specify various option settings in the design in order to fine-tune the configuration settings and the corresponding listings and results. SAS Enterprise Miner is a very easy to learn and very easy to use. You do not even need to know SAS programming and can have very little statistical expertise in designing an Enterprise Miner project in order to develop a completely comprehensive statistical analysis reporting system, whereas an expert statistician can make adjustments to the default settings and run the Enterprise Miner process flow diagram to their own personal specifications. Enterprise Miner takes advantage of the intuitive point-and-click programming within a convenient graphic user interface. The diagram workspace or the process flow diagram has the look and appearance much like the desktop environment in Microsoft Windows. Enterprise Miner is built around various icons or nodes at your disposal that will perform a wide variety of statistical analysis.

Each chapter of the book is organized by the SEMMA acronym based on the various nodes that are a part of the data mining acronym. Each section to the book is arranged by the way in which the node appears within the hierarchical listing that is displayed in the **Project Navigator** within the Enterprise Miner window. Each section of the book will begin by providing the reader with an introduction to the statistics and some of the basic concepts in data mining analysis with regard to the corresponding node. The book will then explain the various tabs along with the associated option settings that are available within each tab of the corresponding node, followed by an explanation of the results that are generated from each one of the nodes.
In the first chapter, the book will begin by explaining the purpose of the various sampling nodes that are a part of the Sample section in the SEMMA design. The first section will introduce the readers to the way in which to

read the data in order to create the analysis data set to be passed on to the subsequent nodes in the process flow diagram. The following section will allow the readers to learn how to both randomly sample and partition the analysis data set within the process flow. This is, sampling the analysis data set into a smaller, more manageable sample size that is an accurate representation of the data that was randomly selected, or even split the analysis data set into separate files that can be used in reducing the bias in the estimates and making an honest assessment in the accuracy of the subsequent predictive model.

The second chapter will focus on the various nodes that are designed to discover various relationships or patterns in the data that are a part of the Explore section in the SEMMA design. The first couple nodes that are presented are designed to generate various frequency bar charts or line graphs in order to visually observe the univariate or multivariate distribution of the variables in the data. In addition, the readers will be introduced to the **Insight** node, which can perform a wide range of statistical analysis through the use of the synchronized windows. The following section will explain the purpose of association analysis that is widely used in market basket analysis in discovering relationships between the different combinations of items that are purchased. The next section will introduce the readers to the variable selection procedure that can be applied within the process flow, which is extremely critical in both predictive or classification modeling designs that are designed to select the best set of input variables among a pool of all possible input variables. The chapter will conclude with the readers getting familiar with link analysis which automatically generates various link graphs in order to view various links or associations between the various class levels that are created from the analysis data set.

The third chapter will introduce readers to the various nodes that are designed to modify the analysis data sets that are a part of the Modify section of the SEMMA design. The chapter will allow the readers to realize how easy it is to modify the various variable roles or level of measurements that are automatically assigned to the analysis data set once the data set is read into the process flow. In addition, the subsequent section will allow readers to transform the variables in the data set in order to meet the various statistical assumptions that must be satisfied in data mining analysis. The following section will allow readers to understand both the importance and the process of filtering, excluding, and removing certain problematic observations from the data set that might otherwise influence the final statistical results. The next section will explain the procedure of imputing missing values and replacing undesirable values in the data set and the subsequent analysis. The following two sections will explain both clustering and SOM/Kohonen analysis, which are designed to group the data into homogenous groups in order to profile or characterize the various groups that are created. The next section will explain the way in which to transform the data set in preparation to repeated measures or time series analysis. And, finally, the chapter will conclude by introducing readers to interactive grouping analysis that is designed to automatically create separate groups from the input variables in the data set based on the class levels of the binary-valued target variable. These input variables may then be used as input variables in subsequent classification modeling designs such as fitting scorecard models.

The fourth chapter will present the readers to the various modeling nodes in Enterprise Miner that are a part of Model section in the SEMMA design. The chapter will begin with traditional least-squares modeling and logistic regression modeling designs that are based on either predicting an interval-valued or a categorically-valued target variable of the model. The following section will introduce the readers to the **Model Manager** which is available in any one of the modeling nodes. The purpose of the **Model Manager** is to store and list the various models that are created within each modeling node. Each model that is displayed is based on the different settings that were previously specified each time you saved the corresponding changes. The **Model Manager** will allow you to select the number of observations from each partitioned data set to be passed along for interactive assessment in evaluating the accuracy of the modeling fit from the **Assessment** node. In addition, the **Model Manager** will allow you to specify if the various diagnostic plots or performance charts will be available for viewing within the **Assessment** node for each partitioned data set. The next section will introduce the readers to decision tree modeling, which is based on a recursive splitting process in which binary splits are automatically performed, and where the average value of the interval-valued target variable is commonly used as the standard cutoff point or the separate class levels of the categorically-valued target variable that are repeatedly divided through the decision tree based on the corresponding range of values or the separate class levels of the input variables in the model. Neural network modeling will then be introduced \. This is essentially nonlinear modeling of the process flow that has the flexibility of interpolating many different functional forms with extreme accuracy or approximating many different classification boundaries with great precision. Neural network modeling is built around a multilayered design in which the linear combination of input variables and weight estimates are transformed through the layers where the weight estimates must be solved by some type of iterative grid search or line search routine. The next section will explain principal components analysis to readers; this is designed to reduce the number of variables in the data based on the linear combination of input variables and components in the model, where the components are selected to explain the largest proportion of the variability in the data. User-defined modeling will be introduced that will allow you to incorporate a wide variety of modeling techniques within the process flow that are unavailable in Enterprise Miner and the various modeling nodes. Furthermore, the node will allow you to create your own scoring code that will generate the standard modeling assessment statistics that can be passed along the process flow in order to compare the accuracy between the other modeling nodes. The book will provide readers with a powerful modeling technique called ensemble modeling that either averages the prediction estimates from various models or averages the prediction estimates based on successive fits from the same predictive model, where the analysis data set is randomly sampled any number of times. The following section will introduce the readers to memory-based reasoning or nearest neighbors modeling that is essentially nonparametric modeling in which there are no distributional assumptions that are assumed in the model in which the fitted values are calculated by averaging the target values or determined by the largest estimated probability based on the most often occurring target category within a predetermined region. The chapter will conclude with two-stage modeling that fits a categorically-valued target variable and an interval-valued target variable in succession, where the categorically-valued target variable that is predicted in the first-stage model will hopefully explain a large majority of the variability in the interval-valued target variable to the second-stage model.

The fifth chapter will explain the various nodes in Enterprise Miner that are a part of Assess section to the SEMMA design. These nodes will allow you to evaluate and assess the results that are generated from the various nodes in the process flow. The chapter will begin by introducing the readers to the **Assessment** node that will allow you to evaluate the accuracy of the prediction estimates from the various modeling nodes based on the listed assessment statistics that are automatically generated and the numerous diagnostic charts and performance plots that are created for each model. The node will allow you to evaluate the accuracy of the modeling estimates by selecting each model separately or any number of models simultaneously. The **Reporter** node will be introduced, which is designed to efficiently organize the various option settings and corresponding results that are generated from the various nodes within the process flow into a HTML file that can be viewed by your favorite Web browser.

The sixth chapter will introduce the readers to the **Score** node that manages, exports, and executes the SAS scoring code in order to generate prediction estimates from previously trained models. In addition, the node will allow you to write your own custom-designed scoring code or scoring formulas that can be applied to an entirely different sample drawn in order to generate your own prediction or classification estimates.

The final chapter of the book will conclude with the remaining nodes that are available in Enterprise Miner. These nodes are listed in the Utility section within the **Project Navigator**. The chapter will begin by explaining the purpose of the **Group Processing** node that will allow you to transform variables by splitting these same variables into separate groups. In addition, the node is used with the **Ensemble** node that determines the way in which the various prediction estimates are formed or combined within the process flow. The subsequent section will explain the purpose of the data mining data set that is designed to accelerate processing time in many of the nodes within the process flow. This is because the data mining data set contains important metadata information to the variables in the analysis, such as the variable roles, level of measurement, formats, labels, range of values, and the target profile information to name a few. The next section will explain the importance of the **SAS Code** node. The node is one of the most powerful nodes in Enterprise Miner. This is because the node will allow you to incorporate SAS programming within the process flow, thereby enabling you to write data step programming within the process flow in order to manipulate the various data sets, write your own scoring code, or access the wide variety of procedures that are available in SAS. The final two sections will explain the purposes of the **Control point** node and the **Subdiagram** node that are designed to manage and maintain the process flow more efficiently.

For anyone looking for a new edge in the field of statistics, *Data Mining Using SAS Enterprise Miner*, offers the inside track to new knowledge in the growing world of technology. I hope that reading the book will help statisticians, researchers, and analysts learn about the new tools that are available in their arsenal making statisticians and programmers aware of this awesome new product that reveals the strength and easy use of Enterprise Miner for creating a process flow diagram in preparation for data mining analysis.

Supervised Training Data Set

Predictive modeling is designed to describe or predict one or more variables based on other variables in the data; it is also called *supervised training*. Predictive models such as traditional linear regression modeling, decision tree modeling, neural network modeling, nearest neighbor modeling, two-stage modeling, and discriminate analysis may be applied in Enterprise Miner. The main idea in predictive modeling is to either minimize the error or maximize the expected profit. In Enterprise Miner, the modeling terms are distinguished by their model roles. For example, the output response variable that you want to predict would be set to a **target** model role and all the predictor variables in the predictive model are assigned a model role of **input**. Time identifier or carryover variables might be passed along in the modeling design that identifies each observation in the data with an **id** model role.

The following is the data set that is used in the various Enterprise Miner nodes based on supervised training in explaining both the configuration settings and the corresponding results. The SAS data set is called HMEQ. The data set is located in the SAMPSIO directory within the folder in which your SAS software is installed. The SAMPSIO directory is automatically available for access once Enterprise Miner is opened. The data consists of applicants granted credit for a certain home equity loan that has 5,960 observations. The categorical target variable that was used in the following examples is a binary-valued variable called BAD that identifies if a client either defaulted or repaid their home equity loan. For interval-valued targets, the variable called DEBTINC, which is the ratio between debt and income, was used in many of the following modeling examples. There are thirteen variables in the data mining data set with nine numeric variables and four categorical variables. The following table displays the variables in the data set, the model role, measurement level, and variable description. The database was used in many of the following predictive modeling designs to determine if the applicant can be approved for a home equity loan.

Name	**Model Role**	**Measurement Level**	**Description**
BAD	Target	Binary	1 = Defaulting on the loan, 0 = Repaid the loan
CLAGE	Input	Interval	Age (in months) of the oldest trade line
CLNO	Input	Interval	Number of trade lines
DEBTINC	Input	Interval	Ratio of debt to income.
DELINQ	Input	Interval	Number of delinquent trade lines
DEROG	Input	Interval	Number of major derogatory reports
JOB	Input	Nominal	Six occupational categories
LOAN	Input	Binary	Amount of the loan request
MORTDUE	Input	Interval	Amount due on the existing mortgage
NINQ	Input	Interval	Number of recent credit inquires
REASON	Input	Binary	DebtCon = debt consolidation, HomeImp = home improvement
VALUE	Input	Interval	Current property value
YOJ	Input	Interval	Number of years at the present job

Unsupervised Training Data Set

Descriptive modeling is designed to identify the underlying patterns in the data without the existence of an outcome or response variable; it is also called *unsupervised training*. That is, observations might be formed into groups without prior knowledge of the data or unknown links and associations might be discovered between certain variables in the underlying data. Cluster analysis, SOM/Kohonen maps, and principal components are considered descriptive modeling techniques. In Enterprise Miner, all the variables in the analysis are set to a variable role of **input** when performing the various unsupervised training techniques.

The following is the data set that is used in the various Enterprise Miner nodes based on unsupervised training for explaining both the configuration settings and the corresponding results. The SAS data set is based on major league baseball hitters during the 2004 season. The data set consists of all baseball hitters in the major leagues that had at least 150 at bats, which resulted in 371 observations of hitters in the major leagues. The data consisted of twenty variables in the data mining data set with fifteen numeric variables and five categorical variables. The following table displays the variable names, model role, measurement level, and variable description of the variables.

Name	**Model Role**	**Measurement Level**	**Description**
TEAM	Input	Nominal	Major League Team
FNAME	Rejected	Nominal	First Name
LNAME	Rejected	Nominal	Last Name
POS1	Input	Nominal	Position
G	Input	Interval	Number of Games Played
OPS	Input	Interval	On-base percentage plus slugging pct.
SLG	Input	Interval	Slugging Percentage *Total Bases Divided by the Number of At Bats*
AVG	Input	Interval	Batting Average *Hits Divided by the Number of At Bats*
OBP	Input	Interval	On-Base Percentage *(H + BB + HBP) / (AB + BB + HBP + SF)*
AB	Input	Interval	At Bats
R	Input	Interval	Runs Scored
H	Input	Interval	Hits
H2	Input	Interval	Doubles
H3	Input	Interval	Triples
HR	Input	Interval	Home Runs
RBI	Input	Interval	Runs Batted In
BB	Input	Interval	Walks
SO	Input	Interval	Strikeouts
SB	Input	Interval	Stolen Bases
LEAGUE	Input	Binary	American League or National League

Chapter 1

Sample Nodes

Chapter Table of Contents

1.1 Input Data Source Node

General Layout of the Enterprise Miner Input Data Source Node

- **Data tab**
- **Variables tab**
- **Interval Variables tab**
- **Class Variables tab**
- **Notes tab**

The purpose of the **Input Data Source** node in Enterprise Miner is to read the source data set to create the input data set. The input data set is then passed on to the subsequent nodes for further processing. The **Input Data Source** node is typically the first node that is used when you create a process flow to read the source data set and create the input data set. The node also creates a data mining data set called the *metadata sample*. The purpose of the metadata sample is to define each variable attribute for later processing in the process flow. The metadata sample is created for each variable in the data set by preprocessing the information from the imported source data set to determine the sample size, model role, type of measurement, formatting assignments, and summary statistics of the variables in order to perform faster processing in the subsequent Enterprise Miner SEMMA analysis. The metadata sample will also provide you with faster processing in the subsequent nodes, such as filtering outliers, replacing missing values, or interactive modeling. The **Input Data Source** node performs the following tasks.

- The node will allow you to import various SAS data sets and data marts to create the input data set. Data marts can be defined using SAS/Warehouse Administrator software by using the Enterprise Miner Warehouse Add-Ins.
- The node creates a metadata sample for each variable in the input data set.
- The node automatically sets the level of measurement and the model role for each variable in the input data set. However, the node will allow you to change both the measurement level and the model role that are automatically assigned to each variable in the input data set.
- The node automatically computes various descriptive statistics that can be viewed within the node for both interval-valued and categorically-valued variables.
- The node will allow you to define target profiles for each target variable in the data set that can be used throughout the entire currently opened Enterprise Miner diagram.
- The node will allow you to view the distribution of each variable by displaying a frequency bar chart of the frequency counts across the range of values placed into separate intervals or separate groups. The metadata sample is used to create the frequency bar charts.
- The node will allow you to import the source data set from many different file formats to create the input data set and the associated metadata sample. Conversely, the node will allow you to export the input data set into many different file formats.
- The node will allow you to refresh the data, assuming that modifications or updates have been performed to the data to ensure you that the current changes or updates have been processed since the last project session. This option might be performed if unexpected results occur within the currently opened process flow diagram.

The node usually assigns all variables in the metadata sample and the associated input data set a model role of **input** that is automatically included in the subsequent analysis. For instance, if the number of values of the variable contains one-third of the sample size, then the variable is assigned the **input** model role. If more than 90% of the values of the variables in the metadata sample are of the same value, then the variables are assigned an **id** model role. Conversely, all variables with one unique value or variables that are formatted dates are automatically set to a level of measurement of **unary** with a model role of **rejected**. Therefore, these rejected variables are removed from the subsequent analysis. However, the node will allow you to assign different type of model roles to the variables in the input data set. For example, you might want to automatically remove certain variables from the subsequent analysis by assigning a model role of **rejected** to these same variables, assigning a model role of **target** to the variable that you want to predict in the subsequent modeling node, assigning a model role of **id** to the variable that will identify the order in which the data set is recorded, or assigning a model role of **predicted** to the variable that consists of the fitted values that are needed in the **User-Defined** modeling node, and so on.

The node will also allow you to change the level of measurement of the variable in the input data set. The possible level of measurements that can be assigned to the variables in the input data set can either be interval, unary, binary, nominal, or ordinal. For example, an ordinal-valued variable with more than ten distinct class levels will be automatically assigned an undesirable measurement level of **interval**. However, the measurement level assignments are quite logical. For example, all variables with more than two distinct categories cannot possibly be assigned a measurement level of **binary**, and all character-valued variables cannot possibly be assigned a level of measurement of **interval**.

The Target Profile

The target profile can only be assigned to the target variable. The purpose of the target profile is designed to assign both the prior probabilities and profit or losses with fixed costs at each decision and each category or class level of the target variable. However, prior probabilities can only be assigned to categorically-valued target variables. The target profile can be assigned to both categorically-valued and interval-valued target variables, where the interval-valued target variable has been split into separate nonoverlapping intervals. The advantage of creating the target profile within the **Input Data Source** node is that the target profile may be shared throughout the currently opened Enterprise Miner process flow diagram.

In statistical modeling, the best model that is selected depends on the modeling criterion that is used in evaluating the usefulness of the model. Often the best model that is selected will accurately predict the target values, although another criterion might be selecting the best model that results in the highest expected profit. However, both of these criterions may result in entirely different models that are selected. In other words, the target profile will allow you to specify the appropriate revenue and costs to select the best model that either maximizes the expected profit or minimizes the expected loss.

Prior Probabilities: The prior probabilities represent the true proportion of the target categories. At times, the sample proportions of the target categories from the input data set might not represent the true proportions in the population of interest. Therefore, prior probabilities should always be applied if the sample proportion of the various target categories from the input data set differs significantly in comparison to the underlying data. In addition, specifying prior probabilities is very effective for target groups that consist of a small number of rare events in order to correct for oversampling. *Oversampling* occurs when the proportion of certain class levels are overrepresented in the data. The purpose of the prior probabilities is that these same prior probabilities are applied to adjust the estimated probabilities to conform more to the true proportions of the underlying data. For instance, an increase or a decrease in the prior probabilities will result in an increase or a decrease in the estimated probabilities. The estimated probabilities are also called the conditional probabilities or posterior probabilities. By default, no prior probabilities are applied. Therefore, Enterprise Miner assumes that the training, validation, and test data sets are actual representations of the underlying data set. Specifying the prior probability, Enterprise Miner assumes that the corresponding prior probabilities represent the true proportions of the target levels in the input data set. As a simple assessment to measure the accuracy of the classification model, the estimated probabilities that are calculated from the model should approximate the given prior probabilities by the class levels of the target variable. Also, the profit or loss summary statistics are affected by the accuracy of the specified prior probabilities since the posterior probabilities are adjusted by the prior probabilities. However, the various goodness-of-fit statistics like the parameter estimates, residuals, error functions, or MSE are unaffected by the prior probabilities that are specified.

Decision Costs: The various modeling nodes not only make predictions for each observation, but can also make a decision for each observation based on the maximum expected profit or the minimum expected loss through the use of the decision matrix. That is, a decision matrix can be specified that represents either profits or losses in which the final model can be selected by the average profit or average loss. A decision matrix can only be specified for one and only one target variable in the data set that you want to predict. For a categorically-valued target variable, the rows of the decision matrix represent the separate categories. For an interval-valued target variable, the rows of the decision matrix represent the separate nonoverlapping numeric intervals. The columns of the decision matrix represent the separate decision levels. At times, the decision matrix might have any number of decision levels. The main diagonal entries of the profit matrix represent the expected profit in accurately identifying the separate decision scenarios and the off-diagonal entries representing the misclassification costs involved in inaccurately identifying the different decision levels. In Enterprise Miner, an identity profit matrix is applied with ones along the main diagonal entries and zeros in all other off-diagonal entries. In other words, the cost of correct classification is equally profitable and the cost of misclassification is equally costly. To better understand the profit matrix, a profit matrix with nonzero entries along the main diagonal and zeros in all other off-diagonal entries is essentially no different than specifying the corresponding prior probabilities to each target category. Similar to the prior probabilities, the decision values do not affect the parameter estimates, the residual values, error functions, and the various goodness-of-fit statistics. In addition, as opposed to the prior probabilities, the estimated probabilities, the classification criteria, and the misclassification rate of the classification model are unaffected by the specified decision values.

One of the most critical steps in increasing the classification performance of the model is correctly identifying the appropriate prior probabilities and decision costs. Therefore, carefully selected prior probabilities or decision costs will lead to a dramatic increase in the classification performance of the model. On the other hand, specifying incorrect prior probabilities and decision costs will lead to erroneous results to the entire classification modeling process. Specifying the correct prior probabilities with accurately specified decision costs will lead to the correct decision results, assuming that the predictive model accurately predicts the variability in the target values and precisely fits the underlying distribution of the target variable. However, you really do not know the true nature of the underlying data that the predictive model is trying to fit. That is, even specifying the correct prior probabilities for each target category or the exact decisions will generally produce incorrect decision results, assuming that the predictive model generates an unsatisfactory fit to the corresponding data. Again, as it often happens that inappropriate decision results are incorporated, which leads to incorrect prediction estimates and inappropriate statistical inferences. Therefore, it is suggested to repeatedly fit the same model any number of times, by fitting the predictive model several times with the same parameter estimates and specifying different decision values for each separate modeling fit in selecting the final predictive model with the largest expected profit or the smallest expected loss.

The Metadata Sample

Enterprise Miner uses a metadata data set to make a preliminary assessment of the way in which to use each variable in the source data set. The default in creating the metadata sample is taking a random sample of 2,000 observations, assuming that the source data set has 2,000 or more observations. Otherwise, the entire source data set is selected to create the metadata sample. Again, the metadata sample is used to speed up processing time during the subsequent analysis by automatically assigning a model role, level of measurement, formats and labels to each variable in the input data set. The metadata procedure is a highly accurate process, but it is always a good idea to check that the variable attributes are correctly assigned to each variable in the input data set. For example, an nominal-valued variable with more than ten distinct levels might be assigned as an undesirable interval-valued variable.

The Model Roles

The following are the possible model roles that can be assigned to the variables within the node:

Model Role	Description
Modeling Model Roles	
• **Target** variables	Dependent, outcome or response variable to the model
• **Input** variables	Independent or predictor variables to the model
• **Rejected** variables	Rejected or omitted variables from the model
• **Id** variables	Record identifier variable like a time, counter or carry-over variables to the predictive model
• **Predict** variables	Predicted values required for the **User-Defined** modeling node

Model Role	Description
• **Column** variables	Column number variable for the matrix used in sparse matrix specification and the **Text Miner** node.
• **Cost** variables	Decision cost variable that represents the amount of cost that can be assigned to a decision level of the decision matrix.
• **Crossid** variables	Identifies levels or groups that is used in cross-sectional time series analysis such as repeated measures modeling in the **Time Series** node.
• **Freq** variables	Number of cases or the frequency of occurrence for each observation.
• **Group** variables	Grouping variable that is used in group processing analysis.
• **Row** variables	Row number variable for the matrix that is used in sparse matrix specification and the **Text Miner** node.
• **Sequence** variables	A variable that identifies the difference in time between successive observations that is used in sequential association analysis and the **Association** node.
• **Timeid** variables	Time identifier variable required for time series analysis that is used in the **Time Series** node.
• **Trial** variables	Contains frequency counts of the binomial target values.

The Measurement Levels

The following are the various measurement levels that can be assigned to the variables based on their own variable attributes and the range of values:

Measurement Level	Description
• **Interval**	Numeric variables that have more than ten distinct levels in the metadata sample are automatically assigned an **interval** measurement level.
• **Binary**	Categorical variables with only two distinct class levels are automatically assigned a **binary** measurement level.
• **Nominal**	Character variables with more than two and less than ten distinct nonmissing levels are automatically assigned a measurement level of **nominal**.
• **Ordinal**	Numerical variables with more than two but no more than ten distinct levels in the metadata sample are automatically assigned a level of measurement of **ordinal**.
• **Unary**	All variables with one nonmissing level or any variables assigned date formats in the metadata sample are assigned a measurement level of **unary**.

The **Data** tab will automatically appear as you open the **Input Data Source** node. The **Data** tab is specifically designed to read the source data set that will then automatically create the input data set and the associated metadata sample. Generally, once the following steps are performed in reading the source data set, then the next step is specifying the appropriate model roles for each variable in the input data set from the **Variables** tab.

To read in the source data set into Enterprise Miner, simply perform the following steps as follows:

- From the **Data** tab, press the **Select...** button or select **File > Select Data ...** from the main menu.
- Select the library reference name from the **Library** pop-up menu items to browse for the appropriate temporary or permanent SAS data set to be read into the currently opened SAS Enterprise Miner session. Press the **OK** button to return back to the **Data** tab.

Data tab

The purpose of the **Data** tab is designed to let you to select the source data set that will then automatically create both the input data set and the metadata sample. By default, Enterprise Miner performs a random sample of 2,000 observations from the source data set to create the metadata sample. However, the tab will allow you to change the sample size of the metadata sample by performing a random sample of the source data set of any desirable number of observations. Generally, it is assumed that the source data originates from an existing SAS data set. However, the tab is designed with the added capability of importing the source data through the use of the SAS **Import Wizard** that might reside in many other file formats. Alternatively, the input data set can be exported from the **Input Data Source** node into many other file formats through the use of the SAS **Export Wizard**.

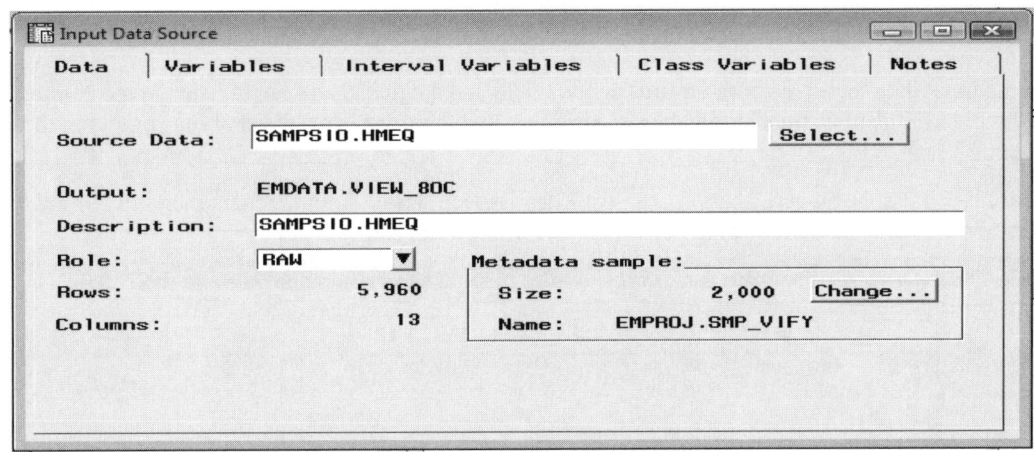

*The **Data** tab used to select the source data set in creating the input data set or the metadata sample that will automatically assign the various model roles and measurement levels to the variables in the metadata sample and the input data set.*

Option Settings

The following are the various data set attributes and options displayed within the **Data** tab:

- **Source Data:** The name of the selected source data set. Simply enter the appropriate libref and data set name to read the source data set within the tab instead of accessing the file through the **SAS Data Set** window.
- **Output:** The Enterprise Miner data set name that is automatically assigned to the input data set that is a member of the EMDATA directory. However, the name of the SEMMA input data set that is automatically created cannot be changed.
- **Description:** The data description label of the incoming source data set that you may be able to change from the corresponding entry field.
- **Role:** Set the role to the input data set that determines the way in which the data set is used throughout the process flow. For example, this option will allow you to perform your own partitioning of the same input data set by identifying the partitioned data sets assuming that the split-sample procedure is applied in evaluating the performance of the model. Conversely, the input data set can be scored by selecting the **Score** role option and connecting the **Input Data Source** node to the **Score** node that will then add prediction information to the data set that it is scoring, assuming that one of the modeling nodes is connected to the **Score** node. The following are the various roles that can be assigned to the input data set.

  RAW: (default) Raw data set.

  TRAIN: Training data set that is used to fit the data for modeling.

  VALIDATE: Validation data set that is used to assess and fine-tune the model.

  TEST: Test data set that is used as an unbiased assessment in measuring the accuracy of the predictions that is not a part of the modeling fit.

  SCORE: Score data set that will allow you to include the prediction information such as the fitted values or clustering assignments within the selected data set with or without the existence of the target variable in the data set.
- **Rows:** Number of rows or records in the input data set.
- **Columns:** Number of columns or fields in the input data set.

The **Metadata Sample** section that is located in the lower right-hand corner of the **Data** tab will allow you to select the number of observations for the metadata sample. The metadata sample is the data set that automatically assigns the model roles and measurement levels to the variables in the input data set that is then passed along to the subsequent nodes for further analysis within the process flow diagram.

- **Size:** By default the **Input Data Source** node creates a metadata sample based on a random sample of 2,000 observations or the entire sample size of the source data set.
- **Name:** The data set name that is automatically assigned to the SEMMA metadata sample.

Note: Both the input data set listed in the **Output** display field and the metadata sample listed in the **Name** display field will have the same number of records, assuming that there are 2,000 observations or less in the source data set.

From the **Data** tab, press the **Select...** button or select **File > Select Data...** from the main menu to browse for the input data set to read with the following **SAS Data Set** window appearing.

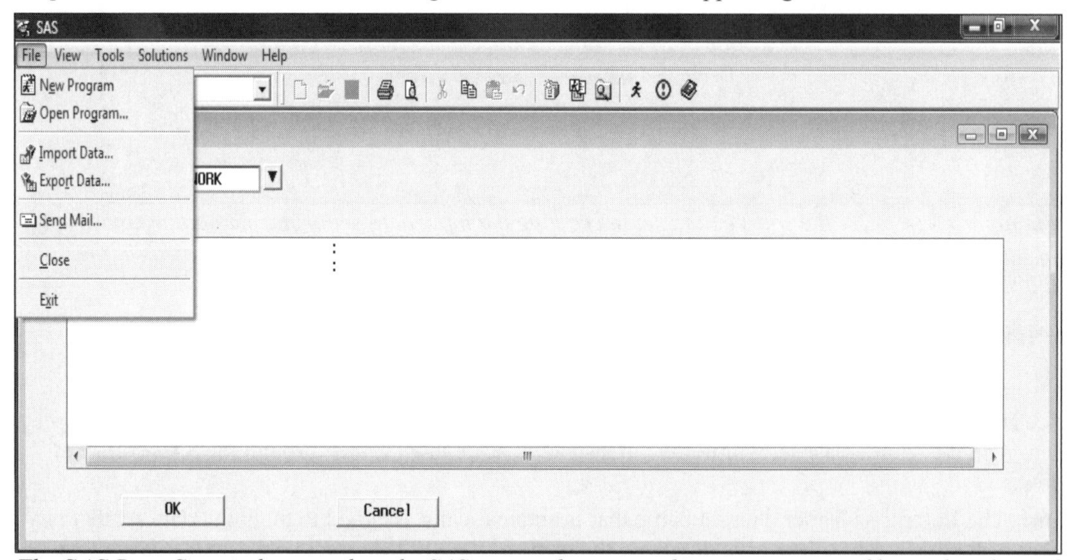

*The **SAS Data Set** window to select the SAS source data set and import or export files within the node.*

Importing Files

The node will allow you to read the source data through the **Import Wizard** that may reside in many different file formats. Simply select **File > Import Data** from the main menu, which will automatically activate the **Import Wizard**. The purpose of the **Import Wizard** is that it will allow you to automatically create the SAS data set by converting the external file into the appropriate SAS file format. The **Import Wizard** will allow you to browse the operating system to import the selected file into the currently opened SAS Enterprise Miner session that will automatically create the input data set.

Exporting SAS Data Sets from the Input Data Source Node

Conversely, select **File > Export Data** from the main menu options to export the input data set into many different file formats by accessing the SAS **Export Wizard**. The purpose of the **Export Wizard** is that it will allow you to automatically convert the SAS data set into a wide variety of file formats.

Creating the Input Data Set

The following are the steps that must be followed in reading the SAS source data set to create the input data set from the **Data** tab within the **Input Data Source** node.

1. From the **Data** tab, press the **Select...** button within the **Source Data** section or select the **File > Select Data** main menu options and the **SAS Data Set** window will appear.
2. SAS data sets are organized or stored in the various SAS libraries. Therefore, click the drop-down arrow button in the **Library** entry field to display the currently available SAS library names that are currently assigned to the associated SAS libraries or folders within the currently opened Enterprise Miner project. For example, select the **WORK** library reference name that represents the SAS work directory. This will allow you to select any one of the temporary SAS data sets that are created within the currently opened SAS session. From the **Tables** listing, select the appropriate temporary SAS data set to read by double clicking the mouse or highlighting the data set row and pressing the **OK** button. The **Input Data Source** window will then reappear.

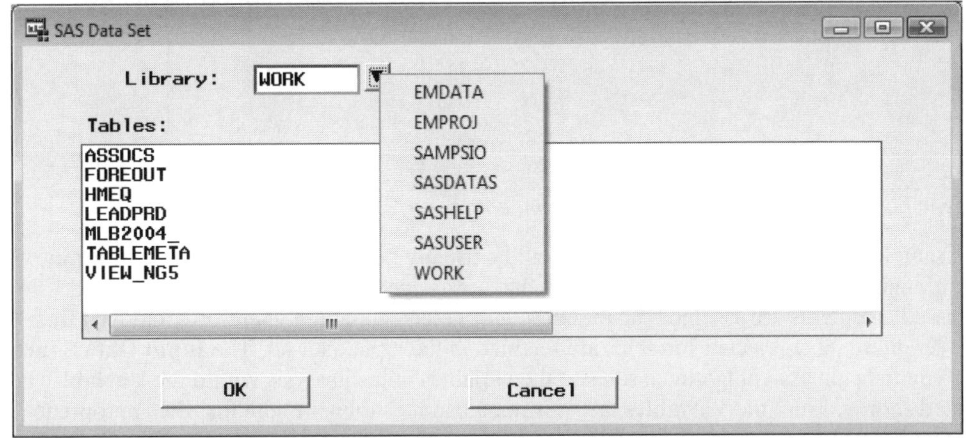

*Selecting the source data set to create the input data set and the associated metadata sample from within the **Input Data Source** node by selecting the **Library** pull-down menu to select the SAS data library or directory of the source data set. Notice the additional SASDATAS data library that is created in the subsequent steps.*

Selecting the SAS Library Reference

- To select a temporary SAS data set, you must first create the SAS source data set within the same SAS session in which you have currently opened the Enterprise Miner session.
- To select a permanent SAS data set, you must first create a library reference name by selecting the **Options > Project > Properties** main menu options, then selecting the **Initialization** tab. From the tab, select the **Edit** button corresponding to the **Run start-up code when project opened** option that will open the following **Edit start-up code** window to enter the LIBNAME statement that will create a library reference name to the directory where the permanent SAS data set is stored in the system folder.

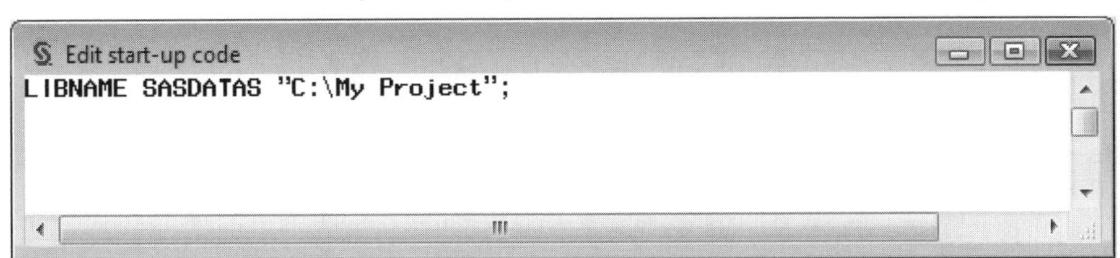

*The **Edit start-up code** window to create the library reference to the system folder.*

An alternative technique in creating a data library is to select the **New Library** toolbar icon within the SAS Program Editor. The **New Library** window will appear for you to specify the libref name within the **Name** entry field. Select the **Enable at startup** check box in order for the library name to be connected to the specified folder each time the SAS session is opened. This will result in Enterprise Miner recognizing the corresponding name of the directory where the permanent data set is located that will be displayed from the previously displayed **Library** pull-down items. Finally, select the **Browse...** button to search your operating system for the appropriate folder to assign to the specified library reference name. After selecting the appropriate SAS data set, then click the **OK** button. This will result in the creation of the input data set and the corresponding metadata sample with the **Input Data Source** window appearing.

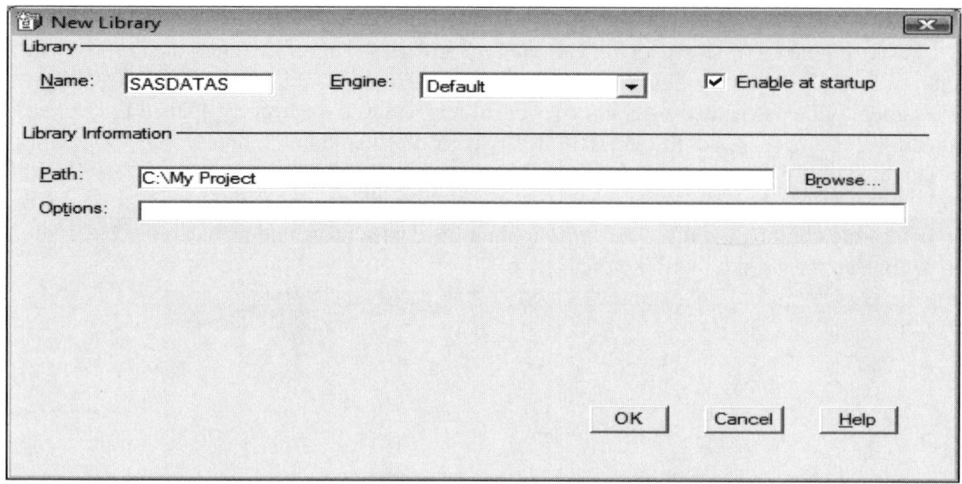

*The **New Library** window to set the library reference name to the corresponding directory.*

From the metadata sample, every variable in the data set will be usually be assigned a model role of **input**, with the exception of unary-valued variables and variables that are assigned date formats. Unary-valued variables and formatted date fields are assigned the model role of **rejected** and are, therefore, removed from the analysis. However, these same rejected variables are retained in the input data set. The **Input Data Source** window will allow you to begin assigning model roles to the variables in the analysis within the **Variables** tab that will be illustrated shortly. From the **Variables** tab, it is not a bad idea to make sure that the appropriate levels of measurement have been correctly assigned to the variables in the input data set, even though the metadata process in the variable assignments is extremely accurate.

Changing the Sample Size of the Metadata Sample

The metadata sample will be automatically created once the source data set is read into Enterprise Miner. By default, Enterprise Miner performs a random sample of 2,000 observations from the source data set. From the **Input Data Source** tab, the **Metadata Sample** section will allow you to change the number of observations of the metadata sample. In other words, the section will allow you to sample the entire data set or a subset of the selected source data set. Press the **Change** button or select the **Edit > Sample Size...** main menu options and the following **Set Sample Size** window will appear.

- Select the **Use complete data as sample** check box to ensure you that the metadata sample contains all of the observations from the source data set. From the **Sample Size** entry field, a sample size can be entered, which determines the number of records to randomly sample from the source data set in creating the metadata sample.

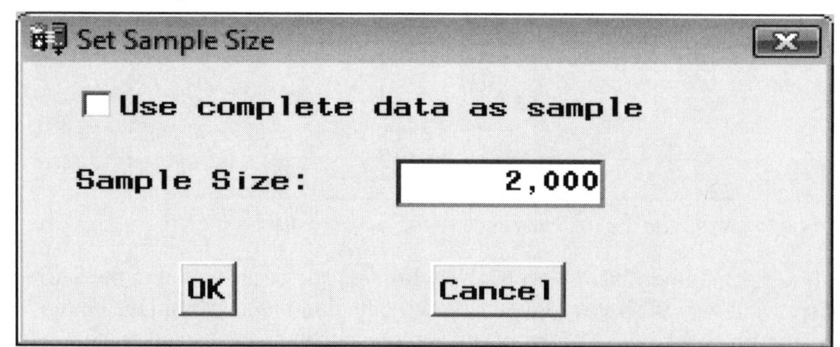

*The **Set Sample Size** window used to specify the sample size of the metadata sample.*

Refreshing the Sample

The node will allow you to refresh the metadata sample to ensure that the current changes or updates have been processed since the last project session. Select **Edit > Refresh Sample** from the main menu options to refresh the metadata sample in order to update the corresponding changes that have been made to the input data set. It is generally recommended to refresh the metadata sample if you have performed updates to the input data set such as redefining the data library of the source data set, adding or removing observations or variables, redefining variables or prior probabilities of the data, updating the decision values in the profit matrix from the target profile, and so on. It is recommended to select this option if you have noticed some unexpected results from the process flow diagram or if all else fails during the construction or execution of the process flow diagram. To refresh the metadata sample, simply select this option then close the **Input Data Source** node to save the current changes and rerun the entire process flow diagram once more.

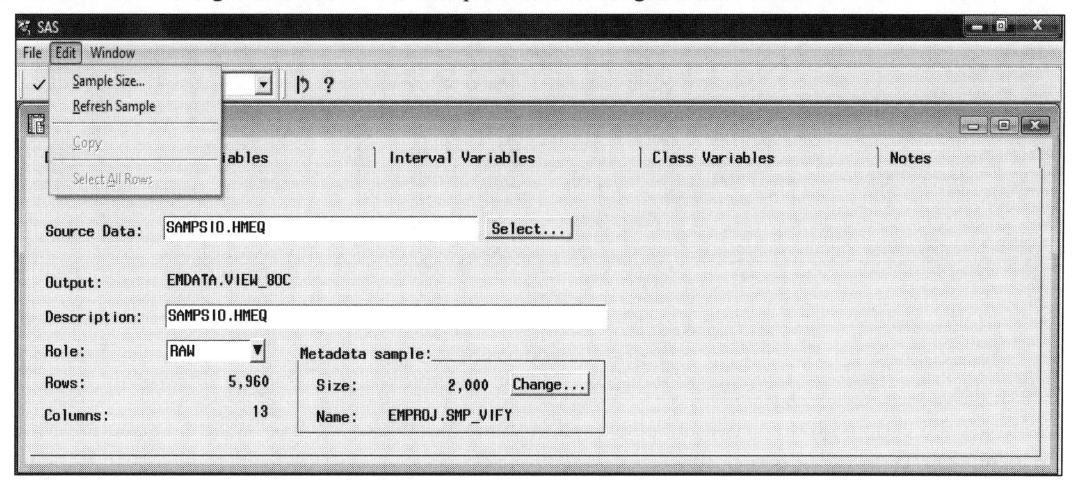

Selecting the menu option to refresh the metadata sample of the current Enterprise Miner diagram.

Browsing the Input Data Set

Select the **File > Details** main menu options to browse the file administration information or the header information, such as the creation date and the number of variables and records of the input data set, or simply view the table listing of the input data set that is illustrated in the following diagram.

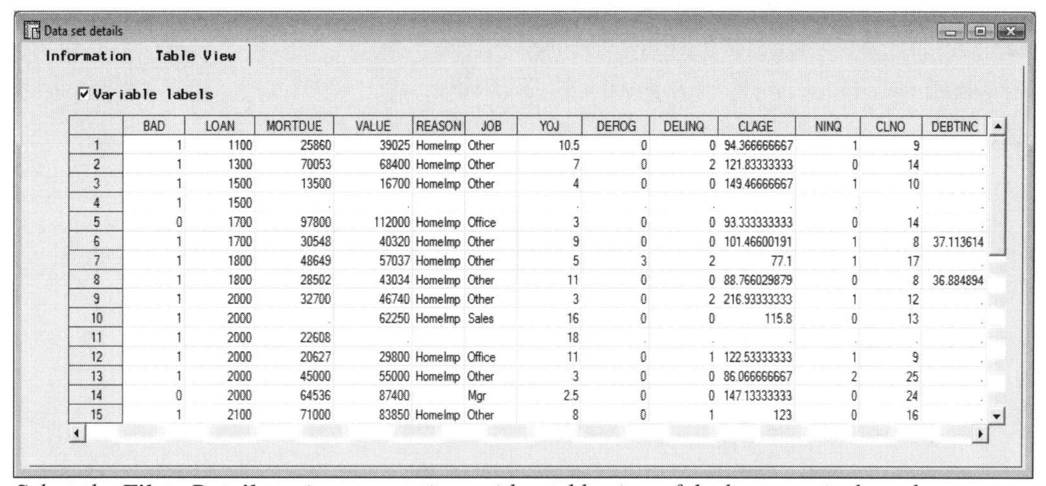

	BAD	LOAN	MORTDUE	VALUE	REASON	JOB	YOJ	DEROG	DELINQ	CLAGE	NINQ	CLNO	DEBTINC
1	1	1100	25860	39025	HomeImp	Other	10.5	0	0	94.366666667	1	9	.
2	1	1300	70053	68400	HomeImp	Other	7	0	2	121.83333333	0	14	.
3	1	1500	13500	16700	HomeImp	Other	4	0	0	149.46666667	1	10	.
4	1	1500	.	.			.	.	.	.	.	.	.
5	0	1700	97800	112000	HomeImp	Office	3	0	0	93.333333333	0	14	.
6	1	1700	30548	40320	HomeImp	Other	9	0	0	101.46600191	1	8	37.113614
7	1	1800	48649	57037	HomeImp	Other	5	3	2	77.1	1	17	.
8	1	1800	28502	43034	HomeImp	Other	11	0	0	88.766029879	0	8	36.884894
9	1	2000	32700	46740	HomeImp	Other	3	0	2	216.93333333	1	12	.
10	1	2000	.	62250	HomeImp	Sales	16	0	0	115.8	0	13	.
11	1	2000	22608	.			18	.	.	.	.	.	.
12	1	2000	20627	29800	HomeImp	Office	11	0	1	122.53333333	1	9	.
13	1	2000	45000	55000	HomeImp	Other	3	0	0	86.066666667	2	25	.
14	0	2000	64536	87400		Mgr	2.5	0	0	147.13333333	0	24	.
15	1	2100	71000	83850	HomeImp	Other	8	0	1	123	0	16	.

*Select the **File > Details** main menu options with a table view of the home equity loan data set.*

Variables tab

The primary purpose of the **Variables** tab is to assign the appropriate model roles to the variables and define the target profile of the target variable in the analysis. In addition, the tab will allow you to reassign a new model role or level of measurement to the selected variable, view the frequency distribution of each variable from the metadata sample, and create, browse, or edit a target profile of the selected target variable. The tab displays a table view of the various properties of the listed variables, such as the model roles, level of measurements, formats, and variable labels that are assigned from the metadata sample. Both the **Name** and the **Type** columns are grayed-out and cannot be changed. In the following tabs that are displayed throughout the book, all columns or cells with white backgrounds will indicate that their attributes may be changed, and all gray columns or cells will indicate that their attributes cannot be changed.

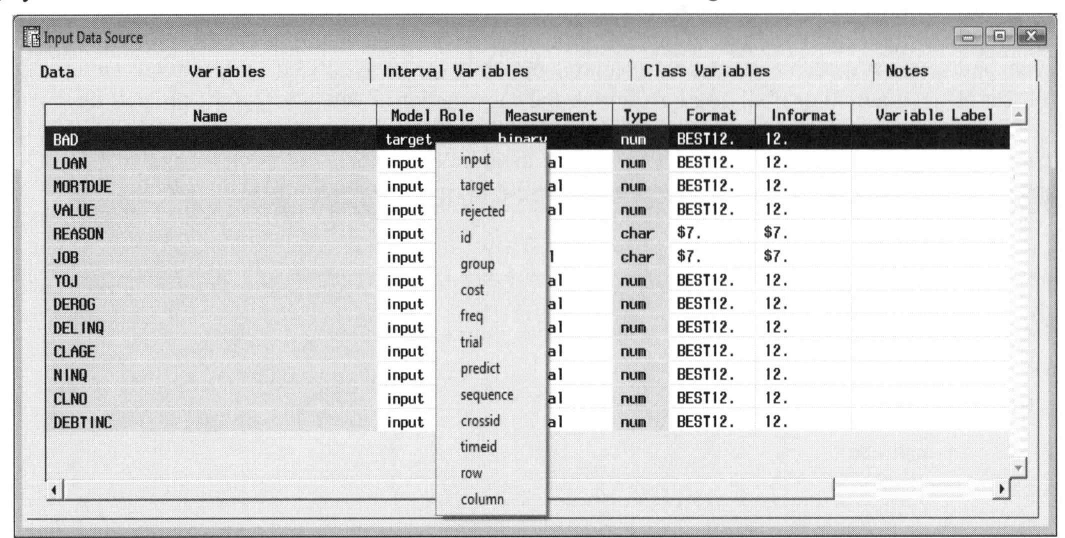

*The **Variables** tab used in setting the target model role to the variable in the classification model.*

The following are the various columns that are displayed from the **Variables** tab that lists the attributes of each variable in the data set:

- **Name:** Variable name assigned to the listed variable.
- **Model Role:** Model role assigned to the listed variable.
- **Measurement:** Level of measurement assigned to the listed variable.
- **Type:** Type of measurement assigned to the listed variable, for example, character or numeric.
- **Format:** SAS format assigned to the listed variable.
- **Informat:** SAS informat assigned to the listed variable.
- **Variable Label:** Description label assigned to the listed variable.

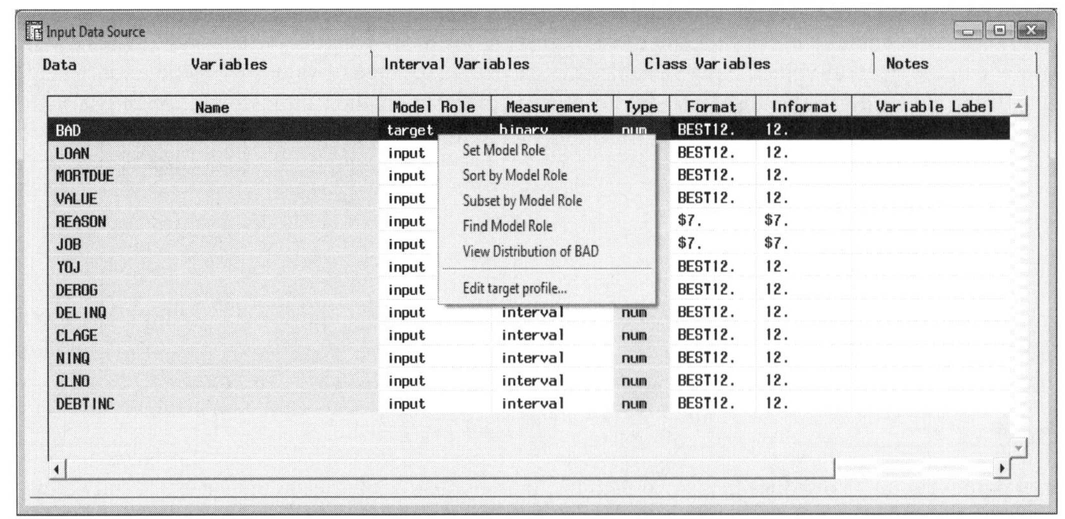

Selecting the pop-up menu items to set the level of measurement, view the frequency bar chart, or access the target profile of the selected target variable.

Assigning the Model Roles

One of the most important purposes of the **Variables** tab is assigning the appropriate model roles to the variables in the input data set. Simply select the appropriate variables to assign the corresponding model roles, then scroll over to the **Model Role** column and right-click the mouse to select from the variety of model roles. For example, in unsupervised training, all variables that are a part of the analysis are assigned a model role of **input**. In predictive modeling or supervised training, all predictor variables are assigned a model role of **input** and the variable that you want to predict is assigned a model role of **target**. Conversely, all variables that are removed from the subsequent analysis are assigned a model role of **rejected**.

Note: If you want to simultaneously assign several variables with the same model role, for example, **input**, simply select any number of noncontiguous rows by left-clicking the mouse and holding down the **Ctrl** key at the same time to highlight any number of input variables or rows, then release the **Ctrl** key. To highlight a block of variable rows or contiguous variable rows to the data set, simply hold down the **Shift** key and select the block of rows. For example, select the first row and last row of the block by left-clicking the mouse, then release the **Shift** key.

Assigning the Measurement Levels

There might be times that you might want to assign different levels of measurement to the variables in the input data set. For instance, a categorically-valued variable with more than ten separate categories might be assigned an undesirable interval level of measurement. To change the level of measurement of the variable, select or highlight the variable row, then scroll over to the **Measurement** column and right-click the mouse to select the **Set Measurement** pop-up menu item that is illustrated in the previous diagram. At times, several variables may be assigned the same level of measurement by selecting any number of variable rows, then assigning the appropriate level of measurement.

Viewing the Frequency Distribution of the Variable

From the **Variables** tab, you may view the distribution of each variable one at a time. It is important to understand that the various descriptive statistics and the range of values that are displayed in the resulting frequency plot are based on the metadata sample that will provide you with a short summary of the data. Simply highlight the row by left-clicking the mouse to select the variable row, then right-click the mouse to select the **View Distribution of <variable name>** option from the pop-up menu items and the **Variable Histogram** window will appear that will display a histogram of the frequency distribution of the selected variable from the metadata sample that is used to create the histogram.

Viewing the distribution of each variable in the data set is usually a preliminary step in statistical modeling, principal component modeling, decision tree modeling, cluster analysis, and overall data mining. This is because it is very important that the variables in the analysis be normally distributed and share the same range of values. In other words, many of the nodes and the corresponding analysis apply the sum-of-squares distance function, in which the interval-valued input variables that have a wide range of values will tend to dominate the final results. If the selected variable displays a highly skewed distribution, then a transformation might be required, such as the standardizing the interval-valued input variables or applying the log transformation, which will usually lead to better results. This is because these small number of outlying data points may have a profound effect to the analysis. For instance, these same outlying data points could possibly shift the prediction line away from the majority of the data points being fitted, create a lop-sided tree design or create its own interactive or cluster groupings. This will then lead to biased estimates and misleading results of the analysis. An alternative to nonnormality in the data is applying the various nonparametric techniques in which the analysis is based on the ranks of the observations. In classification modeling, viewing the separate class levels of the categorically-valued target variable is important since the classification model will have the propensity to predict the class level of the most frequent class level. From the **Replacement** node, specifying the various replacement values of the listed variable in the analysis, it is important that the variable follow a normal distribution when applying the various traditional statistical estimates such as the mean. This is because that many of the traditional statistical estimators can be inefficient due to the existence of outliers in the data. Conversely, when replacing values with the various robust estimates, such as the M-estimators, it is best that the variable follows a symmetric distribution such as a bimodal, trimodal, or other symmetric distributions in generating the most reliable estimates.

The following histogram displays a wide range of values in the ratio between debt and income due to a few outlying data points at the upper end of the chart, therefore, a transformation might be required.

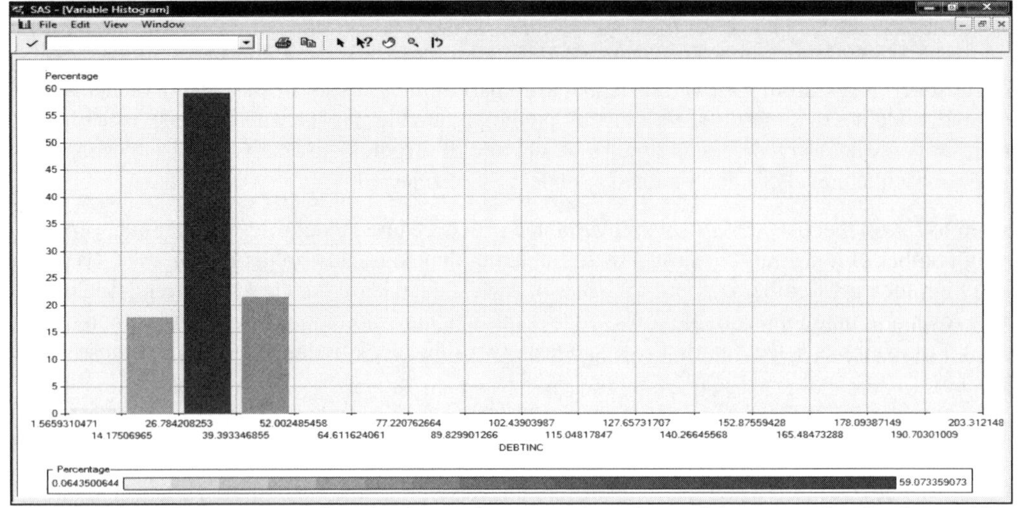

*The **Variable Histogram** window displays a frequency bar chart of the interval-valued variable.*

Interval Variables tab

The **Interval Variables** tab will display the descriptive statistics of all the interval-valued variables in the input data set. The tab will display various descriptive statistics such as the minimum, maximum, mean, standard deviation, percentage of missing values, skewness, and kurtosis of all interval-valued variables from the metadata sample. These statistics will allow you to view the range of values, the overall average value and the overall variability from its own mean, the amount of missing cases, skewness, and peakness of the listed variables. The *skewness* measures the amount of symmetry in the distribution. The *kurtosis* measures the amount of flatness or peakness in the distribution. In SAS, a skewness of zero and a kurtosis of one will indicate a perfectly normal distribution for the listed variable.

Input Data Source

Data | Variables | Interval Variables | Class Variables | Notes

Name	Min	Max	Mean	Std Dev.	Missing %	Skewness	Kurtosis
LOAN	1100	89200	18604	11242	0%	2.0941	7.3916
MORTDUE	2063	399412	74566	45201	8%	2.0341	8.159
VALUE	8800	855909	103769	60783	2%	3.5362	28.747
YOJ	0	36	9.0125	7.5968	8%	0.9486	0.1678
DEROG	0	10	0.2468	0.8354	11%	5.612	42.535
DELINQ	0	11	0.5116	1.1862	9%	3.5159	16.422
CLAGE	0	649.75	179.71	83.363	5%	0.9174	2.2131
NINQ	0	12	1.1198	1.6754	8%	2.5576	8.8353
CLNO	0	71	21.206	9.9067	4%	0.7927	1.4052
DEBTINC	0.7203	143.95	33.672	8.429	21%	1.3837	21.872

*The **Interval Variables** tab displays the descriptive statistics of the interval-valued variables.*

Class Variables tab

The **Class Variables** tab will display the number of levels, percentage of missing values, and the ordering level of all the categorical variables from the metadata sample.

The following tab is important in controlling the level of interest to the categorically-valued target variable to fit in the classification model in estimating the probability of the target event rather than the target nonevent, setting the target event in interpreting the lift values to the performance charts, setting the reference levels to the categorically-valued input variables that plays the role of the intercept term in the logistic regression model, setting the target event in interpreting the development of the tree branching process in the decision tree model, setting the target event to the classification model in the two-stage modeling design, or setting the appropriate target event in interactive grouping.

In Enterprise Miner, the target event is automatically assigned to the last ordered level of the categorically-valued target variable in the classification model. In addition, the intercept term is automatically assigned to the last ordered level of the categorically-valued input variable in the logistic regression model. In the performance charts, in order to make the charts easier to interpret, it is important to make sure that the classification model is fitting the probability of the target event rather than the target nonevent at each decile estimate of the binary-valued target variable. In two-stage modeling, it is important to know the target event that the classification model is fitting where the variable that represents the estimated probabilities of the target event is one of the input variables to the subsequent prediction model. For nominal-valued target variables, the last ordered class level is the input variable that is inserted into the second-stage model. In interactive grouping, it is important to know the target event of the binary-valued target variable in interpreting the response rate between the two separate class levels of the target variable. For binary-valued target variables, its best to sort its values in descending order to easily interpret the modeling results.

From the **Order** column, you may set the class level of the target variable to predict or the reference level of the categorical input variable to the logistic regression model. By default, all the categorically-valued input variables are sorted in ascending order and the categorically-valued target variable is sorted in descending order. Scroll over to the **Order** column, right-click the mouse and select the **Set Order** pull-down menu option to set the ascending or descending ordering level to the selected categorical variable.

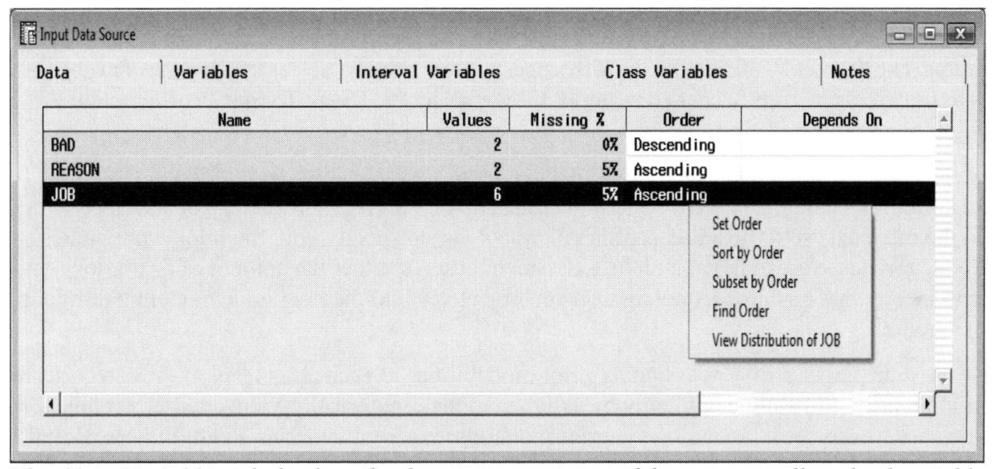

The ***Class Variables*** *tab displays the descriptive statistics of the categorically-valued variables.*

Notes tab

The **Notes** tab is a notepad to enter notes or record any other information to the Enterprise Miner **Input Data Source** node.

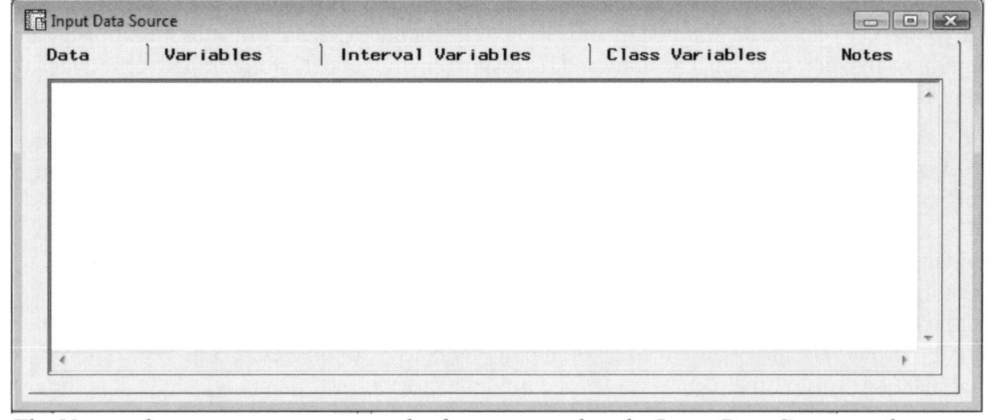

The ***Notes*** *tab to enter notes or record information within the* ***Input Data Source*** *node.*

Editing the Target Profile

The following example is based on the home equity loan data set that estimates the probability of clients defaulting on their home loan.

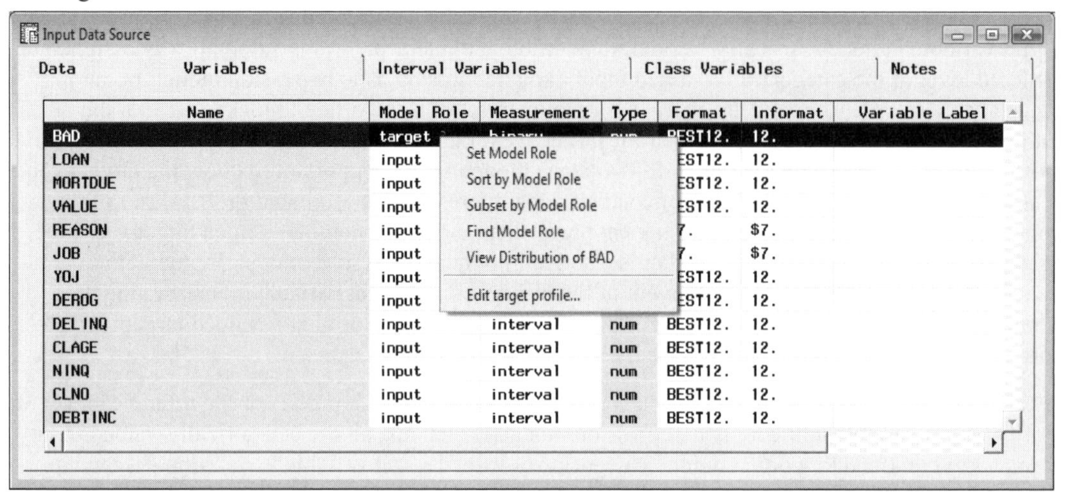

The ***Input Data Source*** *window to either edit, browse, or delete the target profile of the target variable.*

To edit the target profile from the **Variables** tab, highlight the row of the target variable with a model role of **target**, then scroll over to the **Model Role** column and right-click the mouse to select the **Edit target profile...** pop-up menu option. The **Target Profiles** window will appear. Alternatively, select the **Browse Target Profile...** option to browse the currently created target profile or select the **Clear Target Profile** option to delete the existing target profile.

The target profile can only be applied to categorically-valued target variables or interval-valued target variables that are split into separate nonoverlapping intervals. Again, the target profile can be assigned to one target variable in the input data set. The target profile will allow you to specify both the prior probabilities to each level of the categorically-valued target and the assessment costs of either the amount of profit, loss or revenue in order to incorporate business modeling decision objectives into the predictive modeling design in selecting the final model.

One of the purposes of the target profile is assigning prior probabilities to each class level of the categorically-valued target variable. Prior probabilities can only be assigned to the categorically-valued target variable. From the **Priors** tab, the various choices in assigning the prior probabilities are either **Prior Probabilities**, **Equal Probabilities** or **None**. **None** is the default. In other words, the prior probabilities are not applied.

The target profile is also designed to specify the profits, profit with costs, or losses at the various decision levels of the target variable from the **Assessment Information** tab. In the numerous modeling nodes, the final model is selected based on the smallest average error from the validation data set, assuming that you have the luxury of an enormous amount of data to perform the split sample procedure. But by creating a decision matrix and specifying the assessment objective with the corresponding assessment costs, one of the criterions in selecting the final model can then be determined by either maximizing the expected profit or minimizing the expected loss from the validation data set. The drawback in evaluating the model based on both criterions is that they can both lead to selecting entirely different models.

The target profile can ether be edited or browsed from the corresponding **Input Data Source** node, **Data Set Attributes** node, **Assessment** node, and the various modeling nodes. Alternatively, you can define a target profile for a specific data set from the project level. The advantage is that the active profile created from the project level can be accessed from the currently opened Enterprise Miner project. From the project level, you may also specify defaults to the target profile that can be used throughout the currently opened Enterprise Miner project. That is, specifying default settings such as the assessment objective to the decision matrix, defining the prior probabilities, and setting the upper and lower bound to the profit function for the interval-valued target variable that is split into separate nonoverlapping intervals. From the **Diagram Workspace**, you can define the active target profile from the project level within the process flow diagram by selecting the **Options > Project > Data Profiles** main menu options.

Profiles tab

The **Profiles** tab is designed to manage the various target profiles that have been created within the **Input Data Source** node. From the tab, you may either create a target profile, set the active target profile to be shared throughout the currently opened Enterprise Miner project, browse, rename, copy, and delete the selected target profile.

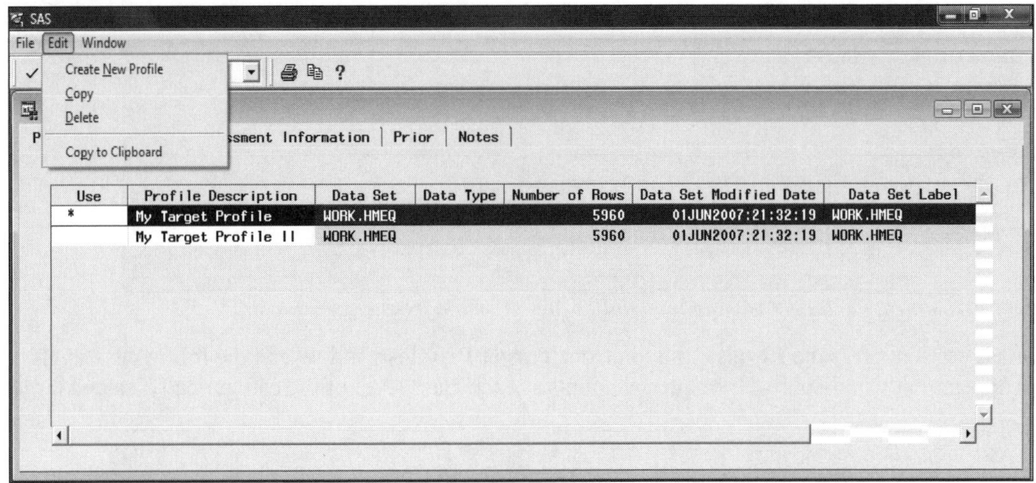

*The **Profiles** tab displays the list of available target profiles of the currently opened project.*

The **Profiles** tab will allow you to create a target profile for the currently selected data set within the **Input Data Source** node. However, the advantage in creating a target profile within the node is that it can be shared throughout the currently opened Enterprise Miner project. By creating the target profile within the **Input Data Source** node, the input data set is used to create the target profile. However, creating a target profile within any one of the modeling nodes means that the target profile cannot be modified outside of the currently opened Enterprise Miner diagram.

In addition to creating a new target profile, you may perform the following options by simply highlighting any one of the currently listed target profiles that have been created:

- To create a new target profile, select the **Edit > Create New Profile** main menu options that will be added to the list of existing target profiles.
- To activate the selected target profile, right-click the mouse and select the **Set to use** pop-up menu item. An asterisk will then be displayed in the **Use** column to indicate that it is the active target profile.
- To rename the description of the selected target profile, select the **Profile Description** column and enter a short description of the target profile.
- To copy the selected target profile, simply select **Edit > Copy** from the main menu items.
- To delete the selected target profile, simply select **Edit > Delete** from the main menu items.

You may also select any one of the listed target profiles in order to browse the various profile tabs and the corresponding target profile information.

Target tab

For a binary-valued target variable, the most important item displayed from this tab is the **Event** display field. It is extremely important to make sure that the value assigned to the event level is correct. This is because it is important that the classification model and lift values are correctly modeling the probability of the target event rather than modeling the target nonevent. Otherwise, it might be quite confusing in interpreting the classification results. The **Target** tab displays the attributes of the target variable such as the variable name, label, format, and level of measurement of the target variable. The **Levels** section displays the ordering level of either ascending or descending order, the number of categorical levels, and the value assigned to the target event of interest.

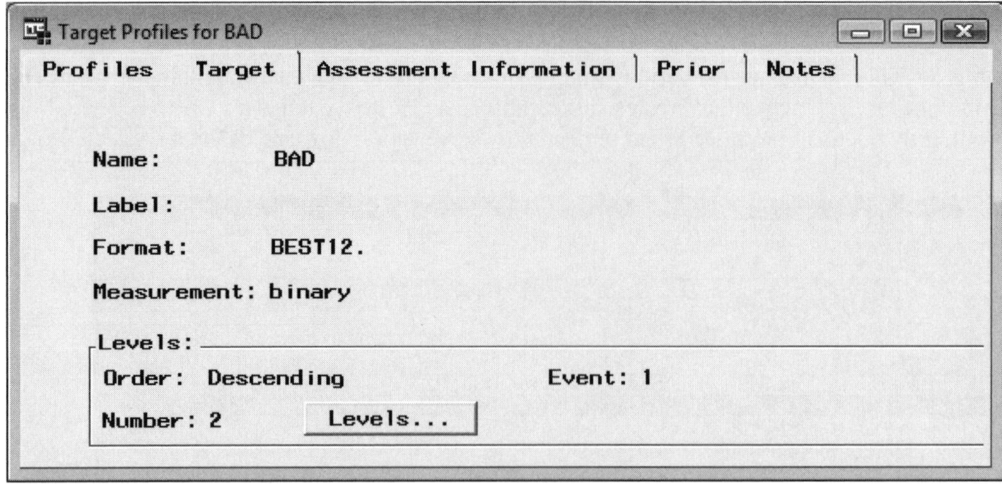

*The **Target** tab displays the target variable attributes and view the frequency counts.*

From the **Levels** section, press the **Levels...** button from **Target Profiles** window and the following **Levels** window will appear that will display the frequency counts at each class level of the categorically-valued target variable.

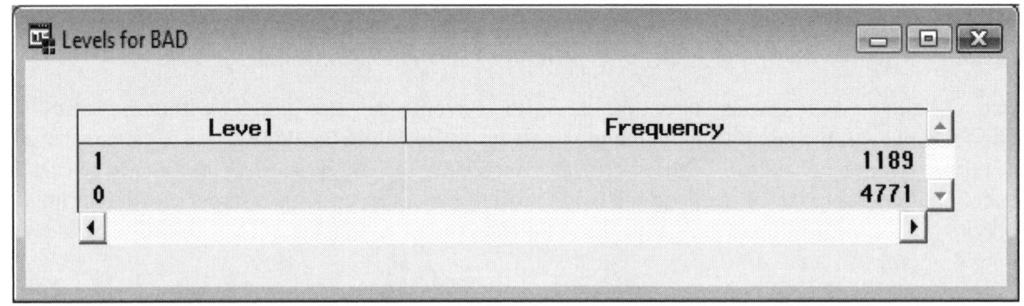

Level	Frequency
1	1189
0	4771

*The **Levels** window displays the frequency counts of each level of the categorically-valued target variable.*

Assessment Information tab

The **Assessment Information** tab is designed to display or edit the default profit vector, loss vector, profit matrix and the loss matrix. The default decision matrices that are displayed from the list of decision matrices cannot be modified. That is, the decision values cannot be modified. Furthermore, you cannot add or delete the decision levels or change the assessment objective to the default decision matrix. Therefore, these same decision matrices are grayed-out. The purpose of the decision matrix is to assign a numerical value or a decision value to each decision level and each class level of the categorically-valued target variable or each separate interval of the stratified interval-valued target variable. In other words, the profit matrix is a table of the expected revenue and the associated fixed costs accrued for each decision level at each class level of the target variable.

There exist three basic decision matrices: a profit matrix, revenue matrix (profit with costs) or a loss matrix. The decision matrices are designed so that the column represents each decision and the rows represent the actual target levels. The default is the default profit matrix. The default profit matrix is a squared matrix that consists of ones along the main diagonal and zeros elsewhere, with the average profit and the correct classification rate sharing the same value. The profit matrix represents the profit amounts; the model that best maximizes the predicted profit is selected. The default profit matrix is set to an identity matrix. Therefore, the profit matrix assigns a profit of one for a correct classification and a profit of zero for misclassifying the target response. Conversely, to calculate the misclassification rate, then you must specify the default loss matrix. The default loss matrix will consist of ones in the off-diagonal entries and zeros in the main diagonal entries of the matrix, where the average loss equals the misclassification rate. The loss matrix represents the losses; the model that best minimizes misclassification rate or predicted loss is selected. The loss matrix assigns a loss of one for incorrect classification and a loss of zero for correct classification.

The profit matrix might also reflect a net revenue assuming that there is a fixed cost incurred in the analysis since the amount of cost cannot be specified in the profit or loss matrix. However, by simply subtracting the fixed costs from the net revenue is one way of specifying the profits (revenue minus costs) of each decision in the profit matrix. The constant costs or fixed costs are strictly assigned to each decision. Therefore, the profit would then be redefined as the difference between net revenue and fixed cost for each target level. That is, the fixed costs are used in computing return on investments, which is defined as the difference between the revenue and cost divided by the fixed cost.

Target Profile Options

The tab will allow you to add a new decision matrix, delete or copy an existing decision matrix that has been created, or set the selected decision matrix as the active profile of the currently opened diagram. Simply highlight the desired decision matrix that is listed to the left of the tab, then right-click the mouse, which will allow you to select from the following menu options:

- **Add:** Adds the selected decision matrix to the bottom of the list of existing decision matrices.
- **Delete:** Deletes the selected decision matrix from the list with the exception of the default decision matrices or the active profile from the existing list of decision matrices.
- **Copy:** Copies the selected decision matrix that will be added to the bottom of the list of the existing decision matrices that are displayed in the list box.
- **Set to use:** Sets the selected decision matrix as the active target profile to the process flow. An asterisk will appear to the left of the listed entry, indicating that the decision matrix is the active profile of the current diagram. Once the added decision matrix is set as the active profile matrix, it cannot be deleted until a separate active profile is specified. In addition, only one target profile can be active in order for the subsequent nodes within the process flow to read the corresponding information from the active target profile.

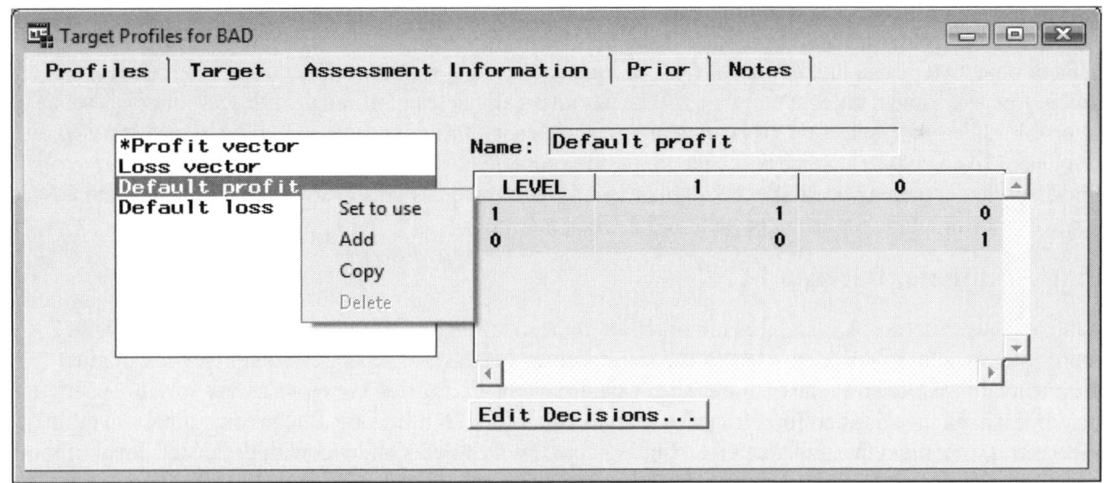

*The **Targets Profiles** window used to view the decision matrix and add, delete or copy the decision matrix.*

Creating the Decision Matrix

Since the predefined decision matrices cannot be changed, we must first begin by selecting from the list of existing default matrices in Enterprise Miner to create our own custom-designed decision matrix. For example, let us add an entirely different profit matrix. Simply highlight the default profit matrix, that is, the matrix called **Default profit**, then right-click the mouse to select the **Add** pop-up menu item. A new decision matrix will be appended to the bottom of the list of existing decision matrices. Select the listed entry with the corresponding decision matrix appearing within the window. The next step is assigning a name to the decision matrix. Therefore, from the **Name** entry field, enter a name for our custom-designed decision matrix. From the **Name** entry field, simply enter "My Decision Matrix" into the field. The next step is assigning this added matrix as the active profile of the process flow as follows:

Setting the Active Profile

To set the current decision matrix as the active profile, highlight the added decision matrix called My Decision Matrix, then right-click the mouse and select the **Set to use** option. This will set the corresponding decision matrix as the default decision matrix in computing either the posterior probabilities, maximum expected profits, or minimum expected losses for the predictive modeling design. Again, only one target profile can be active in order for the subsequent modeling nodes or the **Assessment** node to read the corresponding information from the active target profile.

Decision Values Other than the Profit and Revenue Amounts

If the decision matrix consists of revenue and costs, then the fixed costs for each target level may be specified, where the profit is defined as the difference between revenue and cost for each decision level. Keep in mind that the values that are specified in the decision matrix can, at times, have nothing to do with profits, losses, or revenue, but instead may represent weights or numerical values that are assigned when correctly and incorrectly classifying the target class levels. Likewise, fixed costs may be specified for the target class levels that have other practical interpretations than costs. However, the return on investments might have absolutely no meaning.

Setting the Assessment Objective

The next step is to set the assessment objective to the target profile. The assessment objectives can either be maximizing profits, maximizing revenue and costs, or minimizing losses. The window will allow you to set the assessment objective to the decision matrix of either profits, revenues or losses to each decision. To set the assessment objective to the target profile, simply select the **Edit Decisions...** button and the **Editing Decisions and Utilities** window will appear. The **Editing Decisions and Utilities** window will allow you to enter the assessment objective, and the various fixed costs for each decision level or even add additional decision levels to the active decision matrix. The various assessment objectives to select from are either **Maximize profit**, **Maximize profit with costs**, or **Minimize loss**. By default, the **Maximize profit** assessment objective option is automatically selected since we have specified a default profit matrix as the active target profile, that is, the assessment objective of maximizing profit with no incurred fixed costs for each decision. Conversely, the **Minimize loss** assessment objective option will be automatically selected if the default loss matrix is set as the active profile. Therefore, select the **Maximize profit with costs** option in order to maximize profit based on the associated fixed costs, since we will be entering the separate costs for each decision. The **Maximize profit with costs** option is the only available option that will allow you to assign a cost variable or a constant cost in defining revenue amounts for each decision of the profit matrix.

Adding Additional Decision Levels

For a binary-valued target variable, it is important to understand that we are not limited to the default 2x2 matrix of two rows for the two class levels and two columns for the two separate decision levels. In other words, we are not restricted to the two separate decision levels. Enterprise Miner will allow you to specify any number of decision levels. Therefore, from the decision matrix, select the **Edit Decisions...** button to specify both the assessment objective and the corresponding cost for each decision level with the added capability of including any number of decision levels in the decision matrix. The **Editing Decisions and Utilities** window will then appear. By default, the constant cost vector contains two separate rows or decision levels obtained by fitting the binary-valued target variable. In other words, the default is a squared decision matrix with a separate decision level for each class level of the categorically-valued target variable. Again, it is unlimited to the number of decision levels of the decision matrix. With the exception of decision tree modeling in fitting an ordinal target variable in the **Tree** node, there is no requirement that the number of decision levels must be the same as the number of target levels of the decision matrix.

To add an entirely new decision level, simply select the **Decision** column, right-click the mouse and select **Add Decision** from the pop-up menu items, which will result in a new decision level or row that will be added to the list of existing decisions. To rename the decision level from the decision matrix, simply select the **Decision** column and enter the appropriate decision labels into each one of the decision column entries, that is, overseas, domestic, and over-the-counter. You may also delete any of the decision levels by simply selecting any one of the rows, right-clicking the mouse and selecting **Delete Decision** from the pop-up menu items. To clear the entire table of the various listed cost constant decisions, simply select the **Clear table** pop-up menu item.

Setting the Constant Costs

For example, suppose we want to maximize profit from the associated fixed costs. Constant costs may only be specified if the decision matrix consist of strictly revenue amounts as opposed to profit or loss amounts. This is because revenue is the difference between net revenue and costs. A constant value or a data set variable can be entered for each decision in the matrix. The default is a constant cost of $0 for each decision level. The first row of the cost variable or constant cost represents the fixed cost endured or the production cost at the decision level 1. The second row represents the fixed cost endured at the decision level 2, and so on.

Let us set a fixed cost of $10 with a purchase order made overseas at the decision level of 1. Therefore, scroll over to the **COST** column and left-click the mouse to select the first row and enter the value 10 into the cell. Let us assume there exists a cost of $5 for a domestic purchase order at decision level 2. Therefore, scroll over to the **COST** column and select the second row by left-clicking the mouse and enter the value 5 into the cell, assume that there is no cost at all for an over-the-counter purchase in the last decision level. This is illustrated in the following diagram.

From the table, you may either enter a specified amount for each cost or a cost model role variable in assigning a value to the selected decision level. If you want to enter the fixed cost to the decision level make sure that the option **constant cost** is assigned to the **Cost Variable** cell that is the default. If the **Cost Variable** column is left blank, then right-click the mouse in the **Cost Variable** cell and select the **Set Cost Variable** pop-up menu item. The first option from the pop-up menu items will display every variable in the input data set that has been previously assigned a **Cost** model role to assign to the corresponding decision level. That is, the values from the cost model role variable in the input data set may represent the amount of cost for the decision level. The purpose of the cost variable is that the assessment costs might be quite different for each record in the data set. Therefore, the cost variable will represent the separate costs incurred for each record in the data set. The second option is to select the **constant cost** option to enter an appropriate cost for the decision level. Select the **constant cost** option and enter the appropriate costs located in the adjacent **Cost** column cell. The last option to select from the pop-up menu items is to clear the selected **Cost Variable** cell. If there are no fixed costs assigned to each decision level, then the decision matrix defaults to a profit matrix.

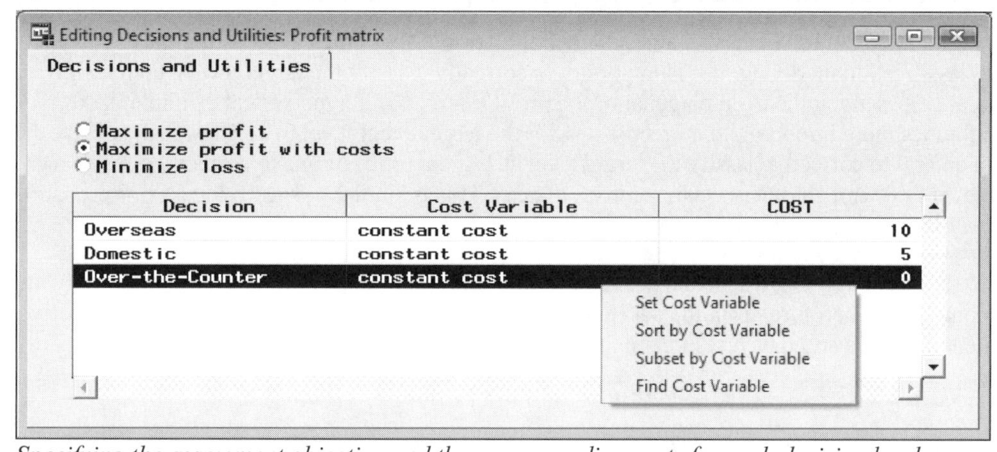

Specifying the assessment objective and the corresponding costs for each decision level.

To save the changes of the assessment objective and the corresponding costs for each decision level, select the upper right-hand button or select the **File > Close** main menu options to close the **Editing and Decisions and Utilities** window. A dialog box will then appear to confirm that you want to save the corresponding changes with the dialog message of "Do you want to Save Decision and Utilities Changes?" Select **Yes** to save all the changes that have been made. The **Editing Decisions and Utilities** window will close and return you back to the **Target Profiles** window. The decision matrix will be updated to reflect the corresponding selections that were performed in the **Editing and Decisions and Utilities** window.

The next step is specifying the various revenue amounts since we have added the additional decision level with the various associated fixed costs of production. From the **Target Profiles** window, we can enter the amount of revenue in production. From the **Target Profiles** window, we can enter the amount of revenue in production to the first row. Let us say that the net revenue of the widget is $90 with an overseas purchase at the first decision level. Therefore, select the first row and first column and enter 90 in the cell. Let us assume that there is a net revenue of the widget of $50 based on a domestic purchase at the second decision level and a net revenue of $10 with a over-the-counter purchase in the last over-the-counter decision level. If the decision values are a part of the input data set, then you might want to calculate the average net revenue or median net revenue or profit for each decision level by selecting the best decision values for the predictive model.

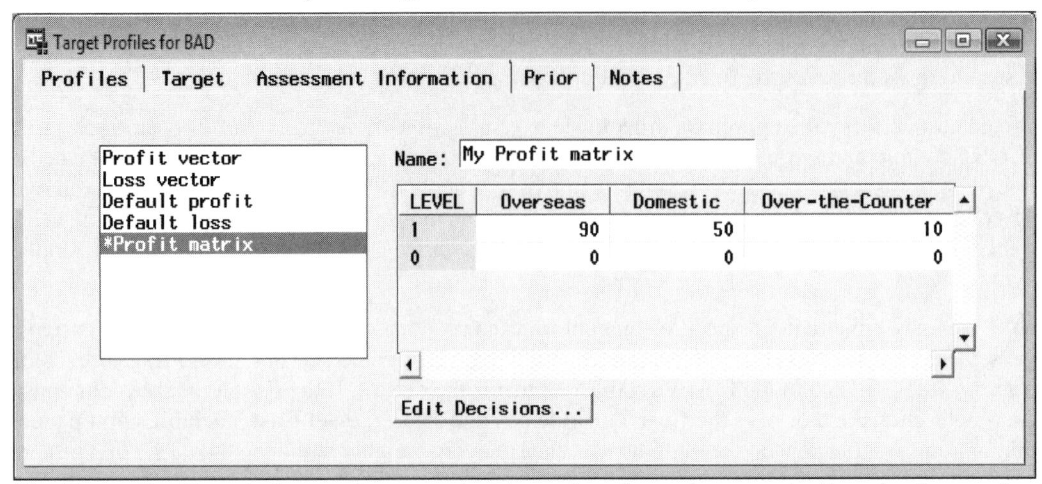

*The **Target Profiles** window used to enter decision values for each categorically-valued target level.*

Saving the Changes to the Decision Entries and Prior Probabilities

To save the changes that have been made to the decision matrix, select the upper right-hand button to close the **Target Profiles** window or select the **File > Close** main menu options and a dialog box will then appear to confirm that you are saving the corresponding changes.

Diagonal Decision Matrix in Adjusting the Estimated Probabilities

A diagonal decision matrix can be used as an alternative to the prior probabilities in adjusting the estimated probabilities. In classification modeling, if it is important to correctly classify the target levels, then simply enter the numerical values along the main diagonal of the profit matrix. As a hypothetical example, let us assume that it is 9 times more important to correctly classify the target event level of 1. Therefore, specify the profit matrix and enter 9 to correctly classify the target level at 1 and a profit of one to correctly classify the target nonevent level of 0 with all other decision entries of zero. This is similar to specifying a prior probability of .9 for the target event and .1 for the target nonevent.

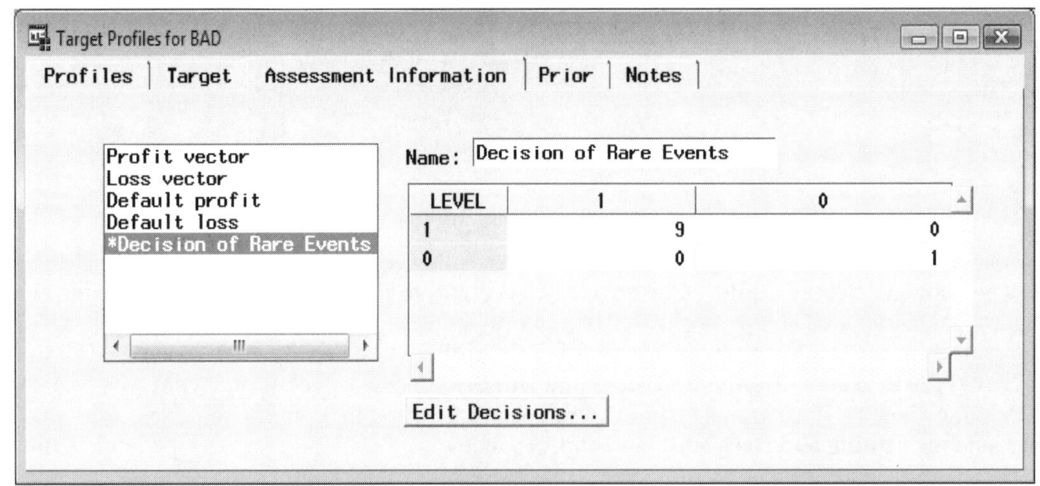

*The **Profit Matrix** used to specify the decision values for the categorically-valued target levels.*

Prior tab

The purpose of the **Prior** tab is to specify the frequency percentages of each class level of the target variable called prior probabilities. By default, prior probabilities are not applied to the modeling design. The prior probabilities represent the expected probability distribution of the true population. In classification modeling, it is not only important to be concerned about the accuracy of the model, but it is also extremely important that the sample that the current model is fitting is an accurate representation of the population. Prior probability should always be specified if the sample proportions from the input data set differ significantly in comparison to the original data set. For example, prior probabilities are highly recommended and very effective when the target variable consists of very rare events or extremely infrequent categories, for example, credit card fraud, defaulting on a loan, purchasing rare items, and so on. It is important that the predicted probabilities be adjusted accordingly by the prior probabilities in achieving predicted values that accurately reflect the true distribution that the model is trying to fit. Usually, the input data set is biased with respect to the original population, since oversampling will often occur within the target categories of rare events. Therefore, prior probabilities are applied to adjust the probability values of each target level back to the probabilities in the underlying data. This will lead to a significant increase in the modeling performance of the classification model assuming that the correct prior probabilities have been specified. The prior probabilities are usually selected by prior knowledge, previous studies, or general intuition.

The prior probability adjusts the corresponding posterior probability at each class level of the target variable as follows:

Adjusted $p_i = \frac{\pi_i}{p_i}$ for some prior probability π_i and posterior probability p_i at the ith target class level.

The three options that are available in specifying the prior probabilities within the **Prior** tab are **Equal probability**, **Proportional to data**, and **None**. The default is **None**. In other words, there are no prior probabilities assigned to the target levels. The **Equal probability** option assigns the prior probabilities of equal proportion to the class level of the target variable by the number of target groups. The **Proportional to data** option assigns the prior probabilities based on the frequency percentages of each target level. The default prior probabilities assigned to the categorically-valued target variable can be specified from the project level. At the project level, the default settings to the prior probabilities may be specified by selecting the **Options > Project > Data Profiles...** main menu options, then selecting the **Target Profile** tab and the **Class** subtab. The baseline probability from the lift charts or performance charts is defined by these same prior probabilities. The purpose of the lift charts is to evaluate the performance of the classification model for binary-valued targets. If there are no prior probabilities specified from the **Prior** tab, then the response rate of the target event from the validation data set is used as the baseline probability of the performance charts called lift charts. The baseline probability is the probability estimate without the existence of the classification model.

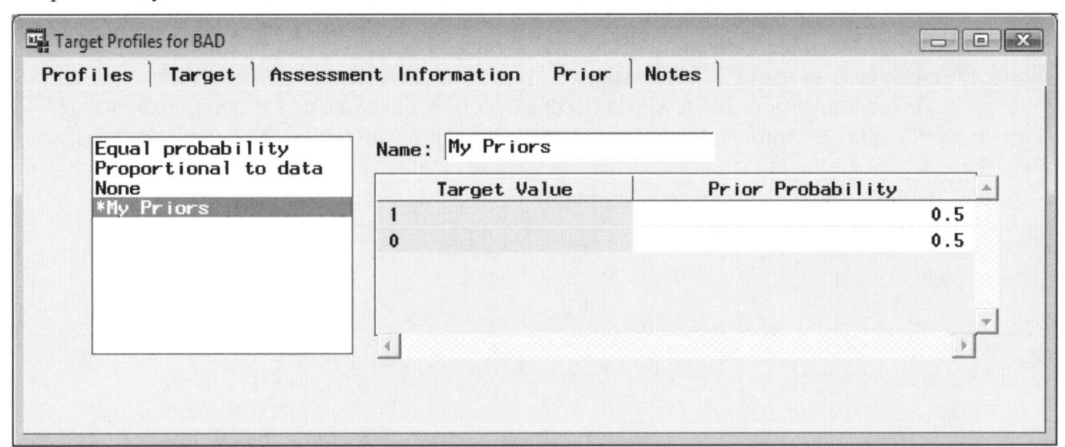

*The **Prior** tab is used to assign prior probabilities to the target classes based on some given probabilities.*

Note: The previous prior probabilities that were specified to adjust the estimated probabilities is equivalent to selecting the **Equal probability** option and specifying an equal probability to each class level of the categorically-valued target variable.

Specifying Prior Probabilities for the Target Levels

To create an additional prior vector to adjust the fitted values by the prior probabilities, highlight the row to select the probability distribution applied to the target variable. Right-click the mouse and the identical pop-up menu items will appear from the **Target Profiles** window of **Add, Delete, Copy** or **Set to use**. In the following example, we will be entering the prior probabilities for each target level. Therefore, select the **Add** option. The next step is to set the probabilities as the default prior probabilities assigned to the categorically-valued target variable by selecting the **Set to use** pop-up menu item. The **Target Profiles** window will remain open until the various prior probabilities that are entered add up to one. To copy an existing prior vector, select any one of the existing prior vectors that are currently displayed, then right-click the mouse and select the **Copy** pop-up menu item. To delete a custom-designed prior vector, simply select any one of the existing prior vectors, then select the **Delete** pop-up menu item. However, the predefined prior vectors in Enterprise Miner or the active prior vector cannot be deleted.

In the following example, let us assume that the sample proportions at each target levels from the given sample are not the same as the original data. For example, suppose there was an average cost of production of $5 per unit with a net revenue of $20. Therefore, the optimum prior probability or the cutoff probability will be at 20%, which is defined by the following theoretical approach:

Decision Matrix		Decision	
		1	0
Actual Response	1	δ_{1c}	δ_{1m}
	0	δ_{2m}	δ_{2c}

Again, one theoretical approach in selecting the best threshold probability for a binary-valued target variable is based on Bayes rule. For binary-valued target variable, the threshold value or cutoff probability θ that is considered to be the best prior probability of the target event is calculated as follows:

$$\theta = \frac{1}{1 + \dfrac{\delta_{1c} - \delta_{1m}}{\delta_{2c} - \delta_{2m}}} \quad \text{for some threshold value } \theta$$

From Bayes rule, the theoretically best threshold probability of the target event for a binary-valued target variable is calculated as follows:

$$\theta = \frac{1}{1 + \dfrac{20 - 0}{0 - (-5)}} = .2$$

Since we will be entering our own prior probabilities, the first step is creating a new prior vector for the list of existing prior probabilities. From the **Prior** tab, select from the list of available prior probabilities, then right-click the mouse to select the **Add** menu item to assign new prior probabilities. A new prior vector will be created to enter the corresponding prior probabilities. In this example, select the added option called **Prior vector** to enter the following prior probabilities. To assign the prior probabilities as the default probability distribution of the categorically-valued target variable, right-click the mouse and select the **Set to use** pop-up menu item from the newly created vector of prior probabilities.

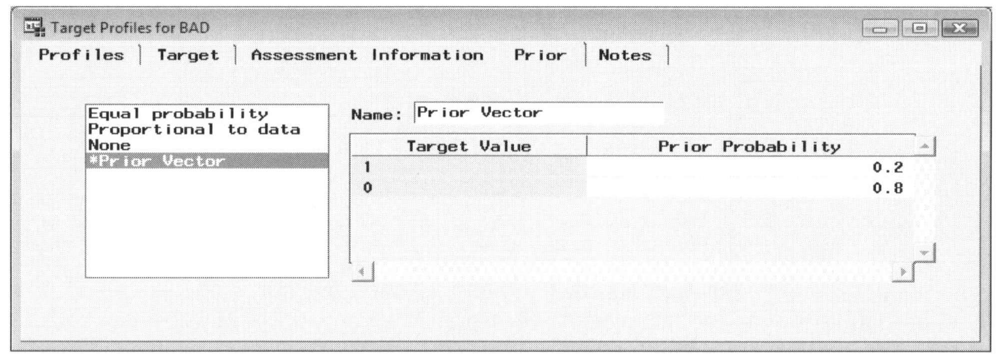

*The **Prior** window used to enter prior probabilities for each level of the binary-valued target variable.*

The next step is to enter the net revenue and fixed costs for the added decision matrix. From the **Assessment Information** tab, select the **Default profit** matrix to create a new decision matrix by right-clicking the mouse and selecting the **Add** pop-up menu item. The next step is entering the revenue and cost amounts in the first column of the decision matrix, that is, the first row and first column represent the net revenue of $20 and the second row and first column represent the average cost of production of $5. Therefore, select the first row and column and enter 20 into the cell and select the first row and second column and enter –5 into the cell that is illustrated in the following diagram.

Note: Alternatively, you could specify the profit amounts of $25 and $0 from the decision matrix and then specify a fixed cost of $5 for Decision 1, that is, Buy, and $0 for Decision 0, or No Buy, from the **Editing Decisions and Utilities** window.

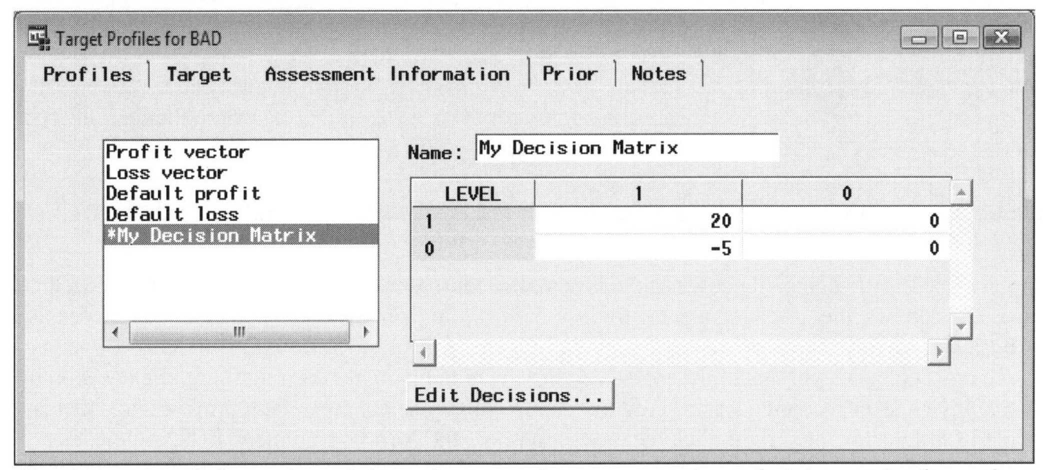

*The **Assessment Information** tab is used to enter the revenue and cost decision entries for each target level to the decision matrix or profit matrix.*

Again, in this hypothetical example, there are no fixed costs involved. Therefore, it is not necessary to open the **Editing Decisions and Utilities** window to set the assessment objective to the decision matrix. Therefore, select the upper right-hand button or select the **File > Close** main menu options to close the **Target Profiles** window to save the changes.

The Loss Matrix

At times, the objective in measuring the accuracy of the predictive model is achieving the lowest misclassification rate. In addition, assuming that there are costs involved to each decision, then the objective to the analysis is minimizing the expected loss. Again, the home equity loan data set will be presented in explaining the following loss matrix. The default loss matrix will result in the correct misclassification rate at each decision with zeros on the main diagonal entries and ones everywhere else. The target variable in this example is based on the binary-valued target variable of either defaulting on a home loan or not with the corresponding business decision of either accepting or rejecting the creditor from this same home loan. In this hypothetical example, let us assume that it is ten times more disastrous to accept a bad creditor than to reject a good creditor. The following loss matrix displays the corresponding decisions that are made about the good or bad creditor at each decision.

The steps in creating the loss matrix are very similar to creating the previous profit matrix. In other words, instead of selecting the **Default profit** matrix from the list box, select the **Default loss** matrix from the list box. The loss matrix will be automatically appended to the list of existing decision matrices. The next step is entering the appropriate decision values into the listed matrix and entering an appropriate name to distinguish between the various listed decision matrices that are displayed in the list box to the left of the tab. The next step is defining the **Edit Decision …** button that will display the **Editing Decisions and Utilities** window to set the assessment objective to the decision matrix. Select the **Minimize loss** radio button to set the assessment objective to the decision matrix in minimizing the expected loss. In addition, you might want to enter appropriate labels for the decision levels of the loss matrix. Close the window and return to the **Assessment**

Information tab. The following illustration displays the corresponding loss matrix in accepting or rejecting good or bad creditors.

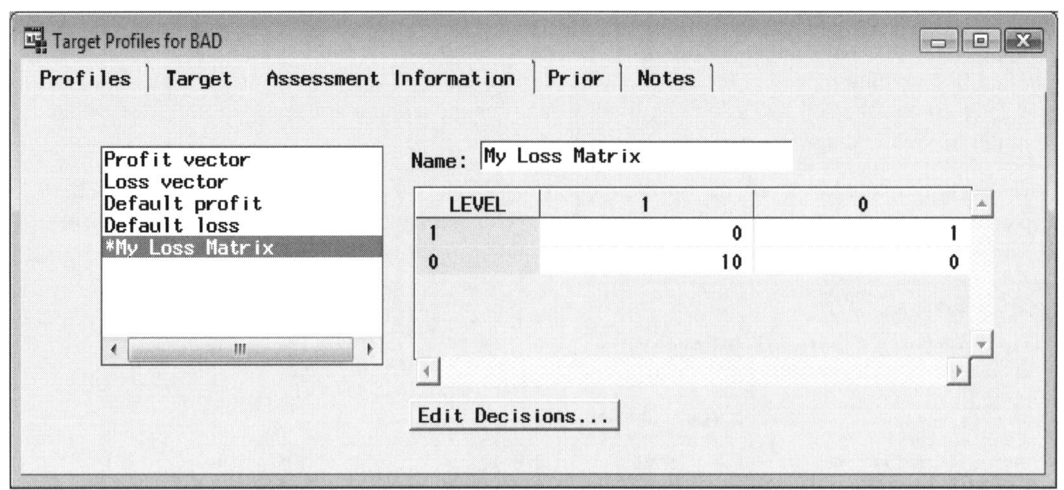

The loss matrix of either accepting or rejecting good or bad creditors from the HMEQ data set.

In our example, since the off-diagonal entries of the loss matrix represent the worse case scenarios at each corresponding decision of either rejecting a good creditor or accepting a bad creditor, this will always result in a loss to the business. Therefore, a more realistic scenario in generating a profit from the model is to specify a loss vector with one decision level that represents accepting the home loan, that is, redefining the decision by specifying a negative loss in accepting a good creditor. Again, simply select the default profit vector, that is, the matrix called **Loss vector**, then right-click the mouse to select the **Add** pop-up menu item. A new loss vector will be created that will allow you to assign this additional loss vector as the active profile to the process flow and enter the appropriate decision values to the corresponding loss vector.

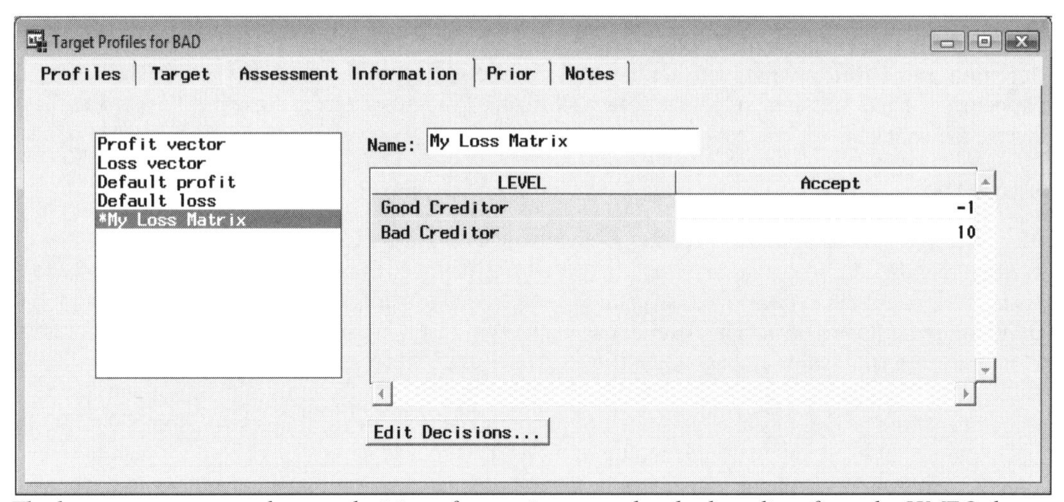

The loss vector to enter the cost decision of accepting a good or bad creditor from the HMEQ data set.

Note: Both loss matrices will result in the same decisions with the same models selected. However, the assessment statistics such as the expected loss will be easier to interpret from the loss vector with a single decision level that is based on all applicants who have been accepted for a home loan.

Defining the Profit Function for Interval-Valued Targets

The *profit function* is based on an interval-valued target variable in the active data set. The default profit function is defined by the number of target intervals associated with a single column of decision values. The default settings of the profit function, such as the minimum and maximum value of the profit function and the number of intervals, can be specified from the project level by selecting the **Options > Project > Data Profiles...** main menu options, then selecting the **Target Profile** tab and the **Interval** subtab. The default settings of the profit function will have two rows obtained by fitting the interval-valued variable that is split into two separate nonoverlapping intervals and a single decision level with decision values of zero and one. The default profit function is grayed-out and, therefore, cannot be changed. In the following profit function, by fitting the interval-valued target variable DEBTINC from the SAMPSIO.HMEQ data set, by default the target values between the interval of [0, .52] are assigned a decision value of zero. Conversely, the target values between the interval of (.52, 203.31] are assigned a decision value of one. This will result in every record assigned a decision value of one since the minimum value is .52 and the maximum value is 203.31. From the **Assessment** node, since the cutoff value is computed by calculating the average of the maximum and minimum value of the first decision level of the decision matrix, this will result in a default cutoff value of 101.92 in the lift plots. For example, all observations with actual profits greater than the cutoff value are assigned as the target event with the remaining observations assigned as the target nonevent. The default profit function will represent profit values. However, the profit function may also represent profit with costs, that is, revenue or losses.

From the **Assessment Information** tab, the steps in creating a new profit function are similar as before. That is, highlight the profit function, then right-click the mouse and select the **Add** pop-up menu item that will display the profit function illustrated in the following diagram in order to redefine the configuration settings of the profit function and the corresponding decision values. For interval-valued targets, there is no active profile. **None** is the default. However, by creating the profit function, the tab will allow you to generate the various performance plots for assessment by setting the default profit function as the active profile. This will result in the creation of the various lift charts that can be viewed within the **Assessment** node.

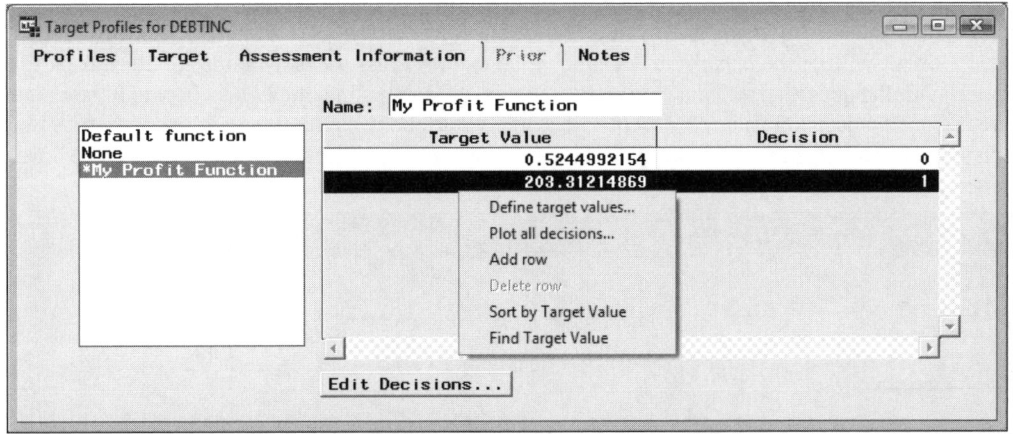

The profit function with two separate nonoverlapping intervals and a single decision level based on the interval-valued target variable.

Redefining the Intervals of the Profit Function

To redefine the upper and lower bounds of the target values or the number of intervals or rows of the profit function, simply select **Target Value** column of the profit function, then right-click the mouse and select the **Define target values...** pop-up menu item with the following **Define Target Values** window appearing. The window is identical to the default settings where you may enter the same entries from the **Interval** subtab at the Enterprise Miner project level. By default, the minimum and maximum values are set at the range of values in the active data set, and the number of points is automatically set to 2. The **Minimum** entry field sets the lower bound to the target values. Conversely, the **Maximum** entry field sets the upper bound to the target values. If the minimum target value is greater than the maximum target value, then the window will remain open. Select the **Number of points** radio button to set the number of intervals or rows to the profit function from the entry field. The default is two. The window will remain open if the number of intervals entered is less than two. Select the **Increment** radio button and enter the incremental value of each interval from the entry field. In other words, the incremental value is added to the lower bound within each interval that defines the range of target values for each interval. In our example, we will apply three separate intervals to the profit function. Select the **Number of points** radio button and set the number of intervals to three. Close the window and the profit function will result in three separate intervals of [0, .52], (.52, 101.91], and (101.91, 203.31] that will not be shown.

Define Target Values

Minimum: 0.5244992154

Maximum: 203.31214869

Number of points 2

Increment

OK Cancel

*The **Define Target Values** window is used to redefine the intervals and the minimum and maximum decision values of the profit function.*

Redefining the Minimum and Maximum Decision Values

To redefine the minimum or maximum decision value of each decision level within the profit function, simply select the appropriate **Decision** column of the profit function, then right-click the mouse and select the **Define function values...** pop-up menu item with the following **Define Function Values for Decision** window appearing. From the **Minimum** entry field, simply enter the minimum decision value of the selected decision level. Conversely, enter the maximum decision value of the selected decision level from the **Maximum** entry field. In our example, we will set the maximum decision value to 100. Press the **OK** button to return back to the profit function with the updates applied accordingly. That is, all debt-to-income values between the range of 0 and .52 are assigned a decision value of 0, debt-to-income values between the range of .52 and 101.91 are assigned a decision value of 50, and all debt-to-income values between the interval of 101.91 and 203.31 are assigned a decision value of 100.

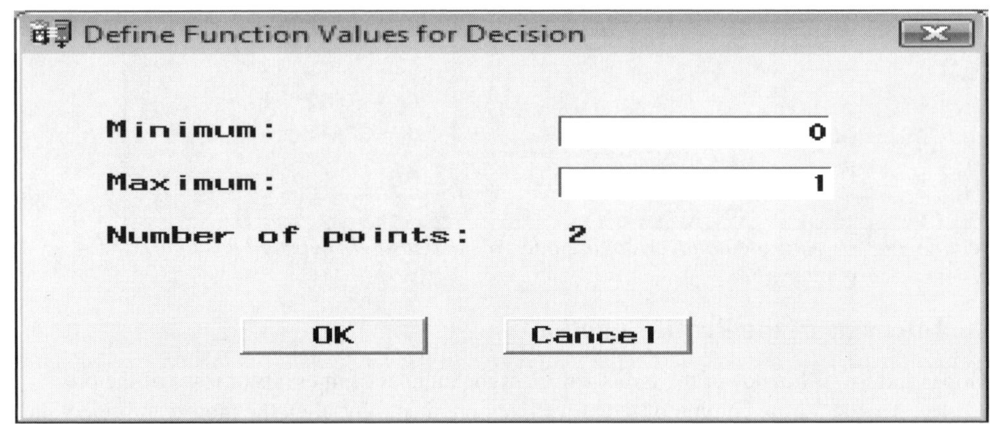

*The **Define Function Values for Decision** to specify the minimum and maximum decision values.*

Entering New Decision Values

From the profit function that we just created, if the minimum or maximum decision values are incorrect, then simply scroll over to the **Decision** column of the profit function and select the appropriate decision cell to enter the corresponding decision value to change.

Adding Additional Intervals to the Profit Function

To add an additional interval to the profit function, simply scroll over to either column of the profit function, right-click the mouse and select the **Add Row** pop-up menu item that will result in a blank row appearing below the selected row. By default, both the **LEVEL** and the **Decision** column will contain missing target values and decision values. Missing target values and decision values are inappropriate in the decision matrix. Therefore, type the appropriate target value and decision value into the corresponding cells.

Deleting Rows from the Profit Function

To delete any interval or row from the profit function, simply select the row to be deleted by right-clicking the mouse and selecting the **Delete Row** pop-up menu item.

Decision Plot of the Decision Values and the Target Values

A line plot can be created that displays the decision values or profits across the range of the target values. To display the plot, select the **Target Value** column, then right-click the mouse and select the **Plot all decisions...** pop-up menu item. The decision plot displays the decision values on the vertical axis with the range of the target values displayed on the horizontal axis from the following **Profit function** window for each decision level. Alternatively, select the appropriate **Decision** column, then right-click the mouse to select the **Plot decision...** pop-up menu item that will then plot each decision separately. The line plot may represent profit (default), profit with costs, or loss. Therefore, the following line plot will allow you to perform linear approximations of the decision values across the range of the various target values for each decision level. In addition, linear extrapolation may be performed by estimating the various decision values outside the range of the target values from the piecewise linear spline function.

Each box that is displayed in the graph represents each decision value from the decision matrix. To adjust the decision values from the decision plot, select any one of the boxes that represents each data point in the decision plot and move the box to a new decision value. Close the graph and the **Profit function for LEVEL** window and the corresponding changes will then be reflected in the profit function. Alternatively, you may make the appropriate changes in the decision matrix, then open the decision plot to view the corresponding changes.

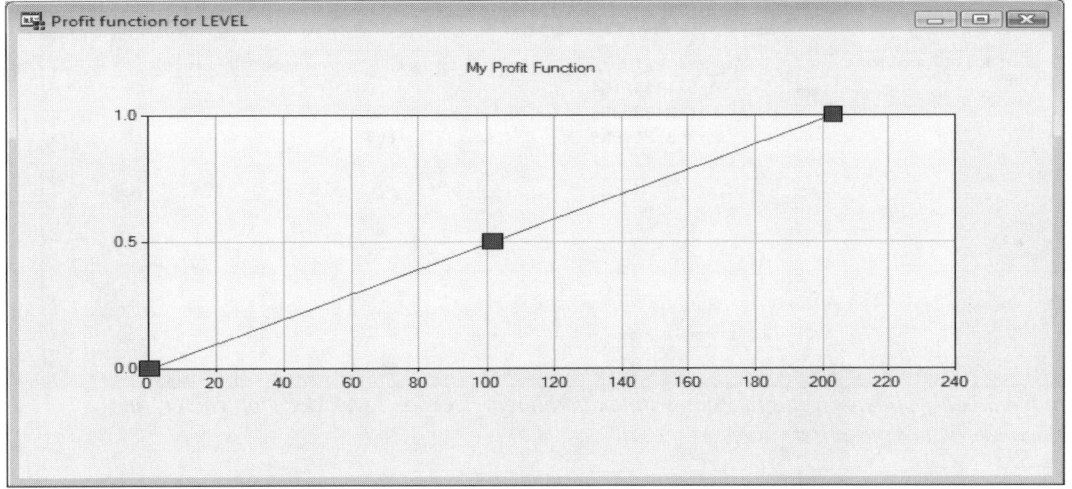

*The **Profit function for LEVEL** window with a line plot of the decision values across the interval-valued target values for each decision level.*

Adding Additional Decision Levels to the Profit Function

The profit function is not just limited to a single decision level, therefore, press the **Edit Decision...** button and the **Editing Decisions and Utilities** window will appear. The steps are the same as before in redefining the profit function and incorporating fixed costs for each decision level, adding additional decision levels, and specifying the assessment objective to the profit vector that has been previously explained. Again, it is important to understand that the profit function is not restricted to a single decision level. In other words, you may redefine the default profit function with an unlimited number of decision levels or columns for the interval-valued target variable that is illustrated as follows:

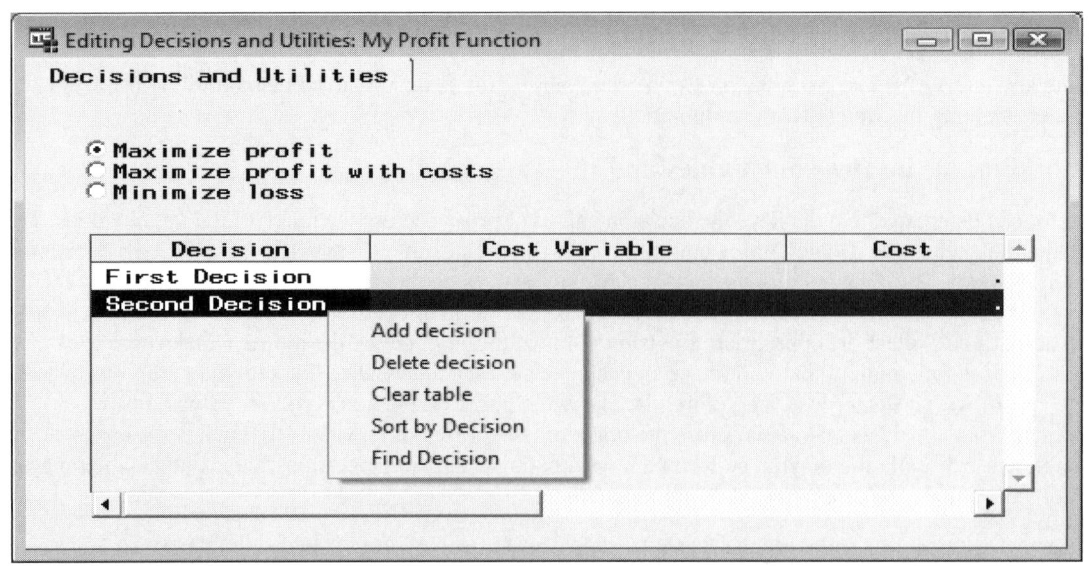

*The **Editing Decisions and Utilities** window to specify the assessment objective, additional decision levels and costs to the decision matrix.*

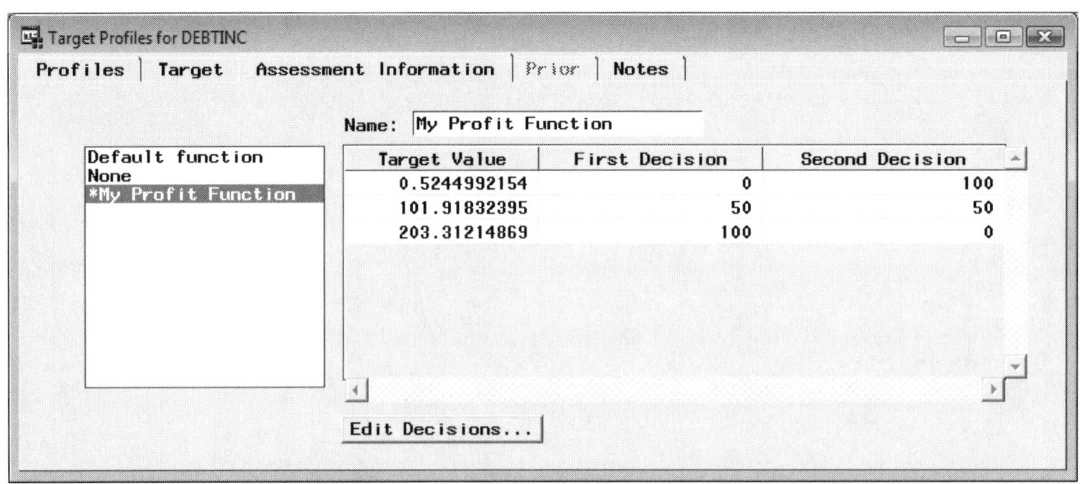

The automatically updated profit function with an additional decision level, decision values, and redefined interval of the target values.

Locked Diagrams

In Enterprise Miner, multiple users can work on the same project simultaneously. However, only one user may open a diagram at one time. When you open an Enterprise Miner diagram, a lock file is created to prevent more than one person from performing modifications to the same diagram at the same time. The purpose of the lock file is to prevent two separate users from opening the same diagram. Therefore, if SAS terminates abnormally, then the lock file for the corresponding project is not deleted and the lock will remain on the diagram. This will prevent you from opening the same diagram in a subsequent SAS session. Therefore, in order to unlock the diagram and gain access to the same locked diagram in Enterprise Miner you must delete the .LCK lock file within the EMPROJ folder, The EMPROJ folder is automatically created below the project folder that is created by Enterprise Miner. The directory structure that is automatically created by Enterprise Miner can be viewed from the **Reporter** node.

Calculating the Expected Profit from the Binary-Valued Target

Since we have now created a profit matrix for the Enterprise Miner diagram, the final parameter estimates can now be determined from the objective function of either a maximum expected profit or minimized expected loss based on the target-specific decision consequences and the prior probabilities at each level of the categorically-valued target variable. The following profit matrix is obtained by fitting the binary-valued target variable with two separate values of Yes or No and three separate decisions, where δ is defined as the net revenue values, p posterior probabilities, and c incurred fixed costs.

Profit Matrix

		Decision		
		1	2	3
Response	Yes	δ_{11}	δ_{12}	δ_{13}
	No	δ_{21}	δ_{22}	δ_{23}

Posterior Probabilities

		Probability
Response	Yes	p_1
	No	p_2

Calculation of the Expected Profit

		Decision		
		1	2	3
Response	Yes	$\delta_{11} \cdot p_1$	$\delta_{12} \cdot p_1$	$\delta_{13} \cdot p_1$
	No	$\delta_{21} \cdot p_2$	$\delta_{22} \cdot p_2$	$\delta_{23} \cdot p_2$
	Sum	$\delta_{11} \cdot p_1 + \delta_{21} \cdot p_2$	$\delta_{12} \cdot p_1 + \delta_{22} \cdot p_2$	$\delta_{13} \cdot p_1 + \delta_{23} \cdot p_2$
	Cost	c_1	c_2	c_3
	Profit	$(\delta_{11} \cdot p_1 + \delta_{21} \cdot p_2) - c_1$	$(\delta_{12} \cdot p_1 + \delta_{22} \cdot p_2) - c_2$	$(\delta_{13} \cdot p_1 + \delta_{23} \cdot p_2) - c_3$

The *expected profit* is basically the difference between the summarized predetermined profit taken from the profit matrix entries, multiplied by the posterior probabilities, along with the reduction to the fixed costs at each decision level. The *maximum expected profit* is then defined as the largest profit among the three separate decision levels that can be determined by the largest value from the last row. Furthermore, if there exist two or more decisions with the same expected profits, then the first decision from the list of various decision levels is selected.

Conversely, the *expected loss*, is defined as the predetermined loss based on the loss matrix entries, multiplied by the posterior probabilities, and then summarized at each decision level. The *minimum expected loss* is then defined as the smallest loss among the separate decisions. Note that the fixed costs are not involved in the calculation of the expected loss since the fixed costs may only be specified for the decision matrix that contains the net revenues as opposed to profits or losses.

Therefore, the classification model calculates the best possible linear combination of parameter estimates that produces some combination of the posterior probabilities at each one of the target levels to determine the largest expected profit or smallest expected loss from the separate decision levels. Conversely, the best predictive model is selected that produces the largest expected profit or smallest expected loss from the separate decision levels.

Note: In Enterprise Miner, calculating the best expected profit or loss from the model will be disabled when fitting multiple target variables from the **Neural Network** node. In other words, the **Neural Network** node is the only modeling node that will fit multiple target variables.

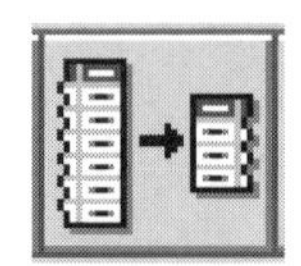

1.2 Sampling Node

General Layout of the Enterprise Miner Sampling Node

- **Data tab**
- **Variables tab**
- **General tab**
- **Stratification tab**
- **Cluster tab**
- **Output tab**
- **Notes tab**

The purpose of the **Sampling** node in Enterprise Miner is to subset the input data set into a smaller, more manageable data set. Subsetting the input data set is recommended by SAS for extremely large data sets in order to significantly reduce processing time and disk space. Assuming that the sampling is performed correctly, then the reduced sample should not have a severe impact on the accuracy of the results in comparison with the original sample. In other words, it is important that the method of drawing the random sample from the input data set will result in partitioned data that properly represent of the population of interest. The goal in sampling is taking a collection of observations from the input data set that accurately covers the entire sample. The default sampling technique that is commonly applied is a simple random sample. Also, many of the other sampling designs that can be applied incorporate a simple random sample in some way or another. The **Sampling** node either performs a simple random sample, systematic sampling, stratified sampling, cluster sampling, or first nth sampling. In addition, you may either specify a certain number of observations or a certain percentage of the input data set in drawing the sample. Again, sampling is recommended for extremely large data sets to reduce both the computational time and memory resources. An output data set is created based on the sample selected from the input data set that is passed forward through the process flow diagram. In the Enterprise Miner process flow diagram, one requirement of the **Sampling** node is that it must proceed the node that creates the input data set that is usually from the **Data Input** node.

Random Sampling

In statistics, the objective is making an inference about a population from a sample. The basic statistical assumption in a large majority of statistical analysis is the independence assumption. The common remedy to independence in the observations can be achieved by performing random sampling from the underlying population. A random sample is defined as selecting a sample of n observations from a population of N items in such a way that each observation has an equal chance of being selected from the underlying population. In random sampling, it is assumed that the sample drawn from the population is absolutely random with absolutely no bias in the responses. A *bias sample* is a sample that is not a true representation of the underlying population from which the sample is drawn. In other words, a bias sample is a sample that is not completely random. For example, a bias sample in predicting the rate of people being arrested results when you sample people who have been arrested are less likely to report the arrest than people who have not been arrested. In predicting personal income, sampling people with high incomes who are less likely to report their income as opposed to people at lower income levels will result in a bias sample. Failure to recognize serious bias or independence in the data can produce bias standard errors and inaccurate test statistics. When the sample is biased, the statistical inferences are questionable, meaningless, and unreliable.

The random seed number determines the sampling. Enterprise Miner uses an identical random seed number to select the sample from the same SAS data set that will create an identical random sample of the data set with the exception when the random seed is set to zero, which sets the random seed number to the computer's clock at run time. From the **Sampling** node, an output data set is created from the random sample selected from the input data set.

Sampling With or Without Replacement

There are two separate sampling procedures used to draw the sample, called "sampling with replacement" and "sampling without replacement". *Sampling with replacement* means that any record that is selected for inclusion in the sample has the chance of being drawn a second time. *Sampling without replacement* means that once the record is drawn, then the observation cannot be drawn again. However, in data mining, due to the existence of an enormous number of records, the difference in the results between the two sampling procedures will usually be minuscule. Typically, the observations selected and the sampling procedure applied should be such that no one observation is selected more than once, that is, without replacement, where each observation has an equal chance of being selected for the sample. In Enterprise Miner, sampling is performed without replacement.

Sampling Error

In statistics, inaccuracies in the statistical estimate may occur in one of two ways from taking a random sample of the population. There are two types of errors that arise, called *sampling error* and *nonsampling error*. Sampling error can be controlled by carefully selecting the appropriate sampling design. Nonsampling error is mainly attributed to nonresponses, inaccurate responses, or bias responses. Nonsampling error is much more difficult to control in comparison to sampling error. The first problem of nonresponses might introduce bias into the sample data. That is, responders who do not respond may not represent the population in which you would like to make an inference. For example, people who have been arrested are less likely to report it as opposed to people who have not been arrested. The second problem of inaccurate responses might arise due to error in recording, miscalculation, or malfunction in the data processing equipment. The third problem is bias responses. In other words, the various statistical estimates from the sample might be biased. For example, performing a study to determine the overall weight of people would lead to a bias sample if the data that was collected were based on the weight of women, that is, the biased estimate of the sample mean would to tend to be too small since women generally weigh less than men. Therefore, in statistical inference, the bias can either be too large or too small.

The key to making valid statistical inferences is that the sample drawn is an accurate representation of the population of interest. In statistical analysis, as a good rule of thumb, it is generally a good idea to obtain a sample as large as possible in order to improve the power of the statistical test, reduce the bias and the variability in the data, and better meet the various distributional assumptions that need to be met in the statistical tests. There are two factors that affect the accuracy of the sample drawn. The first is the size of the sample drawn. More observations drawn from the population will result in a greater chance of replicating the population of interest. The second is the amount of variability in the data, which can be controlled by the sampling technique that is selected. The sampling technique determines the way in which the data is selected.

Sampling Techniques

There are four sampling procedures that are typically applied: a simple random sample, stratified sampling, systematic sampling, and cluster sampling. *Simple random sampling* is the most common sampling procedure. A simple random sample simply selects observations from the data so that each observation has the same chance of being selected. *Stratified sampling* design is applied in obtaining an accurate representation of the population by grouping the data into nonoverlapping groups, for example, gender or income level, then selecting a simple random sample within each group. *Systematic sampling* is designed for hopefully covering the entire population by uniformly sampling the entire population and selecting the observations within an equal number of intervals. *Cluster sampling* is used when the entire population of interest in unobtainable. Therefore, various nonoverlapping cluster groups are created, then a random sample of clusters is taken and all the observations within each cluster selected. In cluster sampling, it is important to achieve an adequate balance in selecting the appropriate number of clusters and the appropriate number of observations within each cluster.

In conclusion, taking a larger sample from the population increases the probability of obtaining an appropriate representation of the population. Also, the second factor that affects the quality of the sample that is selected is based on the sampling method that is applied.

The various sampling techniques are discussed in greater detail in Cochran (1997), Scheaffer, Mendenhall, and Ott (1986), and Thompson (2002).

Data tab

The **Data** tab is designed for you to select the input data set to perform the various sampling techniques. The tab will display the data set library, data set name, data set description, and the table view or table listing of the input data set to perform the random sample. By default, the input data set is randomly sampled.

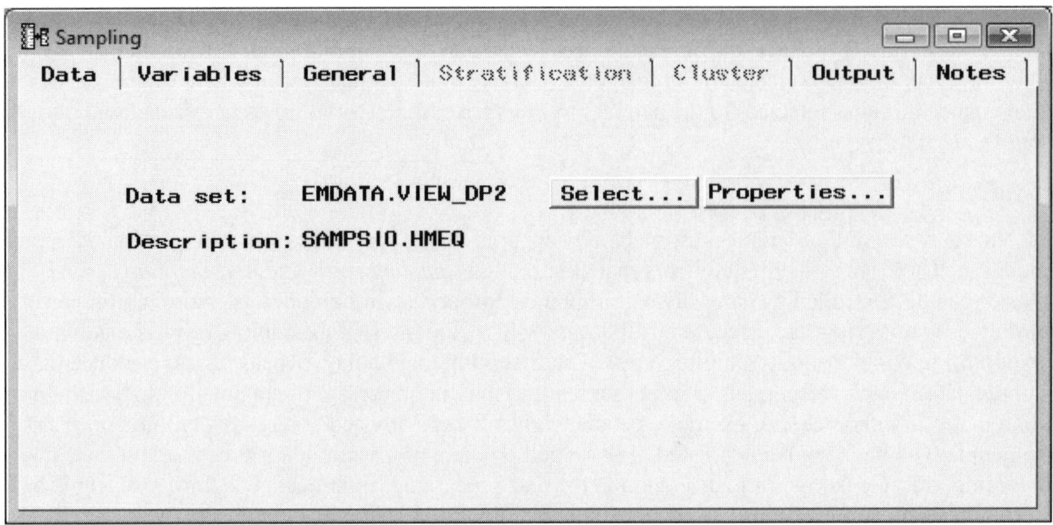

*The **Data** tab displaying the data mining data set name used to perform the sampling techniques.*

Most of the following nodes have a **Data** tab that will allow you to view both the file administrative information and a table view of the selected input data set. The EMDATA library name is automatically assigned to each currently opened project. The folder contains the data mining partitioned data sets. Therefore, when you open an entirely different project, Enterprise Miner will assign the EMDATA library name to the appropriate folder of the currently opened project. The **Description** field will display the data set name and the associated library reference name of the previously created input data set, which is usually created from the **Input Data Source** node.

Imports Map Window

The **Select...** button will be grayed-out if there exists one unique data set that has been previously connected to the **Sampling** node that is usually created from the **Input Data Source** node. However, if there exist two or more previously created input data sets that are connected to the node for sampling, then the node will automatically select one active input data set for sampling in the subsequent nodes. Press the **Select...** button and the following **Imports Map** window will appear that will allow you to select an entirely different active input data set for sampling. Simply select the plus sign (+) that is displayed next to each input data set available within the process flow, click it to expand the tree, then double-click the listed data set. Alternatively, if you are not interested in sampling the training data set, then select the **Clear** button to remove the training data set from the sampling process and the subsequent analysis. The **Role** display that is displayed below the **Imports Map** window will list the role type of the input data set of RAW. The **Selected Data** and **Description** fields will display both the data mining data set name and a short description of the selected input data set. Press the **OK** button to return to the **Sampling** window and the corresponding **Data** tab.

From the **Data** tab, select the **Properties...** button to view the file information and the table view of the input data set that is available for sampling within the node.

Note: Viewing the input data set will result in the data set being locked. This will prevent the currently opened table from being modified or updated by other users until the corresponding table is closed.

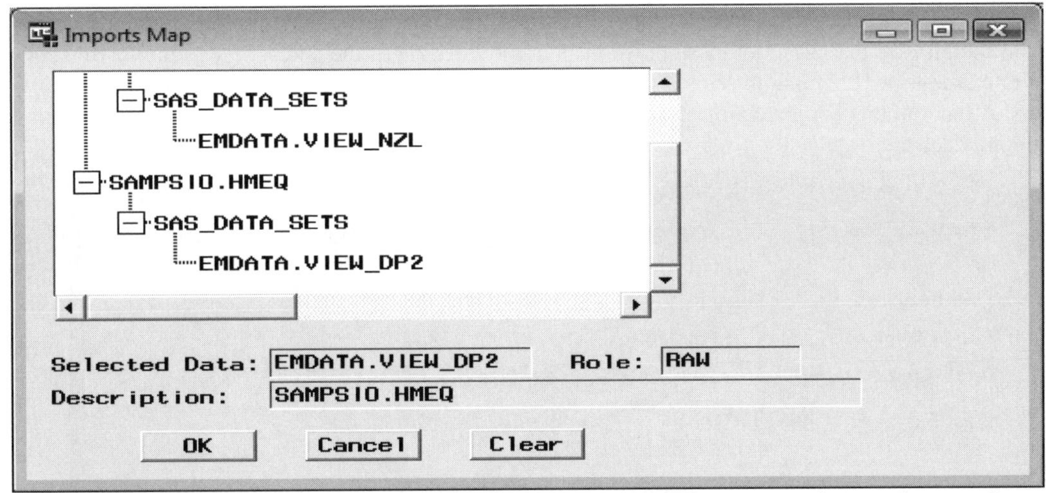

*The **Imports Map** window will allow you to select an entirely different input data set for sampling or remove the selected data set from the analysis.*

Variables tab

The **Variables** tab is designed to display the variables in the input data set that will allow you to view the frequency distribution of the variables in the data set. The tab displays the table listing of the variables in the data mining data set with its associated variable role, level of measurement, variable type, format, and variable label, similar to the **Variables** tab from the previous **Input Data Source** node. To view the distribution of a selected variable, right-click the mouse and select the **View Distribution of <variable name>** pop-up menu item that will display the standard histogram within Enterprise Miner. The metadata sample is used to create the histogram.

Sampling

Data | Variables | General | Stratification | Cluster | Output | Notes

Name	Model Role	Measurement	Type	Format	Label
BAD	input	binary	num	BEST12.	
LOAN	input	interval	num	BEST12.	
MORTDUE	input	interval	num	BEST12.	
VALUE	input	interval	num	BEST12.	
REASON	input	binary	char	$7.	
JOB	input	nominal	char	$7.	
YOJ	input	interval	num	BEST12.	
DEROG	input	interval	num	BEST12.	
DELINQ	input	interval	num	BEST12.	
CLAGE	input	interval	num	BEST12.	
NINQ	input	interval	num	BEST12.	
CLNO	input	interval	num	BEST12.	
DEBTINC	target	interval	num	BEST12.	

Sort by Measurement
Subset by Measurement
Find Measurement
View Distribution of DEBTINC

*The **Variables** tab is used to display the variables and the frequency distribution of the selected variable from the input data set.*

Note: To display the listed column entries in sorted order, simply click the appropriate column heading. To toggle the sorted entries within each column, select the same column heading a second time to reverse the sorted order of the selected column listing. In addition, you can resize any column to the tab by adjusting the column headings in the tab with your mouse to enhance the readability of the numerous column entries.

General tab

The **General** tab will automatically appear as you open the **Sampling** node. The tab is designed for you to specify the sampling method that is performed on the input data set. From the tab, you may specify the sample size by either selecting the number of observations or the percentage of allocation from the input data set. However, the sample size that is randomly sampled is an approximation of the percentage or number that is specified. For example, a 10% random sample of 100 observations may contain either 9, 10, or 11 observations. A random number generator or random seed may be specified that determines the observations that are randomly selected.

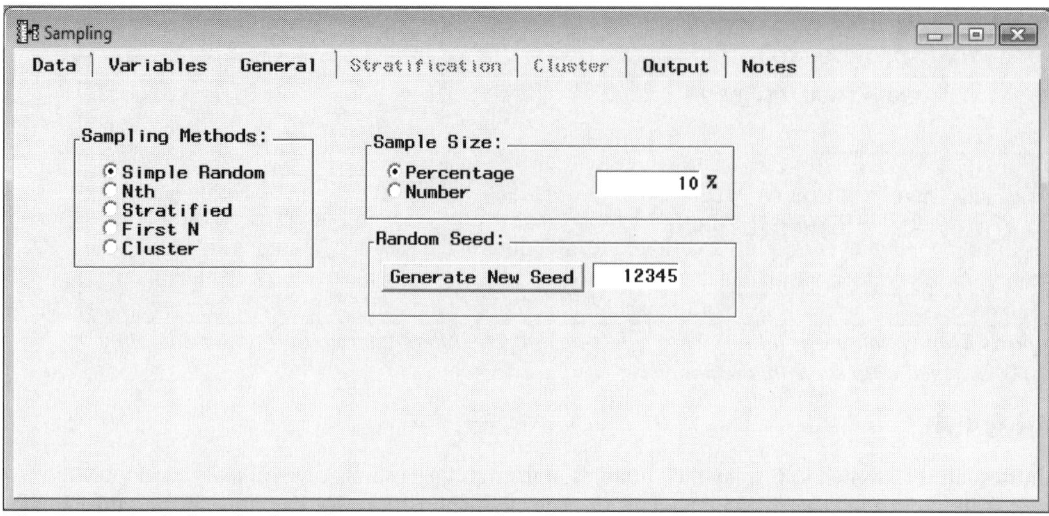

*The **General** tab that displays the various sampling designs to select for the analysis.*

The following are the various sampling techniques that are available within the **Sampling** node:

- **Simple Random** (default)**:** This is the default sampling method. Simple random sample is the most common sampling schema in which any one observation has an equal chance of being selected for the sampling data set.
- **Nth:** Systematic sampling in which every *n*th observation is selected from the input data set. That is, the starting point to begin the sample should be random and then every *n*th observation is taken thereafter. A systemic sample is designed to generally cover more uniformly over the entire input data set with the idea of providing more information about the population of interest as opposed to a simple random sample. As an example, phoning every person in the phone book might be impossible, whereas a more practical approach might be phoning every *n*th person as opposed to drawing random numbers out of a hat. However, if the input data set is designed in an alternating pattern, for instance, male followed by female, then this type of sampling will lead to disastrous results. By default, the node selects 10% of the observations from the input data set. Therefore, the node will select observations that are a multiple of (no. of observations/10%) = *n*, then randomly select a starting point from 1 to *n*, and select every *n*th observation in the data from that point on.
- **Stratified:** Stratification sampling, where the categorically-valued variables control the separation of the observations into nonoverlapping groups or strata, then taking a simple random sample from each group. Stratification sampling is recommended if the data consists of observations of rare events. In classification modeling with an ordinal-valued target variable, stratified sampling might be selected in order to ensure you that the input data set contains all the target class levels with a reasonable number of observations. For example, this type of sampling is effective if the analysis is in regard to weight gain based on sex. Since men generally tend to weigh more than women, therefore you would want to stratify the sample by sex, with weight gain being relatively similar within each sex and different between both sexes. From the input data set, you must specify a categorically-valued variable to create the separate strata or subsets. In the sampling design, each observation within each stratum has an equal chance of being selected. However, observations across the separate subsets will generally not have an equal chance of being selected. By default, this option is disabled if all the variables in the input data set are continuous.
- **First N:** Sequential sampling draws a sample at the beginning and takes the first *n* observations that are based on the number specified from the **Sample Size** option. From the tab, you may either select a percentage or a certain number of observations. By default, the node selects the first 10%

of the observations from the input data set. For instance, if the data set is sorted by zip code, then this sampling technique will result in all records sampled from the eastern United States. This type of sampling can be disastrous if the input data set is not arranged in a random order.

- **Cluster:** In cluster sampling, the observations are selected in clusters or groups. The difference between cluster sampling and stratified sampling is that you select a random sample of distinct groups, and all records that belong to the randomly selected cluster are included in the sample. Cluster sampling is designed for use when each cluster has similar characteristics and different clusters have different characteristics. Cluster sampling is advisable if it is very difficult or impossible to obtain the entire sampling population where the idea is to form various nonoverlapping groups. Cluster sampling is often performed in political elections were the sampling population that is represented by a certain city is divided into various clustering groups or city blocks and a simple random sample is taken within each city block. By default, this option is disabled when every input variable in the data mining data set is continuous since the categorically-valued variable determines the clustering groups. In cluster sampling, an optimum balance needs to be achieved with regard to the number of cluster groups to select and the number of observations within each stratum. When you perform cluster sampling, you must first select the appropriate sampling technique of either a simple random sample, a systematic sample or a sequential sample. The default sampling technique is a random sample. The next step is selecting the number of clusters based on the previously selected sampling technique.

The following are the percentages of allocation that may be specified within the **Sampling** node:

- **Percentage:** This option will allow you to specify the percentage of observations that are selected from the input data set. The default is a random sample of 10% of the input data set to be sampled in creating the output data set. For example, specifying a percentage of 50% would result in selecting half of the observations in the input data set.
- **Number:** This option will allow you to specify the number of observations to select from the input data set. The default is a sample size of 10% of the input data set. In our example, 596 observations are selected from the HMEQ data set, with 5,960 observations in the data set.

Generate New Seed

Enterprise Miner lets you specify a random seed to initialize the randomized sampling process. Randomization is often started by some random seed number that determines the partitioning of the input data set. A different random seed will create a different partitioning of the input data set. In other words, generating a new seed will select an entirely different random sample of the input data set. Therefore, an identical seed number with the same data set will create the same partition of the metadata sample based on repeated runs. The default is a random seed number of 12345. The random number generator is applied to the first three sampling designs, that is, simple random sampling, systematic sampling, and stratified sampling.

Stratification tab

The **Stratification** tab is designed for you to perform a stratified sample of the input data set. The tab will allow you to select the appropriate categorically-valued variable to perform the stratification. The **Variables** subtab will appear as you first open the **Stratification** tab. If the input data set consists of all interval-valued variables, then the **Stratification** tab is grayed-out and unavailable for viewing. However, if the **Stratified** option is selected for stratified sampling, then the following **Stratification** tab will become available for viewing in order for you to specify the various stratification configuration settings.

Variables subtab

The **Variables** subtab lists the various categorically-valued variables that are appropriate for use as stratification variables to subset the input data set. From the tab, you may select the categorically-valued variable to perform the stratification of the various levels or groups of the input data set. By default, every variable in the data set is automatically set to a status of **don't use**. To select the discrete variable to perform the stratification, select the variable row and scroll over to the **Status** column, then right-click the mouse and select the **Set Status** pop-up menu item to assign the selected variable to **use**. All other columns are grayed-out and, therefore, cannot be changed.

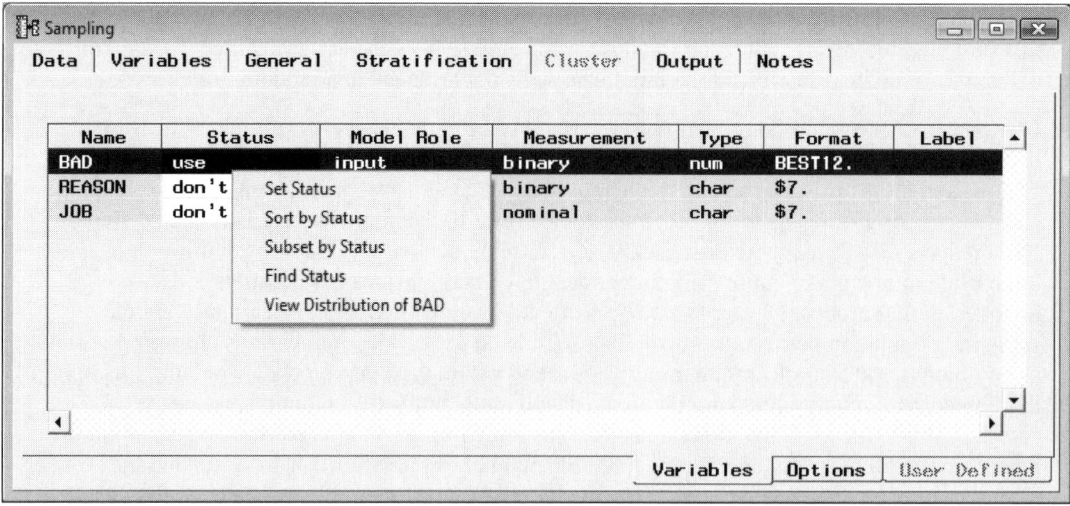

*The **Variables** subtab is used to specify the variable to perform the segmentation.*

Options subtab

The **Options** subtab will display the various stratification sample schemas for the design. These options are designed for you to avoid creating a bias sample from the stratified design. The following options will allow you perform sampling weights to adjust the proportion of the target responses in the biased stratified sample to better represent the actual proportions from the input data set. In order to achieve valid statistical results, it is important that sampling weights be applied to the stratified sample to adjust the proportion of the class levels of the categorically-valued target variable in order to accurately represent the actual target levels from the input data set.

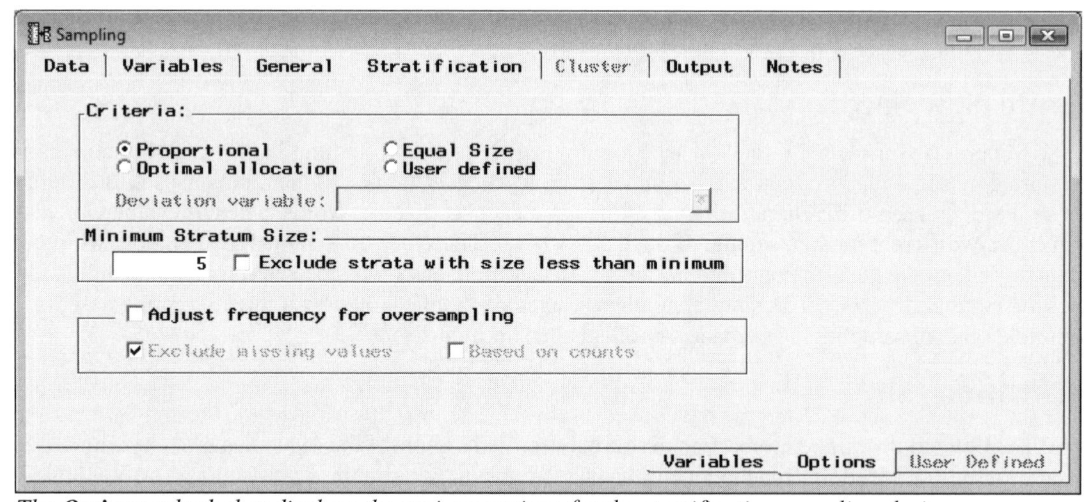

*The **Options** subtab that displays the various settings for the stratification sampling design.*

The following are the various allocation schemas for the stratification sampling design:

- **Proportional** (default)**:** The same proportion of observations are selected within each stratum in order to take a steady fraction throughout the sample.
- **Equal Size:** The same number of observations are selected within each stratum that is defined by the ratio between the total number of observations and the total number of strata.
- **Optimal Allocation:** The same proportion of observations are selected within each stratum, that is relative to the standard deviation of the specified variable based on the interval-valued variable selected from the **Deviation variable** field. Typically, the deviation variable is set to the target variable to perform the analysis that calculates the standard deviation within each stratum. The **Deviation variable** field is available for the **Optimal Allocation** option setting only.

- **User defined:** By default, this option is disabled. This option will be made available by specifying the categorically-valued variable in the data mining data set to perform the stratification or partitioning of the input data set. Selecting this option will result in the **Variable Levels Retrieval** window appearing, as illustrated in the following diagram. The window is designed for you to specify the class levels of the variable and the percentage of allocation that is based on the **Metadata sample** or the **Training data** set. Selecting the appropriate data mining data set will automatically generate the user-defined sample.

- **Minimum Stratum Size:** This option determines the minimum amount of data allocated within each stratum. Selecting the **Exclude strata with size less than minimum** option will result in all strata being removed from the selected sample for which the number of observations within the strata is less than the specified amount. By default, this option is unchecked. The **Minimum Stratum Size** option is associated with both the **Proportional** and the **Equal Size** option settings. Simply enter the minimum number of observations for each stratum from the entry field.

- **Adjust frequency for oversampling:** This option is designed to prevent oversampling that will allow you to specify a frequency variable that represents the sampling weights. By default, the check box is not selected. This will prevent the **Sampling** node from adjusting the active training data set for oversampling. This option is based on the categorically-valued target variable that determines the stratification needed to perform the oversampling adjustments. If the input data set contains a variable with a frequency variable role, then the variable is used to adjust for oversampling in the data. If there are no variables in the active training data set mining data set with a **freq** variable role, then the node will automatically create a frequency variable to perform the oversampling adjustment. If the prior probabilities have been specified from the target profile, then the prior probabilities are used to adjust the frequencies for oversampling. The amount of adjustments that is applied depends on how biased the sample is in comparison to the input data set. In statistical modeling designs, it is important that the biased sample be adjusted in order for the sampling proportions of the target class levels to accurately reflect the target proportions from the original data.

  Selecting the **Exclude missing values** option will result in the node excluding all observations with missing values from the stratified sample and assigning these missing records a weight of zero. When the **Exclude missing values** option is not selected, the weights are adjusted for all observations with missing values and the prior probabilities that are assigned to the stratification variable will be ignored.

  By default, the frequency weights are calculated from the proportion of each class level of the categorically-valued strata variable from the input data set. Selecting the **Based on counts** option will adjust these weights based on the actual frequency counts for each class level.

By selecting the user defined sampling technique from the **Criteria** section, the following **Variable Levels Retrieval** window will appear. The window will allow you to determine whether to sample the metadata sample or the input data set to determine the class levels of the categorically-valued stratification variable and associated percentages that are retrieved from either data set. This selection is used to generate the values that are displayed within the **User Defined** subtab. Both the class levels and the allocation percentages for each class level of the categorically-valued stratification variable are calculated from the selected data set. By default, the metadata sample is automatically selected to determine both the class levels and the allocation percentages of each class level of the stratified categorically-valued variable. However, the training data set might be selected in order to avoid oversampling in the data due to certain class levels of the stratified categorically-valued variable with an insufficient number of records. In other words, the training data set might be selected that will usually result in a better chance that the sample will adequately represent the true distribution of the population of interest such as identifying the rare class levels of the stratified categorically-valued variable. In stratified sampling, it is important that there are a sufficient number of observations within each group of the stratified categorical variable to better meet the various distributional assumptions, obtain stable and consistent estimates, and achieve more accurate estimates in the analysis.

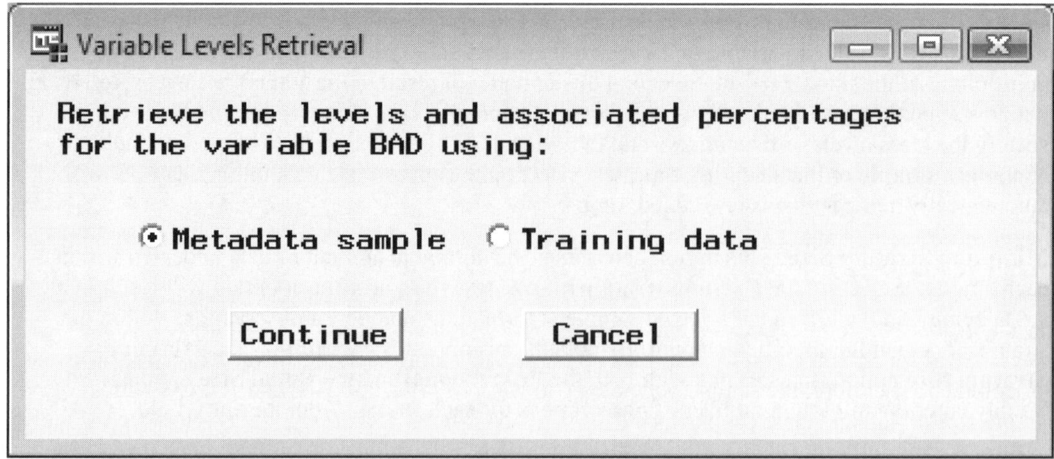

*The **Variables Level Retrieval** window is used to select the training data or metadata sample for sampling.*

User Defined subtab

The **User Defined** subtab is designed for you to specify the percentage of allocation by each class level of the categorical variable that will perform the stratification of the input data set. You may either select the metadata sample or the training data set to apply the stratification sampling technique.

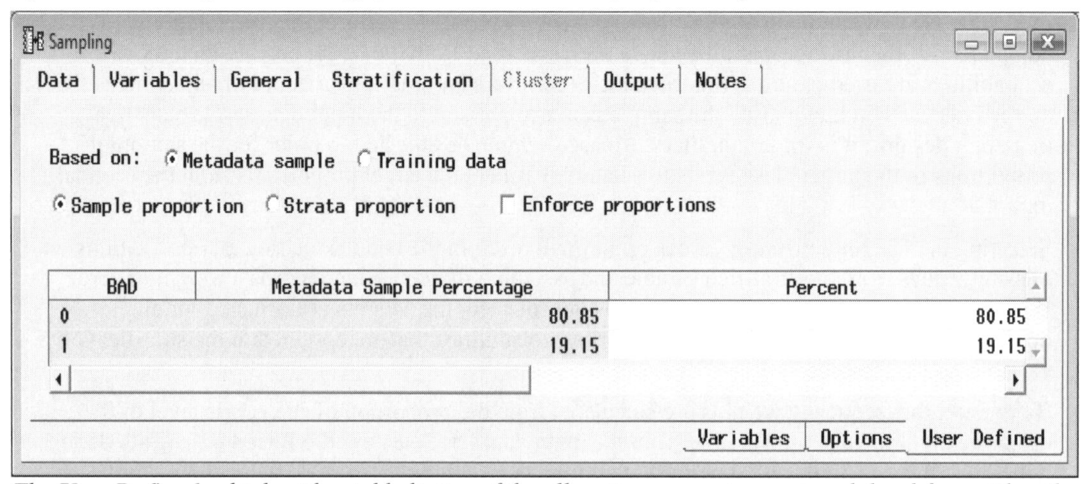

*The **User Defined** subtab with a table listing of the allocation percentages at each level from either the metadata sample or the input data set.*

The following options are based on the allocation proportions of the user-defined sampling technique from the metadata sample or the training data set that you have previously selected:

- **Sample Proportion** (default)**:** This is the default method that is used in determining the allocation percentages of the user defined sampling design. If the **Sample Proportion** check box is selected, then the allocation percentages that are displayed within each class level of the **Percent** column are based on the sample proportions from each class level of the categorically-valued stratified variable. The allocation percentages are calculated from either the previously selected metadata sample or the training data set. From the **Percent** column, you may enter the appropriate proportions for each class level. However, the restriction is that the column percentages that are entered must add up to 100%.
- **Strata Proportion:** If the **Strata Proportion** check box is selected, the percentages that are displayed within each class level of the **Percent** column are determined by the allocation percentage that is specified from the **Sample Size** option setting within the **General** tab. Again, the allocation percentages that are displayed within the **Percentages** column are calculated from either the metadata sample or the training data set. From the **Percent** column, you may enter the separate percentages for each stratum. The restriction is that the percentages must add up to 100%.

- **Enforced Proportions:** This option determines the percentage of allocation to the sampling data set. This option is designed to meet the sample size requirements with regard to the given sampling proportions that are specified from the **Percent** column. At times, the number of observations within certain class levels might fall below the allocation percentages that can be specified from the **Sample Size** option setting. As an example, suppose we have a sample of 1,000 observations with three separate levels A, B and C of 100, 200, and 700 observations. In addition, suppose the class proportions of the stratification variable are set to 50%, 25%, and 25%, respectively. Let us assume that you request a sample size of 600 observations from the previous **General** tab. Therefore, the sample will generate a sample of 300 A's, 150 B's, and 150 C's. However, the problem is that there are only 100 A's in the original sample. In other words, there are not enough observations for the class level A to meet the requested proportion of 50%. Therefore, select the **Enforced Proportions** check box to specify how you want this sample to be created. By default, **Enforced Proportions** is not enforced. Therefore, the sample generated will have 100 A's, 150 B's, and 150 C's with a sample size of 400 and class proportions of 25%, 37.5%, and 37.5%, respectively. However, if the **Enforced Proportions** is selected, then the sample generated will have 100 A's, 50 B's, and 50 C's with a sample size of 200 observations and the proportions enforced at 50%, 25%, and 25%, respectively.

Cluster tab

The **Cluster** tab is designed for you to specify the categorical variable that will create the cluster groupings, the cluster sampling design to determine the clusters that are selected for the sample and the number of clusters that are created for the sampling data set. It is important to understand that the observations that are selected for the sample are based on the randomly selected clusters as opposed to taking a random sample of observations from the entire sample.

From the **Cluster Variable** display field, simply select the drill-down arrow button that displays the list of categorically-valued variables in the input data set that will define the various cluster groupings that are created from each class level. This is illustrated in the following diagram.

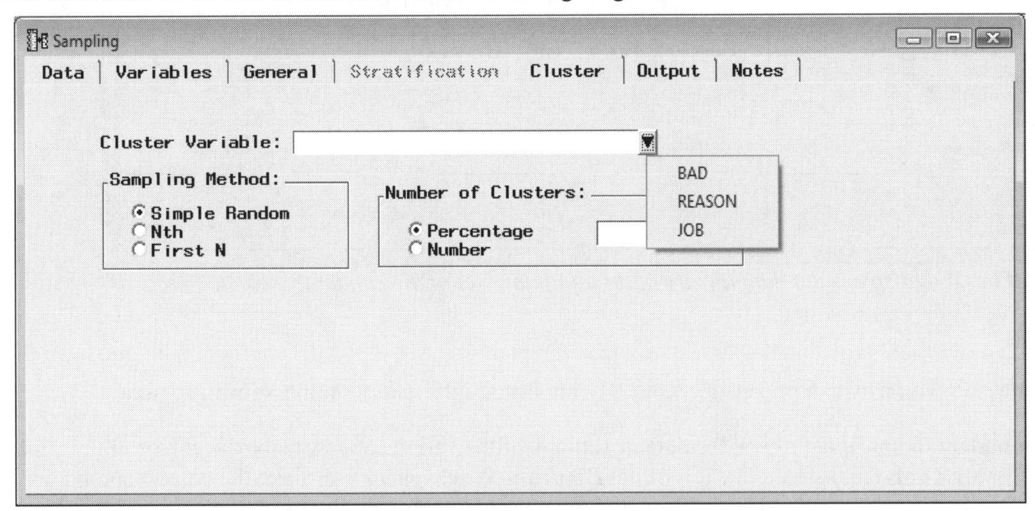

*The **Cluster** tab that displays the various configuration settings for the sampling design.*

The following are the various sampling methods that can be applied to the cluster sampling technique:

- **Simple Random** (default)**:** This is the default cluster sampling technique. Simple random cluster sampling performs a simple random sample of cluster groups, where any one cluster has an equal chance of being selected to the sampling data set.
- **Nth:** Systematic cluster sampling in which every n^{th} cluster is selected from the input data set.
- **First N:** Sequential cluster sampling in which the first *n* clusters are selected based on the number specified from the **Sample Size** option.

The following are the percentages of allocation within each cluster of the cluster sampling method:

- **Percentage** (default)**:** This option will allow you to specify the proportion of clusters to sample from all possible clusters that are created in the sampling schema. By default, Enterprise Miner performs a 10% allocation of all possible clusters that are selected, which is attributed to the sampling schema. The allocation percentage is hidden from the previous diagram.
- **Number:** This option will allow you to specify the number of clusters to sample from all possible clusters that are created in the sampling schema. The number of clusters that are created is defined by the number of class levels of the categorically-valued variable.

Output tab

The **Output** tab is designed to display the label description, library, and name of the sampling output data set that is passed along to the subsequent nodes within the process flow diagram. The output data set is the random sample of observations from either the metadata sample or the input data set that is determined by the sampling method that is selected. The output data set is created once you execute the node. From the **Description** field display, you may enter a separate label description for the sample, as illustrated in the following diagram. After running the node, you may select the **Properties...** button in order to view the file administration information and a table view or table listing of the data sample that is similar to the **Information** tab and **Table View** tab within the **Data** tab. The output data set will be automatically created in the assigned EMDATA project library.

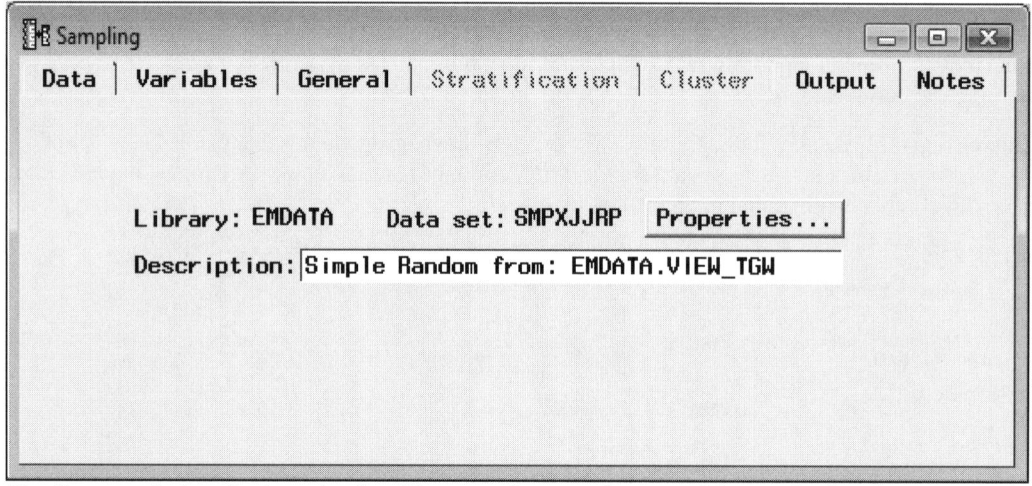

*The **Output** tab that displays the data library and output data set of the randomly selected records.*

Notes tab

From the **Notes** tab, you may enter notes or record any important information in the **Sampling** node.

Note: It is important to note that to view the default settings of the various nodes, place the cursor in the **Project Navigator Tools** tab, listed to the left of the **Diagram Workspace**, and select the corresponding folder of the Enterprise Miner node, then right-click the mouse and select the **Properties** pop-up menu item. Alternatively, place the cursor on to the **Tools Bar** over the desired icon of the Enterprise Miner node, that is, assuming that the respective node is displayed in the **Tools Bar** window. Right-click the mouse and select the **Tool properties** option from the pop-up menu items.

Select the **Edit Default...** button to specify the various default configuration settings for the selected Enterprise Miner node. Select the **OK** button to save the current setting and return to the **Enterprise Miner** window. Select the **Reset** button to reset the current changes that have been made back to the default configuration settings that SAS has carefully selected for the node.

Viewing the Sampling Node Results

General Layout of the Results Browser within the Sampling Node

- **Table View <table name> tab**
- **Strata Information tab**
- **Code tab**
- **Log tab**
- **Output tab**
- **Notes tab**

Table View <table name> tab

The **Table View <table name>** tab displays the output data set based on the sampling method that is selected for the input data set. The following table view displays the first few observations from the output data set. The tab will allow you to export the listed data set to many different file formats by selecting the **File > Export** main menu options, which will result in the standard SAS **Export Wizard** appearing. The listed output data set is passed on to subsequent nodes within the process flow diagram.

Results - Sampling

Table View: EMDATA.SMPXJJRP | Strata Information | Code | Log | Output | Notes

	BAD	LOAN	MORTDUE	VALUE	REASON	JOB	YOJ	DEROG	DELINQ	CLAGE	NINQ	CLNO	DEBTINC
1	1	1800	48649	57037	HomeImp	Other	5	3	2	77.1	1	17	.
2	1	2500	15000	20200	HomeImp		18	0	0	136.06666667	1	19	.
3	1	2800	50795	63100	HomeImp	Self	26	2	15	145.63333333	3	45	.
4	1	3000	7000	20300	HomeImp	Other	3	0	0	50.8	5	9	.
5	1	3100	39589	36100	HomeImp	Other	1.5	0	0	153.16666667	1	14	.
6	1	3600	41900	52490	HomeImp	Mgr	3	0	0	72.033333333	2	4	.
7	1	4000	26572	31960	HomeImp	Office	11	0	0	117.79510387	0	8	41.286491948
8	0	4000	64240	63990	HomeImp	Sales	5	2	0	160.3	1	23	.
9	0	4200	66272	82953	HomeImp	Other	3	0	0	110.27866259	0	15	43.450587911
10	1	4500	57000	.	HomeImp	Other	5	.	.	222.6	1	13	.
11	0	4900	58688	63348	DebtCon	ProfExe	9	0	3	179.16257471	2	31	22.544392837
12	1	4900	25597	23031	HomeImp	Other	5	0	0	101.74676644	0	8	33.243156978
13	0	5000	12457	53448	HomeImp	Self	.	0	0	207.14348133	0	12	44.107951691
14	1	5000	40000	70500	HomeImp	Mgr	0.4	0	0	61.766666667	2	8	.
15	1	5000	123000	157500	HomeImp	Mgr	9	.	.	.	.	.	.

A table view of the output data set obtained by taking a random sample of the input data set.

Strata Information tab

The **Strata Information** tab will only be displayed if stratified sampling is performed. The tab will list the various descriptive statistics such as the frequency count and allocation percentages by each stratum that is created from the categorically-valued stratification variable. The **Stratum Size** and **Stratum Proportion** columns are based on the number of observations in each stratum from the input data set. The **Sample Size** and **Sample Proportion** columns are based on the number of observations in each stratum from the metadata sample.

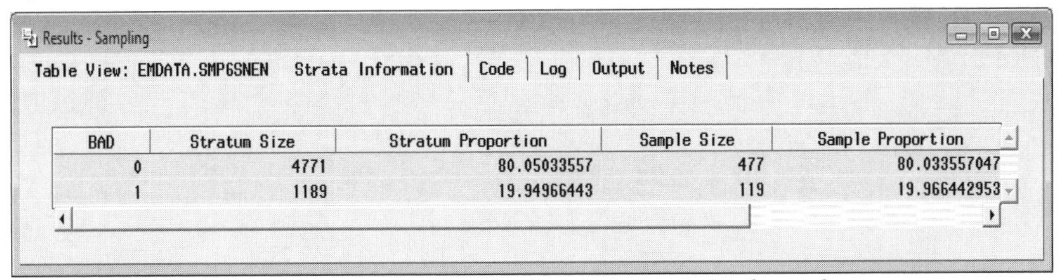

Results - Sampling

Table View: EMDATA.SMP6SNEN | Strata Information | Code | Log | Output | Notes

BAD	Stratum Size	Stratum Proportion	Sample Size	Sample Proportion
0	4771	80.05033557	477	80.033557047
1	1189	19.94966443	119	19.966442953

*The **Strata Information** tab used to display the descriptive statistics for each stratum or group.*

Code tab

The **Code** tab displays the internal SEMMA training code that was executed when running the **Sampling** node to create the output data set by performing a random sample of the input data set.

Results - Sampling

Table View: EMDATA.SMPXJJRP | Strata Information | Code | Log | Output | Notes

Train   Score

```
00001 data EMDATA.SMPXJJRP(label="Sample of EMDATA.VIEW_TGW.");
00002 set EMDATA.VIEW_TGW;
00003 drop _sample_count_;
00004 if _sample_count_ < 596 then do;
00005 if ranuni(12345)*(5961 - _N_) <= (596 - _sample_count_) then do;
00006 _sample_count_ + 1;
00007 output;
00008 end;
00009 end;
00010 run;
00011 quit;
00012 *** END OF FILE ***
```

*The **Code** tab displays the internal SEMMA code used to perform the random sample of the data set.*

Note: To copy the listed code, you must select the **Edit > Copy** menu options.

Log tab

The **Log** tab displays the log listing based on the above listed sampling code that creates the SAS sampling data set. You may view the **Log** tab in order to diagnose any problems from the compiled code that produced the node results and output data set.

Output tab

The **Output** tab displays the procedure output listing from the PROC CONTENTS procedure of the sampling output data set. The procedure listing will not be displayed. The data set is located in the EMDATA folder of the project folder based on the currently opened Enterprise Miner project.

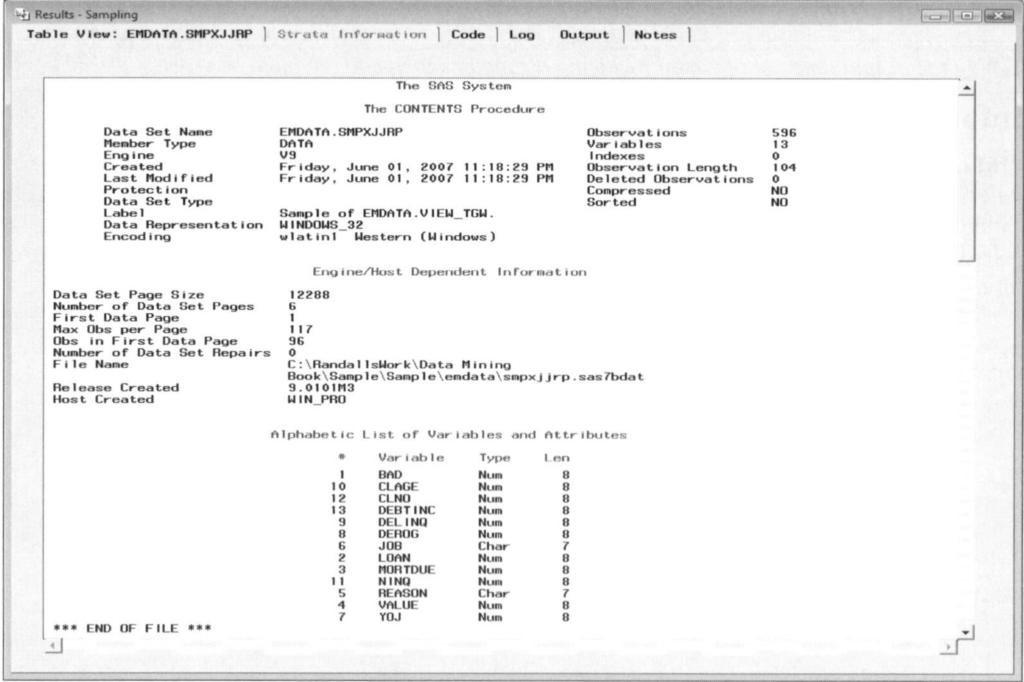


```
                              The SAS System

                          The CONTENTS Procedure

     Data Set Name          EMDATA.SMPXJJRP                     Observations          596
     Member Type            DATA                                Variables             13
     Engine                 V9                                  Indexes               0
     Created                Friday, June 01, 2007 11:18:29 PM   Observation Length    104
     Last Modified          Friday, June 01, 2007 11:18:29 PM   Deleted Observations  0
     Protection                                                 Compressed            NO
     Data Set Type                                              Sorted                NO
     Label                  Sample of EMDATA.VIEW_TGW.
     Data Representation    WINDOWS_32
     Encoding               wlatin1  Western (Windows)

                       Engine/Host Dependent Information

Data Set Page Size          12288
Number of Data Set Pages    6
First Data Page             1
Max Obs per Page            117
Obs in First Data Page      96
Number of Data Set Repairs  0
File Name                   C:\RandallsWork\Data Mining
                            Book\Sample\Sample\emdata\smpxjjrp.sas7bdat
Release Created             9.0101M3
Host Created                WIN_PRO

                 Alphabetic List of Variables and Attributes

                      #    Variable    Type    Len

                      1    BAD         Num       8
                     10    CLAGE       Num       8
                     12    CLNO        Num       8
                     13    DEBTINC     Num       8
                      9    DELINQ      Num       8
                      8    DEROG       Num       8
                      6    JOB         Char      7
                      2    LOAN        Num       8
                      3    MORTDUE     Num       8
                     11    NINQ        Num       8
                      5    REASON      Char      7
                      4    VALUE       Num       8
                      7    YOJ         Num       8
*** END OF FILE ***
```

*The **Output** tab that displays the contents procedure output listing of the sampling data set.*

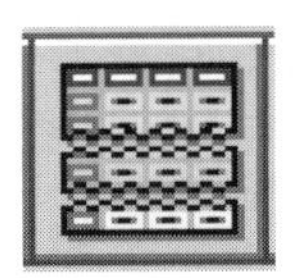

1.3 Data Partition Node

General Layout of the Enterprise Miner Data Partition Node

- **Data tab**
- **Variables tab**
- **Partition tab**
- **Stratification tab**
- **User Defined tab**
- **Output tab**
- **Notes tab**

The purpose of the **Data Partition** node in Enterprise Miner is to randomly divide the input data set, usually created from the **Input Data Source** node, into separate data sets called the training, validation, and test data sets. However, partitioning the input data set should only be performed if you have the luxury of an enormous number of records. This is because data splitting is inefficient with small to moderate sized data sets since reducing the number of observations from the training data set can severely degrade the accuracy of the results in the statistical model. The purpose of partitioning the input data set into mutually exclusive data sets is to assess the generalization of the statistical model and obtain unbiased estimates for predictive modeling or classification modeling. *Generalization* is a major goal of both predictive and classification modeling. In other words, generalization is a process of choosing the appropriate complexity of the predictive or classification model to generate unbiased and accurate predictions or classifications based on data that is entirely separate from the data that was used in fitting the model. The general idea in achieving good generalization is to first split or partition the data into separate parts, then use one part of the data to build your model and use the remaining data points to measure the accuracy of the fitted model.

In predictive modeling, it is important to obtain an unbiased estimate of the MSE or *mean square error* statistic. The MSE statistic is the most commonly used statistic for measuring the accuracy of the model. It is the squared difference between the target values and the predicted values, averaged over the number of observations that the model is fitting. The best models that are selected will have the lowest MSE from the validation data set. This is because the MSE estimate from the training data set that is used to fit the model will almost certainty be overly optimistic since the same data is used to fit the model. It is important to understand that just fitting the model to the training data set does not mean that the model is necessarily correct and that the model will fit well to new data. Therefore, the simplest remedy is splitting the input data set into two separate data sets: a training data set and validation data set. Generalization should be applied so that you can achieve an appropriate balance between the complexity of the predictive model based on an adequate accuracy of the modeling estimates and keeping the model as simple as possible. Again, one of the goals in model building is finding the optimal model complexity that involves a trade-off between the bias and variance. In other words, finding a good fit to the data while at the same time simplifying the complexity of the model. Generally, increasing the complexity of the model, for instance, by adding additional polynomial terms to the multiple linear regression model in fitting the curve, adding more leaves to the decision tree, increasing the number of hidden units in the neural network model, or increasing the smoothing constant in nearest neighbor modeling, will result in *overfitting* that will increase the variance and decrease the squared bias. On the other hand, the opposite will occur by simplifying the model, which will result in *underfitting*.

The *validation data set* is used to reduce the bias and obtain unbiased prediction estimates in validating the accuracy of the fitted model by fine-tuning the model and comparing the accuracy between separate fitted models of different types and complexity. Fine-tuning the model is performed to avoid both underfitting or overfitting. Therefore, the validation data set is applied in order to revert to a simpler predictive model as opposed to fitting the model to the training data set. That is, by fitting the model several times in selecting the appropriate complexity of the model while at the same time avoiding overfitting that leads to poor generalization. This is because the training error can be very low even when the generalization error is very high. In other words, choosing a model based on the training error will result in the most complex models being selected even if the same model generalizes poorly. Therefore, the selected model will result in poor generalization. Again, the validation data set is used in selecting the final predictive model. This data partitioning process is called the *hold-out method*. The process of using the validation data set to assess the stability and consistency of the statistical results is called the *cross-validation method*. As a very naïve approach, it is assumed that the most complex model performs better than any other model. However, this is not true. An overly complex model will lead to overfitting in which the model fits the variability in the data. Therefore, the main objective of model building is to determine a model that has enough flexibility in achieving the best generalization. For example, assuming the target values that you want to fit exhibits extremely high variability, then overfitting the data to achieve a perfect fit is not recommended to achieve good generalization. Therefore, in obtaining good generalization it is best to find a model that averages over the high variability of the data. In the modeling nodes, the validation data set is automatically used to prevent overfitting in the training data set. In the **Regression** node, the validation data set is automatically used in the stepwise regression procedure in order to choose the best subset of input variables from a pool of all possible combinations of input variables in the regression model. In the clustering nodes, the validation data set is used to compare the consistency of the clustering assignments.

If the input data set is sufficiently large enough, then a third data set may be created, called the *test data set*. The test data set is independent of the modeling fit. The test data set is used at the end of the model fit that is used in evaluating the accuracy of the final predictive model. The test data set is used at the end of model fitting in order to obtain a final honest, unbiased assessment of how well the predictive model generalizes by fitting the same model to new data. The test data set should be applied to reduce the bias and obtain unbiased prediction estimates that are entirely separate from the data that generated the prediction estimates in evaluating the performance of the modeling fit. The reason for creating the test data set is that at times the validation data set might generate inaccurate results. Therefore, a test data set might be created in providing an unbiased assessment of the accuracy of the statistical results. The purpose of the validation and test data sets is to fit the model to new data in order to assess the generalization performance of the model. Again, the drawback of the split sample procedure is that it is very inefficient with small or moderate sized samples. The major drawback in creating the test data set for the split sample method is that the predictive model might perform well for the validation data set, but there is no guarantee that the same predictive model will deliver good generalization from the test data set that is not used during model training. In addition, another obvious disadvantage in splitting the input data set into three separate parts is that it will reduce the amount of data that would be allocated to the training and validation data sets. In statistical modeling, reducing the number of observations for the model will result in degrading the fit to the model and reducing the ability to generalize.

It is important to point out that since the input data set is divided into separate data sets, it is critical that the partitioned data sets accurately represent the distribution of the underlying input data set. As in our example, it is important to preserve the ratio of the proportion of clients who either default on a loan or not in the partitioned data sets. If the validation and test data sets are not a proper representation of the input data set, then it will result in the validation statistics being invalid and the test data set producing inaccurate generalization results. For categorically-valued target variables, in order to obtain a proper percentage of the target class levels, you might have to adjust these percentages by specifying appropriate prior probabilities to match the percentages of the target class levels of the underlying data set. These prior probabilities can be specified from the target profile for the target variable.

Sampling Schemas

By default, the partitioned data sets are created from a simple random sample. The node creates a random sample based on a random number generator that follows a uniform distribution between zero and one. However, you may also specify either a stratified random sample or a user-defined sample. Again, a stratified sample is recommended if the target variable is a discrete variable of rare events, to ensure you that there is an adequate representation of each level of the entire data set, since a random sample will result in over-sampling of the data. A user-defined sample is advantageous in time series modeling where you might want to partition the input data set into separate data sets with the main objective of keeping the chronological order of the data structure intact, that is, where each partitioned data set is based on a categorical variable of the separate class levels that determines the partitioning of the input data set.

Data partitioning is discussed in greater detail in Hastie, Tibshirani, and Friedman (2001), Hand, Mannila, and Smyth (2001), and Berry and Linoff (2004).

Data tab

By default, the **Data** tab will first appear as you open the **Data Partition** node. The **Data** tab is designed to display the name and description of the input data set to be partitioned, which usually created from the previous **Input Data Source** node. From the tab, select the **Properties...** button to display the creation date of the active training data set, the number of rows and columns, and the number of deleted rows to the file from the **Information** tab. Alternatively, select the **Table View** tab to display a table view of the input data set.

The Enterprise Miner is designed so that unique data set names are automatically assigned to each partitioned data set. The naming convention for the partitioned data sets is such that the first three letters indicate either the training data set with the first three prefixes of TRN, validation data set with the prefix of VAL, or test data set with the prefix of TRT, along with some random alphanumeric characters, for example, TRNabcde, where abcde are set by some random assignment. In other words, the data mining data set will consist of a maximum of eight alphanumeric characters. However, this naming convention will change in future releases of SAS.

In Enterprise Miner, the naming conversion of the data mining data sets are randomly assigned, which will result in several data mining data sets that will be created within the EMDATA folder. However, by creating a new Enterprise Miner project, this will result in data mining data sets that are created within the currently opened diagram that will be written to the EMDATA folder that is assigned to the newly created Enterprise Miner project.

Note: The table view that displays the records of the partitioned data set must be closed in order to perform updates, or recreate and redefine the source data set that is associated with the corresponding partitioned data set.

Variables tab

The **Variables** tab is designed to display the variables in the input data set and the corresponding variable attributes such as the variable model roles, level of measurement, variable type, format, and labels that are assigned from the metadata sample. The variable attributes may be assigned in either the **Input Data Source** node or the **Data Set Attributes** node. You can change the order of the listed variables of any column by simply selecting the column heading and left-clicking the mouse button.

Partition tab

The **Partition** tab will allow you to specify the sampling schema, the allocation percentages, and initialize the random seed generator to create the partitioned data sets within the node.

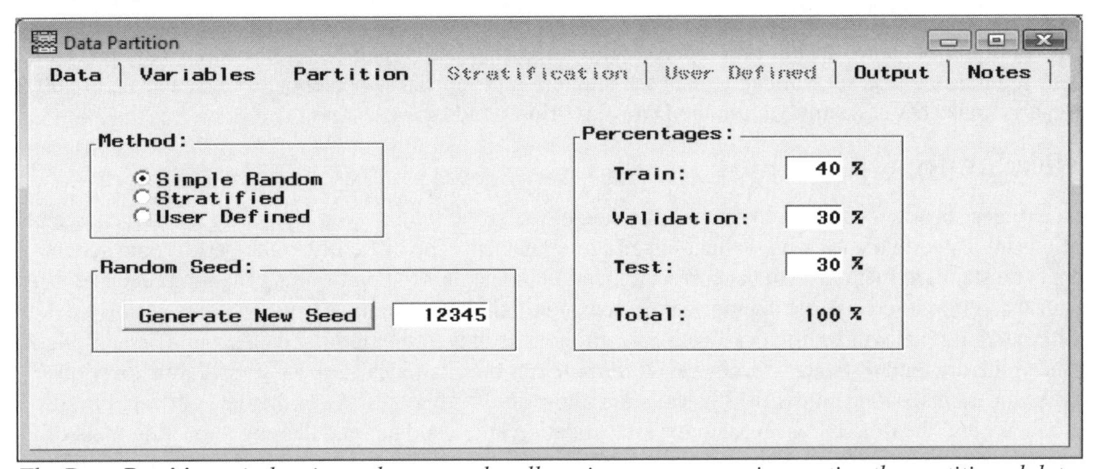

*The **Data Partition** window is used to enter the allocation percentages in creating the partitioned data sets.*

Sampling Methods

The sampling method to partition the input data set may be specified from **Method** section that is located in the upper left-hand corner of the tab. The three separate sampling schemas may be specified from the **Partition** tab, that is, simple random sample, stratified sample or a user-defined sample. By default, the source data set is partitioned by a simple random sample. The observations that are selected from the input data set are based on a random number generator that follows a uniform distribution between the value of zero and one. As the node is executed, a counter variable is created behind the scenes for each data set to regulate the correct number of records to allocate to each partitioned data set that is determined by the allocation percentages.

The following are the sampling methods that may be specified within the **Data Partition** node:

- **Simple Random** (default)**:** This is the default partitioning method. A simple random sample is the most common sampling design; where each observation has an equal chance of being selected. The input data set is randomly partitioned from the **Data Partition** node based on a random seed number that follows a uniform distribution between zero and one.
- **Stratified:** This is a sampling technique that is used when the input data set is randomly partitioned into stratified groups based on the class levels of the categorically-valued input variable. A random stratified sample is a sampling design that partitions data into nonoverlapping groups or strata, then selects a simple random sample within each group or stratum. Selecting this option will allow you to view the following **Stratification** tab. Stratification may be performed by one or more categorically-valued variables in the input data set.
- **User-Defined:** This sampling option will allow you to specify one and only one categorical variable called a *user-defined sample* that defines the partitioning of the training, validation, and test data sets. Selecting this option will allow you to view the **User-Defined** tab. The difference between user-defined sampling and stratified sampling is that each data set that is created for the user-defined sampling design is based on each class level of the categorically-valued partitioning variable for which the data set is not randomly sampled. This method might be used in time series modeling where you might want to partition the input data set into separate data sets and yet retain the chronological order of the data set intact, as is required in time series modeling.

Generate New Seed

Since the various sampling schemas are based on a random sample of the source data set, the node will allow you to specify a random seed from the left-hand portion of the **Partition** tab, to initialize the randomized sampling process. The default random seed is set at 12345. Specifying the same random seed will create the same partitioned data sets. However, resorting the data will result in a different partition of the data. In addition, if the random seed is set to zero, then the random seed generator is based on the computer's clock at run time. Since the node saves the value set to the random seed, it will allow you to replicate the same sample by each successive execution of the node.

From the **Partition** tab, enter the allocation percentages to split the input data set into separate data sets called *roles*, that is, a training data set, validation data set, and test data set. These percentages must add-up to 100%. To save the various changes, simply close the **Data Partition** window.

Allocation Strategy

The **Percentages** section located on the right-hand side of the tab will allow you to specify the percentage of observations allocated to the training, validation, and test data sets. The difficulty in the partitioning schema is that there is no standard method as to the correct allocation percentages in partitioning the input data set. It is under your discretion to choose the appropriate percentage of allocation to the partitioned data sets. By default, the **Data Partition** node will automatically allocate 40% of the data to the training data set and 30% of the data to both the validation and test data sets. However, it might not be a bad idea to specify separate allocation percentages for each separate model fit. Typically, an allocation strategy is 75% of the data allocated to the training data set and the other 25% set aside for the validation and test data sets, thereby allocating most of the input data set to estimate the modeling parameter estimates from the training data set while at the same time leaving enough of the data allocated to the validation data set to determine the smallest average validation error to the split-sample procedure and the test data set in order to create unbiased estimates in evaluating the accuracy of the prediction estimates. The node is designed in such a way that the user must specify allocation percentages that must add up to 100% or the **Data Partition** window will remain open.

In predictive modeling, there is no general rule as to the precise allocation of the input data set. The allocation schema used in partitioning each one of the data sets depends on both the amount of the available cases from the input data set and the noise level in the underlying data set. Also, the amount of data allocated to the training data set depends on both the noise level in the target variable and the amount of complexity in the predictive model that is used in trying to fit the underlying data set. It is also important to point out that there must be more data points than the number of input variables in the model to avoid overfitting. In predictive modeling, as a very conservative approach based on a noise-free distribution in the target values, it is recommended that there be at least ten observations for each input variable in the model, and for classification modeling there should be at least ten observations for each level of the categorically-valued input variable in the predictive model.

Default Allocation Percentages

40% Training	The training data set is used for model fitting.
30% Validation	The validation data set is used for assessment of the estimates in determining the smallest average validation error and to prevent overfitting in the statistical modeling design by fitting the model several times and selecting the most appropriate modeling terms for the statistical model that generates the smallest validation generalization error.
30% Test	The test data set is used at the end of the modeling fit in order to reduce the bias by generating unbiased estimates and obtaining a final honest assessment in evaluating the accuracy of the estimates since the data is entirely separate from the data that is used in fitting the model in order to improve generalization. In other words, the main objective of statistical modeling is to make predictions of the data in which the target values are unknown. Therefore, the role of the test data set can be thought as making predictions of the data with the target values that are not exactly known. However, the drawback is that the target values in the test data set are already known.

Note: The node is designed so that the **Data Partition** window will continue to stay open until you have entered the correct allocation percentages, which must add up to 100%.

Stratification tab

The **Stratification** tab is designed to display the categorical variables that may be used as the stratification variable to partition the input data set, where each class level of the categorically-valued variable will automatically create a separate data set.

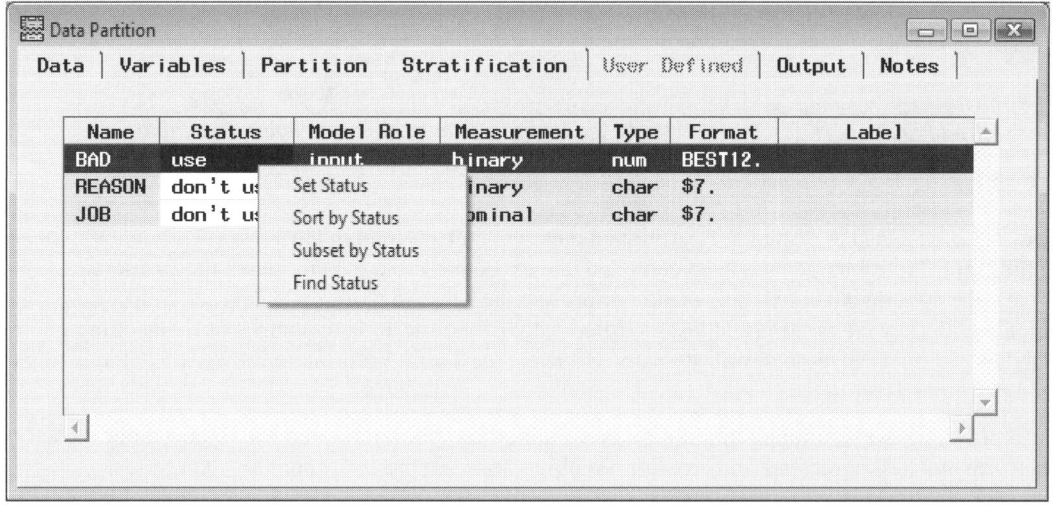

*The **Stratification** tab used in specifying the categorically-valued variables to perform the partitioning.*

The **Stratification** tab can only be viewed if stratified sampling is specified from the previous **Partition** tab. Stratified sampling may be performed by assuming that there exist categorically-valued input variables in the data set. The categorically-valued stratification variable must have a model role of **input** or **group** to perform the partitioning. The **Stratification** tab is identical to the previously displayed **Variables** tab, with the exception that only categorically-valued variables are displayed. Also, the tab has an added **Status** column to specify the categorically-valued variable to perform the stratification. By default, all the variables in the input data set are automatically set to a variable status of **don't use**. Simply select the **Status** cell to assign the selected variable to **use**, as illustrated in the previous diagram, to select the categorically-valued variable that creates the stratified groups. Again, it is important that there are an adequate number of observations in each class level of the stratified categorically-valued variable to better meet the distributional assumptions that are required in the test statistics and achieve more accurate estimates in the analysis.

User Defined tab

The **User Defined** tab is made available for viewing within the **Data Partition** node once you have selected the user-defined partitioning method from the **Partition** tab. The node displays the **Partition variable** field to select the categorically-valued variable that will perform the partitioning of the input data set. Each data set is created by each unique discrete value from the categorically-valued partitioning variable. The categorically-valued partitioning variable must have a model role of **input** or **group** to perform the partitioning. The restriction is that the target variable cannot be selected as the partition variable to perform the user-defined sample. The **Partition Values** section will allow you to specify the class level of each partitioned data set. The **Training** display field will allow you to specify the class level to create the training data set, and the **Validation** display field will allow you to specify the class level to create the validation data set. If you have the luxury of a large amount of data in the input data set, then the **Test** display field will allow you to specify the class level with which to create the test data set. From each display field, simply click the drop-down arrow control button, then select the appropriate value from the drop-down list to set the partitioned values to create each separate data set.

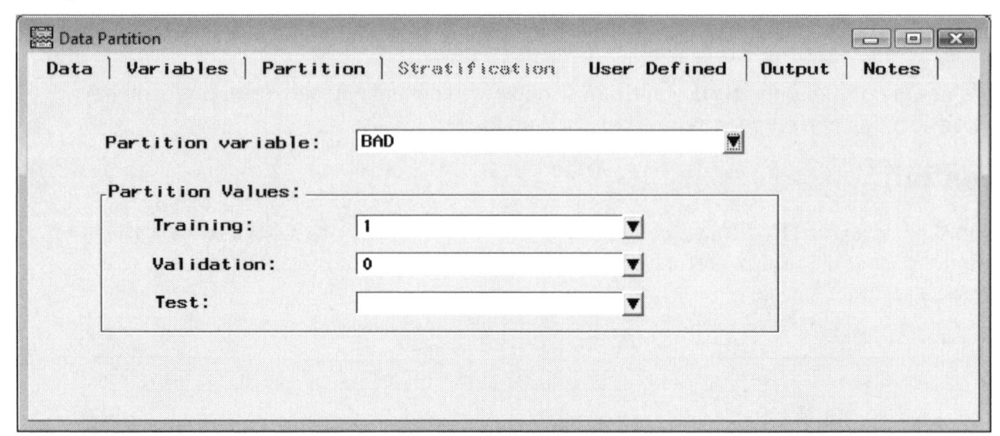

*The **User Defined** tab to partition the data set based on the values of the user-defined variable.*

Output tab

The **Output** tab is designed to display the partitioned data sets that are used in the subsequent modeling nodes assuming the **Data Partition** node has been compiled and executed. From the tab, select the **Properties...** button to view the file administrative information or browse the selected partitioned data set. Select the **Information** tab to view the creation and last modified dates of the file and the number of fields of the partitioned data mining DMDB data set. Alternatively, select the **Table View** tab to browse the table listing of the selected partitioned data set.

Note: When browsing the partitioned data sets from the table view, the partitioned data sets must be closed in order to perform updates, or recreate and redefine the observations in the corresponding data sets.

Notes tab

Enter notes or record any other information in the **Data Partition** node.

Viewing the Data Partition Results

General Layout of the Results Browser within the Data Partition Node

- **Table View tab**
- **Code tab**
- **Log tab**
- **Output tab**
- **Notes tab**

Table View tab

The **Table View** tab will allow you to browse each one of the partitioned data sets that are created in order to verify that the data partitioning routine was performed correctly by executing the node. The following illustration displays the first few records from the training data set that is created by performing a simple random sample of the HMEQ data set.

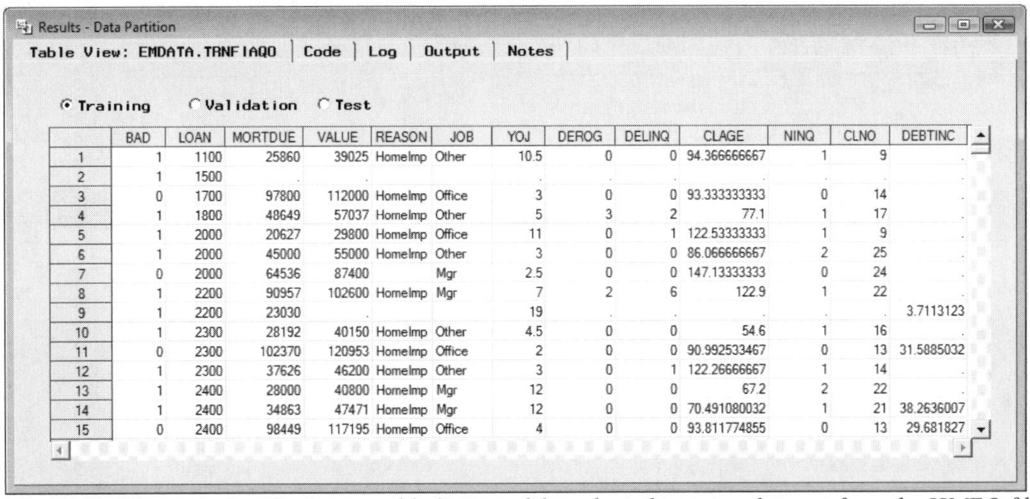

Results - Data Partition

Table View: EMDATA.TRNFIAQ0 | Code | Log | Output | Notes

(•) Training ( ) Validation ( ) Test

	BAD	LOAN	MORTDUE	VALUE	REASON	JOB	YOJ	DEROG	DELINQ	CLAGE	NINQ	CLNO	DEBTINC
1	1	1100	25860	39025	HomeImp	Other	10.5	0	0	94.366666667	1	9	.
2	1	1500	.	.			.	.	.	.	.	.	.
3	0	1700	97800	112000	HomeImp	Office	3	0	0	93.333333333	0	14	.
4	1	1800	48649	57037	HomeImp	Other	5	3	2	77.1	1	17	.
5	1	2000	20627	29800	HomeImp	Office	11	0	1	122.53333333	1	9	.
6	1	2000	45000	55000	HomeImp	Other	3	0	0	86.066666667	2	25	.
7	0	2000	64536	87400		Mgr	2.5	0	0	147.13333333	0	24	.
8	1	2200	90957	102600	HomeImp	Mgr	7	2	6	122.9	1	22	.
9	1	2200	23030	.			19	.	.	.	.	.	3.7113123
10	1	2300	28192	40150	HomeImp	Other	4.5	0	0	54.6	1	16	.
11	0	2300	102370	120953	HomeImp	Office	2	0	0	90.992533467	0	13	31.5885032
12	1	2300	37626	46200	HomeImp	Other	3	0	1	122.26666667	1	14	.
13	1	2400	28000	40800	HomeImp	Mgr	12	0	0	67.2	2	22	.
14	1	2400	34863	47471	HomeImp	Mgr	12	0	0	70.491080032	1	21	38.2636007
15	0	2400	98449	117195	HomeImp	Office	4	0	0	93.811774855	0	13	29.681827

*The **Table View** tab that displays a table listing of the selected training data set from the HMEQ file.*

Note: When browsing the partitioned data sets from the table view, the table listing must be closed in order to perform updates, recreate, and redefine the observations in the selected data set.

Code tab

The **Code** tab will display the internal SAS training code that is compiled within the node to create the partitioned data sets by the sampling technique that was selected. For random sampling, the input data performs a simple random sample based on a random seed number that follows a uniform distribution. A counter variable is automatically created for each partitioned data set to control the number of observations that are assigned to each partitioned data set that is created. For stratified sampling, a random sample is performed within each class level of the categorically-valued stratified variable. The code automatically creates a counter variable within each class level of the categorically-valued stratification variable for each partitioned data set that is created to control the number of observations that are assigned to each partitioned data set. For user-defined sampling, the internal code will automatically generate a PROC SQL procedure for each partitioned data set that is created by each class level that is specified for the partitioned data set from the user-defined, categorically-valued variable.

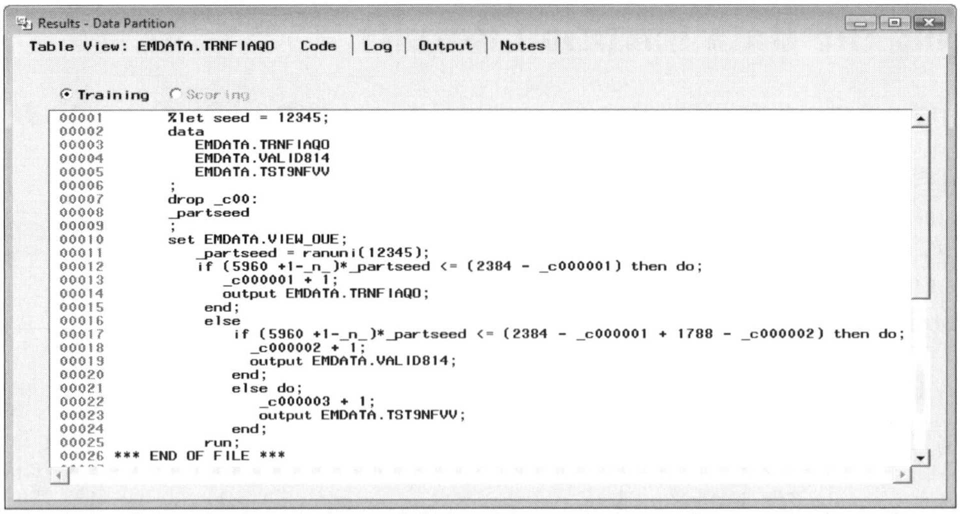

Results - Data Partition

Table View: EMDATA.TRNFIAQ0 | Code | Log | Output | Notes

(•) Training ( ) Scoring

```
00001        %let seed = 12345;
00002        data
00003           EMDATA.TRNFIAQ0
00004           EMDATA.VALID814
00005           EMDATA.TSTSNFVV
00006        ;
00007        drop _c00:
00008        _partseed
00009        ;
00010        set EMDATA.VIEW_OUE;
00011           _partseed = ranuni(12345);
00012           if (5960 +1-_n_)*_partseed <= (2384 - _c000001) then do;
00013              _c000001 + 1;
00014              output EMDATA.TRNFIAQ0;
00015           end;
00016           else
00017              if (5960 +1-_n_)*_partseed <= (2384 - _c000001 + 1788 - _c000002) then do;
00018                 _c000002 + 1;
00019                 output EMDATA.VALID814;
00020              end;
00021              else do;
00022                 _c000003 + 1;
00023                 output EMDATA.TSTSNFVV;
00024              end;
00025           run;
00026 *** END OF FILE ***
```

*The **Code** tab that displays the internal SAS training code that performs a simple random sample.*

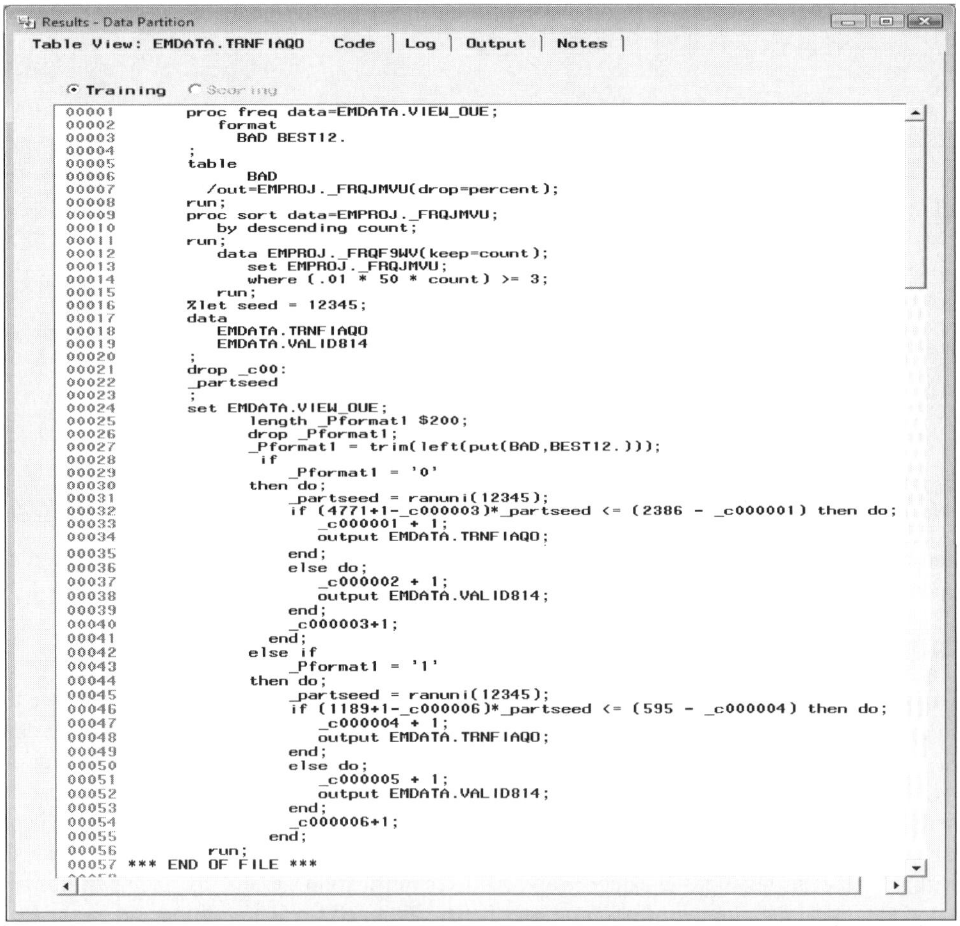

Results - Data Partition

Table View: EMDATA.TRNFIAQ0 | Code | Log | Output | Notes

(•) Training ( ) Scoring

```
00001        proc freq data=EMDATA.VIEW_OUE;
00002           format
00003              BAD BEST12.
00004        ;
00005        table
00006              BAD
00007           /out=EMPROJ._FRQJMVU(drop=percent);
00008        run;
00009        proc sort data=EMPROJ._FRQJMVU;
00010           by descending count;
00011        run;
00012           data EMPROJ._FRQF9WV(keep=count);
00013              set EMPROJ._FRQJMVU;
00014              where (.01 * 50 * count) >= 3;
00015           run;
00016        %let seed = 12345;
00017        data
00018           EMDATA.TRNFIAQ0
00019           EMDATA.VALID814
00020        ;
00021        drop _c00:
00022        _partseed
00023        ;
00024        set EMDATA.VIEW_OUE;
00025              length _Pformat1 $200;
00026              drop _Pformat1;
00027              _Pformat1 = trim(left(put(BAD,BEST12.)));
00028               if
00029                 _Pformat1 = '0'
00030              then do;
00031                 _partseed = ranuni(12345);
00032                 if (4771+1-_c000003)*_partseed <= (2386 - _c000001) then do;
00033                    _c000001 + 1;
00034                    output EMDATA.TRNFIAQ0;
00035                 end;
00036                 else do;
00037                    _c000002 + 1;
00038                    output EMDATA.VALID814;
00039                 end;
00040                 _c000003+1;
00041              end;
00042              else if
00043                 _Pformat1 = '1'
00044              then do;
00045                 _partseed = ranuni(12345);
00046                 if (1189+1-_c000006)*_partseed <= (595 - _c000004) then do;
00047                    _c000004 + 1;
00048                    output EMDATA.TRNFIAQ0;
00049                 end;
00050                 else do;
00051                    _c000005 + 1;
00052                    output EMDATA.VALID814;
00053                 end;
00054                 _c000006+1;
00055              end;
00056           run;
00057 *** END OF FILE ***
```

*The **Code** tab that displays the internal SAS training code that performs a stratified random sample of the HMEQ input data set to create the partitioned data sets based on the binary-valued stratification variable, BAD, by executing the node.*

Results - Data Partition

Table View: EMDATA.TRNQNMZJ | Code | Log | Output | Notes

Training Scoring

```
00001 proc sql;
00002 create view EMDATA.TRNQNMZJ as select * from EMDATA.VIEW_RZA where
00003 trim(left(put(BAD,BEST12.))) = '0';
00004 quit;
00005 proc sql;
00006 create view EMDATA.VALQRPH7 as select * from EMDATA.VIEW_RZA where
00007 trim(left(put(BAD,BEST12.))) = '1';
00008 quit;
00009 *** END OF FILE ***
```

*The **Code** tab that displays the internal SAS training code that performs a user-defined sample.*

Log tab

The **Log** tab will allow you to view the compiled code that is executed once you run the node. The corresponding programming code that is compiled is listed in the previous **Code** tab. The tab will allow you to view any warning or error messages that might have occurred after you execute the **Data Partition** node.

Output tab

The **Output** tab is designed to display the file contents from the PROC CONTENTS procedure listing of each partitioned SAS data set. The following illustration displays the contents listing of the training data set that is created from the simple random sample. For stratified sampling, the procedure output listing will initially display a frequency table listing to view the frequency counts at each class level of the stratified, categorically-valued variable that created the partitioned data sets.

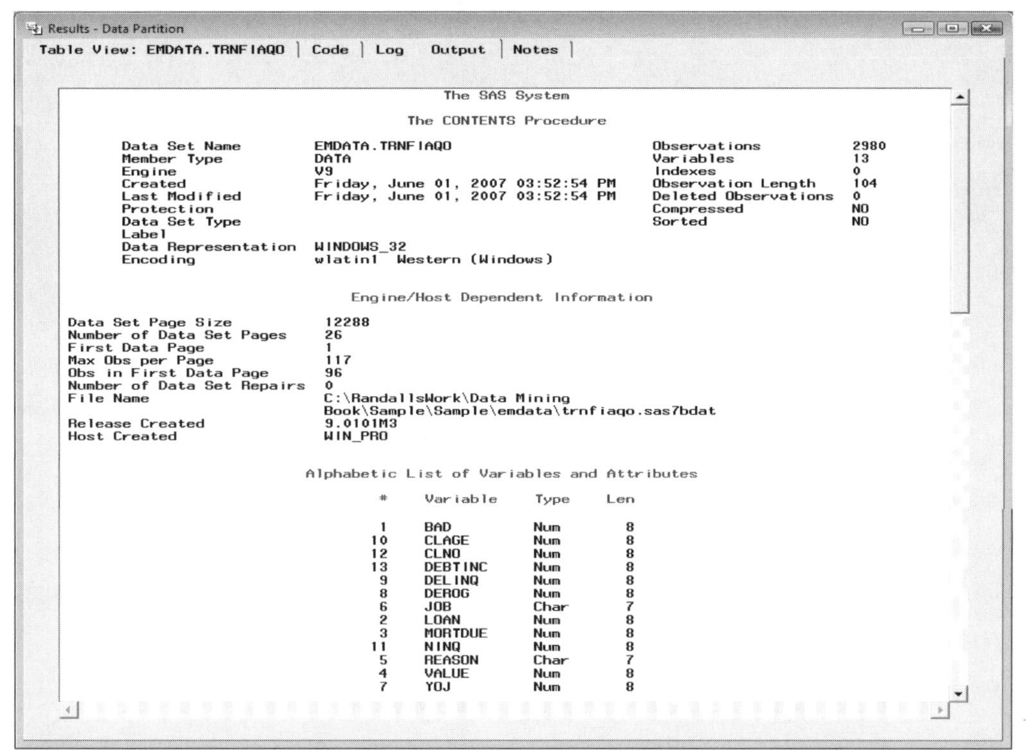

Results - Data Partition

Table View: EMDATA.TRNFIAQO | Code | Log | Output | Notes

```
                                   The SAS System

                              The CONTENTS Procedure

        Data Set Name        EMDATA.TRNFIAQO                      Observations            2980
        Member Type          DATA                                 Variables               13
        Engine               V9                                   Indexes                 0
        Created              Friday, June 01, 2007 03:52:54 PM    Observation Length      104
        Last Modified        Friday, June 01, 2007 03:52:54 PM    Deleted Observations    0
        Protection                                                Compressed              NO
        Data Set Type                                             Sorted                  NO
        Label
        Data Representation  WINDOWS_32
        Encoding             wlatin1  Western (Windows)

                         Engine/Host Dependent Information

Data Set Page Size          12288
Number of Data Set Pages    26
First Data Page             1
Max Obs per Page            117
Obs in First Data Page      96
Number of Data Set Repairs  0
File Name                   C:\RandallsWork\Data Mining
                            Book\Sample\Sample\emdata\trnfiaqo.sas7bdat
Release Created             9.0101M3
Host Created                WIN_PRO

                    Alphabetic List of Variables and Attributes

                         #    Variable    Type    Len

                         1    BAD         Num       8
                        10    CLAGE       Num       8
                        12    CLNO        Num       8
                        13    DEBTINC     Num       8
                         9    DELINQ      Num       8
                         8    DEROG       Num       8
                         6    JOB         Char      7
                         2    LOAN        Num       8
                         3    MORTDUE     Num       8
                        11    NINQ        Num       8
                         5    REASON      Char      7
                         4    VALUE       Num       8
                         7    YOJ         Num       8
```

*The **Output** tab that displays the data set contents procedure output listing of the training data set from the simple random sample.*

Chapter 2

Explore Nodes

Chapter Table of Contents

2.1 Distribution Explorer Node

General Layout of the Enterprise Miner Distribution Explorer Node

- **Data tab**
- **Variables tab**
- **X Axis tab**
- **Y Axis tab**
- **Z Axis tab**
- **Notes tab**

The purpose of the **Distribution Explorer** node in Enterprise Miner is designed to visualize the distribution in the active training data set by producing multidimensional histograms. The node will generate either one, two, or up to three-dimensional frequency bar charts from the metadata sample. By default, a one-dimensional frequency bar chart is created that is identical to the **Variable Histogram** window that displays the standard Enterpriser Miner frequency bar chart from the pop-up menu option within the **Variables** tab. For interval-valued variables, the bars will be displayed by the values divided into separate intervals of equal length. For categorically-valued variables, the bars will be displayed by each class level of the categorical variable. By default, the multidimensional histograms are created from the metadata sample.

The node is an advanced visualization tool. The node creates 3-D plots with up to three separate variables that can be plotted at a time based on the percentage, mean or sum. From the **Variables** tab, simply select the variables to assign the corresponding X, Y, and Z coordinates to the graphs. The node is designed to display a histogram or a frequency bar chart of each variable in the graph. The histogram has the option of transforming interval-valued variables into categorically-valued variables by specifying the number of buckets, bins, or bars for the chart. Cutoff values to the bar charts can be set for interval-valued variables. For interval-valued variables that are plotted, descriptive statistics are created from the **Output** tab. For categorically-valued variables, cross-tabulation statistics are created.

The basic idea of the node is that it will allow you to graphically observe the distributional form of the variables, visualize relationships between the variables, and observe any extreme values or missing values in the data. The node is designed to gain a greater insight in visualizing the interrelationship between the variables in the active training data set. Histograms or frequency charts provide the simplest way of displaying the location, spread, and shape of the distribution based on the range of values in the variable. The frequency charts are designed to view the distribution of the values, such as viewing the spread or variability in the variable and observing whether the range of values are concentrated about its mean or whether the variable has a symmetrical, skewed, multimodal, or unimodal distribution. In other words, if the variable displays an approximately bell-shaped distribution, you may then assume that the variable is normally distributed about its mean, which is extremely important in various statistical modeling assumptions and statistical hypothesis testing. There are two important reasons for the frequency plots. First, it is important to determine the shape of the distribution of the variable. For example, if the variable displays a bimodal distribution, that is, two peaks, then it might suggest that the observations will be normally distributed among two separate groups or populations. Second, many of the statistical tests in addition to the following nodes and corresponding analysis in Enterprise Miner are based on the normality assumption with regard to the distribution of the variables in the analysis. If the variables do not follow a normal distribution, then transformations of the variables might be considered. The reason why it is so important to first view the distribution of each variable in Enterprise Miner is that typically the variables are measured in different units – dollars, pounds, and so on. It is important that the variables in the analysis follow the same range of values since many of the following nodes apply the sum-of-squares distance function in the corresponding analysis. In other words, the variables that have a wide range of values will tend to have a great deal of influence to the results.

As a preliminary step in data mining analysis, you might want to first observe the distribution of each variable in the data set. The next step might be to observe the bivariate relationship between two separate variables and, finally, constructing the multidimensional charts to visualize the distribution and the relationship between the three separate variables. The reason that up to three-dimensional frequency bars are displayed within the node is that the human eye can only visualize up to three-dimensional plots at one time. The advantage to constructing frequency bars of two or three variables at a time is that it might reveal multivariate outliers that would otherwise be unnoticed using the standard frequency bar when viewing one variable only. Since the human eye can only visualize up to three-dimensional plots at one time, principal components analysis might be considered. Principal components analysis is designed to consolidate a majority of the variability in the data into usually at most two separate variables or principal components. In the analysis, each principal component is a linear combination of every input variable in the model, where these same components can be plotted in order to observe multivariate outliers in the data.

For categorical variables, you may specify separate levels from the chart. For interval-valued variables, you may specify a separate range of values. The node is designed to exclude certain class levels from the chart based on the categorically-valued variable. For interval-valued variables, you can set a certain range of values to remove from the analysis. The node is also designed to create several cross-tabulation reports and summary statistical listings based on the specified categorically-valued or interval-valued axes variables from the active training data set.

In statistical modeling, it is not a bad idea to construct various bars charts of the residuals in addition to the modeling terms in order to view any outlying data points that will have a profound effect to the final results. These outlying data points will be located at either end of the bar chart. The bar charts are also important to use in order to visualize normality in the data, which is one of the modeling assumptions that must be satisfied. Also, the error terms must be reasonably normally distributed or have an approximately bell-shaped distribution with a symmetric distribution about its mean of zero. The importance of the error terms in the model having a normal distribution is that the statistical tests used to determine the significance of the effects in the model assumes that the interval target values are normally distributed around its mean. In statistical modeling, a residual is the difference between the observed target value, and the predicted target value where the sum of the residual values equal zero. In analysis of variance, with an interval-valued target variable and one or more categorically-valued input variables, it is important that the target values is approximately normal with equal variability within each group of the categorical input variables in the analysis. Normality can be verified by examining the frequency plot for each group. If the assumptions are not satisfied, then it will result in making invalid inferences to the analysis. In multivariate analysis, it is important that the variables in the analysis are multivariate normally distributed in order to avoid bias and misleading results. This means that not only the variables in the analysis are normally distributed, but also their joint distribution should be normal as well. That is, each pair of variables in the analysis are normally distributed with each other. This can be observed from the bivariate bar charts.

In logistic regression modeling, it is important to avoid both complete and quasi-complete separation that will result in undefined parameter estimates in the model. Quasi-complete separation occurs when none of the observations falls within any one of the class levels of the input variable in the model and the class levels of the target variable that you want to predict. This can be observed from the two-dimensional bar chart displaying any bar within any one of the separate class levels of the input variable and target variable in the classification model.

In decision tree modeling, it is important that there be a balance between the separate class levels of the variables in the model. This is because the model will do better at distinguishing between the separate class levels of equal size when performing the exhaustive split searching routine. In addition, it is important to observe large discrepancies in the frequency counts between the training and validation data sets within each leaf of the tree that might indicate instability and bad generalization to the decision tree model. A remedy would be to remove these splits from the model, which will result in a smaller number of leaves. Therefore, it is important to view the distribution between the target variable and the input variables in the model. The bar chart, might give you an indication of where to best split the range of values or class levels in the input variable. This can be achieved by observing break points in which there are significant changes in the distribution of the target values across the range of values of the input splitting variable. In the analysis, it is important that you select the range of values in the input variable in which most of the data resides in order to achieve consistency and stability in the decision tree estimates.

In discriminant analysis, it is critical to observe both normality and equal variability within each class level of the categorically-valued target variable that you wish to classify from the range of values of the interval-valued input variables in the analysis. If the variability assumption is violated, then this will result in making incorrect inferences with regard to the group differences. In addition, the multivariate tests that are designed to determine equal variability across the separate categories depends on normality in the data within each class level of the target variable.

In cluster analysis, bar charts are helpful to both identify and profile the various cluster groupings, that is, constructing separate bar charts for each input variable in the analysis to distinguish between the various clusters in determining both the similarities and the differences between the clusters. In addition, it is important to distinguish between the various characteristics of the separate clusters that are created. For example, identifying the various characteristics from the clusters, where the first cluster consist of younger people with higher incomes, the second cluster that consist of older people with lower incomes, and so on. Since the node can create bar charts of up to three separate variables, the analysis can be performed by identifying the characteristics of each cluster grouping by stratifying the analysis by using an additional grouping variable.

In interactive grouping analysis, the input variables are transformed into numerous categories in order to create the most optimum groupings in the input variables in the analysis based on a binary-valued target variable. The optimum groupings are determined by certain grouping criterion statistics that are defined by the ratio or log difference between the two separate proportions of the target variable. Therefore, it is important to construct various bar charts to observe normality in the input variables, but also normality in the input variables within each of the two separate groups of the binary-valued target variable, since the outlying data points will result in the interactive grouping results creating their own separate groupings.

The various graphical techniques are discussed in Friendly (1991), and Carpenter and Shipp (2004).

Data tab

The **Data** tab is designed to select the input data set to create the various graphs. The tab displays the selected file administrative information and a table view of the data set. Press the **Select...** button that will open the **Imports Map** window to select an entirely different active training data set that is connected to the node within the process flow diagram.

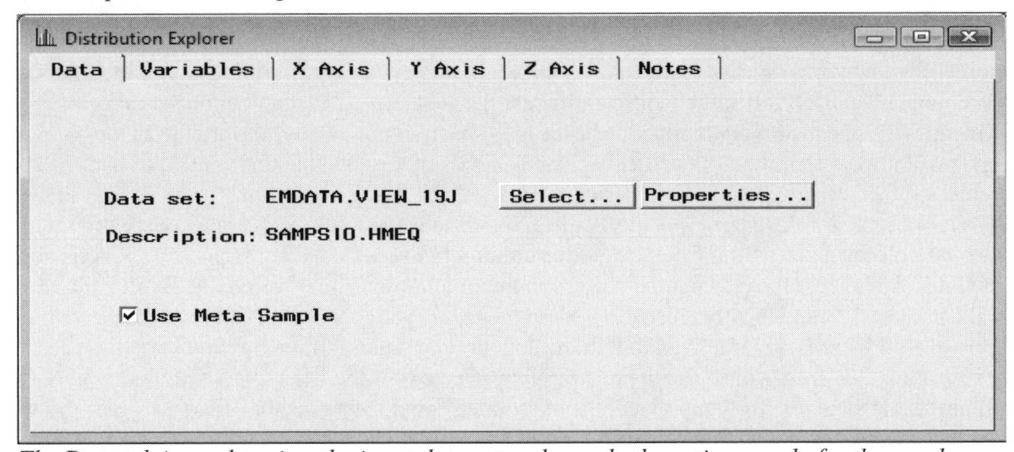

*The **Data** tab is used to view the input data set and sample the entire sample for the graphs.*

- **Use Meta Sample:** By default, this option is checked. The corresponding bar charts are automatically created from the metadata sample. However, clearing the check box will result in Enterprise Miner using the entire input data set to create the corresponding bar charts.

Variables tab

The node is designed so that the **Variables** tab will appear when you open the **Distribution Explorer** node. The tab will allow you to view the various properties of the listed variables in the active training data set. If the previously assigned model roles or measurement levels are incorrect, then they cannot be corrected within the node. Therefore, you must go back to the **Input Data Source** node to make the appropriate corrections. The

Variables tab is designed for you to select the variables for the three X, Y and Z axes variables that create the bar charts and reports. From the tab, up to three separate variables can be selected for the three-dimensional bar charts from the metadata sample. Since the human eye can only view the charts in up to three-dimensions, the node is limited to three separate variables. Although, the **Distribution Explorer** node is capable of creating a bar chart from three separate variables, you may also elect to create a univariate bar chart based on a single axis variable or a bivariate bar chart based on two separate axis variables.

The tab will also allow you to view the standard Enterprise Miner histogram from the selected variable listed within the tab. Select any one of the listed variables, then right-click the mouse to select the **View Distribution of <variable name>** pop-up menu item. The **Variable Histogram** window will appear with the corresponding frequency bar chart of the selected variable from the metadata sample.

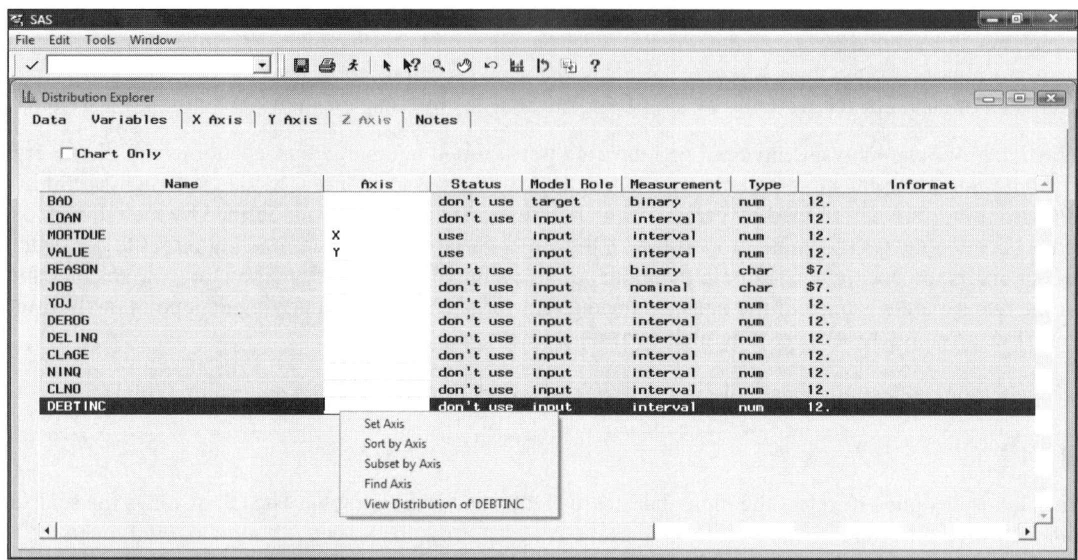

*The **Variables** tab is designed to select the variables for each axis of the three-dimensional bar chart.*

Setting the Axes Variables

The **Axis** column will allow you to specify the variables to the frequency bar chart. Simply scroll the mouse over to the **Axis** column, then right-click the mouse and select the first pop-up menu item of **Set Axis**. A second pop-up menu will appear for you to either clear the axis specification or set the variable as the X, Y, or Z axis. The Y axis will always represent the vertical axis, the X axis represents the horizontal axis and the Z-axis will also represent the horizontal axis. The restriction is that the X and Y axis variables must exist in order to specify the Z axis variable for the chart. Specifying the variable selected to the X, Y, or Z axis setting will change the status variable attribute setting from the **Status** column to **use** with the corresponding **Y Axis** and **Z Axis** tab undimmed and available for viewing the corresponding one-dimensional bar charts. By default, the one-dimensional bar chart is automatically created based on the target variable in the data set with a variable status of **use**. However, if there is no target variable in the data mining data set, then the first variable listed with a variable role other than **freq** will be the default variable selected for the bar chart. By default, the **X Axis** tab is undimmed and available for you to view the one-dimensional frequency bar chart, whereas the **Y Axis** and **Z Axis** tabs are grayed-out and unavailable for viewing.

Specifying the Statistical Listing

By default, the **Chart Only** check box is selected, indicating to you that once the node is executed, then only the charts will be created. However, clearing the check box will instruct the node to create cross-tabulation reports and classification tables from all the categorical variables and calculate descriptive statistics from all the continuous variables with a variable attribute status of **use**. From the **Results Browser**, assuming that the axes variables are all interval-valued variables, the PROC MEANS procedure listing will be displayed within the **Output** tab. For a single categorical axes variable, the PROC MEANS procedure listing will be displayed by each class level. Cross-tabulation reports will be automatically created from all the categorically-valued axes variables in the chart.

X Axis tab

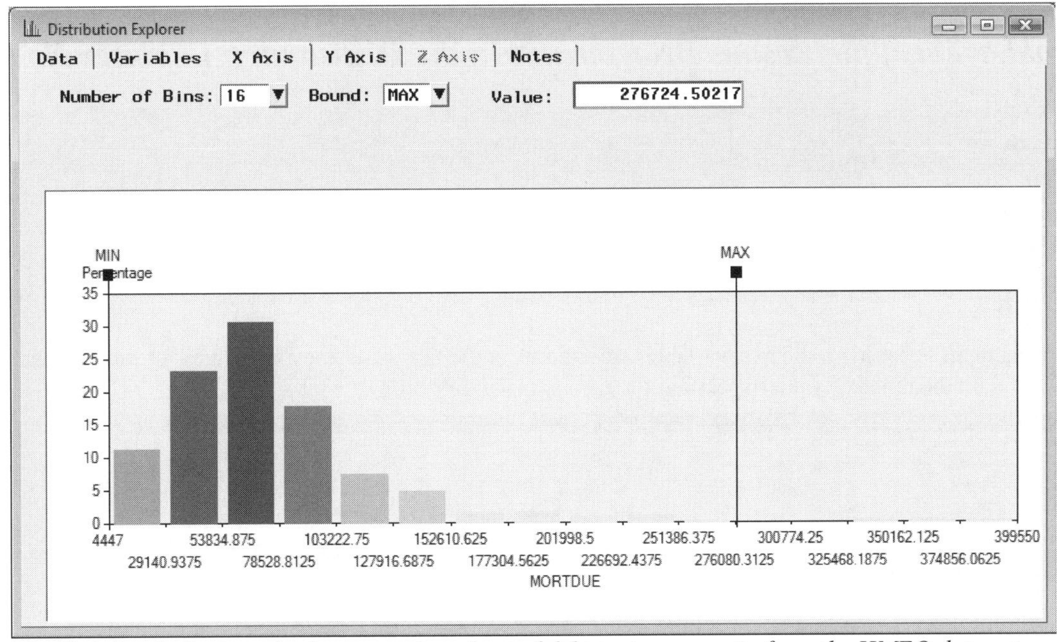

*The **X Axis** tab displays the frequency bar chart of debt-to-income ratio from the HMEQ data set.*

The **X Axis** tab will open the **Variable Histogram** window that will display the standard univariate frequency bar chart that is displayed in many of the other Enterprise Miner nodes within the **Variables** tab. The chart displays the frequency percentages from the interval-valued variables divided into separate intervals of equal length, where each bar is located at the midpoint of each separate interval. By default, 16 separate intervals or bars are displayed. However, you may specify the number of intervals or bars to view from the value entered within the **Number of Bins** entry field. A pop-up menu option will appear for you to enter 2, 4, 8, 16, 32, 64 or up to 128 bars. In the previously displayed bar chart, the red bars represent high values, orange bars represent intermediate values, and yellow bars represent low values. In addition, the reason that the bar chart creates a wide range of values along the horizontal axis is that there are a small number of outlying data points in the variable where there are small yellow bars that are displayed at the upper end of the horizontal axis that is very difficult to view from the previous illustration. For categorically-valued variables, the **X Axis** tab will allow you to view the frequency distribution across the separate class levels.

Adjusting the Axes Boundaries

The node will allow you to set the range of values to the interval-valued axes variables that will be displayed in the following bar chart from the **Result Browser**. Restricting the range of values of the axes variable can only be applied to interval-valued variables in the active training data set. To set the upper limit to the interval-valued variable, select the **Bound** drop-list arrow, and select the **MAX** option, then simply enter the appropriate value from the **Value** entry field. To set the lower limit to the variable, select the **Bound** drop-list arrow, then select the **MIN** option to enter the appropriate value from the **Value** entry field. Otherwise, select the sliders or the black boxes called the *boundary boxes*, which are located on either vertical axis, then hold down the left button on your mouse to scroll across the X axis to set the range of values for the axes variable. The **Value** entry field is synchronized with the slider that displays the corresponding values by scrolling the slider across the range of values from the axis. Therefore, setting the predetermined minimum and maximum values to the axis variable will be reflected in the corresponding charts that are created from the **Result Browser**.

The **Y Axis** and **Z Axis** tabs will not be displayed since the tabs are similar to the previous **X Axis** tab. The only difference in the tabs is that the bar charts are based on the range of values of the selected Y and Z axis variables.

Run the **Distribution Explorer** node to generate the following multidimensional bar charts.

Viewing the Distribution Explorer Results

General Layout of the Results Browser within the Distribution Explorer Node

- **Chart tab**
- **Code tab**
- **Log tab**
- **Output tab**
- **Notes tab**

Chart tab

The following chart is based on the axis variables selected from the tab. Also, the axis values for the 3-D bar chart are based on the range of values specified for each axis variable from the tab.

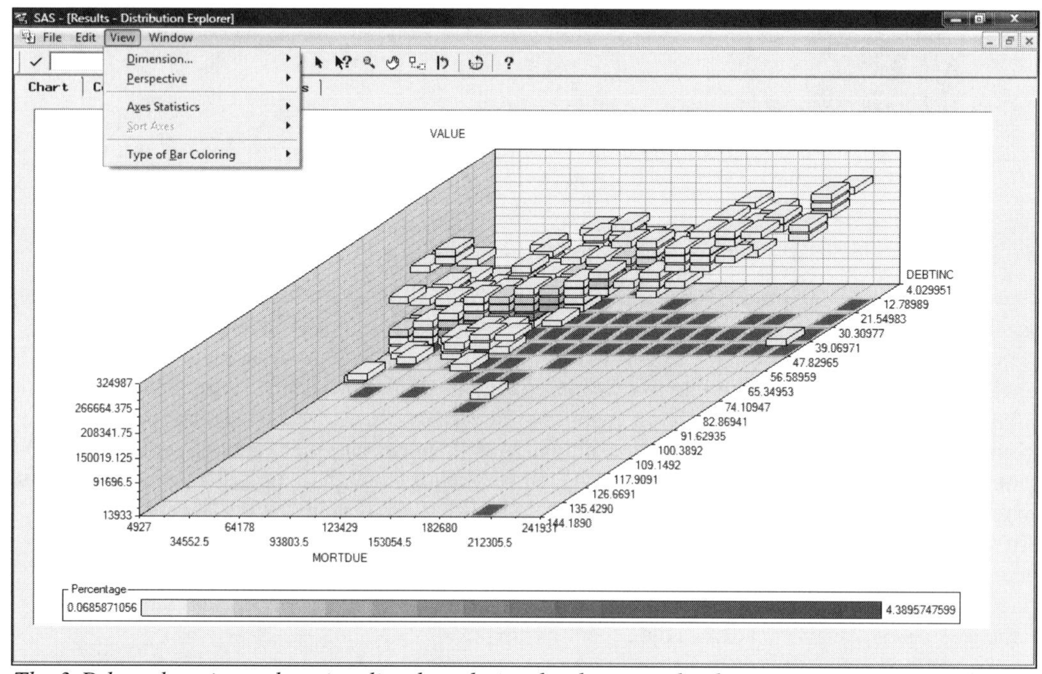

The 3-D bar chart is used to visualize the relationship between the three separate axis variables.

From the **Result Distribution Explorer** window, the various menu options are designed to change the appearance of the frequency bar chart:

- **View > Dimension > Drill Up:** The **Dimension** option will then display the **Drill Up** window that will allow you to increase the number of variables for the chart. However, the **Dimension** menu option is disabled with only one axis variable in the chart.
- **View > Dimension > Drill Down:** The **Dimension** option will then display the **Drill Down** window that will allow you to decrease the number of variables for the chart.

  Note that both dimension options are related to each other. In other words, either option is available in order to add or remove the axes variables from the currently displayed frequency bar chart. By default, the **Drill Down** option will be available that will allow you to reduce the selected variables from the current chart. Conversely, select the **Drill Up** option to instruct the node to restore the axis variables back to the multidimensional chart.
- **Perceptive > 2D View:** The **Perceptive** option sets the mode in viewing the charts to a two-dimensional frequency chart. When there are two separate axis variables in the chart, then the chart will display the top of each bar where you must rely on the color coding schema to determine the size of each bar. This option is not available with three separate axes variables in the chart.

- **Perceptive > 3D View** (default)**:** The **Perceptive** option sets the mode in viewing the charts to a three-dimensional frequency chart.
- **Perceptive > 3D Depth View:** The **Perceptive** option will display an overhead or aerial view of the three-dimensional frequency chart that is designed to make it easier to view the hard-to-see bars that maybe buried among the larger bars.
- **Axes Statistics > Response Axis > Frequency:** The **Axes Statistic** option will allow you to display the frequency values of the corresponding variable within each interval for continuous variables or each group level for categorically-valued variables in the chart.
- **Axes Statistics > Response Axis > Percentage** (default)**:** The **Axes Statistic** option will allow you to display the frequency percentage of the corresponding variable within each interval for continuous variables or each group level for categorically-valued variables in the chart.
- **Sort Axes > Data Order:** The **Sort Axis** menu options are only available for categorically-valued axis variables in the chart. This option sorts the axis values in the original order that is displayed in the data.
- **Sort Axes > Format > Sort** (default)**:** The values across the axis are based on the formatted values assigned to the categorically-valued axis variable in the chart.
- **Sort Axes > Sort > Format:** The values across the axis are based on the raw values of the categorically-valued axis variable, then applying the assigned formats.
- **Type of Bar Coloring > Top:** Changes the coloring settings to the top of the displayed bars.
- **Type of Bar Coloring > All** (default)**:** Changes the coloring settings to the displayed bars.

Code tab

The **Code** tab is designed to display the internal SEMMA training program to the node that generates the listed output in the **Output** tab. Again, if all the axis variables of the chart are categorical, then the **Code** tab will display the listed code based on the execution of the PROC SQL procedure to extract the data. In addition, if the **Chart Only** option is not selected from the **Variables** tab, then the code will display the PROC FREQ procedure that will generate up to a three-dimensional frequency table listing of the categorical axes variables. However, if there are any interval-valued axes variables to the chart, then the PROC MEANS procedure is automatically applied. The PROC MEANS procedure will produce the descriptive statistics based on the range of values of the interval-valued variable by each class level of the categorically-valued axis variable. If the axis variables are all interval-valued responses, then the PROC MEANS procedure listing will display the descriptive statistics for each interval-valued axis variable to the chart. This is illustrated in the following diagram. The **Output** tab will not be illustrated since it simply lists the procedure output listing from either the PROC FREQ or PROC MEANS procedures.

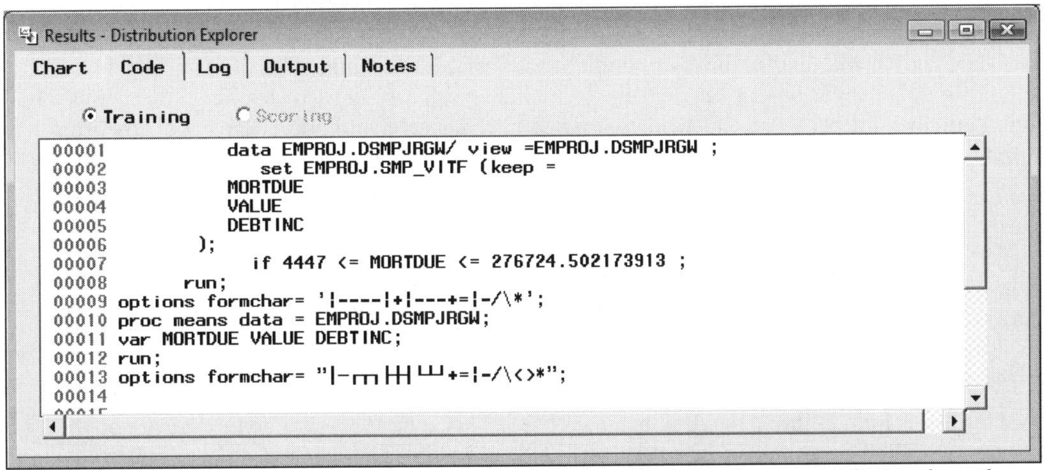

Results - Distribution Explorer

Chart | Code | Log | Output | Notes

Training   Scoring

```
00001            data EMPROJ.DSMPJRGW/ view =EMPROJ.DSMPJRGW ;
00002               set EMPROJ.SMP_VITF (keep =
00003            MORTDUE
00004            VALUE
00005            DEBTINC
00006         );
00007              if 4447 <= MORTDUE <= 276724.502173913 ;
00008         run;
00009 options formchar= '|----|+|---+=|-/\*';
00010 proc means data = EMPROJ.DSMPJRGW;
00011 var MORTDUE VALUE DEBTINC;
00012 run;
00013 options formchar= "|-┌┬┐├┼┤└┴┘+=|-/\<>*";
00014
```

*The **Code** tab that displays the training code that generates the procedure output listing from the interval-valued axis variables.*

2.2 Multiplot Node

General Layout of the Enterprise Miner Multiplot Node

- **Data tab**
- **Variables tab**
- **Output tab**
- **Notes tab**

The purpose of the **Multiplot** node in Enterprise Miner is that it will allow you to graphically view the univariate or bivariate distribution of the variables in the active training data set. The node is designed to create bar charts, stacked bar charts, and scatter plots. The plots are designed to determine the statistical distribution in the variables and to discover outliers, and determine trends and patterns in the input data set. The **Multiplot** node creates both bar charts and scatter plots for the variables in the active training data set. By default, the node is designed to create only bar charts. However, scatter plots can be generated in order to view the functional relationship between the target variable and each input variable in the data set. The assortment of charts and graphs can be viewed from the automated slide show that will automatically scroll through each graph that is created. From the slide show, the node will allow you to quickly view outliers and normality in each input variable in the predictive model and also determine the functional relationship between each input variable and the target variable you want to predict. The node will allow you to manage the charts and graphs that are created by preventing certain charts and graphs from being displayed from the built-in slide show.

Bar Charts

By default, bar charts are generated from the node. In predictive modeling, the slide show that is generated from the node will allow you to view both outliers and normality in the variables in the statistical model. Bar charts are designed to display the distribution of each variable separately by partitioning the interval-valued variable into separate intervals of equal length or to view the distribution of the separate class levels of the categorically-valued variable. The bar chart or histogram is designed to view the distributional form of the selected variable. The bar chart is constructed when the length of the rectangular bars are positioned in the center of the midpoint of each interval for continuous variables. The purpose of the frequency plot is to display either the specific values or a range of values that are most frequent, whether the range of values are concentrated about its mean, whether the distribution is symmetrical or skewed, and whether the distribution is multimodal or unimodal. The bar chart provides the simplest display for viewing the shape of the univariate distribution across the range of values of the variable. Basically, the number of intervals or the number of categories of the variable and the location of the class intervals determines the shape of the bar chart. The drawback to the histograms is that they can be misleading based on a small sample size with the different diagrams created from a wide variety of values and several choices for each end of the intervals. However, for large samples the diagram will display the distribution of the variable. Assuming that the sample size is sufficiently large and there exist several class levels or small intervals within each class level that forms a bell-shaped distribution, then the bar chart can be approximated by a normal distribution. The bar charts listed below are used to either display frequency counts or frequency percentages in the data.

The possible bar charts that are created within the **Multiplot** node are the following:

- Bar chart of each input or target variable
- Bar chart of each input variable across each class level of the target variable
- Bar chart of each input variable grouped by each interval-valued target variable

Stacked Bar Charts

Line plots are far better for presenting the data than bar charts. This is particularly true in comparison to stacked bar charts. Stacked bar charts are bivariate bar charts. In stacked bar charts, each bar in the frequency bar chart represents a certain level of the categorical variable and each bar in the stacked bar chart represents a certain class level of an additional categorical variable. The advantage of the stacked bar charts is that these charts will allow you to view the relationship between two separate categorically-valued variables by viewing the range of values of the interval-valued variable or the various class levels of a categorically-valued variable within each class level of the categorically-valued variable that will allow you to view the differences within the groups by scanning across the vertical bars. The shacked bar charts are very good for visualizing the proportionality between the first grouping that is represented by the bars positioned in the bottom stack. However, the major drawback in interpreting the tiered bar chart is viewing the proportionality between each bar that is positioned above the bottom tier. Also, the limitation of the bar charts is that it is harder to visualize the relationship or interaction between the variables in the data in comparison to line plots.

Scatter Plots

The node is designed to create scatter plots based on the target variable in the active training data set. In predictive modeling, the slide show that is created from the node will allow you to view the functional relationship between the target variable and each input variable in the statistical model. Scatter plots are designed to observe the distributional relationship between two or three separate variables. The most common way to visualize the relationship between two or three separate interval-valued variables is by constructing scatter plots. Scatter plots are typically applied to continuous variables. However, scatter plots may be drawn for all kinds of variables. At times, scatter plots in data mining might not be that useful since the data set can potentially consist of an enormous amount of data. The plot could potentially display a huge cloud of points that might not be that informative. Therefore, contour plots might be constructed. The bivariate plot is also useful in observing bivariate normality, spotting outliers, determining various groupings or showing intercorrelation and trends between two variables at a time.

In cluster analysis, it is important to determine the appropriate number of clusters. Therefore, one of the first steps to the analysis is constructing scatter plots that will provide you with the ability in hopefully viewing a clear separation in the various cluster groupings that exists between two separate interval-valued variables that is defined by the categorically-valued variable that formulates the various cluster groupings.

Scatter plots are designed to display either linear or curvature relationships between the two separate variables. For linear statistical modeling designs, appropriate data transformations might be applied in order to achieve linearity between both variables. In regression modeling, correlation analysis is often applied to examine the linear relationship between two separate interval-valued variables in order to determine the degree of linearity between the variables. However, before performing correlation analysis, it is recommended to construct scatter plots in order to view the functional relationship between both variables. For example, a strong curvature relationship might exist between both variables with a correlation statistic close to zero or outlying data points might influence high correlation that is far beyond the rest of the other data points. Scatter plots provide a diagnostic tool enabling you to determine if your predictive model is an adequate fit to the data in determining both the mathematical form between the input variables and the target variable that you want to predict and the various statistical modeling assumptions that must be satisfied in the model. The scatter plot of the residuals and the predicted values is typically used in determining uniform random variability and nonlinearity in the multiple linear regression model. Scatter plots will allow you to view both trend and variability between one variable across the increasing values of the second variable in the plot. One assumption we must check in traditional regression modeling is that the error terms are chronologically independent or uncorrelated to one another over time. In other words, scatter plots must be constructed to verify that the error terms are randomly distributed, centered about zero across the fitted values, all the input variables in the model, and over time. Since we must have independence in the data, therefore a counter or time identifier variable is used to keep track of the chronological order in the data by checking for cyclical patterns in the residuals against the identifier variable over time from the plot. If the independence assumption is not valid, the error terms are said to be nonnormal or the error terms are autocorrelated with one another over time. If the error terms display a cyclical pattern over time, then it may be due to one of two reasons. Either the incorrect functional form has been applied to the model or there exist a high degree of correlation between successive error terms over time. In time series modeling, it is important to first check for stationarity in the data by constructing scatter plots of the target values across time. In time series modeling in predicting the current target values as a function of its past values, the data must be stationary. If the time series scatter plot display a uniform fluctuating behavior with any increasing or decreasing trend over time, then the time series model can be determined.

In statistical modeling, scatter plots or line plots are used to determine interaction effects between the response variable and two separate input variables in the model. In other words, at times, interaction effects can increase the accuracy in the prediction estimates, however the drawback is that it will make it much harder to interpret the linear relationship between the target variable and the input variables involved in the interaction. Scatter plots or line plots can be used in detecting interacting effects in the statistical model. If there is no interaction between both variables, then each line will be relatively parallel to each other. Assuming that interaction is not present in the statistical model, then it is also important in detecting confounding in the statistical model. Confounding is present in the model when the primary input variable might not explain the variability in the target variable without the existence of an additional confounding input variable in the model.

In logistic regression modeling, one of the first steps is constructing several scatter plots to visually observe the linear relationship between the estimated logits of the target variable and each input variable in the model. Similar to linear regression modeling, it is assumed that there is a linear relationship between the target variable and each input variable in the logistic regression model. In classification modeling, once the model has been fitted, it is important to evaluate the accuracy in the logistic regression model and identifying outlying data points where certain observations fit the model poorly by constructing a scatter plot of the residual values across the estimated probabilities. The accuracy of the model can be determined by overlying a smooth curve between the calculated residuals across the estimated probabilities of the target class levels. The accuracy of the logistic regression model can be determined when the smooth curve does not display a trend across the estimated probabilities with a slope of approximately zero that also crosses the intercept of the residuals at approximately zero. In addition, it is important to determine which observations have a great deal of influence in the final estimated probabilities by constructing various scatter plots of the diagnostic statistics across the estimated probabilities. The diagnostic scatter plots are designed to observe outlying data points that are well separated from the other data points. When you are modeling ordinal-valued target variables, it is important that the "*proportional odds assumption*" is satisfied. The assumption can be visually satisfied by constructing several scatter plots to observe that there is no overlap in the cumulative estimated probabilities at each target class level across each input variable in the classification model.

The possible scatter plots that are created within the **Multiplot** node:

- Scatter plot of each interval-valued input variable across the interval-valued target variable
- Scatter plot of each categorically-valued input variable against the interval-valued target variable

The various graphical techniques are discussed in Friendly (1991), Carpenter and Shipp (2004).

Data tab

The **Data** tab is designed to select the active training data set in order to create the various plots and graphs within the node. Furthermore, the tab will allow you to select a subset of the active training data set. The tab is also designed for you to view the selected training data set. Press the **Select...** button to view the data sets that are available within the process flow diagram. By default, the **Validation** and **Test** radio buttons will be dimmed and unavailable for selection, assuming that a validation or test data sets are not created within the process flow. The tab has the same appearance and functionality as the previously displayed **Data** tab. Therefore, the tab will not be displayed.

From the **Imports Map** window, the tab will allow you to select an entirely different active training data set that is used in producing the various graphs within the node. The listed data sets within the window are based on the various data sets that have been created from previous nodes that are connected to the **Multiplot** node. The **Imports Map** window will provide you with a browsable interface for you to select your favorite data set.

By default, **Use samples of data** is unchecked. However, selecting the **Use samples of data** check box will allow you to select a subset of the active training data set. This will result in the node performing a random sample of the active training data set. The **Sampling options** window will be displayed for you to select the number of observations and the random seed for the random sample.

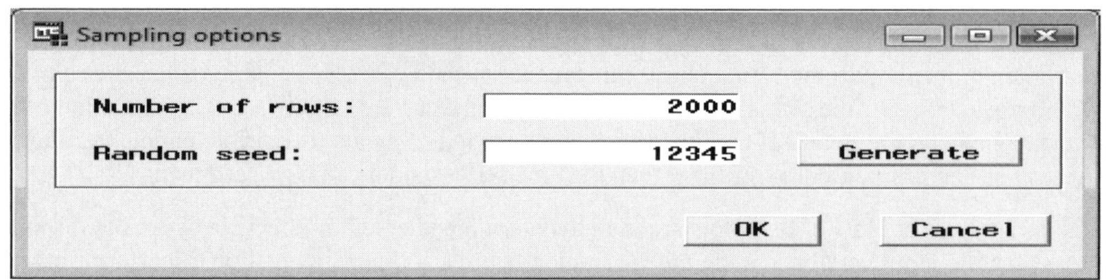

*The **Sampling options** window is used to select the number of records and random seed for the sample.*

Variables tab

When the **Multiplot** node is opened, the **Variables** tab will first appear. The **Variables** tab will allow you to define the variables for the bar charts and scatter plots. You may also view the frequency distribution of each variable in the data mining data set. By default, all the variables are plotted with a variable attribute status of **use.** To prevent the variables from being plotted, simply select the corresponding variable rows and scroll over to the **Status** column, then right-click the mouse and select the **Set Status** pop-up menu item to select the variable status of **don't use**. All variables with a variable role of **rejected** are automatically set to a variable status of **don't use** and are excluded from the following plots. As opposed to many of the following modeling nodes, the **Multiplot** node will allow you to specify any number of target variables within the node in order to view the distribution between all other input variables in the active training data set.

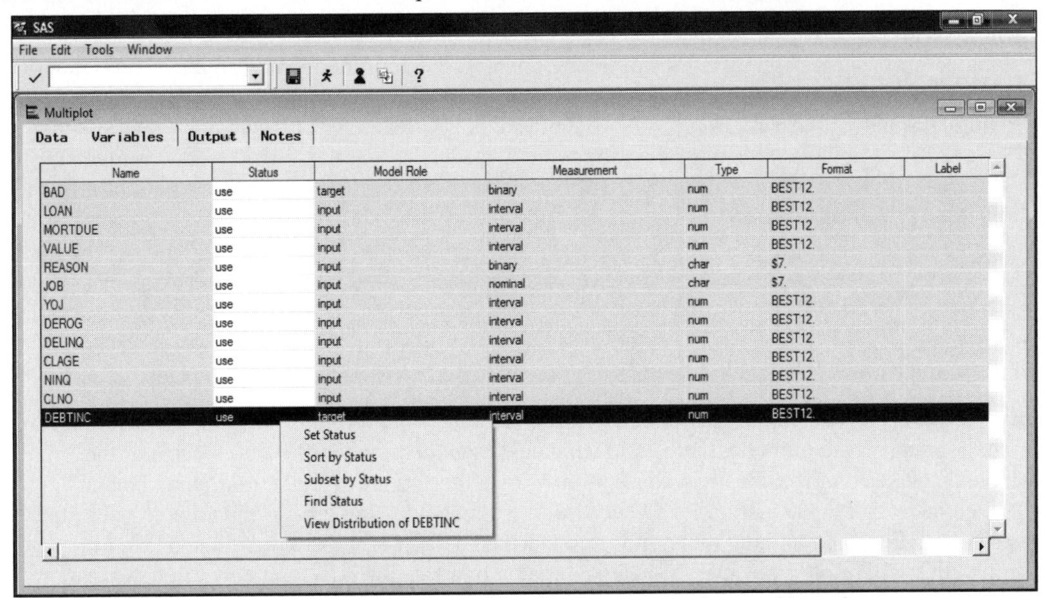

*The **Variables** tab is used to specify the variables for the scatter plots and bar charts.*

To view the various option settings for the bar charts and scatter plots, select the **Tools > Settings** main menu options or select the keyhole **Settings** toolbar icon that is located on the **Tools Bar** menu. The following **Multiplot Settings** window will then appear for you to specify the various option settings for the bar charts and scatter plots. You may also select the default font type for the graphs and plots.

The following are all of the possible bar charts and scatter plots that can be created within the node that depends on the existence of the target variable in the data set or not.

Target Variable	Charts	Description
No Target Variable	Univariate bar charts	Bar charts are created for each variable with a variable role of **input**.
Target Variable	Bivariate bar charts	Bar charts are created for each target variable across the midpoint of each interval-valued or the class level of each categorically-valued input variable with a variable role of **input**.
	Scatter plot	Scatter plots are created for each target variable against each input variable with a variable role of **input** and a variable attribute status of **use.**

Bar charts tab

The **Bar charts** tab is designed for you to specify the various configuration settings for the bar charts.

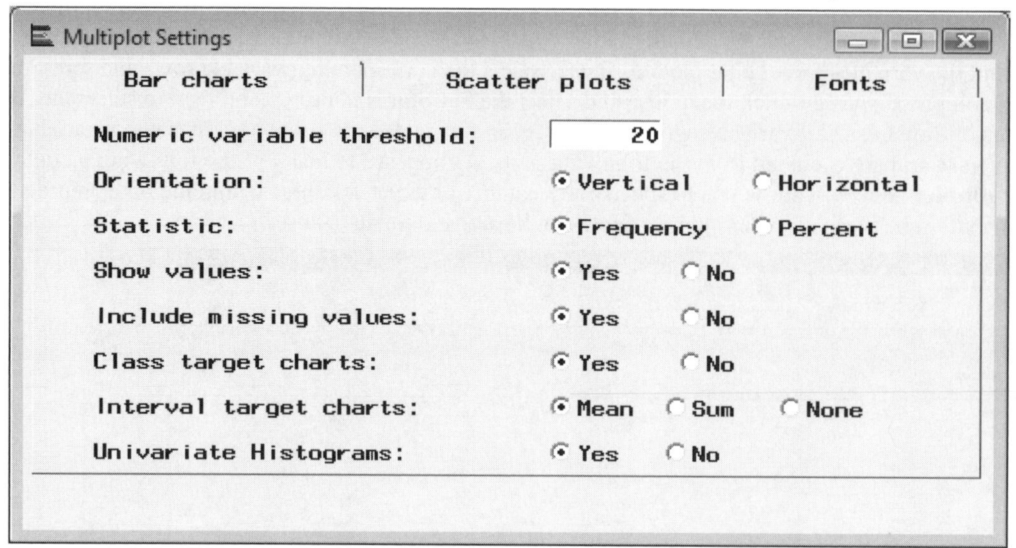

Setting the various configuration settings to the bar charts that are automatically displayed.

The following are the various configuration settings for the bar charts within the **Bar charts** tab:

- **Numeric variable threshold:** This option is design to determine if binning will be applied to the bar chart. Specifying a numerical threshold value less than the number of unique levels of the variable displayed on the x-axis will cause binning to be performed with the midpoints being displayed instead of the actual levels. Otherwise, if a threshold value greater than the number of levels is specified, then binning will not be applied. The default is a threshold value of 20. For example, this will result in an interval-valued variable or a categorically-valued variable with more than 20 unique class levels collapsed into 20 separate levels. In other words, 20 separate bars that will be displayed within the chart.
- **Orientation:** This option sets the orientation of the bar chart to either a vertical or horizontal orientation. The default is a vertical orientation.
- **Statistic:** This option sets the frequency value or percentage to the vertical axis of the bar chart. By default, the charts are displayed by the frequency counts.
- **Show values:** By default, each bar displays the corresponding statistical values that are displayed at the top of each bar. Otherwise, select the **No** radio button to prevent the values from being displayed on the top of each bar that is created.
- **Include missing values:** This option treats missing values as a separate category. By selecting this option, then an additional bar will be created that will represent missing values in the variable.
- **Class target charts:** This option creates frequency bar charts for each class level of the categorically-valued target variable.
- **Interval target charts:** This option creates a bar chart at each interval of the interval-valued target variable by either the **Mean**, **Sum**, or **None**. **None** will display the frequency counts.
- **Univariate Histograms:** This option is designed to create a separate bar chart for each variable with a variable status of **use**. The **Variable Histogram** window will appear that displays the standard univariate frequency bar chart. By default, both the standard univariate histogram and stacked bar charts are created. However, selecting the **No** radio button will create only stacked bar charts based on a target variable with two or more levels. The stack bar charts are based on the class levels of the categorically-valued target variable in the input data set.

Scatter plots tab

The **Scatter plot** tab is designed to specify the various option settings for the scatter plots. The scatter plots are based on the target variable of the data set. If a target variable is not specified in the input data set, then the following scatter plots will not be created and the bar charts with separate class levels of the target variable will not be created. The tab will allow you to generate the bivariate scatter plots or not. By default, the various scatter plots are not automatically generated unless specified otherwise.

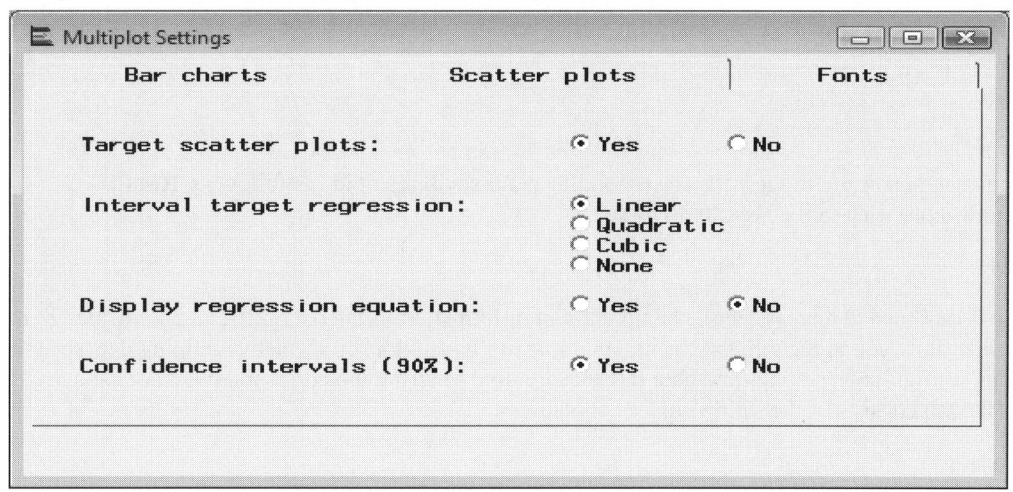

The option settings of the scatter plots are based on the target variable in the input data set.

The following options are the various configuration settings to the scatter plots within the **Scatter plots** tab:

- **Target scatter plots**: The target scatter plots display the distributional relationship between all interval-valued target variables against each interval-valued input variable in the data set. By default, scatter plots are not created based on the target variable in the active training data set.
- **Interval target regression:** This option will allow you to either specify a **Linear**, **Quadratic**, **Cubic** or **No** regression line or trend line to the scatter plot. By default, the options are grayed-out and unavailable for selection unless you specify the corresponding scatter plot of the target variable. In addition, the **Linear** option is automatically selected that will result in the simple linear regression line displayed within the scatter plot.
- **Display regression equation:** Displays the simple linear regression equation with regard to the continuous variables in the simple linear regression model that will be automatically displayed in the lower left-hand corner of the scatter plots.
- **Confidence interval (90%):** Displays the 90% confidence interval of the target mean from the simple linear regression line. The confidence interval is designed to include the most likely values of the target variable with 90% confidence. For instance, with 90% certainty you would conclude that the observed target values will fall within the interval, assuming that there is a linear relationship between the target variable and input variable in the simple linear regression model. By default, the 90% confidence interval is displayed, assuming that the corresponding scatter plots are specified.

Fonts tab

The **Fonts** tab is designed for you to specify the type of fonts that are used in the bar charts and scatter plots. In other words, you may specify the type of font that is used in all titles, axis labels, data points, and the regression equation, assuming that it has been specified.

You may specify a font type other then the available type of fonts that are listed in the display box. Select the last display item, that is, the **User Defined ftext goption** item, and save the changes by closing the node, then from the SAS Program Editor window submit the following SAS statement:

```
goptions ftext=fontname;
```

where the ***fontname*** is the type of font that is applied in the various bar charts and scatter plots.

Viewing the Multiplot Node Results

General Layout of the Results Browser within the Multiplot Node

- **Model tab**
- **Graphs tab**
- **Code tab**
- **Log tab**
- **Output tab**
- **Notes tab**

Running the node and choosing to view the corresponding results will result in the following **Results–Multiplot** window appearing in the **Results Browser**.

Model tab

The **Model** tab is designed to display the file administration information of the data set, a tabular display of the variables that will allow you to reconfigure the variable role of the variables in the active training data set and view the various configuration settings and data sets that are used to create the corresponding bar charts and scatter plots. That tab consists of the following four subtabs:

- **General subtab**
- **Variables subtab**
- **Settings subtab**
- **Status subtab**

General subtab

The **General** subtab displays the file administration information of the plotting data set, such as the creation and modification date of the data set, and the target variable that is assigned to the training data set.

Variables subtab

The **Variables** subtab is similar to the standard **Variables** tab with which you may view the standard frequency bar chart for each variable in the data set. In addition, the tab is also designed for you to reassign new variable roles to the listed variables in the data mining data set.

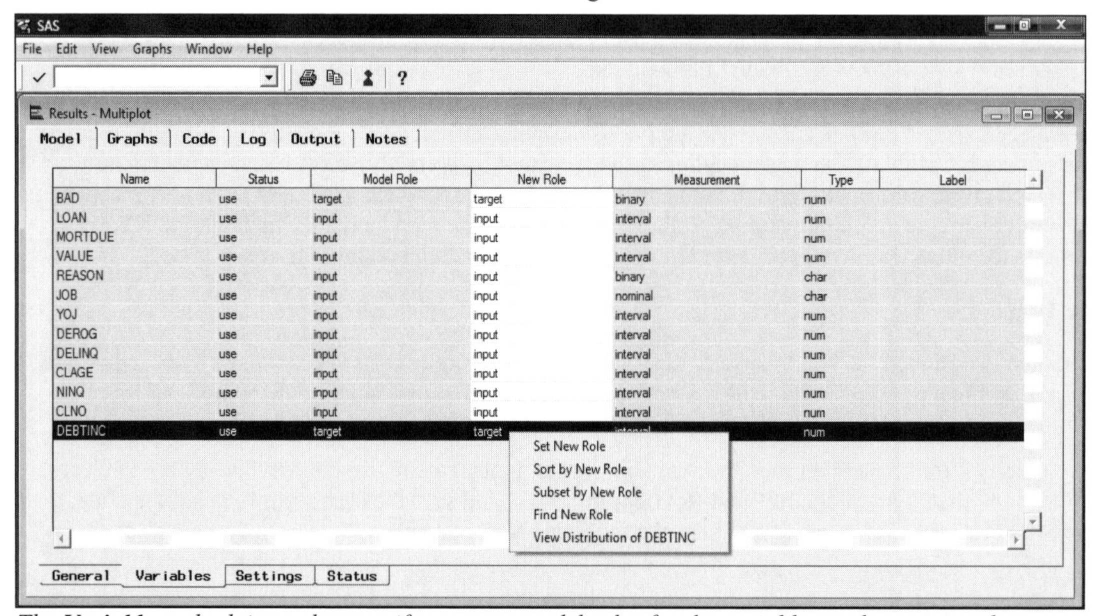

*The **Variables** subtab is used to specify separate model roles for the variables in the training data set.*

The tab will allow you to redefine the variable roles of the variables in the data set. Simply scroll the mouse over to the **New Role** column, then right-click the mouse and select the **Set Variable role** pop-up menu item to redefine the appropriate variable role to the variable. By default, all the variables in the input data set have a variable attribute status automatically set to **use**.

Settings subtab

The **Settings** subtab is designed to display the various configuration settings and the data sets that are used to create the corresponding bar charts and scatter plots. The following listed option settings are specified from the previous tabs within the **Multiplot** node.

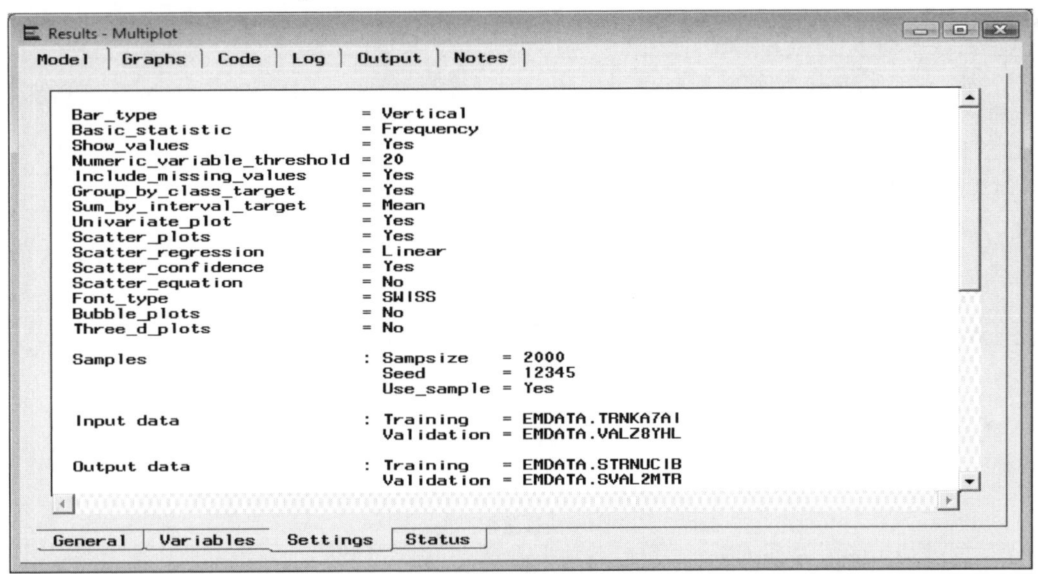

*The **Settings** subtab displays the option settings that are previously specified within the node.*

Status subtab

The **Status** subtab is designed to display the processing status that occurred when executing the node. The subtab displays the creation dates and processing times of the active training and scoring data sets.

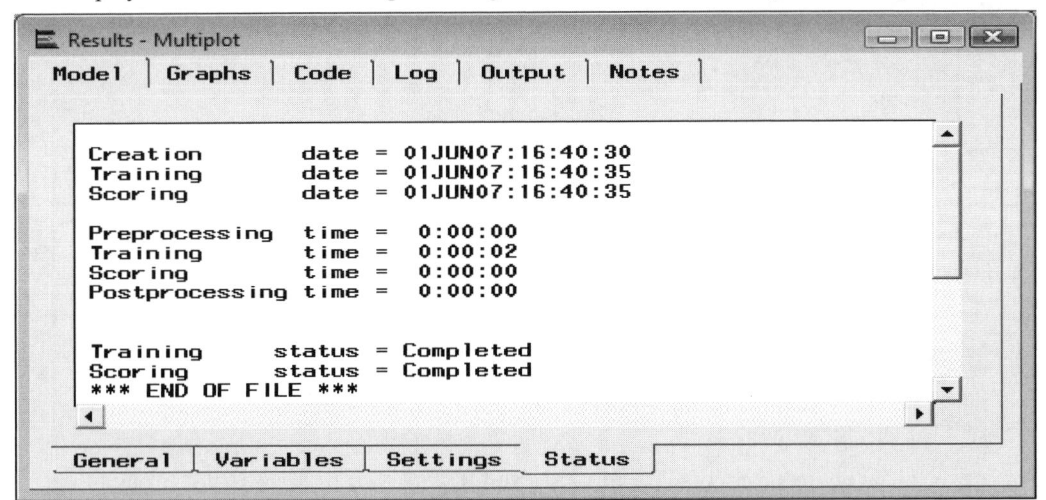

*The **Status** subtab displays various processing information by executing the node.*

Graphs tab

After executing the node, the following **Results–Multiplot** window will appear that will display the various graphs that are generated from the node.

From the **Results–Multiplot** window, there are three separate ways to navigate through the separate graphs in the node:

1. Press the **Page Up** or **Page Down** button from the keyboard to browse backwards or forwards through the list of available graphs that are created within the node.
2. Select the **Graphs > Previous** or **Graphs > Next** to scroll backwards or forwards through the assortment of charts and graphs that are created.
3. From the **Tool Bars** menu, select the left arrow or right arrow icons to go backwards or forwards through the numerous graphs that are created within the node.

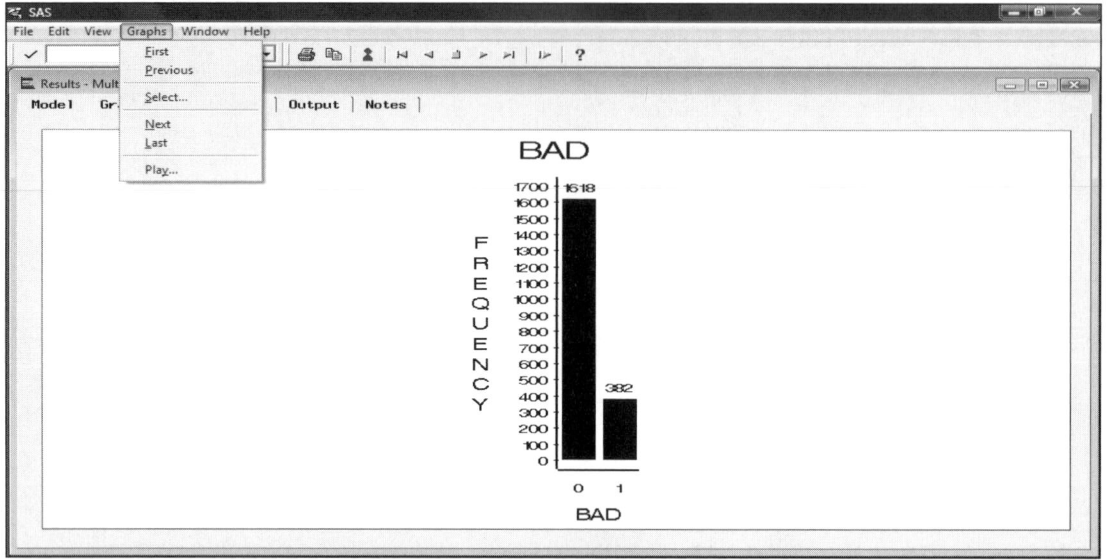

*The **Graphs** tab displays the numerous bar charts and scatter plots of the node.*

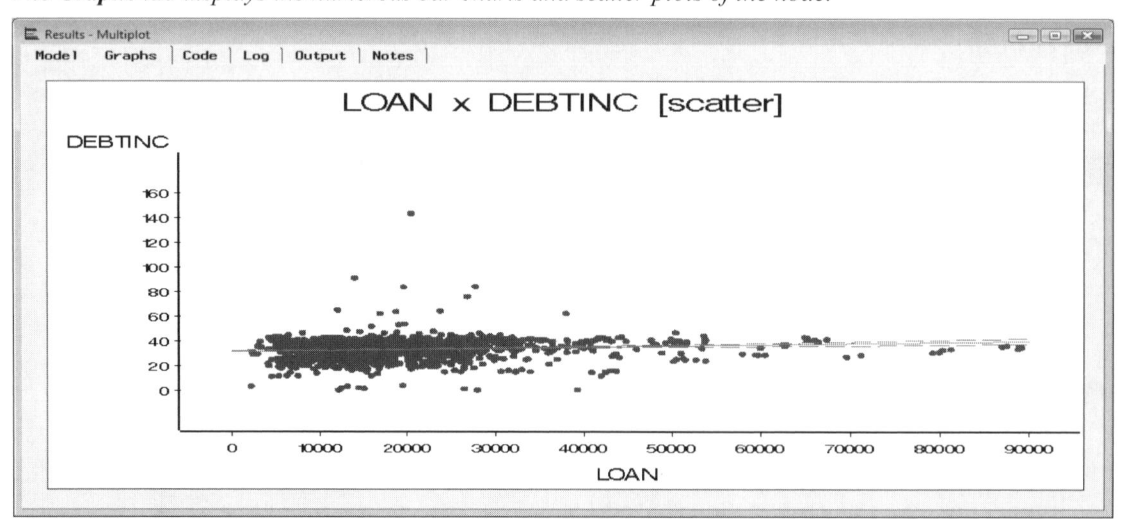

The scatter plot displays the simple linear regression line and the 90% confidence interval by fitting the interval-valued target variable.

The following are the menu options within the **Results–Multiplot** window used to navigate the charts and graphs that are displayed within the tab:

- **View > Detailed Results:** This option will open the **Detailed Results...** window for you to specify the various graphs to be displayed and run within the built-in slide show.
- **Graphs > First:** Navigates to the first graph.
- **Graphs > Previous:** Navigates to the previous graph

- **Graphs > Select:** Opens the **Selection** window that lists the order of every possible bar chart and scatter plot in order to select the corresponding graph to be displayed.
- **Graphs > Next:** Navigates to the next graph within the series of graphs.
- **Graphs > Last:** Navigates to the last graph within the series of graphs.
- **Graphs > Play:** Runs an automated slide show to view the assortment of bar charts and scatter plots by automatically scrolling through each bar chart or scatter plot that is created. To stop the automated slide show from running, simply press the **Enter** key or left-click the mouse on the displayed graph.

Select the **View > Detailed Results** from the main menu options or select the blue keyhole **Settings** toolbar icon to display the following **Detailed Results–Select Graphs** window. The window will allow you to manage all the possible graphs and charts that are automatically generated once you execute the node. By default, every graph is displayed. The slide show will allow you to quickly view the assortment of charts and graphs that are generated from the node. The bar charts will allow you to view the distributional form of each variable in the analysis with a variable status of **use**. The scatter plots of the slide show will allow you to view the functional relationship between the target variable and each input variable in the statistical model. However, you may remove certain graphs from being displayed within the tab. Press the double arrow button to move all the plots from the **Keep** list box to the **Drop** list box to prevent all the plots from being displayed within the **Graphs** tab. This will also prevent the built-in slide show from being displayed. Alternatively, press the single right arrow button to move each plot separately from the **Keep** list box to the **Drop** list box. Conversely, from the **Drop** list box, press the double arrow or the single arrow button to transfer the plots from the **Drop** list box into the **Keep** list box in order to display the plots.

The ***Detailed Result*** *window is used to manage the various bar charts and scatter plots that are created.*

Code tab

The **Code** tab is designed to display the internal SAS SEMMA training code that created the corresponding graphs. The various bar charts are created from the PROC GCHART procedure and the numerous scatter plots are generated from the PROC GPLOT procedure. The code might be used to generate entirely different charts and plots by randomly drawing an entirely different sample from the active training data set.

Output tab

The **Output** tab is designed to display the standard PROC CONTENTS procedure output listing of the plotted data mining data set.

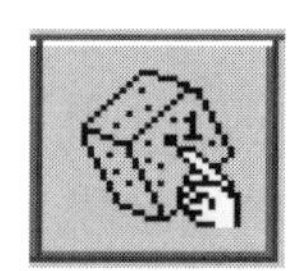

2.3 Insight Node

General Layout of the Enterprise Miner Insight Node

- **Data tab**
- **Variables tab**
- **Notes tab**

The purpose of the **Insight** node in Enterprise Miner is to perform a wide range of statistical analysis along with generating a wide range of graphs and charts. The node opens the SAS/INSIGHT session. SAS/INSIGHT software is an interactive tool for data exploration and statistical analysis. The SAS/INSIGHT interface displays a table view of the input data set that is similar to the SAS/Viewer interface. Several graphs and statistical analyses can be generated based on a number of table views or windows that can be opened within the node. The node is comprised of a data window, graph window, and output window. From the *data window*, you can sort observations, transform variables, perform edits and queries, and save the listed SAS data set. From the *graphics window*, you can create various bar charts, box plots, probability plots, line plots, scatter plots, contour plots, and three-dimensional rotating plots in order to view the distribution of the variables in the analysis. From the *output window*, you can perform a wide variety of analysis from descriptive statistics, both parametric and nonparametric regression modeling and multivariate analysis. The node is designed so that the windows are linked to one another. The windows are linked to each other so that the data is displayed within the data window and then passed on to the subsequent analysis that will be displayed in the output window or the graphics window. In addition, the node is designed for you to create an output data set from the various statistical results. From the node, a random sample of 2,000 observations is performed on the training data set that is automatically used as the analysis data set. The node is discussed in greater detail from my website.

Plots and Graphs

The node creates univariate histograms and bar charts, bivariate box plots, line plots, scatter plots and 3-D contour plots and rotating plots through the SAS/INSIGHT session.

Descriptive Statistics

The node creates various descriptive statistics that can also be stratified into several groups. The node displays various box plots, histograms, and standard univariate results such as means, variance, skewness, kurtosis and the various quartiles and percentile estimates. You may request nonparametric Kolomogorov-Smirnov normality test, student t-test statistics, frequency counts, robust measurement of scale, nonparametric Kolomogorov-Smirnov normality test, trimmed means, and kernal density estimation statistics.

Predictive Modeling

The node can also perform both linear, nonlinear polynomial and logistic regression modeling. Although the node is capable of creating various predictive models, it is recommended that the other modeling nodes be used. This is because the other modeling nodes can be used to fit much larger data sets. By default, the node lists the regression equation, fitted plot, ANOVA table with both the r-square and adjusted r-square statistic, parameter estimates based on Type III testing, and the residual plots. However, you may request the correlation matrix, parameter estimates based on Type I (sequential) testing, confidence intervals of the parameter estimates, collinearity diagnostics, variance–covariance matrix, various leverage plots, probability plot by the residuals, confidence interval plots, spline, kernal, and loess interpolations of the target values.

Multivariate Analysis

Multivariate analysis can be performed from the node. By default, the node generates various univariate statistics in addition to the correlation matrix. However, you may request a wide variety of other multivariate listings such as the covariance matrix, correlation matrix, principal components, canonical correlations, maximum redundancy analysis, canonical discrimination analysis, and the associated scatter plots from the statistical results. The node also creates various scatter plots for you to observe the multivariate distribution between the variables in the analysis in addition to generating elliptical prediction interval in order to observe the various multivariate groupings within the scatter plots.

2.4 Association Node

General Layout of the Enterprise Miner Association Node

- **Data tab**
- **Variables tab**
- **General tab**
- **Sequences tab** (*sequence analysis only*)
- **Time Constraints tab** (*sequence analysis only*)
- **Sort tab**
- **Selected Output tab**

The purpose of the **Association** node in Enterprise Miner is designed to identify or measure the associations and relationships between items or events within the input data set. The node is designed to perform either association discovery analysis or sequential discovery analysis. This type of association or sequence analysis is often called *market basket analysis*. A simple example of market basket analysis is if a customer buys product A, then what are the chances that the same customer will also purchase product B or C ?. The goal to the analysis is basically to determine the buying behavior of customers who purchase a certain combination of items or purchase items in a particular sequence. However, management is usually not only interested in the most popular combination of items that their customers purchase, but, more importantly, the amount of money spent. The association rule of the analysis is defined by the frequency counts or the number of times that the number of combinations or interactions of two or more items occur. The general rule is that if item or event A occurs, then item or event B will occur a certain percentage of the time. Association rules are the most well-known techniques used in discovering associations, interactions, and relationships between the variables in the data. The main goal of the analysis is to determine the strength of the association rules among a set of certain items. The strength and weakness of association analysis is that understanding the calculations and interpreting the results is fairly easy. However, the difficulty in the analysis is selecting the correct set of items that are related to each other.

From the various items that are selected, link analysis might be performed in order to view the separate items that are associated with each other. In addition, the link diagram will allow you to graphically view the strength of each item that is associated with each other based on the color and thickness of the line that connects the two separate nodes. The drawback of the association rules is that they cannot be used to calculate predictions. However, decision trees can be applied to create both classifications or predictions that are based on a set of a certain combination of association rules. These if-then rules are analogous to a backward search of the decision tree, where there are as many rules as there are leaves in the recursive tree branching process.

Data Structure and Data Process of the Association Data Set

The general database layout in market basket analysis must be hierarchically structured such that there are several observations per customer with separate observations for each transaction. The input data set in association analysis must have both a categorically-valued target variable with each class level identifying each transaction and an id variable to identify each item. In sequence discovery, an additional sequence variable must exist in the data set that identifies the sequential order of the numerous transactions.

Initially, the **Association** node computes the single items that have the best support with the largest relative frequency of each separate item. Single items with a support less than the threshold value, that is, those specified from the **General** tab, are discarded. At each subsequent pass through the data, all combination of items are formed based on the single items that are determined from the first pass. During the second pass through the data, all paired items are retained, with a support and confidence probability that is greater than some threshold value, and so on. The process stops when there are no more combinations of items that meet the specified threshold.

Association Discovery

For example, consider the following association rule of A → B. In other words, A preceding B, where A and B might represent either a particular product item or a popular website. The item or event A is called the antecedent, the body of left-hand side of the rule, and B is called the consequence, head, or right-hand side of the rule. A *rule* consists of a condition (body) and a result (head) that is defined in the following logical form.

if *condition* then *result*

To better understand the association or interaction between two separate items or events, it is best to construct various two-way frequency tables. The support (occurrence probability), and the confidence (conditional probability), are the most common type of probabilities used to determine the strength of the association among a set of items. From the following two-way frequency table, the support is the ratio between the frequency count of both events occurring over the total sample size. A low support will indicate to you that there is a small chance of a person purchasing both items. The confidence of the rule is based on the conditional frequency, which is the frequency count of both items occurring divided by the marginal total of the body to the rule. The lift is the ratio between both items of the two marginal frequencies.

It is important to understand that high confidence and support does not necessarily imply cause and effect. Actually, the two separate items might not even be related. For example, from the following table, the association rule of A → B has a support of (4,000/10,000) or 40% and a confidence of (4,000/7,000) or 57%. This would lead you to believe that having item A leads to having item B. However, those not having item A are even more likely to have item B (2,000/3,000) or 66%. This indicates that there is a negative association between both items that can be identified by the lift value.

Frequency Table of Item A by Item B

The FREQ Procedure

Table of A by B

A / Frequency	B: No	Yes	Total
No	1000	2000	3000
Yes	3000	4000	7000
Total	4000	6000	10000

The following support, confidence, and lift statistics are used to measure the strength of an association rule:

- **Support(%):** for A→B is the proportion of times that the rule A and B occur together divided by the number of rules in the data set, that is, if customers purchases item A, then they will also purchase item B which is defined as follows:

$$\text{Support}(A \rightarrow B) = \frac{\text{transactions that contain every item in A and B}}{\text{all transactions}}$$

$$= N_{A \rightarrow B} / N \text{ , that is, } N_{A \rightarrow B} = \text{relative frequency of A then B occurring}$$

$$= 4{,}000 / 10{,}000 = 40\%$$

Note: The support is symmetric since the probability is based on the association or interaction of both items or events divided by the total sample size.

- **Confidence(%):** for A→B is the proportion of times that the rule will contain the left side A that will also contain the right side B. For example, the percentage of customers who bought item B after they have purchased item A that is based on Bayes theorem as follows:

$$\text{Confidence}(A \rightarrow B) = \frac{\text{transactions that contain every item in A and B}}{\text{transactions that contain every item in A}}$$

$$\text{Confidence}(A \rightarrow B) = (N_{A \rightarrow B} / N) / (N_A / N) = (N_{A \rightarrow B} / N_A) = \text{Support}(A \rightarrow B) / \text{Support}(A)$$

$$\text{Confidence}(A \rightarrow B) = 4{,}000 / 7{,}000 = 57.14\%$$

$$\text{Confidence}(B \rightarrow A) = 4{,}000 / 6{,}000 = 66.67\%$$

The confidence is the ratio between the transactions that contain every item in A and B over the transactions that contain A items. The confidence is the most often used measure of an association rule. Although the order of the items is of no concern in association analysis, the order of the items or events does matter with regard to the confidence of the rule, that is, the left-hand side of the rule A preceding the right-hand side of the rule B. The statistic measures the strength between the two items or events. The higher the confidence indicates the greater the chance that the person will buy item A given that they will subsequently purchase item B.

- **Lift** of A→B measures the strength of the association between two items. A lift value of 2 will indicate that the randomly selected customer is twice as likely to purchase item B given that they already purchased item A as opposed to having not purchased item A. Lift values greater than one will indicate a useful rule. In other words, the lift value indicates how much better a rule is in predicting the association in comparison to randomly guessing the association.

$$\text{Lift}(A \rightarrow B) = \text{Confidence}(A \rightarrow B) / \text{Support}(B)$$
$$= \text{Support}(A \rightarrow B) / [\text{Support}(A) \cdot \text{Support}(B)]$$
$$\text{Lift}(A \rightarrow B) = .5714 / (6{,}000 / 10{,}000) = 95.23\%$$
$$\text{Lift}(B \rightarrow A) = .6667 / (7{,}000 / 10{,}000) = 95.23\%$$

Note that the lift is reflexive, that is, the lift value of the rule A → B is the same as the lift value of the rule B → A. The lift value is the ratio that is defined by the relative frequency of both items or events occurring and the relative frequency of the each item or event independently. A lift value greater than one will indicate to you that there is a positive association or correlation, and the significance of the association and the likelihood of the right-hand side increasing given the left-hand side. Conversely, lift values less than one will indicate a negative association. In other words, the association should be interpreted with caution and that randomly guessing the association is better than predicting the association. Therefore, to obtain reliable associations, it is important to retain the rules with the highest lift values. The advantage of the lift value statistic is that you can place an interval bound about this statistic, assuming you have a relatively large active training data set. This interval statistic is analogous to the odds ratio statistic as follows:

$$\log(\text{lift}) \pm z_{1-\alpha/2} \cdot \frac{1}{\sqrt{\text{support}(A \rightarrow B) - 1/N + 1/\text{support}(A) + 1/\text{support}(B)}}$$

Taking the exponential of the statistic will result in the confidence interval of the lift statistic. The strength of the rule can be determined by a relatively large support and confidence probability along with a large lift ratio greater than one. In addition, all rules with a low support probability along with a high confidence probability should be interpreted with caution.

Note: Since the denominator of the lift statistic is based on the relative frequency of each item separately, the lift value statistic is absolute nonsense in sequence discovery.

Association rules might involve various patterns called item sets. So far you have observed items with association rules of order 2, that is, A→B. A simple association rule of order 3 would be $(A \cap B) \rightarrow C$ or if (A and B) then C. In other words, if a customer buys products A and B, then she will also buy product C. Therefore, in order to reduce the number of possible combinations, it may be advisable to reduce the number of combinations to analyze.

Sequence Discovery

The node is also capable of performing sequence discovery. The difference between sequence discovery as opposed to association discovery is that sequence discovery takes into account the order or timing in which the items or events occur for each customer. In Enterprise Miner, a sequence, ordering, or time stamp variable must exist in the data set that identifies the order of occurrence in which the items or events occur for each customer. As an example of sequence analysis, suppose a person purchased item A, then what are the chances that they will then purchase item B, with the added condition that the person has earlier purchased item A, then purchase item B? In sequence discover, the meaning of support is the number of times A precedes B. The confidence is interpreted as the probability of event B occurring conditioned on the fact that event A has already occurred, given that event A occurred earlier in time. In sequence discovery, there are either indirect or direct sequence rules. In indirect sequence discovery, there might be times at which certain events occur between the body A and the head B of the rule that are ignored. Direct sequence looks for occurrences in which A exactly precedes B in succession. The node is designed to define a certain time window for each time sequence, thereby adjusting the window of time in order to observe the trend in the items during the separate time periods.

Transaction Matrix

Since association discovery is typically applied in market basket analysis, the following explanation of the **Association** node and the following results are displayed from association analysis that is based on the data set called ASSOCS that is provided by SAS in the SAMPSIO library. The data set in the following diagram was transformed into a *transaction matrix* in which the rows represent each unique customer and the columns represent the numerous binary indicator variables used in identifying the various items that are purchased, that is, 1-buy or 0-no buy. Since association analysis is based on exploring item-based data to determine whether certain items occur together, the range of values in the **id** and **target** variable to the analysis is usually categorical rather than numeric. The following table listing will allow you to view the various items that have been purchased by the first few customers. Each customer listed can be viewed as a single, large market basket. The **id** variable identifies each customer and the categorically-valued **target** variable, that is, with its values transposed in the following table listing, identifies each item in the analysis.

To visualize the various associations or frequency counts between the *n* items, you would need to collapse the following matrix into an *n* x *n* symmetric matrix that collapses the transaction matrix into two-way associations where the items are comprised of both the rows and columns. The two-way symmetric matrix will indicate the number of times that two items where purchased together. For three-way associations, the interpretation of the various combination of items gets a bit more difficult since the data would then need to be formulated into a symmetric three-dimensional cube, where the same set of items would be represented by the X, Y, and Z coordinates of the three-dimensional cube. The three-way symmetrical cube will indicate the number of times the three separate items where purchased together, and so on.

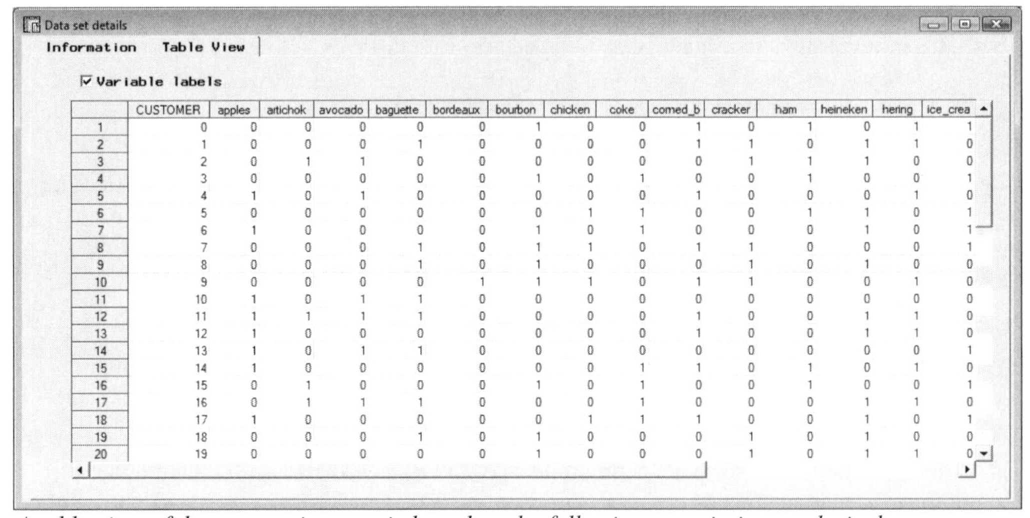

	CUSTOMER	apples	artichok	avocado	baguette	bordeaux	bourbon	chicken	coke	corned_b	cracker	ham	heineken	hering	ice_crea
1	0	0	0	0	0	0	1	0	0	1	0	1	0	1	1
2	1	0	0	0	1	0	0	0	0	1	1	0	1	1	0
3	2	0	1	1	0	0	0	0	0	0	1	1	1	0	0
4	3	0	0	0	0	0	1	0	1	0	0	1	0	0	1
5	4	1	0	1	0	0	0	0	0	1	0	0	0	1	0
6	5	0	0	0	0	0	0	1	1	0	0	1	1	0	1
7	6	1	0	0	0	0	1	0	1	0	0	0	1	0	1
8	7	0	0	0	1	0	1	1	0	1	1	0	0	0	1
9	8	0	0	0	1	0	1	0	0	0	1	0	1	0	0
10	9	0	0	0	0	1	1	1	0	1	1	0	0	1	0
11	10	1	0	1	1	0	0	0	0	0	0	0	0	0	0
12	11	1	1	1	1	0	0	0	0	1	0	0	1	1	0
13	12	1	0	0	0	0	0	0	0	1	0	0	1	1	0
14	13	1	0	1	1	0	0	0	0	0	0	0	0	0	1
15	14	1	0	0	0	0	0	0	1	1	0	1	0	1	0
16	15	0	1	0	0	0	1	0	1	0	0	1	0	0	1
17	16	0	1	1	1	0	0	0	1	0	0	0	1	1	0
18	17	1	0	0	0	0	0	1	1	1	0	0	1	0	1
19	18	0	0	0	1	0	1	0	0	0	1	0	1	0	0
20	19	0	0	0	0	0	1	0	0	0	1	0	1	1	0

A table view of the transaction matrix based on the following association analysis data set.

Missing Values in the Association Node

When there are missing values in the categorically-valued target variable, then the missing values will be assigned as a separate item that will be identified by a period within the listed rules. In association analysis, missing values in the id variable will be assigned as a separate item. In sequence analysis, transactions with missing values in the sequence identifier variable are completely ignored.

Association analysis is discussed in greater detail in Hastie, Tibshirani, and Friedman (2001), Giudici (2003), Berry and Linoff (2004), Hand, Mannila, and Smyth (2001), and Applying Data Mining Techniques Using Enterprise Miner Course Notes, SAS Institute (2002).

Data tab

The **Data** tab is designed to select the input data set to perform association analysis. Otherwise, the active training data set is automatically selected for the subsequent analysis, assuming that the partitioned data sets have been previously created within the process flow. Association analysis is performed on the active training data set only. Similar to the other **Data** tabs, simply press the **Select...** button to display the **Imports Map** window, assuming that there are several **Input Data Source** nodes within the process flow. The **Imports Map** window displays all the previously created data sets that are currently connected to the **Association** node. Click the directory tree check box to interactively find and select the desired data set that will be used in the subsequent analysis. Press the **Close** button to return back to the **Data** tab. Select the **Properties...** button to view both the file administrative information and the table listing of the selected training data set. The following diagram displays a table view of the association data set.

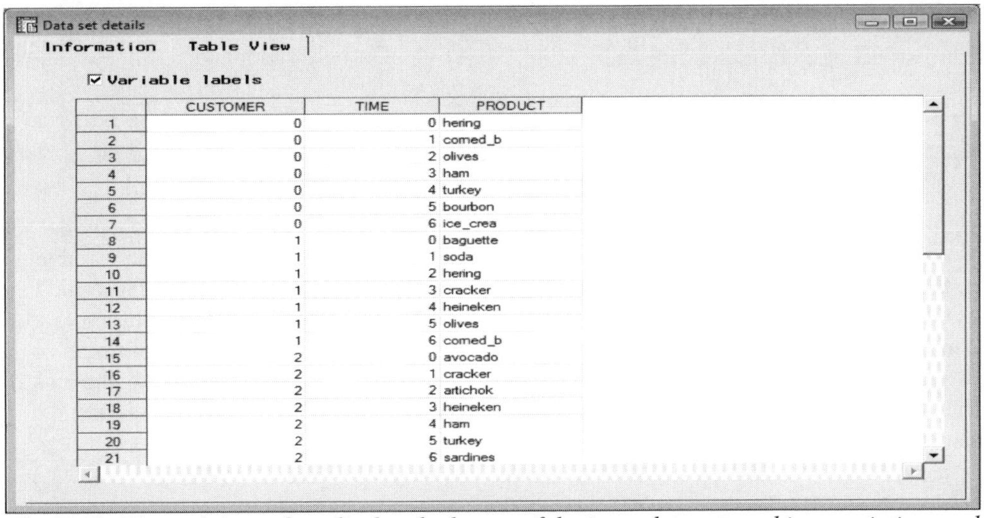

Data set details

Information   Table View

☑ Variable labels

	CUSTOMER	TIME	PRODUCT
1	0	0	hering
2	0	1	corned_b
3	0	2	olives
4	0	3	ham
5	0	4	turkey
6	0	5	bourbon
7	0	6	ice_crea
8	1	0	baguette
9	1	1	soda
10	1	2	hering
11	1	3	cracker
12	1	4	heineken
13	1	5	olives
14	1	6	corned_b
15	2	0	avocado
16	2	1	cracker
17	2	2	artichok
18	2	3	heineken
19	2	4	ham
20	2	5	turkey
21	2	6	sardines

*The **Table View** tab is used to display the layout of the input data set used in association analysis.*

From the above data set listing, notice that the customer identified as zero had first purchased herring, then corned beef, and, finally, some ice cream. This same data set may also be used in sequence discovery with the TIME variable that would be identified as the **sequence** variable in the analysis. The table listing must be closed in order to perform updates to the selected data set.

Variables tab

Initially, the **Variables** tab will appear as you first open the **Association** node. The **Variables** tab is designed to view the variables in the active training data set. The tab lists the various properties of the listed variables that are assigned from the metadata sample. In association analysis, the active training data set must have variables with **id** and **target** model roles. In association discovery, the input variable called TIME that is defined from the **Input Data Source** node is automatically removed from the variable listing and excluded in the subsequent association discovery analysis. The model role assigned to the listed variables must be specified from the previously connected **Input Data Source** node or the **Data Set Attributes** node. In association discovery or sequence discovery, one and only one target variable can be specified in the active training data set. If you define more than one target variable to the data set, usually from the **Input Data Source** node, then the **Target Selector** window will appear once you open the node, which will force you to select the unique target variable from the active training data set. The same restriction applies to sequence variables. In sequence discovery, only one sequence variable can be assigned in the input data set.

From this tab, you have the option to remove certain **id** variables from the analysis by selecting the **Status** column and setting the variable status attribute to **don't use**. Similar to the other **Variables** tabs, you may also view the distribution of each variable separately, by simply selecting the variable row, right-clicking the mouse and selecting the **View Distribution of <variable name>** pop-up menu item in order to view the frequency bar chart of the selected variable. Initially, the histogram might be a good indication of an adequate cutoff value to exclude certain customers or items to remove from the analysis. The metadata sample is used to view the distribution of the variables from the histogram.

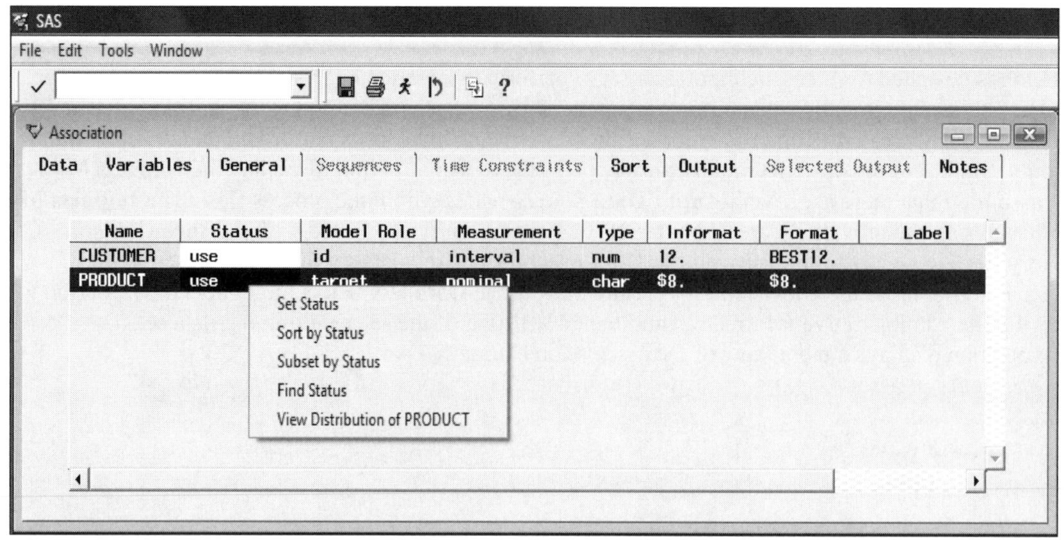

*The **Variables** tab that displays a table view of the variables for association discovery.*

To perform association analysis, there must exist a categorically-valued variable with a **target** role model and a categorically-valued variable in the data with an **id** model role. The data may have more than one variable with an **id** model role, that is, you might have an additional id variable for which the association model would have a separate **id** model role in identifying the various attributes of one variable within another. For example, suppose you are interested in the buying habits of certain customers buying items at particular stores. Therefore, you would assign a primary **id** model role to the variable that identifies the various stores you are interested in selling the items and a secondary **id** model role assigned to a separate variable identifying customers who purchase items at these same stores.

General tab

The **General** tab is designed to set the mode of the association analysis and control how many association rules that will be generated within the **Association** node.

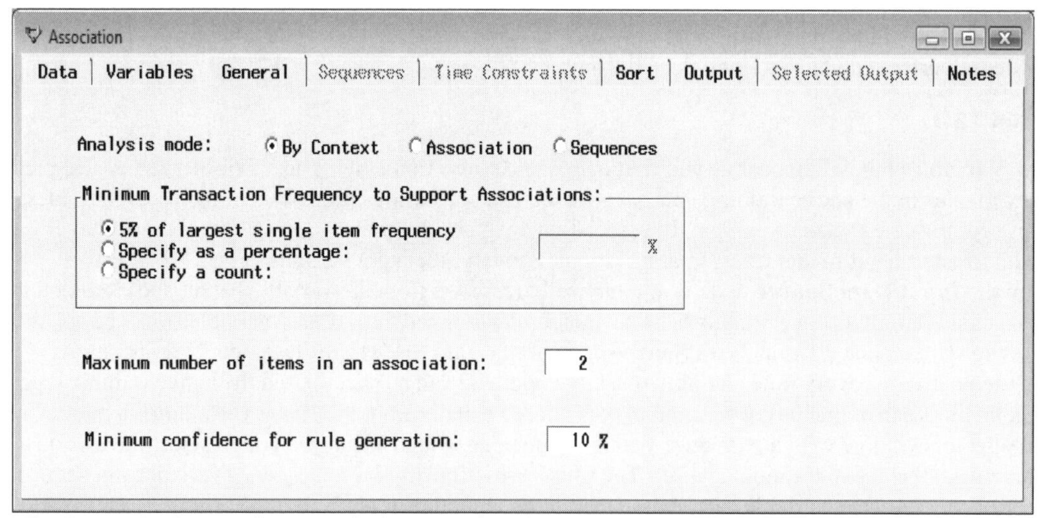

*The **General** tab is used to specify the type of analysis and cutoff values of the association rules.*

By default, the mode of the association analysis is By Context. The mode of the analysis is related to the model roles specified for the variables in the **Input Data Source** node.

The following are the various modes of analysis to select from the **General** tab:

- **By Context** (default): This option is based on the way in which the variables in the data are identified model roles within the **Input Data Source** node.
- **Association:** This type of association analysis is automatically performed if there exists both a categorically-valued **target** variable and an **id** variable in the input data set. Association analysis ignores the order in which the data is recorded.
- **Sequences:** This type of association analysis is performed assuming that a **sequence** variable has an attribute status of **use**. Again, sequence analysis takes into account the order or time in which the items were purchased or the order in which the Web pages were browsed. Therefore, a variable is required in the training data set with a variable role of **sequence**.

The **Minimum Transaction Frequency to Support Associations** option is based on the minimum level of support needed to claim that the items are associated with each other in the association analysis. In other words, you may specify either a frequency rate or a frequency count. Processing an extremely large number of items in the active training data set can potentially cause your system to run out of disk space and reduce memory resources. Therefore, these options will allow you to reduce the number of items that are initially created in the analysis. Conversely, if you interested in associations of extremely rare items to your analysis, then you should consider reducing the minimum number of transactions.

- **5% of the largest single item frequency** (default): This option will allow you to specify the minimum level of support to claim that the items are related to each other. The default is a 5% rate of occurrence. A higher percentage will result in a fewer number of single items, or conversely, a fewer number of combinations of items that will be considered in the analysis.
- **Specify a count:** This option will allow you to specify the minimum number of times that a single item occurs or a combination of two items occurs together. A higher count will result in fewer combinations of items being considered in the association analysis.
- **Maximum number of items in an association:** This option will allow you to specify the maximum number of order n associations or n-way associations performed in the analysis. The default is order 4 associations. The node will create the combinations or associations of up to four separate items. If you are interested in associations involving fairly rare items, then you should consider smaller n-way associations. Conversely, if you have created too many rules that are practically useless, then you should consider higher n-way associations. This option is available in both association and sequence discovery.
- **Minimum confidence for rule generation:** This option will allow you to specify the minimum confidence needed to generate an association rule. The default confidence level is automatically set at 10%. If you are interested in rules that have a higher level of confidence, such as 50%, then it is to your advantage to set the minimum confidence to this level. Otherwise, the node can generate too many rules. This option is designed so that only the rules that meet the minimum confidence value are outputted. This option will not appear if you are performing sequence discovery analysis.

Sequence tab

The **Sequence** tab is designed for you to specify the various configuration settings from the sequence rules in sequence analysis. The tab will allow you to remove the sequence of items from the output data set that either occur too frequency, or infrequently. If you are performing association analysis, then **Sequence** tab is grayed-out and unavailable for viewing. In other words, the tab may be accessed if there exists a variable in the input data set with a **sequence** variable role. Again, in order to perform sequential analysis, there must exist a sequential or time stamp variable that distinguishes between the duration of time or dates between the successive observations.

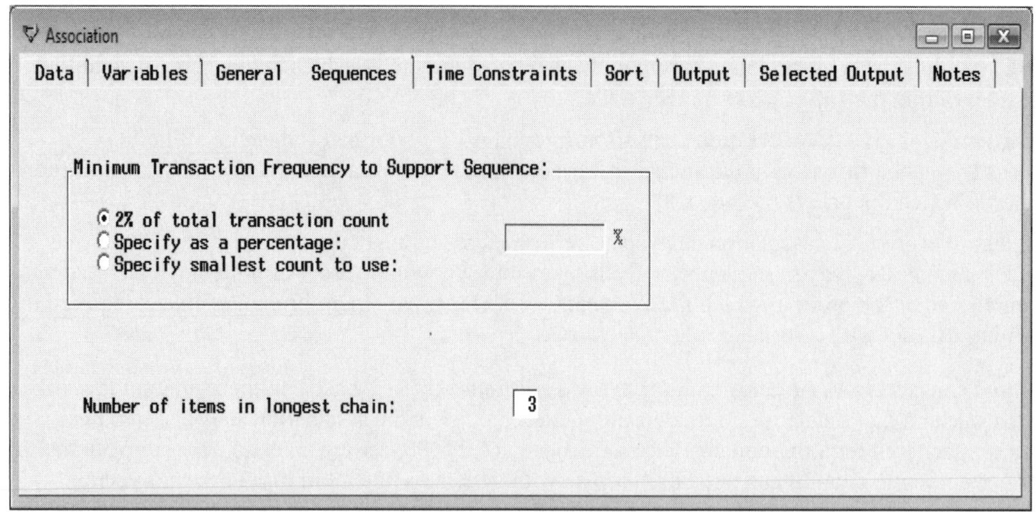

*The **Sequence** tab is used to specify the various cutoff values to the sequence rules.*

The **Sequence** tab has the following options for sequence discovery:

- **Minimum Transaction Frequency to Support Sequence:** This option is designed to exclude sequences from the output data set that occur too infrequently in the input data set. The default is set at 2%, that is, the sequence of items that occur less than 2% of the time in the data are automatically excluded from the output data set. Similar to the minimum transaction frequency from the **General** tab, you may select the corresponding radio button to change the default settings by specifying either a minimum rate of occurrence from the **Specify as a percentage** entry field or a minimum frequency of occurrence from the **Specify smallest count to use** entry field.
- **Number of items in longest chain:** This option is designed to remove sequences that occur too frequently. By default, the maximum number is set at 3. The maximum value that may be specified is 10. This option limits the number of sequences in the data mining analysis. If the number of items specified is larger than the number of items found in the data, then the chain length in the results will be the actual number of sequences that are found.

Time Constraints tab

The **Time Constraints** tab is designed to specify the length of the time sequence in the analysis. The **Time Constraints** tab is based on sequence discovery. If you are performing association analysis, then the **Time Constraints** tab is grayed-out and unavailable for viewing.

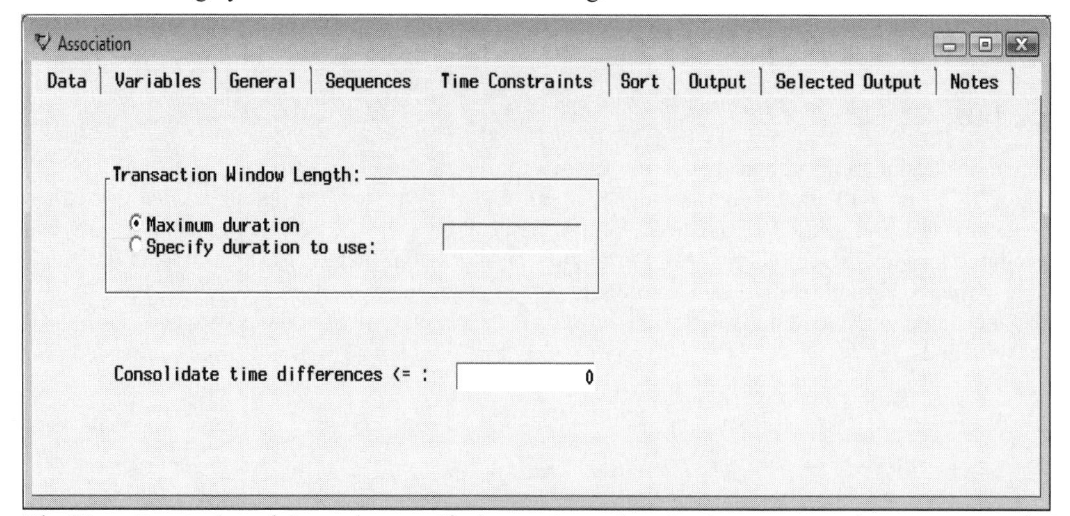

*The **Time Constraint** tab is used to specify the maximum duration in a sequence in sequence analysis.*

- **Transaction Window Length:** This option defines the time interval, that is, the time window length, of the time sequence. Select the **Specify duration to use** radio button to specify the time window of the desired numeric time range of the time sequence. The default is the **Maximum duration**, where the node automatically determines the maximum time range based on the values of the **sequence** variable. The node is designed so that any difference between successive observations in the sequence variable that is less than or equal to the value entered in the **Transaction Window Length** entry field is considered the same time and the same transaction. Conversely, any time difference greater than the window length value does not constitute as a sequence and is, therefore, removed from the analysis.
- **Consolidate time differences:** This option is designed to collapse sequences that occur at different times. For example, consider a customer arriving at a restaurant in four separate visits. At the beginning, the customer goes to the restaurant to buy breakfast in the morning, then the same customer returns in the afternoon to purchase lunch, and, finally, returns in the evening to buy dinner. This same customer returns two days later to buy soda and ice cream. In this example, the sequence variable is defined as visits with four separate entries. Therefore, this option is designed for you to perform sequence discovery on a daily basis by consolidating the multiple visits into a single visit. If you want to perform sequence discovery on a daily basis, then you would need to enter 24 (hours) in the **Consolidate time differences** entry field. Again, this option would instruct the node to consolidate these multiple visits on the same day into a single visit.

Sort tab

The **Sort** tab is designed to specify the primary and secondary sort keys for the **id** variable in the analysis. The output data set will be sorted by the primary sort key within each corresponding secondary key.

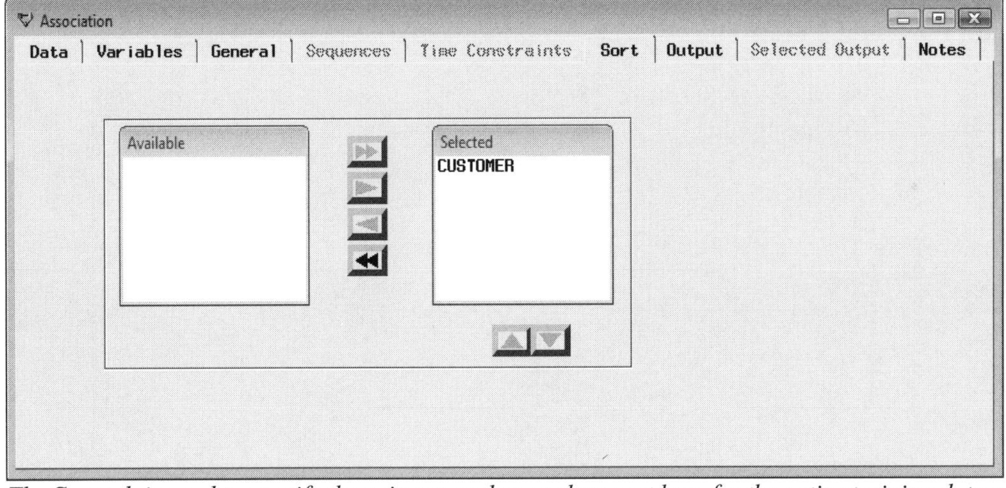

*The **Sort** tab is used to specify the primary and secondary sort keys for the active training data set.*

The tab displays both an **Available** list box and a **Selected** list box. The variables listed in the **Available** list box are all the variables in the active training data set that have been assigned a variable role of **id** with a variable status of **use**. Simply select the variables listed in the **Available** list box to the **Selected** list box in order to assign primary and secondary sort keys to the data mining analysis by clicking the right arrow control button to move the selected variables from one list box to the other. In first step, you must select the **id** variable that represents the primary sort key for the data set. The primary sort key will be the first **id** variable displayed in the **Selected** list box. All other **id** variables listed in the **Selected** list box are assigned as the secondary sort keys. The arrow button located at the bottom of the **Selected** list box will allow you to change the sort order of the selected variables one at a time. You must set at least one of the **id** variables to a variable status of **use** if you have defined several **id** variables in the **Input Data Source** node. By default, the node sets all nonselected **id** variables to variable attribute status of **don't use**.

Output tab

The **Output** tab is designed to browse the output scored data set from the PROC ASSOC, SEQUENCE, and RULEGEN data mining procedure output listing. Which procedures are executed depends on the type of analysis that is specified, that is, association discovery or sequence discovery. The PROC ASSOC data mining procedure is applied in both types of analysis. However, association discovery applies the PROC RULEGEN data mining procedure in creating the rules, whereas sequence discovery uses the PROC SEQUENCE data mining procedure in creating the rules within each sequence.

The PROC ASSOC procedure determines the various items that are related to each other. In other words, the PROC RULEGEN procedure generates the association rules. The PROC SEQUENCE procedure uses the time stamp variable to construct the sequence rules. The output is saved from these procedures as SAS data sets after you execute the node. The data sets will allow you to observe the various evaluation criteria statistics from the various *if-then* rules that are listed.

Select the association **Properties...** button to view the **Information** tab that displays the administrative information about the data set, such as the name, type, created date, date last modified, columns, rows, and deleted rows. Alternatively, select the **Table View** tab to browse the output data set that is created from the PROC ASSOC procedure by running the node. The output data set displays a table view of the frequency count by all the possible *n*-way associations or interactions of the items or events. The scored data set will display separate columns called ITEM1, ITEM2, and so on, based on the number of associations that are specified from the **Maximum number of items in an association** option within the **General** tab. The frequency listing to association discovery is displayed in the following diagram.

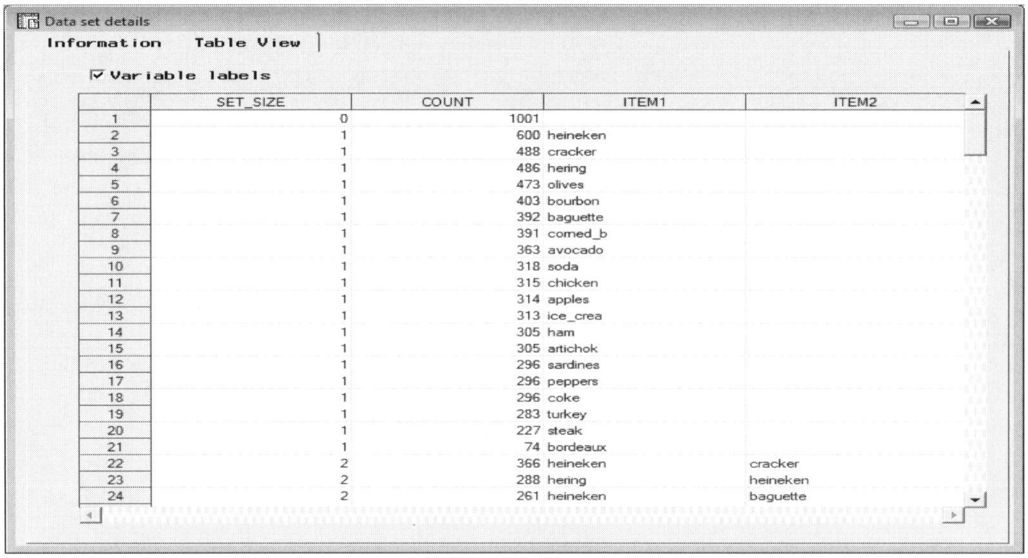

Data set details

Information   Table View

☑ Variable labels

	SET_SIZE	COUNT	ITEM1	ITEM2
1	0	1001		
2	1	600	heineken	
3	1	488	cracker	
4	1	486	hering	
5	1	473	olives	
6	1	403	bourbon	
7	1	392	baguette	
8	1	391	corned_b	
9	1	363	avocado	
10	1	318	soda	
11	1	315	chicken	
12	1	314	apples	
13	1	313	ice_crea	
14	1	305	ham	
15	1	305	artichok	
16	1	296	sardines	
17	1	296	peppers	
18	1	296	coke	
19	1	283	turkey	
20	1	227	steak	
21	1	74	bordeaux	
22	2	366	heineken	cracker
23	2	288	hering	heineken
24	2	261	heineken	baguette

The ***Table View*** *tab displays the unique frequency counts from all possible association of items.*

Select the rules **Properties...** button to display the file administrative information and a table view of the rules output data set. The options are similar to the previous association options. From the tab, you may view the file information of the rules data set. However, the file layout of the rules data set is a bit more detailed. Select the **Table View** tab to view the rules data set that contains the evaluation criterion statistics such as the frequency count, support, confidence, and lift values, based on the various rules that were specified from the **General** tab. The table listing is sorted in descending order by the support probability between each item or event in the association analysis. From the following listing, focusing on the combination of Heineken beer and crackers then purchased, of the 1,001 customers purchasing items approximately 36.56% of these customers purchased both Heineken beer and crackers, customers purchasing Heineken beer resulted in 61.00% of these same customers then purchasing crackers, with a lift value of 1.25 indicates the reliability of the association and that the rule is better at predicting the association between both items as opposed to randomly guessing the association of both items purchased. The table listing will provide you with the list of all combinations of items with a low support and a high confidence probability that should be interpreted with caution. Although it is not shown in the following table listing, there were four separate combinations of items where customers first purchased Bordeaux, then olives, herring, crackers or Heineken beer, with a low support ranging between 3.2 and 4.4 and a high confidence probability ranging between 43.25 and 59.36.

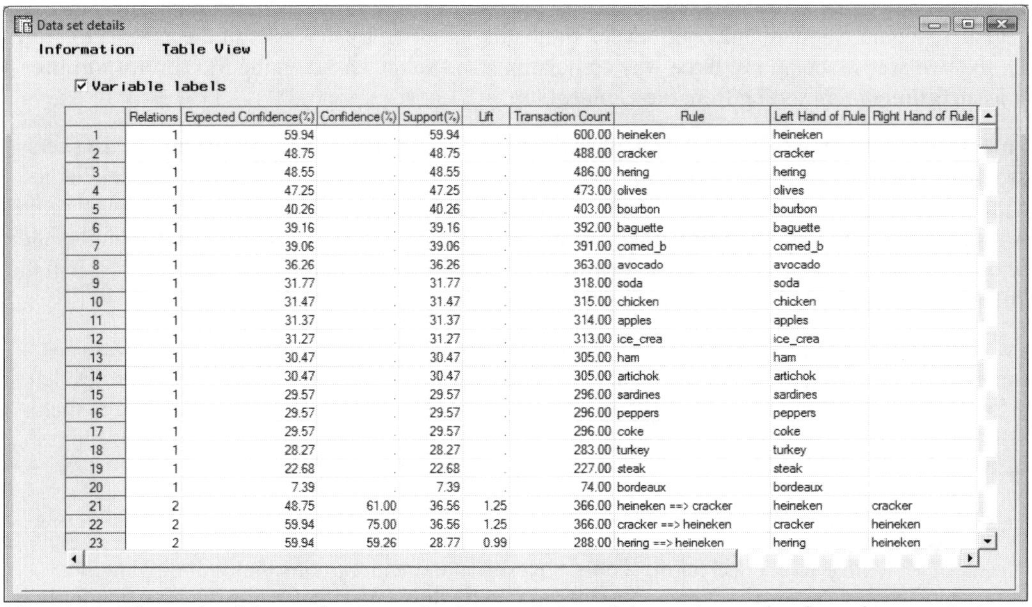

Data set details

Information | Table View

☑ Variable labels

	Relations	Expected Confidence(%)	Confidence(%)	Support(%)	Lift	Transaction Count	Rule	Left Hand of Rule	Right Hand of Rule
1	1	59.94	.	59.94	.	600.00	heineken	heineken	
2	1	48.75	.	48.75	.	488.00	cracker	cracker	
3	1	48.55	.	48.55	.	486.00	hering	hering	
4	1	47.25	.	47.25	.	473.00	olives	olives	
5	1	40.26	.	40.26	.	403.00	bourbon	bourbon	
6	1	39.16	.	39.16	.	392.00	baguette	baguette	
7	1	39.06	.	39.06	.	391.00	corned_b	corned_b	
8	1	36.26	.	36.26	.	363.00	avocado	avocado	
9	1	31.77	.	31.77	.	318.00	soda	soda	
10	1	31.47	.	31.47	.	315.00	chicken	chicken	
11	1	31.37	.	31.37	.	314.00	apples	apples	
12	1	31.27	.	31.27	.	313.00	ice_crea	ice_crea	
13	1	30.47	.	30.47	.	305.00	ham	ham	
14	1	30.47	.	30.47	.	305.00	artichok	artichok	
15	1	29.57	.	29.57	.	296.00	sardines	sardines	
16	1	29.57	.	29.57	.	296.00	peppers	peppers	
17	1	29.57	.	29.57	.	296.00	coke	coke	
18	1	28.27	.	28.27	.	283.00	turkey	turkey	
19	1	22.68	.	22.68	.	227.00	steak	steak	
20	1	7.39	.	7.39	.	74.00	bordeaux	bordeaux	
21	2	48.75	61.00	36.56	1.25	366.00	heineken ==> cracker	heineken	cracker
22	2	59.94	75.00	36.56	1.25	366.00	cracker ==> heineken	cracker	heineken
23	2	59.94	59.26	28.77	0.99	288.00	hering ==> heineken	hering	heineken

*The **Table View** tab of the evaluation criterion statistics of the various rules from the association analysis.*

Conversely, selecting the sequence rule will display the procedure output listing from the PROC SEQUENCE data mining procedure. From the following diagram, the same combination of Heineken beer and crackers are the most popular items purchased from the sequence analysis results. Notice that the support probability is smaller in the following table listing from sequence analysis among customers purchasing crackers and Heineken beer since the order in which these items are purchased are taken into account.

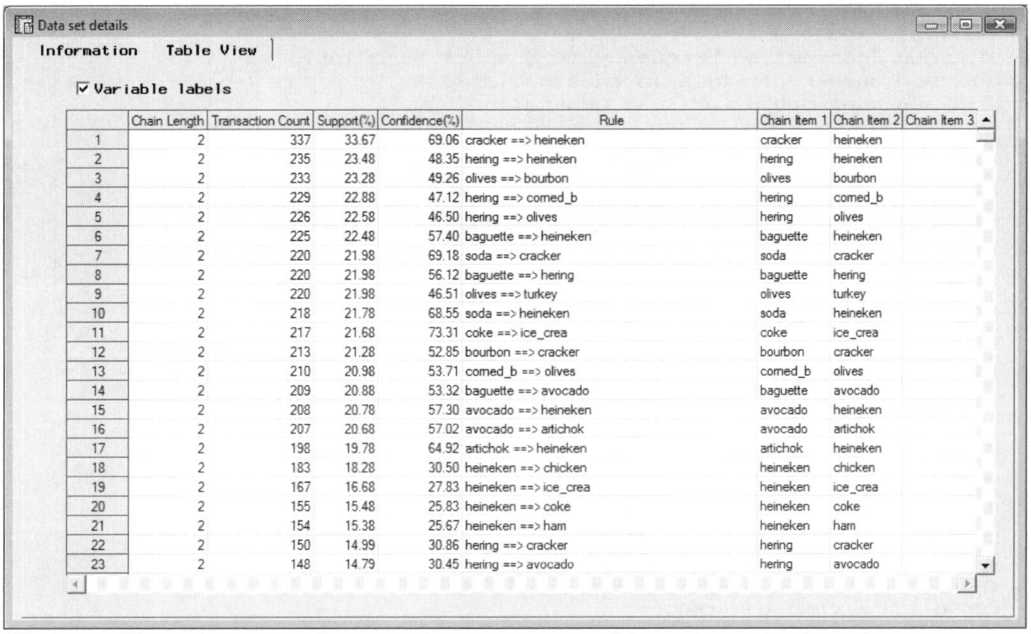

Data set details

Information | Table View

☑ Variable labels

	Chain Length	Transaction Count	Support(%)	Confidence(%)	Rule	Chain Item 1	Chain Item 2	Chain Item 3
1	2	337	33.67	69.06	cracker ==> heineken	cracker	heineken	
2	2	235	23.48	48.35	hering ==> heineken	hering	heineken	
3	2	233	23.28	49.26	olives ==> bourbon	olives	bourbon	
4	2	229	22.88	47.12	hering ==> corned_b	hering	corned_b	
5	2	226	22.58	46.50	hering ==> olives	hering	olives	
6	2	225	22.48	57.40	baguette ==> heineken	baguette	heineken	
7	2	220	21.98	69.18	soda ==> cracker	soda	cracker	
8	2	220	21.98	56.12	baguette ==> hering	baguette	hering	
9	2	220	21.98	46.51	olives ==> turkey	olives	turkey	
10	2	218	21.78	68.55	soda ==> heineken	soda	heineken	
11	2	217	21.68	73.31	coke ==> ice_crea	coke	ice_crea	
12	2	213	21.28	52.85	bourbon ==> cracker	bourbon	cracker	
13	2	210	20.98	53.71	corned_b ==> olives	corned_b	olives	
14	2	209	20.88	53.32	baguette ==> avocado	baguette	avocado	
15	2	208	20.78	57.30	avocado ==> heineken	avocado	heineken	
16	2	207	20.68	57.02	avocado ==> artichok	avocado	artichok	
17	2	198	19.78	64.92	artichok ==> heineken	artichok	heineken	
18	2	183	18.28	30.50	heineken ==> chicken	heineken	chicken	
19	2	167	16.68	27.83	heineken ==> ice_crea	heineken	ice_crea	
20	2	155	15.48	25.83	heineken ==> coke	heineken	coke	
21	2	154	15.38	25.67	heineken ==> ham	heineken	ham	
22	2	150	14.99	30.86	hering ==> cracker	hering	cracker	
23	2	148	14.79	30.45	hering ==> avocado	hering	avocado	

*The **Table View** tab of the evaluation criterion statistics of the various rules from the sequence analysis.*

Selected Output tab

The **Selected Output** tab is designed to create separate output SAS data sets for each one of the association and sequence rules selected from the **Rules** tab. The tab will allow you to browse the listed rules data sets that have been created from the compiled node. The **Association** node assigns a data role to each output set from the analysis that is selected. For instance, the data roles are automatically assigned **Assoc** and **Rulegen** based on association analysis and **Assoc** and **Sequence** based on sequence analysis. The data set lists the corresponding rules, the order of the association rule, the frequency count for each association rule, and the three rule statistics: the lift, support, and confidence. The listing is sorted by the order of the association rules, that is, by all the two-way associations, three-way associations and so on, based on the **Maximum number of items in an association** option setting from the **General** tab.

From the table, you may either retain or delete the listed output data set. By default, all the output data sets from the associated rules are saved with a variable attribute status automatically set to **Keep**. To delete the selected output data sets, select the corresponding rows and set the **Status** column to **Delete**. To select multiple variable rows, either select the Shift key or Ctrl key or drag the mouse over the desired rows. Rerun the node, to delete the output data sets from the **Selected Output** tab. Deleting all the output data sets will result in the **Selected Output** tab being unavailable for viewing when you reopen the node.

If you intend to fit a model to the output data set from a modeling node, then you should use the **Data Set Attributes** node. From the node, you can set the data set role to **training** and define any variable roles that are appropriate for the subsequent modeling node. The output data sets do not have any rules assigned to them.

Resetting to the Default Settings

At times, it might be advantageous to revert back to the basic default configuration settings within the node.

Select the **Reset** toolbar icon or select the **Tools > Reset** main menu options that will display the following **Reset to Shipped Defaults** window. The window will allow you to select any one option or all of the listed options in order to reset the original association discovery or sequence discovery configuration settings back to the default settings that SAS has carefully selected.

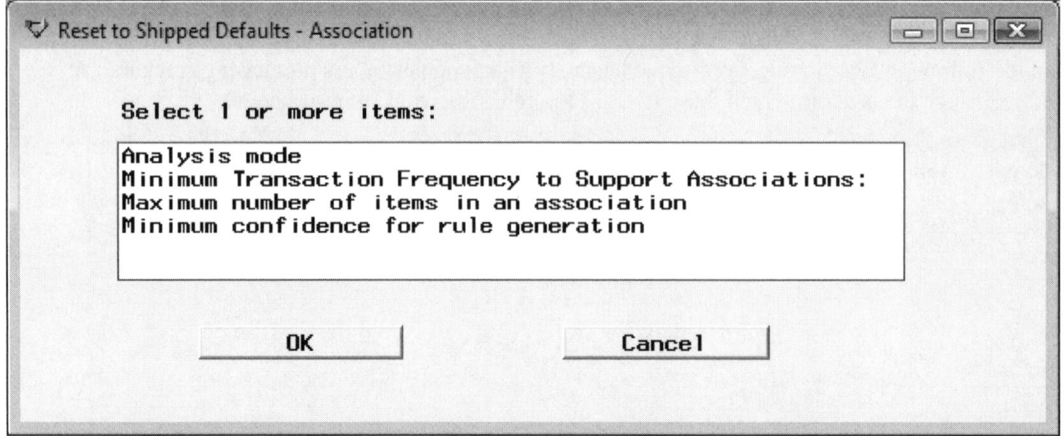

*The **Reset to Shipped Defaults** window is used to reset the association options back to the default settings.*

Reset to Shipped Defaults - Association
Select 1 or more items:
Analysis mode
Minimum Transaction Frequency to Support Associations:
Maximum number of items in an association
Minimum Transaction Frequency to Support Sequence:
Number of items in longest chain
Transaction Window Length
Consolidate time differences
OK
Cancel

*The **Reset to Shipped Defaults** window is used to reset the sequence options back to the default settings.*

Viewing the Association Node Results

General Layout of the Results Browser within the Association Node

- **Rules tab**
- **Frequencies tab**
- **Code tab**
- **Log tab**
- **Notes tab**

Association Analysis

The following tabs from the **Results Browser** are based on the association analysis results that will automatically open the **Results–Association** window.

Rules tab

The **Rules** tab is designed to display a table listing of the association rule, the order of the association rules and the evaluation criterion statistics (frequency count, support, confidence, and lift values) in the following **Results–Association** window. The tab will allow you to specify a variety of queries in reducing the number of associations or interactions between the items under analysis. You may either save the following listing to a SAS data set by selecting the **File > Save as Data Set** from the main menu or select the **File > Export** main menu option that opens the SAS Export Wizard to save the following table listing to an external file for further processing.

The table listing is sorted by the highest support probabilities based on order 2 rules of association. Many of the following listed associations are listed next to each other since the support probability is a symmetric probability. It is also important to keep in mind that the list of if-then association rules are based on a minimum support rate at 5%, which is the default or whatever support rate is specified from the **Minimum Transaction Frequency to Support Associations** option from the **General** tab. The following table listing displays some of the best two-way association rules that were generated from the **Association** node. The most desirable associations will have high support and confidence probabilities along with lift values greater than one that will indicate to you of the strength and reliability of the corresponding associations. In addition, it might be of interest to determine what items that should be interpreted with caution. Since the table listing displays the support probabilities in descending order, scroll to the bottom of the table listing to observe these suspicious rules with a low support and a high confidence probability.

Results - Association

Rules | Frequencies | Code | Log | Notes

	Relations	Lift	Support(%)	Confidence(%)	Transaction Count	Rule
1	2	1.25	36.56	61.00	366.00	heineken ==> cracker
2	2	1.25	36.56	75.00	366.00	cracker ==> heineken
3	2	1.11	26.07	43.50	261.00	heineken ==> baguette
4	2	1.11	26.07	66.58	261.00	baguette ==> heineken
5	2	1.35	25.67	80.82	257.00	soda ==> heineken
6	2	1.35	25.67	42.83	257.00	heineken ==> soda
7	2	1.11	25.57	54.12	256.00	olives ==> hering
8	2	1.11	25.57	52.67	256.00	hering ==> olives
9	2	1.38	25.17	42.00	252.00	heineken ==> artichok
10	2	1.38	25.17	82.62	252.00	artichok ==> heineken
11	2	1.62	25.07	78.93	251.00	soda ==> cracker
12	2	1.62	25.07	51.43	251.00	cracker ==> soda
13	2	1.31	24.88	51.23	249.00	hering ==> baguette
14	2	1.31	24.88	63.52	249.00	baguette ==> hering
15	2	1.14	24.88	41.50	249.00	heineken ==> avocado

*The **Rules** tab displays the evaluation criteria and the various association rules.*

From the **Results Browser**, the following are the various menu options that are available from the **View** main menu option within the **Rules** tab.

- **View > Table:** This option displays a table view of the various rules and the corresponding evaluation criterion statistics such as the frequency counts, the support and confidence probabilities and lift ratio.
- **View > Graph:** This option displays a grid plot of the *if-then* association rules and the corresponding evaluation criterions.
- **View > Relations > 1:** This option displays a table listing of all associations of order two or more, that is, all order 2 associations, order 3 associations, and so on. Otherwise, the table displays all single item association rules.
- **When Confidence > Expected Confidence:** This option displays a table listing of all of the association or sequence rules that have a lift ratio greater than one, that is, all positive associations.
- **Subset Table:** This option displays a frequency table listing of all of the possible rules and frequency bar charts from the evaluation criterion statistics such as the confidence, support, and lift values. The tabs do not only view the distribution of the various *n*-way association of items, but also remove certain *n*-way association of items from the table listing.
- **Reset Table:** This option resets the table to the original settings that SAS has carefully selected.
- **Swap Stats:** By default, the symbols in the scatter plot graph correspond to the magnitude of the confidence probability and the color of the graph indicates the magnitude of the support to the rule. This option will allow you to switch the meaning of both the symbols and the color of the graph. However, the size of the symbols will stay the same, representing the magnitude of the lift values.
- **Grid:** By default, the rules graph does not display grid lines. This option will allow you to add grid lines to the symbol scatter plot.

Grid Plot

Select the **View > Graph** main menu options that will graphically display both the support and confidence probabilities from the combination of items. The grid plot is designed to display all the association rules of order two, that is, based on order 2 associations between the set of items under analysis that have been previously specified from the **General** tab. The horizontal axis represents the left-hand side of the rule and the vertical axis represents the right-hand side of the rule. The size of the symbols in the graph indicates the magnitude of the lift value.

If there exist too many association rules in the grid plot, then simply select the desirable associations from the table listing in the **Rules** tab and replot the graph by selecting the **View > Graph** menu option to display the desired association rules. Otherwise, select the **Create Viewport into Data** toolbar icon and drag the mouse over the plot to resize the listed symbols. You may also select the **View Info** toolbar icon that will display a text box of the body and head of the rule along with the associated evaluation criterion probabilities.

Interpretation of the Grid Plot

From the graphical display, the head of the rule is displayed on horizontal axis and the body of the rule is displayed on the vertical axis. The support probability of each rule is identified by the color of the symbols and the confidence probability of each rule is identified by the shape of the symbols. The items that have the highest confidence are Heineken beer, crackers, chicken, and peppers. In other words, customers purchasing Heineken beer will also buy crackers and customers buying chicken or peppers will also purchase Heineken beer. Also, there exists a relatively small probability that customers purchasing corned beef will also buy steak, or people buying an avocado will also buy sardines. Conversely, there are low support probabilities to certain items, like apples, that can be distinguished by the light blue symbols indicating that there is a small probability that customers will purchase both apples and steak or apples and Bordeaux. Interpreting both probabilities, it is important that the confidence probability be read from the items displayed on the vertical axis that represent the left-hand side of the rule, since it is a conditional probability. However, it makes no difference which axis is read from in interpreting the support probability, which is basically the relative frequency of both items combined. The larger symbols that are in reference to the lift values will indicate the strength and reliability of the associated items with a higher occurrence of customers purchasing the items displayed on the horizontal axis given that they have already purchased the items displayed on the vertical axis as opposed to not buying items displayed on the same axis. On the other hand, smaller symbols will indicate less reliable associations. From the following graph, the association between customers purchasing Bordeaux then buying olives displays a small light blue circle, indicating that the relationship should be interpreted with caution since the items generated a low support and a high confidence probability.

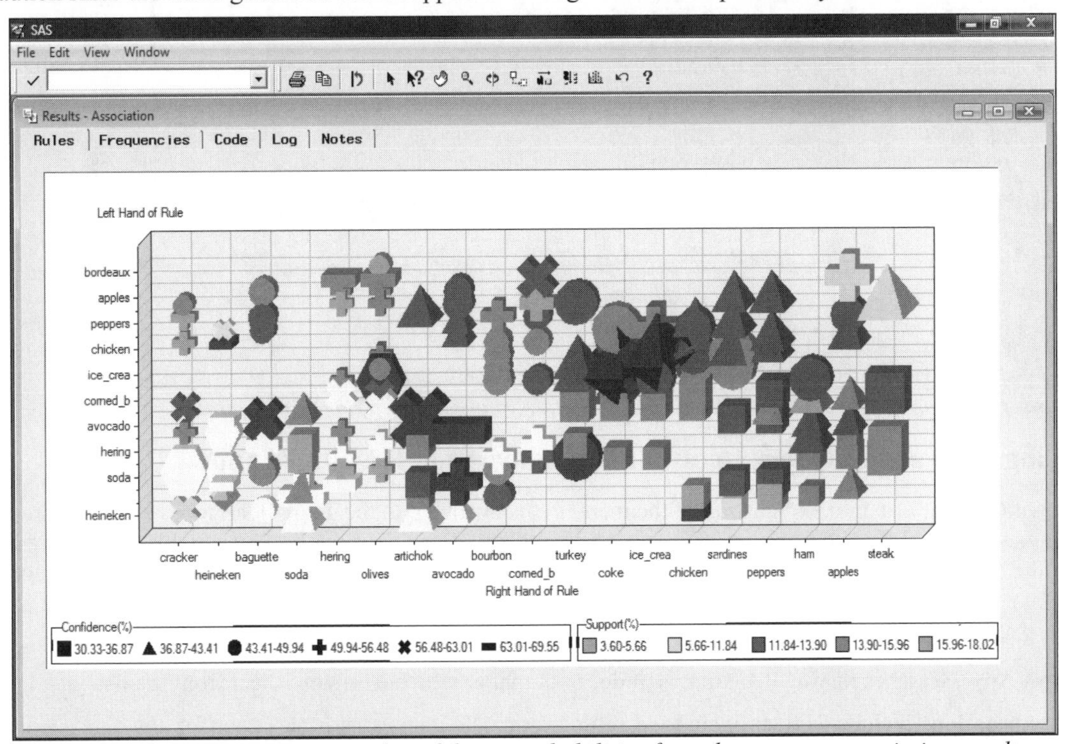

A graphical view of the support and confidence probabilities from the two-way association results.

Subsetting the Association Rules

Select the **View > Subset Table** main menu option to view the **Subset Table** window based on association discovery. The window displays the frequency count of each item based on the body and head of the *if-then* association rule. The window also displays a frequency bar chart of the confidence, support, and lift of the association items.

Predecessor → Successor tab

The **Predecessor → Successor** tab is designed to display the evaluation criterion statistics such as the frequency count of all items under analysis. In other words, the frequency count of the body to the rule is listed on the left-hand side of the tab and the frequency count of the head of each association rule is listed on the right-hand side of the tab.

The tab is designed for you to subset the list of all the desired combination of items that have been determined from the various criterions specified within the **General** tab. The tab will allow you to build your own queries from the list of the if-then rules in order to subset the list of items into various combinations or interactions. The items listed in the left-hand side of the tab are based on the body of the association rules and the similar items listed to the right of the tab represent the items of the head of the association rule.

Simply press the Shift or Ctrl key to select several items at one time. Alternatively, right-click the mouse and select the **Select All** pop-up menu item to select all items from either the left-hand or right-hand side. Conversely, select the **Deselect All** pop-up menu item to clear all items from either the left-hand or right-hand side.

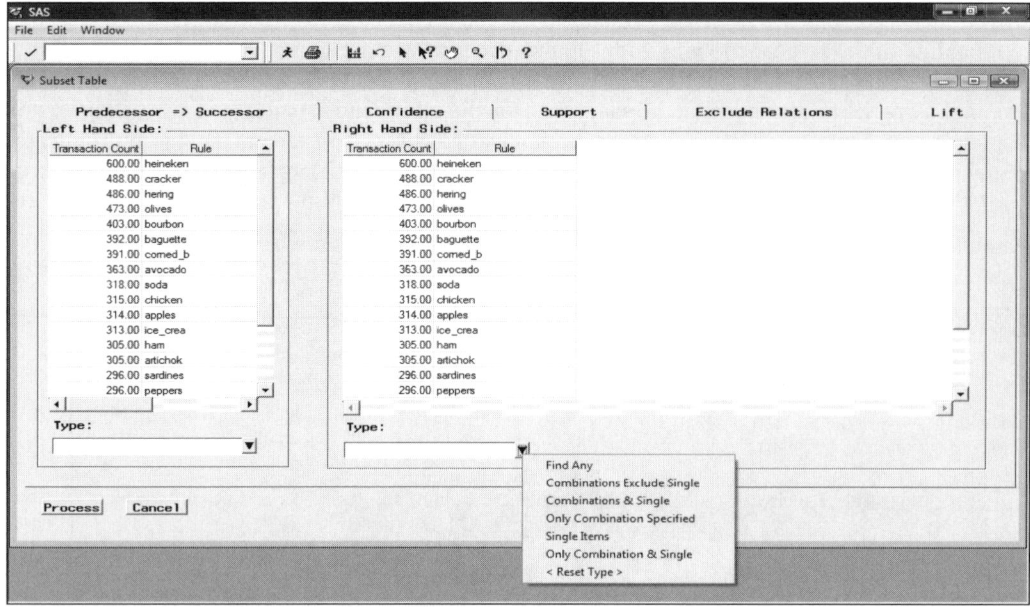

*The **Subset Table** option is used to query the combination of items based on the association rules.*

Specifying the Various Query Items for the Association Results Window

The following listing is a brief explanation of the menu items available in the **Type** field in order to query the combination of items from either the left-hand or right-hand side. The following list of available queries are based on the combination of items desired either from the left-hand or right-hand side of the association rule. For example, if you want to query a combination of items, then you must first select two or more items from the list.

- **Find Any:** Searches for any item, that is, it does not matter which item you select from the list.
- **Combination Exclude Single:** Exclude searching for single items, that is, searches for any combination of two or more items, for example, (AB, AC and ABC).
- **Combination & Single:** Searches for both single items and any combination of items of two or more items, for example, (A, B, C, AB, AC, BC and ABC).
- **Only Combination Specified:** Searches for a specific combination of items.
- **Single Items:** Searches for single items, for example, (A, B and C).
- **Only Combination and Single:** Searches for a specific combination of items and also searches for single items, for example, (A, B, C and ABC).
- **<Reset Type>:** Sets the **Type** field to blank that is the default, that is, displays all possible combinations of items.

Select the **Process** button to execute the specified query. The **Association Discovery Subset** window will appear, which displays the listing of the subset of association rules that met the corresponding query. You may display the grid plot of the resulting combination of association rules by selecting **View > Graph** from the main menu. From the **Subset Table** window, you may also save the desired subset of association rules to a SAS data set by selecting the **File > Save as Data Set** from the main menu. The node will also allow you to export the list of association rules and the corresponding evaluation criterion statistics through the SAS Export Wizard by selecting the **File > Export** main menu option.

If your subset request did not find any valid matches, then a message window will open asking you to modify the subset request. To modify the subset request, click **Yes** and make your changes in the **Association Discovery Subset** window. If you do not want to modify the subset request, then select **No** to return to the **Associations Results** table.

Confidence tab

The **Confidence** tab is designed to display a frequency bar chart of the confidence probabilities that have been automatically grouped into separate intervals. Large confidence probabilities that are displayed within the graph will indicate to you the strength of association between the separate items.

In Enterprise Miner, there exists a color coding schema of the bars used to identify the frequency counts or the frequency percentages of the range of values in the following evaluation criteria probabilities. All bars colored in red will indicate a relatively large frequency percentage of the associated items, yellow bars will indicate a relatively low frequency of the associated items, and orange bars will indicate relatively moderate frequency of the associated items. However, the height of each bar will be consistent with the color coding that is assigned to each bar. Large bars will be colored in red and small bars will be colored in yellow.

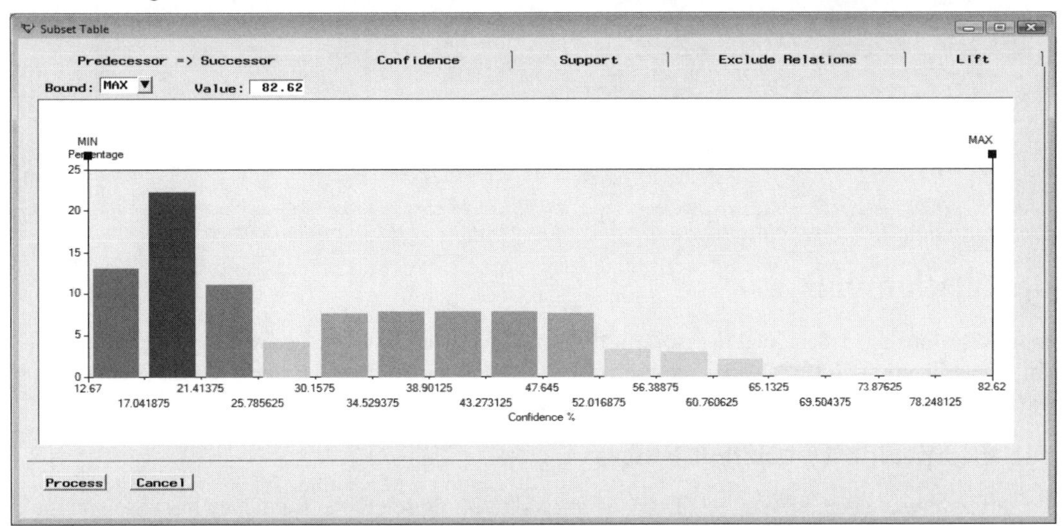

*The **Confidence** tab that displays the distribution and range of values of the confidence probability.*

Setting the Cutoff Ranges to the Confidence, Support, and Lift Values

The charts will not only allow you to view the distribution of the association statistics, but you may also remove certain items from the analysis. For instance, the charts will allow you to set certain cutoff values to remove items that have a weak association between one another. From the chart, simply select the slider or boundary box to a desired location along the horizontal axis to retain all items that are within the specified range of values. Alternatively, select the **Bound** drop list to select either the **Min** or **Max** menu items. Simply enter a value into the **Value** entry field and the slider will automatically be positioned to the corresponding cutoff value of the confidence probability.

By pressing the **Process** button, the **Results Associations** window will open with the **Rules** tab being displayed. The tab will display the resulting rules that have been selected from the graph. From the tab, you may save the rules to a data set by simply selecting the **File** > **Save as Data Set** from the main menu.

Support tab

The **Support** tab is designed to display a frequency bar chart of the support probabilities that are automatically grouped into separate intervals. Similar to the previous **Confidence** tab, you may also remove certain associations from the analysis and the corresponding output data set. Simply select the desired cutoff values based on the support probability between the interaction of the various items by selecting the slider along the horizontal axis or entering the desired upper or lower bounds from the **Value** entry field. Since items with a low support and a high confidence probability should be interpreted with caution. It might not be a bad idea to remove all support probabilities that are located at the lower end of the graph with a low support probability. From the following plot, the minimum support probability was entered in the **Value** entry field to remove all items that seemed suspicious (with a low support and high confidence probability) that were observed from the previous association rules scored output table listing.

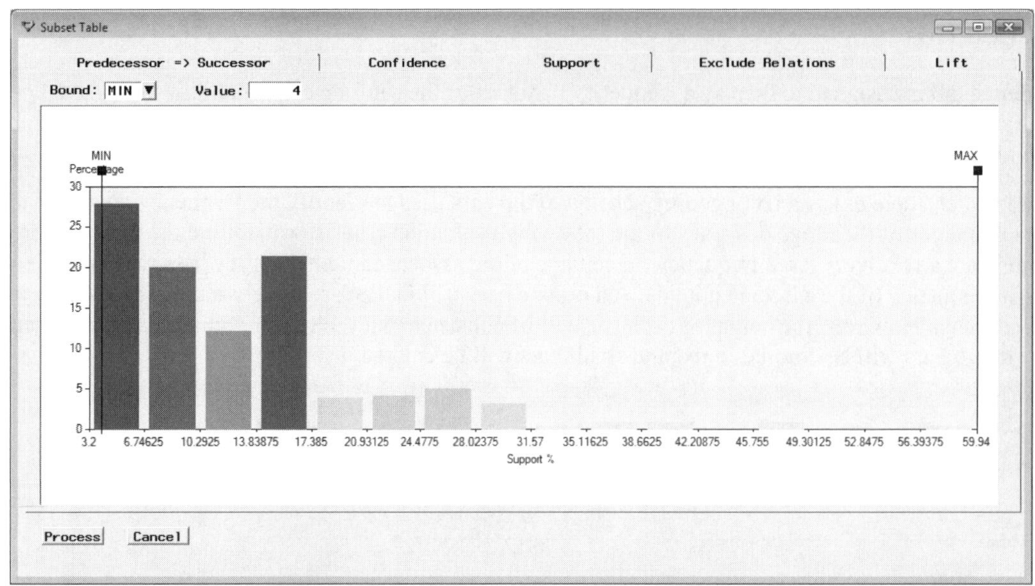

*The **Support** tab that displays the distribution and range of values of the support probability.*

Exclude Relations tab

The **Exclude Relations** tab is designed to display a frequency bar chart to view the frequency distribution as to the order of the association rules. Each bar represents the *n*-way association of the items from the number of associations specified from **Maximum number of items in an association** options within the **General** tab.

Excluding the Specific Association Rules

To exclude certain *n*-way interaction effects from the analysis, simply select the frequency bar based on the relation number that you want to remove from the list of *n*-way association rules. To remove more than one *n*-way associations, simply press the Ctrl key while at the same time selecting more than one bar. The selected bars will turn to the color gray to indicate to you that the corresponding *n*-way associations have been removed from the analysis. Once the desired order of associations is selected, then the results will be reflected in the **Rules** tab. Select the **Process** button and the **Rules** tab will appear with the desired *n*-way order of associations.

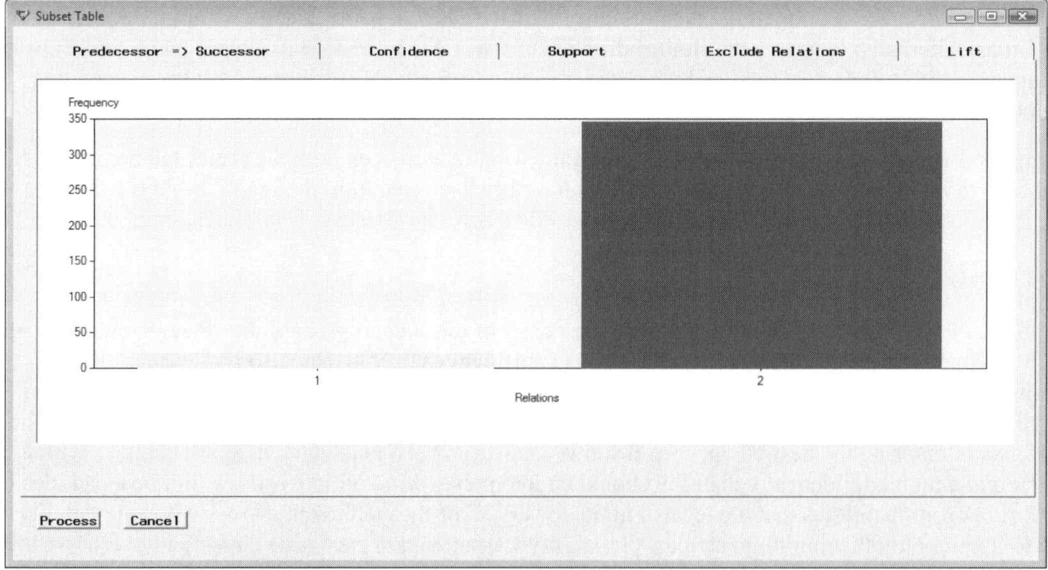

*The **Exclude Relation** tab used to exclude certain association of items or events from the analysis.*

Lift tab

The **Lift** tab is designed to display a frequency bar chart of the lift values. Since the bar chart groups the statistic into corresponding intervals, the lift chart will allow you to view both the positive and negative associations among the various items under analysis. Lift values greater than one are the most desirable. A positive association or correlation between the corresponding rules will indicate how much better the rules is in predicting the association as opposed to a random guess. Conversely, lifts below the value of one will indicate that the associations should be interpreted with caution and that randomly guessing the association is better than predicting the association. To retain all items that are positively associated with each other, reposition the slider to the left, that is, the minimum slider, along the X-axis to the value of one. Alternatively, to remove all items that are positively associated with each other, then reposition the slider to the right, that is, the maximum slider, along the X-axis to the value of one.

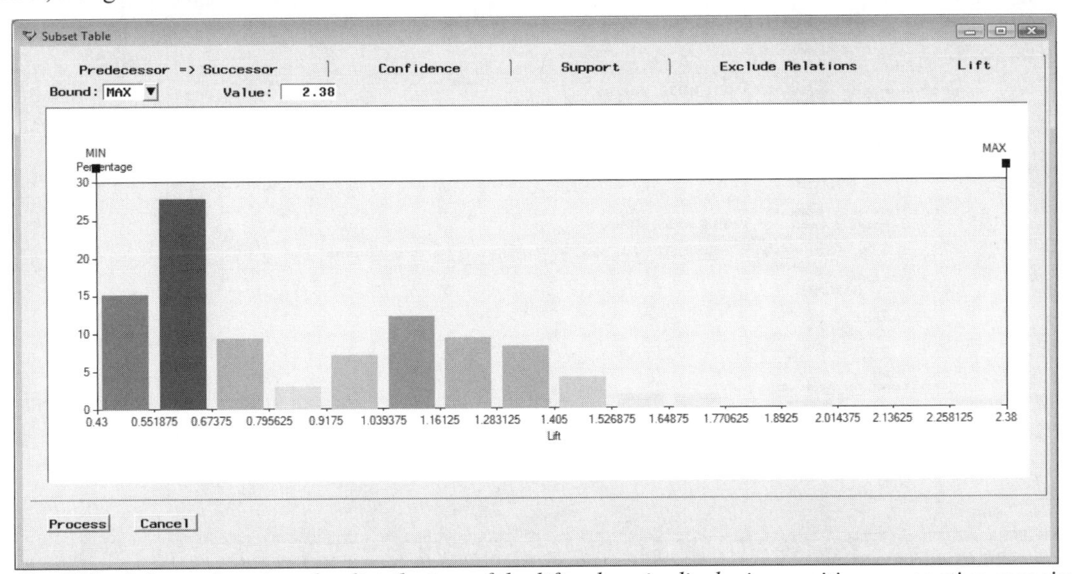

*The **Lift** tab is used to view the distribution of the lift values in displaying positive or negative associations.*

Frequencies tab

The **Frequencies** tab is designed to display the frequency count of each item or event. For example, the frequency count for Heineken beer can be interpreted as 600 different customers who have purchased Heineken beer at least once. That is, the actual frequency count in the active training data set is actually 602. In other words, there were two customers – customer id 380 and 719 – who purchased Heineken beer more than once. The listed items display the variable assigned the **target** variable role with the associated frequency counts that are displayed in descending order by the **id** variable role in the training data set.

Results - Association

Rules | Frequencies | Code | Log | Notes

	Count	Item
1	600	heineken
2	488	cracker
3	486	hering
4	473	olives
5	403	bourbon
6	392	baguette
7	391	corned_b
8	363	avocado
9	318	soda
10	315	chicken
11	314	apples
12	313	ice_crea

*The **Frequencies** tab that displays a table listing of the frequency counts of all single items or events.*

Code tab

The **Code** tab is designed to display the internal SEMMA training code that generates the evaluation statistics from the corresponding rules. In Enterprise Miner, the PROC ASSOC and RULEGEN data mining procedures are performed for association discovery. The process used in generating the association results is designed so that the internal SAS code must initially create a DMDB data mining data set. The DMDB data set is then read into the PROC ASSOC procedure that determines the items that are related to each other. The procedures create an output data set based on all items that meet the various criteria that are then used as inputs to the PROC RULEGEN procedure that creates the various support, confidence, and lift statistics. These statistics are written to the rules output data set that can be viewed when making your own interpretation of the strength and relationship between the related items.

Results - Association

Rules | Frequencies | Code | Log | Notes

(•) Training ( ) Scoring

```
00001     proc sort data=EMDATA.VIEW_KPT( keep=
00002                PRODUCT
00003                CUSTOMER
00004     )
00005     out=_emtrain ;
00006     by
00007        CUSTOMER
00008     ;
00009     run ;
00010       options nocleanup;
00011       Proc Assoc dmdbcat= EMPROJ.dm_DGM00004
00012                  data= _emtrain
00013                  out=EMDATA.ASCOK6GX (label = "Output from Proc Assoc")
00014                    items=2;
00015          customer
00016              CUSTOMER
00017          ;
00018          target
00019              PRODUCT
00020        ;
00021            run;
00022            quit;
00023       options nocleanup;
00024
00025        Proc Rulegen in = EMDATA.ASCOK6GX
00026           out      = EMDATA.RLAJTIVL (label = "Output from Proc Rulegen")
00027           minconf = 10;
00028        run;
00029        quit;
```

*The **Code** tab that displays the SEMMA training code that created the association rules results.*

Log tab

The **Log** tab is designed to display the procedure log listing of the SEMMA programming code that is executed once you run the node. The corresponding programming code is listed in the previous **Code** tab. The **Log** tab will display the number of records that are processed for each single items and all combination of items based on the single items that are determined from the first pass.

Notes tab

Enter notes or record any other information to the **Association** node.

Sequence Analysis

The following tabs from **Results Browser** are based on the sequence analysis results that will open the following **Results - Sequence** window. The **Frequencies** and **Log** tab will not be displayed since it is essentially identical to the previously illustrated tabs that have been explained in association discovery.

Rules tab

The following listing displays the most frequent sequence of items purchased. Notice that the lift values are not displayed since they are based on the relative frequency of each separate item. The listing is sorted in descending order by the most frequent sequence of items purchased. The tab displays the number of chained sequences of items or events, the support probability, confidence probability, and the frequency of occurrence of each listed sequence of items. Similar to the previous **Association Rules** tab, you may save the following table listing to either a SAS data set or to an external file by selecting the **File > Save as Data Set** or **File > Export** main menu options.

Results - Sequence

Rules | Frequencies | Code | Log | Notes

	Chain Length	Support(%)	Confidence(%)	Transaction Count	Rule
1	2	33.67	69.06	337	cracker ==> heineken
2	2	23.48	48.35	235	hering ==> heineken
3	2	23.28	49.26	233	olives ==> bourbon
4	2	22.88	47.12	229	hering ==> corned_b
5	2	22.58	46.50	226	hering ==> olives
6	2	22.48	57.40	225	baguette ==> heineken
7	2	21.98	69.18	220	soda ==> cracker
8	2	21.98	56.12	220	baguette ==> hering
9	2	21.98	46.51	220	olives ==> turkey
10	2	21.78	68.55	218	soda ==> heineken
11	2	21.68	73.31	217	coke ==> ice_crea
12	2	21.28	52.85	213	bourbon ==> cracker
13	2	20.98	53.71	210	corned_b ==> olives
14	2	20.88	53.32	209	baguette ==> avocado
15	2	20.78	57.30	208	avocado ==> heineken
16	2	20.68	57.02	207	avocado ==> artichok
17	2	19.78	64.92	198	artichok ==> heineken
18	2	18.28	30.50	183	heineken ==> chicken

*The **Rules** tab displaying the evaluation criteria statistics from sequence analysis.*

From the **Rules** tab, the table listing displays the highest confidence. Although Heineken beer is the most popular single item purchased, it seems that the buying behavior of customers is to either first purchase crackers, herring, baguette, soda, avocados, and artichokes, then subsequently purchase Heineken beer.

Subset Table

The **Subset Table** option is designed to subset the sequence rules that are automatically generated. Simply select the **View > Subset Table** main menu option and the following **Subset Table** window will appear. The **Subset Table** window is designed for you to specify a built-in operator of the requested chain of sequences based on a simple WHERE expression. To begin building the WHERE expression, simply select any one of the evaluation criteria listed in the **Available Columns** list box and the following list of operators will appear. This is illustrated in the following diagram. The next step is to select from the listed evaluation criteria statistics to complete the corresponding WHERE expression that is listed in the **Where** container box. Execute the specified WHERE query expression by selecting the **OK** button. The **Rules** tab will then appear that will list the various rules based on the query condition specified by the WHERE statement.

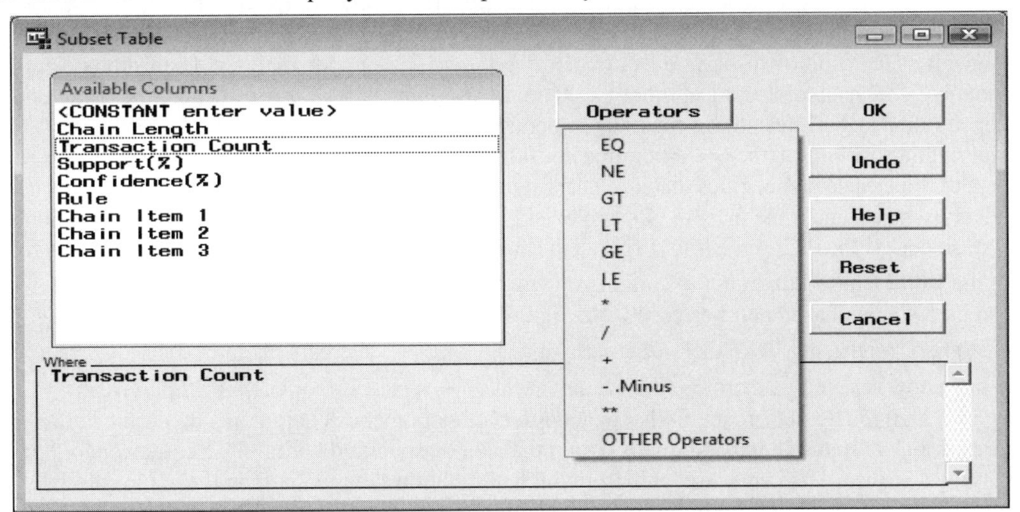

*The **Subset Table** window is used in building your query from sequence discovery.*

The following are the various operands to select from within the **Available Columns** list box.

- **<CONSTANT enter value>:** This option will open an additional dialog box that will appear for you to enter the appropriate constants or text in building your WHERE expression.
- **Chain Length:** This option will allow you to specify the number of events in a sequence.
- **Transaction Count:** This option will allow you to specify the number of occurrences.
- **Support (%):** This option will allow you to subset the listed rules from the relative frequency criteria.
- **Confidence (%):** This option will allow you to subset the listed rules from the confidence probability criteria.
- **Rule:** This option will allow you to enter an expression for the rule that can be up to 200 characters.
- **Chain Item 1-n:** The number of chains that are listed based on the **Number of items in longest chain** specified from the **Sequence** tab.
- **<LOOKUP distinct values>:** This option opens the **Lookup Distinct Values** window that will allow you to select an item name of interest.

For example, suppose you are interested in a listing of all ordered sequences of two or more based on all customers purchasing Heineken beer. Therefore, simply perform the following steps.

1. Select **View > Subset Table**.
2. From the **Available Columns** list, select **Chain Item 1**.
3. Select the **Operators** button, then select the **EQ** pull-down menu item.
4. From the **Available Columns** list, select **<LOOKUP distinct values>**.
5. Select Heineken from the **Lookup Distinct Values** window.
6. Select the **Operators** button, then select the **OR** pull-down menu item.
7. From the **Available Columns** list, select **Chain Item 2**.
8. Select Heineken from the **Lookup Distinct Values** window and press the **OK** button.

Event Chain

Selecting the **View > Event Chain** main menu options will display the following **Event Chains** window. The window is designed for you to build a sequence of rules to query from the active training data set. The list box positioned to the left of the window displays every two-way sequence rule under analysis. The list box displays the order 2 sequence of items in alphabetical order based on the left-hand side of the sequence rule. Only one two-way sequence rule may be selected at a time. By selecting each sequence rule from the **Possible** list box and pressing the right arrow button, the corresponding confidence and support probabilities will be displayed that is based on the selected sequence rules that are listed in the **Selected** list box positioned to the right. Continue to select each sequence rule from the **Possible** list box until the desirable chain of sequence of items is reached. Next, select the **Find Chain** tab to display the chain of sequence of items, that is, the sequence rule, specified from the **Build Chain** tab. In our example, you will first select the sequence of bourbon→Coke to query all customers who purchased bourbon then Coke. From the **Possible** list box, you will next select the sequence of customers purchasing Coke then ice cream, that is, Coke → ice cream chain of the sequence of items that you have now selected is, therefore, bourbon => Coke → ice cream, or customers purchasing bourbon then Coke, and, finally, ice cream, in that exact order. The number of chain items that can be queried is based on the **Number of items in longest chain** option that can be specified within the **Sequence** tab. For instance, specifying a sequence of items or events from the **Build Chain** tab greater than the value entered from the **Sequence** tab will result in the **Find Chain** tab being grayed-out and unavailable for viewing.

Press the left-hand double arrow control button to start all over again by moving all the two-way sequence rules from the **Selected** list box back to the **Possible** list box.

*The **Build Chain** window that lists all the possible sequence rules to query the desired sequence rule.*

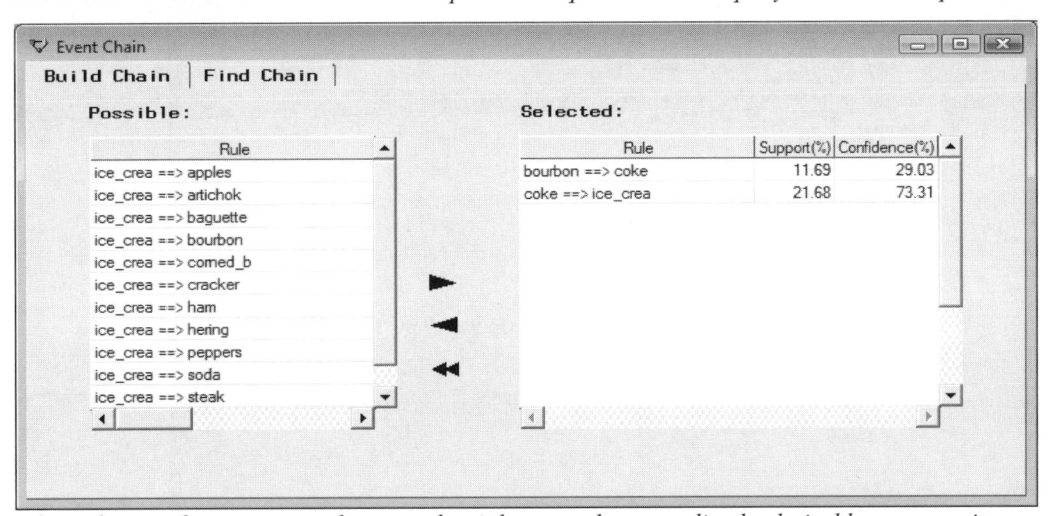

After selecting the sequence rule, press the right arrow button to list the desirable sequence items.

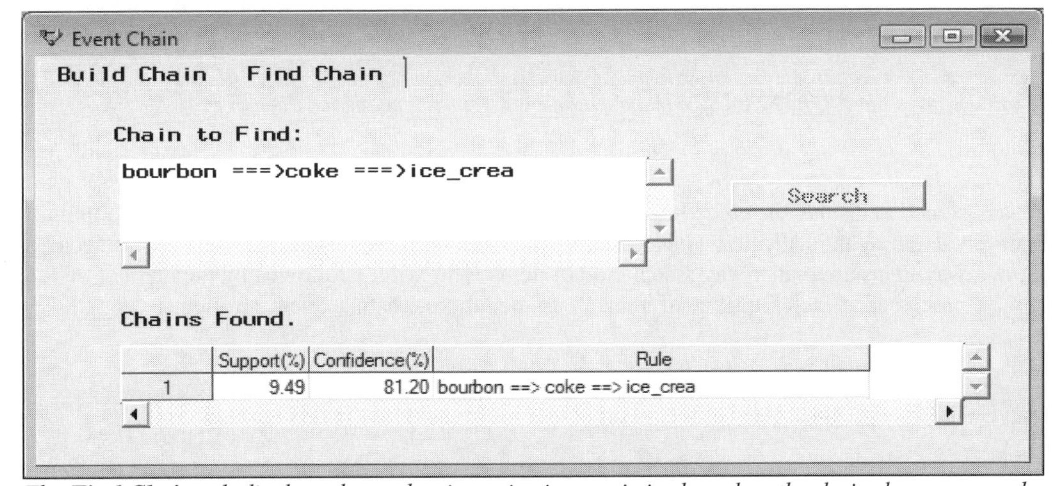

*The **Find Chain** tab displays the evaluation criterion statistics based on the desired sequence rule.*

From the **Find Chain** tab, you may save the resulting sequence of events to a SAS data set by selecting **File > Save As Data Set** main menu options. The query results can also be exported to an external file by selecting the **File > Export** main menu option with the SAS Import Wizard appearing.

Code tab

The **Code** tab displays the internal SEMMA training code that is applied to sequence analysis. Initially, the training code ensures that the active training data set is structured as a sequence analysis database layout by sorting the file by the customer identifier variable within the time identifier variable. The sorted file is then read into the PROC ASSOC procedure to determine the best set of items that are related to each other. The PROC SEQUENCE procedure is then applied to determine the best sequence of related items from the rules output data set that is generated from the previous PROC ASSOC procedure.

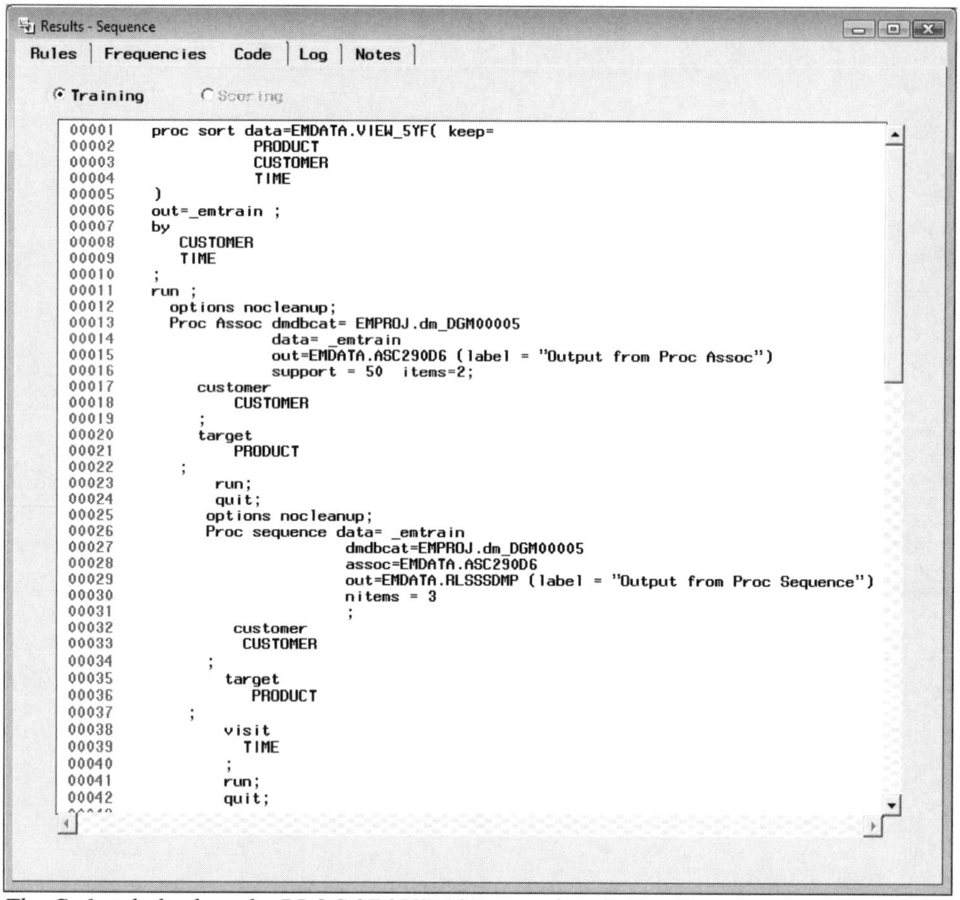

Results - Sequence

Rules | Frequencies | Code | Log | Notes

(•) Training ( ) Scoring

```
00001   proc sort data=EMDATA.VIEW_5YF( keep=
00002              PRODUCT
00003              CUSTOMER
00004              TIME
00005   )
00006   out=_emtrain ;
00007   by
00008      CUSTOMER
00009      TIME
00010   ;
00011   run ;
00012     options nocleanup;
00013     Proc Assoc dmdbcat= EMPROJ.dm_DGM00005
00014                data= _emtrain
00015                out=EMDATA.ASC290D6 (label = "Output from Proc Assoc")
00016                support = 50  items=2;
00017        customer
00018            CUSTOMER
00019        ;
00020        target
00021            PRODUCT
00022      ;
00023          run;
00024          quit;
00025          options nocleanup;
00026          Proc sequence data= _emtrain
00027                         dmdbcat=EMPROJ.dm_DGM00005
00028                         assoc=EMDATA.ASC290D6
00029                         out=EMDATA.RLSSSDMP (label = "Output from Proc Sequence")
00030                         nitems = 3
00031                         ;
00032             customer
00033              CUSTOMER
00034         ;
00035            target
00036               PRODUCT
00037        ;
00038            visit
00039              TIME
00040            ;
00041            run;
00042            quit;
```

*The **Code** tab displays the PROC SEQUENCE procedure listing from sequence discovery.*

Log tab

The **Log** tab is designed to display the procedure log listing from the programming code that is listed in the previous **Code** tab. The **Log** tab will allow you to observe the number of records processed at each stage of all single items, two-way items up to all *n*-way association of items. This will be followed by the number of records that were processed at each sequence of items up to the longest chain sequence of items.

Notes tab

The **Notes** tab is a notepad to enter notes or record any other information to the **Association** node.

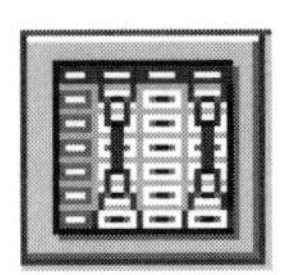

2.5 Variable Selection Node

General Layout of the Enterprise Miner Variable Selection Node

- **Data tab**
- **Variables tab**
- **Manual Selection tab**
- **Target Associations tab**
- **General tab**
- **Output tab**
- **Notes tab**

The purpose of the **Variable Selection** node in Enterprise Miner is to perform a variable selection procedure in determining the best set of input variables for the predictive model from a pool of all possible input variables that best predicts the variability of the unique target variable from the training data set. The following variable selection routines are advantageous when there are a large number of input variables in the model. By default, the node performs stepwise regression to the process flow. The node performs the variable selection routine for either an interval-valued or binary-valued target variable. Therefore, if the target variable is ordinal or nominal with more than two class levels, you may want to use the **Regression** node to perform the variable selection routine in selecting the important input variables for the model. In model building, it is not only very important to construct a model with the smallest error, but it is also important to find a model with the fewest number of parameters. The reason is because there are a larger number of degrees of freedom for the error term that leads to a more powerful test, since there are fewer terms in the model to predict the target variable. Failure to reduce the number of input variables for the model will result in overfitting and bad generalization, a tremendous increase in computational time, a high probability of removing observations from the analysis due to missing values in any one of the input variables and a high probability that certain input variables will be related to each other, which will result in unstable estimates and invalid test statistics. *Unstable estimates* means that the estimates are quite different by fitting the same model to the same data that is resampled any number of times. Before performing the following variable selection routines in order to develop a well-designed predictive or classification model, it is advisable to perform the various preprocessing routines such as filtering the extreme values from the training data set, transforming all the variables that are not normally distributed, and imputing missing values and replacing incorrect nonmissing values. The node is designed to remove input variables from the model based on a certain percentage of missing values, and the number of unique values of the categorical input variables, and removing variables based on model hierarchy. Model hierarchy is defined by the level of generalization or the interacting effect between the two separate input variables in the predictive model. The metadata sample is used to determine variable hierarchies in the **Variable Selection** node.

In predictive modeling, there are two reasons for eliminating variables from the analysis: *redundancy* and *irrelevancy*. In other words, most of the modeling selection routines are designed to minimize input redundancy and maximize input relevancy. At times, the statistical model can potentially consist of an enormous number of input variables in the model to predict the target variable. Therefore, irrelevancy in some of the input variables might not provide a sufficient amount of information in describing or predicting the target variable. Redundancy in the input variables suggests that a particular input variable does not provide any added information in explaining the variability in the target variable that has not already been explained by some other input variables already in the model. For example, state and zip code, weight in pounds and weight in ounces, or height in feet and height in inches provide the same redundant information in predicting the target variable. In eliminating both the redundant and irrelevant variables from the predictive model, it is easier to first eliminate the redundant input variables and then consider the removal of the irrelevant variables. Redundant input variables can be identified by a high correlation between both variables or basically observing the same range of values between both variables. Conversely, irrelevancy in the input variables can be determined by prior knowledge, previous studies, or general intuition. Otherwise, irrelevancy in the input variables can be determined by insignificant p-values in its associated parameter estimates.

It is assumed that the input variables in the model are not highly correlated to one another. The reason in removing the redundant input variables from the model is to avoid the multicollinearity phenomenon. *Multicollinearity* occurs when certain input variables are highly correlated with one another, providing redundant or identical information in explaining or predicting the underlying effect of the target variable. For example, a variable that is the total sum or an accumulation of separate input variables in the model will provide redundant information about the model that will result in the inversed correlation matrix that cannot be computed. It is important to understand that the problem of multicollinearity has to do with the data and not with the predictive model. The reason that multicollinearity is so problematic is because there are a certain number of input variables in the model that are identical or proportional to one another, which will result in the system of linear equations of input variables that best describes the relationship in the target variable that cannot be computed algebraically. Highly collinear input variables in the statistical model will result in singularity matrices that cannot be inverted. Again, the existence of multicollinearity in the input variables can cause the model to be very unstable, which can also make the interpretation of the linear relationship between the input variables and the target variable nearly impossible. Multicollinearity can cause some related input variables to appear less important than they actually are, that is, multicollinearity can hide the significance of certain input variables in the model. In extreme cases of multicollinearity, the t-statistic and the associated *p*-value of a combination of related input variables can make every input variable in the model nonsignificant, and when actually taken together accurately describe and predict the target variable. In predictive modeling, multicollinearity will contribute to an increase in the variance of the parameter estimates, which will result in unstable and less precise parameter estimates that will contribute to increased error in the predicted values. Examples of pairs of variables that are related to one another are height and weight, store sales and number of employees, and household income and mortgage rates.

The **Variable Selection** node typically performs two steps in selecting the best set of input variables to the predictive model. Initially, the node performs correlation analysis from the simple linear relationship between each input variable and the target variable. From the analysis, the input variables with the highest correlations are retained in the model. However, the shortcoming of this technique is that correlation analysis does not account for the partial correlation between the other input variables in the model. Therefore, inputs could be mistakenly added to or removed from the predictive model. *Partial correlation* measures the association between each input variable and the target variable in the model, that is, as the values of the input variable change, they will result in the change in values of the target variable in the model, or vice versa. Therefore, this is the reason that the forward stepwise regression procedure is performed in the subsequent step. Stepwise regression takes into account the intercorrelation between the input variables in the predictive model. However, at times two variables might be uncorrelated with each other unless you take into account a third variable. As an example, height generally would not contribute that much in predicting hair length unless you consider including females in the regression model. In other words, the partial correlation between height and hair length is approximately zero as opposed to an increase to the multiple correlation in the model with females included in the model in predicting hair length. The reason is because women tend to wear long hair.

At times, data mining data sets can potentially amount to an enormous number of input variables in the predictive model. Therefore, the **Variable Selection** node is designed to eliminate a certain number of input variables from the predictive model based on the simple linear relationship between each input variable and the target variable from the training data set. Variable selection routines are designed to eliminate the *curse of dimensionality*. However, removing the number of variables or reducing the dimensionality to the modeling design can also result in disregarding important information in predicting the target variable. The main idea in the following modeling selection procedure is to eliminate the unimportant variables from the predictive model and produce a smaller combination of input variables that contain as much information as possible from the original data set in predicting the target variable. In the variable selection routine, there is this trade-off in reducing the number of input variables in the model while at the same time making sure that the variable selection routine does not remove important input variables from the model that accurately explains the variability in the target variable. In nonlinear predictive modeling, such as neural network modeling, eliminating the input variables from the design and reducing the dimensionality of the extremely nonlinear error surface will increase the generalization performance of the modeling design that is the main goal of any statistical modeling design. The **Tree** node and the **Regression** node are the other two Enterprise Miner nodes that perform variable selection routines within the process flow. In addition, the **Princomp/Dmneural** node is another dimension reduction routine that can be used in determining the best set of principal components that best explains the variability in the data. In principal component analysis, instead of dropping input variables from the model, you remove the linear combination of independent variables, that is, principal components from the model. Stepwise regression is probably the most well-known modeling selection technique used in predictive model building. However, there is no universal variable selection technique that is considered superior. Therefore, it is recommended to apply different modeling selection criterions in evaluating the importance of the input variables in the predictive model and also look for consistency in the input variables that are selected for the final model from the various modeling selection routines. Selecting from many different criteria can result in many different models. Therefore, as a general modeling selection strategy, it is not a bad idea to specify many different modeling selection criteria, will result in many different models to select from in order to select the most reliable model.

The Key to a Well-Designed Model

The reason that the modeling selection procedure is so important is because including enough polynomial terms in the multiple linear regression model, constructing a tree large enough in decision tree modeling or selecting enough hidden layer units in neural network modeling can result in an absolutely perfect fit with absolutely no training error. On the other hand, ignoring important input variables from the model that predicts the target responses extremely well will result in poor generalization. Adding too many terms in the regression model, too many hidden layer units in the neural network model, or constructing too small of decision tree will result in overfitting and poor generalization. Therefore, the goal in the modeling selection routine is finding a balance in fitting the model that is not too simple and yet not too complex.

Determining the correct combination of input variables for the model is the key to an effective predictive and classification modeling design. There are two reasons why it is important to reduce the number of input variables in the predictive model. The first reason is that it will result in an increase in the predictive accuracy of the model with a reduction in the variance and a small sacrifice of an increase in the bias. The second reason is that reducing the numerous input variables in the predictive modeling design will lead to an easier understanding in the interpretation of the modeling terms based on the linear relationship between the input variables in the predictive model that best explains or predicts the target variable. Typically, input variables with little or no correlation to the target variable are excluded from the predictive model.

Preprocessing the data in statistical modeling, like performing the modeling selection procedure in order to reduce the number of input variables in the predictive model, is one of the most important steps in a well-designed predictive modeling design. In statistical modeling, reducing the number of input variables decreases the chance that the **X'X** correlation matrix will result in an ill-conditioned matrix and instability in the parameter estimates in the model. Also, reducing the number of input variables from the statistical model will reduce the chance that certain observations will be removed from the modeling fit since Enterprise Miner will automatically remove all cases from the model with missing values in any one of the input variables. From the **General** tab, the node will allow you to remove certain input variables from the model based on a specified proportion of missing cases. In logistic regression modeling, reducing the number of categorically-valued input variables to the model will reduce the chance that the model will consist of certain class levels with zero cell counts, which will lead to quasi-complete separation in the model. Quasi-complete separation will result in an undefined parameter estimate in the categorically-valued input variable. In nearest neighbor modeling, reducing the number of input variables of the model will increase the chance that each hypersphere will adequately encompass the number of specified observations. In decision tree modeling, eliminating certain input variables from the model will reduce the number of splits to consider where the algorithm first determines the best split based on each input variable in the model, then performs an exhaustive search in selecting the best splits from a multitude of possible splits at each node. In principal component analysis, reducing the number of input variables, where each principal component is a linear combination of every input variable in the model, will reduce the complexity in interpreting the variability or correlation in each principal component that is selected.

An Overview of the Node

The **Variable Selection** node facilitates the variable modeling selection procedure for both ordinary least-squares regression and logistic modeling designs. The node supports both binary-valued or interval-valued target variables. Therefore, if the target variable consists of more than three separate class levels, then you must create separate binary dummy target variables to fit, instead of fitting the original nominal or ordinal-valued target variable and observing the consistency in the final input variables selected from the different classification models. Otherwise, assume it is an interval-valued target variable by fitting the least-squares model. The input variables that are selected for the statistical model are determined by one of two variable selection criteria of either the r-square criterion or chi-square model selection criterion. For an interval-valued target variable, the node applies an r-square criterion. The r-square statistic is based on the linear relationship between each input variable in the predictive model and the target variable to predict. The r-square statistic is calculated in determining how well the input variables in the predictive model explain the variability in the target variable. The node uses the stepwise method in selecting the best linear combination of input variables to the model with the modeling selection procedure terminating when the improvement in the r-square is less than .0005. By default, the method rejects input variables whose contribution to the target variability is less than .0005. For a categorically-valued target variable, the node applies a chi-square criterion. This variable selection routine produces a series of 2 x 2 frequency tables in determining the most appropriate input variables to the logistic regression model based on the highest chi-square statistic, where each nominal or ordinal-valued input variable is remapped into binary-valued dummy variables and each interval-valued input variable is remapped into several categories of equally sized intervals.

For each interval-valued input variable in the model, the **Variable Selection** node performs simple linear regression to calculate the r-square value by either fitting the interval-valued or binary-valued target variable to the model. For each categorically-valued input variable, the node performs one-way analysis of variance to calculate the chi-square statistic by fitting the interval-valued target variable. In addition, from the **Target Associations** tab, the node is designed to create a scored data set from the final input variables that are selected from the variable selection routine. For interval-valued targets, the node automatically performs linear regression modeling. For binary-valued targets, the node automatically performs logistic regression modeling.

However, for binary-valued target variables, the **Variable Selection** node will allow you to select the r-square criterion. In other words, the node has the added flexibility in performing the variable selection routine by applying the multiple linear regression model when fitting the binary-valued target variable. The reason for applying the least-squares model is to avoid computational difficulties and long lengthy runs due to the iterative maximum likelihood algorithm that is applied in calculating the parameter estimates and standard errors in the logistic regression model which may take many passes of the data to reach stable parameter estimates when performing the variable selection routine. The algorithm can result in a tremendous amount of computational time when fitting an enormous amount of data to the logistic regression model and generating a large number of classification models from a pool of all the possible combination of input variables by assessing the significance of each input variable added or removed from the classification model one at a time, which is determined by the difference in the likelihood functions. In addition, memory limitations with regard to the Hessian matrix might occur when fitting several categorically-valued input variables or an enormous number of input variables in the logistic regression model. The Hessian matrix is the matrix of second derivatives of the error function with respect to the parameter estimates. The standard errors from the logistic regression model are determined by the diagonal elements of the negative of the inverted Hessian that can be computationally demanding with several categorically-valued input variables in the logistic regression model. This is compounded by the fact that algebraic problems might occur when inverting the Hessian matrix due to certain input variables in the model that are highly correlated to each other that will result in undesirable bad local minimums in the error function. The drawback to this approach is that, at times, the least-squares model might generate fitted values less than zero or greater than one. The reason is because the linear regression line might go below zero and above one when fitting the responses of the binary-valued target variable with values of either zero or one.

To generate assessment statistics from the scored data set that is created from the node, simply connect the **Variable Selection** node to the **User Defined Model** node. During the variable selection procedure, missing values in the target variable are removed from the analysis. Missing values in any one of the interval-valued input variables are replaced by their own mean and missing values in the categorically-valued input variable are treated as a separate class level.

The Three Steps to the Variable Selection Node

The node performs a three-step process in selecting the best set of input variables from the R^2 variable selection routine where the input variables are added or removed from the predictive model one at a time by fitting the interval-valued or binary-valued target variable in the model as follows:

First Step: In the first step, the node performs correlation analysis between each input variable and the target variable. All input variables are retained to the model with a squared correlation r-square statistic greater than the default criterion value of .005. However, the drawback to this approach is that it ignores the partial correlations between the other input variables in the model that might inadvertently add or remove input variables to the model. Therefore, setting the default criterion value of zero will result in all input variables retained to the model, which will prevent the correlation analysis procedure from being performed by removing certain input variables from the model.

Second Step: In the second step, the node performs forward stepwise r-square regression to the input variables that are not rejected from the previous step. The input variables that have the largest squared correlation coefficient value to the target variable are first entered into the regression model. Input variables with an improvement in the R^2 statistic less than the threshold criterion value, that is, with a default of.0005, are removed from the model. Setting the default threshold criterion value of zero will result in all input variables retained to the model that will prevent the stepwise regression procedure from being performed in removing the input variables from the model.

Third Step: For binary-valued targets, an additional third step is performed that fits a logistic regression model based on the predicted values that are outputted from the previous forward stepwise regression routine that are used as the only input variable for the model. Once the final predictive model is selected with the best set of linear combination of input variables, the **Variable Selection** node will then automatically create an output scored data set that can be passed along to the subsequent modeling node.

From the node, input variables with a model role set to **input** are automatically included in the variable selection routine. Conversely, the input variables removed from the predictive model due to variable selection routine are automatically set to a model role of **rejected**. However, these same input variables are passed along to the subsequent modeling nodes. The node will allow you to initially remove certain input variables from the variable selection routine that might be considered unimportant to the modeling fit by setting the variable status of these same input variables to **don't use** from the **Variables** tab. Furthermore, the input variables that are automatically dropped from the predictive model can be included back into the predictive model by reassigning the input model role to these same variables in any of the subsequent modeling nodes.

The various modeling selection techniques are discussed in greater detail in Neter, Wasserman, Kutner and Nachtsheim (2004), Freund and Littell (2000), Bowerman and O'Connell (1990), and Applying Data Mining Techniques Using Enterprise Miner Course Notes, SAS Institute (2002).

Data tab

The **Data** tab is designed to perform the variable selection routine on the training data set only. In other words, the node is not designed to perform the variable selection routine on the validation, test, or scored data sets. However, the validation and test data sets can be applied within the subsequent modeling nodes to compare the consistency in the results based on the best set of input variables that are selected for the predictive model from the variable selection routine that is performed on the training data set. The **Select...** button is grayed-out assuming that the node is connected to a single data set to be processed, which is usually created from the **Input Data Source** node. Otherwise, press the **Select...** button to choose among the various previously created data sets that are connected to the node in order to select the active training data set to perform the subsequent variable selection routine. Press the **Properties...** button to either view the file administrative information or a table view of the active training data set.

Variables tab

By default, opening the **Variable Selection** node will result in the following **Variables** tab appearing. The **Variables** tab displays a table view of the various properties of the variables such as the variable roles, level of measurement, variable type, and the variable formats that are assigned from the metadata sample. The tab is designed for you to manually add or remove the input variables from the predictive model. Simply select the input variable rows of interest, then scroll the mouse over to the **Status** column and right-click the mouse to select the **Set Status** pop-up menu item that will allow you to select either the **use** or **don't use** option in order to add or delete the corresponding input variable from the predictive model. By default, all the variables in the modeling selection process will have a variable status set to **use**. However, setting the target variable to **don't use** will prevent the modeling selection procedure from being performed. Again, it is important to point out that removing the input variable from the analysis will not result in the variable being removed from the active training data set. Notice that the other columns are grayed-out and cannot be changed.

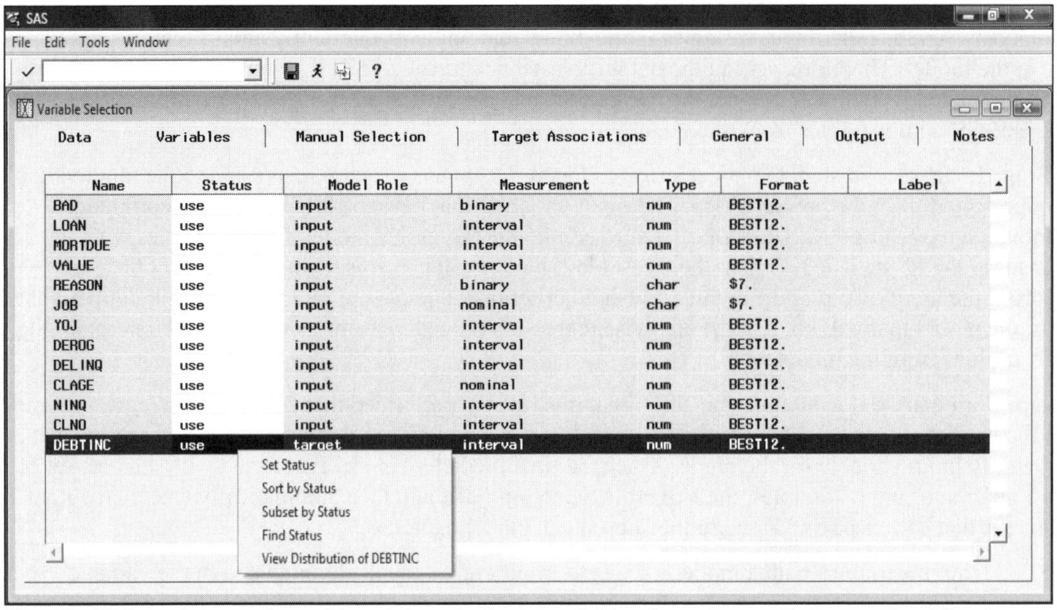

SAS — File Edit Tools Window

Variable Selection

Data | Variables | Manual Selection | Target Associations | General | Output | Notes

Name	Status	Model Role	Measurement	Type	Format	Label
BAD	use	input	binary	num	BEST12.	
LOAN	use	input	interval	num	BEST12.	
MORTDUE	use	input	interval	num	BEST12.	
VALUE	use	input	interval	num	BEST12.	
REASON	use	input	binary	char	$7.	
JOB	use	input	nominal	char	$7.	
YOJ	use	input	interval	num	BEST12.	
DEROG	use	input	interval	num	BEST12.	
DELINQ	use	input	interval	num	BEST12.	
CLAGE	use	input	nominal	num	BEST12.	
NINQ	use	input	interval	num	BEST12.	
CLNO	use	input	interval	num	BEST12.	
DEBTINC	use	target	interval	num	BEST12.	

Set Status
Sort by Status
Subset by Status
Find Status
View Distribution of DEBTINC

*The **Variables** tab is used to add or delete the input variables from the predictive or classification model.*

Manual Selection tab

The **Manual Selection** tab is designed for you to automatically add or remove input variables from the variable selection procedure based on the **Role Assignment** column setting. By default, input variables are set to **<automatic>**, indicating to you that the input variable will be part of the modeling selection process in which the automatic selection criterion that is selected from the following the **Target Associations** tab is used to determine which inputs are added or removed from the model. However, you may automatically include the input variable in the predictive model by setting the **Role Assignment** setting to **input**. If you already know that an input variable is not important in predicting the target variable, then you may automatically drop the variable from the predictive model by assigning the input variable a **Role Assignment** setting of **rejected**. Input variables with a variable status of **don't use** will not appear in the **Manual Selection** tab. The **Reason** column is designed for you to enter a short description of the listed variables in the active training data set.

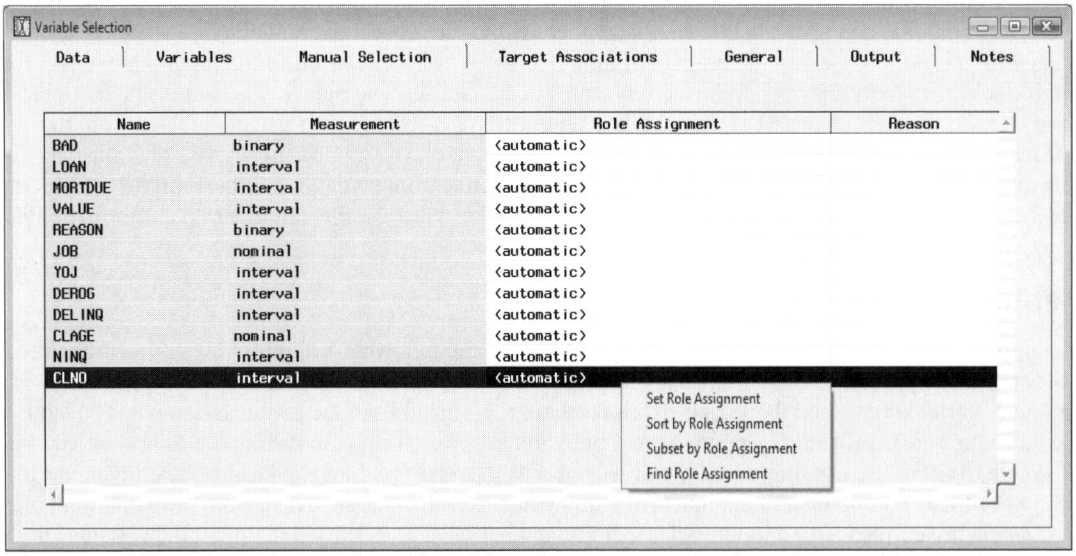

Variable Selection

Data | Variables | Manual Selection | Target Associations | General | Output | Notes

Name	Measurement	Role Assignment	Reason
BAD	binary	<automatic>	
LOAN	interval	<automatic>	
MORTDUE	interval	<automatic>	
VALUE	interval	<automatic>	
REASON	binary	<automatic>	
JOB	nominal	<automatic>	
YOJ	interval	<automatic>	
DEROG	interval	<automatic>	
DELINQ	interval	<automatic>	
CLAGE	nominal	<automatic>	
NINQ	interval	<automatic>	
CLNO	interval	<automatic>	

Set Role Assignment
Sort by Role Assignment
Subset by Role Assignment
Find Role Assignment

*The **Manual Selection** tab is used to add or delete the input variables from the variable selection procedure.*

Target Associations tab

The **Target Associations** tab is designed to specify the modeling selection criterion statistic that are used in determining the appropriate input variables of the predictive modeling design, prevent the creation of the output scored data set by training the regression model with the selected input variables, and setting the cutoff probability to the binary-valued target response. If a target variable does not exist in the input data set or if the categorically-valued target variable has more than two class levels, then the tab is grayed-out and unavailable for viewing.

Again, the tab will allow you to select from the two separate model selection criteria that are available within the node. By default, the r-square selection criterion is automatically selected.

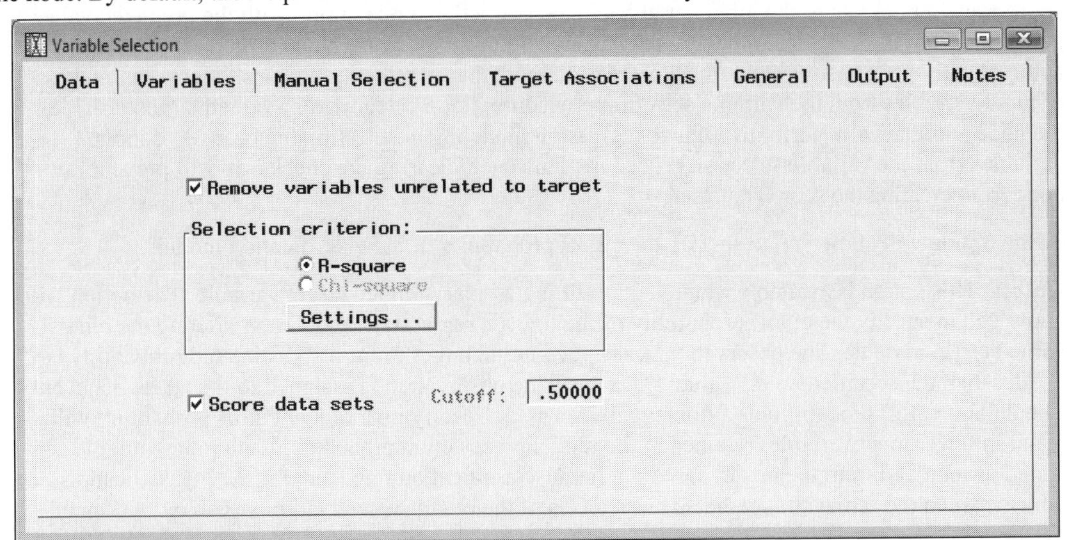

*The **Target Associations** tab is used to specify the criterion statistic for the modeling selection routine by fitting the interval-valued target variable.*

- **Remove variables unrelated to target:** By default, this option is checked. The node automatically removes input variables from the model that are unrelated to the target variable based on the various modeling selection criteria. It is a preliminary assessment for quickly determining the input variables in the model that best predict the variability in the target variable based on a linear relationship between each input variable. To prevent the node from using the selection criterion value that is used in removing variables from the predictive model, clear the check box. Clearing the check box also will prevent the node from creating the scored data set with the following **Selection criterion** and **Score data sets** options unavailable for selection. This will result in the node selecting or rejecting the input variables in the model from the settings you defined in the **Manual Selection** and **General** tabs.

The following are the two separate modeling selection statistics that you may select from the **Target Associations** tab:

- **R-squared:** By default, the r-square or the coefficient of determination statistic is used in determining the appropriate input variables added or removed from the predictive model. The r-square statistic is the proportion of variability of the target variable explained by the input variable in the model, that is, with values ranging between zero and one. A r-square statistic of zero indicates that there is absolutely no linear relationship between both the input variable and the target variable in the model, and a r-square statistic of one indicates a perfect linear relationship between both variables in the predictive model. For interval-valued input variables, the node performs a simple linear regression model by fitting either the interval-valued or binary-valued target variable. For categorically-valued input variables in the predictive model, the node performs one-way analysis of variance by fitting the interval-valued target variable. Selecting the **Settings...** button will result in the following **R-square** window appearing. Input variables are removed from the regression model with an r-square statistic less than the value specified from the **Squared correlation** value.

- **Chi-Square:** This modeling selection criterion is available for binary-valued target variables only. Otherwise, this option will be automatically unavailable if there exits an interval-valued target variable in the regression model. The variable selection process evaluates the importance of the input variable to the model by performing various splits in maximizing the chi-square statistic from the 2x2 frequency table. For nominal or ordinal-valued input variables, the class levels are decomposed into dummy variables. The hypothesis is rejected for large values of the chi-square statistic, indicating to you that the input variable is a good predictor of the classification model.

The following option controls the creation of the output scored data set:

- **Score data sets**: By default, the **Variable Selection** node creates an output data set that assigns the appropriate model role to the input variables in the predictive model along with the predicted values generated from the statistical model within the node. For an interval-valued target variable, the node automatically performs multiple linear regression modeling in selecting the best linear combination of input variables from the variable selection procedure. For a categorically-valued target variable, the node automatically performs logistic regression modeling in selecting the best set of input variables from the variable selection procedure. However, clearing the check box will prevent the node from creating the scored data set.

The following option will allow you to specify the cutoff probability of the classification model:

- **Cutoff:** This option is available when you are fitting a binary-valued target variable. The option will allow you to specify the cutoff probability of the logistic regression model in predicting the binary-valued target variable. The observation is assigned to the target event if its estimated probability is greater than the specified cutoff value. Otherwise, the observation is assigned to the target nonevent. The default cutoff probability is automatically set at .5. Specifying a higher cutoff probability will result in fewer input variables retained to the modeling selection procedure. Each input variable is added or removed from the model based on the misclassification rate from the 2x2 classification table between the actual and predicted class levels of the binary-valued target variable.

R-Square Modeling Selection Criterion

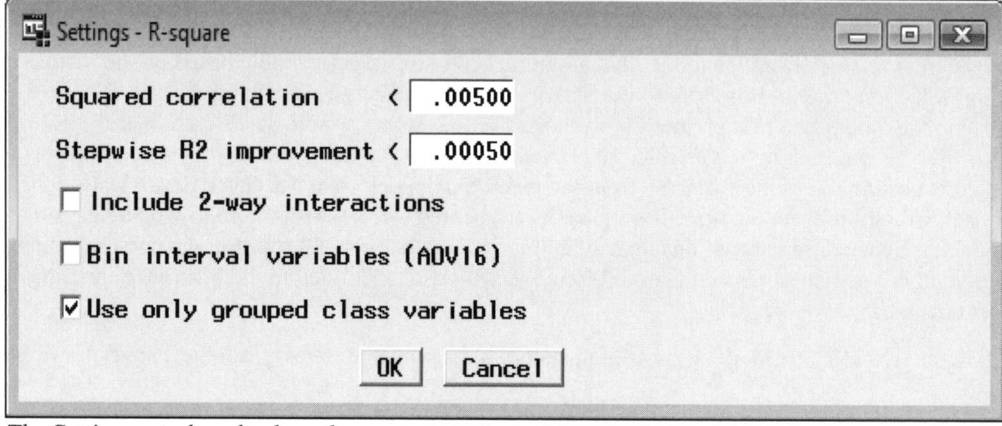

*The **Settings** window displays the r-square option settings to the variable selection procedure.*

The following are the various option settings for the R-square modeling criterion statistic:

- **Squared correlation:** In the first step, correlation analysis is performed between each input variable and the target responses. Input variables are entered into the predictive model in which the node sequentially evaluates each input variable in the predictive model based on the squared correlation coefficient value. From the **R-square** window, you may specify an appropriate r-square statistic used in assessing the strength of the linear relationship between each input variable and the target variable in the model. The node sets each input variable in the model to a model role of **rejected** if the target correlation for each input variable falls below the specified threshold value. By default, the squared correlation p-value is set to .005. Setting the r-square value higher will tend to exclude more input variables from the model. Conversely, setting the value in the entry field too low will tend to include more input variables in the model. Also, you can elect to retain all input variables to the model by setting the squared correlation value to zero. This will allow the node to select the best input variables for the model from the following forward stepwise modeling selection routine and avoid the possibility of eliminating important intercorrelated input variables from the predictive model. For categorically-valued input variables, the node performs one-way ANOVA in computing the squared correlation based on the grouped target means and the overall target mean.

- **Stepwise fit improvement:** This option will allow you to control the termination criteria that is used in the forward stepwise regression procedure to the significant input variables that failed to be rejected from the previous correlation analysis. Setting a minimum value to the improvement of the global r-square value from the overall predictive model is used in stopping the stepwise regression process. By default, the p-value to this r-square criterion is set to .0005. Setting a higher threshold value will tend to exclude more input variables from the model. Conversely, setting the value in the entry field too low will tend to include more input variables in the predictive model.
- **Include 2-way interactions:** Includes all two-way interaction effects in the categorically-valued input variables. Not selecting this option, will result in the node performing the variable selection routine on the main effects in the model only and preventing the interaction terms from being included in the final model. Typically, it is important to first determine if the various interaction effects might have a significant effect in predicting the target response. However, the big drawback in introducing interaction into the predictive model is that it makes the relationship between the target response variable and the input variables in the model much harder to interpret. By default, the interaction terms are automatically removed from the statistical model.
- **Bin interval variables (AOV16):** This option will create 16 equally spaced intervals for all interval-valued input variables in the linear regression model. Intervals with zero frequency counts are automatically removed. Therefore, at times there can be less than 16 intervals. The AOV16 method is designed to bin the input variables in the model to determine the nonlinear relationship between the input and target variable. By default, this method is not selected since it can be computationally intensive.
- **Use only group class variables:** This option is designed to reduce memory and processing time. In other words, the option is designed to reduce the number of group levels to the categorically-valued input variable using only the transformed categorical variable to evaluate the variable's importance. By default, this option is selected, indicating to you that if a class variable can be reduced into a group variable with fewer class levels, then only the transformed grouping variable is applied to the variable selection process. However, clearing this check box will result in the node using both the grouped class variable and the original class variable in the subsequent variable selection procedure.

Grouping the Categorical Input Variables

The following example is based on a categorical input variable in the predictive model in predicting the interval-valued target variable. The group class variables method can be viewed as an analysis of the ordered cell means to determine if specific class levels of the categorical input variable can be collapsed into a single group. The first step in the group class variable method is to order the target means in descending order across the separate class levels of the categorical input variable that explains a predetermined percentage of variability in the target variable that is determined by the r-square criterion value. The method then determines which class levels can be combined or collapsed into a single group. The separate levels of the input variable are combined from top to bottom with the class levels ordered by the largest target mean ranked at the top, that is, (A), to bottom, (C), in determining the best groups to combine. Again, the node always works from top to bottom when considering the levels that can be collapsed. If the top two input groups (A) and (B) are combined, leading to a reduction in the r-square statistic at .05 or lower, then the input variable categories are collapsed. Otherwise, the procedure then determines the next adjacent input categories to combine, that is, (B) and (C). The node stops combining levels when the r-square threshold is not met for all possible class levels that are adjacent to each other or when the original variable is reduced to two groups.

Note: The **Code** tab will display the internal SEMMA scoring code from a series of *if-then* statements that perform the group class variable method in determining the process in which the class levels of the categorical-value input variable are combined.

Explanation of the AOV16 Variables in the Predictive Model

The following code will briefly explain the meaning of the various AOV16 variables that identify the nonlinear relationship in the target variable. The main idea of the AOV16 variables is to bin the interval-valued input variables in the model into equally sized intervals or intervals of equal length. Typically, the average values of the target variable will display a nonlinear trend across the separate class levels of the nonlinear transformed AOV16 input variable in the model. Therefore, the reason for the **Variable Selection** node to perform this transformation in an attempt to increase the r-square statistic by discovering nonlinear relationships between the interval-valued input and target variable in the model with the drawback being an increase in the degrees of freedom to the multiple linear regression model.

As an example, income will typically have a nonlinear relationship with age since income will generally increase over time, then decline during the retirement years. In the following example, suppose the target variable represents income and the transformed categorically-valued input variable represents age, which has been split into three separate class levels or age groups. The following SAS programming code is provided to better understand the meaning of the AOV16 variable in the linear regression model and the variable selection procedure. In the following example, the GLM procedure is typically used in SAS for unbalanced one-way analysis of variable designs to explain the variability in the interval-valued target variable from the categorically-valued input variables of the model.

```
proc glm data=work.sasdata;
  class x;
  model y = x;
run;
```

where the input variable has two degrees of freedom. However, the same modeling results can be generated from the PROC GLM procedure by specifying a second-degree polynomial regression model as follows:

```
proc glm data=work.sasdata;
  model y = x x*x;
run;
```

In the **Variable Selection** node, the AOV16 variable will account for at most 15 degrees of freedom. In other words, the AOV16 variable represents the transformed categorical input variable of up to 16 separate class levels that will account for up to 15 separate polynomial terms in the nonlinear regression model in order to explain the nonlinear relationship of the interval-valued target variable to the model.

Chi-Square Modeling Selection Criterion

The following are the various options that are available from the chi-square modeling criterion statistic by fitting the binary-valued target variable.

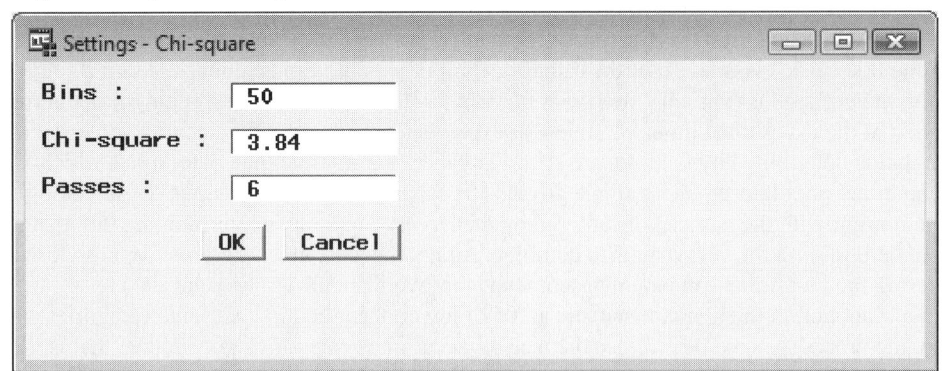

*The **Settings – Chi-Square** window is used to set the criterion values to the chi-square statistic.*

The following chi-square modeling selection criteria can only be applied to binary-valued target variables in the model. The node is designed to perform a series of two-way, that is, 2x2, frequency tables in maximizing the chi-square value by performing a succession of binary splits of the categorical input variables in the classification model.

For nominal or ordinal-valued input variables, each level is decomposed into binary dummy variables. For interval-valued input variables, the range of values for each variable is split into equally sized intervals that are binned into 50 separate levels.

- **Bins:** You may specify the number of bins of equal-sized intervals that are created from the range of values of the interval-valued input variables in the model. By default, interval-valued input variables are partitioned into 50 separate class levels, intervals, or categories. Increasing the number of bins will increase the accuracy of the chi-square statistic with the drawback of requiring more memory resources and computational time.
- **Chi-Square:** You may specify the chi-square modeling selection criterion value. The statistic is designed to determine the minimum bound in deciding whether the input variable is eligible in making a binary split and identifying the input variables that are useful for predicting the binary-valued target variable. The default is a chi-square value of 3.84. Specifying higher chi-square values will result in fewer binary splits performed and fewer input variables added to the model.
- **Passes:** This option is designed for you to specify the maximum number of passes that the procedure must make through the data as the variable selection process performs optimal partitioning of the input variable into the corresponding binary groups. The default is 6 passes. You may specify between 1 to 200 passes. Increasing the number of passes will increase the precision of the partitioning schema applied to the interval-valued input variable. However, the obvious drawback is the increased processing time.

General tab

The **General** tab is designed to remove the input variables from the predictive model with a certain percentage of missing cases. In addition, the node will allow you to remove input variables from the predictive model which share the same range of values in the other input variables in the selected data set.

Since predictive and classification modeling uses complete cases only, this means that if there are any missing values in any one of the input variables in the model, it will result in the observation removed from the analysis and the subsequent modeling fit. Therefore, assuming there are several input variables to the model, then this could result in several cases being removed from the analysis. Therefore, imputing missing values in the input variables should be performed prior to training the model. However, an alternative is removing input variable altogether from the model that have a large percentage of observations with missing values. The first option will allow you to remove input variables from the model by a specified percentage of cases with missing values.

The subsequent option is designed to remove redundant variables from the modeling fit that might lead to algebraic problems and computational difficulties in computing the regression estimates. A redundant input variable does not give any new information that has not been already explained in the model. The option keeps input variables to the model based on the amount of information in the variable, which will be explained shortly.

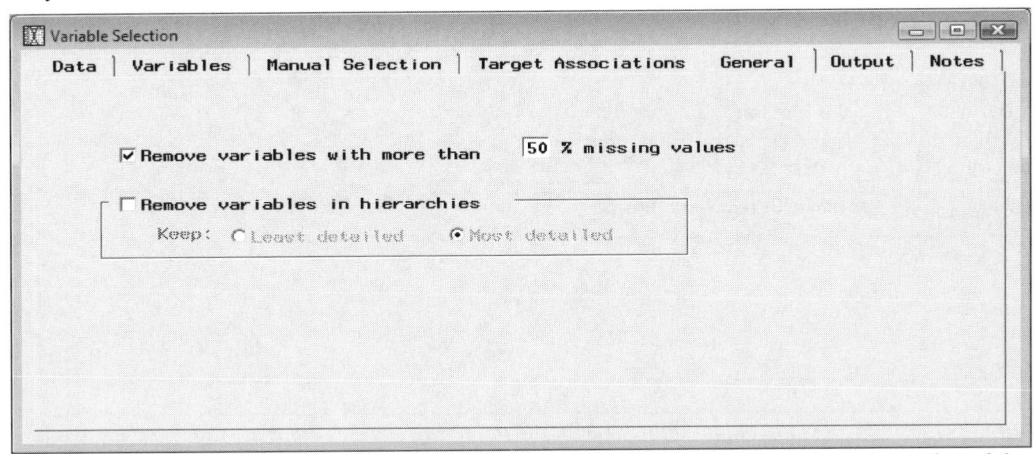

*The **General** tab is used to remove input variables with missing values from the statistical model.*

The following are the two separate options for removing input variables from the modeling selection process:

- **Remove variables with more than % missing values:** This option is designed so that you may remove input variables from the predictive model based on the specified proportion of missing cases. The default in removing input variable from the analysis if half of its values are missing.
- **Remove variables in hierarchies:** This option is designed to determine the hierarchal relationship, interdependency, or generalization between the input variables in the predictive model. Variable hierarchy analysis is performed on the metadata sample. This option determines the level of generalization between the input variables in the predictive model. For example, the zip code of an area generalizes to its corresponding state, which is associated with the region of the country. In other words, you may not want to use both the input variables since they might provide redundant information. Initially, you may want to determine which input variable to include in the predictive model based on the variable's information. By default, the node will keep the variable with the **Keep: Most detailed** information as in our example zip code because it contains more detailed information with regard to geographical location as opposed to the state identifier variable. Conversely, selecting **Keep: Least detailed** option will result in the variable state being retained since the variable contains fewer values.

  The drawback to this option is that it can dramatically increase memory resources and processing time of the **Variable Selection** node.

Output tab

The **Variable Selection** node will automatically create an output data set for each partitioned data set that is then passed to the subsequent modeling nodes. The node must first be executed with the corresponding variable selection routine performed in order for the various output data sets to be created. Since the variable selection routine is performed on the training data set, the variable role assignments to the input variables that are selected for the following modeling node and the corresponding output data set will be the same for the various partitioned data sets. The **Output** tab will allow you to view the output data set with the input variables, target variable, and the corresponding predicted values. The output or score data set is then passed along to the subsequent modeling nodes within the process flow diagram. From the **Show details of:** section, simply select either the **Training**, **Validation**, or **Test** radio button, then select the **Properties...** button to view both the file administrative information of the selected data set and the table view of the corresponding data. The output data set will be located within the EMDATA folder of the project library. From the **Description** field, you may enter an identifiable short descriptive label for the output data set that is created for each partitioned data set by selecting the corresponding radio button.

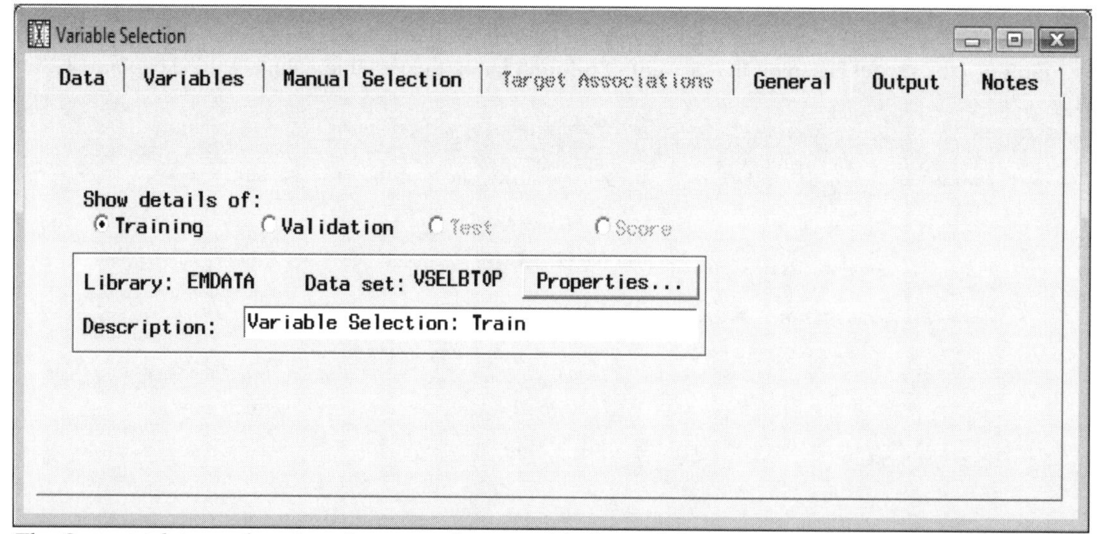

*The **Output** tab is used to view the output data set with the modeling terms and fitted values for the model.*

Viewing the Variable Selection Node Results

General Layout of the Results Browser within the Variable Selection Node

- **Variables tab**
- **R-square tab**
- **Effects tab**
- **Code tab**
- **Log tab**
- **Output tab**
- **Notes tab**

The following procedure output listing is based on the various preprocessing routines that were applied to the least-squares modeling design in the following **Regression** node. In other words, outliers were removed from the fit and some of the interval-valued input variables were transformed into categorical variables to predict the interval-valued target variable, DEBTINC, the ratio of debt to income. After running the node and opening the **Results Browser**, the following **Results – Variable Selection** window will appear.

Variables tab

The **Variables** tab will be displayed once you view the results. The tab is designed for you to determine which input variables have been added or removed from the given predictive model based on the modeling selection procedure. From the tab, you may view the input variables that have been currently included in the predictive model with a variable role of **input** displayed under the **Role** column. Conversely, input variables that are removed from the model will have a variable role attribute of **rejected** listed under the **Role** column. In other words, the corresponding input variables will be assigned the model roles in the subsequent nodes within the process flow diagram. The **Rejection Reason** column will display a short description explaining the reason why the input variable was removed from the predictive model that is displayed in the following diagram. The subsequent output listing that is displayed on the following page is based on the r-square modeling criterion statistic. The **Variable Selection** node eliminated nearly half of the input variables. The input variables of VALUE, YOJ, CLAGE, and both indicator variables, INDEROG and INDELINQ, were all removed from the multiple linear regression model. In addition, the input variable JOB was replaced with its grouped variable G_JOB. If there are still too many input variables in the final regression model, then simply increase the corresponding criteria values. That is, simply increase the squared correlation or the stepwise R2 improvement values from the **Settings** window. The **Variables** tab also displays the proportion of missing observations and the number of distinct levels for each input variable in the active training data set.

The following are possible explanations why the input variable is removed from the least-squares model:

Reason	Description
Overridden	Rejected manually Included manually
Missing Values	Missing % greater than the *n* percent value, which specified in the **General** tab.
Hierarchies	Duplicate Info, which occurs when a variable is removed because of being in a hierarchy or providing redundant information to the statistical model.
R-square Selection Criterion	Low R2 w/ target Group variable <input variable> preferred AOV16 variable <input variable> preferred
Chi-square Selection Criterion	Small chi-square

Note: In Enterprise Miner, if a categorical input variable has been incorrectly assigned a variable role of **interval**, then the node will generate an additional reason for rejection of **Levels exceed set maximum** with the input variable automatically removed from the final regression model.

The **Variables** tab is designed for you to reset certain model roles for the input variable in the data set and the predictive modeling design. That is, input variables may be forced back into the predictive model by specifying a model role of **input** that has been set to a model role of **rejected** from the modeling selection procedure. Conversely, input variables may be deleted from the analysis by setting a model role of **rejected**. Some analysts might include all main effects in the modeling design even though they have been removed from the predictive model, assuming that the predictive model adequately predicts the target values. From the modeling selection results, the **File > Reset** main menu option will allow you to reset all the changes that have been performed on the model roles back to their initial settings. The **Dependencies** column will list the input variable that provides redundant information for the respective listed input variable that has been removed from the model. The column will display the hierarchical relationship between both input variables. The **% Missing** column will display the percent of missing cases for the listed input variable. The **% of Levels** column displays the number of unique values of the listed input variable in the active training data set.

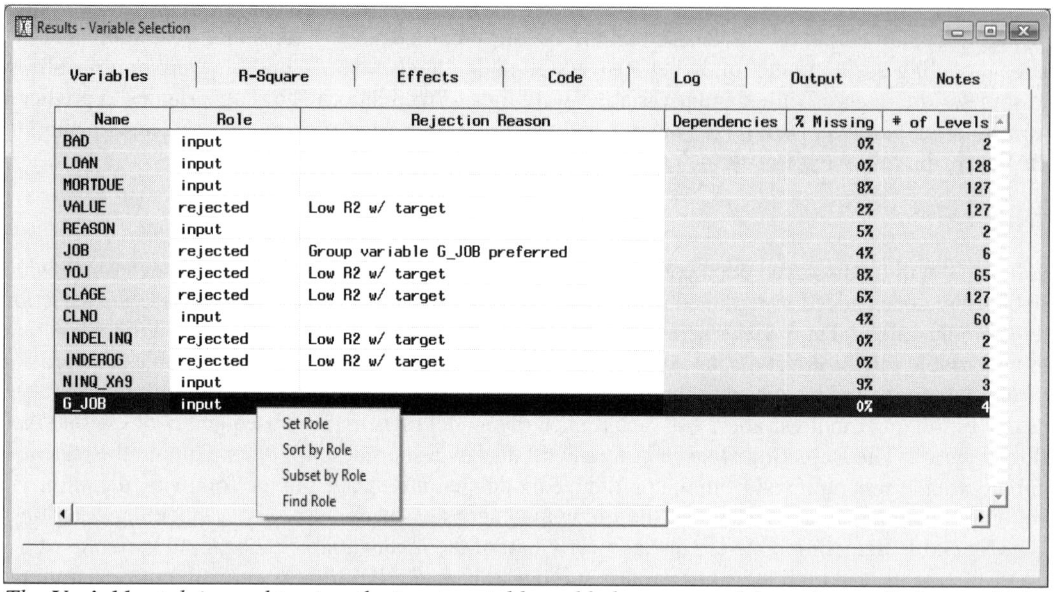

Name	Role	Rejection Reason	Dependencies	% Missing	# of Levels
BAD	input			0%	2
LOAN	input			0%	128
MORTDUE	input			8%	127
VALUE	rejected	Low R2 w/ target		2%	127
REASON	input			5%	2
JOB	rejected	Group variable G_JOB preferred		4%	6
YOJ	rejected	Low R2 w/ target		8%	65
CLAGE	rejected	Low R2 w/ target		6%	127
CLNO	input			4%	60
INDELINQ	rejected	Low R2 w/ target		0%	2
INDEROG	rejected	Low R2 w/ target		0%	2
NINQ_XA9	input			9%	3
G_JOB	input			0%	4

The ***Variables*** *tab is used to view the input variables added or removed from the predictive model based on the r-square modeling selection criterion.*

Note: By fitting the binary-valued target variable, the **Rejection Reason** column will display all the input variables that have been rejected from the model based on the variable selection procedure due to small chi-square values, assuming that the chi-square selection criterion option is selected. The window will be displayed shortly.

R-Square tab

By default, both the **R-Square** and the **Effects** tab can be viewed, assuming that the default r-square modeling selection criterion is specified. Both tabs are not available if you have previously selected the chi-square modeling selection criterion by fitting the binary-valued target variable. The **R-square** tab displays the r-square statistic for each input variable in the predictive model. The r-square statistic is computed by fitting a simple linear regression model between each input variable. This is the first step in the modeling selection routine obtained by fitting an interval-valued target variable. The variables listed from the chart are ranked by the largest contributions to the r-square statistic that best explain or predict the target variable. Simply slide the vertical scroll bar that is located on the right-hand side of the tab to view each modeling effect and its associated r-square statistic.

The diagram displays the r-square statistic for each modeling effect in predicting the ratio of debt to income. The two-way interaction effects are automatically displayed in the following plot even if the **Include 2-way interaction** option is not selected from the previous **Settings-R-square** tab. Since the **Include 2-way interaction** option is not selected, the interaction terms that are displayed in the following **R-Square** tab have been removed by the **Variable Selection** node.

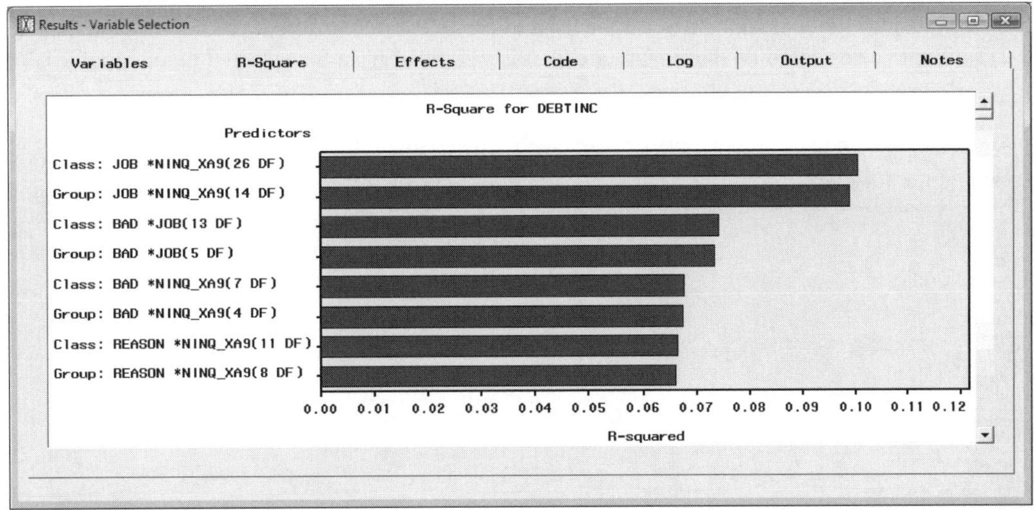

*The **R-square** tab is used to graphically view the r-square statistic by the forward selection method.*

Effects tab

The **Effects** tab is designed to display the incremental increase in the r-square statistic as each input variable is sequentially entered into the predictive model from the forward stepwise regression procedure. The plot displays the modeling terms that have been retained in the statistical model. From the variable selection routine, the grouped job occupation input variable JOB contributed the most to the overall r-square statistic. Therefore, the input variable was first entered into the regression model. This was followed by the input variable of number of trade lines, CLNO, and, finally, the amount of the loan, LOAN, that was the last input variable entered into the multiple linear regression model, which contributes the least to the overall r-square statistic and the total variability in the target variable. The overall r-square statistic is 14.6%. Again, the following main effects that are shown in the plot do not display a reason for rejection from the **Rejection Reason** column within the previously displayed **Variables** tab.

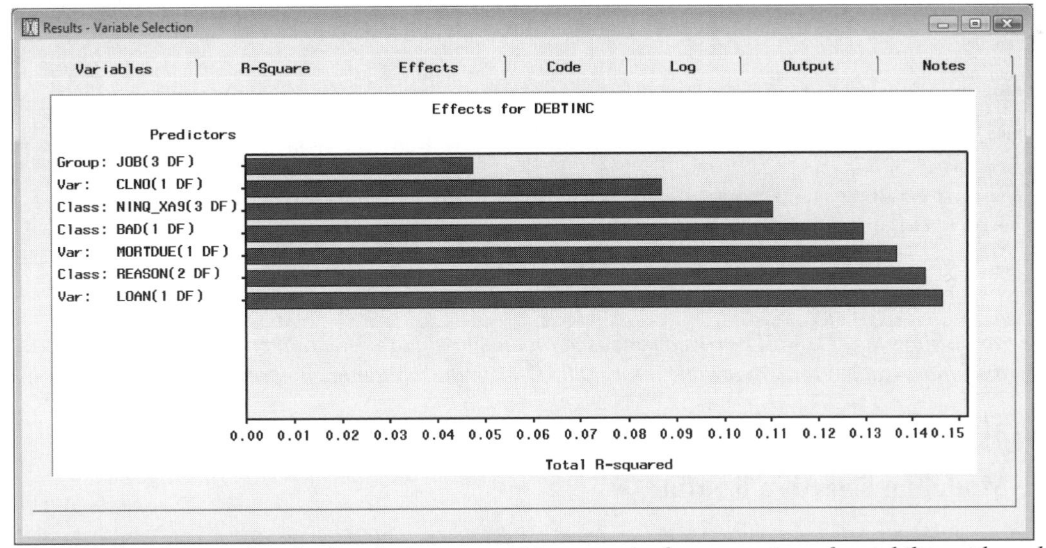

*The **Effects** tab is used to display the incremental increase in the proportion of variability with each input variable sequentially entered into the regression model.*

Code tab

The following **Code** tab displays the training code from the PROC DMINE data mining procedure that created the variable selection results from the forward stepwise modeling selection criterion. Note that many of the configuration settings specified from the node are listed in the data mining procedure option statement. The

Scoring option will display the score code that can be used to create entirely different prediction estimates from a new set of input values based on the final model that is generated from the selected input variables.

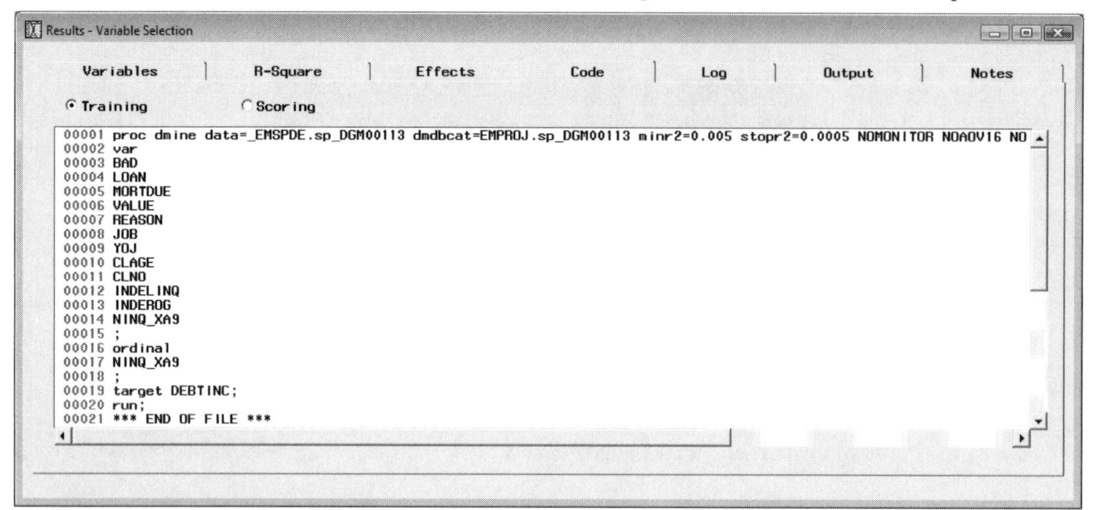

*The **Code** tab displays the DMINE procedure based on the r-square selection criterion statistic in predicting the interval-valued target variable, that is, DEBTINC, from the least-squares model.*

For a categorically-valued target variable, the PROC DMSPLIT data mining procedure performs the modeling selection technique based on the chi-square modeling selection criterion.

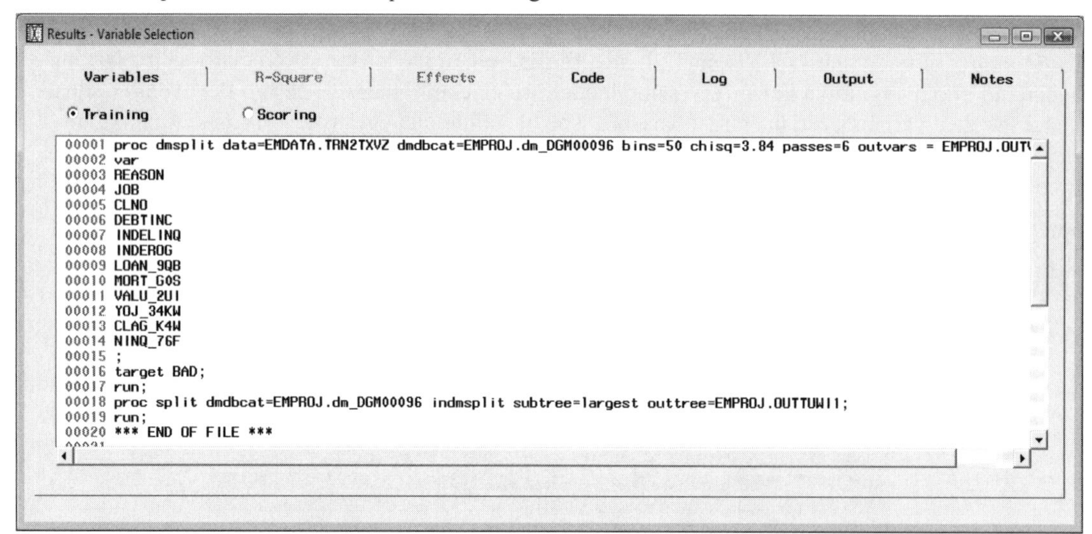

*The **Code** tab displays the DMSPLIT procedure based on the chi-square selection criterion statistic in predicting the binary-valued target variable, that is, BAD from the modeling comparison procedure.*

Output tab

R-Square Modeling Selection Routine

The **Output** tab displays the output listing from the PROC DMINE data mining procedure based on the r-square modeling selection criterion statistic in predicting the interval-valued target variable, that is, DEBTINC. The first section of the output listing displays the various modeling effects and their associated r-square statistic to the stepwise regression modeling selection process. The following section displays the overall r-square statistic, F-value, sum-of-squares and mean square error for each modeling effect. Note that the AOV16 variable CLNO has the largest r-square statistic. One reason is because the r-square statistic is calculated from the target variable and the CLNO input variable in the model in addition to the binned AOV16 CLNO input variable. Therefore, the r-square statistic measures the linear relationship between the target variable and the CLNO input variable along with the nonlinear relationship between the target variable and the AOV16 CLNO input variable. If the modeling effect r-square statistic is less than the specified minimum r-square statistic, then it will be listed in the table. For example, the CLAGE and YOJ input variables are removed from the multiple linear regression model along with both indicator variables, INDELINQ and INDEROG, whose r-square statistic are less than the specified minimum R^2 statistic of .005.

```
                         The DMINE Procedure

              R-Squares for Target Variable: DEBTINC

Effect                       DF          R-Square

AOV16: CLNO                  14          0.058318
AOV16: MORTDUE               14          0.049075
Class: JOB                    6          0.047699
Group: JOB                    3          0.047306
Class: NINQ_XA9               3          0.043602
AOV16: VALUE                 14          0.041251
Var:   CLNO                   1          0.038460
Class: BAD                    1          0.025652
Class: REASON                 2          0.024785
AOV16: LOAN                  15          0.022657
AOV16: CLAGE                 15          0.019075
Var:   MORTDUE                1          0.013120
AOV16: YOJ                   14          0.011972
Var:   VALUE                  1          0.009215
Var:   LOAN                   1          0.006445
AOV16: INDELINQ               1          0.002790    R2 < MINR2
Var:   INDELINQ               1          0.002790    R2 < MINR2
Var:   CLAGE                  1          0.000964    R2 < MINR2
Var:   YOJ                    1          0.000327    R2 < MINR2
AOV16: INDEROG                1       0.000068872    R2 < MINR2
Var:   INDEROG                1       0.000068872    R2 < MINR2
```

The following is the four possible modeling effects to the predictive model.

Effect	**Description**
Var	Identifies the interval-valued input variables in the regression model. The R^2 statistic is based on the simple linear relationship between the input variable and the target variable in the simple linear regression model. **Note:** The degrees of freedom is always equal to one.
Class	Identifies the categorically-valued input variables in the model that is estimated for all categorically-valued input variables and all two-way interaction modeling effects. A one-way ANOVA is performed for each categorically-valued input variable to calculate the R^2 statistic. Two-factor interaction effects are constructed by combining all possible class levels into one modeling term as follows: degrees of freedom – one-factor effects: (# of class levels) – 1 two-factor effects: (# of class levels of factor 1) · (# of class levels of factor 2) – (the number of nonzero cells in the two-factor cross-tabulation) – 1
Group	The transformed categorically-valued input variable that is collapsed into smaller groups. The degrees of freedom for each class effect is equal to the number of class levels.
AOV16	Transforming the interval-valued input variables by grouping these same input variables into 16 equally spaced buckets or bins. The AOV16 modeling effects explain the nonlinear relationship of the target means across the separate class levels of the input variable. The degrees of freedom is determined by the number of groups that are created.

```
                      Effects Chosen for Target: DEBTINC

                                                                Sum of   Error Mean
Effect             DF     R-Square      F Value  p-Value       Squares       Square

Group: JOB          3     0.047306    38.929730   <.0001   6434.629106    55.096102
Var:   CLNO         1     0.039316   101.198120   <.0001   5347.799107    52.844846
Class: NINQ_XA9     3     0.023483    20.653024   <.0001   3194.114628    51.552010
Class: BAD          1     0.019344    52.152360   <.0001   2631.236167    50.452869
Var:   MORTDUE      1     0.007212    19.597902   <.0001    980.996860    50.056217
Class: REASON       2     0.006090     8.326029   0.0002    828.365415    49.745529
Var:   LOAN         1     0.003603     9.889751   0.0017    490.112136    49.557579
```

The following explains the statistics listed from the previous procedure output listing.

Label	Description
Effect	The modeling effects to the predictive model that are sequentially entered into the regression model based on the r-square statistics that are listed in descending order.
DF	The degrees of freedom of the modeling effects.
R-Square	The modeling selection criterion statistic is based on the sequential improvement in the r-square statistic of the modeling effects sequentially entered into the predictive model. That is, 3.93% of the variability in the target response, that is, the ratio of debt to income or DEBTINC, is explained by the linear relationship with the interval-valued input variable CLNO. From the table listing, you are looking for a large r-square statistic indicating to you that the modeling effect is a really good predictor. The r-square statistic is calculated by dividing the sum-of-squares of the modeling effect by the total sum-of-squares. A pictorial explanation of the r-square statistic will be displayed in the subsequent **Regression** node.
F Value	The F value tests the variability between the target variable and each input variable in the predictive model. The F-statistic is based on the ratio between the mean square error of the respective input variable in the predictive model and the mean square error in the residuals. A large F-value will indicate that the input variable is a good predictor of the predictive model.
p-Value	The *p*-value is based on the F-test statistic that tests the significance of the simple linear regression of each input variable in the model and the target variable. Typically, a *p*-value below a predetermined significance level, usually set at .05, indicates that the input variable is a good predictor in the linear regression model.
Sum of Squares	The sum of squares error of the modeling effects.
Error Mean Square	The mean square error that measures the variability due to either random error in the target variable or due to the other input variables not in the predictive model. The mean square error should get smaller as important input variables are added to the regression model. For instance, the input variable, that is, LOAN, was the last modeling term entered into the final multiple linear regression model. The accumulated sum-of-squares is the Model Sum of Squares that is listed in the subsequent ANOVA table listing. The mean square error is calculated by dividing the sum-of-squares error by its associated degrees of freedom.

```
                      The DMINE Procedure

          The Final ANOVA Table for Target: DEBTINC

                                                      Sum of
     Effect              DF          R-Square        Squares

     Model               12          0.146355          19907
     Error             2343                 .         116113
     Total             2355                 .         136021
```

The following explains the column listings that are displayed in the ANOVA table that is illustrated in the previous diagram.

Label	Description
Effect	Source of variation
DF	The degrees of freedom from the source of variation.

R-Square	The r-square statistic measures the overall fit to the model that is based on the overall variability in the predictive model. From the table listing, 14.63% of the variability in the target variable DEBTINC is explained by the final input variables selected for the multiple linear regression model. A r-square statistic close to one will indicate an extremely good fit. The r-square statistic is calculated by the ratio between the model sum-of-squares and the total sum-of-squares.
Sum of Squares	The sum-of-squares is the total sum-of-squares that is partitioned in both the regression sum-of-squares and the error sum-of-squares. In predictive modeling, you want the error sum-of-squares to be as small as possible indicating a very good fit. The Total sum-of-squares is the squared difference between each target value and the overall target mean. The Error sum-of-squares is the squared difference between each target value and the predicted values and the Model sum-of-squares is the difference between the two sum-of-squares. The mean square error that is the error sum-of-squares divided by its degrees of freedom is often called the variance of the model.

The Effects Not Chosen for Target: <target variable> section lists the input variables that do not meet the various selection criterion values of the stepwise regression procedure. The table listing displays the input variables that are removed from the model and are not passed along to the subsequent modeling nodes with a model role set to **rejected**. The following section displays all input variables with extremely low correlations to the target response variable. The following listing displays a very low r-square statistic, indicating to you that the ratio of debt to income stays the same as the property value increases.

The DMINE Procedure

Effects Not Chosen for Target: DEBTINC

Effect	DF	R-Square	F Value	p-Value	Sum of Squares
Var: VALUE	1	0.000111	0.304704	0.5810	15.104862

Chi-Square Modeling Selection Routine

From the following output listing, logistic regression modeling was performed in predicting the binary-valued target variable BAD, that is, clients defaulting on a home loan. The following results are based on the chi-square modeling selection procedure that is performed to the logistic regression model that is one of the models under assessment from the modeling comparison procedure that is displayed in the **Assessment** node.

The following displays the **Variables** tab listing of the input variables that have been removed from the logistic regression model. From the table listing, the input variables REASON, JOB, DEROG, MORT, and NINQ were removed from the final model due to low chi-square values and a weak association with the target class levels.

Results - Variable Selection

Variables | R-Square | Effects | Code | Log | Output | Notes

Name	Role	Rejection Reason	Dependencies	% Missing	# of Levels	Label
REASON	rejected	Small chi-square		0%	2	
JOB	rejected	Small chi-square		0%	6	
CLNO	input			0%	57	
DEBTINC	input			0%	128	
INDELINQ	input			0%	2	DELINQ > 0
INDEROG	rejected	Small chi-square		0%	2	DEROG > 0
LOAN_9QB	input			0%	128	log(LOAN)
MORT_G0S	rejected	Small chi-square		8%	127	log(MORTDUE)
VALU_2UI	input			2%	127	log(VALUE)
YOJ_34KW	input			16%	64	log(YOJ)
CLAG_K4W	input			6%	127	log(CLAGE)
NINQ_76F	rejected	Small chi-square		0%	3	Bucket(NINQ)

*The **Variables** tab that lists the modeling terms added to the regression model and an explanation to the input variables removed from the classification model based on the chi-square model selection criterion.*

The following PROC DMSPLIT procedure output listing is based on the chi-square modeling selection criterion for predicting the binary-valued target variable in the logistic regression model. The output listing is analogous to the recursive decision tree branching process. The following table listing will indicate the importance of the input variables for the classification model. The input variables that are displayed at or near the beginning of the splitting procedure listing will indicate the variables that have the strongest split for the binary target values. In this example, DEBTINC seems to be the input variable that performs best, with the strongest split in the tree branching process to the binary-valued target variable.

```
                        The DMSPLIT Procedure

                       History of Node Splits

Node     Parent        ChiSqu     Split              Value     ---Levels---

   1          0    205.201253     DEBTINC        32.970523

                       History of Node Splits

Node     Parent        ChiSqu     Split              Value     ---Levels---

   2          1     80.900262     DEBTINC         6.364784
   3          1    203.121285     DEBTINC        36.377356

                       History of Node Splits

Node     Parent        ChiSqu     Split              Value     ---Levels---

   4          2     11.484375     CLNO            9.100000
   5          2     24.620593     VALU_2UI       12.631920
   6          3    207.117635     DEBTINC        33.720026
   7          3    230.485404     DEBTINC        46.393443
                       History of Node Splits

Node     Parent        ChiSqu     Split              Value     ---Levels---

   8          4      8.000000     INDELINQ        0.020000
  10          5     17.490344     INDELINQ        0.020000
  12          6     93.577749     DEBTINC        33.705036
  13          6     41.158107     LOAN_9QB        8.323748
  14          7     57.723903     DEBTINC        44.590547

                       History of Node Splits

Node     Parent        ChiSqu     Split              Value     ---Levels---

  18         10     22.235275     YOJ_34KW        1.115171
  19         10     34.814280     CLAG_K4W        4.416915
  20         12     16.841961     CLNO            1.300000
  21         12     39.650710     INDELINQ        0.020000
  23         13     35.514189     CLAG_K4W        3.794249
  24         14     31.002770     VALU_2UI       13.099193
  25         14     10.000000     CLNO           18.200000
```

The following explains the statistics listed from the previous procedure output listing.

Column Label	Description
Node	The node number of the corresponding split. The table listing displays the node number in descending order based on the recursive binary splits that are automatically performed. The first node that is displayed at the top of the table listing represents the root of the tree where the first split is performed.
Parent	The parent node is ranked by the construction of the tree branching process. For example, a parent node of 0 identifies the root of the tree.
ChiSqu	Chi-square value for the split with large values will indicate a significant input variable of the logistic regression model.
Split	The splitting variable or the input variable of the classification model.
Value	The average value of the interval-valued splitting variable or the number of class levels of the categorically-valued splitting variable.
Levels	The number of class levels of the categorically-valued input variable that has been partitioned to the left-hand or right-hand side of the listed splitting node.

The following listing displays the input variables in the logistic regression classification model. The output listing indicates that the input variable DEBTINC, that is, the ratio between debt and income, performs the strongest split to the binary-valued target variable, BAD, by performing its split at the root of the decision tree. In addition, the same input variable was involved in seven separate binary splits of the target variable in the recursive tree splitting process.

```
                    The DMSPLIT Procedure

                       Effect Summary

                             Node        Total
                              1st        Times
            Effect          Split        Split

            DEBTINC             1            7
            CLNO                4            3
            VALU_2UI            5            2
            INDELINQ            8            3
            LOAN_9QB           13            1
            YOJ_34KW           18            1
            CLAG_K4W           19            2
```

The following explains the statistics listed from the previous procedure output listing.

Column Label	Description
Effect	The modeling effect or the input variable of the classification model.
Node 1st Split	The ordering sequence of the recursive tree branching splitting procedure. Low values indicate the strongest splits in measuring the importance of the input variable of the tree model.
Total Time Split	Number of times that the input variable was involved throughout the entire tree branching process that partitioned the binary-valued target variable that will also indicate the importance of the input variable for the tree model.

Generally, the modeling assessment statistic that is used in measuring the overall accuracy of the classification model is determined by the accuracy rate of the actual target values and its predicted values at each class level. The posterior probabilities are assigned to the corresponding target levels by the prior probability of the target event that is determined by the corresponding cutoff value. The cutoff probability can be specified from the **cutoff** entry field within the **Target Associations** tab.

At the bottom of the **Output** tab, the following output listing will display the classification table by selecting the r-square modeling selection criterion by fitting the binary-valued target variable. The following classification table that is displayed is based on fitting the logistic regression model to the binary-valued target variable, BAD, from the HMEQ data set that is one of the classification models under assessment in the **Assessment** node. The purpose of the following classification table is to view the accuracy at each target level of the categorically-valued target variable, BAD, in the classification model. The cutoff value is automatically set at .5 for the binary-valued target variable, indicating to you that an estimated probability greater than .5 is identified as the target event that you want to predict. Otherwise, the observation is classified as the target nonevent. A classification table with all off-diagonal entries of zero will indicate an absolutely perfect fit to the classification model. From the procedure output listing, there were 176 clients who were correctly classified in defaulting on their loan and 2297 clients who were correctly classified as paying their home loan from the training data set with an accuracy rate of 82.99%.

```
                 The DMINE Procedure

       Classification Table for CUTOFF = 0.5000

                 Accuracy =   82.99

                            Predicted
         Observed               1            0

         1                    176          415
         0                     92         2297
         Missing                0            0
```

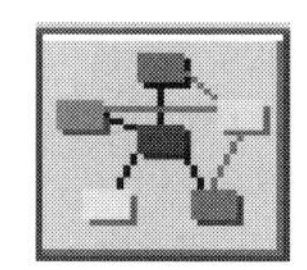

2.6 Link Analysis Node

General Layout of the Enterprise Miner Link Analysis Node

- **Data tab**
- **Variables tab**
- **Output tab**
- **Notes tab**

The purpose of the **Link Analysis** node in Enterprise Miner is to visualize the relationships and associations between the variables in defining patterns or characteristics in the active training data set. Link analysis can be used as a visualization tool to both association discovery and sequence discovery in determining the most significant associations between the items under analysis. Also, link analysis can be applied to clustering analysis, which generates various cluster scores from the training data set that can be used for data reduction and segmentation. Although link analysis does not perform predictive modeling, models can be developed by the patterns in the relationship between the variables that can be discovered from the analysis.

The **Link Analysis** node automatically creates a link graph. The *link graph* consists of two distinct components: the nodes and the lines connected to these same nodes, called *links*. Certain nodes are connected to each other to visually display the relationship between the variables in the active training data set. A fully connected graph will consist of every node connected to each other. The purpose of link analysis might be to analyze fraud detection, when visiting various Web pages, calling patterns of various phone users, or purchasing behaviors of customers. That is, link analysis can be used as a visualization tool for market basket analysis. In other words, the node would represent the various items that the customers will purchase and the linked lines would indicate the combination of items that these same customers will purchase. Link analysis may also be applied in viewing a sequence of events, that is, sequence discovery, where the nodes represent the more popular Web pages and the linked lines indicate the sequence of Web pages that you might view in successive order. Link analysis might also be used in finding the distance between two nodes by placing weights to the links that connect the nodes.

The basic goal of link analysis is to recognize the connection between the nodes and maybe even visualize certain groupings from the fully connected nodes forming separate groups. By default, the link graph displays a circular pattern of nodes in which various lines are connected to a certain number of the nodes. However, the **Link Analysis** node is designed to view the relationship between the variables in many different graphical layouts. These lines are of different thicknesses that measure the number of connections between the nodes. The size of each node indicates the number of occurrences of the node. Therefore, the linked graph will indicate to you which nodes are connected to each other and which nodes are not. The values of the nodes can be determined by the marginal frequency counts, frequency percentages, or the weighted centrality measurements of the assigned class levels of the node. Conversely, the value of the links is based on the joint distribution of the two separate nodes such as the relative frequency counts of the two separate class levels from the two-way frequency table, the frequency percentages, or the partial chi-square value of the class levels of the two separate nodes that are connected to each other.

Each node is represented by the level of measurement of the variables in the input data set. For categorically-valued variables, each node represents each class level of the corresponding variable. For interval-valued variables, the **Link Analysis** node will automatically split its values into three separate buckets or bins of approximately equal intervals. At times, the link graph can potentially amount to hundreds of nodes and links based on the number of variables in the active training data set. Therefore, the **Link Analysis** node is designed to delete certain nodes from the analysis. The **Link Analysis** node will allow you to randomly sample a smaller subset of the active training data set, remove certain interaction effects from the analysis, and exclude certain combination of items from association analysis or a certain number of sequences from sequence analysis. On the other hand, additional nodes can be added to the analysis. Conversely, certain links can be removed that are connected to the various nodes, in order to view the more significant links in the active training data set. In addition, the **Link Analysis** node will allow you to display the various links or connections within the link graph from a specific node of interest. This will be illustrated in the following diagrams within this section.

The Centrality Measurements

From the link analysis results, there are two separate centrality measurements called the 1st and 2nd order undirected weighted centrality. The centrality measurements are calculated to measure the importance of each node to the analysis. The *first centrality measurement* measures the importance of the node based on the number of nodes that are connected to it. The *second centrality measurement* determines the cumulative importance of the nodes that are connected to the corresponding node. The node calculates both the weighted and unweighted centrality measurements. Since the nodes and links are each ranked by their importance in the diagram, the weighted centrality measurements are automatically calculated. Conversely, the unweighted centrality measurements are calculated when all the nodes are assumed to be equal in importance to all other nodes and all links are assumed to be equal in importance to all other links. In other words, the unweighted centrality measurements are computed without the value of the links where every link is assigned a value of one. The unweighted centrality value might be important if you believe that a few strong links are undesirably biasing your results. The weighted first-order centrality value is calculated in order to assign numerical values to the various links within the graph. For example, suppose you are interested in the importance of an item, then you would want to look at all items that are linked to it and sort them by the centrality measurements.

The centrality measurements would be useful in cross-selling if you were analyzing market basket items. If a customer buys or expresses an interest in a certain item, then should you sell a second item? A third item? Is the second item more likely lead to buying a third item? Can you chain the items optimally? Unweighted means that the centrality is computed without the value of the links, that is, every link has a value of one. Again, you might want to look at the unweighted measurements if there are a few strong links that are undesirably biasing your results. Maybe you are interested in low-volume, high-profit links. Remember that centrality is a node property, not a link property, so the centrality measure will not affect path finding or layouts. So if there is item 1, and you want to know what is the most important item that you could co-market with item 1, then you would look at the items linked to item 1 and sort them by a centrality measure. You might select the top 10 candidates, and then sort the top items by an external measure of profitability. The next step is that you might then select the next 10 items by centrality measurements that are linked to item 2, and so on. The weighted first-order and second-order undirected centrality measurements are called C1 and C2 and are a part of the NODES data set.

The Data Structure

From the **Link Analysis** node, you can create a database layout from either coordinate data or transaction data. Transaction data is used in association or sequence analysis. For transaction data, link analysis may be interpreted as indirect sequence in describing the relationship between the sequence of items or events. Since each observation represents a particular sequence in sequence discovery, then the number of observations that include a certain sequence is the fundamental measurement of link analysis.

The link graph is based on two separate data sets that are automatically created once the node is executed, called the NODES and LINKS data sets. The file layout of the NODES data set will contain a row for each node in the link graph that is identified by a unique ID variable. The output data set will also display various columns of descriptive statistics such as the frequency counts and centrality measurement statistics used in determining the importance of the various nodes within the link analysis diagram. The file layout of the LINKS data set will contain a row for each link between two separate nodes that are connected to each other in the link graph. Each row of the LINKS data set must have two separate identifier variables for both nodes that are connected to one another, called ID1 and ID2, with the corresponding frequency counts of each connected node along with the two-way frequency link count between both levels and the link identifier variable that distinguishes between the various links or connections between the two separate nodes. For coordinate data, the links data set will also generate the partial chi-square statistic, the expected values, and the deviation. The Pearson chi-square statistic is used to determine whether there is an association between the class levels of the two separate nodes. The *chi-square test* measures the difference between the observed cell frequencies and the expected cell frequencies. The observed cell frequency is the individual cell frequency count between the class levels of the two separate nodes that are connected. The expected cell frequency is calculated by the ratio between the marginal row and column totals, divided by the total number of observations from the active training data set. The deviance is the difference between the observed and the expected cell frequencies. The deviation will enable you to determine the systematic patterns in the class levels between the connected nodes. Interpreting the chi-square results, deviations greater than zero will indicate the observed frequency count is higher than the expected count and deviations less than zero will indicate that the observed frequency count is smaller than the expected count. Large partial chi-square values will indicate to you that there is a difference between the observed count and what is expected. In other words, large partial chi-square values will indicate that there is strong evidence that there exists an association between the two separate class levels.

Link analysis is discussed in greater detail in Berry and Linoff (2004), and Giudici (2003).

Data tab

The **Data** tab is designed for you to select the active training data set. The tab displays the name and description of the previously created data set that is connected to the node. From the tab, select the **Properties...** button to view the file information of the active training data set or browse the table view listing of the active training data set. Press the **Select...** button that opens the **Imports Map** window. The window will allow you to interactively select from a list of the currently available data mining data sets that have been created within the currently opened Enterprise Miner diagram based on the nodes that are connected to the **Link Analysis** node. For instance, the process flow diagram might have created several previously created data sets that have the same model role in comparison to the currently selected training data set.

By default, the **Use sample of data sets** is unchecked. Select the check box and press the **Configure...** button to perform a random sample of the active training data set. The **Sampling options** window will appear. The **Number of rows** entry field will allow you to specify the number of observations to randomly select from the training data set. The **Random seed** entry field will allow you to specify a seed for the random sample.

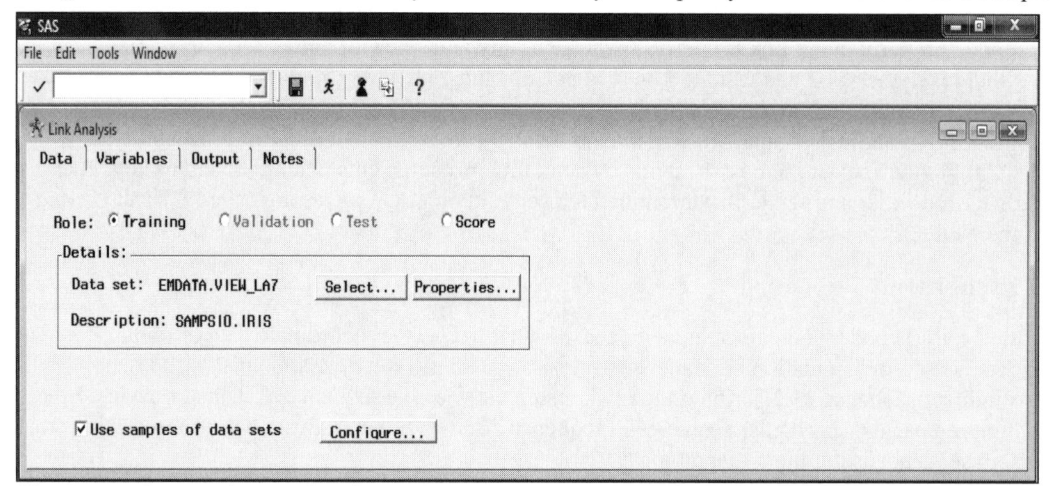

*The **Data** tab is used to read in the analysis data set to create the subsequent link analysis diagrams.*

Variables tab

The **Variables** tab will automatically appear as you open the **Link Analysis** node. The **Variables** tab is designed to display a table view of the variables in the active training data set. The tab will display the name, status, model role, level of measurement, type format, and label for each variable in the analysis. The model roles and the variable attributes may be changed from either the **Input Data Source** node or the **Data Attributes** node. The tab will allow you to remove certain variables from the analysis based on the **Status** column. By default, all variables are included in the analysis with a variable status automatically set to **use**, which are used in creating the LINKS and NODES data sets. However, you can also remove certain variables from the data sets and the subsequent analysis by simply selecting the variable rows, then scrolling to the **Status** column and setting the variable status to **don't use**. In addition, the tab will allow you to view the frequency distribution of each listed variable within the tab by selecting the **View Distribution of <variable name>** pop-up menu option.

Output tab

The **Output** tab will allow you to browse the selected data set that will be passed along to the subsequent nodes within the process flow diagram. Select the **Properties...** button to either view the file information or view the table listing of the selected output data sets to be passed along within the process flow. The node must be compiled before viewing the output data set. By performing the various clustering techniques from the link graph, the scored data set will display the numeric cluster score variable for each cluster along with a segment identifier variable and all other variables in the analysis data set. The segment identifier variable can be used for further group processing within the process flow.

Detailed Setting

Select the keyhole **Settings** toolbar icon from the tools bar menu, which will open the following **Detailed Settings** window. The **Detailed Setting** window contains the following four tabs.

- **General tab**
- **Interactions tab**
- **Transactions tab**
- **Post Processing tab**

General tab

The **General** tab is designed for you to specify the type of data structure of the active training data set that is used in link analysis, specify the interaction terms for the analysis, the association transactions or defining the sequences of the analysis or post-processing of the output data sets by controlling the size of the data set, configuring the sorted order of the frequency counts of the number of nodes connected, or saving the output data set into matrix form for subsequent analysis such as clustering analysis.

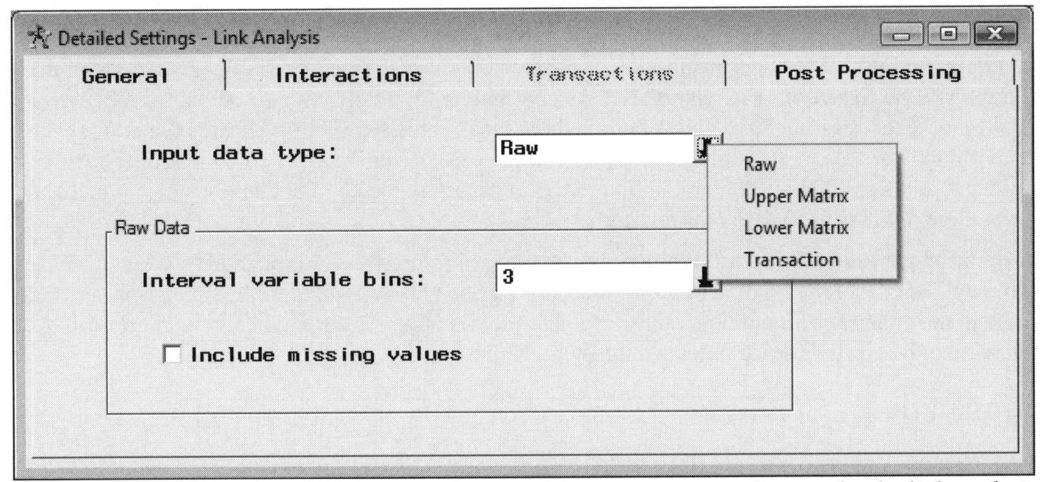

*The **General** tab is used to select the data layout of the active training data set for the link analysis.*

From the **Input data type** field, the following data structures may be selected from the active training data set:

- **Raw:** The input data source that is considered to be coordinate data. The data structure is such that each row represents each observation and each column representing each variable in the analysis. This option will be automatically displayed within the **Input data type** field when the input variables or the target variable is either continuous or discrete. The statistics that are listed within the NODES and LINKS data sets are generated from the PROC FREQ procedure.
- **Upper Matrix:** The input source data set is composed of a symmetric matrix. The nodes and links are created from the upper diagonal entries of the symmetric matrix. However, it is important to note that although this option is available within the **General** tab, the option is experimental in Enterprise Miner v4.1 and should be removed from Enterprise Miner v4.3 since this input data type is not currently supported by SAS. In other words, specifying this option will result in Enterprise Miner

automatically converting the symmetric matrix into a raw data type that will result in each node being created for each nonzero cell in the matrix.

- **Lower Matrix:** The input source data set is composed of a symmetric matrix. The nodes and links are created from the lower diagonal entries of the symmetric matrix. However, similar to the previous **Upper Matrix** option, this option is invalid and should not be specified in Enterprise Miner v4.3 since this input data type is not currently supported by SAS.
- **Transaction:** The input source data set that is used in association discovery contains both a target variable and an id variable. The file layout has a hierarchical data structure with several target values that are usually represented as items within each id variable that are usually represented as customers. The nodes and links data sets are created from the PROC ASSOC data mining procedure. Each node in the link graph represents each value of the target variable. The **Transactions** tab can be viewed with the **Associations** section available for you to reduce the number of items to view in the analysis. Since each node in the link graph is based on each single item, reducing the number of items displayed in the graphs will reduce the clutter of the various connected nodes within the link graph.

  In addition, this option will be displayed when the data structure is based on sequence discovery. That is, the input data set is identical to the association discovery data structure with an added sequence variable or time stamp variable to identify the sequence of items or events. The active training data set must consist of a target variable, id variable, and a sequence variable. Again, the **Transactions** tab can be viewed. However, the **Associations** section will be dimmed and grayed-out, and the **Sequences** section will be available for you to define the number of sequences, the length of the sequences, and filtering repeating or duplicating sequences. In the link graph, a single node will indicate a single sequence in the active training data set.

The **Raw Data** section will be available for you to select the following options if you have selected the **Raw** option from the **Input data type** field that is based on the interval-valued variables in the data set.

- **Interval variable bins:** This drop-down menu is designed for you to specify the number of nonoverlapping intervals or bins to subset the transformed interval-valued variable. Each node in the link graph will be created for each interval of the continuous variable. You may select from either three, five or seven separate bins. The default is three bins. Three separate nodes are created for each interval-valued variable in the analysis by creating three separate intervals of equal length. The three separate cutpoints are defined, where the first two groups are defined by the cutpoints of $\mu \pm \sigma/2$ and the remaining data assigned to the third group.
- **Include missing values:** By selecting the check box, the missing values of the interval-valued variable will be part of the binning process. A separate node will be created in the link graph for each variable in the analysis with missing values. By default, this option is automatically unchecked and nodes will not be created for variables with missing values.

Interactions tab

The **Interactions** tab is designed for you to specify the various interactions or associations to add or remove from the analysis. At times, the link graph can potentially amount to several links, which might inhibit your ability to visualize the various links. Therefore, this tab will allow you to reduce the number of interactions in the analysis. By default, all interactions are included in the analysis. The various interactions are displayed in the **Include** list box. To remove certain interactions from the analysis, simply select any number of the listed interactions from the **Include** list box by pressing the Ctrl key and selecting the right-arrow keys located between each list box. The tab will allow you to add or remove interaction effects between the target variables or the input variables in the analysis. Select the **Interaction between targets** check box to include all possible interactions between the input variables and the target variables in the analysis data set. Selecting the **Interaction between inputs** check box will allow you to include all interactions between the input variables in the active training data set.

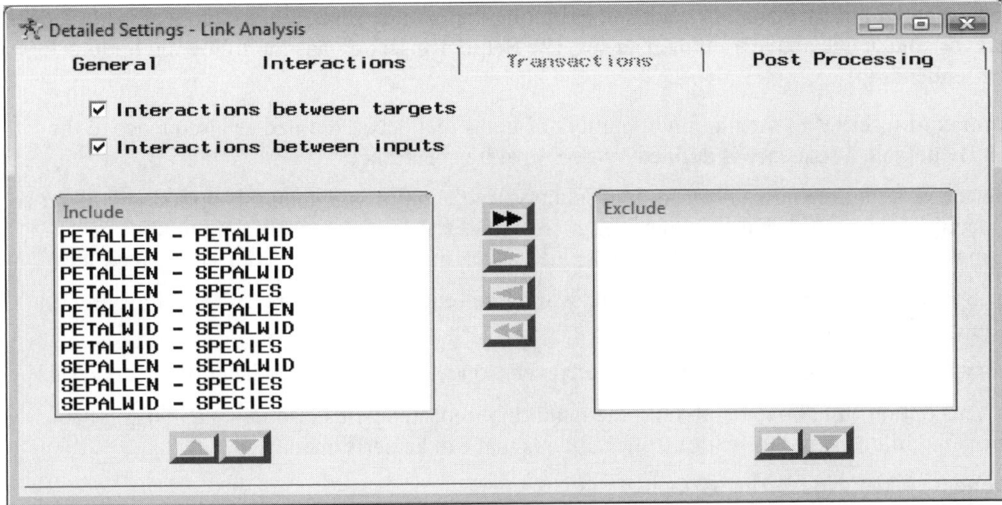

*The **Interactions** tab is used to select the various interactions or links for the link analysis diagram.*

Transactions tab

The **Transactions** tab is designed for you to configure the way in which the associations or the sequences are defined in the link analysis. The tab is designed to limit the number of links in the analysis based on the various configuration settings from either the association or sequence data structures that are specified from the previous **General** tab. The **Transactions** tab will become available for viewing assuming that you have selected the **Transaction** option from **Input data type** field. In Enterprise Miner, either the **Association** or the **Sequence** section will be available for you to specify the various options based on the input source data set having either an association or sequence data structure.

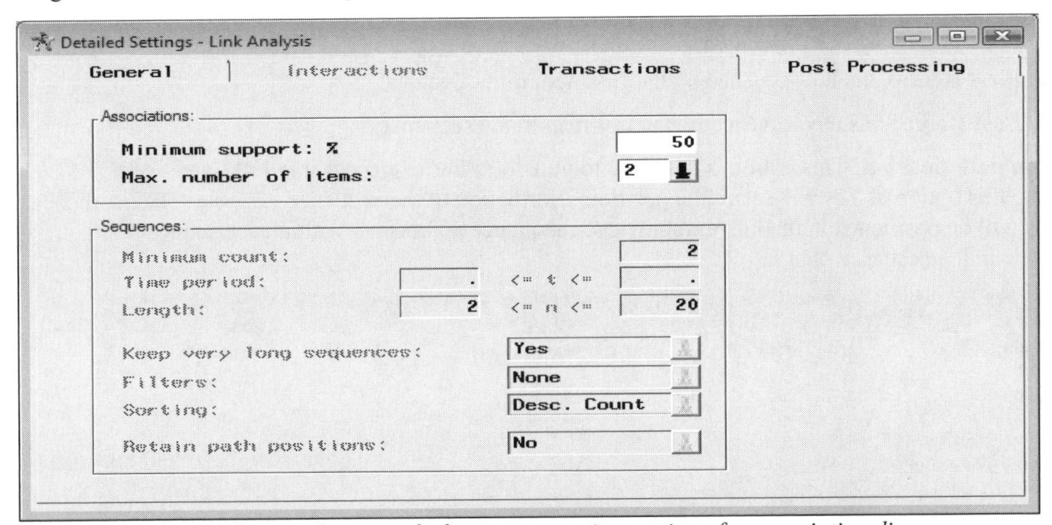

*The **Transactions** tab is used to specify the various option settings for association discovery.*

Again, the **Association** section will be available to you assuming that the active training data set matches the expected data structure to association analysis. The **Association** section has the following two options:

- **Minimum Support: %:** This option will limit the number of associations or connections based on the relative frequency of the two separate variables or nodes in the analysis. The option will limit the number of combinations of items that occur too infrequently. By default, associations or interactions are removed from the analysis with a support probability of 5% or less.
- **Max. number of items:** This option will limit the number of combination of items that occur in the association data set. The default is up to four-way interactions that will be displayed within the link graph.

The **Sequence** section will be available for selection assuming that you have sequence data, that is, a sequence or a time stamp variable, in the active training data set. The **Sequence** section has the following options for configuring the sequence:

- **Minimum count:** Specifies the minimum number of items that occur, defined as a sequence to the analysis. By default, a sequence is defined by two separate occurrences.
- **Time period:** Assuming that you have a time stamp variable in the data, then this option will allow you to specify the minimum and maximum length of time for the sequence. By default, the entries are missing. Therefore, the time period is determined at run time from the active training data set.
- **Length:** Specifies the numerical range of items within the sequence. The default is a range between 2 and 20 items in the sequence.
- **Keep very long sequences:** The default is to keep very long sequences in the data mining analysis.
- **Filters:** This option will allow you to limit the number of repeated items or duplicate items in the sequence. The following are the various filter options that can be performed:
  - **None** (default)**:** No filtering is applied to the sequence.
  - **Repeats:** Removes all repeated sequences, for example, A→A→B→C→A to A→B→C→A
  - **Duplicates:** Removes all duplicate sequences, where A→A→B→C→C→A to A→B→C
- **Sorting:** Defines the sequence of the link analysis sequence based on the sorted order of the sequence that can be specified from the following options:
  - **Count:** Sorts the active training data set by the most frequent sequences that are determined during processing.
  - **Desc. Count** (default)**:** Sorts the sequence of items in descending order. Selecting this option will sort the sequence by reversing the order of the sequence by the most frequently occurring items followed by the least frequent items.
  - **N items:** Sorts the first n items in the sequence.
  - **First Item:** Sorts the sequence by the first item in the sequence.
  - **Last Item:** Sorts the sequence by the last item in the sequence.
- **Retain path position:** This option is designed to transform the sequences into links and nodes. Setting this option to **Yes** will retain the position information of the sequence variable. That is, the nodes will be positioned in the link graph by each sequence that occurs within each sequence variable in the sequence data set.

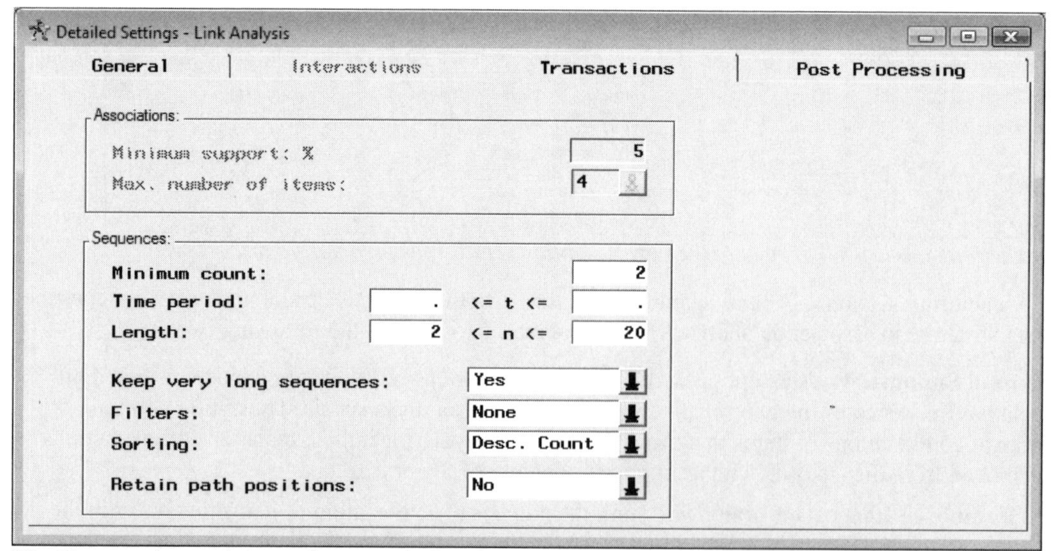

*The **Transactions** tab is used to specify the various option settings in sequence discovery.*

Post Processing tab

The **Post Processing** tab will allow you to parse variables and control the way in which the output data set is created. Once the node is executed and the NODES and LINKS data sets are created, then a post-processing routine is performed to determine the way in which the link analysis data sets are constructed. The following are the various options in configuring the post-processing routine.

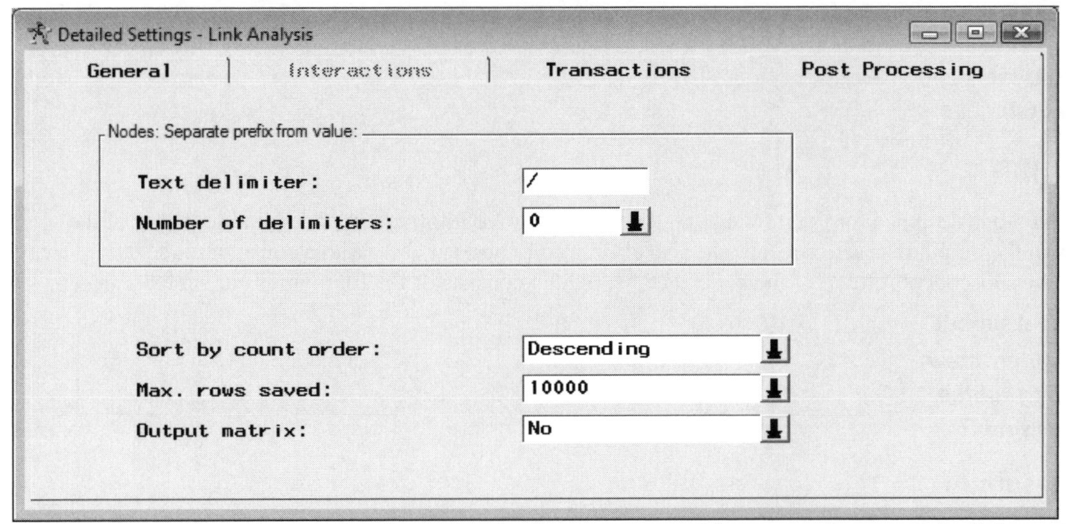

*The **Post Processing** tab is used to determine the data structure of the output data set of the link graph.*

The **Nodes: Separate prefix from value** section is designed to handle input variables that contain text strings. This option is analogous to the SCAN function that can be used in standard SAS programming code to parse words from sentences. The following two options are identical to the options used in the SCAN function. For example, if the input variables are Web URL text strings, then this section will allow you to create a sequence of words based on the URL path names, in which each word will represent a particular node that is created in the link analysis diagram.

- **Text delimiter:** The text delimiter is used to distinguish between the separate words in the sentence. The default is a delimiter of /. The option will allow you to specify other delimiters such as commas, periods, or backslashes by typing them in the **Text delimiter** input field.
- **Number of delimiters:** Limits the number of parses that are performed, that is, the number of words to parse from the given sentence. The option will allow you to specify between zero to ten delimiters.

From the **Post Processing** tab, the following options will allow you to configure the output data set used to create the link graphs such as specifying the sort order, the number of rows, and save the output data in a matrix form for further processing.

- **Sort by count order:** Specify the sorted order of the listed items to the output data set based on the frequency count. The default is descending order.
- **Max. rows saved:** This option controls the size of the output data set by specifying the number of rows written to the output data set. You may specify either 100, 500, 1000, 5000, 10000, 25000, 50000, 75000, or 100000 rows written to the output data set. The default is 10000 rows.
- **Output matrix:** This option will allow you to save the output data set in matrix form. For example, a matrix form that is the square form of the LINKS data set where the rows and columns are based on the co-occurrences of each link in the link analysis graph. However, the matrix is not used by Enterprise Miner.

Viewing the Link Analysis Node Results

General Layout of the Results Browser within the Link Analysis Node

- **Model tab**
- **Code tab**
- **Log tab**
- **Output tab**
- **Notes tab**

Model tab

The **Model** tab is designed for you to view the file administrative information and the table listing of the variables in the output data set. The tab will also allow you to observe the various configuration settings that generate the subsequent results for the node. The **Model** tab consists of the following four subtabs.

- **General subtab**
- **Variables subtab**
- **Settings subtab**
- **Status subtab**

General subtab

The **General** subtab is designed for you to view the file administrative information of the output data set. File administrative information such as the name of the node, creation date, and the last modification date with an indication that a target variable was specified in the analysis.

Variables subtab

The **Variables** subtab is designed for you to view the variables in the output data set. The subtab displays the standard **Variables** tab with a tabular view of the variables and the various attributes assigned to each variable in the output data set. The tab will allow you to change the model role of the listed variables.

Settings subtab

The **Settings** subtab is designed for you to display the various configuration settings that have been specified within the **Link Analysis** node that generate the results that follow. The subtab is illustrated in the following diagram.

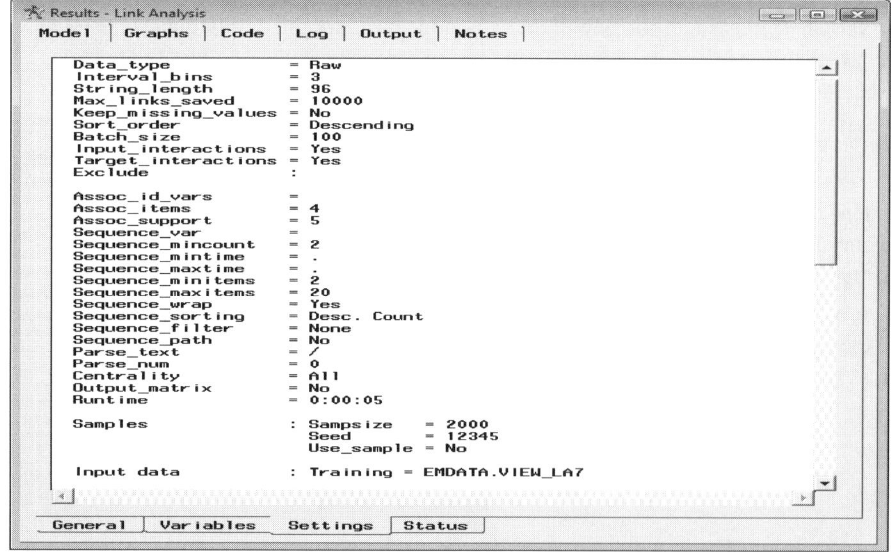

*The **Setting subtab** displaying the configuration settings that generated the results.*

Status subtab

The **Status** subtab is designed for you to display the processing status that occurs when executing the node. The subtab displays the creation dates and processing times of the active training and scoring data sets.

Graphs tab

The **Graphs** tab is designed for you to display various frequency bar charts. The arrow buttons displayed on the toolbar will allow you to scroll thorough the various bar charts that are created. The histograms will display the frequency distribution of the various links and nodes that are created. The first few bar charts will display the frequency distribution of the number of links between the connected nodes, with the subsequent charts displaying the distribution of the number of nodes that are created within the link analysis diagram. In addition, the bar charts will display the importance of the nodes that are created from the first, and second-order weight and unweighted centrality measurements. The following bar chart is generated from the iris data set. The first bar indicates that there is one node, PETALLEN = 37.58, with a frequency count of 40. That is, there were forty occurrences where the petal length was within the highest cutoff value of 37.58, and so on.

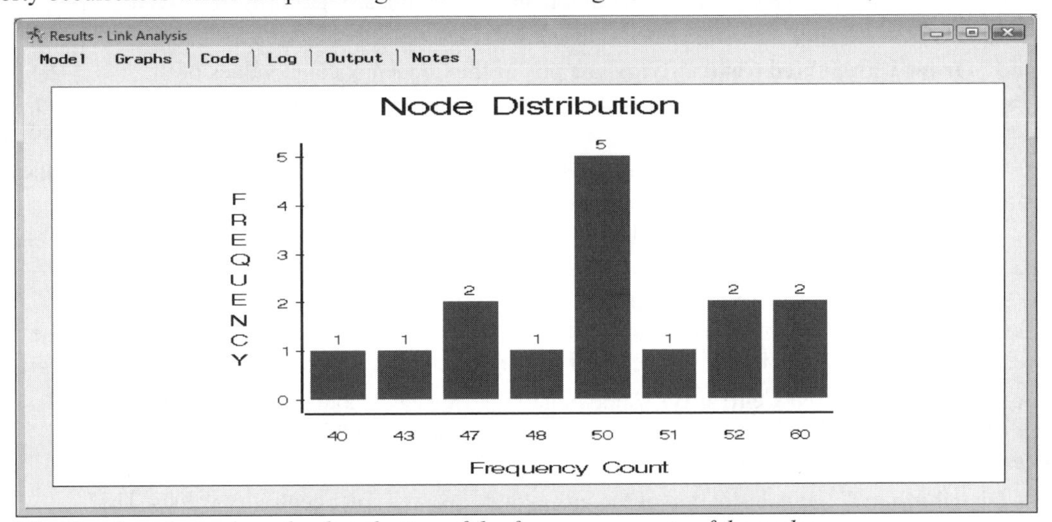

*The **Graphs** tab displays the distribution of the frequency counts of the nodes.*

The following is the sequence of graphs that are displayed within the node:

- **Link Chi Squared Distribution:** A histogram of the partial chi-square values. The chi-square values are computed from the two separate class levels of the connected nodes. The chi-square tests the association between the binary-valued categories of the two nodes that are connected. The values that are displayed on the X-axis are the midpoints of the bins of the partial chi-square values.
- **Link Chi Squared:** A line plot that displays the partial chi-square values and the frequency counts that are displayed in descending order along the X-axis. The chi-square value is the partial chi-square value between the two separate categories occurring together, where the two-way frequency table is formulated from the two separate connected nodes.
- **Link Chi Squared by Count:** A scatter plot of each unique partial chi-square value across the frequency counts of the link. A linear regression line and the 95% upper and lower confidence limits are displayed in the scatter plot to predict the linear relationship between the various chi-square values by the frequency count between the links.
- **Link Distribution:** A histogram of the frequency count for each link. That is, the frequency counts between the class levels of the two separate nodes that are connected to each other.
- **Link Count:** A line plot of the frequency counts for each link. The frequency count index values are sorted in descending order along the X-axis.
- **Node Distribution: C1:** A histogram of the weighted first-order centrality values that displays the importance of each node based on the number of nodes that are connected to it. The midpoints of the binned centrality values are displayed along the X-axis.

- **Node 1st Order Weighted Centrality by Count:** A scatter plot of the frequency counts of the weighted first-order centrality values of each node that is created. The weighted first-order undirected centrality value uses the importance calculations by ranking the various links connected to the node in order to assign numeric values to these same links. A linear regression line and the 95% upper and lower confidence limits are displayed in the scatter plot to predict the linear relationship between the weighted first-order centrality values of each node by the frequency counts that are displayed along the X-axis.
- **Node 1st Order Weighted Centrality:** A line plot of the frequency counts of the weighted first-order centrality value of each node that is created. The frequency count index values are sorted in descending order along the X-axis.
- **Node Distribution: C1U:** A histogram of the unweighted first-order centrality values. The midpoints of the binned first-order unweighted centrality values are displayed along the X-axis.
- **Node 1st Order Unweighted Centrality by Count:** A scatter plot of the frequency counts of the unweighted first-order centrality values of each node. A regression line and 95% upper and lower confidence limits are displayed in the scatter plot to predict the linear relationship between the first-order unweighted centrality values by the frequency counts of each node.
- **Node 1st Order Unweighted Centrality:** A line plot of the frequency count values of the unweighted first-order centrality value of each node. The frequency count index values are sorted in descending order along the X-axis.
- **Node Distribution: C2:** A histogram of the weighted second-order centrality values that displays the importance of each node based on the number of links it has. The X-axis displays the midpoints of the binned second-order weighted centrality values.
- **Node 2nd Order Weighted Centrality by Count:** A scatter plot of the frequency counts of the node's weighted second order centrality values. The linear regression line and the 95% upper and lower confidence limits are displayed in the scatter plot to predict the linear relationship between the unweighted second-order centrality values based on the frequency counts of each node.
- **Node 2nd Order Weighted Centrality:** A line plot of the frequency counts of the weighted first-order centrality value of each node that is created. The frequency count index values are sorted in descending order along the X-axis.
- **Node Distribution: C2U:** A histogram of the unweighted second-order centrality values. The midpoints of the binned, unweighted second-order centrality values are displayed along the X-axis.
- **Node 2nd Order Unweighted Centrality by Count:** A scatter plot of the frequency counts of the unweighted second-order centrality values of each node. A regression line and 95% upper and lower confidence limits are shown in the scatter plot to display the linear relationship between the frequency counts of the second-order unweighted centrality values across each node.
- **Node Distribution: C2U:** A line plot of the frequency counts of the unweighted second-order centrality value of each node that is created. The frequency count index values are sorted in descending order along the X-axis.
- **Node Distribution:** Displays a histogram of the sorted frequency counts at each class level of the target variable or binned interval of the input variable with regard to each node that is created. The plot is displayed in the previous diagram.
- **Node Count:** A line plot of the marginal frequency value of the class level that represents the node. The frequency count index values are sorted in descending order along the X-axis.

Code tab

The **Code** tab is designed for you to display the internal SEMMA training code that produces the following link analysis results.

Log tab

The **Log** tab is designed for you to display the log listing based on the internal SEMMA data mining SAS programming code that was compiled to generate the various results. The **Log** tab is designed for you to diagnose any compiling problems that might have occurred in executing the node.

Output tab

The **Output** tab is designed for you to display the data structure of the NODES and LINKS data set followed by data listings of both data sets.

The output data set that is displayed within the **Output** tab depends on the type of data that was selected from the **General** tab. In addition, the size of the output data set that is created depends on the **Max. rows saved** option specified from the **Post Processing** tab according to how the output data set is internally sorted, as specified by the **Sort by count order** option setting. By default, the output data set is sorted in descending order by the frequency counts.

- **Transaction data for association:** The various data sets that are displayed within the **Output** tab are the result of specifying the **Transaction** option setting within the **Data** tab from association analysis. In the following listing, the default setting of up to four-way associations between the items is displayed.
  - **Associations by product:** By default, the tab will display the items that occur most frequently in descending order by listing the frequency count of all possible unique combination of items of up to four-way associations.

Associations - max rows=50
Sorted by descending size and count

Obs	COUNT	ITEM001	ITEM002	ITEM003	ITEM004
1	90	apples	avocado	peppers	sardines
2	90	apples	baguette	peppers	sardines
3	91	avocado	baguette	peppers	sardines
4	92	apples	avocado	baguette	peppers
5	92	apples	avocado	baguette	sardines
6	92	bourbon	chicken	corned_b	peppers
7	92	chicken	corned_b	cracker	peppers
8	93	bourbon	chicken	corned_b	cracker
9	93	bourbon	corned_b	cracker	peppers
10	94	bourbon	chicken	cracker	peppers
11	95	bourbon	coke	olives	turkey
12	95	coke	ice_crea	olives	turkey
13	96	bourbon	coke	ice_crea	turkey
14	96	bourbon	ice_crea	olives	turkey
15	97	apples	corned_b	hering	steak
16	97	apples	corned_b	olives	steak
17	97	apples	hering	olives	steak
18	97	bourbon	coke	ice_crea	olives
19	99	artichok	avocado	baguette	hering
20	99	artichok	avocado	baguette	ham

- **Link Analysis nodes:** By default, the tab will display the nodes that have the most frequently occurring unique number of items along with a listing of the first and second order undirected weighted centrality variables, frequency count, node label, and the node identifier.

Nodes - maximum 100 rows

Obs	1st order unweighted centrality	2nd order unweighted centrality	1st order weighted centrality	2nd order weighted centrality	Count	Value	Node Identifier
1	.	.	.	.	74	bordeaux	A20
2	0.60	10.70	12	214	227	steak	A19
3	0.75	13.20	15	264	283	turkey	A18
4	0.80	13.85	16	277	296	coke	A15
5	0.75	13.15	15	263	296	peppers	A16
6	0.85	14.65	17	293	296	sardines	A17
7	0.85	14.50	17	290	305	artichok	A13
8	0.90	15.20	18	304	305	ham	A14
9	0.85	14.65	17	293	313	ice_crea	A12
10	0.90	15.20	18	304	314	apples	A11
11	0.80	13.95	16	279	315	chicken	A10
12	0.90	15.20	18	304	318	soda	A9
13	0.85	14.45	17	289	363	avocado	A8
14	0.90	15.20	18	304	391	corned_b	A7
15	0.90	15.20	18	304	392	baguette	A6
16	0.90	15.20	18	304	403	bourbon	A5
17	0.90	15.20	18	304	473	olives	A4
18	0.90	15.20	18	304	486	hering	A3
19	0.90	15.20	18	304	488	cracker	A2
20	0.90	15.20	18	304	600	heineken	A1

- **Link Analysis links:** By default, the tab will display the most frequent links in descending order by listing the relative frequency count of the number of links between the two separate nodes that are displayed along with the link identifier.

Links - maximum 100 rows

Obs	Link Count	First Node Identity	Second Node Identity	Link Identifier
1	50	A17	A18	L162
2	50	A13	A19	L138
3	51	A9	A18	L163
4	51	A9	A19	L148
5	52	A14	A17	L115
6	52	A14	A19	L144
7	52	A15	A9	L127
8	53	A11	A14	L38
9	53	A13	A15	L19
10	53	A14	A16	L101
11	54	A8	A12	L70
12	54	A17	A9	L136
13	55	A15	A16	L98
14	55	A13	A17	L107
15	55	A13	A18	L150
16	55	A14	A9	L130
17	56	A8	A10	L15
18	56	A6	A14	L41
19	56	A8	A18	L151
20	56	A12	A9	L133

- **Transaction data for sequence:** The various data sets that are displayed within the **Output** tab are the result of specifying the **Transaction** option setting within the **Data** tab based on sequence analysis.

  - **Sequences of Items:** By default, the tab will display the most frequent items that occur within a sequence for a specific customer. The listing will display the customer identifier, number of sequences, the minimum and maximum length of the sequence, the average length of the sequence, and the total number of items in a defined sequence. By default, the tab will display the most frequent sequence of items along with a listing of the frequency counts, the number of items within the listed sequence, and the sequence of items.

PRODUCT: Most Frequent CUSTOMER - max 50 rows
Count)= 2 : 2 (= length (= 20 : 0 (= time period (= 1E64

CUSTOMER	Number of sequences	Min sequence length	Max sequence length	Mean sequence length	Total number of items in sequences	count	nitems	Sequences
0	1	7	7	7	7	5	7	baguette -) sardines -) apples -) peppers -) avocado -) bourbon -) bourbon
1	1	7	7	7	7	4	7	olives -) bourbon -) coke -) turkey -) ice_crea -) cracker -) baguette
10	1	7	7	7	7	4	7	soda -) olives -) bourbon -) cracker -) heineken -) steak -) hering
100	1	7	7	7	7	3	7	avocado -) cracker -) artichok -) heineken -) ham -) steak -) steak
1000	1	7	7	7	7	3	7	baguette -) hering -) avocado -) artichok -) heineken -) bordeaux -) soda
101	1	7	7	7	7	3	7	baguette -) hering -) avocado -) artichok -) heineken -) bourbon -) corned_b
102	1	7	7	7	7	3	7	baguette -) soda -) hering -) cracker -) heineken -) artichok -) turkey
103	1	7	7	7	7	3	7	baguette -) soda -) hering -) cracker -) heineken -) avocado -) steak
104	1	7	7	7	7	3	7	baguette -) soda -) hering -) cracker -) heineken -) bourbon -) peppers
105	1	7	7	7	7	3	7	baguette -) soda -) hering -) cracker -) heineken -) corned_b -) ham

  - **Link Analysis nodes:** The table listing is similar to the previous listed **Link Analysis nodes**.

Nodes - maximum 100 rows

Obs	1st order unweighted centrality	2nd order unweighted centrality	1st order weighted centrality	2nd order weighted centrality	Value	Count	Node Identifier
1	1.60	33.35	32	667	heineken	249	A12
2	0.90	20.00	18	400	cracker	189	A10
3	0.80	17.55	16	351	hering	175	A13
4	1.00	21.95	20	439	bourbon	173	A6
5	0.80	18.55	16	371	olives	167	A15
6	0.95	20.40	19	408	corned_b	149	A9
7	0.85	20.05	17	401	baguette	147	A4
8	1.10	25.10	22	502	artichok	131	A2
9	1.00	21.65	20	433	avocado	126	A3
10	1.20	26.05	24	521	chicken	124	A7
11	1.00	21.25	20	425	soda	115	A18
12	1.35	28.50	27	570	apples	114	A1
13	1.30	28.30	26	566	ice_crea	114	A14
14	1.05	23.90	21	478	peppers	112	A16
15	1.40	30.50	28	610	ham	111	A11
16	0.70	15.85	14	317	coke	110	A8
17	1.05	23.65	21	473	turkey	107	A20
18	0.70	16.70	14	334	sardines	96	A17
19	1.30	28.70	26	574	steak	76	A19
20	0.95	21.50	19	430	bordeaux	33	A5

- **Link Analysis links:** The table listing is similar to the previous listed **Link Analysis links** with an additional column that lists the direction of the links that are displayed.

Links - maximum 100 rows

Obs	Link Count	First Node Identity	Second Node Identity	Link Identifier	Direction
1	85	A10	A12	L126	1
2	79	A6	A10	L98	1
3	77	A2	A12	L121	1
4	76	A15	A6	L61	1
5	63	A13	A9	L91	1
6	58	A12	A7	L71	1
7	49	A7	A8	L79	1
8	48	A13	A10	L101	1
9	48	A18	A13	L138	1
10	48	A4	A18	L182	1
11	47	A8	A14	L143	1
12	45	A17	A12	L130	1
13	45	A16	A6	L62	1
14	44	A20	A14	L150	1
15	44	A9	A16	L164	1
16	43	A18	A15	L158	1
17	42	A11	A20	L207	1
18	42	A10	A7	L69	1
19	41	A8	A20	L206	1
20	40	A4	A13	L133	1

- **Raw data for interaction:** The various data sets that are displayed within the **Output** tab are the result of specifying the **Raw** option setting within the **Data** tab. By specifying raw data for interaction data layout, the **Output** tab will display the following NODES and LINKS data sets.
  - **Link analysis nodes:** By default, the tab will display the marginal frequencies within each node in descending order in addition to the first and second order weighted and unweighted centrality measurements, and along with the interval-valued variables cutoff point values assigned to the node that are split into three separate groups or the various categories of the categorically-valued variables, the variable label of the node, and its associated variable role.

Nodes - maximum 100 rows

Obs	1st order unweighted centrality	2nd order unweighted centrality	1st order weighted centrality	2nd order weighted centrality	Value	Variable Name	Role	Count	Percent of Total Frequency
1	0.80000	7.40000	12	111	30.57	SEPALWID	input	60	40.0000
2	0.60000	6.33333	9	95	55.23	PETALLEN	input	60	40.0000
3	0.66667	7.06667	10	106	19.612	PETALWID	input	52	34.6667
4	0.73333	7.00000	11	105	50.15	SEPALLEN	input	52	34.6667
5	0.60000	6.20000	9	93	66.711	SEPALLEN	input	51	34.0000
6	0.46667	4.80000	7	72	19.93	PETALLEN	input	50	33.3333
7	0.46667	4.80000	7	72	SETOSA	SPECIES	target	50	33.3333
8	0.66667	7.00000	10	105	VERSICOLOR	SPECIES	target	50	33.3333
9	0.66667	7.00000	10	105	VIRGINICA	SPECIES	target	50	33.3333
10	0.46667	4.80000	7	72	4.368	PETALWID	input	50	33.3333
11	0.60000	6.33333	9	95	11.99	PETALWID	input	48	32.0000
12	0.80000	7.40000	12	111	26.21	SEPALWID	input	47	31.3333
13	0.80000	7.60000	12	114	58.43	SEPALLEN	input	47	31.3333
14	0.73333	6.80000	11	102	34.929	SEPALWID	input	43	28.6667
15	0.66667	7.06667	10	106	37.58	PETALLEN	input	40	26.6667

  - **Link analysis links:** The tab will display the frequency percent of the listed links, the marginal frequency counts of each class level of the two linked nodes, the link identifier variable, the expected frequency count, the deviation (observed frequency count – expected frequency count), and the partial chi-square value between the separate class levels of the connected nodes.

Links - maximum 100 rows

Obs	Link Count	First Node Identity	Second Node Identity	Link Identifier	Percent of Total Frequency	First Node Count	Second Node Count	Expected	Deviation	Chi Squar
1	50	N1	N13	L50	33.3333	50	50	16.6667	33.3333	66.66
2	50	N4	N13	L51	33.3333	50	50	16.6667	33.3333	66.66
3	50	N1	N4	L8	33.3333	50	50	16.6667	33.3333	66.66
4	50	N3	N6	L7	33.3333	60	52	20.8000	29.2000	40.99
5	49	N3	N15	L68	32.6667	60	50	20.0000	29.0000	42.05
6	47	N6	N15	L70	31.3333	52	50	17.3333	29.6667	50.77
7	45	N7	N13	L52	30.0000	52	50	17.3333	27.6667	44.16
8	45	N5	N14	L59	30.0000	48	50	16.0000	29.0000	52.56
9	45	N1	N7	L9	30.0000	50	52	17.3333	27.6667	44.16
10	45	N4	N7	L13	30.0000	50	52	17.3333	27.6667	44.16
11	44	N3	N9	L21	29.3333	60	51	20.4000	23.6000	27.30
12	39	N2	N14	L57	26.0000	40	50	13.3333	25.6667	49.40
13	38	N2	N5	L4	25.3333	40	48	12.8000	25.2000	49.61
14	38	N6	N9	L23	25.3333	52	51	17.6800	20.3200	23.35
15	37	N9	N15	L73	24.6667	51	50	17.0000	20.0000	23.52
16	33	N1	N12	L42	22.0000	50	43	14.3333	18.6667	24.31
17	33	N4	N12	L46	22.0000	50	43	14.3333	18.6667	24.31
18	33	N12	N13	L56	22.0000	43	50	14.3333	18.6667	24.31
19	30	N8	N14	L62	20.0000	47	50	15.6667	14.3333	13.11
20	29	N5	N10	L27	19.3333	48	47	15.0400	13.9600	12.95

General Layout of the Detailed Results

- **Link Graph tab**
- **Plots tab**
- **Nodes tab**
- **Links tab**

The four tabs are displayed within the **Detailed Results** window. From the **Result Browser**, select the **Detailed Results...** toolbar icon or select the **View > Detailed Results...** main menu options that will allow you to view the link analysis diagrams that follow. The **Detailed Results** window is designed for you to view the results from the link analysis by both a graphical display and a table view listing of the nodes and links that are created. An initial message window will appear to inform you as to the number of nodes and links that are created within the link graph. Simply press the **OK** button that will then automatically display the circular link graph.

Link Graph tab

The **Link Graph** tab is designed for you to view the link graph. There are several different graphical layouts that you may select from in order to view the results from the link analysis. The following are the six different graphical layout designs available within the **Link Analysis** node. By default, the numerous nodes that are created within the link graph are displayed in a circle.

- **Circle** (default)**:** Displays the association between the variables in the active data set by positioning the nodes within the linked graph in a circular pattern.
- **Grid:** Displays the nodes within the graph as separate grids based on a group variable in the active training data set or when your data has created separate cluster groupings.
- **MDS:** The link analysis results are based on the PROC MDS procedure that calculates the multidimensional scaling that is designed to reduce the dimensionality in the data to two dimensions based on the distance between the pair of data points while at the same time preserving the ranked-order distance between the original data points.
- **Parallel Axis:** Displays the results within the graph on a parallel axis where the nodes are positioned in either a column or a row based on the order of importance that is determined by the unweighted centrality measurements. The configuration layout of the graph is such that each column represents each variable in the active training data set and each row represents the class levels of the corresponding variable.
- **Swap:** The Swap layout works on the x/y coordinates of the nodes and scales the inter-node distance measures by the selected link metric, that is, frequency count, partial chi-square, and so on, based on whatever is in the link table that has been selected from the **Links** tab. The swap technique iteratively repositions the nodes, in order to maximize the energy state of the graphical layout. The energy state is calculated by the value of the link divided by the distance between two separate nodes.
- **Tree:** The tree technique is useful when you have a hierarchical data structure in which the nodes and links are displayed in a tree branching layout. For example, hierarchical data structures such as association or sequence data sets in which the most significant associations between the different items or nodes are displayed within the graph based on association or sequence analysis.

The following link graphs were created from the iris data set. The iris data set is one of the most popular data sets used in introducing a number of multivariate statistical techniques to separate the different varieties of wildflowers by the sepal and petal length and width of the wildflower. In the following link analysis, the data set will be used in explaining the various links and connections that are created from the various wildflower groups and corresponding petal length and width and sepal length and width, which are divided into separate intervals.

Circle Layout

By default, the nodes to the graph are based on the variables in the active training data set. For categorically-valued variables, each separate class level represents each node in the link graph. For interval-valued variables, the values are automatically split into three separate buckets or bins for each node that is displayed in the link graph. Since the following graph will display every node that is connected, however it is very difficult to view the 73 different links. Therefore, the only links that are displayed in the following graph will originate from the Setosa wildflower through the use of the **Filter** option. Limiting the number of links to the graph will be discussed shortly. In the following diagrams, we will focus our attention on the various nodes that are connected to the Setosa wildflower. The graph indicates that the Setosa wildflower can be mainly characterized by its sepal length and width, with highest cutoff values of 50.15 and 34.929, respectively, and the petal length and width, with highest cutoff values of 19.93 and 4.368, based on the color and thickness of the lines.

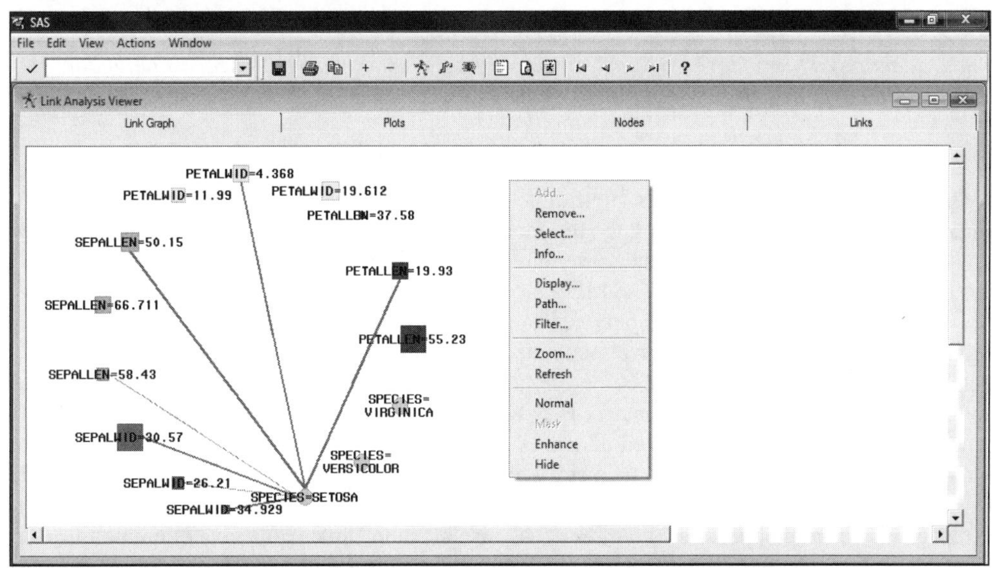

The default circular layout of the link analysis diagram from the iris data set.

The following configurations are the default settings of the link graph layouts:

- **Symbol shape:** From the link graph, the circular nodes represent the target variables in the active training data set and the square nodes identify the input variables.
- **Symbol color:** Each node of the same variable in the graph will have the same color.
- **Symbol size:** The size of the node will indicate the frequency count at each class level of the variable. Larger nodes will indicate larger frequency counts than smaller nodes.
- **Link width:** The width or thickness of each line or link will indicate the count of each node connected to it. The thicker lines will indicate to you that the node has more links or connections.
- **Link color:** The color of the links is identified by three separate colors. Red links will indicate the highest count values, blue links will indicate intermediate count values, and green links will indicate the lowest count values based on the number of links to the corresponding node.

Grid Layout

The nodes to the link graph are formed into a rectangular grid. This type of graphical layout is useful in cluster analysis. Each variable in the analysis is positioned within each grid of the graph. The nodes are formed into separate groups by the clustering segment identifier variable. The link graph displays the connection between the input variable that is connected to the corresponding target categories. For example, the node that represents the sepal width at the highest cutoff point of 34.929, that is, SEPALWID = 34.929, is more closely associated with the Setosa wildflower, that is, SPECIES = SETOSA within the link analysis diagram. The Setosa cluster grouping is located at the left-hand portion of the link graph.

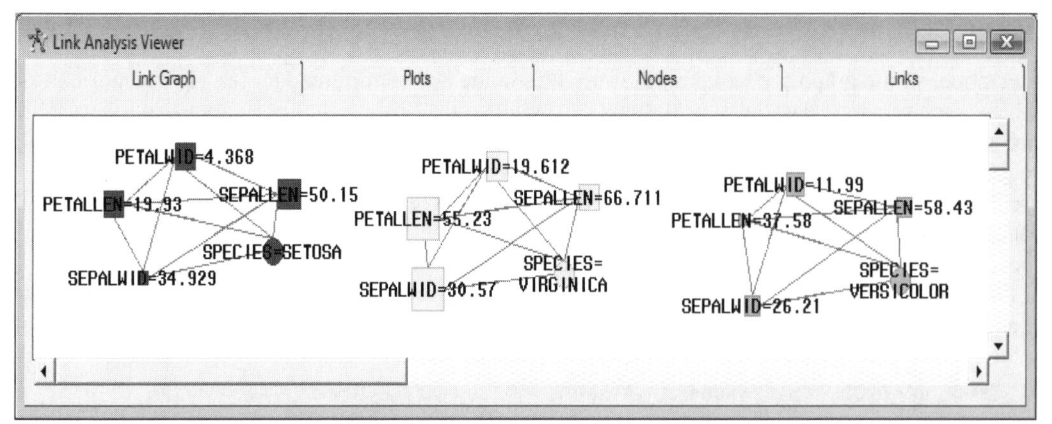

*The **Grid** layout that arranges the nodes into a rectangular grid of clusters based on nearest neighbor clustering with a smoothing constant of five.*

MDS Layout

The graphical layout is based on the MDS, that is, multidimensional scaling, procedure that is performed on the active training data set. The procedure is designed to reduce the data with several input variables into a two dimensional layout in cluster analysis, while at the same time preserving the original ranked-order distance between the various input variables in the data. The MDS model iteratively fits the data by reducing the dimensionality in the data based on the distances between each data point. The procedure rearranges the data points in an efficient manner in order to best approximate the original distance while preserving the ranked-order distance between the data points. The most common measure of goodness-of-fit in determining how well the procedure reproduces the observed distance matrix is as follows: $\sum[d_{ij} - f(\delta_{ij})]^2$ where d_{ij} stands for the reproduced distance and δ_{ij} represents the observed distances in the input data set with $f(\delta_{ij})$ that represents the nonmetric, monotone transformation of the input data set. This will result in the reduction in the dimensionality in the data, while at the same time preserving the observed distance between the data points. In the following diagram, the iris data set is clustered by nearest neighbor analysis since the **Link Analysis** node will allow you to visualize the clusters that are created. The nodes are grouped by the cluster variable CLUSTER_K_5, that is, a smoothing constant of 5, with the default MDS option settings that are applied with the grouping bias option that was set to –2 to better visualize the separate target groupings. In addition, the only links that are displayed within the link graph are based on the nodes that are grouped within each cluster that was obtained by selecting the **CLUSTER_MASK** pull-down menu item from the **Mask** option and selecting the corresponding **Mask** option from the **Mode** section within the **Links** tab. The **CLUSTER_MASK** pull-down option was selected to display all links within each cluster grouping that was created within the graph. From the following MDS layout, the sepal length and width at the highest cutoff points of 50.15 and 34.929, respectfully, and the petal length and width at the highest cutoff points of 19.93 and 4.368 are grouped within the Setosa wildflower.

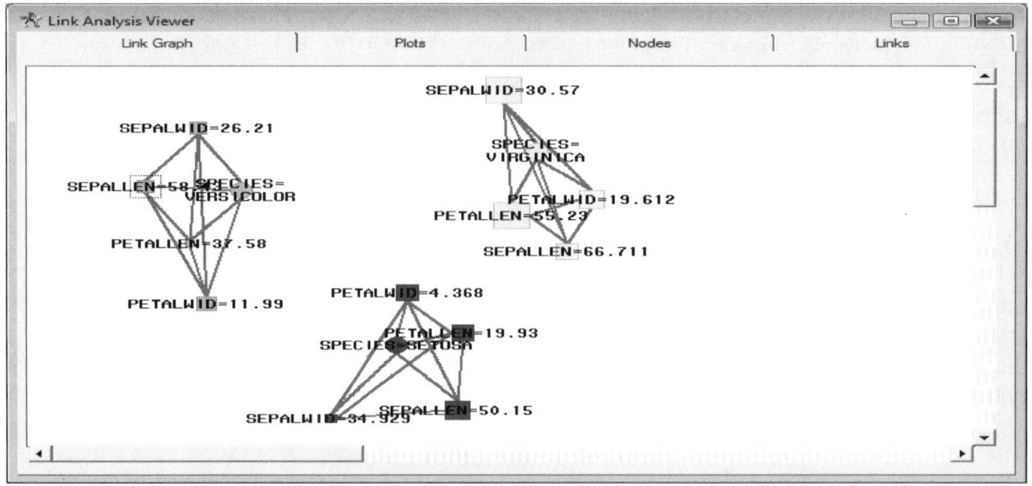

*The **MDS** layout of the iris data set with cluster groupings from nearest neighbors at k = 5.*

From the previous MDS design, three separate cluster groupings were created from the three different varieties of iris wildflowers. The sepal and petal length and width at the same cutpoints from the previous non-parametric clustering technique were grouped in the Setosa variety of wildflower. The previous cluster display was created by selecting the MDS technique and the VAR pop-up menu item from the **Grouping** option within the **Layout** tab. The labels of the nodes were created by selecting the TEXT pop-up menu item from the **Label** option. A further step might be to create a scored data set to check the accuracy of the clustering assignments that are created from the three separate target class levels.

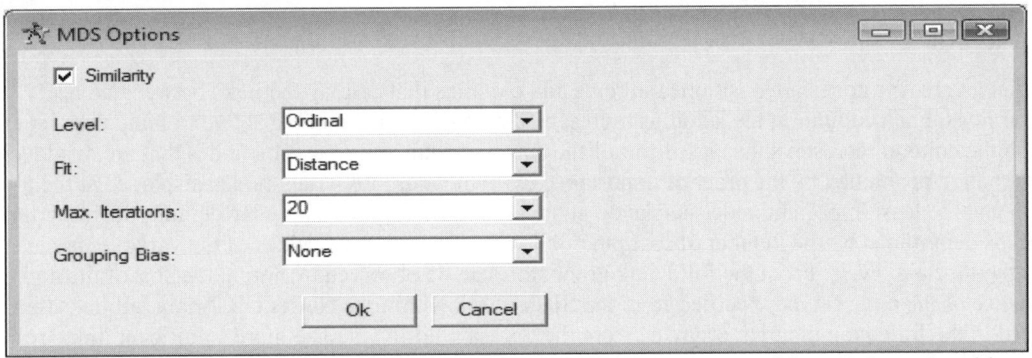

*The **MDS Options** window is used to specify the various option settings of the MDS graphical layout.*

The following are the various option settings available for modifying the MDS graphical layout:

- **Similarity** (default)**:** The similarity check box is automatically selected, indicting that the training data set is treated as similarity measurements, assuming that the data is formulated in a matrix form. Similarity measurements are based on differences between two separate variables, such as distances between different cities. Clearing the check box will result in the node treating all the data points as dissimilarity measurements.
- **Level:** Specify the measurement level of the variables and the type of transformation that is applied to the data or distances. The default measurement level is **Ordinal**. However, the other options you may select from are **Absolute, Ratio, Interval**, and **Log** measurement levels as follows:
  - **Ordinal** (default)**:** This option specifies a nonmetric analysis, while all other **Level** options specify metric analyses. This option fits a measurement model in which a least-squares monotone increasing transformation is applied to the data in each partition or interval. At the ordinal measurement level, the regression and measurement models differ.
  - **Absolute:** This option allows for no optimal transformations. Hence, the distinction between regression and measurement models is irrelevant. In other words, this option would be selected with no transformation that is applied to distance data.
  - **Ratio:** This option fits a regression model in which the distances are multiplied by a slope parameter in each partition, that is, a linear transformation. In this case, the regression model is equivalent to the measurement model with the slope parameter reciprocated.
  - **Interval:** This option fits a regression model in which the distances are multiplied by a slope parameter and added to an intercept parameter in each partition, that is, an affine transformation. In this case, the regression and measurement models differ if there is more than one partition.
  - **Log:** This option fits a regression model in which the distances are raised to a power and multiplied by a slope parameter in each partition, that is, a power transformation.
- **Fit:** Transforms the data to both the target values and the input variables in the MDS procedure. The following are the various transformations that can be applied to the model.
  - **Distance** (default)**:** Fits the MDS model by transforming the data into Euclidean distance measurements between all possible pairs of data points. This is the most common technique.
  - **Squared:** Fits the MDS model to the squared distances; large data and distances are considered more important to the MDS model as opposed to small data and distances. The drawback to this option is that the model is very sensitive to extreme values in the active training data set.

- **Max. Iterations:** Specifies the maximum number of iterations in determining the optimum distances between the pair of data points. The default is 20 iterations.
- **Grouping Bias:** This option controls how the nodes are grouped when they are initially displayed. Changing this value may or may not change the resulting graph depending on the data set and links. You can tell the impact on any given data set by running the plot using different values. Select a positive number to move the nodes within each group closer together or select a negative number to move the nodes within each group further apart.

Parallel Layout

In the parallel layout, the nodes are positioned in rows and columns that display the links between adjacent columns and rows. Each column in the layout is represented by each variable in the active training data set and each node in the column represents the class level of the corresponding variable. The nodes that are displayed in each column are positioned by the order of importance, with the more important nodes displayed at the top of the graph and the least important nodes displayed at the bottom of the link graph. By default, the importance of the nodes is determined by the number of cases that fall within the range of values of the variable that is split into separate class levels. From the following graph, this can be observed by noting the size of the nodes. The importance of the node can be specified from the **Size** setting within the **Nodes** tab. Again, all links that are displayed in the following diagram originate from the Setosa wildflower. The more significant links from the Setosa node are consistent with the previous link diagrams.

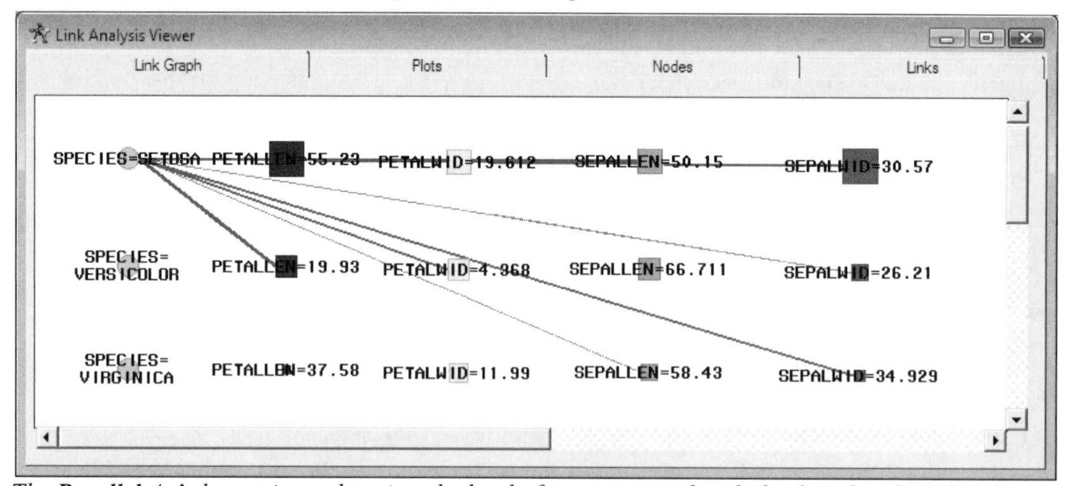

*The **Parallel Axis** layout is used to view the level of importance of each displayed node within the graph.*

From the **Parallel Axis Setup** window, you may select the **Orientation** field to change the orientation of the nodes to the link graph from left to right or top to bottom. The input variables and target variable to the analysis are displayed in the **Axis** field. From the **Axis** field, you may select the order of each axis by changing the ordering level of the **Axis** field. Select the listed node to change the order of the column by using the arrow buttons. From the window, the variable that is listed at the top of the **Axis** list box represents the nodes located in the first column of the graph and the last input variable that is listed at the bottom of the **Axis** list box represents the nodes located in the last column of the link graph, as illustrated in the following diagram. From the previous diagram, the SPECIES variable was repositioned to the top of the **Axis** list box and is, therefore, displayed in the first column of the listed nodes.

*The **Parallel Axis** window is used to reposition the nodes into each column of the link graph.*

Swap Layout

The **Swap Layout** is designed to change the positioning of the nodes based on the energy state of the linked nodes. This option is designed to interchange the nodes through an iterative process in order to maximize the overall energy state of the linked nodes. The energy state of the graph is the ratio of the value of the link to the distance between two separate nodes. In Enterprise Miner, nodes that are highly linked to each other are moved closer together based on an increased energy state of the graph. Conversely, nodes are moved further apart from each other that are not highly linked, based on a decreased energy state of the graph. From the swap layout, the three separate target categories along with the separate class levels of the same input variable are well separated from one another within the graph. The swap layout works on the x/y coordinates of the nodes and scales the internode distance measures by the selected link metric, that is, frequency count, partial chi-square and so on, which can be specified from the **Links** tab. From the following link graph, the only links that are displayed in the diagram originate from the Setosa wildflower. The associated nodes that are highly linked to the Setosa node are positioned closest to the node.

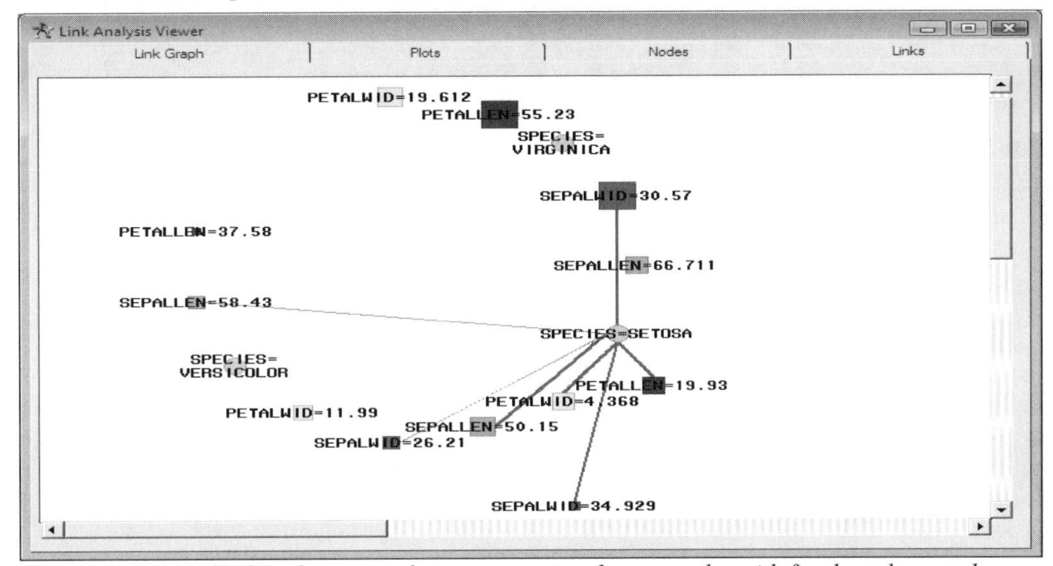

*The **Swap** layout displayed in a circular pattern using dummy nodes with fixed random seeds.*

From the **Swap Options** window, you can specify the various configuration settings of the link graph based on the swap layout. To maximize the change in network display, choose random initialization, dummy nodes, and random seed options. However, the selection of a square grid or circular disk of random initialization is just a visual preference and has no computational significance.

- **Initialization:** You may either select the current x/y coordinates or randomly assigned coordinates. Select random coordinates to force the network into an entirely new configuration. You may also select a square grid or a circular pattern of the random initialization.
  - **Current** (default)**:** The displayed nodes will be positioned in the current x/y coordinates.
  - **Circle:** The displayed nodes will be positioned in a circular pattern within the link graph.
  - **Square:** The displayed nodes will be positioned in a square grid within the link graph.
- **Iterations:** The algorithm works by establishing a global energy metric as a function of all node positions and links and then swaps the positions of two nodes to try to achieve a lower energy level. This is a standard entropy algorithm. The number of iterations is simply the number of swaps. The default number of iterations is a function of the number of nodes. The maximum number of iterations is 40,000. The default is 1,500 iterations.
- **Add dummy nodes:** To give the network more flexibility, a number of dummy nodes are uniformly added to span the diagram with unit linkage to all other nodes, but not to each other. When the algorithm runs, real nodes can swap with dummy nodes, which will result in achieving a new position not previously occupied. However, run time is increased since you need to perform more iterations to achieve the same swap coverage. By default, the check box is automatically selected.

- **Fixed Seed:** This option is based on random initialization. If you want the process to repeat, then select the **Fixed Seed** check box. If you want the process to have more variability, for instance in a very unstable network to find entirely different layouts, then uncheck the **Fixed Seed** check box to select random seeds. By default, the check box is automatically selected.

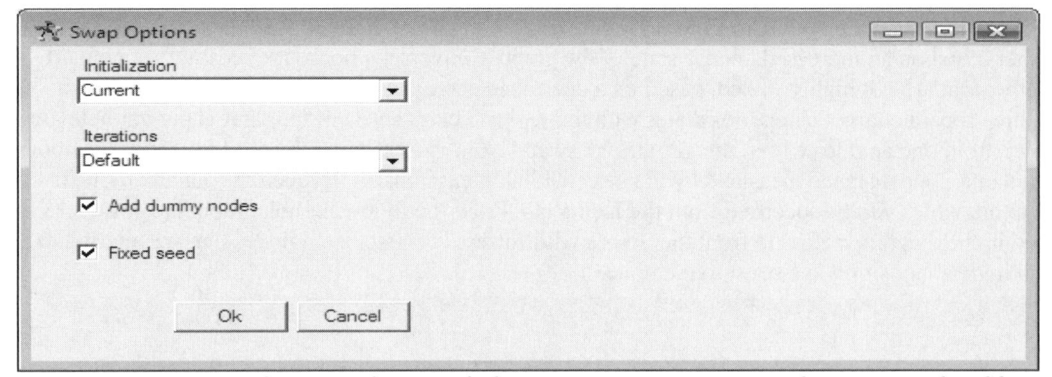

*The **Swap Options** window is used to specify the various option settings to the swap graphical layout.*

Tree Layout

The graphical layout is displayed in a tree-like structure. The tree structure that is created in the link graph is based on the PROC NETDRAW procedure results. This tree-like structured layout is useful when the data is in a hierarchical structure such as sequence data sets. The layout of the tree is designed to display the most significant associations or sequences between different items or nodes. By default, the nodes displayed at the top of the layout represent the left-hand side of the if-then rule and the nodes that are connected just below each associated node represent the right-hand side of the rule. The links that are displayed within the graph represent the most frequently associated items. The size of the nodes that are displayed as boxes will indicate the number of occurrences of each item. The following link graph was created by selecting a minimum support of 50% with up to four-way associations. In addition, the frequency counts from the output data set are sorted in ascending order with a maximum number of rows saved of 100. Therefore, all nodes that are displayed within the tree have some of the other nodes linked to it with at least 100 occurrences. The reason that the Bordeaux item is not part of the tree is that there weren't any nodes that were significantly linked to it. Also, the **LAYOUT_MASK** mask option was selected to prevent several links from cluttering the following diagram.

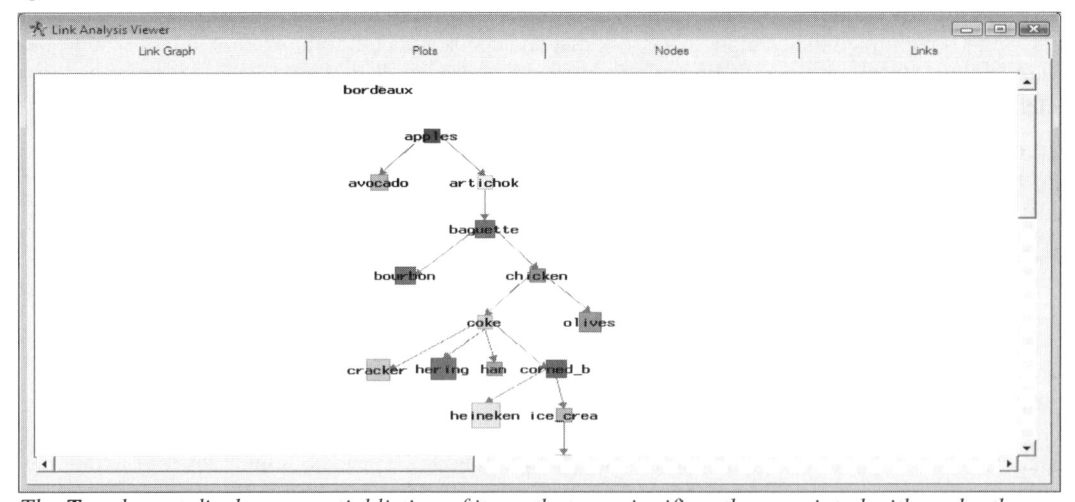

*The **Tree** layout displays a partial listing of items that are significantly associated with each other.*

From the **Tree Options** window, you may change the orientation of the link graph. By default, the link graph displays a vertical tree structure orientation from top to bottom based on the **Top → Down** orientation setting. However, you may change the layout of the tree structure to a horizontal tree structure layout by selecting the **Left → Right** orientation option. By selecting the **Polar** option, you may change the orientation of the tree structure to a circular layout.

Link Graph Pop-Up Menu Items

There are various pop-up menu items that are designed for use within the link graph to control the display of the graphs. Simply right-click the mouse button to view the various pop-up menu items from the **Link Graph** pop-up menu. The **Link Graph** pop-up menu is separated into five separate display options: node modification, node and link display options, link graph display options and modifications, view properties, and graph view settings.

Many of the modifications that you make to the link graph result in a mask. A *mask* is a column in the LINKS data set that indicates which links appear in the layout, cluster, path, or filter display. For example, if you cluster a link graph, the resulting links that are displayed in the clustered link graph would be to the cluster results. All the original links that are displayed in the clustered link graph would have a **CLUSTER_MASK** value of 1 and all of the original links that are not displayed in the graph would have a **CLUSTER_MASK** of 0. In the previous **Tree** graphical layout, all significant links that are displayed within the graph will have a **LAYOUT_MASK** value of 1 in the LINKS data set. That is, a value of 1 indicates that the link is included in the layout mask and displayed within the link graph. Conversely, a value of 0 indicates that the link is excluded from the layout mask and removed from the link graph. In our examples, all filtered nodes and corresponding links that are connected to the node or nodes that can be selected from the **FILTER-Text** Window will have a **FILTER_MASK** value of 1.

At times, the number of nodes and links can potentially amount to an unmanageable size based on the number of variables in the active training data set, which will inhibit your ability to view the various links in the graph. Therefore, the various pop-up menu items are designed to control the visualization of the nodes and links to the link analysis diagram. The following options are based on the various modifications that can be performed to the nodes that are displayed within the link graph. The following menu items are displayed from the pop-up menu items that are displayed within the link analysis diagram:

- **Add:** Nodes can be added to the graph that have been previously removed. An additional pop-up menu will appear for you to select the specific values of the added node with the following options: **Value, Var, Role, Id**, and **Text**. Selecting the **All** option will automatically include all previously removed nodes back into the link graph. The various **Value, Var, Role, Id**, and **Text** option settings will be discussed in greater detail in the following **Grouping** option. In general, this option will add all the nodes that have been previously removed back into the graph based on the selected variable attribute.
- **Remove:** Removes the select nodes from the link graph based on the following five options that can be selected:
    - **Selected:** Removes all the selected nodes from the graph. Select the node to be removed from the graph, then right-click the mouse to display the corresponding pop-up menu by selecting the **Remove > Selected** menu items. To select several nodes from the graph, press the Ctrl key, then select each one of the nodes to be removed from the graph or drag the mouse over several nodes in the graph to remove them.
    - **Deselected:** Removes all nodes that are not currently selected from the link graph.
    - **Value, Var, Role, Id, Text, User**, or **Cluster:** Removes all nodes from the link graph based on the selected value of the node.
    - **All linked:** Removes all nodes that are connected to the selected node from the link graph.
    - **All:** Removes all the nodes from the link graph.
- **Select:** Select the various nodes based on a certain attribute that you can specify. The **Select** pop-up menu will appear, enabling you to select various nodes from the graph based on a certain variable by selecting the **Var** menu item, or selecting the nodes based on the variable role associated with the node by selecting the **Role** menu item. By selecting the **Var** menu option, a separate pop-up menu will be displayed to select the names of all the variables in the link graph. By selecting the **Role** menu item, a separate pop-up menu will be displayed to select the various variable roles. This option is advantageous if there exist several nodes that are displayed within the link graph.
- **Info:** Displays information about the link graph or the selected nodes and links.
    - **Link graph information:** Displays information about the currently opened link graph and the selected nodes or links with the **Link Analysis Information** window appearing. The

window will display information about both the NODES and LINKS data sets and the various options applied to the currently active link graph that is followed by information about the selected link, such as the two separate nodes that are connected, the frequency count, and the chi-square statistic that tests the proportions between the class levels of the two separate categorical variables.

Note: The chi-square statistic is only valid for large cell counts. One of the requirements of this statistic is that there must be at least 10 observations in each group in order for the statistic to follow a chi-square distribution. An additional sample size requirement of the statistic is that there must be at least 80% of the cells counts with an expected cell count of greater than 5. If the sample size is too small in any one of the groups, then the chi-square test might not be valid. This is because the asymptotic *p*-values are based on the assumption that the chi-square test statistic follows a particular distribution when the sample size is sufficiently large.

- **Node information:** By selecting any one of the nodes, the **Link Analysis Information** window will appear that displays information about an individually selected node. The table listing, which is similar to the NODES data set, will appear, displaying the node ID, node role, number of observations of the class level to the selected node, and centrality measurements that are associated with the selected node, followed by a table listing of the various nodes that are connected to the selected node. The link analysis information window is displayed in the following diagram. Note the large frequency counts that are based on all nodes, with the thickest lines within the graph connected to the selected SETOSA node. From the larger frequency counts, the Setosa iris wildflower can be mainly identified by having a petal length and width of no more than 19.93 and 4.368 centimeters along with a sepal length and width of no greater than 50.15 and 34.929 centimeters.

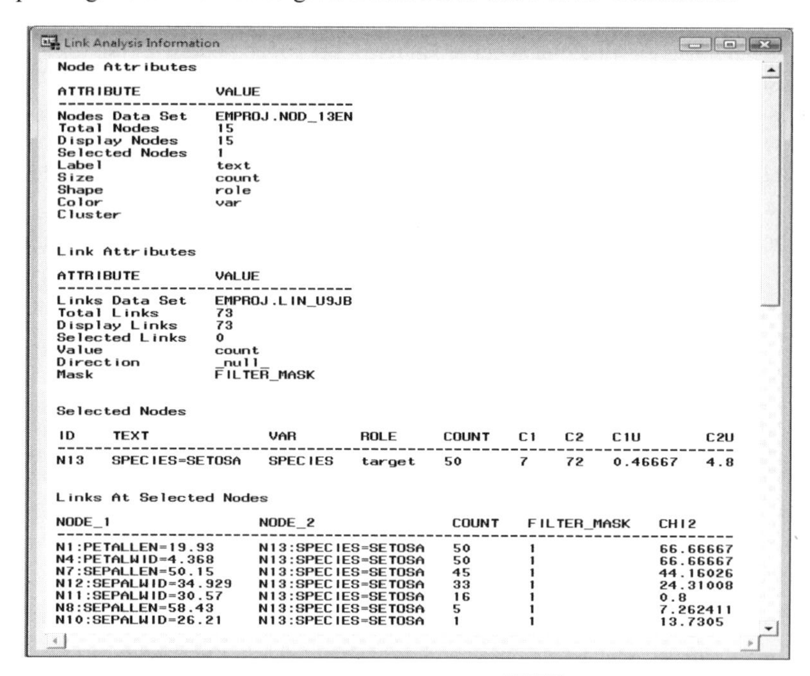

Link Analysis Information

Node Attributes

ATTRIBUTE	VALUE
Nodes Data Set	EMPROJ.NOD_13EN
Total Nodes	15
Display Nodes	15
Selected Nodes	1
Label	text
Size	count
Shape	role
Color	var
Cluster	

Link Attributes

ATTRIBUTE	VALUE
Links Data Set	EMPROJ.LIN_U9JB
Total Links	73
Display Links	73
Selected Links	0
Value	count
Direction	_null_
Mask	FILTER_MASK

Selected Nodes

ID	TEXT	VAR	ROLE	COUNT	C1	C2	C1U	C2U
N13	SPECIES=SETOSA	SPECIES	target	50	7	72	0.46667	4.8

Links At Selected Nodes

NODE_1	NODE_2	COUNT	FILTER_MASK	CHI2
N1:PETALLEN=19.93	N13:SPECIES=SETOSA	50	1	66.66667
N4:PETALWID=4.368	N13:SPECIES=SETOSA	50	1	66.66667
N7:SEPALLEN=50.15	N13:SPECIES=SETOSA	45	1	44.16026
N12:SEPALWID=34.929	N13:SPECIES=SETOSA	33	1	24.31008
N11:SEPALWID=30.57	N13:SPECIES=SETOSA	16	1	0.8
N8:SEPALLEN=58.43	N13:SPECIES=SETOSA	5	1	7.262411
N10:SEPALWID=26.21	N13:SPECIES=SETOSA	1	1	13.7305

By selecting the **Actions > Paths** main menu options, the **Paths** toolbar icon or the **Path...** pop-up menu item will result in the following **Find Paths...** window appearing.

- **Path:** This option will allow you to find the shortest path from one node to another that uses the SAS/OR PROC NEWFLOW procedure. From the graph, you must select two or more nodes from the link graph to access this option from the **Link Graph** pop-up menu that will then open the following **Find Paths...** window.

  The **Find Paths...** window will display two separate columns. The **Source** column is positioned to the left of the window and the **Sink** column is positioned to the right. The nodes listed in the **Source** column will display all links originating from the nodes selected within the corresponding column. The nodes selected in the **Sink** column will display the significant links that originate from the selected nodes displayed in the **Source** column along with the corresponding nodes selected in the same **Sink** column. From the **Links Values** display, you may select from the following three options to set the value of the links:

- **Similarity:** Count values of the links are calculated in PROC NETFLOW as follows:
  2 • (maximum count value of all the links) – (count of the specific link).
- **Distance:** Count values that are the actual frequency count of the number of links that are connected to the node.
- **Unary:** All count values of the links are assigned a value of one.

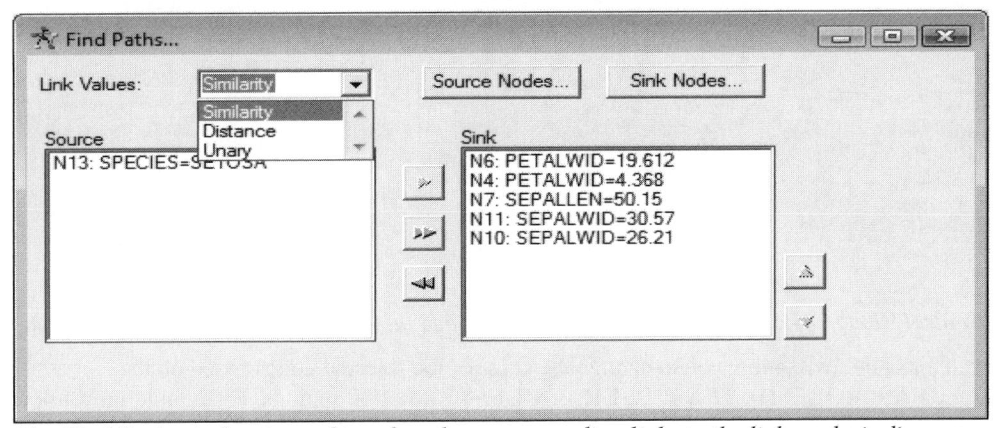

*The **Find Path** window is used to select the corresponding links to the link analysis diagram.*

Selecting the **Filter...** pop-up menu item will allow you to filter the various links connected to the displayed nodes within the link analysis diagram. The various links within the previously illustrated link analysis diagrams that were connected to the SPECIES=SETOSA node were displayed by selecting the **TEXT** pop-up option setting and selecting the SPECIES=SETOSA item from the **FILTER – Text** window.

- **Filter:** This option will result in the link graph only displaying the links that are connected to the selected nodes. When you select this option, you can filter the various links that are connected to the corresponding nodes by the following attributes of **VALUE, VAR, ROLE, ID, TEXT, USER**, or **CLUSTER**.

*The **FILTER-Text** window is used to display all links from the SETOSA node in the link graph.*

Selecting the **Actions > Display** main menu options, the **Display** toolbar icon or the **Display** pop-up menu item from the link graph will result in the subsequent **Display Control** window appearing. The window will allow you to change the layout of the link graph, set the display options for the selected nodes and links within the link graph, or create various cluster groupings for the displayed nodes.

Nodes tab

The **Nodes** tab will allow you to change the appearance of the selected node. This option will allow you to change the color, shape, size, and label of the selected nodes. The following are the various options that may be selected from the **Nodes** pop-up menu:

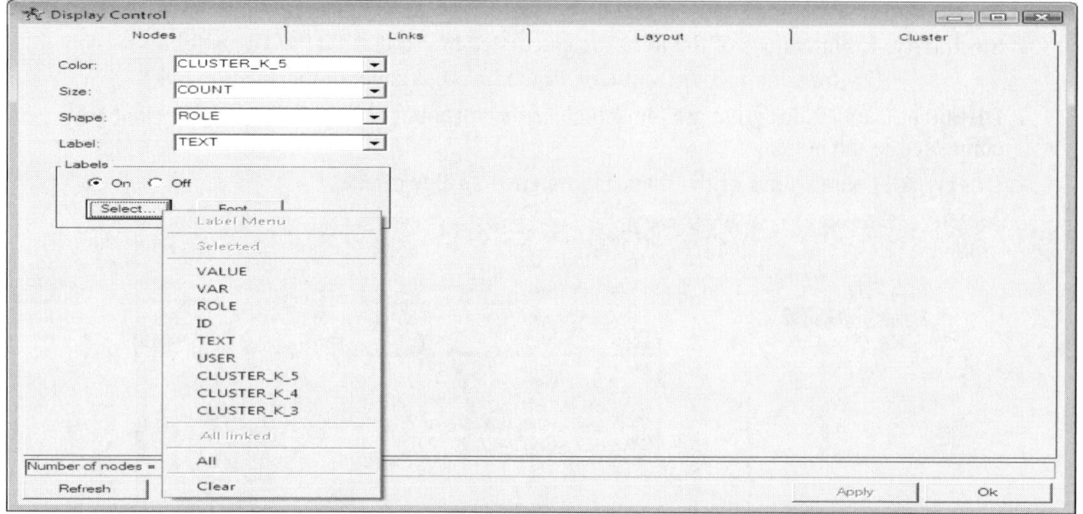

*The **Display Control** window and the **Nodes** tab are used to change the attributes of the nodes in the graph.*

- **Color:** This option will allow you to change the color of the selected nodes based on the **VALUE, VAR, ROLE, ID, TEXT, USER**, or **CLUSTER** option settings. By default, the **Link Analysis** node will automatically assign unique colors to each level of the variable that is specified. Selecting the **None** option will automatically change the color of all the nodes within the link graph to the same color. Again, the various attributes will be discussed in greater detail in the following **Grouping** option.
- **Shape:** This option will allow you to assign different shapes to the selected nodes based on the **VALUE, VAR, ROLE, ID, TEXT, USER**, or **CLUSTER** option settings. By default, the **Link Analysis** node will automatically assign different shapes to the nodes based on each unique level of the variable that is specified. By default, the nodes that represent input variables are displayed as boxes and the nodes that represent the target variable are displayed as circles. Selecting the **None** option will automatically change the nodes to all the same shape.
- **Size:** This option will allow you to assign different sizes to the nodes based on the **VALUE, VAR, ROLE, ID, TEXT, USER**, or **CLUSTER** item. The size of the node is based on either the first centrality measurement, **C1**; the second centrality measurement, **C2**; frequency count, **COUNT**; frequency percentages, **PERCENT**; and the (**X, Y**) coordinates. Select the **Uniform** option to change the nodes to the same size. The **None** option will change the size of the nodes into points. The default is the frequency count of the class level of the node.
- **Label:** This option will allow you to assign text or numeric labels to the selected nodes. The other options that can be applied are based on the **Label Text** option. The specific label that is applied is determined from the following **Label Text** option. Select the **All** option to label every node in the link graph. To clear all labels to the nodes, select the **Clear** option. From the graph, you can also select a node or any number of nodes by holding down the Ctrl button or dragging over the corresponding nodes to label all nodes that are connected to the selected node by selecting the **All linked** pop-up menu option.
- **Label Text:** This option will allow you to select a specific label to assign to the selected nodes based on the **VALUE, VAR, ROLE, ID, TEXT, USER**, or **CLUSTER** option settings that will be discussed in the following **Links tab** section.
- **Labels Section:** The **On** radio button is automatically selected that will allow you to assign a specific type of font to the label of the selected nodes. Selecting the **Font** button will open the **Font** window for you to assign the font, font style, and font size to the labels. Select the **Off** radio button to turn off the font attributes within the graph.

Links tab

From the **Links** tab, the following links options will allow you to change the attributes of the links between the displayed nodes, such as the number of links that are displayed, the value of the links, direction of the links and the mask.

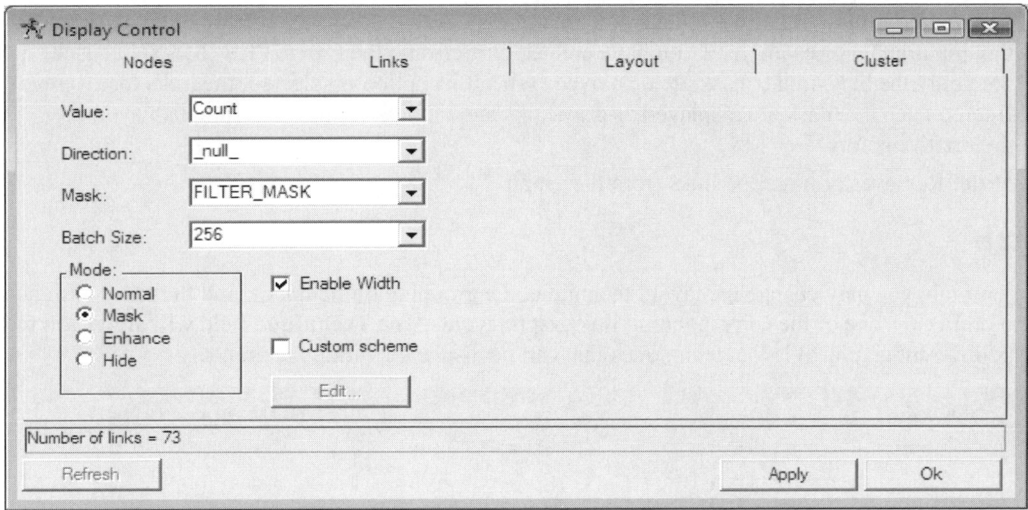

*The **Links** tab is used to set the various options for the displayed links within the link graph with the FILTER_MASK option setting selected along with the **Mask** option setting selected from the **Mode** section to display all links originating from the Setosa wildflower in the iris data set.*

- **Value:** This option will allow you to change the value of the selected links. In other words, this option will allow you to set the value that will be used in identifying the importance of the link. By default, the value of the links is based on the frequency counts, **Count**, which determines both the color and thickness of the links. This option will allow you to set the value of the link by using either the **Percent**, **Count1**, **Count2**, **Expect**, **Dev**, or **Chi2** option settings.
- **Direction:** This option will allow you to specify the direction of the links. By default, the links do not display any arrows to identify the direction of the links or connections between the nodes called *undirected connections*. However, this option will allow you to create *directed connections* by specifying the direction of the links based on the larger nodes that are connected to the smaller nodes. Therefore, this will allow you to visualize cyclical patterns that might be occurring in the link analysis diagram. The various directions to select from are the following: **Percent**, **Count1**, **Count2**, **Expect**, **Dev,** or **Chi2** option settings.
- **Mask:** This option will allow you to apply a mask to the selected variable. By default, no mask is applied. In other words, the **_null_** option is automatically selected, that is, no mask is applied. Assuming that you have created clusters, a path or a filter for the graph, then you may apply the **CLUSTER_MASK**, **PATH_MASK**, or **FILTER_MASK** option settings.
- **Batch Size:** Specifies the maximum number of links that are displayed at one time. You can choose from a maximum of 128, 256, 512, 1024, or 2048 links to be displayed in each graph.
- **Enable Width:** This option will allow you to display the width or thickness of the links from the **Value** option that you have selected. By default, the thickness of the links is based on the frequency count of the two separate levels of the connected nodes.
- **Custom Scheme:** By default, the check box is unselected. However, selecting the check box, then the **Edit** button that is displayed at the bottom of the tab, will open the **Link value** window. The option is designed to modify the attributes of the link such as the link cutoff values, colors, and widths of the corresponding link.

The **Mode** section is designed to define the way in which the links are displayed within the link graph. The following are the four separate ways to display the links within the link graph:

- **Normal:** Displays the normal links to the connected nodes within the graph.
- **Mask** (default)**:** Applies the link mask that you can specify from the previous **Mask** option.
- **Enhance:** Displays the filtered links that are not displayed when you have applied a mask to the link graph. For example, if you have created clusters and the **CLUSTER_MASK** is used, then only the links to the mask are displayed. When this option is selected, the links that were filtered from the mask are displayed as gray links and the links in the mask are displayed in their native colors.
- **Hide:** Removes or hides the links from the graph.

Layout tab

From the **Layout** tab, you may change the layout technique, the grouping of the nodes, and the value assigned to each node within each one of the corresponding link graph layouts. The **Technique** field will allow you to specify the six different graphical layout techniques that can be displayed within the graph.

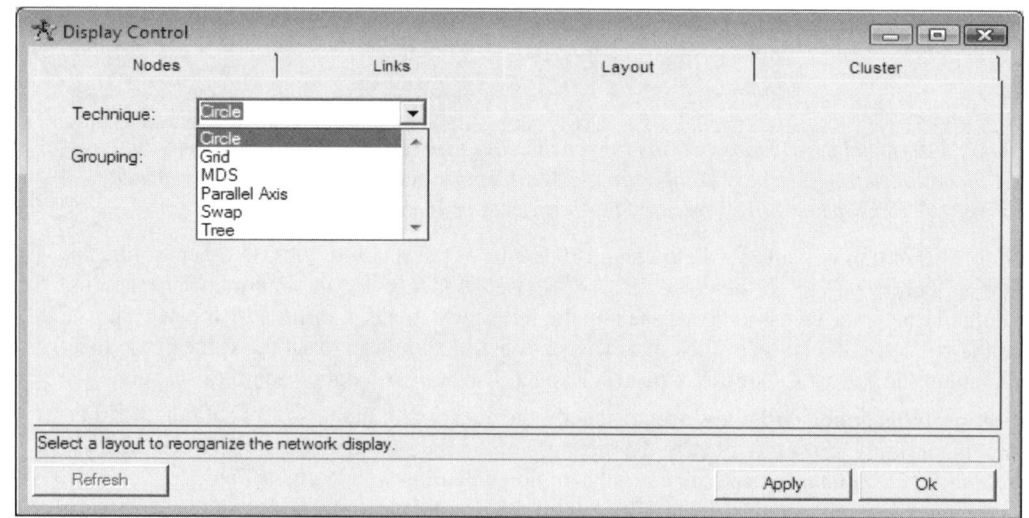

*The **Layout** tab is used to change the graphical layout of the displayed link analysis diagram.*

- **Technique:** This option will allow you to change the graphical layout of the link analysis diagram of either the **Circle**, **Grid**, **MDS**, **Parallel Axis**, **Swap**, or **Tree** layout designs.
- **Grouping:** This option is designed for you to change the way in which the displayed nodes are grouped together within the link graph. You may group the various nodes by the following attributes:
  - **VALUE:** The nodes are grouped within the link graph by its own values. For interval-valued variables, the nodes are grouped in ascending order in a counter clockwise direction within the link graph. For categorically-valued input, the nodes are grouped in alphabetical order within the graph.
  - **VAR** (default)**:** The nodes are grouped by their variable names for all the layout techniques.
  - **ROLE:** The nodes are separately grouped together within the link graph by their model role. When this option is selected, then the input variables are grouped together and the target variables are grouped together, and so on.
  - **ID:** The nodes are grouped together by the node ID identifier variable. The node id of *N#* is assigned to the node that is automatically created in the **Link Analysis** node.
  - **TEXT:** Nodes are grouped together by the text, that is, VAR = VALUE.

Cluster tab

The **Cluster** tab will allow you to cluster the NODES data set either interactively or using the following two nonparametric clustering techniques. The following are the various options that you may select from within the **Cluster** tab:

- **Select Clusters:** The drop-down menu option will list the cluster variable that you want to display in your link graph, either **VALUE**, **VAR**, **ROLE**, **ID**, **TEXT**, or **_NULL_**. **_NULL_** is the default. In Enterprise Miner, the nodes are not automatically clustered within the graph.
- **Create Clusters:** This section will allow you to cluster the NODES data set either interactively or using following two nonparametric techniques. Each node will represent a single degree of freedom. The following are the various clustering options that you may select from based on the **Create Clusters** option:
  - **Interactive:** This option will allow you to define the clusters by selecting the nodes that you want to include in the cluster that is created. The maximum number of clusters that can be created is the total number of nodes in the linked graph. Select the **Interactive** pop-up menu item, then select the **Run** button that will result in the **Interactive Clustering** window being displayed. The window is illustrated in the following page. From the nodes displayed within the **Unassigned** list box, select the corresponding nodes to create a unique clustering grouping within the link graph by selecting the arrow button that will result in the selected listed nodes displayed in the **User_1** list box. Selecting the **Create** button will result in the list box at the right being automatically cleared, which will allow you to create an entirely different set of nodes to create an entirely different cluster grouping within the link graph.

    Select the **Interactive > Copy** pop-up menu item that will open an additional pop-up menu in order to select the variables that you want to create the cluster. From this option, you may select the following pop-up menu items of either **Value**, **Var**, **Role**, **Id**, **Text**, or **User** to create a cluster for each level of the **Value**, **Var**, **Role**, **Id**, **Text**, or **User** variable.
  - **Modeclus:** This option performs nonparametric clustering that will enable you to cluster the selected nodes based on the following two separate nonparametric clustering methods. The following nonparametric clustering methods are calculated from the PROC MODECLUS procedure. Press the **Run** button that will result in a subsequent **Modeclus Options** window appearing in order for you to select the following two nonparametric clustering techniques.
    - **Kernal Density** (default)**:** This nonparametric technique determines the cluster groupings based on spherical kernals with a fixed radius. By default, the **Link Analysis** node initially defines two fixed radii to determine the optimal number of clusters. The results are generated from the PROC MODECLUS procedure. The procedure output listing will be displayed within the SAS system output window. The **Grouping** option setting will allow you to select the radii of the nonparametric clustering routine based on the listed **CLUSTER_R_#** variable, where # denotes the number of radii from the kernal density clustering technique. The **Link Analysis** node uses ten fixed radii to find the optimal number of clusters. From the iris data set, the **Grouping** option will allow you to select radii of either one or two by selecting the listed variable of **CLUSTER_R_1** or **CLUSTER_R_2**.
    - **Nearest Neighbor:** This nonparametric technique determines the cluster groupings that are created by the number of neighbors. The **Grouping** option setting will allow you to select the smoothing constant for the nonparametric technique based on the listed **CLUSTER_K_#** variable, where # denotes the number of nearest neighbors in the nearest neighbor clustering technique. The *k*-nearest neighbor clustering technique performs well in classifying observations into elongated elliptical clusters. The **Link Analysis** node uses ten different neighbor values to find the optimal number of clusters. From the iris data set, the **Grouping** option will allow you to select the number of nearest neighbors between three and five, that is, **CLUSTER_K_3**, **CLUSTER_K_4**, or **CLUSTER_K_5**.

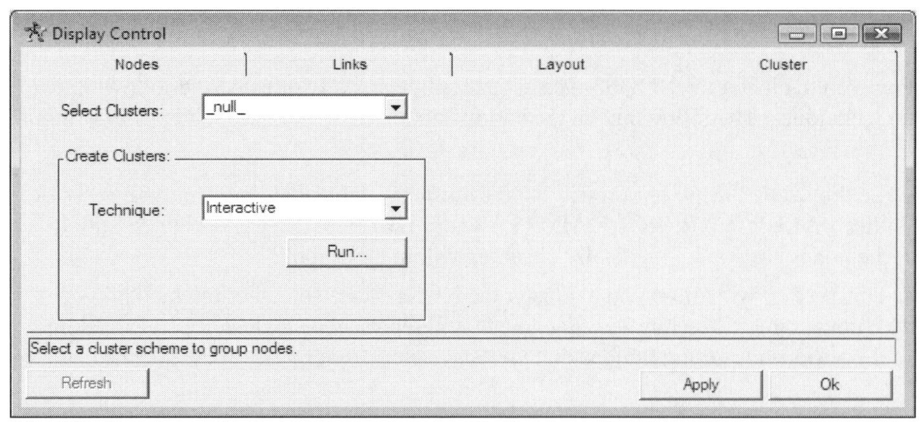

*The **Display Control** window is used to specify the clustering identifier and clustering technique.*

Selecting the **Interactive** clustering technique and pressing the **Run** button will display the following **Interactive Clustering** window to create the selected cluster groupings within the graph.

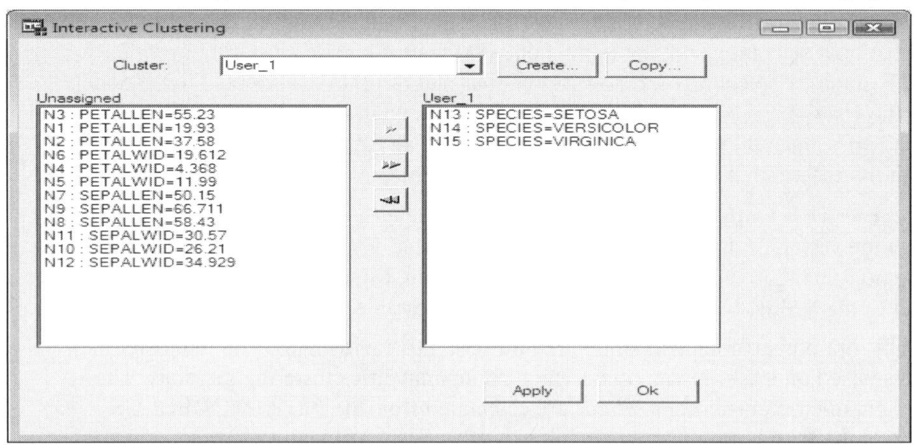

*The **Interactive Clustering** window creates cluster groups for the three iris wildflowers.*

Plots tab

The **Plots** tab is designed to create frequency bar charts that are based on the variables in the NODES or LINKS data sets. By default, the bar chart displays the frequency count of each input variable or the separate class levels of the categorically-valued target variable in the NODES data set. The frequency counts are displayed in descending order. The following bar chart displays the frequency distribution in the grouped nodes from the iris data set. For association analysis, the bar chart plot will display the number of unique items in descending order based on the various nodes that are displayed within the link graph.

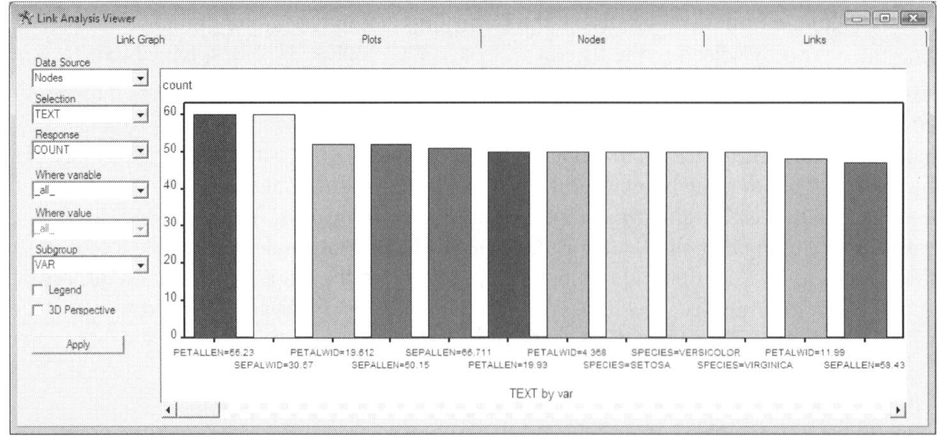

*The **Plots** tab displays the frequency bar chart of the displayed nodes in the link graph.*

The following are the various pull-down menu options to select from within the **Plots** tab:

- **Data Source:** This option will allow you to specify the NODES or LINKS data set to create the bar chart.
- **Selection:** This option will allow you to set the X-axis variable to the bar chart based on the following values of the **VALUE, VAR, ROLE, ID**, or **TEXT** option settings. The default is **VAR**.
- **Response:** This option will allow you to specify the Y-axis variable to the bar chart based on the following values of the **C1**, **C2**, **COUNT**, or **PERCENT** variable. The default is **COUNT**.
- **Where Variable:** This option will allow you to subset the attribute displayed along the X-axis of the bar chart based on either the **VALUE, VAR, ROLE, ID**, or **TEXT** option settings.
- **Where Value:** This option will allow you to set the values displayed within the bar chart based on the variable selected for the bar chart from the previous **Where Variable** option setting.
- **Subgroup:** This option will allow to create a stacked bar chart based on the following grouping option settings of **VALUE, VAR, ROLE, ID**, or **TEXT**. By default, no stack bar chart is displayed.
- **Legend:** By default, the check box is unselected. However, selecting the **Legend** check box and pressing the **Apply** button will result in the bar chart automatically displaying a legend at the bottom of the chart.
- **3D Perceptive:** Selecting this check box and pressing the **Apply** button will result in the bar chart automatically displaying a 3-D bar chart.

Nodes tab

The **Nodes** tab is designed to display a table view of the NODES data set. The tab will list the various nodes that are displayed within the link graph. By default, the table is sorted in alphabetical order by the variable name. In addition, the table listing is sorted in descending order by the frequency counts. The table displays the first-ordered weighted and unweighted centrality measurements, **C1** and **C1U**; the second ordered weighted and unweighted centrality measurements, **C2** and **C2U**; the value of the class variable or the cutoff point of the interval-valued variable that is split into three separate intervals, **VALUE**; the variable name of the node, **VAR**; the variable role, **ROLE**; the frequency count based on the level of the variable, **COUNT**; the frequency percentage of the node, **PERCENT**; the node identifier, **ID**; the variable text based on the **VAR=VALUE** option, **TEXT**; the X coordinate, **X**; and the Y coordinate, **Y**; of the node that is displayed within the link graph. Both the **Nodes** tab and the subsequent **Links** tab are designed so that you may click the selected column that will result in tab sorting of the selected column in ascending or descending order.

Link Analysis Viewer

Link Graph | Plots | Nodes | Links

C1U	C2U	C1	C2	VALUE	VAR	ROLE	COUNT	PERCENT	ID	TEXT	X	Y	PLOT_SHOWLABEL	FILTER_MASK	USER
0.6	6.3333333333	9	95	55.23	PETALLEN	input	60	40	N3	PETALLEN=55.23	374	199	1	0	Unassigne
0.4666666667	4.8	7	72	19.93	PETALLEN	input	50	33.333333333	N1	PETALLEN=19.93	360	132	1	0	Unassigne
0.6666666667	7.0666666667	10	106	37.58	PETALLEN	input	40	26.666666667	N2	PETALLEN=37.58	320	77	1	0	Unassigne
0.6666666667	7.0666666667	10	106	19.612	PETALWID	input	52	34.666666667	N6	PETALWID=19.612	261	43	1	0	Unassigne
0.4666666667	4.8	7	72	4.368	PETALWID	input	50	33.333333333	N4	PETALWID=4.368	194	36	1	0	Unassigne
0.6	6.3333333333	9	95	11.99	PETALWID	input	48	32	N5	PETALWID=11.99	129	57	1	0	Unassigne
0.7333333333	7	11	105	50.15	SEPALLEN	input	52	34.666666667	N7	SEPALLEN=50.15	79	103	1	0	Unassigne
0.6	6.2	9	93	66.711	SEPALLEN	input	51	34	N9	SEPALLEN=66.711	51	165	1	0	Unassigne
0.8	7.6	12	114	58.43	SEPALLEN	input	47	31.333333333	N8	SEPALLEN=58.43	51	233	1	0	Unassigne
0.8	7.4	12	111	30.57	SEPALWID	input	60	40	N11	SEPALWID=30.57	79	295	1	0	Unassigne
0.8	7.4	12	111	26.21	SEPALWID	input	47	31.333333333	N10	SEPALWID=26.21	129	340	1	0	Unassigne
0.7333333333	6.8	11	102	34.929	SEPALWID	input	43	28.666666667	N12	SEPALWID=34.929	194	361	1	0	Unassigne
0.4666666667	4.8	7	72	SETOSA	SPECIES	target	50	33.333333333	N13	SPECIES=SETOSA	261	354	1	1	Unassigne
0.6666666667	7	10	105	VERSICOLOR	SPECIES	target	50	33.333333333	N14	SPECIES=VERSICOLOR	320	320	1	0	Unassigne
0.6666666667	7	10	105	VIRGINICA	SPECIES	target	50	33.333333333	N15	SPECIES=VIRGINICA	360	265	1	0	Unassigne

*The **Nodes** tab displaying the table listing of the various nodes created within the NODES data set.*

Links tab

The **Links** tab is designed to display a table view of the LINKS data set. The tab will list the various links between the separate nodes that are displayed within the link graph. By default, the table is displayed in descending order by the number of links that occur within each node. The **Links** tab displays the number of times that the two separate nodes are connected, **COUNT**; the percentage of link counts, **PERCENT**; the marginal frequency counts of the class level from the first node, **COUNT1**; the identifier of the first node connected, **ID1**; the marginal frequency counts of the class level from the second node, **COUNT2**; the node identifier for the second node connected, **ID2**; the link identifier, **LINKID**; and the expected frequency count between the two separate class levels of the connected nodes, **EXPECTED**; based on the two-way contingency table, the deviation from the expected count, **DEV**; the partial chi-square value between the two separate class levels of the connected nodes, **CHI2**; and the layout mask that is applied to the link, **LAYOUT_MASK**.

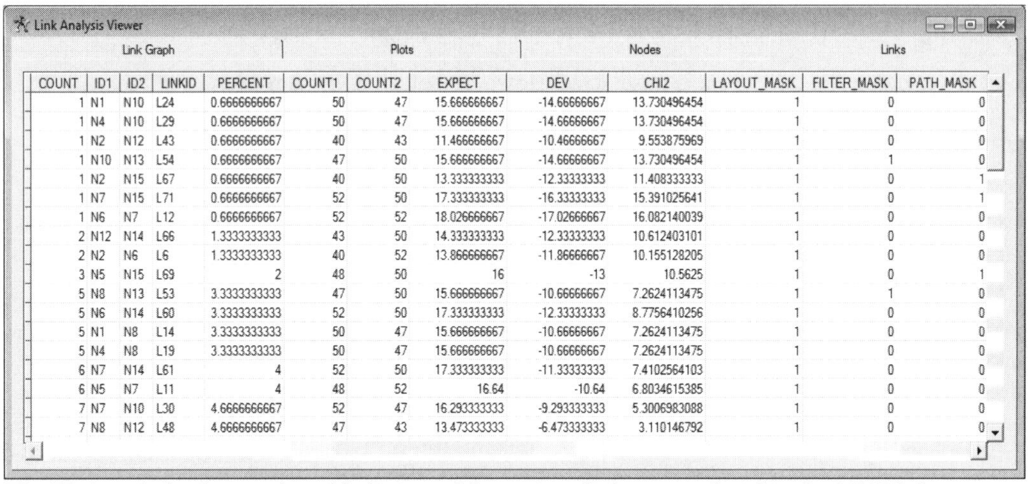

Link Analysis Viewer

Link Graph | Plots | Nodes | Links

COUNT	ID1	ID2	LINKID	PERCENT	COUNT1	COUNT2	EXPECT	DEV	CHI2	LAYOUT_MASK	FILTER_MASK	PATH_MASK
1	N1	N10	L24	0.6666666667	50	47	15.666666667	-14.66666667	13.730496454	1	0	0
1	N4	N10	L29	0.6666666667	50	47	15.666666667	-14.66666667	13.730496454	1	0	0
1	N2	N12	L43	0.6666666667	40	43	11.466666667	-10.46666667	9.553875969	1	0	0
1	N10	N13	L54	0.6666666667	47	50	15.666666667	-14.66666667	13.730496454	1	1	0
1	N2	N15	L67	0.6666666667	40	50	13.333333333	-12.33333333	11.408333333	1	0	1
1	N7	N15	L71	0.6666666667	52	50	17.333333333	-16.33333333	15.391025641	1	0	1
1	N6	N7	L12	0.6666666667	52	52	18.026666667	-17.02666667	16.082140039	1	0	0
2	N12	N14	L66	1.3333333333	43	50	14.333333333	-12.33333333	10.612403101	1	0	0
2	N2	N6	L6	1.3333333333	40	52	13.866666667	-11.86666667	10.155128205	1	0	0
3	N5	N15	L69	2	48	50	16	-13	10.5625	1	0	1
5	N8	N13	L53	3.3333333333	47	50	15.666666667	-10.66666667	7.2624113475	1	1	0
5	N6	N14	L60	3.3333333333	52	50	17.333333333	-12.33333333	8.7756410256	1	0	0
5	N1	N8	L14	3.3333333333	50	47	15.666666667	-10.66666667	7.2624113475	1	0	0
5	N4	N8	L19	3.3333333333	50	47	15.666666667	-10.66666667	7.2624113475	1	0	0
6	N7	N14	L61	4	52	50	17.333333333	-11.33333333	7.4102564103	1	0	0
6	N5	N7	L11	4	48	52	16.64	-10.64	6.8034615385	1	0	0
7	N7	N10	L30	4.6666666667	52	47	16.293333333	-9.293333333	5.3006983088	1	0	0
7	N8	N12	L48	4.6666666667	47	43	13.473333333	-6.473333333	3.110146792	1	0	0

*The **Links** tab displaying the table listing of the various links created within the LINKS data set.*

Scoring the Clustering Link Results

The **Scoring Options** window is designed to create score code by selecting the interactive or nonparametric clustering options that can be specified from the **Clusters** tab. The scored data set can only be created by performing interactive or nonparametric clustering within the tab. Select the **Action > Create Scorecode** main menu options or select the **Create Scorecode**  toolbar icon from the tools bar menu and the following **Scoring Options** window will appear.

Select the drop-down arrow button from the **Scores** display field that will create the score code based on the nodes that are created within the clusters, the links that are created within the clusters, or the nodes and links that are created within the clusters. The scoring code generates a numeric cluster score variable for each cluster along with a segment identifier variable with values equal to the name of the cluster that is determined by the highest numeric cluster score. These cluster indicator variables can then be used in subsequent nodes within the process flow diagram as categorically-valued input variables within the subsequent modeling nodes. The clustering scores from the nodes are the proportion of times that the range of input values fall within the target class level that it has been grouped within the **Link Analysis** node. In other words, the cluster scores from the nodes are calculated by the accumulated frequency counts of the nodes in which the range of input values fall within the target class level that it has been grouped within the **Link Analysis** node, divided by the total frequency counts of the nodes that are assigned within each target group. The clustering scores from the links are calculated by the accumulated linked frequency counts in which the range of input values between the two separate linked nodes fall within the target class level that it has been grouped within the **Link Analysis** node, divided by the total frequency count of the linked nodes within each target category. And finally, the clustering scores from the nodes and links are calculated by the accumulated frequency counts from the nodes and links in which the range of values of the input variables between the separate nodes and two separate linked nodes fall within the target class level that it has been grouped, divided by the total frequency count of the nodes and linked nodes within each target category. By default, the clustering assignments are determined by all links that creates the segmentation identifier variable in the scored data set. The various score code can be viewed from my website.

- **Include masked links:** The generated score code will include all masked links. This option is grayed-out and unavailable when you select the **Nodes** drop-down menu option.
- **Create segmentation:** This option creates the segmentation identifier variable within the output scored data sets. This option must be selected to check the accuracy of the clusters that are created within the node and the various target groups by constructing two-way frequency tables.
- **Include target:** Includes the target variable to the output scored data sets.

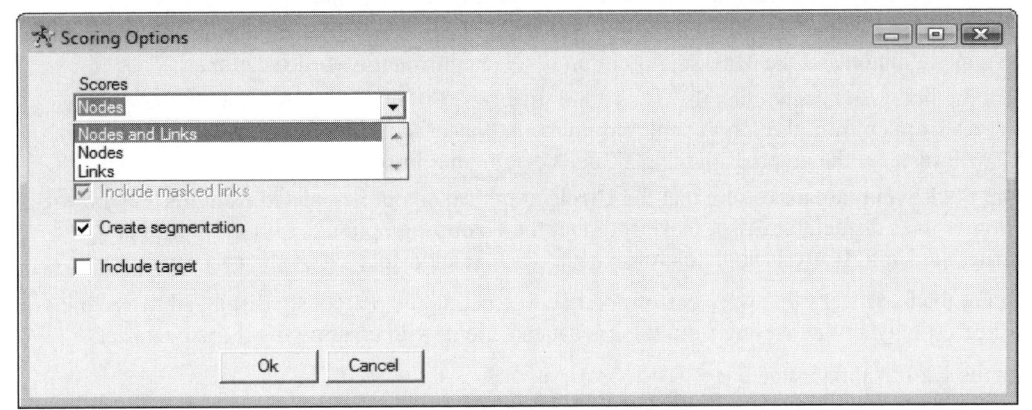

*The **Scoring Options** window is used to specify the way in which to score the partitioned data sets from the clustering nodes that are created within the link graph.*

Select the **Actions > Run Scorecode** main menu options or select the **Run Scorecode** toolbar icon that will result in scoring the corresponding partitioned data sets. A **Message** window will appear that will display a message confirming that you will score the output data set. Select the **Yes** button to score the output data sets. A subsequent **Message** window will appear to inform you that the output data sets have been scored, then select the **OK** button. The output data set will be passed along to subsequent nodes within the process flow. The **Insight** node can be used to view the scored output data set. The scored data set can be used in evaluating the accuracy of the results by creating a two-way frequency table between the categorically-valued target variable, Species, and the segmentation identifier variable, Segmentation. The validation data set might be applied in order to generate an honest assessment of the accuracy of the clustering assignments. Furthermore, the scored data sets from the nodes and links might be compared with the actual target values to determine which clustering assignments performed the best.

Data set details

Information Table View

Variable labels

	Sepal Length (mm.)	Sepal Width (mm.)	Petal Length (mm.)	Petal Width (mm.)	Species	Cluster M2	Cluster M1	Cluster M3	Segmentation
1	50	33	14	2	Setosa	0	1	0	M1
2	64	28	56	22	Virginica	0.730941704	0	0.2582417582	M2
3	65	28	46	15	Versicolor	0.2286995516	0	0.7417582418	M3
4	67	31	56	24	Virginica	1	0	0	M2
5	63	28	51	15	Virginica	0.4977578475	0	0.521978022	M3
6	46	34	14	3	Setosa	0	1	0	M1
7	69	31	51	23	Virginica	1	0	0	M2
8	62	22	45	15	Versicolor	0	0	1	M3
9	59	32	48	18	Versicolor	0.7713004484	0	0.2582417582	M2
10	46	36	10	2	Setosa	0	1	0	M1
11	61	30	46	14	Versicolor	0.269058296	0	0.7417582418	M3
12	60	27	51	16	Versicolor	0.5022421525	0	0.5164835165	M3
13	65	30	52	20	Virginica	1	0	0	M2
14	56	25	39	11	Versicolor	0	0	1	M3
15	65	30	55	18	Virginica	1	0	0	M2

The scored training data set with the clustering assignments from the MDS layout that are determined by the displayed nodes with the number of nearest neighbors set to five in the iris data set. Notice that observation number 5 and 9 have misclassified the wildflower variety from the cluster scores.

The Steps in Creating the Previous Link Graphs

The following are the basic steps that were performed in constructing the previously displayed link graphs:

1. Run the **Link Analysis** node.
2. From the **Result Browser**, select the **View > Detailed Results** pull-down menu options.
3. From the link graph, right-click the mouse and select the **Display** pop-up menu item.
4. From the **Nodes** tab, select the TEXT pull-down menu item from the **Label** option. From the **Labels** section, press the **Select…** button and make sure that the **On** radio button is selected. Press the **Select…** button, then select the TEXT pop-up menu item. From the **LABEL** window, select all the listed variables that will label each node in the link graph.
5. From the **Link** tab, make sure the FILTER_MASK option setting is selected from the **Mask** pull-down menu option and the **Mask** radio button is selected from the **Mode** section.
6. From the link graph, right-click the mouse and select the **Filter** pop-up menu item, then select the TEXT option from the subsequent pop-up menu. Select the SPECIES=SETOSA menu item. This will result in the graph displaying all links originating from the Setosa wildflower only.
7. From the **Layout** tab, make sure that the **Circle** graphical layout is selected from the **Technique** option. This is the default. Also, make sure that the **Grouping** option is set to VAR. This will instruct the **Link Analysis** node to group the nodes that are displayed by their variable names.

The following are the basic steps that were performed in constructing the previously displayed MDS link graph based on five-nearest neighbor clustering from the iris data set along with creating the scored data set.

1. Run the **Link Analysis** node.
2. From the **Result Browser**, select the **View > Detailed Results** pull-down menu options.
3. Right-click the mouse and select the **Display** pop-up menu item.
4. Select the **Nodes** tab. From the **Label** option, select the TEXT pull-down menu item to create labels to the displayed nodes by the text labeling of VAR = VALUE.
5. From the **Layout** tab, select the **MDS** graphical layout from the **Technique** option. Also, make sure that the **Grouping** option is set to VAR, that is, a cluster grouping scheme of VAR. This will instruct the node to group together the nodes by their variable names. Select the **Run** button.
6. From the **MDS Options** window, set the **Grouping Bias** option to –2. This will instruct Enterprise Miner to group the nodes further apart.
7. View the graph and the various target groupings that are created.
8. Right-click the mouse and select the **Display** pop-up item. Select the **Cluster** tab. From the **Create Clusters** section, select the **Modeclus** pop-up menu item from the **Technique** option.
9. Select the **Run** button. This will open the **Modeclus** option window to select the nonparametric cluster technique. Select the **Nearest Neighbor** pop-up menu item. Press the **OK** button. A separate window will appear to confirm to you that five-nearest neighbor clustering was performed with the CLUSTER_K_5 grouping variable created. Press the **OK** button that will then display the previous **Display Control** window.
10. From the **Select Clusters** option, select the smoothing constant of either 3, 4 or 5, that is, CLUSTER_K_3, CLUSTER_K_4, or CLUSTER_K_5 pop-up menu items. A message window will appear that will display the name of the optimal cluster variable.
11. From the **Link** tab, make sure the CLUSTER_MASK option setting is selected from the **Mask** pull-down menu option and the **Mask** radio button is selected from the **Mode** section to display all links within each cluster grouping.
12. To score the clustering results, select the **Create Scorecode** toolbar icon or select the **Action > Create Scorecode** main menu options that will display the **Scoring Option** window.
13. From the **Scoring Option** window, select the cluster score variable based on either the displayed links or nodes. By default, the link identifiers are written to the scored data set. Press the **OK** button that will create the score code based on the grouping variable of CLUSTER_K_5.
14. From the link graph, select the **Run Scorecode** toolbar icon that will create the scored data set.
15. Open the **Link Analysis** node, then select the **Output** tab to view the scored data set.

Chapter 3

Modify Nodes

Chapter Table of Contents

3.1 Data Set Attributes Node

General Layout of the Enterprise Miner Data Set Attributes Node

- **Data tab**
- **Variables tab**
- **Class Variables tab**
- **Notes tab**

The purpose of the **Data Set Attributes** node in Enterprise Miner is designed to modify the attributes of the metadata sample that is associated with the input data set such as the data set name, description, and the role to the corresponding data mining data set. The node is also designed for you to modify the variable role, and variable labels, measurement levels of the variables in the data mining data set, and set the ordering level of the categorical variables in the data set. In Enterprise Miner, the **Data Set Attributes** node is one of the few nodes that will allow you to assign variable roles to the variables in the process flow diagram. Typically, the **Input Data Source** node is used to create the input data set for the process flow. Alternatively, the **SAS Code** node can be used to create the input data set by creating the exported raw data set, followed by the **Data Set Attributes** node that is used to assign the appropriate roles to the variables in the exported raw data set. From the node, you may also prevent the data mining data set or certain variables in the data set from being passed along to the subsequent nodes in the Enterprise Miner process flow diagram. The node can also set the target profiles for the categorically-valued target variable in the input data set. In addition, the node can also delete variables from the Enterprise Miner process flow diagram. From the **Data** tab, the node will allow you to export the SAS data mining data set into many other file formats using the SAS Export Wizard. The restriction to the **Data Set Attributes** node is that it must follow the node that creates the data set to modify its metadata information. In addition, one and only one node can be connected to the **Data Set Attributes** node in the process flow diagram.

Data tab

Opening the **Data Set Attributes** node will result in the following **Data** tab appearing. The **Data** tab controls the exportation, role, and description assigned to the output data set. The node is designed for you to control the status of exportation of the data set to be passed along to the subsequent Enterprise Miner nodes. By default, all data sets are exported to the subsequent nodes in the process flow. To prevent the selected data set from being passed along to the process flow diagram, select the data set row and scroll over to the **Export** column, then right-click the mouse and select the **Set Export** pop-up menu item to select the **No** pop-up menu option. If several data sets are set with an exportation status of **Yes**, then the node defaults to the first listed data set.

From the SAS **Export Wizard**, you may also export the input data set into different file formats. Select the **File > Export** main menu option to export the data mining data set. You may also specify an entirely different role for the selected data set from the **New Role** column. Simply select the **New Role** cell to set the appropriate role to the active training data set from the **Set New Role** pop-up menu item. The various roles to assign to the training data set are listed below:

- **Raw**: The input data set for the data mining analysis, that is, the input data set used to score new data from the various modeling nodes, the input data set used to assign the various clustering assignments from the **Clustering** node, the input data set used to transform variables from the **Transform Variables** node, the input data set used to replace or impute data from the **Replacement** node, and so on.
- **Train**: The data set that is used in data mining analysis, that is, the data set used to fit the predictive model.
- **Validate**: Since the training data set will often be overly optimistic in determining the accuracy of its own results in comparison to a new sample drawn, the validation data set is applied that will provide an unbiased estimate in determining the accuracy of the corresponding results, analogous to performing the analysis on a new set of values. For example, the **Tree** and **Neural Network** nodes have the propensity of overfitting the training data set. Therefore, the validation data set is automatically used during modeling

assessment in order to prevent these nodes from overfitting the training data set.

- **Test**: This data set might be used to validate the assessment results since the validation data set can, at times, generate inaccurate generalization results. A test data set might be used in comparing the consistency in the results between the separate data sets and obtaining unbiased estimates in measuring the accuracy in the data mining results. For example, this additional data set can be used during the modeling assessment process in order to create unbiased prediction estimates, where the data set is entirely separate from the data that generated the prediction estimates in evaluating the performance of the predictive model.
- **Score**: Used to score a new set of values from the results that may or may not contain the target variable.
- **Predict**: Output data set that contains the fitted values from the predictive model that is required in user-defined modeling.
- **Rulegen**: Output data set generated from the data mining PROC RULEGEN procedure and the **Association** node by performing either association discovery analysis or sequence discovery analysis that lists the various items that are related to each other along with the various evaluation criterion statistics such as the frequency count, support, confidence and lift values.
- **Sequence**: Output data set of the various of items that are related to each other within a sequence along with the corresponding frequency counts generated from the data mining PROC SEQUENCE procedure and the **Association** node by performing sequence discovery analysis.
- **Seloutput**: Output data set that contains the outputted rules from the **Association** node results browser.
- **Assoc**: Output data set of the various items that are related to each other and the corresponding frequency counts from the PROC ASSOC procedure and the **Association** node by performing association discovery.
- **Result**: The data set of the results.
- **Estimate**: Output data set that contains the parameter estimates from the predictive model.
- **Statistic**: Output data set that contains the goodness-of-fit statistics from the predictive model.
- **Train_Sparse:** Contains information, that is, parsed term, term number, document number, and frequency of a term in a document of the parsed text from the training data set in the text-mining application.
- **Validate_Sparse**: Contains information, that is, parsed term, term number, document number, and frequency of a term in a document of the parsed text from the validation data set in the text-mining application.
- **Test_Sparse:** Contains information, that is, parsed term, term number, document number, and frequency of a term in a document, of the parsed text from the test data set in the text-mining application.

In order to access the following tab within the node, you must first select or highlight the appropriate data set. Assuming that there is more than one data set listed within the following **Data** tab, then you may select the appropriate data set and view the listed variables within the following **Variables** tab in order to set the following roles to the variables in the active data set. The data set description or the data set label may be change within the tab under the **New Description** column.

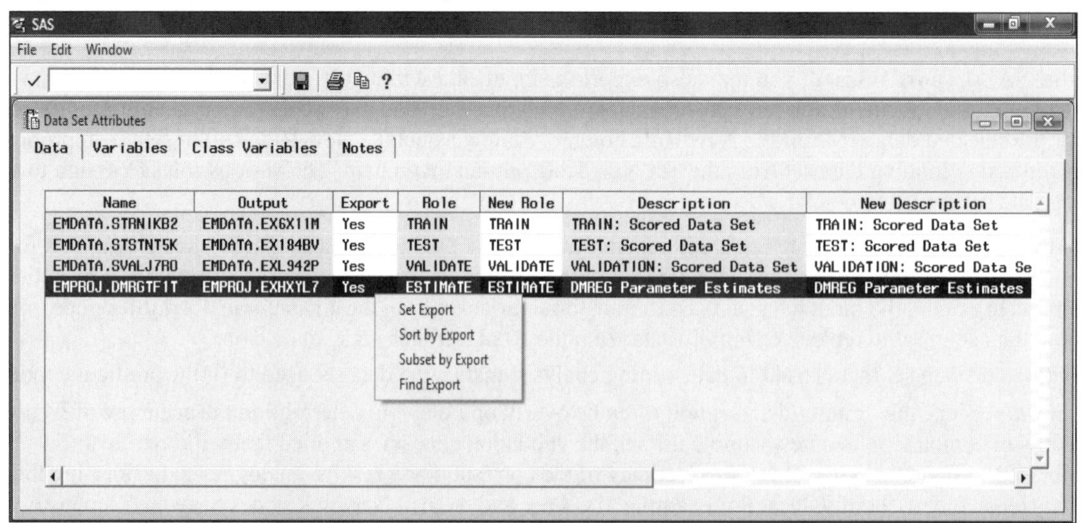

*The **Data** tab is used to control the exportation of the selected data to be passed to the subsequent nodes.*

Variables tab

The **Variables** tab is designed to view the corresponding variables based on the selected data set specified from the previous **Data** tab. From the tab, you may change the variable roles, level of measurements, and variable labels of the list of existing variables in the data mining data set. The node is also designed for you to prevent certain variables from being passed along to the subsequent nodes in the process flow design. In addition, the tab will also allow you to edit the target profile of the target variable and view the frequency distribution of the listed variable from the metadata sample.

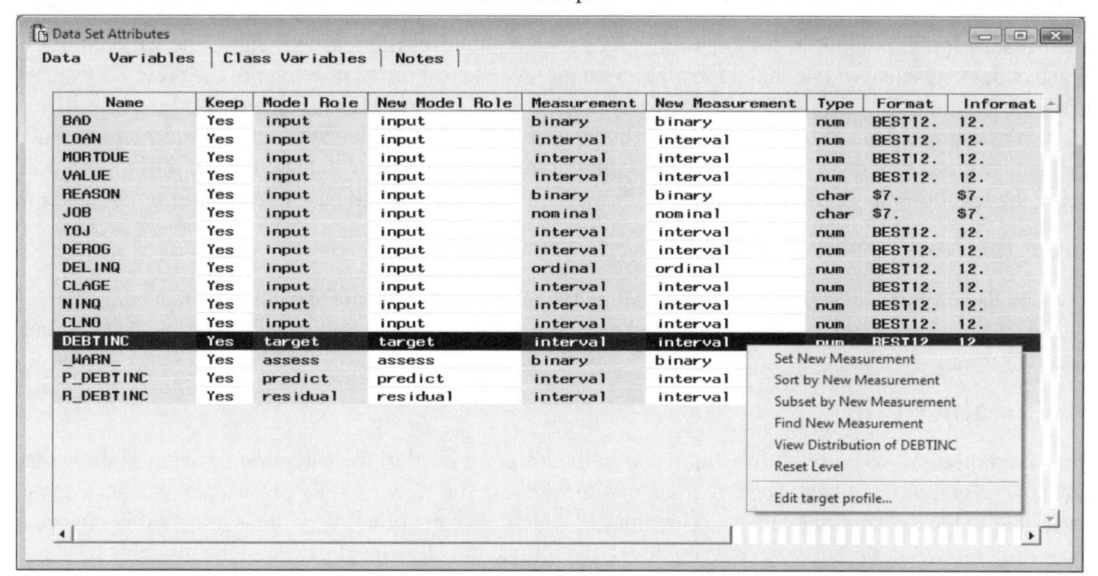

Name	Keep	Model Role	New Model Role	Measurement	New Measurement	Type	Format	Informat
BAD	Yes	input	input	binary	binary	num	BEST12.	12.
LOAN	Yes	input	input	interval	interval	num	BEST12.	12.
MORTDUE	Yes	input	input	interval	interval	num	BEST12.	12.
VALUE	Yes	input	input	interval	interval	num	BEST12.	12.
REASON	Yes	input	input	binary	binary	char	$7.	$7.
JOB	Yes	input	input	nominal	nominal	char	$7.	$7.
YOJ	Yes	input	input	interval	interval	num	BEST12.	12.
DEROG	Yes	input	input	interval	interval	num	BEST12.	12.
DELINQ	Yes	input	input	ordinal	ordinal	num	BEST12.	12.
CLAGE	Yes	input	input	interval	interval	num	BEST12.	12.
NINQ	Yes	input	input	interval	interval	num	BEST12.	12.
CLNO	Yes	input	input	interval	interval	num	BEST12.	12.
DEBTINC	Yes	target	target	interval	interval	num	BEST12.	12.
WARN	Yes	assess	assess	binary	binary			
P_DEBTINC	Yes	predict	predict	interval	interval			
R_DEBTINC	Yes	residual	residual	interval	interval			

*The scored data set from the multiple linear regression model that consists of the input variables with the predicted values and residuals from the interval-valued target variable. The **Variables** tab will allow you to set variable roles and level of measurements of the variables in the data set.*

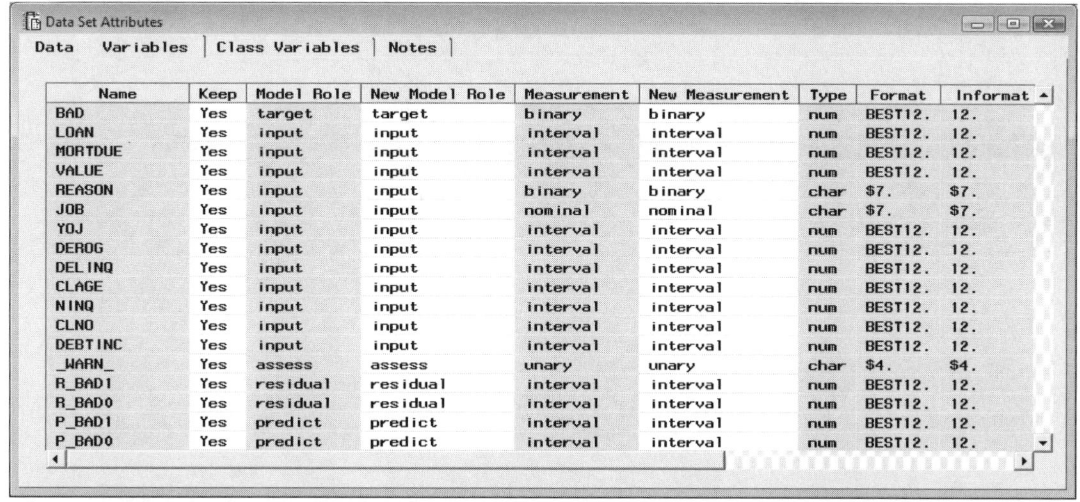

Name	Keep	Model Role	New Model Role	Measurement	New Measurement	Type	Format	Informat
BAD	Yes	target	target	binary	binary	num	BEST12.	12.
LOAN	Yes	input	input	interval	interval	num	BEST12.	12.
MORTDUE	Yes	input	input	interval	interval	num	BEST12.	12.
VALUE	Yes	input	input	interval	interval	num	BEST12.	12.
REASON	Yes	input	input	binary	binary	char	$7.	$7.
JOB	Yes	input	input	nominal	nominal	char	$7.	$7.
YOJ	Yes	input	input	interval	interval	num	BEST12.	12.
DEROG	Yes	input	input	interval	interval	num	BEST12.	12.
DELINQ	Yes	input	input	interval	interval	num	BEST12.	12.
CLAGE	Yes	input	input	interval	interval	num	BEST12.	12.
NINQ	Yes	input	input	interval	interval	num	BEST12.	12.
CLNO	Yes	input	input	interval	interval	num	BEST12.	12.
DEBTINC	Yes	input	input	interval	interval	num	BEST12.	12.
WARN	Yes	assess	assess	unary	unary	char	$4.	$4.
R_BAD1	Yes	residual	residual	interval	interval	num	BEST12.	12.
R_BAD0	Yes	residual	residual	interval	interval	num	BEST12.	12.
P_BAD1	Yes	predict	predict	interval	interval	num	BEST12.	12.
P_BAD0	Yes	predict	predict	interval	interval	num	BEST12.	12.

The scored data set from the logistic regression model that consists of the input variables and the predicted probabilities and residuals at each class level of the categorically-valued target variable.

Retaining the Variables into the Subsequent Nodes

The **Keep** column is designed for you to prevent certain variables from being passed along to the subsequent nodes in the process flow. Select the variable row to set the corresponding status, then scroll over to the **Keep** column and right-click the mouse to select the **No** pop-up item to remove the variable from the analysis. However, several variables can be selected simultaneously by dragging the mouse over the adjacent variable rows or by selecting either the Shift key or the Ctrl key.

Assigning the New Variable Role

The **New Variable Role** column is designed for you to specify an entirely different variable role. From the tab, select the variable row to change the variable role, then scroll over to the **New Variable role** column and right-click the mouse to select the **Set New Role Model** pop-up menu item. You may also reset and undo the specified variable role of the variable by selecting the **Reset Role** pop-up menu item. The numerous model roles that can be assigned to the variables are documented in the previous **Input Data Source** node.

Assigning the New Measurement Level

The **New Measurement** column is designed for you to change the level of measurement of the variable. From the tab, select the variable row to change the measurement level, scroll the mouse over to the **New Measurement** column, then right-click the mouse and select the **Set New Measurement** pop-up menu item that will display a pop-up list of measurement attributes to select from in order to change the current level of measurement assigned to the variable. Similar to changing the variable role, you have the flexibility of undoing or resetting the level of measurement assigned to the variable by selecting the **Reset Level** pop-up menu item.

Assigning the New Variable Labels

The **New Labels** column is designed for you to change the label of the variable. To change the label of the variable, select the corresponding variable row and enter the appropriate variable label within the **New Label** cell.

Class Variables tab

The **Class Variables** tab is designed for you to assign the ordering level of the categorical variables in the data mining data set. By default, the tab is grayed-out and unavailable for viewing if there isn't any categorically-valued variable in the data mining data set. The order in which the categorical variable is assigned is important in classification modeling because the ordering level determines the class level in which the model is trying to fit. By default, the categorically-valued target variable is sorted in descending order and the corresponding categorically-valued input variables are sorted in ascending order. The **Values** column will allow you to view the number of class levels of each categorically-valued variable in the data set.

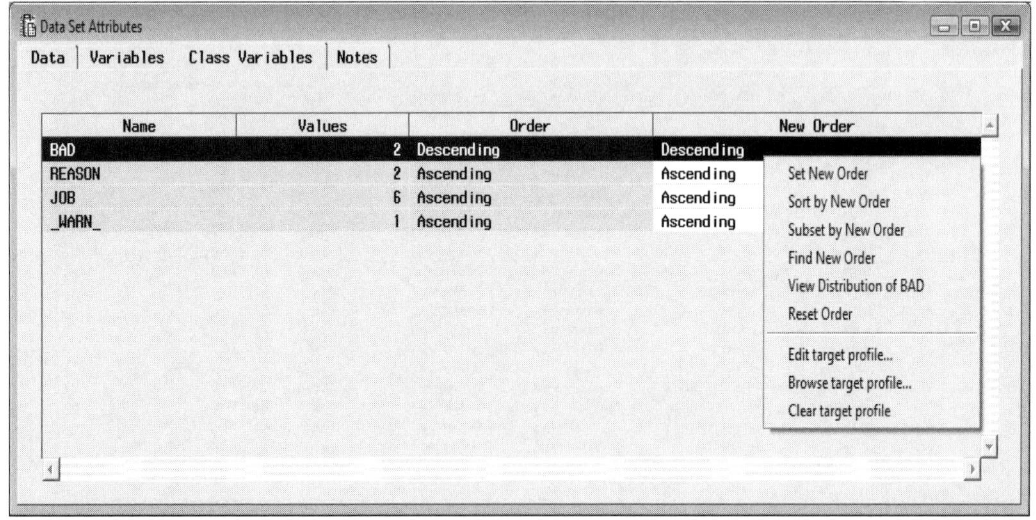

*The **Class Variables** tab is used to set the ordering level of the categorical variables in the data set.*

Assigning the Ordering Level

By default, the ordering level of the target variable is sorted in descending order as opposed to the other input variables in the training data set that are sorted in ascending order. In classification modeling, the ordering level of the target variable is important in interpreting the class level that the model is predicting. Furthermore, the ordering level of the categorically-valued input variable of the logistic regression model is important in determining the reference class level to the categorically-valued input variable that plays the role of the intercept term for the model. This will result in logistic regression model comparing the odds of the target event for each class level of the categorically-valued input variable against the assigned reference class level that is the last class level. In modeling assessment, knowing the correct ordering level is important in determining the class level in which the classification model is trying to fit, which will make it much easier for you to interpret the various lift charts that are created from the **Assessment** node. The reason is because the lift charts are based on the estimated probabilities of the target event that are sorted in descending order across various fixed intervals. In decision modeling, it is important to know the target event in interpreting the development of the tree branching process. In two-stage modeling, it is important to correctly set the target event of the categorically-valued target variable that the classification model is predicting, where the variable that represents the fitted values of the target event is used as one of the input variables for the subsequent prediction model. In interactive grouping, it is important to know the event level of the binary-valued target variable in interpreting the criterion statistics that are used in measuring the predictive power of the input variable.

The ordering level of the variables may be changed by selecting the variable row, then scrolling the mouse over to the **New Order** column and right-clicking the mouse to select the **Set New Order** pop-up menu item. A subsequent pop-up menu will appear for you to select the following ordering levels to the listed variable:

- **Ascending:** Sorts the class levels of the categorical variable by the lowest level first. Ascending order is the default for the input variables in the statistical modeling design. This option will allow you to determine the reference level of the categorical input variable that is a part of the intercept term of the classification model.
- **Descending:** Sorts the class levels of the categorical variable by the highest level first. Descending order is the default for the target variable in the classification modeling design. For binary-valued target variables with zero or one values, the target event is usually set at one and typically assigned to the class level of the target event with the target nonevent typically defined as zero.
- **Formatted ascending:** Sets the lowest level of the categorical variable based on the formatted value that is important in certain categorical analysis statistics such as logistic regression modeling in predicting the target event or the Cochran Mantel Haenszel chi-square test that depends on the ordering level of the categorical variables in the analysis.
- **Formatted descending:** Sets the highest level of the categorical variable from the formatted values.

Note: Similar to the other option settings within the node, you have an added option in resetting the ordering level of the categorical variables to their original order by selecting the **Reset Order** pop-up menu item.

Notes tab

The **Notes** tab is designed for you to enter notes within the **Data Set Attribute** node.

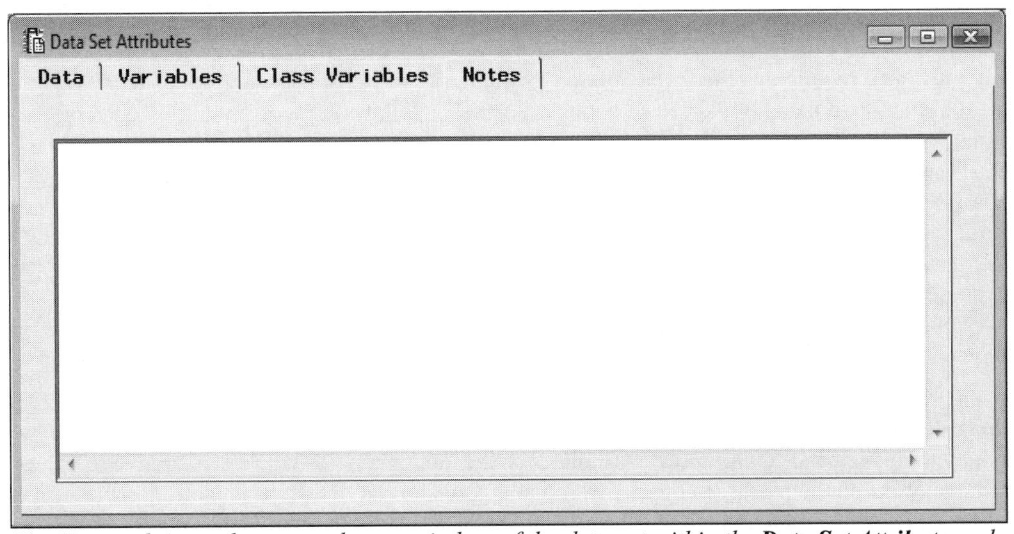

*The **Notes** tab is used to enter short reminders of the data set within the **Data Set Attribute** node.*

3.2 Transform Variables Node

General Layout of the Enterprise Miner Transform Variables Node

- **Data tab**
- **Variables tab**
- **Output tab**
- **Notes tab**

The purpose of the **Transform Variables** node is designed to create new variables by transforming the interval-valued variables that exist in the active training data set. The node supports user-defined formulas from numerous built-in functions that are available in SAS to transform the variable. The node will allow you to select from some of the most common transformations such as the log, square root, inverse, squared, exponential, and standardization of a variable. The purpose of the various transformations that are provided within the node are designed to achieve linearity, normality, or to stabilize the variability in the variables. The node also provides various power transformations that are a subset of Box-Cox transformations that are designed to find the best transformation for the variables in the predictive model. In addition, the node has the option of transforming interval variables into ordinal-valued variables by placing the interval variables into buckets or quartiles to view the distributional form of the variable. For example, there might be more interest in grouping ages into separate nonoverlapping age groups or personal income into separate income groups, and so on. Grouping variables may, at times, lead to better estimates from the predictive model. The **Variables** tab displays various univariate statistics like the mean, standard deviation, skewness, kurtosis, and coefficient of variation to statistically determine normality in the distribution of the interval variables. From the metadata sample, you may also view the distribution of the selected variable from the standard frequency bar chart. For instance, if the chart displays a highly skewed distribution in the variable, then an appropriate transformation might be considered, such as standardizing or taking the log, in order to achieve normality in the interval-valued variable that is assumed in many of the statistical tests, or generating a better fitting model to achieve more stable and consistent results. However, interpretation of the results and the estimates should be performed with caution since the results are based on the transformed variables as opposed to the original values.

The node is designed so that if the range of values does not follow the distribution of the transformation that is applied to the variable, then the node will automatically set the values to the variable to conform to the transformation that is applied to the variable. For example, applying a logarithmic transformation to a variable within the node will result in the node automatically adding the appropriate constant value to the variable, which will prevent the node from performing an illegal transformation.

Advantages of Transforming the Data

Many of the statistical tests require that the variable have a symmetric, bell-shaped or normal distribution. In other words, the distribution of many of the test statistics is normal or follows some form that is based on the normal distribution. Therefore, if the data follows a normal distribution, then various statistical tests may be used. However, if the data is not normal, then a certain transformation should be applied so that the transformed data follows a normal distribution. Applying the correct transformation to the data will achieve the correct asymptotic significance levels with an increased power in the normal distribution-based tests such as the two-sample t-tests, linear regression t-tests, and in several analysis-of-variance tests. *Asymptotic significance level* means that as the sample size increases, then the significance level does not change that much. The power of a statistical test of some hypothesis is the probability that it rejects the null hypothesis when the hypothesis is false. The *statistical power* is the probability that you have made the right decision.

In many of the nodes that apply the sum-of-squares distance function to the analysis, it is important that the **Transform Variables** node be used to standardize interval-valued input variables in the model in order to interpret the input variables as one common scale or unit since the input variables with a wide range of values will have a profound effect on the results. In predictive modeling, one of the first steps is plotting the relationship between the input variables and the target variable that you want to predict. The purpose in creating the bivariate scatter plot is to determine the functional relationship between both variables. For example, if the functional relationship displays a curvature pattern, then an additional squared input variable might be included in the model to increase the accuracy of the modeling fit. Therefore, the node enables you to perform transformations to the input variables in the data set in which both the original input variable and the transformed input variable are included to the data set that can then be passed along to the subsequent modeling node to increase the precision of the prediction estimates.

In statistical modeling, the purpose of transforming the continuous variables is to hopefully achieve linearity in the relationship between separate variables in the model, stabilize the variability, and achieve normality in the data. In addition, transformations might be a remedy to solve autocorrelation or cyclical patterns in the residual values over time since there must be independence in the data that the traditional regression model is trying to fit. At times, one transformation might achieve linearity whereas a different transformation might achieve constant variability. Conversely, one transformation might achieve normality while a different transformation might meet the equal variance assumption, or vice versa. For example, if the normal assumption is met, but there still exists unequal variability in the model, then transforming the target variable to achieve equal variability can cause nonnormality in the data, or vice versa. Residual plots are usually used in detecting the best transformation of the target values. Typically, the variance is a function of the mean that can be observed by creating a scatter plot of the residuals in the model against the fitted values. The appropriate transformation should be selected by observing the rate in which the variability changes as the fitted values increase. The distributional properties of the error terms in the predictive model can usually be detected from the residual plots. Typically, the logarithmic function is the most widely used transformation applied in stabilizing the variability and achieving normality in the residual values. The functional form of the residual values will indicate the type of transformation to apply to the target variable since transformations applied to the input variable do not affect the variability in the residual values in the regression model. Selecting the most appropriate transformation to apply to the model will usually involve theoretical knowledge, general intuition, or trial and error in trying many different transformations to meet the various modeling assumptions.

Variable transformation is one of the more important steps in preprocessing the data in statistical modeling designs. In analysis of variance, the advantage of applying transformations is to remove the relationship between the range of values and central tendency. For example, if you are interested in comparing the difference in central tendency, that is, mean or medians, between the separate groups, then the groups with larger variability will make it harder to compare the group means. An effective transformation will result in a range of values in each group nearly equal and uncorrelated with the group means. In statistical modeling, the fitted values are restricted to the range of the target values. For binary-valued target variables, the predicted values must fall within a range of zero and one. Therefore, transformation or link functions are introduced in order to limit the range of the target values. Again, the importance of transforming the variables in the statistical modeling design is to satisfy the various modeling assumptions such as normality in the data, but more importantly, stability in the variability in the data. Transformations play a key role in linear statistical modeling designs, like least-squares regression and logistic regression modeling, when there exists a nonlinear relationship between the modeling terms with an appropriate transformation that is applied to achieve a linear relationship between the input variable and the target variable that you want to predict. Furthermore, it is assumed that the main effects do not interact in the statistical model. If the main effects do in fact interact, then applying the correct transformation to the data will reduce the effect of interaction in the main effects to the statistical model. In addition, when you transform the target variable in the model, it is not uncommon to have the functional relationship change in the input variable. Therefore, it is important that the various modeling selection routines be performed once more, since a different set of input variables might be selected for the transformed model. Finally, the error distribution from the predictive model should be consistent with the numerical range of the corresponding model being fitted, which can be achieved by applying the correct transformation. The error distribution should account for the relationship between the fitted values and the target variance. Usually, the error distribution should conform to an approximately normal distribution. However, at times the error distribution might actually have a skewed distribution.

In predictive modeling, transformations are applied to the target response variable to satisfy the various modeling assumptions like stabilizing the variance. However, the predicted values are typically retransformed since the predicted values are usually not as interesting or informative as the original data points that are calculated as follows:

Functional form:	$f(\mathrm{x}) = \beta_0(\beta_1^{\mathrm{x}})$	original form of the nonlinear model
Transform function:	$g(\mathrm{x}) = \ln(f(\mathrm{x})) = \ln(\beta_0) + \ln(\beta_1)\cdot \mathrm{x}$	transform Y to achieve linearity with X
Inverse function:	$g(\mathrm{x}) = g^{-1}(\mathrm{x}) = \exp(g(\mathrm{x})) = f(\mathrm{x})$	back transform to its original form

Note that applying the wrong transformation to the variables in the model can result in an inadequate model. Therefore, after linearizing the given model, it is important that diagnostic analysis be performed, that is, checking for constant variability and normality in the residuals, to ensure you that the appropriate transformation has been applied to the model, which will result in the best prediction estimates. If the various statistical assumptions are still not satisfied, then continued transformations might be considered. For data mining analysis, the listed transformations that are suggested depends on the degree of skewness in the distribution of the variable in order to meet the normality assumption to the various statistical tests.

Log Transformation

- If there is large positive skewness, then the log transformation, that is, LOG(Y) or LN(Y), may be applied to remedy the nonnormality in the distribution of the Y variable.
- If there is large negative skewness, then the log transformation, that is, LOG(K – Y) or LN(K – Y) for some constant K where K = MAX(Y) + 1, may be applied to remedy the non-normality in the distribution of the Y variable.

Square Root Transformation

- If there is a moderate positive skewness, then the square root transformation, that is, $\sqrt{Y}$, may be applied to remedy the nonnormality in the distribution of the Y variable.
- If there is a moderate negative skewness, then the square root transformation, $\sqrt{K - Y}$ for some constant K where K = MAX(Y) + 1, may be applied to remedy the nonnormality in the distribution of the Y variable.

Reciprocal Transformation

- For an extreme L-shaped distribution, reciprocal transformation, 1 / Y, may be applied to remedy the nonnormality in the distribution of the Y variable.
- For an extreme J-shaped distribution, reciprocal transformation, 1 / (K – Y) for some constant K where K = MAX(Y) + 1, may be applied to remedy the nonnormality in the distribution of the Y variable.

In predictive modeling, the variance is a function of its mean residual that can be observed by constructing scatter plots of the residual values against the fitted values. The functional form of the residual values will indicate the appropriate transformation to apply to the target variable, since transformations applied to the input variables do not affect the variability in the residual values in the predictive model. The following are the various transformations to apply that are designed to achieve normality and uniform variability in the predictive model that is based on the relationship between the target variable and input variable in the model.

Log Transformation

- The functional relationship between the target variable and the input variable displays a curvature pattern.
- If the variance increases in a fanned-out pattern across the predicted values squared, then transform the target variable with the log transformation.

Square Root Transformation

- The functional relationship between the target variable and input variable displays a curvature pattern.
- This transformation is typically applied to data that consist of nonnegative integer values or counts of rare events, such as number of accidents, failures, injuries, incidence of certain diseases or insurance claims over time, called a Poisson distribution. The distribution has the property that the sample variance is approximately equal to its sample mean. However, if the variance seems to be increasing in a fanned-out pattern across the predicted values, then transform the target variable with the square root transformation to stabilize the variability and achieve normality in the model.

Reciprocal Transformation

- The functional relationship between the target variable and input variable displays a curvature pattern.
- If the standard deviation increases in a fanned-out pattern across the predicted values squared, then transform the target variable with the reciprocal transformation.

In predictive modeling, the following variability stabilizing transformations are arranged by the amount of curvature in the target variable so that σ^2 is proportional to μ^{2k}, then the appropriate transformation is $y' = y^{1-k}$ with the exception when k = 1 then y' = log(y).

K	0	1/2	2/3	1	3/2	2	5/2	3
$\sigma\ \alpha$	μ^0	$\mu^{1/2}$	$\mu^{2/3}$	μ	$\mu^{3/2}$	μ^2	$\mu^{5/2}$	μ^3
Y'	y	$\sqrt{y}$	$\sqrt[3]{y}$	log(y)	$1/\sqrt{y}$	$1/y$	$1/\sqrt{y^3}$	$1/y^2$

Data tab

The **Data** tab is designed for you to specify the partitioned or scored data sets in performing transformations of the variables and the corresponding data sets. The tab will allow you to select from the various partitioned training, validation and test data sets along with the scored data set. Select the radio button to view the various details of the partitioned data sets. By default, the training data is selected. However, any transformations that you create in the **Variables** tab are applied to the active training data set, validation, test, and scored data sets. The scored data set may be used in transforming new data to be scored. Press the **Select...** button and the **Imports Map** window will appear. The window is designed for you to select the active training data set in which you would like to perform the various transformations of the variables in the analysis.

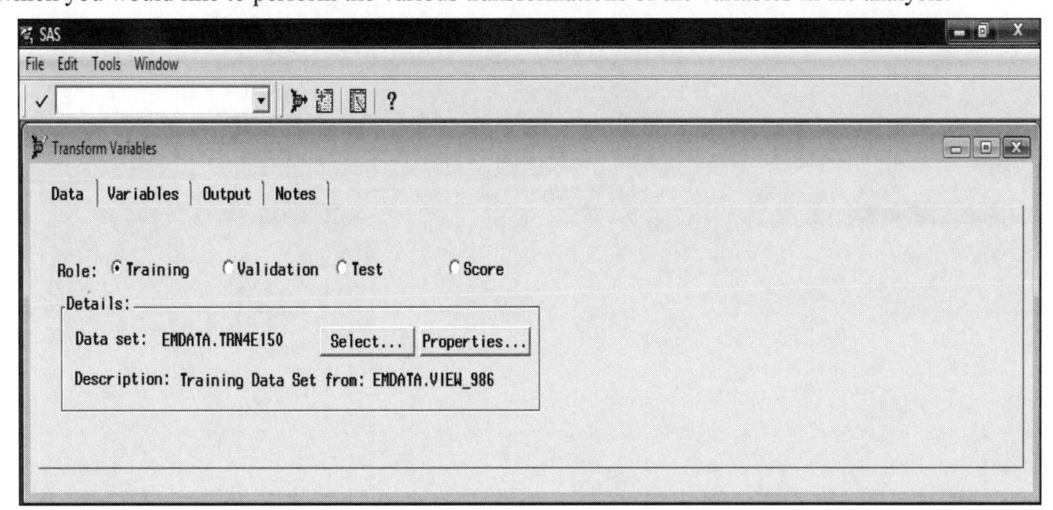

*The **Data** tab to select the active data set to apply various transformations to the variables.*

Variables tab

By default, the **Variables** tab is automatically displayed as you open the **Transform Variables** node. The tab is designed for you to create separate variables to the node based on an appropriate transformation applied to the existing interval-valued variables within the active training data set. The tab displays various descriptive statistics such as the mean, standard deviation, skewness, kurtosis, and coefficient of variation of each interval-valued variable from the metadata sample. The skewness and kurtosis statistics measure the shape of the distribution in the variable. The skewness statistic measures the distribution of the data being spread out on either one side or the other. Positive skewness indicates that values located to the right of the mean are more spread out in comparison to values that are located to the left of the mean. Negative skewness indicates the opposite. The kurtosis statistic measures the tendency of the distribution of the data being distributed toward its tails. Large kurtosis values will indicate that the data contain some values that are very distant from the mean relative to the standard deviation. In addition, the tab will also allow you to view the distribution of the variable from the metadata sample by selecting the **View Distribution of <variable>** pop-up menu item that will display the standard frequency bar chart within the **Variable Histogram** window. The purpose of viewing the distribution of the variable is to perhaps give you a clue as to the correct transformation to apply to the variable. Conversely, this option will allow you to view the distribution of the transformed variable in order to ensure you that the variable is normally distributed. For example, if the variable has a wide range of values, then the logarithmic transformation might be appropriate in achieving normality of the variable.

Transforming the Variable

In order to redefine or transform the listed variables within the tab, select the variable that you would like to transform, then scroll over to the **Formula** column and right-click the mouse to select the **Transform** pop-up menu item. The node is designed so that after a transformation has been applied to the variable, then the corresponding transformed variable will be appended to the list of existing variables that are displayed within the tab. In addition, you may also select the variable to be transformed and select the **Transform Variables** toolbar icon that is located next to the command line, which will then display a list of all the transformations that are available within the **Transform Variables** node. Alternatively, highlight the variable row that you would like to transform by selecting the variable row, then left-click the mouse and select the **Tools > Transform Variables** main menu options.

The naming convention within the node assigns the newly created variable the identical variable name that is used in creating the transformed variable along with some alphanumeric random assignment of the last four letters to the transformed variable.

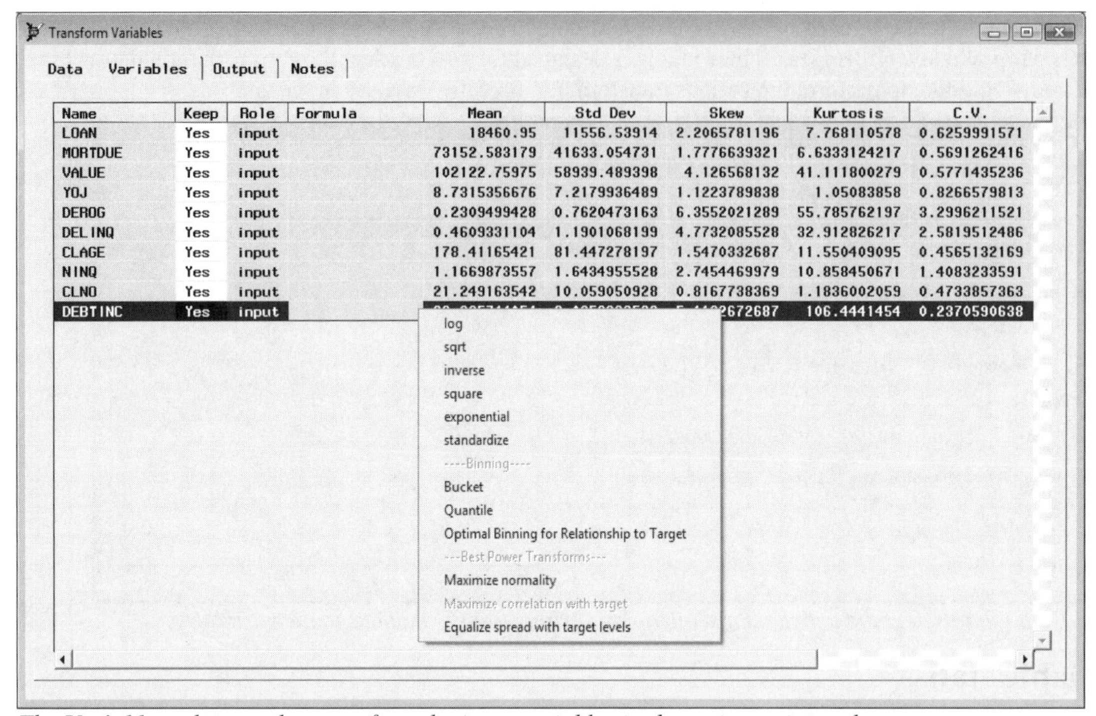

*The **Variables** tab is used to transform the input variables in the active training data set.*

The following is the list of the standard transformations that may be specified from the list.

- **log:** Performs a logarithmic transformation that is one of the most common transformations applied in order to achieve normality in the distribution of the interval-valued variable, with the restriction that the variable consists of positive values.
- **sqrt:** Performs a square root transformation to the interval-valued variable with the restriction that the variable consists of positive values.
- **inverse:** Performs the reciprocal transformation, that is, one over the value of the variable. This transformation will change the values of the variable into rates or proportions assuming that the values of the variable are greater than one.
- **square:** Performs a squared transformation. This transformation is appropriate if the variable consist of both positive and negative values, in which case this transformation will change the values to the variable into strictly positive values.
- **exponential:** Performs an exponential transformation of the variable. This transformation is appropriate if the target variable displays a logarithmic distribution across the values of the interval-valued variable, in which case selecting this transformation will create a linear relationship between both variables in the predictive model.
- **standardize:** This transformation converts the interval-valued variable into z-scores by subtracting its mean and dividing by its standard deviation. In many of the nodes, this transformation is automatically applied to the interval-valued input variables to the model in which the squared distance function is applied. The main idea in standardizing the input variables is to interpret the corresponding variables to a common scale with approximately the same variability, assuming that the variables are measured in entirely different units with a wide range of values.

--- **Binning** ---

- **Bucket:** Transforms the variable by dividing the variable into separate class levels or buckets of equal intervals based on the difference between the minimum and maximum values. This binning option will transform an interval-valued variable into a ordinal-valued variable. The default is four separate buckets or bins. However, the **Input Number** window will appear for you to enter the appropriate number of buckets or bins. You may enter up to 16 separate buckets. The **Variable Histogram** window will display the standard frequency bar chart for you to observe the variable's transformation into separate intervals. Typically, the number of observations are unequally distributed within each interval as opposed to the following quartiles technique.
- **Quartile:** This transformation is similar to the previous bucket technique. However, the values of the variable's frequency distribution are subdivided by the quartiles or the percentiles of the variable distribution, where the observations are uniformly distributed within each quartile. This option partitions the variable into intervals with approximately the same number of observations. The **Input Number** window will appear for you to enter the appropriate number of quartiles or percentile estimates. By default, the variable is partitioned into the four separate quartile estimates. Similar to the previous binning transformation, the standard histogram will appear for you to observe the transformation of the variable that has been split into separate intervals.
- **Optimal Binning for Relationship to Target:** This transformation partitions the values of the variable by the class levels of the binary-valued target variable. This option determines the *n* optimal groups by performing a recursive process of splitting the input variable into groups that maximizes the association with the binary-valued target variable based on the PROC DMSPLIT data mining procedure. This procedure is advantageous when there exists a nonlinear relationship between the interval-valued input variable and the binary-valued target variable. In the first step, recursive binary splits are performed, which maximizes the chi-square statistic based on the 2 x 2 tables, with the rows representing the two separate class levels of the binary-valued target variable and the columns representing the bins that are consolidated into two separate groups. Initially, the inputs are grouped into 64 different bins that are grouped into two separate groups, which maximizes the chi-square value. The recursive grouping procedure stops and the input variable is not transformed if the chi-square value for the initial split does not exceed the cutoff value. In the second iteration, this recursive process then forms a maximum of four separate groups by applying this same group processing procedure in transforming the range of values of the input variable. In the third iteration, this recursive process then forms a maximum of eight groups, and so on. Connecting the **Tree** node after binning a variable may result in the tree model performing worse since binning reduces the number of splits for the tree to search for unless the original variable is added to the model. The metadata sample is used to find the optimum groupings in order to speed up this recursive grouping process.

--- **Best Power Transforms** ---

- **Maximize normality:** The node automatically applies a transformation to the variable that yields sample quantiles that are closest to the theoretical quantiles of a normal distribution. The most common transformation that achieves normality is the logarithmic transformation based on the interval-valued variables in the data set. This option is designed to achieve normality in the variable, assuming that the distribution of the variable is highly skewed, peaked or flat.
- **Maximize correlation with target:** The node automatically applies a transformation to the variable in order to achieve the best squared correlation of the target variable. The transformation that is applied is designed to linearize the relationship between the interval-valued input variable and the interval-valued target variable in addition to stabilizing variability of the target variable. In statistical modeling, applying the appropriate transformation in making both variables linear to one another is called creating *intrinsically linear models*.
- **Equalize spread with target levels:** The node applies a transformation to the variable that reduces the spread in the range of values across the categories of the target variable. This transformation helps stabilize the variability in the input variables across the target levels. The transformation partitions the values of the interval-valued variable by the class levels of the categorically-valued target variable in the data set to achieve the smallest variance within each target class level.

Removing Variables from the Output Data Set

You may also remove certain interval-valued variables in the data set from the subsequent nodes by selecting the variable row, then scrolling over to the **Keep** column and right-clicking the mouse in order to select the pop-up menu option of **No**. By default, all the variables are kept in the output data set with a **Keep** option setting of **Yes**. However, transforming the variable will result in the original variable being removed from the output data set with a keep variable status automatically set to **No** that is displayed in the following diagram.

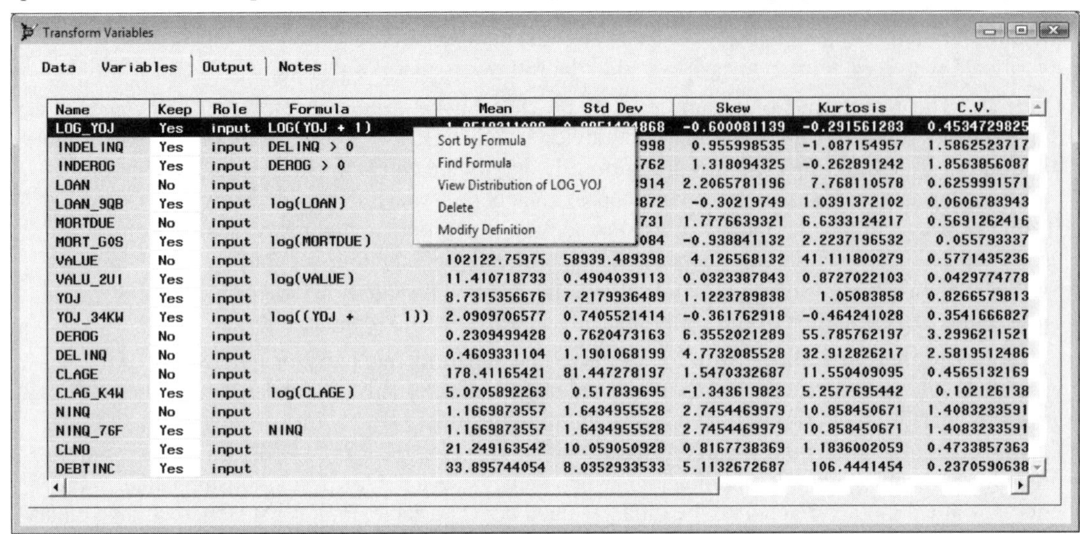

*The various options used to either delete or modify the transformed variable or redefine the variables by fitting the logistic regression model in the modeling comparison procedure from the **Assessment** node.*

Deleting the Transformed Variable from the Data Set

If the transformed variable is not what you expected, then the node is designed for you to delete the transformed variable from the corresponding data set. From the tab, select the variable that has been created, scroll over to the **Formula** column and right-click the mouse to select the **Delete** pop-up option in order to delete the variable from the list of variables in the data mining data set. In addition, you may select the variable to delete from the file by selecting the **Delete Variables** toolbar icon that is located to the left of the question mark icon. Alternatively, highlight the row of the newly created variable that you would like to delete and select the **Tools > Delete Variables** main menu options. However, the node is designed so that the existing variables in the input data set cannot be deleted from the data mining data set.

User-Defined Transformations

The **Transform Variables** node is also designed for you to apply user-defined transformations. The advantage of providing your own user-defined transformation is that the values of the transformed variable will not be set to missing whenever the values of the transformed variable do not comply with some of the user-defined transformations, that is, like the logarithmic transformation. From the tab, select the **Create Variable** toolbar icon from the tools bar menu or select the **Tools > Create Variable** main menu options with the following **Create Variable** window that will appear for you to modify the variable's definition by specifying an appropriate format to the newly created variable. Alternatively, you may modify the existing transformation by selecting the **Modify Definition** pop-up option item. This is displayed in the previous illustration.

- **Name:** This option will allow you to specify a variable name for the transformed variable.
- **Format:** This option will allow you to specify a format for the newly created variable. The default is a format of BEST12., that is, SAS determines the best notation for the numeric variable. Press the drop-list arrow and the **Formats** window will appear that will allow you to select from a list of available SAS formats. The **Width** entry field will allow you to specify the variable length of the newly created variable in the data mining data set or the number of decimal places with a **Decimal** entry field appearing for interval-valued formats, for example, *w.d* variable formats.

- **Label:** This option will allow you to specify a variable label for the newly created variable.
- **Formula:** This option displays the current transformation applied to the variable. Otherwise, enter the appropriate formula from the wide variety of built-in functions that are available in SAS.
- **Type:** This option will allow you to specify either a numeric or character variable attribute.

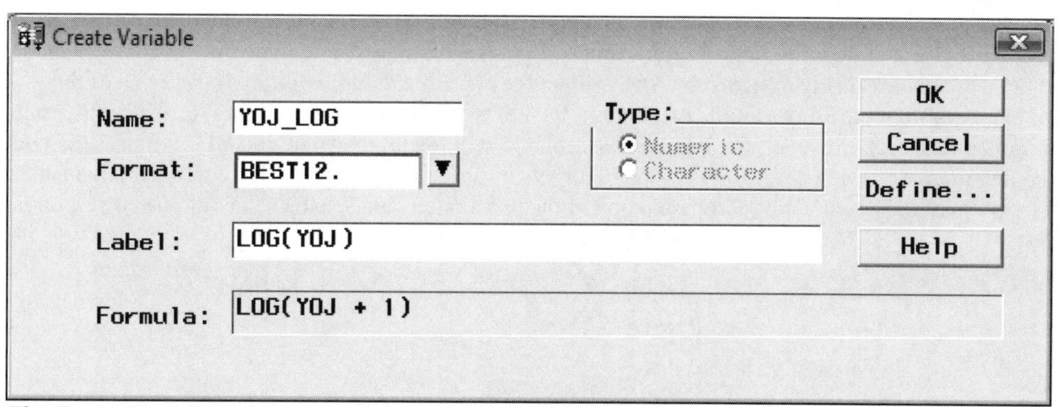

*The **Create Variable** window is used to modify the attributes of the transformed variable that is created.*

Press the **Define...** button that will display the following **Customize** window. The **Customize** window is designed to create custom-designed transformations based on the continuous variables in the data set. In addition, the window will allow you to create interaction terms for the active training data set. At times, adding interaction terms to the statistical model will increase the accuracy of the model; however, the drawback is that it will make the interpretation of the relationship between that target variable and the input variables that interact with each other much harder. The window contains a column list box, a number pad, an operator pad, and a list box of several built-in functions that are available in SAS in order to build an expression in the equation box that is listed at the bottom of the window.

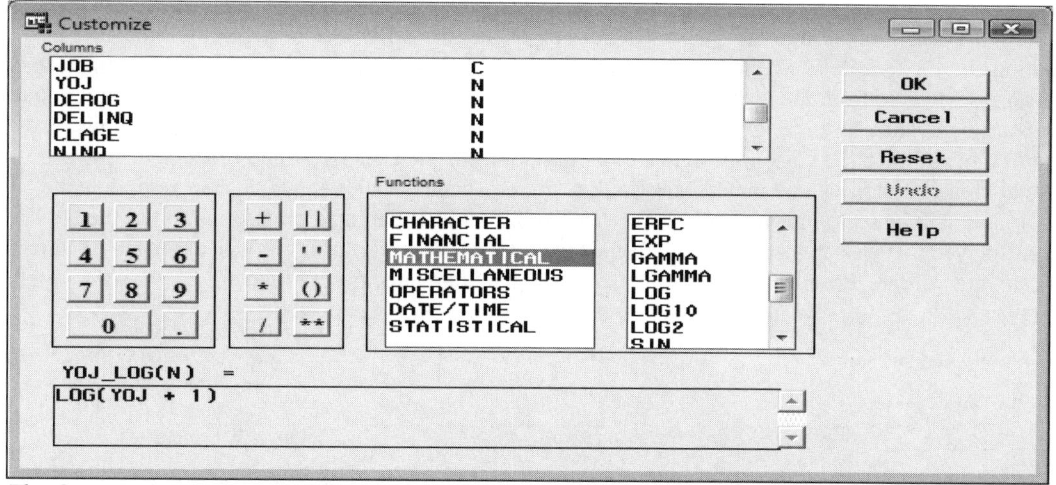

*The **Customize** window is used to perform custom transformations of the variables in the data set.*

First press the **Reset** button to clear the notation field display. The next step is scrolling through the list of available functions that are grouped by the function type. The display field is located to the left of the list of available functions that can be applied to the selected variable. Left-click the mouse to select the appropriate function and the function will then appear in the notation field display. The list of functions that are available to select from within the tab is based on several built-in functions that are a part of the SAS software. Finally, select the variable that you would like to transform from the list of variables that are displayed in the **Columns** list box. You might need to edit the notation field in order for the selected variable to be correctly positioned within the parentheses of the selected function as shown in the **Customize** window. Press the **Undo** key to undo any changes. Otherwise, select the **OK** button and the **Transform Variables** window will reappear with the newly created user-defined transformed variable appended to the list of existing variables in the data set.

Output tab

The **Transformed Variables** node will automatically create an output data set for each partitioned data set that is passed along to the subsequent nodes within the process flow diagram. By default, the output data set will consist of the all variables that have not been transformed along with all variables that have been transformed. The transformed variables that are created within the node will not be written to the output data set until the node has been executed. In addition, the variable names of these same transformed variables that are created within the node are automatically assigned by SAS. Select the appropriate radio button of each one of the partitioned data sets from the **Show details of:** section, then press the **Properties...** button to view either one of the partitioned data sets. However, the **Properties...** button will be dimmed and grayed-out unless the node is first executed in order to view the various output data sets. From the **Description** entry field, you may enter a short description of the output data set for each one of the partitioned data sets by selecting the corresponding radio button.

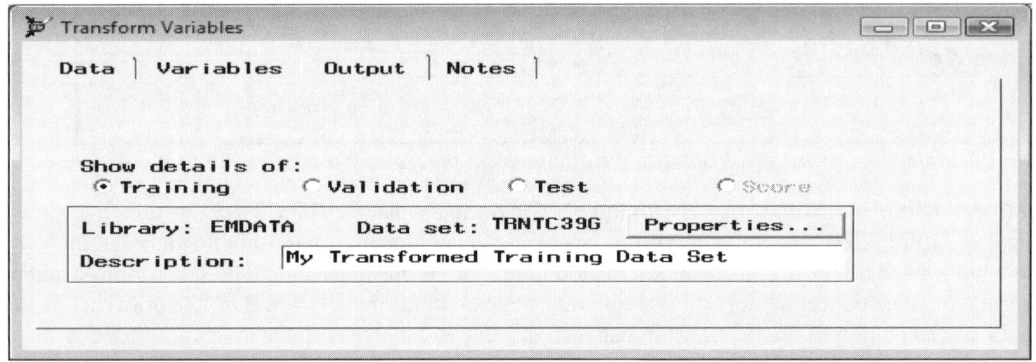

*The **Output** tab is used to view the transformed variables along with all other variables in the data set.*

The following illustration displays a partial listing of the output data set with the transformed variables that was performed to the logistic regression model in the following modeling assessment exercise. The purpose of applying the same transformation to the YOJ variable is to display the difference between the standard transformation and the user-defined transformation. From the table listing, notice that some of the transformed YOJ values are set to zero from the user-defined transformation as oppose to the standard transformation that sets the transformed values to missing with regard to the values that are undefined from the logarithmic transformation. In addition, the custom-designed transformations of the DELINQ and DEROG variables are displayed that creates binary-valued indicator variables where its values that are greater than zero, then the indicator variable will be set to one, and zero otherwise. And, finally, the binned transformation of the NINQ variable is displayed that was created by selecting the **Bucket** option setting from the **Variables** tab in which three separate intervals are created with approximately an equal number of observations.

Data set details

Information | Table View

☑ Variable labels

	BAD	REASON	JOB	YOJ	LOG(YOJ + 1)	DELINQ > 0	DEROG > 0	log(YOJ)	Bucket(NINQ)
49	1	HomeImp	Other	0	0	0	0	.	0001:low-0.5
50	1	HomeImp	Other	8	2.1972245773	1	1	2.1972245773	0003:1.5-high
51	0	DebtCon	ProfExe	7	2.0794415417	1	0	2.0794415417	0002:0.5-1.5
52	0	HomeImp	Office	19	2.9957322736	0	0	2.9957322736	0002:0.5-1.5
53	0	HomeImp	Other	3	1.3862943611	0	0	1.3862943611	0001:low-0.5
54	0	HomeImp	Office	19	2.9957322736	0	0	2.9957322736	0001:low-0.5
55	0	DebtCon	Sales	3	1.3862943611	0	0	1.3862943611	0001:low-0.5
56	0	HomeImp	Other	6	1.9459101491	0	0	1.9459101491	0001:low-0.5
57	0	HomeImp	Other	4	1.6094379124	0	1	1.6094379124	0001:low-0.5
58	0	DebtCon	Other	8.6672752044	2.2687464915	1	1	2.2687464915	0002:0.5-1.5
59	0	HomeImp	Other	4	1.6094379124	0	0	1.6094379124	0001:low-0.5
60	0	HomeImp	Other	0	0	0	0	.	0001:low-0.5
61	1	HomeImp	Other	5	1.7917594692	1	1	1.7917594692	0002:0.5-1.5

The table view of the active training data set with the added transformed variables from the standard logarithmic transformation and the user-defined logarithmic transformation.

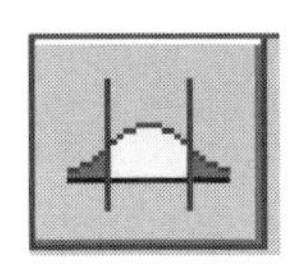

3.3 Filter Outliers Node

General Layout of the Enterprise Miner Filter Outliers Node

- **Data tab**
- **Settings tab**
- **Class Vars tab**
- **Interval Vars tab**
- **Output tab**
- **Notes tab**

The purpose of the **Filter Outliers** node in Enterprise Miner is designed to identify and remove outliers or extreme values from the active training data set only. The node is not designed to filter observations from the validation, test, or scored data sets. The reason is because there should not be any observations removed from the validation and the test data sets since these data sets are specifically used for modeling assessment. Conversely, the scored data set is used to calculate new predictions. Therefore, we need every possible observation to be scored. An *outlier* or *extreme value* is an observation that is well separated from the rest of the other data points. However, outliers might not be errors and may or may not be that influential in the data mining analysis. One strategy is performing the analysis with and without the outlying data points. If the outliers make little difference in the results, then leave these same data points in. Outliers can be detected by performing various data listings, constructing various exploratory plots and graphs, or constructing various descriptive statistical listings such as frequency counts or listings of the range of values, that is, maximum and minimum value. Outliers can be identified as impossible values, an impossible combination of values, coding mistakes, technical difficulties in the automated system or just common knowledge.

From the node, outliers can be determined from the standard frequency chart based on the metadata sample. However, selecting the **Use entire data** option from the **Settings** tab will allow you to determine outliers within the tab for every observation in the training data set. The node will allow you to filter or remove rare values from the active training data set based on the categorically-valued variables with less than 25 unique class levels, class levels occurring a certain number of times, extreme values in the interval-valued variable based on various statistical dispersion criteria or it will allow you to remove missing values altogether. From the node, you have the option of filtering or removing rare values from the process flow and keeping missing values in the analysis. The node has the option of including variables in the analysis that fall within a specified range. Extreme values are eliminated for categorical variables by a specified number of different values or values occurring a certain number of times. For interval-valued variables, rare values can be removed from the analysis by specifying a restricted range of values for each variable in the analysis such as the median absolute deviance, modal center, standard deviation from the mean, and extreme percentiles. The node will create two separate data sets. A SAS data set is created for all records that are deleted and a SAS data set is created for all records that are retained from the training data set.

Missing Values

The node will also allow you to remove observations from the following analysis due to missing values in any one of the variables in the active training data set. In Enterprise Miner, when there are missing values in any one of the input variables in the predictive model, then the target variable is automatically estimated by its own mean. However, the node will allow you to remove these same observations from the modeling fit that is performed in the modeling procedures in SAS. Removing observations due to missing values in the target variables is not recommended since its values are estimated by the given statistical model.

Reasons for Removing Extreme Values

Outliers might have a profound effect on the analysis that can lead to misleading results and biased estimates. In predictive modeling, influential data points affect the modeling statistics such as the parameter estimates, standard errors of the parameter estimates, predicted values, studentized residuals, and so on when they are removed from the data. Outliers are observations that are well separated from the general trend of the other data points that are consistent with the functional relationship between the input and target variable. However, *influential data points* are well separated far beyond the range of the rest of the input values that are also inconsistent with the functional relationship between the input and target variables. Outlying data points that are influential values will substantially change the fit of the regression line. This is illustrated in the following diagrams. An outlier observation might or might not be an influential data point, or vice versa. The purpose of removing extreme data points from the input data set is that it will result in stable and consistent estimates. *Stable estimates* mean that if you were to resample the data set any number of times and fit it to the same model, it will result in comparable estimates of the true values. The reason is because the prediction line or the principal components line will not be influenced by a large shift in the line because the prediction line or the principal components line adjusts to the outlying data points. In neural network modeling, eliminating extreme values from the fit will result in better initial weight estimates that will result in faster convergence in delivering the best fit with the smallest error by seeking the error function at a minimum through the use of the iterative grid search procedure. In clustering analysis, extreme data point tends to form their own clusters. In decision tree modeling, outlying observations tend to lop-side the construction of the tree that form their own separate branches. Since many of the modeling nodes apply the sum-of-squares distance function between the data points, outlying data points or input variables with a wide range of values will have a profound effect to the fitted values. In data analysis, even one outlying data point can result in biased estimates. A common remedy in dealing with extreme values is to either delete the observation from the analysis, apply an appropriate transformation to the variable or redefine the existing model. In predictive modeling, one remedy in eliminating extreme values in the data is simply collecting more data, with the idea that the added data points will follow the same trend as the outlying data points. Another approach is adding additional input variables that are related to the target variable in order to improve the modeling fit due to these outlying data points. Outlying data points can be extremely difficult to detect if there are several input variables in the model. Therefore, there are a wide variety of diagnostic statistics that are designed to identify influential observations. In predictive modeling, these diagnostic statistics are calculated by the removal of each observation from the data, then measuring the change in the residuals, (RSTUDENT residuals), measuring the change in the parameter estimates, (Cook's D statistic), measuring the change in each one of the parameter estimates, (DFBETA statistic), measuring the change in the parameter estimates from the estimated variance–covariance matrix, (COVRATIO statistic), or measuring the change in the predicted values, (DFFITS statistic). These same diagnostic statistics are standardized in order to make unusual observations stand out more noticeably. In addition, these standardized statistics are also useful when the input variables have a wide range of values or are measured in entirely different units.

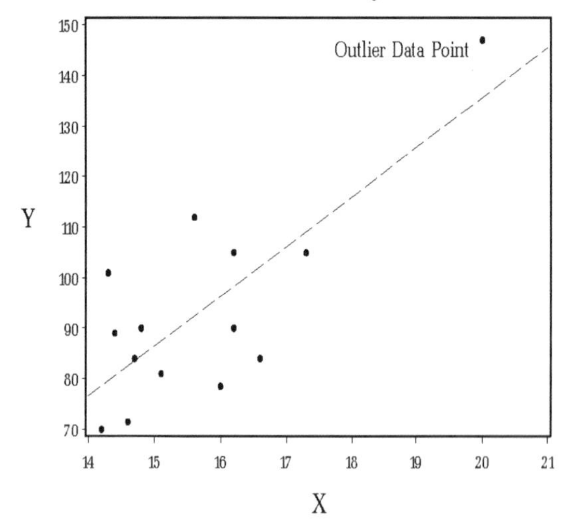

An outlier data point that is well separated from the rest of the other data points. However, the data point is consistent with the general trend in the other data points that does not result in a large shift in the regression line.

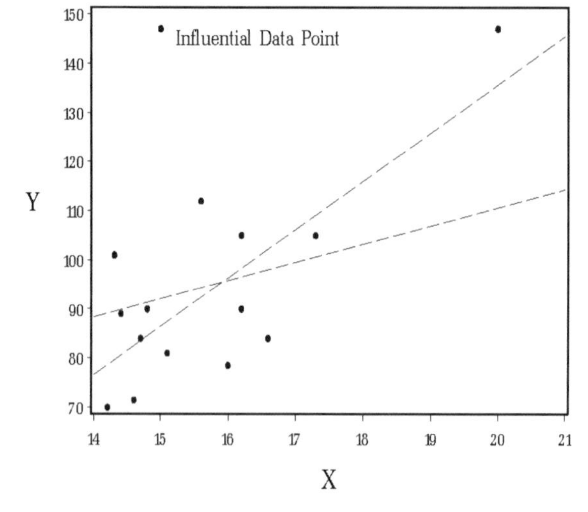

In least-squares modeling, the outlying data point that is an influential data point will result in a large shift in the regression line, with a dramatic change in the parameter estimates, since the observation is not consistent with the general trend for the rest of the other data points.

Typically, outliers might be due to an error in measurement or a malfunction in the data processing equipment. However, outliers might actually contain extremely important information for the analysis, requiring further investigation, for example, a fraudulent claim, an error in a standard procedure, or a fluke of nature. An outlier might also indicate to you that important input variables, interaction or nonlinear effects have been excluded from the predictive model, indicating to you that the observation might turn out to be the most important data point to the analysis. When the outlying influential observation is correctly recorded, the inaccuracy of the modeling fit might be due to the incorrect functional form in the current model. In general, it is not always a good idea to remove observations from the modeling fit. Therefore, outlying influential observations should not be automatically removed from the data because these unusual observations might actually represent an unlikely event. In addition, automatically disregarding valid outlying data points can lead to an undesirable reduction in the accuracy of the fitted model.

The histograms or the frequency bar charts can detect unusual observations with values either too high or too low. These same outliers can be observed at either end of the bar chart. However, the difficulty with the histogram is that these same outlying data points are very difficult to observe since the histogram will display extremely small frequency bars. In addition, these same frequency bars will be difficult to view in Enterprise Miner since they will be displayed in the color yellow. From the **Insight** node, schematic box plots can be constructed by observing the data points that are beyond either end of the whiskers. The whiskers extend at either end of the box out to the 1.5 interquartile range. Probability plots are the best way to determine normality in the variables. Outliers to the plot can be viewed at either end of the tails. From the probability plot, the first and/or last data points that are well separated from the other data points might be considered outliers or extreme values in the data. Scatter plots can be used in observing the bivariate relationship between two separate variables. Extreme values in the scatter plots can be observed in the data with certain data points that are well separated from the rest of the other data points. In predictive modeling, outliers can be observed by creating a scatter plot of the target variable against the input variables. However, outliers can be observed from the scatter plot of the residual values or, even better, the standardized residual values against the input variables or the fitted values in the model and observing any residual value that might be too large or too small in comparison to the other residual values. Again, the reason for standardizing the residuals in predictive modeling is that it will make unusual observations stand out more noticeably.

Filtering Data in the Process Flow

There are two separate techniques for eliminating outliers based on interval-valued variables. The first method is performing a transformation of the variable that can be performed from the previous **Transform Variables** node. The second idea is to trim or truncate the tails of the variable's distribution by deleting the extreme values that are above or below some predetermined threshold or cutoff value. In Enterprise Miner, the **Filter Outliers** node is designed to remove observations from the analysis based on some specified cutoff value.

The following process flow diagram was constructed to generate the least-squares modeling results that is displayed in the following **Regression** node. The **Filter Outliers** node was incorporated into the process flow to remove the extreme values in the interval-valued target variable of DEBTINC, that is, debt to income, from the home equity loan data set along with removing observations in some of the interval-valued input variables in the model that were far beyond the rest of the other data points. The data points that were removed from the regression model are displayed in the following **Interval Vars** tab. The subsequent **Transform Variables** node was used to transform the interval-valued input variables, DEROG, DELINQ, and NINQ into categorical variables in order to increase the predictive accuracy of the modeling results. The transformations that were applied to bin these same input variables into separate categories are displayed in the **Transform Variables** node. The **Regression** node was used to fit the least-squares regression model. In addition, the same preprocessing of the data was performed in many of the following modeling node by fitting the interval-valued variable, DEBTINC.

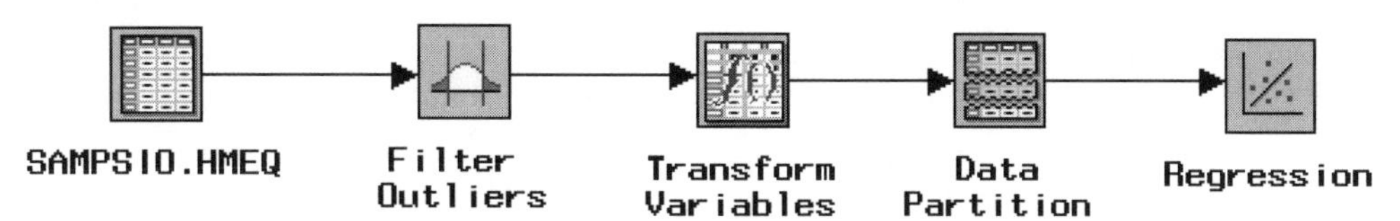

*The process flow diagram that is constructed to generate the least-squares modeling results that are displayed in the following **Regression** node.*

Data tab

The **Data** tab will automatically appear as you first open the **Filter Outliers** node. The **Data** tab is designed for you to specify the active training data set to apply the following filter techniques in removing extreme values from the output data set. Again, the node does not filter observations from the validation, test, or scored data sets. The **Filter Outliers** node is designed to remove observations from the training data set. The reason is because the validation and test data sets are mutually exclusive data sets that are used strictly for modeling assessment. Therefore, you should not remove observations from these data sets. Also, the scored data set is designed to score every observation of some outcome variable. Therefore, you should not filter the scored data set. The **Data set** and **Description** fields will display the active training data set and a description that is usually created from the **Input Data Source** node.

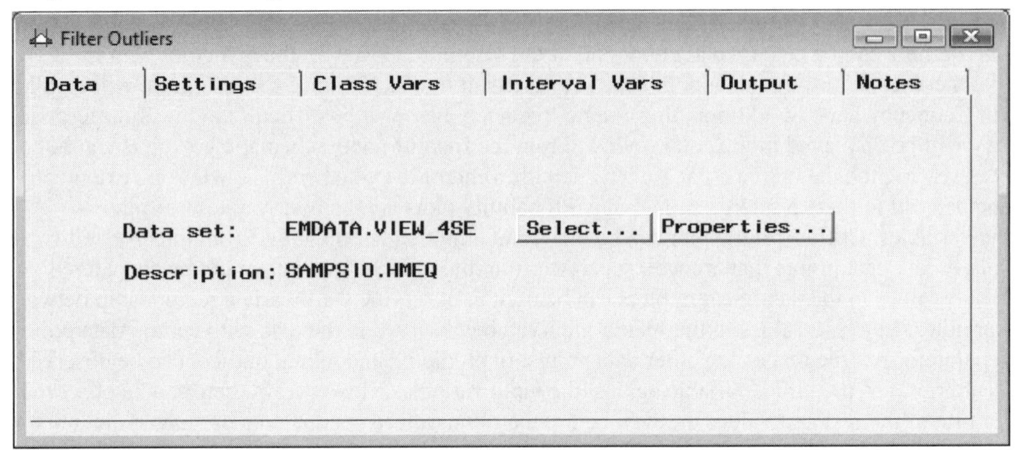

*The **Data** tab lists the active training data set to perform automatic filtering to the data.*

Settings tab

The **Settings** tab is designed for you to specify the various configuration settings in automatically filtering or removing extreme values from the output data set based on either the interval-valued or categorically-valued variables in the data mining data set. By default, the metadata sample is used to determine the range of values in the active training data set. However, selecting the **Use entire data** option, it will allow you to view the range of values of the entire training data set. By default, the **Eliminate rare values** and **Eliminate extreme values in interval vars** option settings are turned off. However, the **Keep Missing?** check box is selected indicating to you that missing values are retained in the data mining data set. Therefore, clear the check box to remove the missing values from the output data set.

By default, both the **Eliminate rare values** and the **Eliminate extreme values in interval var** check boxes are unchecked and the **Keep Missing?** check box selected. This will prevent the node from removing any of the data points from the active training data set. Again, selecting the **Keep Missing?** check box will result the node retaining all observations with missing values for any one of the variables in the active training data set.

To apply the automatic filter settings to all variables in the active training data set, select the **Apply these filters to all vars** button. To eliminate values that you have not already set to be removed from the output data set from the filter settings specified in either the **Class Vars** or **Interval Vars** tabs, simply select the **Apply only to vars without existing filters** button. In other words, all filter settings that are specified from the tab will apply to each variable in the active training data set with the exception to the existing filters that have been assigned to the variables from the following **Class Vars** and **Interval Vars** tabs.

Once the default filter settings have been specified from the **Settings** tab, you can view the filter values to each variable in the active training data set from the following **Class Vars** and **Interval Vars** tabs.

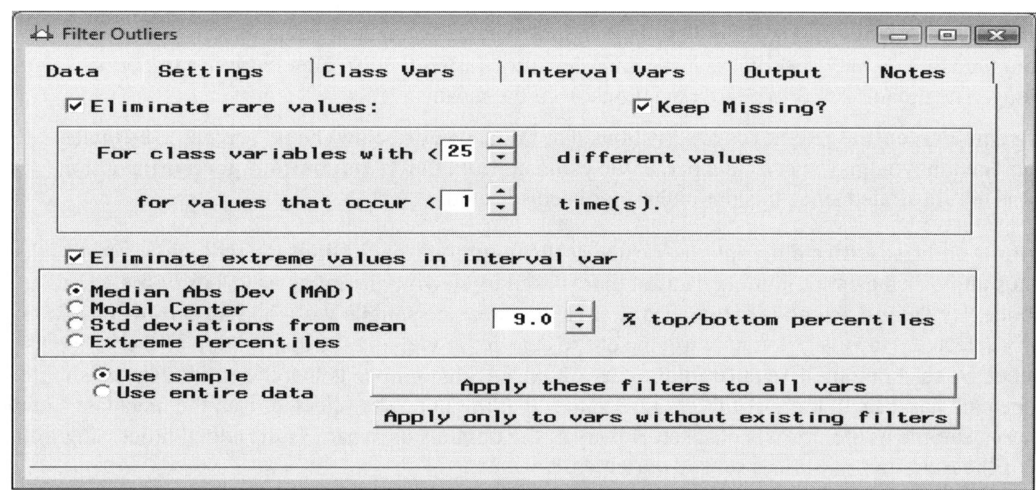

*The **Settings** tab with the automatic filter settings is used to remove extreme values from the data set.*

The following options are the various automatic filter settings used to examine, adjust, and remove extreme values from the output data set based on the categorically-valued variables in the active training data set that can be performed by selecting the **Eliminate rare values** check box:

- **For class variables with <:** The most often occurring class levels are written to the output data set. By default, observations are considered for automatic filtering, that is, they are included in the output data set, from the top 25 most often occurring class levels of the categorically-valued variable. However, any integer between 3 and 99 can be specified from the up and down arrow buttons.
- **for values that occur <**: This option is associated with the previous option. Observations are written to the output data set based on each class level that occurs *n* number of times in the training data set. By default, observations are retained in the analysis if they occur at least once within each class level of the categorical variable. Any positive integer less than 100 may be specified. The value that is selected will be listed in the **Min Freq** column of the following **Class Vars** tab.

The following automatic filtering options are based on interval-valued variables in the active training data set. From the tab, you may select only one automatic filtering method by selecting the **Eliminate extreme values in interval var** check box**.** In addition, the range of values of the corresponding interval-valued input variables to include in the output data set will be automatically listed in the **Range to include** column from the **Interval Vars** tab.

- **Median Abs Dev (MAD)** (default)**:** Eliminates observations from the data set that are more than *n* deviations from the median. This nonparametric statistic is not affected by extreme values in the active training data set. The statistic is calculated by computing the absolute difference from the median, then calculating the median value, that is, Median[$|X_i - \text{Median}(X_i)|$]. You may specify the *n* deviations from the **Deviations from median** option setting that is located to the right of this radio button. This is illustrated in the previous diagram. The default is 9.0 deviations from the median.
- **Modal Center:** Eliminates observations that are *n* spaces from the modal center. You may specify the *n* spaces from the **Spacings from modal center** option located to the right of the radio button. The default is 9.0 spaces from the modal center. The modal center statistic is calculated as follows:
    1. Sort the data in ascending order.
    2. Count the number of observations and divide the data into two halves.
    3. Calculate the range of the top half of the data.
    4. Find the second modal spacing by dropping the first observation from the top half and adding the first observation in the bottom half of the remaining observations, then calculate the range of the values in the bottom half.
    5. Iteratively calculate the range of the modal spacing from the bottom half of the data.
    6. Select the modal spacing with the smallest range.
    7. Calculate the modal center as the middle value of the minimum modal spacing.

- **Std deviations from mean:** Removes observations that are *n* standard deviations from the mean. From the tab, you may specify the *n* standard deviations from the **Std Dev from Mean** option setting. The default is 9.0 standard deviations from the mean.
- **Extreme Percentiles:** Removes observations that fall above or below the pth percentile estimate. From the tab, you may specify the percentile estimate from the **% top/bottom percentile** option that is not illustrated since this option has not been selected.

The **Use sample** and **Use entire data** options located at the bottom of the **Settings** tab will allow you to select the metadata sample or the entire training data set that is used in viewing the range of values of the variables within the node. By default, the node automatically uses the metadata sample with the **Use sample** check box automatically selected. However, you may use the entire data set in viewing the distribution of the variables within the node by selecting the **Use entire data** check box. SAS recommends that the entire data set should be selected if the data has a small number of very rare values that might not be selected from the metadata sample that is a random sample of the original data set. However, the obvious drawback is the added processing time, particularly if the training data set has several variables.

Class Vars tab

The **Class Vars** tab is designed for you to override the automatic filter settings previously mentioned in the **Settings** tab and gives you more control in removing extreme values from the output data set based on each categorically-valued variable in the training data set. From the tab, select the listed categorically-valued variable and specify the class levels to remove the corresponding records from the output data set. Initially, the class levels for each categorically-valued variable that are automatically filtered are based on the automatic filtering method that you selected from the previous **Setting** tab.

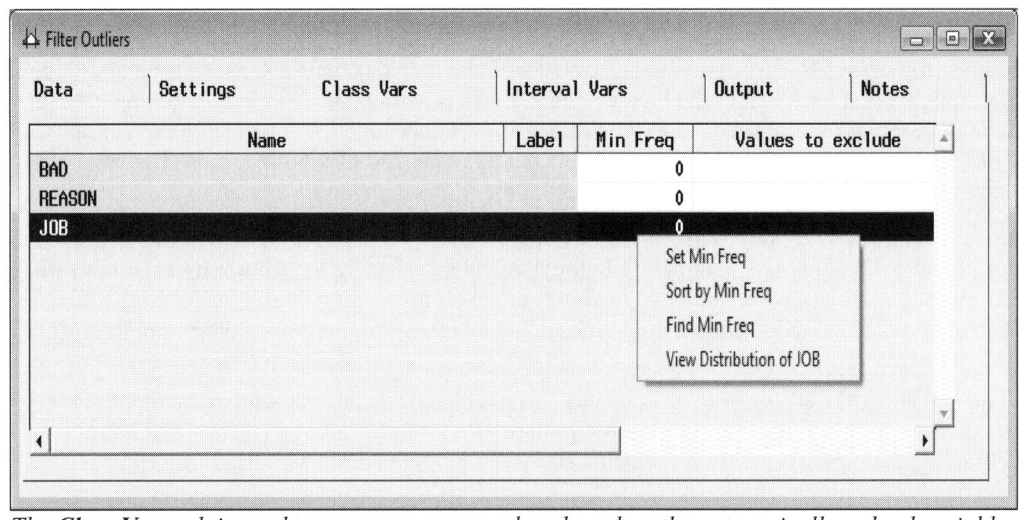

*The **Class Vars** tab is used to remove extreme values based on the categorically-valued variables.*

The **Min Freq** column is designed for you to specify the minimum number of occurrences of each class level of the selected categorically-valued variable to remove from the output data set. The values displayed in the **Min Freq** column are based on the values from the metadata sample. Simply enter the minimum frequency count of the selected categorically-valued variable within the **Min Freq** column. Otherwise, select the variable row to set the smallest frequency value, scroll over to the **Min Freq** column, then right-click the mouse and select the **Set Min Freq** pop-up menu item. The **Select values** window will open and the frequency bar chart will appear, as illustrated in the following diagram. Select the **MIN FREQ** slider bar to drag along the axis to visually adjust the threshold cutoff value with the value automatically updated within the **Value** entry field. However, you may also enter the cutoff value from the **Value** entry field. In the following example, all records that lie above the displayed horizontal line are written to the output data set, that is, all records with a job occupation of missing, sales representative, or self-employed are removed from the output data set. Once the **Select values** window is closed and the appropriate minimum cutoff value has been set, then the corresponding value will be displayed within the **Min Freq** column.

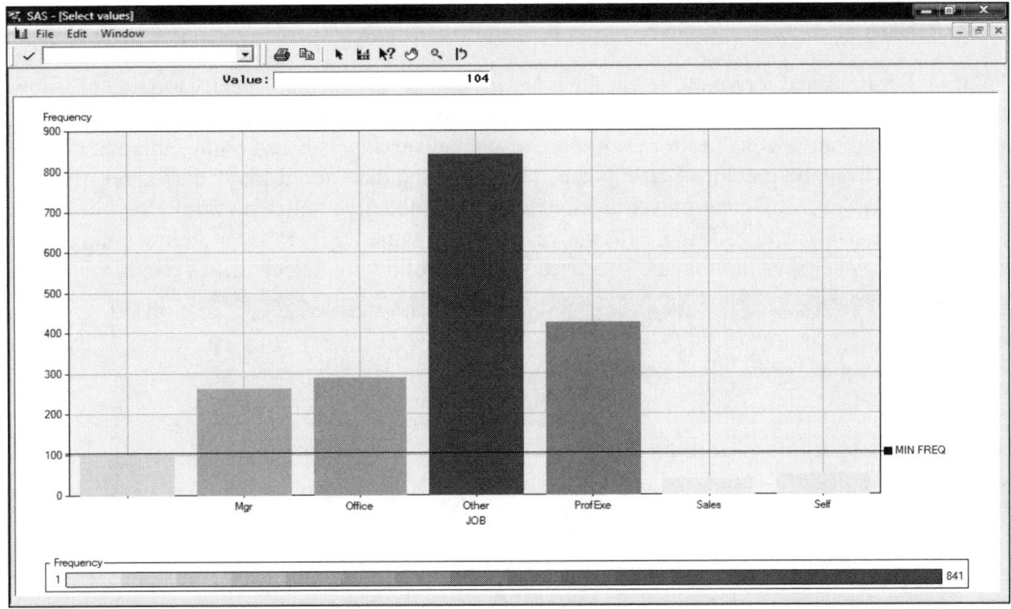

*The **Select values** window is used to specify the cutoff values within each categorical level.*

The **Values to exclude** column within the **Class Vars** tab is designed for you to specify values for the corresponding variables to be removed from the output data set. From the **Values to exclude** column, select the **Set values to exclude** pop-up menu item, which will result in the **Variable Histogram** window appearing for you to select the frequency bar and the corresponding class level to remove from the analysis. Additionally, hold down the Ctrl key to select several bars in order to remove more than one class level from the selected categorically-valued variable in the analysis. The observation will be removed from the output data set if the class level of the selected categorically-valued variable equals the values specified in the **Values to exclude** column. However, selecting every class level for removal will result in every observation being deleted from the output data set. Again, once the **Variable Histogram** window is closed and the appropriate class levels have been selected, then the corresponding class levels to be removed will be displayed within the **Values to exclude** column.

From within the **Select values** window, to remove class levels from the output data set based on the selected categorically-valued variable, select the **View > Frequency table** main menu option that will display the following frequency table. From the window, select the class level or hold down the Ctrl key to select several rows or class levels to be removed from the output data set, then close the window. The frequency bar chart will reappear that will display some of the bars that will be displayed in the color of gray indicating to you which class levels will be removed from the output data set.

Variable Value	Frequency
	96
Mgr	262
Office	288
Other	841
ProfExe	424
Sales	39
Self	50

The frequency table is used to select the class levels to remove from the subsequent analysis.

Interval Vars tab

The **Interval Vars** tab is designed for you to select the interval-valued variable and specify a range of values to remove from the output data set instead of applying the default filter to all interval-valued variables in the active training data set. The range of values for each interval-valued variable is based on the automatic filtering method that you selected from the previous **Settings** tab. The following diagram displays the values of the interval-valued input variables that were removed from many of the modeling nodes by fitting the interval-valued target variable, DEBTINC. In addition, extreme values in the same DEBTINC variable were removed from the fit based on the range of values that are specified from the following **Select values** window.

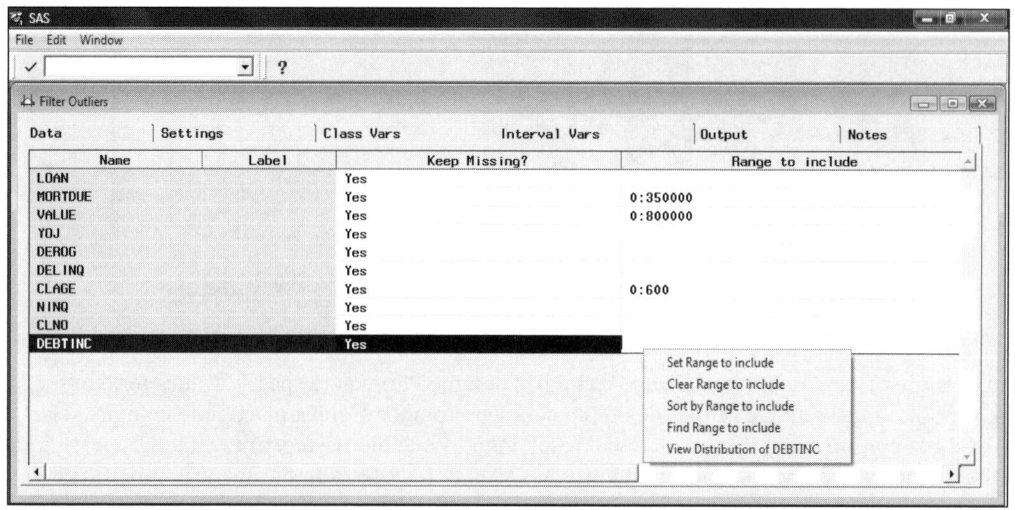

*The **Interval Vars** tab is used to remove extreme values from the interval-valued variables.*

By default, the missing values in the data set are retained in the output data set with a **Keep Missing?** attribute of **Yes**. To remove missing values based on the corresponding variable from the output data set, select the variable row, scroll over to the **Keep Missing?** column, then right-click the mouse and select the **Set Keep Missing?** menu option to select the **No** pop-up menu item. Selecting **Keep Missing?** attribute of **No** for all the input variables in the active training data set is actually what is done in predictive modeling in which all observations are deleted from the modeling fit if there are any missing values in any one of the input variables in the model.

From the tab, you may specify the minimum or maximum cutoff values for the selected variable. This will limit the range of values to the selected variables that are written to the output data set. From the **Range to include** column, simply enter the lower and upper bound to the selected interval-valued variable. Otherwise, select the **Range to include** column, then right-click the mouse and select the **Set Range to include** pop-up menu item. This will result in the **Select values** window appearing that is illustrated in the following diagram. Select the **Max** or **Min** slider bars to set the upper and lower limits and the range of values for the variable that are written to the output data set. Alternatively, select the **Bound** drop-list arrow to specify the maximum or minimum cutoff values from the **Value** entry field. Observations that are above or below the specified interval of the selected interval-valued variable are excluded from the output data set. If the maximum or minimum values are specified beyond the range of the actual data points of the selected variable, then Enterprise Miner will automatically reset the value to the actual maximum or minimum value in the data set. As a review, it is important that the **Use entire data** from the **Settings** tab be selected in order to view every value of the selected variable from the frequency charts. The **Use sample** option is automatically selected that will result in a subset of the original data set to be viewed.

To sort the output data set by the range of values included to the data mining analysis, select the variable to sort, then, from the same **Range to include** column, right-click the mouse and select the **Sort by Range to include** pop-up menu item.

Note: To undo all changes and filter settings that have been specified from the **Class Vars** and **Interval Vars** tabs and revert back to the default filter settings, then simply select the **Settings** tab and press the **Apply these filters to all vars** button.

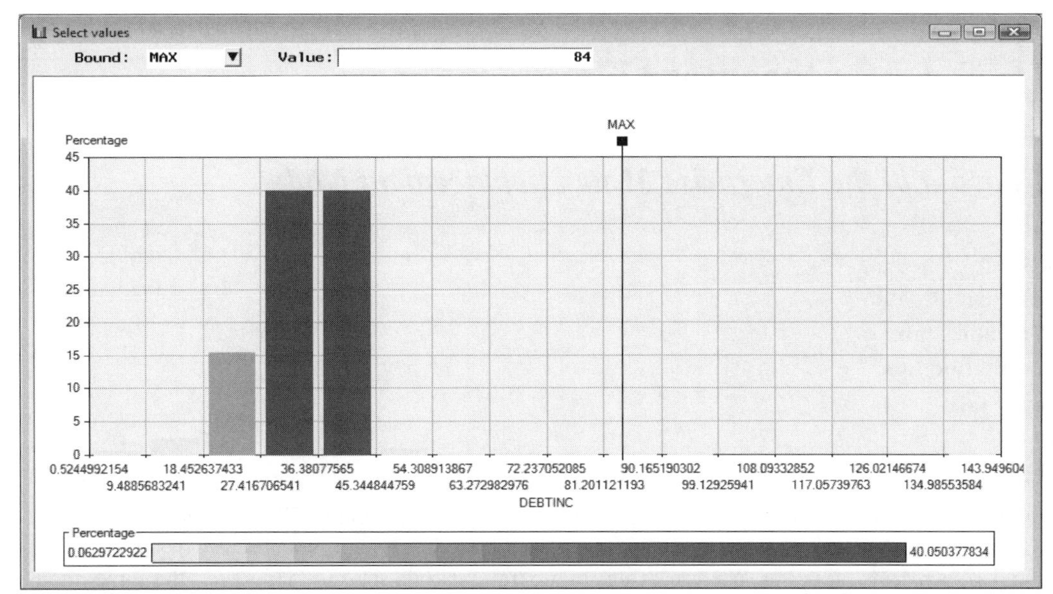

*The **Select values** window is used to set the cutoff values for the interval-valued variables.*

Output tab

The **Output** tab is designed for you to view both output data sets with the retained values and the removed values that are generated from the previously specified option settings. The data set with the filtered values that have been retained are passed along to the subsequent nodes within the process flow diagram for further analysis. The data set with the retained values can be viewed from the **Included Observations** section. The data set with the values that have been removed from the following analysis within the process flow can be viewed from the **Excluded Observations** section. The tab displays the standard table view of each partitioned data set in order for you to browse the filtered data sets. Both **Description** entry fields will allow you to enter a short description to each data set. The restriction is that the node must be executed in order to view both output data sets.

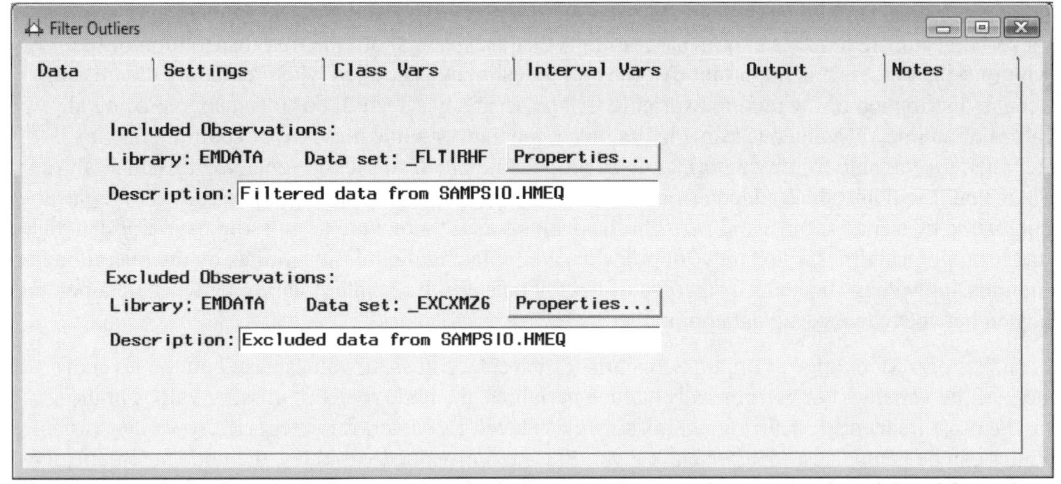

*The **Output** tab displays both data sets in which the records are either retained or deleted from the active training data set.*

Notes tab

The **Notes** tab is designed for you to document any modifications that have been performed on the active training data set with the extreme values or missing values that have been removed from the analysis.

3.4 Replacement Node

General Layout of the Enterprise Miner Replacement Node

- **Data tab**
- **Defaults tab**
- **Interval Variables tab**
- **Class Variables tab**
- **Tree Imputation tab**
- **Frequency tab**
- **Output tab**
- **Notes tab**

The purpose of the **Replacement** node in Enterprise Miner is designed to impute, fill in, or replace missing values based on one or more variables in the active training data set. The node is specifically designed to replace missing values, trim values, or redefine values in the training data set only. However, the same imputed values in the training data set will replace missing values in the other two partitioned data sets. The **Replacement** node might be considered if your data has several missing values or several incorrectly recorded values to achieve normality in the data. The node is designed for you to specify your own imputed values and estimate the missing values for each input variable in the training data set. In other words, the node will allow you to replace missing values by some reasonable estimate or redefine the range of values for every input variable in the active training data set or each input variable separately. For example, you might want to replace all records coded 999 that represent unknown values. Conversely, you might want to achieve normality and remove outliers in the data by replacing all values that have been incorrectly recorded as less than zero to zero or values that have been recorded as greater than 100 to 100. By default, the node gives you the added flexibility of imputing missing values, then replacing these same imputed values by a specified interval. In general, the node gives you the capability of replacing values in order to remove outliers and achieve normality in the data and replace missing values by some reasonable estimate in order to obtain a complete data set.

There is no one universal method or estimate that is used in replacing missing values. Replacing missing values by a certain estimate requires certain assumptions that are made about the true distribution of the variable with missing values. It is important that the correct estimate is applied when replacing the missing values since the distribution of the imputed variable will be seriously affected. For instance, replacing all missing values of an interval-valued variable by its mean, median, or some measure of central tendency is appropriate when the variable follows a normal distribution. The big drawback in replacing missing values in the variable is that it will introduce added error into the analysis. The basic idea of imputation is to replace each missing value by some estimate and perform the analysis as if there were no missing data. For data that is recorded in chronological order, some naïve approaches are replacing the missing values by the overall mean of all the nonmissing values, the mean or median of nearby data points, or interpolating between data points that are located between the missing data point over time.

The node consists of a wide range of imputation statistics to replace missing values based on the level of measurement of the variable. For categorically-valued variables, the node replaces missing values in the variable by the most frequently occurring nonmissing class level. However, for categorically-valued variables, missing values can be treated as a separate class level. For interval-valued variables, the node automatically replaces the missing values by its own mean. However, you may also select from a wide variety of other estimates such as the median, midrange, tree imputation, trimmed-mean, and various robust estimators.

For interval-valued variables, the robust estimators such as the Tukey's biweight, Huber's, and Andrew's wave are effective in larger sample sizes with several outliers. The purpose of the robust estimators is to try to reduce the effect of outliers by using substitute functions that are symmetric with a unique minimum at zero and do not dramatically increase in comparison to the sum-of-squares function. M-estimators are resistant, indicating that these estimators are not seriously affected by outliers or by rounding and grouping errors in the data. In addition, M-estimators have robustness of efficiency, meaning that the estimator is good when samples are repeatedly drawn from a distribution that is not precisely known. However, M-estimators perform best when the data has a symmetric distribution. The drawback to many of the traditional statistical estimators is that the performance of maximum likelihood estimators depends heavily on the assumption that the data belongs to the exponential family of distributions. In other words, maximum likelihood estimators can be inefficient and biased when the data is not from the assumed distribution, especially when outliers are present in the data.

Reasons for Estimating Missing Values

The reason for estimating missing values is to increase the number of records to the data mining analysis. This will result in increased power and better meet the distributional assumptions of the various statistical tests. Typically, as in the case with Enterprise Miner, the solution to handling observations with certain variables recorded with missing values is to completely discard the incomplete observation and analyze only those records that consist of complete data. This is called *listwise deletion* or the complete case method. In data analysis, there are two ways to handle missing values–either remove the observation from the analysis or impute the missing values. At times, excluding cases with missing values in a certain number of variables might lead to discarding information that is quite useful because of the other nonmissing values. Removing observations might result in a bias sample since the records with missing values might have some underlying characteristic associated with the nonrespondents, assuming that the missing values do not occur at random. If the missing values are related to either the input variables or the target variable, then ignoring the missing values can bias the results. As an example, a biased sample results when you sample people who have been arrested are less likely to report it than people who have not been arrested in analyzing the rate of people being arrested. When analyzing personal income, people with high incomes are less likely to report their income. Failure to recognize serious bias or independence in the data can create biased standard errors and inaccurate test statistics. That is, the statistical inferences are questionable and meaningless with a biased sample.

For unsupervised training, Enterprise Miner will automatically remove observations from the analysis with any missing values in the input variables. In Enterprise Miner, imputation is important in the various modeling nodes since missing values in any one of the input variables will result in observations being removed from the modeling fit. Therefore, these observations will not be scored. This suggests that as the number of input variables increases in the analysis, this will result in a greater chance that certain observations will be removed from the analysis. In supervised training, if there are missing values in any one of the input variables, then the interval-valued target variable will be estimated by its own average value. For the categorically-valued target variable, if there are missing values in any one of the input variables, then Enterprise Miner will impute the estimated probabilities from the prior probabilities for each target category. If the prior probabilities are not specified, then the sample proportions of the various class levels of the target variable will be applied.

The main idea of imputation or filling in missing values is to achieve a complete data set that will allow you to apply the standard statistical methods. In experimental designs that are applied compare the means of more than two target groups, an appealing approach is to fill in the missing values to restore the balance in the design and proceed with the standard analysis by estimating the mean between the groups in achieving an equal sample size within each group. The importance of a balanced design is because the ANOVA model is robust against unequal variability between the treatment cells. In cluster analysis, even if the data set consists of a majority of the observations with nonmissing cases, with the exception to one or two variables that have a large proportion of missing values, the node will automatically remove these same observations from the analysis, which will lead to disastrous results since a large majority of the data will be removed from the analysis.

There are various imputation methods used in estimating missing values. There exist two separate scenarios in dealing with missing data: data that are missing completely at random, (MCAR), or missing at random, (MAR). Most of the imputation methods depend on the MCAR assumption. MCAR assumption is such that the missingness of the missing variable is unrelated to missingness of the other variables, that is, the probability of the missing value is not related to the observed value and the value that would have been available had it not been missing. A simple check of the MCAR assumption is to divide the data set into two separate groups where one data set consists of the complete data set with all nonmissing observations and the other data set containing the intermittent observations with missing values. The MCAR assumption can be satisfied by performing a simple t-test statistic in testing the means between both groups or more general tests for equality of distributions between the two groups. An example of the MAR assumption is one in which the probability of missing data on age is related to a person's sex. However, the probability of missing age is unrelated within each sex. MAR is assumed when the missing values of the variable are explained by the other variables in the data set that are observed.

If the missing values occur randomly along with nonmissing values, they are called *intermittent missing values*. In other words, these missing values are unrelated to the sampling process, so these same missing values can be assumed to be missing at random that will result in valid statistical inference. However, if the missing values occur after a certain time period, then the missing values are called drop-outs or monotone missing data. *Drop-out observations* present a much bigger problem in estimating missing values in comparison to intermittent missing values. The reason is because the observations that are missing from the study are directly or indirectly related to the sampling process. Therefore, assuming that the drop-out observations occur at random during the sampling process is misleading and will result in biased estimates.

Imputation Methods

There are three separate methods of imputation based on the number of imputed values: single imputation, multiple imputation, and EM imputation. *Single imputation* replaces each missing value with a single value. *Multiple imputation* replaces the missing values with more than one value. Multiple imputation replaces each missing value in the first pass to create the first complete data set, then replaces each missing value in the second pass by the second imputed estimate, and so on. The standard complete-data methods are used to analyze each data set. The resulting complete-data analysis can then be combined to reflect both the variability of the separate imputation estimates by filling in the missing values and the separate imputation methods that are applied. One disadvantage to multiple imputation as opposed to single imputation is the extra work that is involved in creating the separate imputation estimates.

The most common imputation procedures in estimating missing values are as follows:

Hot-Deck Imputation: The missing values are replaced by values that exist in the data set that share some common characteristics or similar responding units. This method replaces missing values of incomplete records using values from similar, but complete records of the same data set. One advantage of this method is that it does not require any of the analytical calculations of the missing value estimates. This method can involve numerous imputation schemas in determining units that are similar for imputation.

Mean Imputation: The missing values are replaced by the mean of the recorded values.

Regression Imputation: The missing values are replaced by the predicted values from the regression model by every other input variable in the data set with recorded values. The model is designed so that the input variable with the missing values plays the role of the target variable for the predictive model. The advantages to tree imputation is its tolerance to missing values in the data, its flexibility in handling different types of variables and its robustness in the distributional assumptions in the input variables in the model.

Alternative imputation schemas that might be applied due to an insufficient sample size consist of repeatedly resampling the original data. These iterative techniques are called the bootstrap and the jackknife methods. In other words, they involve resampling and subsampling the original data and calculating the imputed estimate, that is, the sample mean of all nonmissing observations, of the data which can be performed any number of times.

Bootstrapping estimates: Bootstrapping estimates fill in missing values by calculating the average of the bootstrapping estimates, that is, the summation of each estimate such as the sample mean, then dividing by the number of bootstrap samples.

Jackknifing estimates: An alternative iterative technique that is similar to bootstrapping is called *jackknifing*. This technique also performs an iterative routine in calculating an estimate from every other data point with the exception that an arbitrary point is removed from the original sample, then calculates an estimate, the sample mean, from all other observations in the data set. This process is repeated any number of times. Therefore, similar to the bootstrapping estimate, the jackknife estimate is the average of the various estimates.

Imputing Values Using the Enterprise Miner Nodes

The **Tree**, **Clustering**, and **Princomp/Dmneural** nodes are the other Enterprise Miner nodes that are designed to impute missing values. The **Tree** node will allow you to specify surrogate rules in order to use other variables with similar characteristics or attributes in replacing missing values of the corresponding primary splitting variable. The **Clustering** node will allow you to create clustering groupings by replacing missing values by the corresponding value of the closest cluster seed or cluster mean in which the observation is assigned and the **Princomp/Dmneural** node that imputes missing values in any one of the input variables.

The various imputation techniques are discussed in greater detail in Rubin and Little (2002).

Data tab

The **Data** tab is designed to select the active training data set to impute missing values. Again, by default, imputation is performed on all input variables in the active training data set. In other words, missing values in the target variable are left unchanged. The tab contains the following two subtabs:

- **Inputs subtab**
- **Imputation Methods subtab**

Inputs subtab

By default, the **Inputs** subtab is displayed once you select the tab. That subtab will allow you to view the separate partitioned data sets or select an entirely different data set for the data mining analysis. The subtab has the same appearance and functionality as the previously displayed **Data** tabs. Therefore, the subtab will not be displayed.

Training subtab

The **Training** subtab is illustrated in the following diagram. The node is designed to replace the missing values by some reasonable estimate of the active training data set only. Therefore, the subtab will allow you to either select a random sample of the training data set or the entire sample. By default, a random sample of 2,000 observations is used to impute the missing values from the training data set. Otherwise, select the **Entire data set** option that will result in the node using the entire training data set in calculating the various imputation estimates. Select the **Generate new seed** button in order for the node to generate a different random seed that will create an entirely different random sample from the active training data set. However, the corresponding random seed value that is specified is saved by the node and will result in the node creating exactly the same partitioning of the active training data set based on each successive execution of the node.

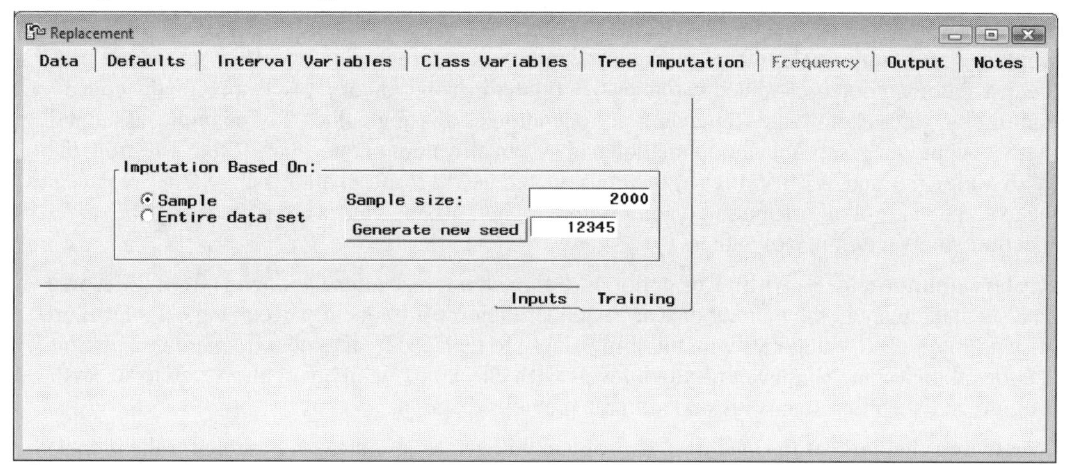

*The **Training** subtab is used to select a random sample of the training data set to impute missing values.*

Defaults tab

The **Defaults** tab will be displayed as you first open the **Replacement** node. The **Defaults** tab is designed for you to replace the missing values before imputation, create identifiers for the output data set to identify all observations that have been imputed that can be used as a categorically-valued input variable for the various modeling procedures, specify the type of imputation for both the interval-valued and categorically-valued variables, specify a range of values for the interval-valued variables, or specify a constant value for all missing values for both the interval-valued and categorically-valued variables in the active training data set.

The **Defaults** tab consists of the following three subtabs.

- **General subtab**
- **Imputation Methods subtab**
- **Constant Values subtab**

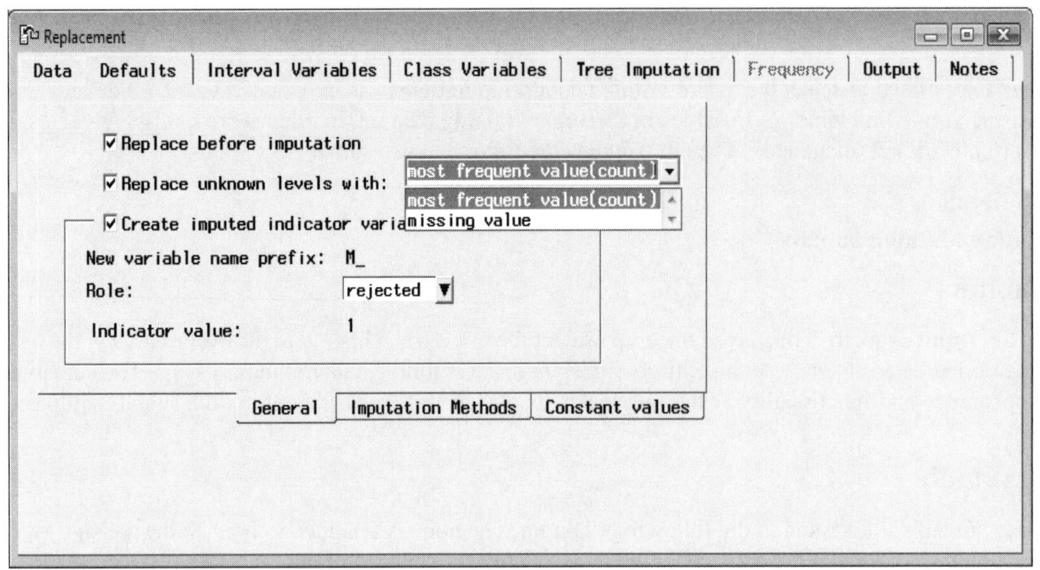

*The **Defaults** window is used to specify the default settings for the imputed estimates of the analysis.*

General subtab

Enterprise Miner is designed to replace the missing values or redefine missing values before performing the various imputation methods. The following options are designed for you to specify the method of replacement of the missing values before the imputation techniques are applied. By default, the check boxes of the following options are clear. Therefore, the following imputation settings are not applied unless otherwise specified. The node is designed for you to select any or all of the following listed imputation options:

- **Replace before imputation:** Replace the missing values before imputation. This option will trim the distribution of interval-valued variables to a tighter distribution or replace specifically coded nonmissing values before the **Replacement** node imputes missing values. For example, assuming that you apply the mean imputation method and assign all values greater than 100 to 100 from the **With Value (<)** and **With Value (>)** columns, then this will result in all missing values replaced by the average value of all nonmissing values before all nonmissing values are replaced by the specified interval or range of values.
- **Replace unknown levels with:** This option is designed to replace the class levels from the scored data set that are not in the training data set by either their most frequently occurring class level or missing. By default, categories with missing values are replaced by the most frequent nonmissing class level. Select the **Replace unknown levels with** check box, which will allow you to select the drop-down list options that is illustrated in the previous diagram.
- **Create imputed indicator variables:** This option will create an indicator variable for the output data sets that identifies each observation that has been imputed based on the input variables in the training data set. The imputed indicator variable name is called M_*variable name*. At times, this binary-valued indicator variable might be an important predictor variable in describing the target values of the statistical modeling design. That is, there might exist some underlying nature that the observations are missing that might contribute to explaining the variability in the target variable. By default, the indicator variable is excluded from the modeling design with the model role set to **rejected**. In order to include the indicator variable to the modeling design, select the **Role** field drop-down arrow in order to select the **input** model role.

Imputation Methods subtab

The **Imputation Methods** tab is designed for you to specify the imputation statistic that is used to fill in the missing values of all the interval-valued or categorically-valued input variables in the active training data set.

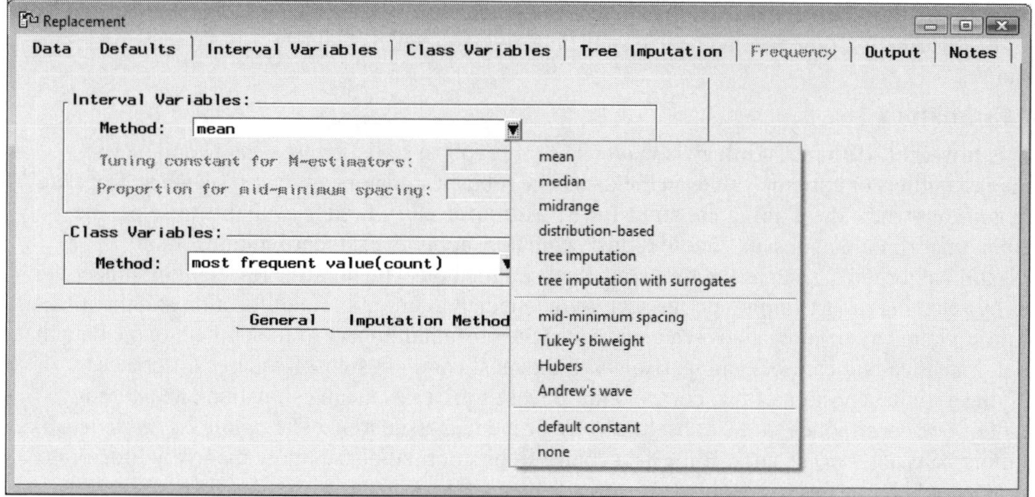

*The **Imputation Method** subtab is used to specify the various imputation estimates for the analysis.*

The following are the various imputation statistics that you may select from within the **Imputation Method** subtab based on the interval-valued variables in the active training data set. Select the drop-down arrow button from the **Method** display field in order to select from the following imputation estimates:

- **mean** (default)**:** Replaces missing values in the variable by its own mean from the training data set that is the default. The mean is the sum of all values divided by the number of nonmissing observations. The mean is a measure of central tendency that is recommended as the best estimate assuming that the data follows an approximately normal distribution.
- **median:** Replaces missing values in the variable by its own median or the 50th percentile, that is, the data point located in the middle or the average of the two separate data points located in the middle. The median is preferable when you want to impute missing values for variables that have skewed distributions. The advantage of imputing missing values by the median as opposed to the mean is that the median is less sensitive to extreme values in the data.
- **midrange:** Replaces missing values in the variable with the average of the maximum and minimum value, that is, the ratio between the sum of the maximum and minimum values over two.
- **distribution-based:** Replaces missing values in the variable by the probability distribution of the nonmissing values of the variable. This option replaces missing values by the random percentile estimates of the variable.
- **tree imputation:** Replaces missing values in the variable from the PROC SPLIT procedure using decision tree modeling. The procedure fits a separate predictive model based on the input variable with missing values playing the role of the target variable regressed against all other input and rejected variables with nonmissing values in the data set. The restriction is that variables that have a model role of target cannot be imputed. This method is analogous to the regression imputation procedure. Again, the advantage of imputing values from decision tree modeling is its tolerance to missing values in the data and its robustness to the distributional assumptions in the input variables in the model. However, it is suggested that when the tree imputation option is selected, then it is important that you use the entire training data set for more consistent results and reliable estimates.
- **tree imputation with surrogates:** Replaces missing values in the variable by using the previous technique with additional surrogate splitting rules such that other variables with similar characteristics, attributes or same range of values be used in replacing missing values of the corresponding variable. The surrogate splitting rules are recursive – if they rely on the input variable whose values are missing, then the next surrogate rule is applied. If the missing value prevents the main rule and all surrogates from being applied to the observation, then the main rule assigns the observation to the branch that is assigned to the missing value.

- **mid-minimum spacing:** Replaces missing values in the variable by the trimmed mean statistic that is based on a specified percentage of the data. This statistic is considered a robust estimate of the mean. Specifying this option, then the **Proportion for mid-minimum spacing** will become available, which will allow you to specify the proportion of the data to calculate the arithmetic mean. The default is 90% of the data that is used to calculate the mean with 10% of the data removed at both tails of the distribution. The trimmed mean statistic is recommended if there exists several outliers in the data.

Robust Estimators

- **Tukey's biweight, Huber's, Andrew's wave:** The three robust statistics are designed to be less sensitive to outliers or extreme values in the data. The robust estimators are based on the appropriate tuning constant where the **Tuning constant for M-estimator** entry field is available to select the corresponding trimming constant. These robust estimators apply the standard minimization optimization algorithms, such as the Newton – Raphson method. The iterative reweighting procedure begins by selecting robust starting points such as the median as an estimate of location and the median absolute deviation as an estimate of scale, to reduce the profound impact of the outliers in the data that will not dramatically increase in comparison with the most common sum-of-squares differences. Specifying a tuning constant of two corresponds to least-squares or mean estimation and a tuning constant of one corresponds to least absolute value or median estimation. The *tuning constant* acts like a trimming constant. For instance, if the data comes from a normal distribution, then very little or no trimming should be performed. If the data comes from a heavy-tailed distribution, then a heavier amount of trimming should be performed at both ends. For M-estimators, the degree of trimming to the data is determined by the tuning constant.

Constants

- **default constant:** Replaces the missing values with a default constant value or a single value.
- **None:** Prevents Enterprise Miner from replacing missing values in the interval-valued variables.

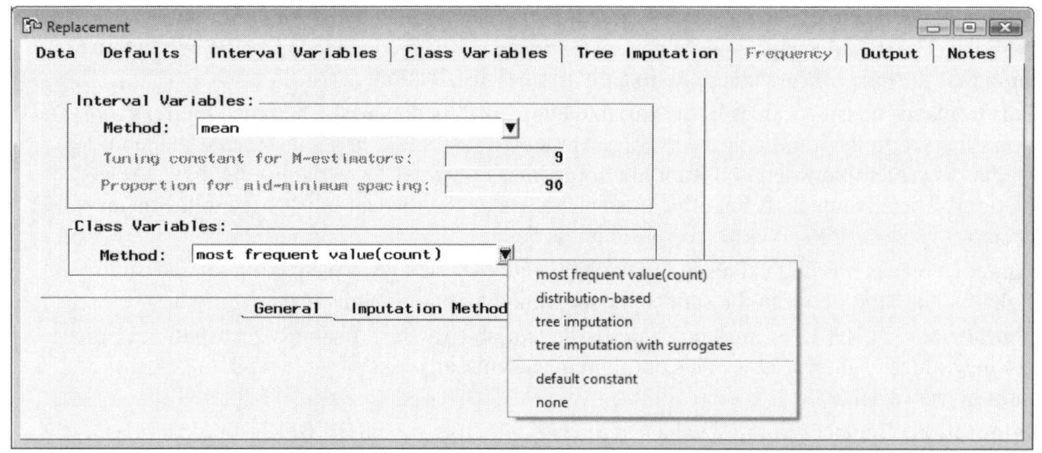

*The **Imputation Method** subtab selects missing value estimates of the categorically-valued variable.*

The following are the various imputation statistics that you may select from the **Imputation Method** subtab that is based on the categorically-valued variables in the active training data set.

- **Most frequent value (count)** (default)**:** This is the default method used that replaces missing class levels with the most frequently occurring nonmissing class level. If there exist several class levels that have the same frequency, then the smallest value is applied. If the most frequently occurring value is missing, then the next most frequently occurring value is used, and so on.
- **distribution-based:** See above.
- **tree imputation:** See above.
- **tree imputation with surrogates:** See above.
- **default constant:** Replaces the missing values with a constant value that can be specified from the following **Constant Values** subtab.
- **None:** Prevents Enterprise Miner from replacing the missing values with the most frequent categorical level. Imputation is not performed to the categorical variables in the active training data set.

Constant values subtab

The **Constant values** subtab is designed for you to estimate missing values by a specified constant or specified value to all of the input variables in the active training data set. For numeric variables, the subtab will allow you to redefine the range of values or replace extreme values by a specified constant or a single value. For character variables, missing values are replaced by a specified label before imputing is performed. However, it is important to understand that setting the default values in the subtab does not automatically result in trimming the values of the interval-valued variables or impute numeric or categorically-valued variables. In other words, imputation takes place by specifying the various methods for each variable in the active training data set from the subsequent **Interval Variables** and **Class Variables** tabs.

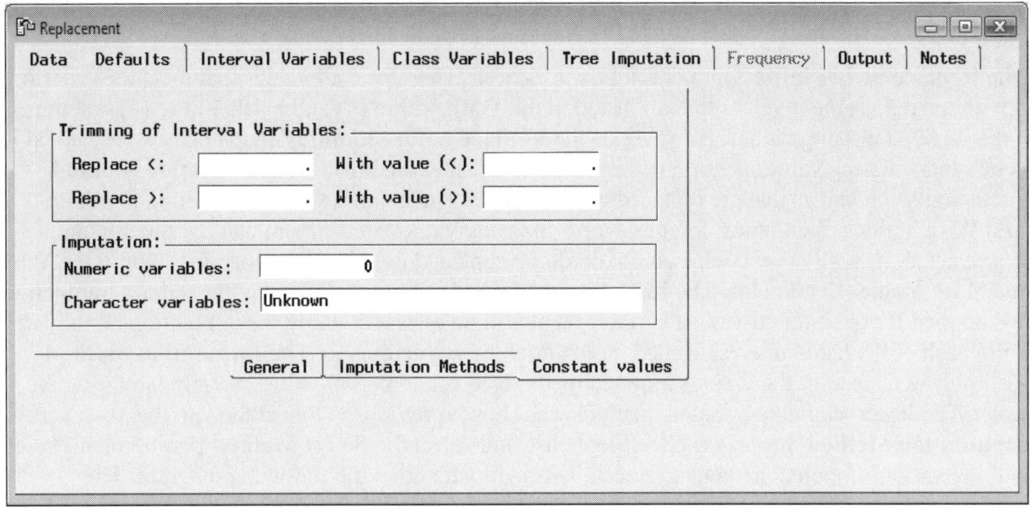

*The **Constant values** subtab is used to specify constant values to replace the missing values for the interval-valued or categorically-valued variables in the training data set.*

- **Trimming of Interval Variables:** This section will allow you to place a default numeric bound about the distribution of the variables in the training data set in order to replace incorrect nonmissing values. From the **Replace <** entry field, you may specify a default lower bound with a constant value that may be specified from the **With value (<)** entry field. For example, from the previously listed diagram, you might want to replace values less than zero with the value of zero.

  From the **Replace >** entry field, you may specify a default upper bound with a constant value that may be specified from the **With value (>)** entry field. For example, you might want to replace values greater than 1000 with the value of 1000 in order to remove outliers and achieve normality in the data.

- **Imputation:** This section is in relationship with the **default constant** imputation option setting. In other words, the section will allow you to specify a default value for all missing values in the interval-valued or categorically-valued input variables in the training data set by specifying the previous **default constant** imputation option setting. From the **Numeric variables** entry field, you may specify a default numerical value to replace all missing values in the interval-valued variables or categorically-valued variables with numeric values from the training data set. For example, you might want to replace all numeric missing values with the value of zero, as illustrated in the previous diagram. The value that is entered will be carried over to the **Interval Variables** tab when the **default constant** imputation method is selected for anyone of the interval-valued variables that are listed within the tab.

  From the **Character variables** entry field, you may specify a default character value to replace all missing values in the categorically-valued variables of the training data set. For example, you might want to replace all missing values in the categorically-valued variables with the value of Unknown, as illustrated in the previous diagram. The character value that is entered will be displayed in the **Class Variables** tab when the **default constant** imputation method is selected for anyone of the categorically-valued variables.

Interval Variables tab

The **Interval Variables** tab is designed to override the default imputation techniques that have been specified from the previous **Defaults** tab in order for you to specify the various imputed estimates to apply to each interval-valued variable in the training data set with missing values. The imputation method that is assigned to the interval-valued variables is associated with the option settings specified from the previous **Defaults** tab.

The **Status** column indicates that the imputed values will replace the missing values. By default, the missing values from the target response are left unchanged, with all input variables with missing values imputed. To prevent Enterprise Miner from automatically replacing missing values of certain variables with imputed values, select the variables row and scroll over to the **Status** column, then right-click the mouse to select the **don't use** pop-up menu item.

By default, the **Replace before imputation** check box is cleared. Therefore, all values that meet the criterion are replaced with the replacement values that are listed in the **With Value (<)** and **With Value (>)** columns, that is, even the imputed missing values. However, if the **Replace before imputation** check box is selected in the **Defaults** tab, then missing values are imputed by the method specified from the **Imputation Method** column before nonmissing values that are replaced with the replacement values that are listed in the **With Value (<)** and **With Value (>)** columns. In other words, missing values are first imputed by the imputation method that is selected, then all other nonmissing records are replaced by the values specified within the **With Value (<)** and **With Value (>)** columns. The **Imputation Method** column will display the default imputation method that is applied for the interval-valued variables that will be over written by the imputation method that is selected from each of the listed interval-valued input variables within the tab. The **Imputation Method** column will allow you to specify the various imputed methods to each interval-valued variable in the active training data set. The list of available imputed methods was shown previously. Select the variable row, scroll over to the **Imputation Method**, then right-click the mouse and select the **Select Method** pop-up menu item with the list of the various imputed methods to select. This is illustrated in the following diagram. The imputation method that is selected will automatically overwrite the default imputation method that is selected from the previous **Imputation Methods** subtab.

From the tab, you may want to replace the missing values in the variable by a default constant. Therefore, select the **Imputation Method** column and right-click the mouse to select the **default constant** pop-up menu option. By default, the constant value is set to missing. However from the **Imputation Method** column, you may right-click the mouse and select the **Set Value** pop-up menu item with the **Specify Numeric Value** window appearing for you to enter the corresponding value to replace all missing values of the selected interval-valued variable.

From the tab, you may want to trim the distribution of the selected input variable by specifying a predetermined upper and lower bound with the idea in replacing the incorrect nonmissing values that fall above or below a given range. Both the **Replace <** and **With Value (<)** columns are designed for you to replace values that fall below the lower limit bound by the value that is specified in the **With Value (<)** column. Similarly, the upper limit bound to the variable may be specified from both the **Replace >** and the **With Value (>)** columns. For example, all the interval-valued input variables with values less than zero are replaced with the value of zero. Conversely, all records with the number of years at the present job, YOJ, that are over 60 years are replaced with 60 years, as illustrated in the following diagram. By selecting the **default constant** option setting, all missing values in the DEROG, DELINQ, and NINQ input variables, are set to zero that is defined by the value entered from the previous **Defaults** tab. As an alternative, the **set value...** option setting may be selected to perform the same imputation. It is important to keep in mind that the imputation method that is selected to replace missing values is calculated from the range of values from the incoming active training data set.

By default, the replacement value is set to either the **With Value (<)** or the **With Value (>)** data value, depending on the selected column. The idea is to remove the extreme values that are either too small or too large in comparison to the other values of the corresponding variable, thereby creating a more normal distribution of the selected variable. Simply enter the imputed values from the **With Value (<)** or the **With Value (>)** columns and enter the corresponding value. Otherwise, right-click the mouse and select the **Specify Value** pop-up menu item that will display the **Specify Numeric Value** window in which you may enter the imputed value from the **New Value** entry field.

For each variable that is listed within the tab, select the **Clear** pull-down menu option within the **Replace <**, **With Value (<)**, **Replace >**, and **With Value (>)** columns to clear the corresponding changes that are displayed within the column.

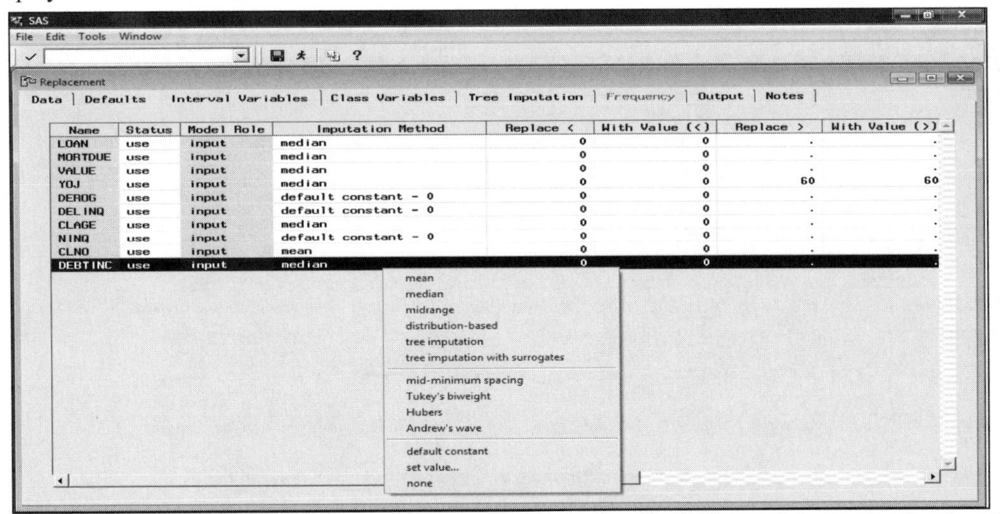

*The **Interval Variables** tab is used to specify the imputed methods for each interval-valued variable.*

Class Variables tab

The **Class Variables** tab is designed similarly to the **Interval Variables** tab that customizes the default settings from the **Imputation Methods** subtab by replacing missing class levels of each categorically-valued variable in the training data set. As long as the **Replace unknown level with** check box is not selected in the **General** subtab, then nonmissing class levels are replaced with the new replacement value that is listed in the **Replace Value** column before missing values are imputed by the method shown in the **Imputation Method** column. From the node, missing values of the listed categorically-valued variable are automatically imputed by the method that is selected from the **Replace unknown levels with** check box within the **General** subtab. The default method that is selected will be automatically listed within the **Imputation Method** column. However, the **Class Variables** tab is designed to override the default imputation techniques that can be specified from the previous **Defaults** tab in order for you to select the various imputed estimates to apply to each categorically-valued variable in the training data set with missing values. Similar to the **Interval Variables** tab, you may select any one of the **Imputation Method** cells to select the various imputation methods that are available within the node for the listed categorically-valued variables in the active training data set.

The **Replace Value** column is designed for you to specify or redefine values of the categorically-valued variable in the output data set. Select the variable, then select the **Replace Value** cell and right-click the mouse to select the **Specify Values** pop-up menu item. The **Specify Values** window will appear, as illustrated in the following diagram. The window will allow you to redefine and replace either character or numerical values with discrete levels. From within the **Specify Values** window, the **Character Variables** tab is available for viewing if you specify the replacement values of the character variable from the **Replacement Value** column. For example, all entries listed in the **Replacement Value** column will replace all observations in the output data set with the existing class levels that are listed in the **Value** column. Conversely, the **Numeric Variables** tab is available for you to specify the replacement values of the discrete values of the selected categorically-valued variable with numeric class levels. In statistical modeling, there are times when a unique class level will be the only class level that significantly predicts the target variable. Therefore, this might suggest that you might consider redefining the class levels of the categorical variable by collapsing the other nonsignificant class levels. This option will allow you to collapse the nonsignificant class levels into one unique class level.

The **Status** column will allow you to control the imputation method that is performed for all categorically-valued input variables in the active training data set. Notice in the following diagram that the variable status attribute of the categorically-valued target variable BAD (defaulting on a home loan) is automatically set to **don't use**. That is, missing values in the target variable are left unchanged.

Note: For character values, you can enter the automatic SAS variable of _BLANK_ or _blank_ to replace all nonmissing class levels to missing within the **Replacement Value** column.

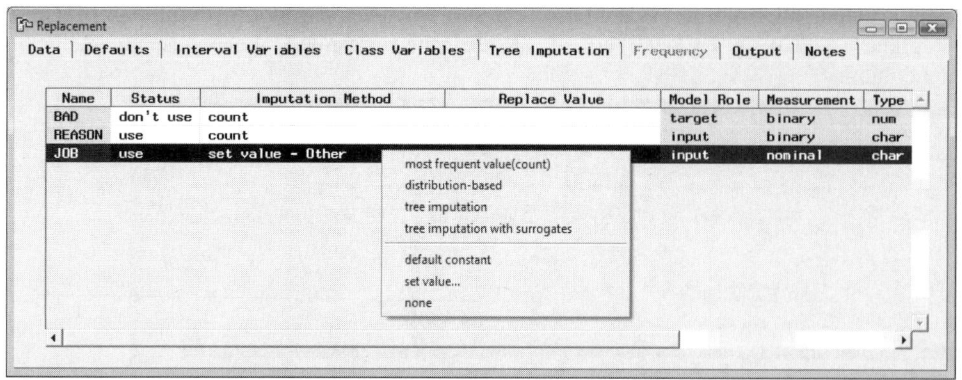

*The **Class Variables** tab is used to specify the imputation method for each categorical variable.*

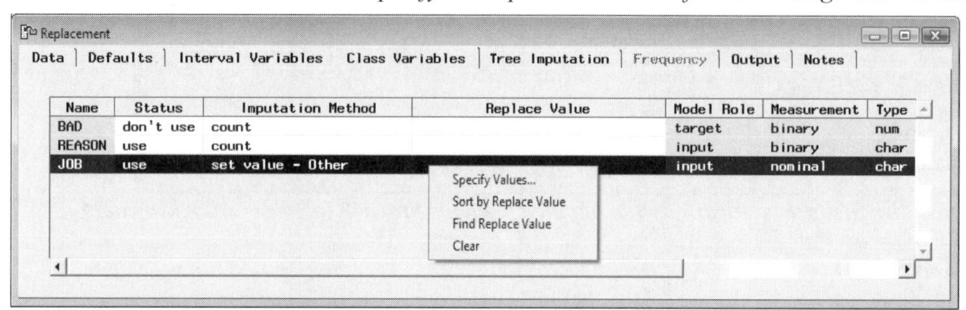

*The **Class Variables** window is used to replace values for each category of the categorical variable.*

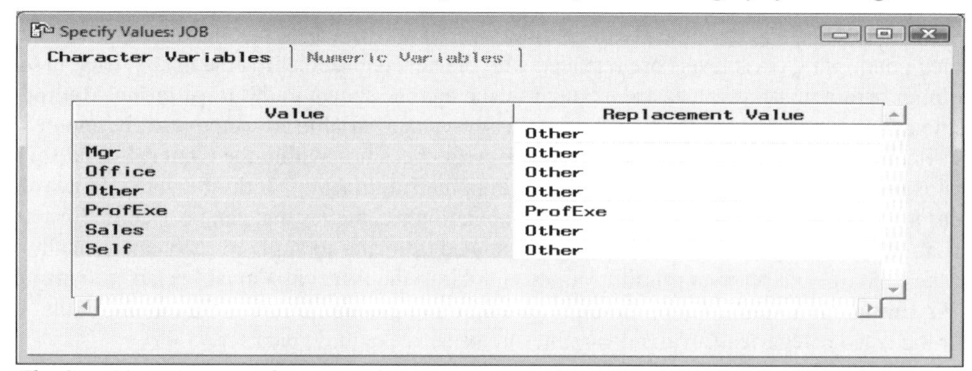

*The **Specify Values** window is used to redefine the character values in the output data set.*

Tree Imputation tab

The tab will allow you to select the input variables that will be included in the tree imputation model. From **Tree Imputation** tab, select the **Status** column to specify the input variables that you want to use during tree imputation. The tree imputation estimates are used when you specify the tree imputation method from the previous tabs. By default, all the input variables are initially set to a default status of **use**. This indicates that the input variable with missing values plays the role of the target variable in the model based on all the other input and rejected variables that are automatically included into the tree imputation model. However, if you believe that certain input variables are not good predictor variables in the imputation model, then set the status of the selected input variable to **don't use** from the **Status** column to remove the input variable from the tree imputation model.

For imputing interval-valued variables, a regression decision tree is performed. In the predictive tree model, the average values at each leaf is used to replace missing values in the input variable that is defined by the recursive tree branching process from the various range of values of the other input variables in the model. For imputing categorically-valued input variables, a classification decision tree is performed. In the classification tree model, the most frequent class level at each leaf is used to replace the missing class levels of the input variable that is defined by the various range of values of the other input variables in the model. Since the imputation values are estimated from the other input variables in the data, then the imputation technique may be more accurate than simply replacing missing values in the interval-valued input variables by its mean or median. Although, tree imputation is robust to the distributional assumptions in the input variables, however, the tree imputation method is most effective when the input variables are approximately normally distributed. And finally, the tree imputation method is tolerant to missing values in the data, since the recursive tree splitting routine is performed to each input variable one at a time from the range of values in the corresponding variable with comparison to models such as regression modeling that combines several input variables in the model. One drawback to this imputation method is that if there exist missing values in the other input variables in the model, then their missing values must be imputed to avoid these same observations from being removed from the fit.

Replacement

Data | Defaults | Interval Variables | Class Variables | Tree Imputation | Frequency | Output | Notes

Inputs for tree imputation:

Name	Status	Model Role	Measurement	Type	Format
LOAN	use	input	interval	num	BEST12.
MORTDUE	use	input	interval	num	BEST12.
VALUE	use	input	interval	num	BEST12.
YOJ	use	input	interval	num	BEST12.
DEROG	use	input	interval	num	BEST12.
DELINQ	use	input	interval	num	BEST12.
CLAGE	use	input	interval	num	BEST12.
NINQ	use	input	interval	num	BEST12.
CLNO	use	input	interval	num	BEST12.
DEBTINC	use	input	interval	num	BEST12.
REASON	use	input	binary	char	$7.
JOB	use	input	nominal	char	$7.

*The **Tree Imputation** tab is used to specify the various input variables for tree imputation.*

Frequency Tab

The **Frequency** tab is designed to display all the variables with a variable role of **freq**. Therefore, the tab is grayed-out and unavailable for viewing unless a variable in the training data set has a **freq** variable role. By default, the frequency variable is automatically used to calculate the imputed statistics that are indicated from the **Status** column of **use**. To prevent the frequency variable from calculating the imputed statistics, simply set the variable role of the frequency variable to **reject**, usually from the previous **Input Data Source** tab. The data set structure is assumed to have each observation with *n* observations, where *n* is the value of the frequency variable. For example, if the mean imputation method is selected for the interval-valued variables, then the imputed values are calculated by multiplying their values by the *n* frequency counts divided by the total number of frequency counts. Therefore, the restriction is that only one frequency variable can be specified in the active training data set. If the value of the frequency variable is less than one or missing, then the observation is not used in the calculations. In addition, the node does not replace missing values in the frequency variable.

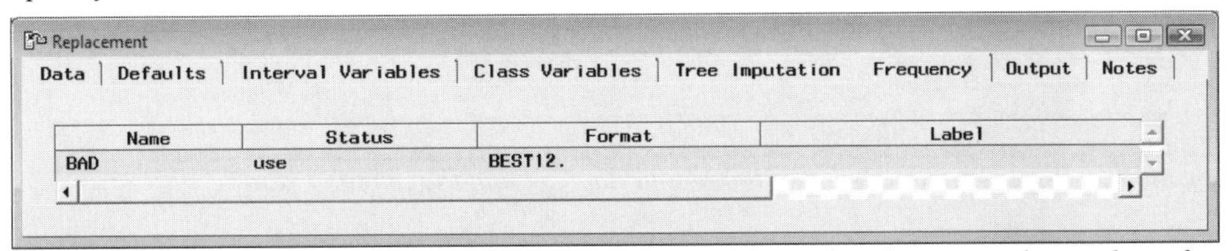
Replacement

Data | Defaults | Interval Variables | Class Variables | Tree Imputation | Frequency | Output | Notes

Name	Status	Format	Label
BAD	use	BEST12.	

*The **Frequency** tab displays the unique frequency variable used to determine the number of records.*

Output Tab

The **Output** tab will allow you to view the imputed values from the active training data set. The **Output** tab displays the detailed listing of the training, validation, test, and scored output data sets. By default, the name of the training data set is listed in the **Output** tab. To view the name and description of the other partitioned or scored data sets, select the **Validation**, **Test**, or **Score** radio button from the **Show details of:** section. However, the one restriction is that the node must first be executed in order to select the **Properties...** button to view the selected output data sets. Since the various imputation methods are calculated from the active training data set, the validation and test data sets will have their missing values replaced by the same estimates that are imputed in the active training data set. These data sets are passed on to the subsequent nodes in the process flow diagram.

Viewing the Replacement Node Results

General Layout of the Results Browser within the Replacement Node

- **Table View <data set name> tab**
- **Interval Variables tab**
- **Class Variables tab**
- **Code tab**
- **Log tab**
- **Output tab**
- **Notes tab**

The following results are based on the **Replacement** node that was applied to the modeling comparison procedure that is displayed in the following **Assessment** node. The median value was selected to replace missing values in most of the interval-valued input variables in the classification models since these same input variables in the active training data set displayed a highly skewed distribution.

Table View <data set name> tab

The **Table View <data set name>** tab will be displayed as you view the results from the **Replacement** node. The tab will allow you to display a table view of the output data set with a listing of the various imputed values that estimated the missing data points, along with the rest of other data from the active training data set. By selecting the **File > Export** main menu options, the tab will allow you to export the output data set into many other file formats.

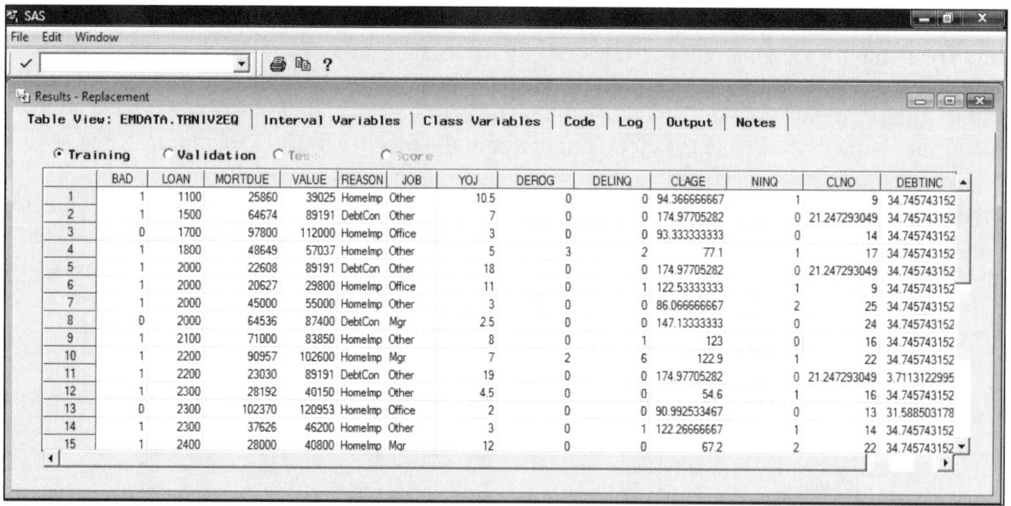

SAS
File Edit Window
Results - Replacement
Table View: EMDATA.TRNIV2EQ | Interval Variables | Class Variables | Code | Log | Output | Notes
Training Validation Test Score

	BAD	LOAN	MORTDUE	VALUE	REASON	JOB	YOJ	DEROG	DELINQ	CLAGE	NINQ	CLNO	DEBTINC
1	1	1100	25860	39025	HomeImp	Other	10.5	0	0	94.366666667	1	9	34.745743152
2	1	1500	64674	89191	DebtCon	Other	7	0	0	174.97705282	0	21.247293049	34.745743152
3	0	1700	97800	112000	HomeImp	Office	3	0	0	93.333333333	0	14	34.745743152
4	1	1800	48649	57037	HomeImp	Other	5	3	2	77.1	1	17	34.745743152
5	1	2000	22608	89191	DebtCon	Other	18	0	0	174.97705282	0	21.247293049	34.745743152
6	1	2000	20627	29800	HomeImp	Office	11	0	1	122.53333333	1	9	34.745743152
7	1	2000	45000	55000	HomeImp	Other	3	0	0	86.066666667	2	25	34.745743152
8	0	2000	64536	87400	DebtCon	Mgr	2.5	0	0	147.13333333	0	24	34.745743152
9	1	2100	71000	83850	HomeImp	Other	8	0	1	123	0	16	34.745743152
10	1	2200	90957	102600	HomeImp	Mgr	7	2	6	122.9	1	22	34.745743152
11	1	2200	23030	89191	DebtCon	Other	19	0	0	174.97705282	0	21.247293049	3.7113122995
12	1	2300	28192	40150	HomeImp	Other	4.5	0	0	54.6	1	16	34.745743152
13	0	2300	102370	120953	HomeImp	Office	2	0	0	90.992533467	0	13	31.588503178
14	1	2300	37626	46200	HomeImp	Other	3	0	1	122.26666667	1	14	34.745743152
15	1	2400	28000	40800	HomeImp	Mgr	12	0	0	67.2	2	22	34.745743152

*The **Table View** tab that displays a table view of the output data set and the imputed values.*

Interval Variables tab

The **Interval Variables** tab will allow you to display the various input variables that were imputed or replaced in the training data set. The **Imputation Method** column displays the method of imputation that was applied in estimating the missing values. Again, missing values in many the following listed interval-valued variables were estimated by their own median value. The **Number Imputed** column will display the number of data points or the number of missing values that were imputed. The **Number Replaced (<)** and the **Number Replaced (>)** columns will display the number of observations that were replaced at both ends of the distribution of each input variable.

From the following table listing, the variable with the largest portion of missing values was DEBTINC, (ratio of debt-to-income), with 617 missing cases in the training data set. Conversely, the amount of the requested home loan, LOAN, has absolutely no missing values.

Results - Replacement

Table View: EMDATA.TRNIV2EQ | Interval Variables | Class Variables | Code | Log | Output | Notes

Name	Model Role	Imputation Method	Imputation Value	Number Imputed	Number Replaced (<)	Number Replaced (>)
LOAN	input	median	16200	0	0	0
MORTDUE	input	median	64674	253	0	0
VALUE	input	median	89191	56	0	0
YOJ	input	median	7	250	0	0
DEROG	input	default constant	0	352	0	0
DELINQ	input	default constant	0	292	0	0
CLAGE	input	median	174.97705282	159	0	0
NINQ	input	default constant	0	260	0	0
CLNO	input	mean	21.247293049	117	0	0
DEBTINC	input	median	34.745743152	617	0	0

*The **Interval Variables** tab is used to view the imputed statistics of the interval-valued input variables in the training data set.*

Class Variables tab

The **Class Variables** tab is designed to list all of the categorically-valued variables of the active training data set in which its missing values were imputed or replaced. The **Imputation Method** column will display the method of imputation. The missing values of the following categorically-valued variable REASON was replaced by its most frequently occurring class level. On the other hand, missing values in the categorical variable JOB was replaced by the specified class level of Other from the **set value...** option setting. The **Number Replaced** column displays the number of observations replaced by the replacement values that are specified from the **Specify Value** window. These same imputed values are listed in the **Imputation Value** column. The **Number Imputed** column displays the number of times that the missing values were replaced by each categorically-valued variable. By default, the categorically-valued target variable, BAD, is not listed in the following tab since the variable is not involved in the imputation process.

Results - Replacement

Table View: EMDATA.RTRNJCM6 | Interval Variables | Class Variables | Code | Log | Output | Notes

Name	Model Role	Imputation Method	Imputation Value	Number Imputed	Number Replaced
REASON	input	count	DebtCon	131	0
JOB	input	set value	Other	141	0

*The **Class Variables** tab is used to view the imputed statistics of the categorically-valued input variables in the training data set.*

The **Code** tab will not be displayed since it simply displays the internal SEMMA score code that generates the imputed values based on several *if-then* statements that can be used to replace values for new data. In other words, the scoring code lists the *if-then* code to replace missing values for each categorically-valued input variable by their most frequent class level that was specified from the **General** subtab. This is followed by *if-then* code to replace missing values for each interval-valued input variable by their average values that were specified from the **Imputation Methods** subtab, followed by *if-then* SAS programming code to redefine the range of values of the interval-valued input variables that were specified from the **Interval Variables** tab.

In addition, the **Log** tab will display the log listing of the *if-then* programming logic that is used to impute missing values in the input variables. Also, the **Output** tab will not be displayed since it simply displays the procedure output listing from the PROC CONTENTS procedure of the output scored data set with the imputed values for each partitioned data set.

3.5 Clustering Node

General Layout of the Enterprise Miner Clustering Node

- **Data tab**
- **Variables tab**
- **Clusters tab**
- **Seeds tab**
- **Missing Values tab**
- **Output tab**
- **Notes tab**

The purpose of the **Clustering** node in Enterprise Miner is to perform cluster analysis. *Clustering* is a process used in dividing the data set into mutually exclusive or nonoverlapping groups with the data points within each group as close as possible to one another and different groups that are separated as far apart from one another as possible. In the clustering assignments, each observation is assigned to one and only one cluster that is created. The most obvious way to identify clusters is to construct scatter plots to graphically visualize the characteristics or patterns of each cluster grouping and observe the differences between each cluster group. Cluster analysis is designed so that the data points within each cluster will generally tend to be similar to each other and data points in separate clusters are usually quite different. It is often important to observe both the formulation and the meaning of the cluster groupings from the variables in the analysis. Examples of cluster analysis are grouping the best baseball hitters in the game from various hitting statistics to grouping big cities in the United States based on the distance between each city, that is, the division of the big cities in the country that can be divided into two separate clusters, (West, East), three separate clusters, (West, Mid-West, and East) or four separate clusters, (North, South, East, and West), and so on.

In cluster analysis, the data set layout that can be applied to the analysis can either be numeric coordinates or numeric distances. The data matrix with regard to the numeric coordinates is such that the rows of the matrix represent the corresponding observations, and the columns represent each variable in the analysis. In the following example, the data set consists of flying mileages between five big cities in the United States, that is, numeric distances. The clustering algorithm attempts to identify these cities into separate groups or regions based on the flying mileage between them.

One of the most critical decisions in the preliminary stage of cluster analysis is to determine the number of clusters in the data. This can be done by constructing scatter plots and interpreting the various clustering statistics in determining the number of clusters. The goal of cluster analysis is finding the number of cluster groupings that are meaningful to you, which will make it easier for you to determine if the clusters that are created make common sense to you. It is also important to determine both the shapes and sizes of the various clusters since many of the clustering methods do not perform well with elongated elliptical clusters. Since the human eye can only visualize plots up of to three dimensions, the dimension reduction routines such as principal component plots, canonical discriminant plots, and multidimensional scaling plots should be considered, assuming that you have more than three input variables in the analysis. The next compelling decision for the analysis is deciding on which clustering method to use. However, once the final clusters are created, then the next step is to both identify and characterize the various cluster groupings that are formed by interpreting the various characteristics and dissimilarities between the clusters that are created. This can be done by generating various box plots or *profile plots*, and observing the normality within each cluster grouping and then comparing the means between the separate cluster groupings from the various input variables in the analysis with the same level of measurement and the same range of values. It is also important to make sure that there are no outliers or extreme values that will have a profound effect on the final values of the various cluster seeds. The reason that outliers should be removed before performing cluster analysis is because outliers tend to form their own clusters. The final step might be turning the descriptive clustering procedure into a predictive modeling procedure by predicting the clustering assignments from a new set of observations.

Actually, in some analysis clustering might be the first stage of your analysis. This is because you might want to perform further analysis within each group based on a certain attribute of interest, for example, defaulting on a loan or acquiring a certain illness and then performing, for example, predictive modeling from the various clustering groups that are created. Since data points might be inadvertently misclassified in the initial stages of the clustering algorithm, it is recommended to carefully examine the configuration of the cluster groupings that make common sense to you.

The node performs both hierarchical and partitive clustering. In Enterprise Miner, hierarchical clustering techniques are performed, such as the average, centroid, and Wald's methods, and partitive clustering technique are performed, such as the popular *k*-means clustering criterion technique. One reason that these three hierarchical methods are available in the **Clustering** node is because they are the only clustering techniques that can handle both distance and coordinate data structures. In addition, the three hierarchical methods do not require a distance matrix, therefore, reducing both the computational time and memory resources. The various clustering techniques differ in the way in which the distance between the observations and the cluster centers are computed. Typically, the distance matrix is computed by the Euclidean distance between the data points and their cluster mean. It is important to understand that a *cluster mean* or *cluster centroid* is not an actual data point, but a calculated value that represents the average value or the center of the cloud of points in the cluster and the *cluster seeds* are the initial data points which form the temporary cluster groupings. For partitive nonhierarchical clustering, the node automatically determines the number of clusters created from a random sample of 2,000 observations using the cubic clustering criterion statistic. The node is also designed to impute or fill in missing values from the cluster seed or cluster mean of the nearest cluster, since the initial clustering seed assignments are determined by the first few nonmissing observations.

Hierarchical Clustering and Partitive Clustering Algorithms

There are two major clustering methods: *hierarchical clustering* and *partitive clustering*. Hierarchical clustering involves two separate methods called agglomerative and divisive clustering. Agglomerative is the most common clustering method. However, if there are an enormous number of observations, then the partitive methods are usually applied. Agglomerative clustering bottom-up design begins with each observation assigned to its own cluster, then joining clusters together at each step until there is only one cluster with all the observations. Conversely, the divisive top-down design takes the opposite approach that begins with all observations assigned to a single cluster, then dividing the cluster into two separate clusters at each step until every cluster has a single observation. Although, the hierarchical clustering methods are the most common clustering techniques, there is not a common hierarchical clustering technique that is universally applied. Also, the hierarchical clustering methods do not perform well with several observations, and these methods suffer from creating poor partitions in previous steps that cannot be backtracked and corrected. However, the advantage to hierarchical clustering is that it is not affected by the *ordering effect*. The ordering effect is the order in which the observations are read to assign the initial seeds. In addition, the method does not depend upon you making an initial guess as to the number of clusters to create which is a huge drawback to the partitive clustering methods.

The agglomerative clustering method is performed as follows:

1. Initially, all observations are assigned to their own clusters.
2. Compute the similarity between each of the clusters, that is, usually the Euclidean distance.
3. Merge the clusters that are closest together. The clusters that are joined together are determined by the smallest Euclidean distance.
4. Repeat steps 2 and 3 until there exists one and only one cluster.

The divisive clustering method is performed as follows:

1. Initially, all observations are assigned to one cluster.
2. Compute the similarity between each of the clusters, that is, usually the Euclidean distance.
3. Partition two separate clusters that are most distant from each other.
4. Repeat steps 2 and 3 until each observation is assigned to its own cluster.

Partitive clustering creates various clusters from the data based on some error function. Initially, the number of clusters or centroids must be specified so that the observations are assigned to different clusters, as determined by some error function. The partitive clustering methods depend on the adequate selection of the initial cluster seeds. In addition, partitive clustering methods are affected by the existence of outliers and the order in which the observations are read and assigned as the initial seeds. Therefore, it is advisable to run the various partitive clustering techniques several times with the data set sorted several different ways each time, in order to achieve stability in the clustering assignments across the separate sorted data sets. Also, the partitive nonhierarchical methods will make assumptions about the shapes of the clusters that are created and depend on an initial guess as to the number of clusters. Therefore, hierarchical methods are most commonly applied. One of the most intuitive assessments in measuring the accuracy in creating the clusters is calculating the maximum squared difference between the observations in different clusters, that is, between-cluster sum-of-squares, and minimizing the squared difference between observations in the same cluster, that is, within-cluster sum-of-squares. The partitive clustering techniques are built around an iterative process of repeatedly moving the cluster means until the clusters are stable and the cluster means stop moving with the general idea of internal cohesion and external separation in the creation of the final cluster groupings.

Partitive clustering runs an iterative heuristic searching algorithm in creating the clusters as follows:

1. Initially, the data is partitioned into g groups from n observations.
2. Each observation is reassigned to different clusters based on the Euclidean distance function.
3. Repeat step 2 until there is no improvement in the clustering criterion.

Note: In step 1, the initial g groups that must be specified can be determined by prior knowledge, previous studies, or general intuition. Another technique that is typically applied is to create an initial split of the data based on the results that are generated from hierarchical clustering. The number of initial clusters can even be randomly selected. However, the choice of the initial clusters is critical for the correct number of clusters that are created. Since the different choices of the initial number of clusters can lead to many different solutions, it is advised to perform the partitive clustering technique any number of times with several different initial clusters to achieve consistency in the final clustering assignments that are created.

The Basic Steps to the k-means Clustering Algorithm

The most common partitive clustering technique is called the k-means clustering technique. k-means clustering can be specified from the **General** subtab by selecting the **Least Squares (Fast)** clustering criterion. The following are the basic steps in the k-means clustering algorithm. The iterative steps are performed until each observation is assigned to the most appropriate cluster.

1. Select the k data points that are the first guess of the initial k cluster seeds.
2. Each observation is then assigned to the nearest cluster seed in creating the temporary clusters. By default, each observation is assigned to the closest seed based on the smallest Euclidean distance.
3. The cluster seeds are then recalculated by the mean of the temporary cluster, that is, the cluster means are recalculated for each and every temporary cluster that is created.

   Repeat steps 2 and 3 until there is no significant change in the clustering criterion.

The big drawback to both hierarchical and partitive clustering techniques is that many of the clustering methods are based on minimizing the variability within each cluster by the squared distances that will result in good performance in creating spherical shaped clusters, but performs rather poorly in creating elongated elliptical clusters. Therefore, transformations might be considered in changing the elliptical clusters into more desirable spherical shaped clusters. The PROC ACECLUS procedure is specifically designed for performing linear transformations of the input variables in the analysis that result in spherical clusters. Another alternative is the nonparametric clustering methods, for example, k-nearest neighbors clustering, that perform well in classifying observations into highly elongated elliptical clusters or clusters with irregular shapes.

Supervised versus Unsupervised Classification

There are two separate types of classification: *supervised classification* and *unsupervised classification*. Cluster analysis performs unsupervised classification. Unsupervised classification is based on only the input variables in the analysis that determines the mutually exclusive cluster groupings. In unsupervised classification, there is no target variable or a supervisor variable in the descriptive model that defines the various groups. In supervised classification, each observation consists of a target variable along with any number of input variables. Examples of supervised classification are logistic modeling, discriminant analysis and k-nearest neighbors, where the values of the categorically-valued target variable may be explained by some linear combination of the input variables.

Similarity Measurements

Since the objective of the analysis is to determine a certain grouping of the variables, the choice of the measure of similarity or distance is important. Although there are many ways to compute the distance between two clusters, the squared Euclidean distance is the most common technique that is applied with interval-valued input variables in the data set. However, the Euclidean distance similarity measure tends to form spherical clusters. *City block distance* or *Manhattan distance* is another similarity measure; the name is derived from the idea in the two variable case in which the two separate variables can be plotted on a squared grid that can be compared to city streets, with the distance between two points measuring the number of blocks. The similarity measurement calculates the distances between the two points by simply summing the distances between the two joining line segments at a right angle as opposed to the Euclidean distance obtained by calculating the square root of the sum-of-squares based on the length of the line that joins the two points, that is, Pythagorean theorem. The advantage of the city block similarity measurement is that it is relatively insensitive to outlying data points and tends to form more cubical shaped clusters. Correlation can also measure similarity in which the correlation between the pair of variables might be regarded as large positive correlations indicating very similar items. Conversely, large negative correlation will indicate dissimilar items. The drawback to the correlation statistic is that it does not take into account the differences in the mean between the observations. For example, two groups of observations might have a perfect correlation and yet the observations in both groups might have very different magnitudes. That is, the correlation statistic measures the difference between each pair of observations, but it does not take into account the absolute closeness or the distance between each observation. Therefore, it is possible to have observations that are very different assigned to the same cluster. For categorically-valued input variables, the data may be formed by various contingency tables in which the groups might be created by the entries displayed along the diagonal representing the measure of similarity. For binary-valued input variables, the node automatically creates one dummy variable with the values of 0 or 1. For nominal-valued input variables, the node automatically creates one dummy variable for each class level with the values of 0 or 1, that is, the GLM coding schema. For ordinal-valued input variables, one dummy variable is created with the smallest ordered value mapped to one, the next smallest ordered value mapped to two, and so on. Another similarity measure is the distance between two clusters that can be determined by forming pairwise comparisons and calculating the absolute difference between the two separate class levels from the binary-valued input variables as follows:

	1	**2**	**3**	**4**	**5**	**6**	**7**	**8**	**9**	**10**	
Cluster 1	0	1	1	0	1	0	1	1	1	1	
Cluster 2	0	0	1	1	0	0	1	0	0	1	
$\lvert x_1 - x_2 \rvert$	0	1	0	1	1	0	0	1	1	0	$\sum \lvert x_1 - x_2 \rvert = 5$

Data Preparation for Cluster Analysis

Similar to predictive modeling, the selection of the best combination of input variables that determines the best clustering groups is a very critical preliminary step in a good clustering design. Therefore, it is recommended that principal components might be applied to reduce the number of input variables in the cluster analysis. Actually, the PROC VARCLUS procedure is specifically designed to create mutually exclusive cluster groupings based on the hierarchical method that applies the divisive clustering routine. Initially, all variables are assigned to one cluster for each variable, then ultimately assigned to their own unique clusters. The PROC VARCLUS procedure is based on the principal component method, where each cluster is a linear combination of the input variables and the first principal component explains a majority of the variability of the data. The next step is selecting the most important input variables from each cluster. The $1 - R^2$ ratio is the ratio of one minus two correlation measurements within and between the clusters. The statistic is used to select the best input variables within each cluster that is created, where small values will indicate a strong correlation with its own cluster and a weak correlation between the separate clusters.

$$1 - R^2 \text{ ratio} = \frac{1 - R^2 \text{ own cluster}}{1 - R^2 \text{ next closest cluster}}$$

It is also important to first graphically, visualize, and interpret the inputs separately. One reason is because outliers may have a profound effect on the results. Therefore, it is highly recommended to perform preliminary analysis of the variables in the data before performing cluster analysis. The reason that extreme values in the data will distort the analysis is because these same outlying observations will create their own cluster groups.

For example, performing cluster analysis based on the distances between big cities – Fairbanks, Alaska or Honolulu, Hawaii – might distort the analysis by forming their own geographical cluster groupings since the states of Alaska and Hawaii are far distant from the rest of the other cities within the continental United States.

The Distance Matrix

The best way to interpret cluster analysis is illustrated by the following two numerical examples. The following examples demonstrate the two separate data structures that may be applied in cluster analysis. The first example is based on a data structure that is in the form of a distance matrix and the second example is based on coordinate data.

In the following example, cluster analysis is performed in determining the various groupings or geographical regions that are formed by the distance between cities in the United States. The distances between the cities may be expressed by a pxp symmetric matrix of pairwise comparisons, where the p cities denote both the number of rows and columns. The distance matrix would then consist of zeros along the main diagonal entries and the difference in flying miles between different cities indicated by the off-diagonal entries. Therefore, the different cluster groupings may be determined by the difference in the flying mileage between each city. In the following example, note that there is no need to convert the data into squared distances since the flying mileages are very good approximations of the Euclidean distances.

Average Linkage Method

The following iterative hierarchical algorithm is called the *average linkage method.* The method is applied in determining the most appropriate number of clusters based on the distance matrix. The algorithm basically determines the average values from the similarity matrix. Since the matrix is based on pairwise comparisons, the smallest average value determines the two separate items to be merged. A new distance matrix is obtained with the two separate rows replaced with a new row, based on the distance between the merged cluster and the remaining clusters. This process is repeated until the algorithm determines the best possible cluster groupings.

To best understand the average linkage algorithm, consider the following distance matrix based on the difference between the flying mileage between the five listed cities in the United States. The reason that the average linkage method is being presented is that it is one of the easiest clustering algorithms to apply and interpret.

City	San Francisco	Los Angeles	Chicago	Washington D.C.	New York
1. San Francisco	0	347	1858	2442	2571
2. Los Angeles	347	0	1745	2300	2451
3. Chicago	1858	1745	0	597	713
4. Washington D.C.	2442	2300	597	0	205
5. New York	2571	2451	713	205	0

In the first step, merge the two closest cities in the United States, that is, New York and Washington D.C. with the shortest flying distance of 205 miles. Both cities are merged together to form the cluster (45).

$d_{(45)1} = \text{avg}(d_{41}, d_{51}) = \text{avg}(2442, 2571) = 2506.5$
$d_{(45)2} = \text{avg}(d_{42}, d_{52}) = \text{avg}(2300, 2451) = 2375.5$
$d_{(45)3} = \text{avg}(d_{43}, d_{53}) = \text{avg}(597, 713) = 655$

Therefore, applying the average linkage method, the following distance matrix will reduce to four separate clusters that are essentially the flying distances from the other three cities and the two separate cities of Washington D.C and New York combined.

City	San Francisco	Los Angeles	Chicago	N.Y./D.C.
San Francisco	0	347	1858	2506.5
Los Angeles	347	0	1745	2375.5
Chicago	1858	1745	0	655
NY/D.C.	2506.5	2375.5	655	0

Now, the shortest distance between the pair of merged clusters is the city of Chicago, that is, $d_{(45)3} = 655$. Therefore, merging the city of Chicago with cluster (45) will form cluster (345), which will result in three separate clusters being created. Since we are applying the average linkage method, we would then calculate the average distance between both cluster (345) and the cities of San Francisco and Los Angeles as follows:

$d_{(345)1} = \text{avg}(d_{(45)1}, d_{31}) = \text{avg}(2506.5, 1858) = 2182.25$
$d_{(345)2} = \text{avg}(d_{(45)2}, d_{32}) = \text{avg}(2375.5, 1745) = 2060.25$

City	San Francisco	Los Angeles	N.Y./D.C./Chicago
San Francisco	0	347	2182.25
Los Angeles	347	0	2060.25
NY/D.C./Chi.	2182.25	2060.25	0

If we want to create two separate clusters, the shortest distance between the pair of clusters is now the distance between the cities of San Francisco and Los Angeles. Therefore, these two cities will now form their own cluster (12). Hence, by requesting the number of clusters of two, the algorithm resulted in two separate clusters or regions of East versus West as follows: cluster (345) of the eastern cities of New York, Washington D.C., and Chicago and cluster (12) of the western cities of San Francisco and Los Angeles.

K-Means Clustering Method

The following is the *k*-means partitive clustering technique based on a coordinate data set with two separate input variables, Variable 1 and Variable 2, that we want to group into two separate clusters. The SAS programming code that performs *k*-means clustering is displayed in the **SAS Code** node. The procedure is designed so that each observation is assigned to the most appropriate cluster that is determined by the shortest Euclidean distance, that is, the smallest sum-of-squares difference, between the two separate clusters that are initially created. In Enterprise Miner, *k*-means clustering is the default method that is automatically applied.

Observation Number	Variable 1	Variable 2
1	2.0	3.0
2	4.5	1.0
3	5.0	4.5
4	5.0	7.0
5	2.5	5.0
6	5.5	4.0
7	6.5	3.5
8	3.0	8.0
9	1.5	5.5

From the partitive *k*-means clustering technique, the initial number of clusters must be specified. Therefore at the beginning, two separate clusters are created. One of the simplest criteria in creating the two initial clusters is calculating the two separate observations furthest apart, based on the Euclidean squared distance.

Cluster Number	Observation	Mean Vector
Cluster 1	3	(5.0, 4.5)
Cluster 2	8	(3.0, 8.0)

Note: The performance of the partitive *k*-means clustering technique depends on the initial seeds being a close approximation of the true cluster means. The initial seeds represent the cluster means at the beginning of this iterative process. Therefore, it is suggested to first perform the hierarchical clustering technique by calculating the cluster means from the final clusters that are created. These same cluster means can then be used as the initial seeds for *k*-means clustering.

The next step is calculating the squared distances between each pair of observations. The observations are assigned to the corresponding cluster based on the smallest distance between the two separate cluster distances as follows:

Cluster Number	Observation	Mean Vector	Total Sum-of-Squares
Cluster 1	1, 2, 3, 5, 6, 7	(4.33, 3.50)	25.33
Cluster 2	4, 8, 9	(3.16, 6.83)	9.33

From the iterative clustering assignment procedure, the next step is calculating the squared distances between the pair of observations for each cluster grouping based on the updated cluster means. The observations are assigned to the corresponding cluster with the smallest distance between the two separate cluster distances as follows:

Observation Number	Clustering Assignment	Distance to Cluster 1 Mean	Distance to Cluster 2 Mean
1	1	2.38630	4.00694
2	1	2.50555	5.98377
3	1	1.20185	2.96742
4	2	3.56293	1.84089
5	2	2.36878	1.95078
6	1	1.26930	3.67045
7	1	2.16667	4.71405
8	2	4.69338	1.17851
9	2	3.46811	2.13437

From the new clustering assignments that are listed in the previous table, the next step is calculating the cluster means and the corresponding squared distances between the pair of observations as follows:

Cluster Number	Observation	Mean Vector	Total Sum-of-Squares
Cluster 1	1, 2, 3, 6, 7	(4.7, 3.2)	18.6
Cluster 2	4, 5, 8, 9	(3.0, 6.375)	12.1

This iterative clustering assignment routine can continue any number of times in calculating the cluster means and the corresponding squared distances between the pair of observations until the observations are assigned to the appropriate clusters. However, the previous listing displays the final cluster assignments that are assigned to the corresponding observations from the iterative *k*-means clustering procedure with the cluster means for each input variable remaining stable from the second iteration and beyond.

Cluster analysis is discussed in greater detail in Johnson and Wichern (2002), Hastie, Tibshirani, and Friedman (2001), Stevens (1982), Berry and Linoff (2004), Lattin, Carroll, and Green (2003), and Applied Clustering Techniques Course Notes, SAS Institute (2003).

Data tab

The **Data** tab is designed to select the training, validation, test, and scored data sets. For coordinate data, clustering analysis can be performed for each one of the partitioned data sets when you run the process flow. By default, the training data set is selected for cluster analysis. However, the clustering algorithm will be performed for each one of the partitioned data sets once the node is executed, which will enable you to validate the consistency between the clustering assignments in comparison to the active training data set. The scored data set can be used in creating new cluster groupings from a new set of values of the input variables. Since clustering analysis is unsupervised training, variables assigned a **target** variable role from the previous nodes are automatically removed from the analysis. The **Data** tab consists of the following two subtabs that either select the previously created data sets that are connected to the node for training, validating and scoring, or set aside a small portion of the training data set in order to perform preliminary training to determine the appropriate number of initial clusters from the randomly selected subset of observations. The **Data** tab consists of the following two subtabs:

- **Inputs subtab**
- **Preliminary Training and Profiles subtab**

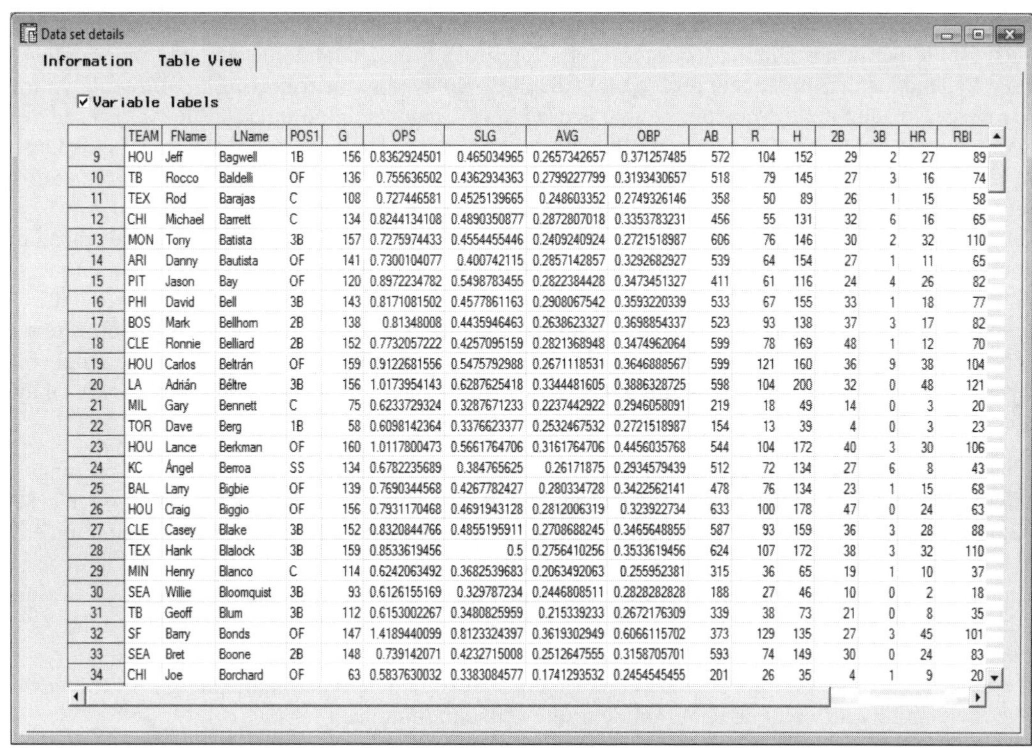

Data set details

Information | Table View

☑ Variable labels

	TEAM	FName	LName	POS1	G	OPS	SLG	AVG	OBP	AB	R	H	2B	3B	HR	RBI
9	HOU	Jeff	Bagwell	1B	156	0.8362924501	0.465034965	0.2657342657	0.371257485	572	104	152	29	2	27	89
10	TB	Rocco	Baldelli	OF	136	0.755636502	0.4362934363	0.2799227799	0.3193430657	518	79	145	27	3	16	74
11	TEX	Rod	Barajas	C	108	0.727446581	0.4525139665	0.248603352	0.2749326146	358	50	89	26	1	15	58
12	CHI	Michael	Barrett	C	134	0.8244134108	0.4890350877	0.2872807018	0.3353783231	456	55	131	32	6	16	65
13	MON	Tony	Batista	3B	157	0.7275974433	0.4554455446	0.2409240924	0.2721518987	606	76	146	30	2	32	110
14	ARI	Danny	Bautista	OF	141	0.7300104077	0.400742115	0.2857142857	0.3292682927	539	64	154	27	1	11	65
15	PIT	Jason	Bay	OF	120	0.8972234782	0.5498783455	0.2822384428	0.3473451327	411	61	116	24	4	26	82
16	PHI	David	Bell	3B	143	0.8171081502	0.4577861163	0.2908067542	0.3593220339	533	67	155	33	1	18	77
17	BOS	Mark	Bellhorn	2B	138	0.81348008	0.4435946463	0.2638623327	0.3698854337	523	93	138	37	3	17	82
18	CLE	Ronnie	Belliard	2B	152	0.7732057222	0.4257095159	0.2821368948	0.3474962064	599	78	169	48	1	12	70
19	HOU	Carlos	Beltrán	OF	159	0.9122681556	0.5475792988	0.2671118531	0.3646888567	599	121	160	36	9	38	104
20	LA	Adrián	Béltre	3B	156	1.0173954143	0.6287625418	0.3344481605	0.3886328725	598	104	200	32	0	48	121
21	MIL	Gary	Bennett	C	75	0.6233729324	0.3287671233	0.2237442922	0.2946058091	219	18	49	14	0	3	20
22	TOR	Dave	Berg	1B	58	0.6098142364	0.3376623377	0.2532467532	0.2721518987	154	13	39	4	0	3	23
23	HOU	Lance	Berkman	OF	160	1.0117800473	0.5661764706	0.3161764706	0.4456035768	544	104	172	40	3	30	106
24	KC	Ángel	Berroa	SS	134	0.6782235689	0.384765625	0.26171875	0.2934579439	512	72	134	27	6	8	43
25	BAL	Larry	Bigbie	OF	139	0.7690344568	0.4267782427	0.280334728	0.3422562141	478	76	134	23	1	15	68
26	HOU	Craig	Biggio	OF	156	0.7931170468	0.4691943128	0.2812006319	0.323922734	633	100	178	47	0	24	63
27	CLE	Casey	Blake	3B	152	0.8320844766	0.4855195911	0.2708688245	0.3465648855	587	93	159	36	3	28	88
28	TEX	Hank	Blalock	3B	159	0.8533619456	0.5	0.2756410256	0.3533619456	624	107	172	38	3	32	110
29	MIN	Henry	Blanco	C	114	0.6242063492	0.3682539683	0.2063492063	0.255952381	315	36	65	19	1	10	37
30	SEA	Willie	Bloomquist	3B	93	0.6126155169	0.329787234	0.2446808511	0.2828282828	188	27	46	10	0	2	18
31	TB	Geoff	Blum	3B	112	0.6153002267	0.3480825959	0.215339233	0.2672176309	339	38	73	21	0	8	35
32	SF	Barry	Bonds	OF	147	1.4189440099	0.8123324397	0.3619302949	0.6066115702	373	129	135	27	3	45	101
33	SEA	Bret	Boone	2B	148	0.739142071	0.4232715008	0.2512647555	0.3158705701	593	74	149	30	0	24	83
34	CHI	Joe	Borchard	OF	63	0.5837630032	0.3383084577	0.1741293532	0.2454545455	201	26	35	4	1	9	20

*The **Table View** tab displaying the training data set of 2004 major league baseball hitters.*

Inputs subtab

The **Inputs** subtab is designed to select the previously created data sets that are connected to the node for training, validating, testing, and scoring the cluster analysis results. Similar to the standard **Data** tabs, you may press the **Select...** button to either view the file administrative information or browse the table view of the selected data set. To train the model and perform the initial cluster analysis, the **Training** role radio button must be selected. Assuming that the sample size is sufficiently large, then both the validation and the test data sets can be used to cross-validate the accuracy and the stability of the clustering assignments from the training data set. Select the **Properties...** button to open the **Imports Map** window that will allow you to select any of the available data mining data sets created within the currently opened Enterprise Miner project that are connected to the node in order to select an entirely different partitioned data set for the process flow or remove the selected data set from the analysis.

Clustering
Data | Variables | Clusters | Seeds | Missing Values | Output | Notes
Role: Training Validation Test Score
Details:
Data set: EMDATA.TRN82HU8 Select... Properties...
Description: Training Data Set from: EMDATA.VIEW_69F
Inputs | Preliminary Training and Profiles

*The **Inputs** tab is used to specify the analysis data set and browse the selected partitioned data set.*

Preliminary Training and Profiles subtab

The **Preliminary Training and Profiles** subtab is designed to allow you to randomly select a small portion of the training data set to perform preliminary clustering. In Enterprise Miner, a random sample of the training data set is used to generate the subsequent profile plots that are displayed in the following **Profiles** tab within the **Result Browser**. In addition, the random sample is used to calculate the importance value that is used in determining the input variables that best separates the cluster groupings. The importance statistics is displayed in the following **Variables** tab within the **Result Browser**. Although, all of the data is used for cluster analysis, however, only 2,000 observations are used for the cluster profile presentation. From the subtab, you may change the number of observations that are randomly sampled from the active training data set. One reason that 2,000 observations are automatically selected to generate the following profile plots is that if there are too many points to be used, it will increase both time and memory. This can degrade the performance. Like the metadata sample, 2,000 observations is a good sample to see what the data looks like. Another reason is that if you have a couple hundred thousand observations and you want to graph all of them, there is a much higher chance that you will be unable to get meaningful graphs to view. In other words, the entire graphics area may be compressed with so many points that poor separation would be seen and observed.

From the subtab, you may specify the amount of data to set aside to create the preliminary cluster groupings with the random seed number that determines the records that are randomly selected from the training data set. By default, 2,000 records are randomly selected from the training data set based on the random seed number that is automatically set to value of 12345. In our example, this will result in all observations that will be used to create the preliminary cluster groupings and the corresponding profile pie charts, bar charts, and CCC plot. By default, the **Entire data set** radio button is dimmed and unavailable to process every record from the training data set. To change the number of records allocated from the training data set, enter the number of records in the **Sample Size** entry field. Press the **Generate new seed** button to automatically generate a new seed that creates an entirely different random sample of the active training data set.

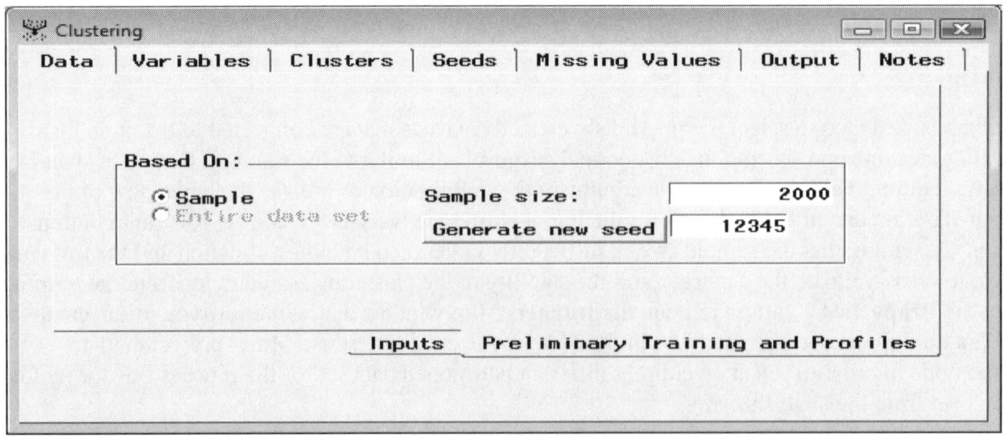

*The **Preliminary Training and Profiles** subtab is used in the allocation of the training data set.*

Variables tab

The **Variables** tab will be displayed as you first open the **Clustering** node. The **Variables** tab is designed for you to browse the variable attributes, that is, the variable status, model role, measurement level, format and label. All input variables that have a measurement level of interval, ordinal, binary, or nominal are displayed, as well as the frequency, id, and rejected variables that were specified in the **Input Data Source** node or the **Data Set Attributes** node. Cluster analysis performs unsupervised classification that does not require a target variable to the analysis. The **Status** and **Model Role** columns will allow you to remove certain input variables from the cluster analysis. By default, the input variables are set to a variable status of **use** with an associated model role of **input**. To remove the input variables from the analysis, set the variable attribute status from the **Status** column to **don't use**. Again, either hold down the Shift key or the Ctrl key to select several variable rows to remove the corresponding input variables from the analysis. For instance, the input variables called on-base and slugging percentage, OPS, and the combination of both on-base percentage, OBP, and slugging percentage, SLG, was set to a variable role of **rejected** and, therefore, was removed from the analysis. The reason is because it will reduce the redundancy in the analysis by removing the input variables on-base and slugging percentage from the analysis that does not provide any added information in explaining the variability between the data points that has not already been explained by the other two input variables.

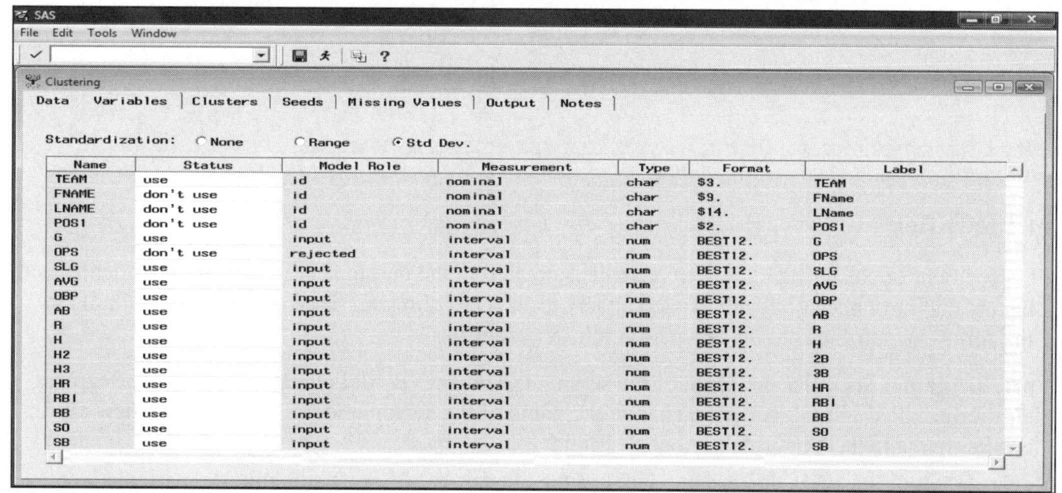

Name	Status	Model Role	Measurement	Type	Format	Label
TEAM	use	id	nominal	char	$3.	TEAM
FNAME	don't use	id	nominal	char	$9.	FName
LNAME	don't use	id	nominal	char	$14.	LName
POS1	don't use	id	nominal	char	$2.	POS1
G	use	input	interval	num	BEST12.	G
OPS	don't use	rejected	interval	num	BEST12.	OPS
SLG	use	input	interval	num	BEST12.	SLG
AVG	use	input	interval	num	BEST12.	AVG
OBP	use	input	interval	num	BEST12.	OBP
AB	use	input	interval	num	BEST12.	AB
R	use	input	interval	num	BEST12.	R
H	use	input	interval	num	BEST12.	H
H2	use	input	interval	num	BEST12.	2B
H3	use	input	interval	num	BEST12.	3B
HR	use	input	interval	num	BEST12.	HR
RBI	use	input	interval	num	BEST12.	RBI
BB	use	input	interval	num	BEST12.	BB
SO	use	input	interval	num	BEST12.	SO
SB	use	input	interval	num	BEST12.	SB

*The **Variables** tab is used to the control the input variables for the analysis and perform standardization.*

The **Standardization** button that is located in the tab will allow you to specify the method of standardization before the clustering algorithm is performed. By default, standardization is not applied. Standardization is strongly recommended if the interval-valued input variables in the analysis have a wide range of values or are measured in entirely different units such as pounds, kilograms, and so on. For example, the range of values in millions of dollars would have a much bigger impact on the variability in the data than the range of values in cents. *Standardization* is applied in order to interpret the variables with a common unit or the same range of values. Standardization might be extremely critical at times since the algorithm depends on the difference between all pairs of Euclidean distances based on the separate input variables during the formulation of the various cluster groupings. In cluster analysis, variables with large variability tend to have a greater influence on the resulting clusters that are created in comparison to variables with smaller variability. Therefore, standardization should be applied even if the input variables are measured with the same unit. For the various clustering techniques, different standardization methods can lead to different clusters since the methods are based on minimizing the sum-of-squares Euclidean distances between each observation and the group mean.

- **None** (default)**:** No standardization is performed prior to performing the clustering procedure.
- **Range:** This option standardizes the variable by subtracting its minimum value, then dividing by its range of values, which transforms its values into a [0, 1] interval. This standardization method is not as sensitive to outliers in the data in comparison to the following standard deviation technique.
- **Standard Deviation:** This is the most common method of standardization that divides all variables in the analysis by its standard deviation, that is, the mean is not subtracted. However, since the standard deviation is calculated by squaring the difference between each observation and its own mean, therefore this technique is very sensitive to extreme values in the data.

Clusters tab

The **Clusters** tab is designed to specify the clustering identifier, the model role of the clustering identifier, and the number of clusters. The **Number of Clusters** section will allow you to specify either the hierarchical or partitive clustering technique. Note that in the following diagram, partitive clustering was performed in the subsequent analysis by specifying the number of clusters, which was set to three within the **Number of Clusters** section to classify the greatest hitters, good hitters, and the rest of the other hitters in the game of baseball.

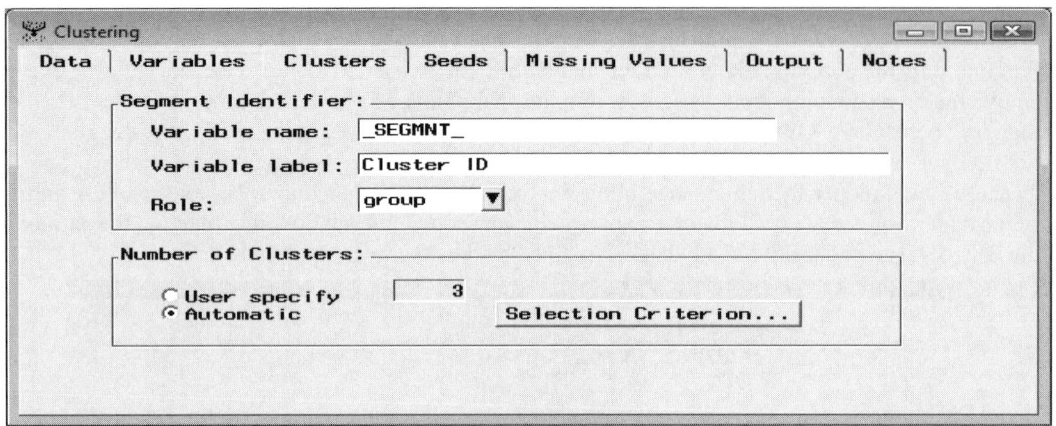

*The **Clusters** tab is used to specify the number of clusters and the attributes of the segment id variable.*

Segment Identifier

By default, a segment or clustering identifier variable is created in the output scored data set that identifies the separate clusters that have been assigned to each observation. The following three options are available in assigning the following attributes to the segment or cluster identifier:

- **Variable name:** Specifies the name of the segment identifier. By default, the name of the segment identifier is called _SEGMNT_. To change the name of the segment identifier, enter the new cluster identifier name in the **Variable name** entry field.
- **Variable label:** Specifies the variable label of the cluster identifier. By default, the variable label name is called Cluster ID. From the **Variable label** entry field, you may enter a new label for the segment identifier variable.
- **Role:** By default, the variable role used as the cluster identifier has a **group** variable role. This option will allow you to select an entirely different variable role that will be assigned to the cluster identifier. The variable roles that will be listed from the pop-up menu items are either the **group**, **id**, **input**, or **target** model roles. The model role of **group** is useful for by-group processing. In addition, assigning a model role of **target** will result in the segment identifier that will be used as the classification target variable that you want to predict in the subsequent classification modeling node. Conversely, assigning a model role of **input** will result in the segment identifier that can be used as an categorically-valued input variable in the subsequent modeling nodes. The variable role assigned to the segment or cluster identifier can be passed along to the subsequent nodes within the process flow diagram for further analysis.

Number of Clusters

The **Number of Cluster** option actually determines if you would like to perform either hierarchical or partitive clustering. Hierarchical clustering can be performed by selecting the **Automatic** option. Hierarchical clustering is the most common clustering technique. Conversely, selecting the **User specify** option performs partitive clustering. In partitive clustering, the selection of the number of cluster groups is extremely important to the *k*-means clustering analysis. The number of clusters must be initially specified before the clustering procedure is performed. The number of clustering groups are created where the observations are very similar within each cluster while at the same time the cluster groupings are well separated from each other. By default, the number of clusters that are created is automatically set to the number of input variables in the input data set. It is important to understand that you may either enter the number of clusters to create based on general knowledge, as in our example the group of hitters in the game, that constitute the number of clusters or by preliminary cluster analysis. By default, the **Automatic** option is automatically selected. The option applies the hierarchical clustering technique in which you may change the minimum and maximum number of clusters to create by selecting the **Selection Criterion** button that opens the **Selection Criterion** window. The window displays the various clustering criteria used in determining the number of clusters to create. It is recommended to repeatedly perform cluster analysis any number of times by specifying a different number of clusters or a different number of observations assigned to any one cluster.

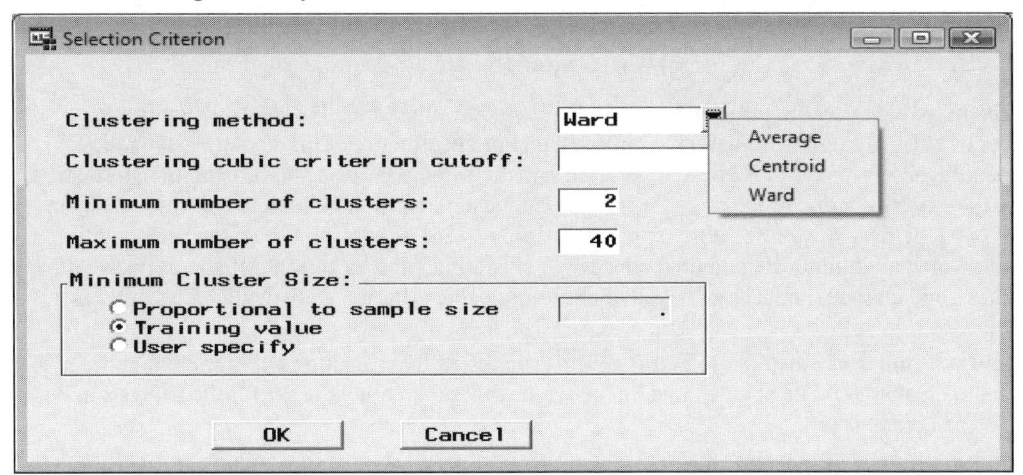

*The **Selection Criterion** window is used to specify the various clustering option settings for the hierarchical clustering method.*

The following are the available options that are displayed within the **Selection Criterion** window.

- **Clustering method:** Specify the cluster method, that is, the **Average**, **Centroid**, and **Wald** options. Since there is no standard cluster method that is generally applied, then it is recommended to try the different clustering methods and compare the results to determine the ideal partitioning of the data.
  - **Average:** The average linkage method forms cluster groupings by computing the average Euclidean distance between two clusters based on the average distance between the pair of observations. Average linkage tends to be less influenced by extreme values than Wald's method and is slightly biased in creating clusters with the same variability. The distance function for the average linkage method is computed as follows:

    $D_{KL} = \frac{1}{n_K n_L} \cdot \sum\sum d(x_i, x_j)$ where $d(x_i, x_j)$ is the distance between the ith and j^{th} observations

  - **Centroid:** The distance matrix is based on the distance between two clusters that are determined by the squared distance between the cluster centroids or the cluster means. The difference between this method and the average linkage method is that the centroids for each cluster are first calculated, then the distance between the cluster centroids are computed. This method is also robust to outliers in the data since the clustering assignments are based on the comparison of the cluster means. However, this method may not perform as well as the average linkage or Wald's method. The distance function for the centroid linkage method is computed as follows:

    $D_{KL} = \| \bar{X}_K - \bar{X}_L \|^2$ where $\bar{X}_K$ and $\bar{X}_L$ are the centroids at cluster K and cluster L

  - **Wald** (default)**:** Wald's method is analogous to predictive modeling where the total variability of the target variable is separated into two different components by using the ANOVA sum-of-squares between the two clusters for every input variable in the analysis. The different variability components are the Total deviance (T) of the p input variables that is based on the variability in the data between both the within-cluster groups (W) and the between-cluster groups (B). The within deviance is the sum-of-squares of the data within each cluster group, the between deviance is the sum-of-squares in the data between the cluster groups and the total deviance is the sum of both sources of variability, that is, T=B+W, based on the p input variables in the data. The basic idea is to have a high between deviance (B) with a low within deviance (W), that is, until the cluster analysis achieves a minimized variability within the clusters and maximized variability between the clusters with the measurement of assessment of the following form: $|W| / (|B + W|)$. The one drawback to this method is that it is very sensitive to extreme values since the estimates are computed from the squared distances. In addition, although this method is considered very efficient, the method tends to form clusters based on a small number of observations and tends to be biased by forming spherical clusters with approximately the same number of observations. The distance function for the Wald's method is computed as follows:

$$D_{KL} = ||\bar{X}_K - \bar{X}_L||^2 / (\frac{1}{n_K} + \frac{1}{n_L})$$ where $\bar{X}_K$ and $\bar{X}_L$ are the centroids at cluster K, L

- **Clustering cubic criterion cutoff** (default is 3)**:** The node automatically selects the optimum number of clusters from the **Clustering cubic criterion cutoff** value. This statistic is the ratio between the observed and the expected r-square statistic using a variance-stabilizing transformation, where the expected value of the r-square statistic can be entered in the **Clustering cubic criterion cutoff** entry field. A clustering cubic criterion statistic of zero or two should be interpreted with caution, which might indicate potential clusters. A clustering cubic criterion statistic of two or three indicates good clusters, and a large negative clustering cubic criterion statistic indicates outliers in the data.
- **Minimum number of clusters** (default is 2)**:** It is suggested that selecting a minimum number of clusters to create should be any positive integer value other than one in order for the analysis to be useful. The default is two.
- **Maximum number of clusters** (default is 40)**:** This option sets the maximum number of clusters to the iterative clustering procedure. It might not be a bad idea to repeatedly perform the clustering procedure and specify an entirely different maximum number of clusters to create each time in achieving consistent clustering assignments. The default is set at 40.
- **Minimum Cluster Size:** Specify the minimum number of data points assigned to any one cluster that is created. This option is designed to avoid instability in the clustering assignments by creating clusters with a reasonable number of observations.
  - **Proportional to sample size:** The proportional sample size assigned to any one cluster is related to the **Minimum Cluster Size** specified from the following **Final** subtab within the **Seeds** tab. The ratio is based on the minimum number of observations assigned to any one cluster from the **Minimum Cluster Size** option divided by the number of observations from the training data set. For example, suppose your training data set had 10,000 observations and you specified a **Minimum Cluster Size** of 500 to any one cluster from the **Final** subtab within the **Seeds** tab. Selecting **Proportional to sample size** option will result in the fact that if your sample size is 2,000, then **Minimum Cluster Size** is set to $(2000 \cdot [500/10000])$, or 100.
  - **Training value:** This option is based on the value specified in the **Minimum Cluster Size** entry field from the training data set. In our example, the **Minimum Cluster Size** would be set to 500.
  - **User specify:** By selecting this option you can enter any positive integer value to set the minimum number of observations within any one cluster that is created. Selecting this option will result in the entry field that will be available for viewing in order to specify an appropriate positive integer value.

Seeds tab

The **Seeds** tab is designed to specify the clustering criterion that determines the process in the way in which the observations are assigned to the corresponding clusters and the way in which the cluster seeds or cluster centers are updated or replaced. The cluster seeds are the n observations that determine the initial g cluster groups in the data, that is, where $g < n$. The clustering criterion also controls the stopping criterion in creating the temporary clusters and the corresponding cluster seeds. The **Seeds** tab consists of three separate subtabs that are used to configure the various clustering criteria:

- **General subtab**
- **Initial subtab**
- **Final subtab**

General subtab

The **General** subtab is designed to specify the clustering criterion that determines the distance between each data point within each cluster and the distance between each separate cluster. The following methods are iteratively performed to determine the most appropriate observations to assign within each nonoverlapping cluster group. The following criteria are generally an iterative process in determining the best set of *g* clusters from the *n* observations in the data set.

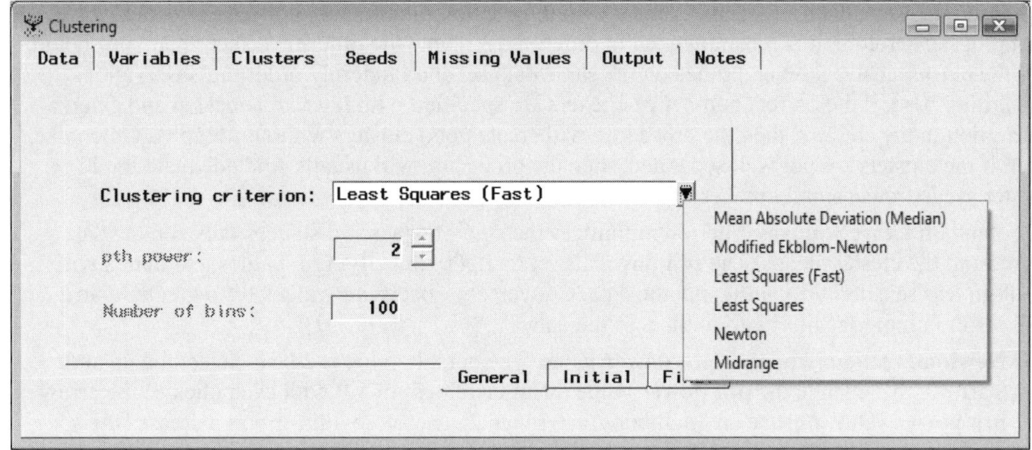

The ***General*** *subtab is used in selecting the clustering criterion used to create the cluster groupings.*

For both interval-valued and categorically-valued input variables in the analysis, with the exception of the **Midrange** clustering criterion method, the clusters are determined by the minimum sum-of-squares Euclidean distance between each observation within the corresponding cluster and its cluster mean. The default clustering criterion is the **Least Squares** (Fast) method that is based on the minimization of the squared distance of the observations to the cluster mean. However, the drawback to this method is that only one iteration is performed.

To specify a different clustering criterion, click the drop-down arrow to open the **Clustering criterion** drop-down list to select the appropriate criterion from the list. Some of the clustering criteria require additional option settings such as the **pth power** and **Number of bins** option settings. Unless additional specifications are required, these options are grayed-out and unavailable.

Clustering Criterion

From the **Clustering criterion** display field, press the drop-down arrow button to select from the various clustering criteria as follows:

- The **Mean Absolute Deviation (Median)** criterion has a default of 100 bins. To specify a different value, select the **Number of bins** entry field to enter an appropriate value. The **pth power** is set to 1.000. The clusters are determined by the absolute difference between the data and the median value of the respective cluster. This criterion might be specified since the following methods apply a squared distance that would have a profound effect on the extreme values in the data. The closest distance between point *a* and point *b* is determined by the shortest absolute distance in the two separate line segments as follows: $D_x = \sum | a_i - b_i |$.
- The **Modified Ekblom–Newton** criterion requires a **pth power** value. The default value is 1.500. Therefore, this option reduces the effect of outliers in calculating the cluster seeds. To specify a different value, use the up and down arrows, where each click of the arrow changes the value by .001. A range from 1 to 2 is allowed for the **pth power**. By default, a maximum of 20 iterations are performed.
- **Least Squares (Fast)** (default): This is the default clustering criterion method used. This option actually performs the most well-known partitive clustering algorithm, called *k*-means clustering. *K*-means clustering is based on the PROC FASTCLUS procedure. The PROC FASTCLUS procedure is designed to determine an adequate number of nonoverlapping clusters based on two or three passes through the data. It is suggested by SAS to perform a preliminary PROC FASTCLUS run with a large number of clusters, that is, 20 to 100, assuming that the data set is

large enough or contains several outliers. The criterion minimizes the sum of squared distances of data points from the cluster means within each cluster that is created. This method is restricted to only one iteration. By default, a maximum of one iteration is performed. The **pth power** is set to 2.000, that is, the squared difference between each data point within the cluster and its center. The drawback to the procedure is that it generally requires that there be at least 100 observations in the data set. Otherwise, the procedure will be highly affected by the ordering effect since the PROC FASTCLUS procedure is highly sensitive to the order of the observations in small data sets. Applying the same data set and the same clustering method might lead to different cluster groupings. Therefore, it is recommended in order to overcome the ordering effect, then you should run the cluster analysis several times on the same data set and randomly order the observations each time. Also, if the correct number of clusters are specified with internal cohesion and external separation in the clusters, then the procedure will create good clusters without iterating. Otherwise, even if the clusters are not well separated, then the procedure will usually find adequate initial cluster seeds from a small number of iterations.

- The standard **Least Squares** criterion minimizes the sum of squared distances between the data points and the cluster means. The **pth power** is set to 2.000. Specifying *p* values less than 2 will result in less sensitivity to outliers in the data. Conversely, specifying values for *p* greater than 2 will result in more sensitivity to outliers in the data.
- The **Newton** criterion requires a **pth power** value. The default value is 2.001. Select the up and down arrows that change the **pth power** value by an increment of 1.0 with each click of the arrow. The **pth power** value must be any number greater than 2. Therefore, this option increases the effect of outliers in the seed assignments. By default, a maximum of 20 iterations are performed.
- The **Midrange** criterion minimizes the midrange distance of the data points from the cluster means. This method is more robust to extreme values in comparison to the other clustering criterion methods.

With the exception to both least squares methods, that is, the **Least Squares** and **Least Squares (Fast)** methods, the node will automatically perform a maximum of 20 iterations to the above-mentioned clustering criterions with the cluster seeds replaced at each iteration based on the **Seed replacement** option set at **Full** and the **Convergence Criterion** set at .02.

Note: The maximum number of iterations for the **Clustering Criterion** can be specified from the **Final** subtab within the **Seeds** tab, with the exception to the **Least Squares (Fast)** method.

Initial subtab

The **Initial** subtab is designed to determine the process in which the cluster seeds are updated or replaced. By default, the node automatically sets the seed replacement option to **Full**, which will result in the node selecting the starting values of the initial cluster seeds to be as far apart as possible from each other. However, the initial seed configuration setting may also be set to **Partial**, **None**, or **Random**.

The initial cluster seeds are selected to be as far apart as possible. Each observation is assigned to the nearest cluster seed to form the temporary cluster groupings. The cluster mean is calculated for each temporary cluster grouping that is created and the iterative process is repeated until no further changes occur in the clusters that are created. The first seed is selected from the first complete case with no missing values in the input variables. Each subsequent observation is selected as a new seed, accordingly. There are two criteria to determine if an observation is a candidate for a new seed. Otherwise, the procedure continues to the next observation.

- First, the old seed is replaced if the distance between the observation and the closest seed is greater than the minimum distance between the existing seeds. The seed is replaced by the two seeds that are closest to each other. Between these two seeds, the seed that is replaced is the one of the two seeds with the shortest distance to the closest of the remaining seeds, where the other seed is replaced by the current observation.
- If the observations fail the first criterion test, then a second criterion is applied as follows:

  The observation replaces the nearest seed if the smallest distance from the observation to all seeds other than the nearest seed is greater than the shortest distance from the nearest seed to all other seeds. If the observation fails to meet this criterion, then the PROC FASTCLUS procedure moves on to the next observation.

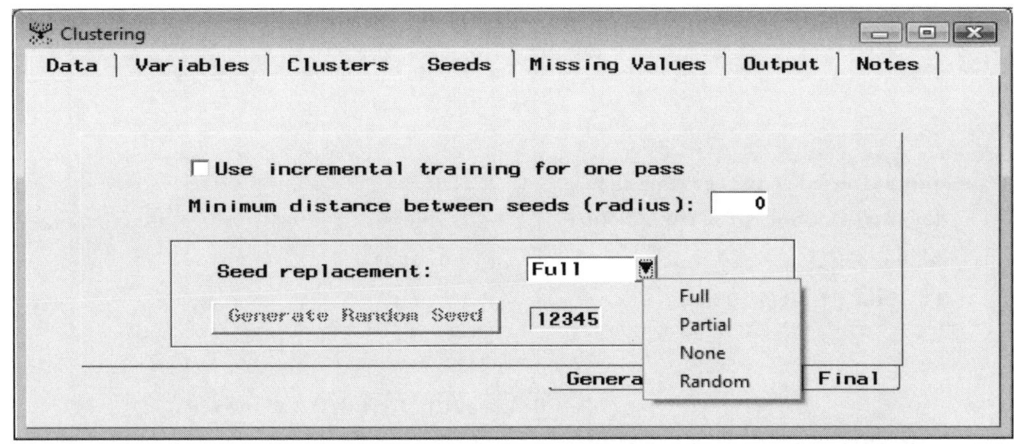

*The **Initial** subtab is used to specify the method in updating the cluster seed to the analysis.*

The **Use incremental training for one pass** check box allows the node to determine the cluster seeds to drift as the algorithm selects the initial seeds. When this option is selected, the **Cluster** node uses incremental training in one pass. In the algorithm, the cluster seed is updated within the corresponding cluster in which the new observation is assigned, with the other seeds staying the same. If this option is selected, then the clustering algorithm performs MacQueen's on-line *k*-means algorithm that is essentially Kohonen's network training clustering. By default, this option is not selected. Therefore, the difference between the drift algorithm as opposed to the default method is that the default method will assign the observation to its unique temporary cluster grouping, then adjust the cluster seeds of every cluster grouping that is created at the same time.

The initial seeds must be observations with no missing values, that is, complete cases, that are separated by the distance at least as large as the value specified from the **Minimum distance between cluster seeds (radius)** option.

The following are the various cluster replacement options for determining the process of modification or replacement of the cluster seeds within the cluster groupings:

- **Full** (default)**:** The initial clusters are determined by the starting values of the cluster seeds that are separated as far apart from each other as possible.
- **Partial:** This option omits the second criterion test for seed replacement. This will result in faster processing time in the seed assignments. However, the final seed assignments that are created may not be as clearly separated in comparison to the previous default initialization method.
- **None:** The initial seeds for the *k* clusters are the first *k* complete observations in the data set. This option will lead to much faster computational time in comparison to the other options. However, to obtain reliable cluster groupings, you must first specify an appropriate value for the **Minimum distance between seeds (radius)**, that is, the minimum distance between the most remote data point and its seed, called the *cluster radius*.

  The first cluster seed is initialized by the first record, where all the input variables are nonmissing. The second seed is based on the next complete case, in which the observations are separated from the first cluster seed by at least the value entered from the **Minimum distance between cluster seeds (radius)** entry field. All subsequent cases are then selected as new seeds if the maximum distance between the previously selected seeds is greater than the specified amount and so on.
- **Random:** The initial cluster seeds are randomly selected from the training cases in which all the input variables are nonmissing. Selecting this option will result in the **Generate Random Seed** button becoming available for selection. Press the **Generate Random Seed** button to determine the initial cluster seeds, where the initial seeds are randomly selected values of complete cases from the training data set.

Final subtab

The **Final** subtab is designed to determine the stopping criterion for the iterative clustering process in producing the final cluster seeds.

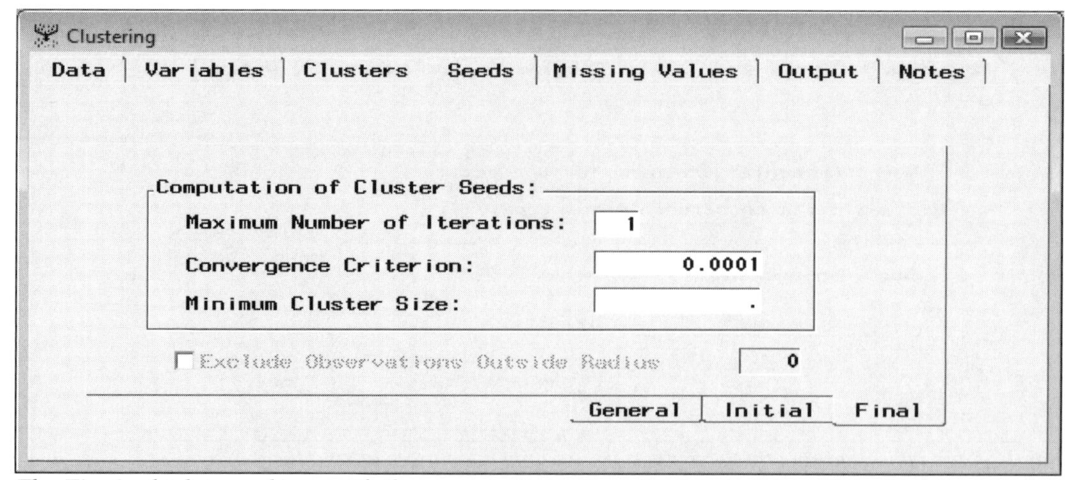

*The **Final** subtab is used to specify the computation options for the cluster seeds in the iterative process.*

- **Maximum Number of Iterations:** Specify the maximum number of iterations in replacing and updating the cluster seeds until the clusters are stable in achieving internal cohesion and external separation in the creation of the final cluster groupings. The default is a maximum of one iteration. However, setting the maximum number of iterations to one will result in the procedure terminating before convergence is achieved. This will usually result in final cluster seeds that will be quite different from the cluster means or cluster centers. Therefore, if you want complete convergence, then you should specify a large value for the **Maximum Number of Iterations** option.
- **Convergence Criterion:** This option is based on the relative change in the cluster seeds. Any nonnegative number can be specified. The default is .0001. The iterative procedure in determining the initial cluster seeds stops when the maximum relative change in the cluster seeds is less than or equal to the convergence criterion value. The relative change in the cluster seed is the distance between the old seed and the new seed divided by the scaling factor. The scaling factor is the determined by the **pth power** option setting. The convergence criterion value should be specified if the maximum number of iterations is greater than one.
- **Minimum Cluster Size:** This option is similar to the minimum cluster size specified from the **Cluster** tab in the **Selection Criterion** window. This option sets the smallest number of observations assigned to any one cluster that is created. By default, the cluster seeds are updated from a single iteration. A large minimum cluster size might be considered in order to obtain reliable cluster means that will result in more consistent clustering assignments.
- **Exclude Observations Outside Radius**: This option will become available if you set a value for the **Minimum distance between seeds (radius)** option from the **Initial** subtab. From the previous illustration, this option is grayed-out and unavailable for viewing. The option removes all observations from the analysis that fall outside the specified distance that is set. In other words, you can only analyze the observations falling within the radius that you have specified from the previous **Initial** subtab.

Missing Values tab

The **Missing Values** tab will allow you to specify the way in which the node handles missing observations in the training data set. In cluster analysis, if any of the input variables contain missing values, then the observation cannot be used to compute the cluster centroids or cluster seeds. Conversely, if all the input variables have missing values, then the observation is automatically removed from the analysis. There are two basic strategies for handling missing values in some of the input variables: remove the observations with missing values during the cluster initialization, or replace missing values in the output data set. Replacing the missing values will occur after the various cluster groupings are created. Select the **Exclude incomplete observations** check box to automatically delete observations from the output data set with missing values in any one of the input variables. By default, the **Imputation** check box is cleared, indicating to you that imputation of the data is not performed. Otherwise, select the **Imputation** check box to impute the missing values by the following imputation methods.

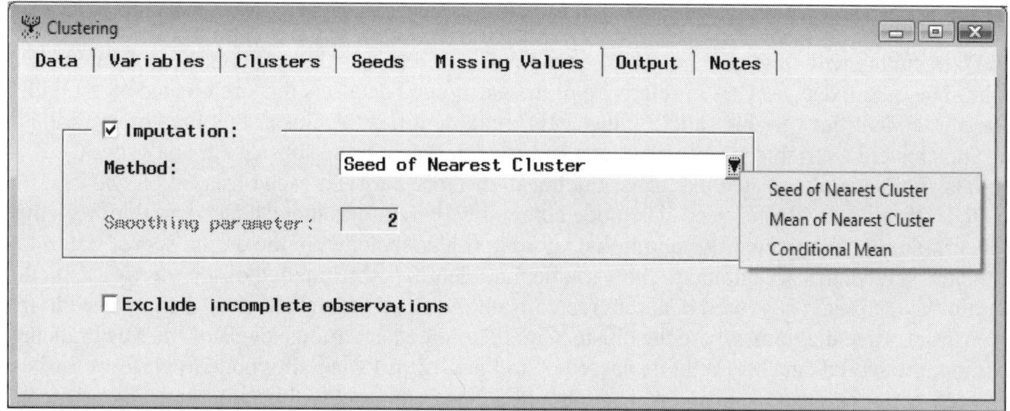

*The **Missing Values** tab is used to specify the type of imputation for the missing observations.*

The following are the various imputation methods to select from the **Missing Values** tab:

- **Seed of Nearest Cluster** (default)**:** This is the default imputation method. This option replaces the missing observation with the cluster seed that is closest to it. In other words, if the observation has a missing value in any one of the input variables, then the missing values will be replaced by the closest cluster seed in which the observation is assigned. The cluster seeds are actual data points as opposed to the cluster means that are the average value of the clusters.
- **Mean of Nearest Cluster:** This option replaces the missing observation by the average value of the nearest cluster closest to the data point. In other words, if the observation has a missing value in any one of the input variables, then the missing values will be replaced by cluster mean in which the observation is assigned. This method may only be used if the least-squares method is selected.
- **Conditional Mean:** This option may only be used if the least-squares method is specified. In addition, the option is also associated with the following **Smoothing parameter**. The smoothing value is multiplied by the within-cluster standard deviation. The default value is 2, but all positive numbers are acceptable. The conditional mean of the missing variable is computed by integrating the estimated mixture density (based on the nonmissing variables) with a within-cluster standard deviation that is increased by a factor that is specified by the smoothing parameter. The larger the smoothing parameter, the smoother the density estimate.
- **Smoothing parameter:** The smoothing parameter is the default number of clusters in the preliminary cluster analysis. The smoothing parameter will increase as the number of clusters in the preliminary analysis increases. If the smoothing parameter is applied, then it is advisable to try different smoothing parameters in each run since there is no general rule of thumb for selecting the correct value of the smoothing parameter. Imputation happens after clustering, so it does not affect the clusters that are created. During the computation of the imputed values, the smoothing parameter is used to smooth out the cluster boundaries.
- **Exclude incomplete observations:** By default, this option is not selected. Therefore, the observation is not removed from the analysis if some of the values in the input variables are missing. Selecting this option will instruct the node to remove the observation from the clustering algorithm if there are missing values in any one of the input variables. However, selecting this option will not prevent the node from replacing missing values and removing the observation from the output data set.

Output tab

The **Output** tab is designed for you to configure the output listing created from the cluster analysis results that are generated from the **Clustering** node and view the various cluster scored data sets that are created when you execute the node. The **Output** tab consists of the following three subtabs.

- **Clustered subtab**
- **Statistics Data Sets subtab**
- **Print subtab**

Clustered Data subtab

The **Clustered Data** subtab will allow you to view the output scored data sets that contain the generated clustering results. The output data sets are in reference to the partitioned data sets that are created along with the following segment identifier variable called "Cluster ID" that identifies the cluster that the observation has been assigned. The cluster id variable may be used as a stratified variable in the subsequent analysis. A distance variable is automatically created that is the Euclidean distance between each observation and its cluster seed, called Distance to Cluster Seed. The node automatically removes the distance variable from the analysis with a variable role of **rejected**. An additional variable is also included in the output scored data set, assuming that the missing values are imputed. The variable that identifies the imputed observations is called _IMPUTE_ within the data set. The scored data set created from the clustering results may then be used in the **Tree** node to construct a tree diagram, where the clusters are determined at various levels of the tree branching process. In addition, the scored data set might be passed on to the **Group Processing** node to perform further analysis within each group or cluster, and so on. The outputted cluster data set will be automatically created in the EMDATA project library. By selecting the appropriate partitioned data set from the **Show details of:** section and pressing the **Properties...** button, you will be able to view the file information or browse the selected data set. The **Description** entry field will allow you to enter a short description to the selected data set.

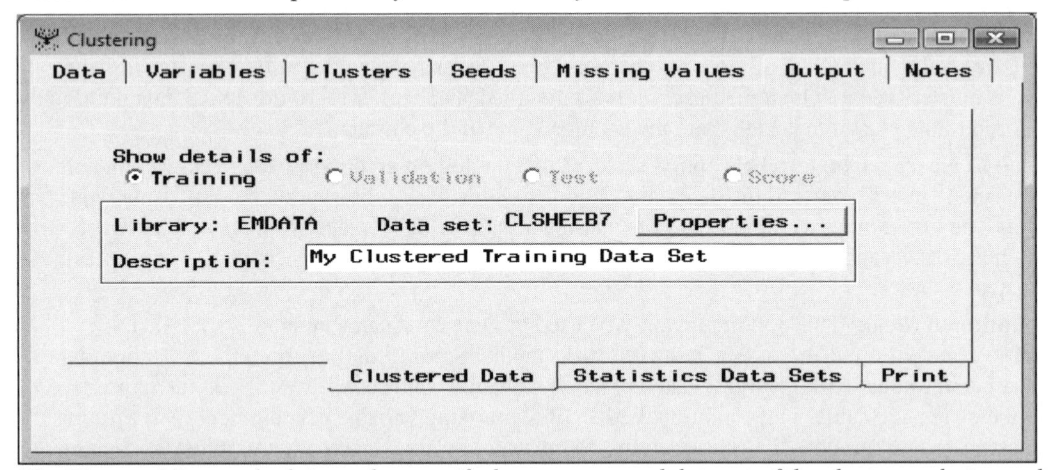

*The **Clustered Data** subtab is used to specify the output scored data set of the cluster analysis results.*

Data set details

Information | Table View

☑ Variable labels

	TEAM	FName	LName	POS1	SLG	AVG	OBP	AB	R	H	HR	RBI	SB	Cluster ID	Distance to Cluster Seed
11	TEX	Rod	Barajas	C	0.452514	0.248603	0.2749326146	358	50	89	15	58	0	3	1.882963021
12	CHI	Michael	Barrett	C	0.489035	0.287281	0.3353783231	456	55	131	16	65	1	3	2.8789413157
13	MON	Tony	Batista	3B	0.455446	0.240924	0.2721518987	606	76	146	32	110	14	1	3.674487106
14	ARI	Danny	Bautista	OF	0.400742	0.285714	0.3292682927	539	64	154	11	65	6	3	2.4308729209
15	PIT	Jason	Bay	OF	0.549878	0.282238	0.3473451327	411	61	116	26	82	4	1	3.4346130779
16	PHI	David	Bell	3B	0.457786	0.290807	0.3593220339	533	67	155	18	77	1	1	2.9574902206
17	BOS	Mark	Bellhorn	2B	0.443595	0.263862	0.3698854337	523	93	138	17	82	6	1	3.5735570952
18	CLE	Ronnie	Belliard	2B	0.42571	0.282137	0.3474962064	599	78	169	12	70	3	1	3.1189557495
19	HOU	Carlos	Beltrán	OF	0.547579	0.267112	0.3646888567	599	121	160	38	104	42	1	4.1113735211
20	LA	Adrián	Béltre	3B	0.628763	0.334448	0.3886328725	598	104	200	48	121	7	2	2.7318409642
21	MIL	Gary	Bennett	C	0.328767	0.223744	0.2946058091	219	18	49	3	20	1	3	3.3497080939
22	TOR	Dave	Berg	1B	0.337662	0.253247	0.2721518987	154	13	39	3	23	0	3	4.1201305171
23	HOU	Lance	Berkman	OF	0.566176	0.316176	0.4456035768	544	104	172	30	106	9	2	1.6736354727
24	KC	Ángel	Berroa	SS	0.384766	0.261719	0.2934579439	512	72	134	8	43	14	3	3.0169168105
25	BAL	Larry	Bigbie	OF	0.426778	0.280335	0.3422562141	478	76	134	15	68	8	3	2.6354656063
26	HOU	Craig	Biggio	OF	0.469194	0.281201	0.323922734	633	100	178	24	63	7	1	3.1107535425
27	CLE	Casey	Blake	3B	0.48552	0.270869	0.3465648855	587	93	159	28	88	5	1	2.5377265616
28	TEX	Hank	Blalock	3B	0.5	0.275641	0.3533619456	624	107	172	32	110	2	2	3.0725782533
29	MIN	Henry	Blanco	C	0.368254	0.206349	0.255952381	315	36	65	10	37	0	3	2.8419175055
30	SEA	Willie	Bloomquist	3B	0.329787	0.244681	0.2828282828	188	27	46	2	18	13	3	3.3210610966
31	TB	Geoff	Blum	3B	0.348083	0.215339	0.2672176309	339	38	73	8	35	2	3	2.5728271351
32	SF	Barry	Bonds	OF	0.812332	0.36193	0.6066115702	373	129	135	45	101	6	2	8.2019238391
33	SEA	Bret	Boone	2B	0.423272	0.251265	0.3158705701	593	74	149	24	83	10	1	3.2807026986
34	CHI	Joe	Borchard	OF	0.338308	0.174129	0.2454545455	201	26	35	9	20	1	3	4.7052552508
35	LA	Milton	Bradley	OF	0.424419	0.267442	0.3560477002	516	72	138	19	67	15	1	3.0215971317
36	MIL	Russell	Branyan	3B	0.525316	0.234177	0.3202247191	158	21	37	11	27	1	3	3.4543464831

The scored data set with the various clustering assignments and distance measurements.

Statistics Data Sets subtab

The **Cluster Statistics** data set will consist of the clustering statistics of each cluster that is created.

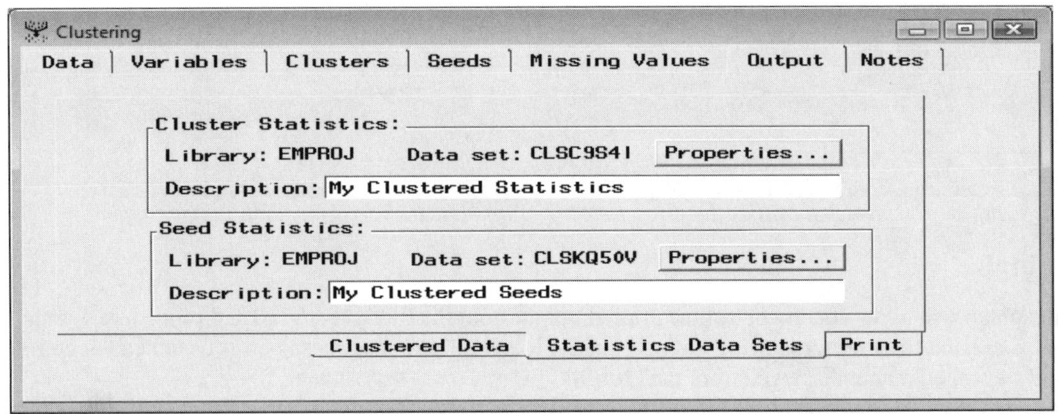

The ***Statistics Data Sets*** *subtab that displays the table listing of the various clustering analysis results.*

The **Cluster Statistics** data set lists the following clustering statistics for both the overall data set and each cluster that is created. The table listing displays various clustering statistics such as the cluster means, dispersion from the mean, frequency count, and the initial seed value of each cluster based on each input variable in the training data set. The data set also displays the various cluster criterions such as the CCC, average value, Pseudo F, and r-square statistics based on each input variable in the training data set. By checking the distance values, you can determine if any observations are unusually far from their own cluster seed. Conversely, small frequency counts might indicate outliers in the data. The data set might be used to create your own customized profile plots in order to characterize the input variables from the various clustering statistics.

Data set details

Information | Table View

☑ Variable labels

	Type of Observation	Cluster	Statistic Applying Over All Variables	G	SLG	AVG	OBP	AB	R	H	2B	3B	HR
1	ACV	.	3336.8828021	212.9525	260.8281	324.323	297.772	180.888	152.464	177.912	198.88	357.5163	186.90655
2	CCC	.	135.20114025	.	.	.	.	.	.	.	.	.	.
3	CRITERION	.	0.7800435232	.	.	.	.	.	.	.	.	.	.
4	ERSQ	.	0.1043947036	.	.	.	.	.	.	.	.	.	.
5	LEAST	.	2	.	.	.	.	.	.	.	.	.	.
6	MEAN	.	.	3.65546	5.827941	8.71391	8.15862	2.634957	1.9935	2.29028	2.0774	0.888149	1.3265434
7	PSEUDO_F	.	145.27793225	.	.	.	.	.	.	.	.	.	.
8	RSQ	.	0.4412015444	0.483387	0.373626	0.19965	0.29351	0.580682	0.67752	0.59354	0.5357	0.164573	0.5645451
9	RSQ_RATIO	.	0.7895539796	0.935685	0.596491	0.24945	0.41544	1.384822	2.101	1.46026	1.1536	0.196993	1.2964491
10	SMOOTH	.	.	.	.	.	.	.	.	.	.	.	.
11	STD	.	1	1	1	1	1	1	1	1	1	1	1
12	STDIZE	.	.	32.10572	0.074538	0.03086	0.04083	150.9888	28.9094	47.6712	10.611	2.528048	10.403383
13	WITHIN_STD	.	0.7495568049	0.720708	0.793586	0.89705	0.84281	0.649305	0.56941	0.63927	0.6833	0.916497	0.6616808
14	CENTER	1	.	4.61782	6.086039	9.12583	8.44671	3.737012	2.97987	3.34171	2.9819	1.608916	1.8058084
15	DISPERSION	1	0.7977879933	0.277022	0.612754	0.77929	0.58979	0.39366	0.49755	0.52658	0.6648	1.456807	0.7957501
16	FREQ	1	89	89	89	89	89	89	89	89	89	89	89
17	INITIAL	1	.	4.734359	6.043582	9.57771	8.15174	4.146002	3.59745	3.88075	2.4504	7.515681	1.0573484
18	SEED	1	52	4.710399	6.287177	9.38685	8.62317	3.90821	3.31473	3.59353	3.0919	2.160378	1.9612703
19	CENTER	2	.	4.679184	7.553235	9.74641	9.64076	3.635463	3.53222	3.46061	3.3928	0.70071	3.3093631
20	DISPERSION	2	0.7872278563	0.287369	0.845905	0.8227	1.14975	0.37242	0.51078	0.44715	0.7435	0.658317	0.7141156
21	FREQ	2	35	35	35	35	35	35	35	35	35	35	35
22	INITIAL	2	.	4.578623	10.89817	11.7298	14.8573	2.470381	4.46222	2.8319	2.5446	1.186686	4.325516
23	SEED	2	18	4.701481	8.035872	10.1148	10.2349	3.571272	3.78193	3.52181	3.5342	0.8131	3.5672157
24	CENTER	3	.	3.163637	5.490467	8.41918	7.8448	2.096088	1.42005	1.74559	1.5651	0.655	0.8728864
25	DISPERSION	3	0.7260420232	0.859159	0.841959	0.94498	0.86918	0.745706	0.60034	0.69607	0.6811	0.661397	0.5982917
26	FREQ	3	247	247	247	247	247	247	247	247	247	247	247

The clustering statistics data set with the various clustering statistic for each input variable.

The **Seed Statistics** data set contains seed information that can be applied to cluster new data when you use the PROC FASTCLUS procedure and the SEED option setting. The data set lists the following statistics based on each cluster that is created: the number of observations within each cluster, root mean square error, maximum distance from the cluster seed, nearest cluster, distance to the nearest cluster, and the cluster center of each input variable in the training data set.

Data set details

Information | Table View

☑ Variable labels

	Optimization Criterion	Cluster	Frequency of Cluster	Root-Mean-Square Standard Deviation	Maximum Distance from Cluster Seed	Nearest Cluster	Distance to Nearest Cluster
1	0.7800435232	1	89	0.7977879933	7.2098780304	2	3.3517372327
2	0.7800435232	2	35	0.7872278563	8.2019238386	1	3.3517372327
3	0.7800435232	3	247	0.7260420232	5.2777219671	1	4.4684401321

The clustering seed statistics that lists the cluster statistics and cluster seeds for each input variable.

Print subtab

The **Print** subtab will allow you to control the printed output from the PROC FASTCLUS procedure output listing that is created from the **Clustering** node. From the **Results Browser**, the subtab is designed to control the amount of printed output displayed from the **Output** tab.

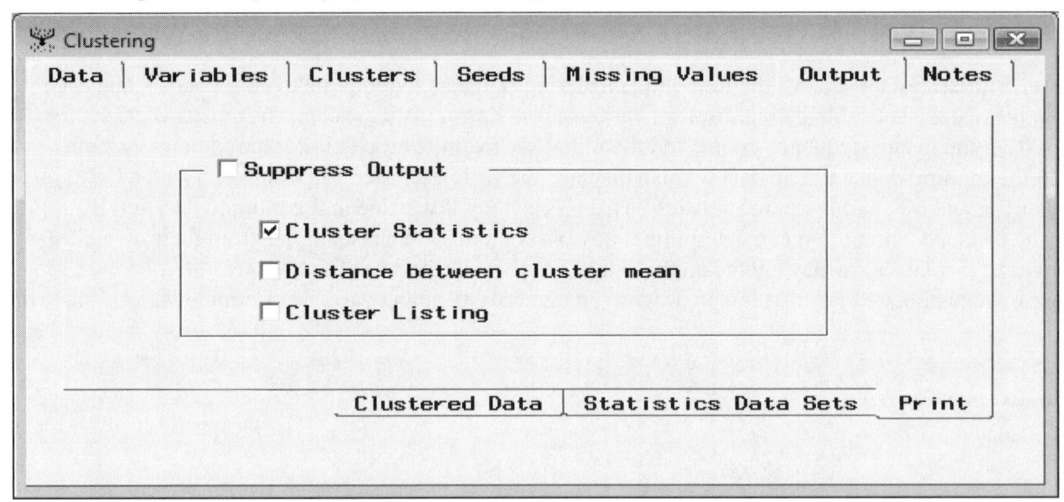

*The **Print** subtab is used to control the printed output of the clustering results from the **Output** tab.*

- **Suppress Output:** Suppresses the printed output from the node. Selecting the check box will result in the **Output** tab being grayed-out and unavailable for viewing from the **Results Browser**, with the following options grayed-out and unavailable for selection. By default, the **Suppress Output** check box is unselected, which will result in the default procedure output listings being available for viewing from the **Results Browser**.
- **Cluster Statistics** (default)**:** Prints the clustering statistics of each cluster. This is the default option. The cluster statistics that are printed are the initial seeds of each cluster and cluster summary statistics from the final cluster groupings, that is, the number of observations within each cluster, standard deviation, maximum distance between the cluster seed and the observation, nearest cluster and the distance between clusters, descriptive statistics, cubic clustering criterion statistic, and cluster means for each input variable in the analysis.
- **Distance between cluster mean:** Prints the distance statistics that are the squared distances between the cluster means based on the final cluster groupings from the PROC FASTCLUS procedure output listing.
- **Cluster Listing:** Prints the Cluster Listing table within the **Output** tab from the **Results Browser**. The Cluster Listing table displays the cluster assignments of each observation in the data along with the distance between the observation and the cluster seed of the assigned cluster.

Viewing the Clustering Node Results

General Layout of the Results Browser within the Clustering Node

- **Partition tab**
- **Variables tab**
- **Distance tab**
- **Profiles tab**
- **Statistics tab**
- **CCC Plot tab** *(hierarchical clustering only)*
- **Code tab**
- **Log tab**
- **Output tab**
- **Notes tab**

It is important to understand that the results that are listed from the various Enterprise Miner nodes are based on their associated data mining procedures. Executing the node will result in the PROC FASTCLUS and PROC CLUSTER procedures running behind the scenes, which creates the following results from the node. However, it is important to note that not all the options that are used in the PROC FASTCLUS and PROC CLUSTER procedures are incorporated in the node, and vice versa.

In the following analysis, the objective was to classify the best hitters in major league baseball by various hitting statistics in the 2004 season. Hitting statistics are batting average, on-base percentage, doubles, triples, home runs, and runs batted in just to name a few. The *k*-means partitive clustering method was performed with three separate cluster groups initially selected. The reason for initially selecting the three separate groups is because we want to determine the best hitters, good hitters, and the rest of the other hitters in major league baseball. Standardization was performed on the variables in the analysis since the various hitting categories displayed a wide range of values that are measured in entirely different units.

Partition tab

The **Partition** tab is designed to display both the proportion of Euclidean squared distance among the final clusters that the node automatically creates and the normalized means of the input variables in the analysis. The charts and plots will allow you to graphically compare the clustering assignments. In addition, you may also select a subset of the listed clusters by saving or exporting the scored data set from the clustering results.

In the following example, three separate clusters were selected that were specified from the **Number of Clusters** option within the **Cluster** tab. One reason for setting the number of clusters to three is that it resulted in a clear separation between the different clusters that were created from the various hitting categories.

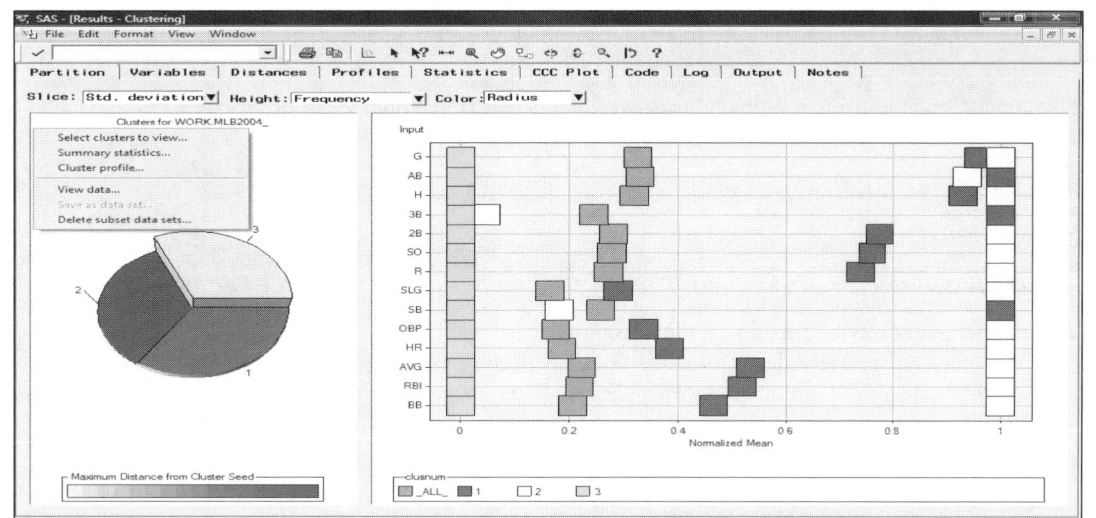

*The **Partition** tab that displays the various hitting categories of the grouped hitters in the game.*

3-D Pie Chart

From the **Partition** tab, a three-dimensional pie chart is displayed on the left-hand portion of the window. Each cluster is identified by each piece of the pie. The pie chart graphically displays the proportion of the maximum Euclidean distance of the observations from the cluster seeds by each cluster grouping. By default, the slices of the pie represent the standard deviation. The slices of the pie are the square root of the difference between the summarized distance between each data point and the cluster mean within each cluster. The three statistics that are used in the 3-D pie may be changed. In other words, you may change the meaning of the slices of the pie from the **Slice** pull-down menu options to **Std. Deviation** (default), **Frequency** or **Radius**. The height of each pie slice displays the number of observations within each cluster that is the default. The **Height** pull-down menu determines the height of each slice. The height represents the number of observations assigned to the cluster. The colors of the pie slices, that is, the radius, represent the distance of the farthest data point from the center. You can use the **View Info** toolbar icon from the tools bar menu to select a pie segment and holding down the right-click button that will result in a text box displaying the cluster id, root mean square error, frequency, and maximum distance. Select the **Rotate** icon or the **Tilt** icon to rotate or tilt the pie chart to various angles.

To create customized labels for each cluster that is created, simply select the **Edit > Cluster Variable Format** main menu options that will open the **Cluster Variable Format** window. From the **Description** column, simply enter the corresponding labels for each cluster that is listed.

Since the input data set consists of numeric coordinates, the frequency counts are based on the number of players assigned to each cluster. From the pie chart, the third cluster has the most baseball hitters in major league baseball, which can be viewed from the height of the pie chart. From the size of each slice of the pie chart, all three clusters display the same variability in the squared distance between each observation and its cluster mean.

From within the pie chart window, you may right-click the mouse and select the following options that are displayed from the previous diagram.

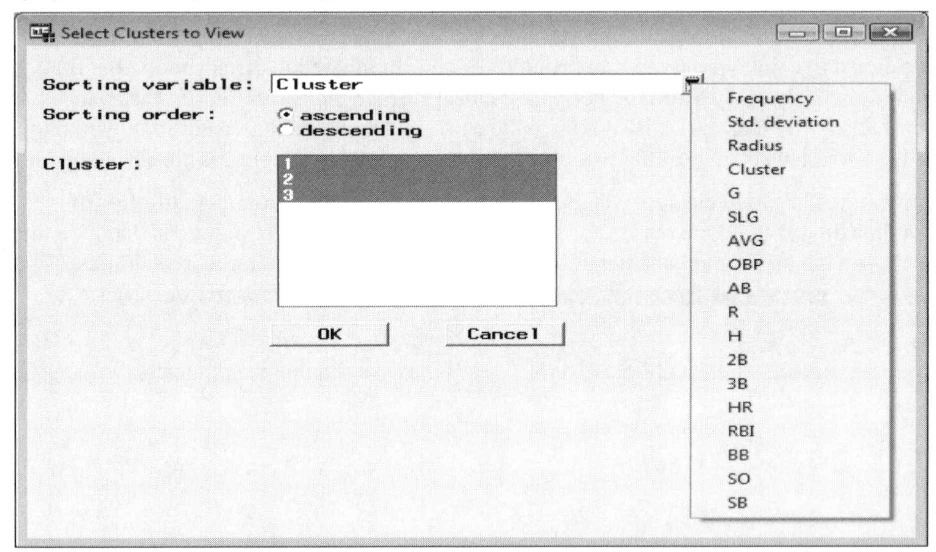

*The **Select Clusters to View** window is used to select the clusters to view within the **Partition** tab.*

- **Select clusters to view:** At times, the pie chart might be difficult to interpret, since the data might create too many clusters. Therefore, this option is designed for you to select the number of clusters that will be displayed within the tab. The **Select Clusters to View** window will appear for you to select the various clusters to view within the **Partition** tab. From the **Sorting variable** display field, click the down-arrow button and select a new variable from the drop-down list. The clusters will be sorted accordingly from the list box by the sorting variable that is selected. Sorting among clusters is performed by using the mean of the selected analysis variable.

- **Summary Statistics:** The **Cluster Profile: All Clusters** window will appear with two separate tabs. The **Interval** tab displays the various descriptive statistics such as the frequency count, mean, standard deviation, median, and minimum and maximum value of each cluster based on the interval-valued input variables in the analysis. The **Class** tab displays the various percentages based on each class level of the categorically-valued variables in the analysis. The summary statistics based on all observations are combined and segregated by each cluster grouping. The table listing that displays a period in the **Cluster** column indicates the overall cluster statistics. The contingency tables are generated by all the class levels of the categorically-valued input variables.
- **Cluster profile:** The **Cluster Profile Tree** window will appear in order for you to view the classification decision tree that displays both the frequency count and the percentages from the binary tree splits that are performed by each cluster that is created. The decision tree will allow you to determine which input variables form the strongest splits or the input variables that are most effective in forming the various clustering groups. The splits performed at or near the root of the tree will indicate the range of values or class levels that best divides the various cluster groupings that are created. This decision tree is created from the **Tree** node that is executed in the background when running the **Clustering** node. The clustering profile statistics are created when you run the **Clustering** node with the **Tree** node executing in the background.
- **View Data:** The **Browsing Clusters: <Cluster #>** window will appear that is a table view of the output score data set of the selected cluster. The table listing is extremely important since it displays the clustering assignments of the input variables based on the distance matrix or the values of the input variables from the coordinate data. This option will display a table listing of all the observations from the training data set with the automatically generated segment or cluster identifier and the associated Euclidean distances to the corresponding cluster seed value. There is an added option within the window of either saving the listed data set or exporting the data set through the standard SAS Export Wizard into many different file formats.
- **Delete subset data sets:** The **Delete Subset Data Sets** window will appear to select the data sets that you want to delete from the list.

Normalized Plot

The plot displayed on the right-hand side of the **Partition** tab shows the average values of each input variable that have been normalized using a scale-transformation function. The *normalized means* are computed by the mean divided by the maximum value of each input variable that will result in its values varying between the value of zero and one. The plot is useful if it is assumed that the various interval-valued inputs are measured in entirely different units, for example, variables measured in pounds as opposed to variables measured in dollars. Thereby, the ratio of change between the variables is comparable, that is, doubling the weight will have the same effect as doubling the income. The purpose of the normalized plot is to identify the input variables that are significantly different from the overall mean, indicating to you the input variables that best characterize the corresponding cluster. Conversely, input variables that are very close to the overall mean will not help us very much in describing the underlying characteristic of the corresponding cluster or clusters. The plot displays the relative difference in the input means of each cluster and the overall mean. From the plot, the node will display each class level of the categorically-valued variable. By selecting the **Select Points** toolbar icon and selecting any one of the slices from the pie chart, you may then select the **Refresh input means plot** toolbar icon that displays the difference between the selected cluster and the overall mean of all the clusters from the various input variables in the analysis. Select any slice or cluster from the pie chart, then select the toolbar icon that will display the cluster means with comparison to the overall average. To select more than one cluster, select the Ctrl key and the corresponding **Refresh input means plot** toolbar icon to display any number of clusters within the input means plot, as illustrated in the previous normalized plot. From the tools bar menu, you can select the **Scroll data** toolbar icon to see any additional input variables in the analysis that have been truncated off the bottom vertical axis of the plot. This can be accomplished by dragging the mouse along the vertical axis of the graph with the open hand.

From the normalized plot, the second cluster grouping is the most noticeable. The reason is because this grouping of players are the most devastating hitters in the game, far better than the overall average in most all hitting categories with the exception of hitting triples or stealing bases. The first cluster grouping consists of mainly base stealers and triple hitters with batting average, home runs, and on-base percentages that are above the norm. The third cluster grouping of baseball players contains a majority of the hitters in the game who are some of the weakest hitters in the game, with a cluster average in all hitting categories lower than the overall average.

Note: If any of the input variables are categorical and the **Standard Deviation** standardization method is applied, then the normalized mean plot will be incorrect.

Variables tab

The **Variables** tab is designed to display a table view of the various input variables in the analysis along with the level of measurement and the variable labels. An *importance value* or ratio is also displayed that measures the importance of the input variable in forming the clusters. The importance value is based on a tree splitting rule with a value of zero indicating to you that the variable was not used as a splitting variable. The statistic indicates how well the input variable splits the data into separate groups. Input variables with an importance statistic close to one will indicate that the variables are good splitting variables and, therefore, make the greatest contribution in creating the various clusters. The listed statistic is determined by the number of observations that are randomly sampled from the training data set that can be specified from the **Preliminary Training and Profiles** subtab. From the following table listing, the number of runs scored and slugging percentage appear to be the most important input variables in creating the clusters. The importance statistic is calculated as follows:

For categorical input variables, the importance statistic is based on the chi-square statistic that is calculated from each class level of the categorically-valued input variable and the various cluster groupings as follows:

Importance Statistic for the Categorically-valued Input Variable = $-\log[(\text{Chi-Square}) \cdot p\text{-value}]$

For interval input variables, the importance statistic is based on the F-test statistic that is calculated by the group variability of the interval-valued input variable and the various cluster groupings as follows:

Importance Statistic for the Interval-Valued Input Variable = $-\log[(\text{F-test}) \cdot p\text{-value}]$

Results - Clustering

Partition | Variables | Distances | Profiles | Statistics | CCC Plot | Code | Log | Output | Notes

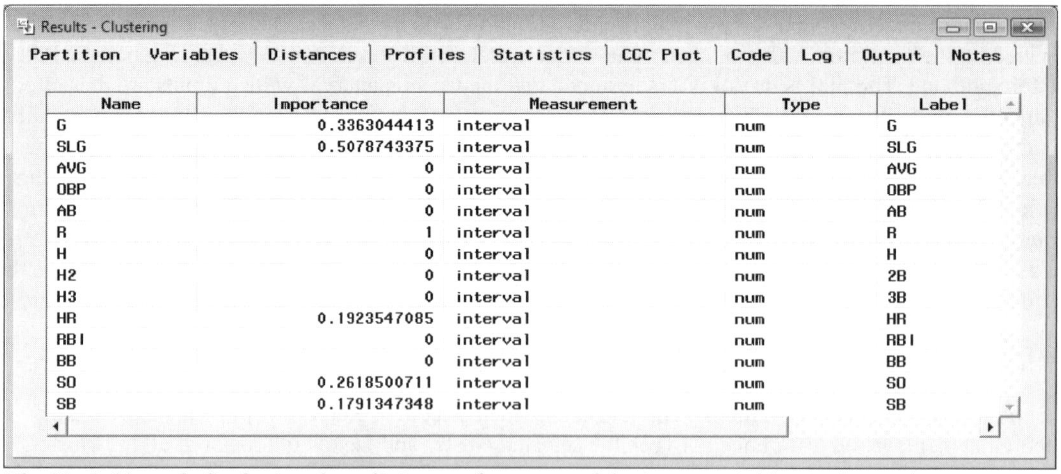

Name	Importance	Measurement	Type	Label
G	0.3363044413	interval	num	G
SLG	0.5078743375	interval	num	SLG
AVG	0	interval	num	AVG
OBP	0	interval	num	OBP
AB	0	interval	num	AB
R	1	interval	num	R
H	0	interval	num	H
H2	0	interval	num	2B
H3	0	interval	num	3B
HR	0.1923547085	interval	num	HR
RBI	0	interval	num	RBI
BB	0	interval	num	BB
SO	0.2618500711	interval	num	SO
SB	0.1791347348	interval	num	SB

*The **Variables** tab displaying the splitting performance of the input variables in the analysis.*

Distances tab

For hierarchical clustering, the **Distances** tab is designed to display a graphical representation of the Euclidean distance within each cluster seed and the most extreme data point based on the squared distance between each cluster that is created. This will allow you to graphically observe the size of each cluster and the relationship between the clusters. The asterisks will indicate the cluster centers and the circles indicate the cluster radius. Clusters that contain a single data point will be displayed as an asterisk without a drawn circle. The tab also displays the distance matrix that is the basis for forming the various clusters that are created. The measurement used in determining the Euclidean distances between the clusters is determined by the **Clustering Criterion** option settings from the previous **Seeds** tab. The following distance plot displays the flying miles between the various cities in the United States. The reason for using this example in explaining the distance plot and the MDS procedure is that there is no need for the procedure to convert the data into squared distances since the flying mileages are a very good approximation of the Euclidean distances. From the plot, the MDS procedure seems to capture the relative position of the cities; however, the geographical locations of the cities have not been retained.

Distance Plot

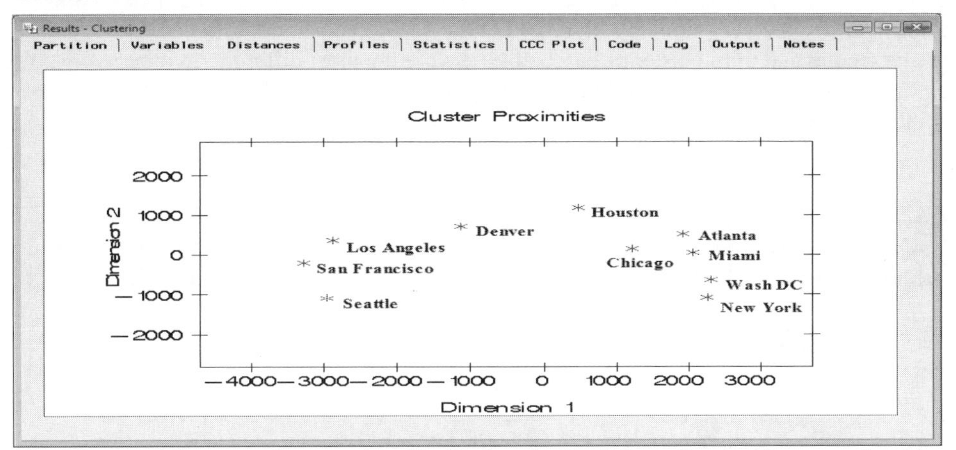

*The **Distance Plot** is used to graphically display the Euclidean distance between the cities.*

The axes of the graph are determined by the two separate dimensions that are calculated from multidimensional scaling analysis based on the distance matrix from the cluster means. The purpose of *multidimensional scaling* is to scale the $n \times n$ distance matrix down to two separate dimensions while at the same time preserving the ranked-order distance between the separate cluster means. The PROC MDS procedure will allow you to map the distances between the points that are in a high dimensional space reduced down into a two-dimensional space. The first step is standardizing all the interval-valued variables in the data set in order to ensure that the data is expressed in comparable units. The next step is calculating the Euclidean distances between the observations. The data set is then formulated into a squared distance matrix based on the Euclidean distances, where each row and column is determined by the unique class levels of the categorically-valued variable, that is, the segment identifier variable in the output scored data set. The PROC MDS procedure is then applied. From the PROC MDS procedure, these Euclidean distances are then normalized to calculate a minimization function that is based on the ratio between the difference in the original distances between the pair of points and the same pair of points in the reduced two-dimensional space over the original distances between the pair of data points so that the paired distances are preserved as closely as possible to the true ordered distances between the observed data points. The minimization function is displayed in the following formula. Again, the main idea of the two-dimensional plot is to reduce the complexity of the data in terms of fewer dimensions based on the squared distance matrix, while at the same time preserving the original ranked-order distances between the pair of data points. As in our example, multidimensional scaling is designed to make it easier for you to visualize the cluster groupings of the hitters in the game of baseball while at the same time preserving the distances between the various hitting categories from the standardized interval-valued variables in the distance matrix.

The following is the minimization function that is used in multidimensional scaling to evaluate how well a particular configuration reproduces the observed distance matrix.

$$E = \frac{\sum_{i=1}^{n} \sum_{j=1}^{i-1} (D_{ij} - d_{ij})^2 / D_{ij}}{\sum_{i=1}^{n} \sum_{j=1}^{i-1} D_{ij}}$$

where E is the minimization formula we want at a minimum, D_{ij} is the original distance between the ith and j^{th} observations, and d_{ij} is the distance in the reduced space between the ith and j^{th} observations.

The radius of each cluster from the plot is based on the squared difference, between the most distant data point and the seed within each cluster. At times, the displayed clusters, circles, or bubbles may appear to overlap in the plot. However, it is important to realize that each data point is assigned to exactly one cluster. For nonhierarchical clustering, the **Distances** tab is designed to display the distance matrix that lists the Euclidean distance between each cluster in the analysis. The matrix consists of zeros along the main diagonal and the distance between each separate cluster in the analysis that are displayed in the off-diagonal entries. The distance matrix is useful in determining if any cluster is distant from the other clusters that are most desirable. The distance among the clusters is based on the criteria that are specified to construct the clusters. The clustering criteria can be specified from the **General** subtab. From the following table listing, the first and second clusters of baseball hitters are very similar with regard to the squared distances between the various hitting categories and the second and third cluster of baseball hitters are the most different. In other words, the best hitters in the game are comparable to the good hitters in the game, but are very different in comparison to the rest of the other hitters in the game.

Distance Matrix

Results - Clustering

Partition | Variables | Distances | Profiles | Statistics | CCC Plot | Code | Log | Output | Notes

CLUSTER	Cluster 1	Cluster 2	Cluster 3
1	0	54.415576426	274.11799271
2	54.415576426	0	279.20297007
3	274.11799271	279.20297007	0

*The **Distance Matrix** is used to view the Euclidean distances between each cluster in the analysis.*

Profiles tab

The **Profiles** tab is designed to display the graphical representation of both the categorically-valued and interval-valued input variables in the analysis. The tab is designed for you to determine certain profiles or characteristics between the clusters that are created. The following profile pie chart and bar chart are created from the random sample of the training data set that is selected from the previous **Preliminary Training and Profiles** subtab. In our example, every observation from the training data set was selected to create the plots.

Categorical Variable Plots

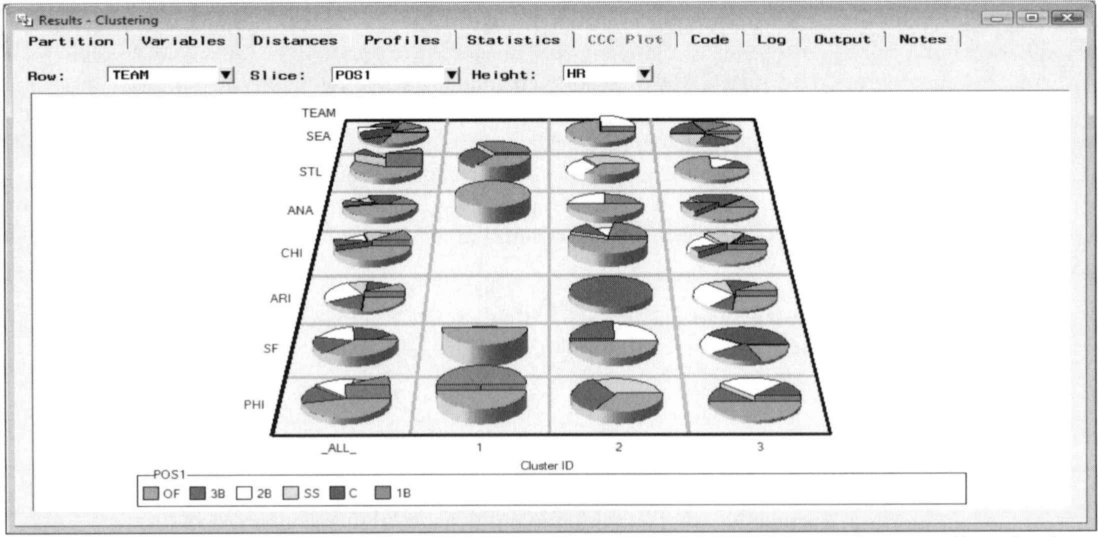

*The **Profile** tab displays the profile plot to identify various characteristics of the categorically-valued variables based on a random sample of the active training data set.*

Selecting the **View > Categorical Variables** main menu option, then this will result in the profile plot appearing. The profile plot is a three-dimensional pie chart displaying the relative frequencies between the two separate categorically-valued variables across the various clusters. Since the node automatically creates separate class levels for each categorically-valued input variable in the analysis, the rows of the profile chart display the relative frequencies based on the class levels of the first categorical input variable with the slices of the pie chart representing the class levels of the second categorical input variable. The heights of the pie charts represent the values within each class level from the interval-valued input variable in the training data set. By default, the height of the pie charts is set to **None**. Therefore, in our example, the pie charts display the various teams in the major leagues with the slices of each pie chart representing the different positions and the height of each piece of the pie displaying the number of home runs that are hit from the three separate cluster groupings that are created. In the example, the pie charts indicate that the second cluster is comprised of the most devastating hitters in the game, with the best hitting outfielders from San Francisco, Chicago, Anaheim, and Seattle hitting at least half of the team's overall home runs.

Select the **Row** pull-down menu item to select a different categorical row variable for the profile plot. The slices of the pie chart may be changed by selecting a separate categorical variable from the **Slices** pull-down menu items. The appearance of the pie chart will change into a 3-D pie chart by selecting any of the available interval-valued input variables in the data from the **Height** pull-down menu item. Select the **Tilt** toolbar icon to make it easier to view the various heights and the corresponding frequencies of the displayed pie charts. Selecting the **Format > Set Grid dimensions** main menu option will allow you to specify the number of columns or clusters to view from the profile plots. By default, the first three clusters are automatically displayed when you select the tab that is illustrated in the previous diagram. Select the **Scroll data** toolbar icon to see any additional class levels that have been truncated off the bottom vertical axis of the chart by dragging the mouse along the vertical axis of the graph with the open hand. Conversely, scroll the horizontal axis to the right if there are any additional clusters to be viewed.

Interval Variable Plots

Selecting the **View > Interval Variables** main menu item will allow you to view the three-dimensional frequency bar chart for every interval-valued input variable in the training data set.

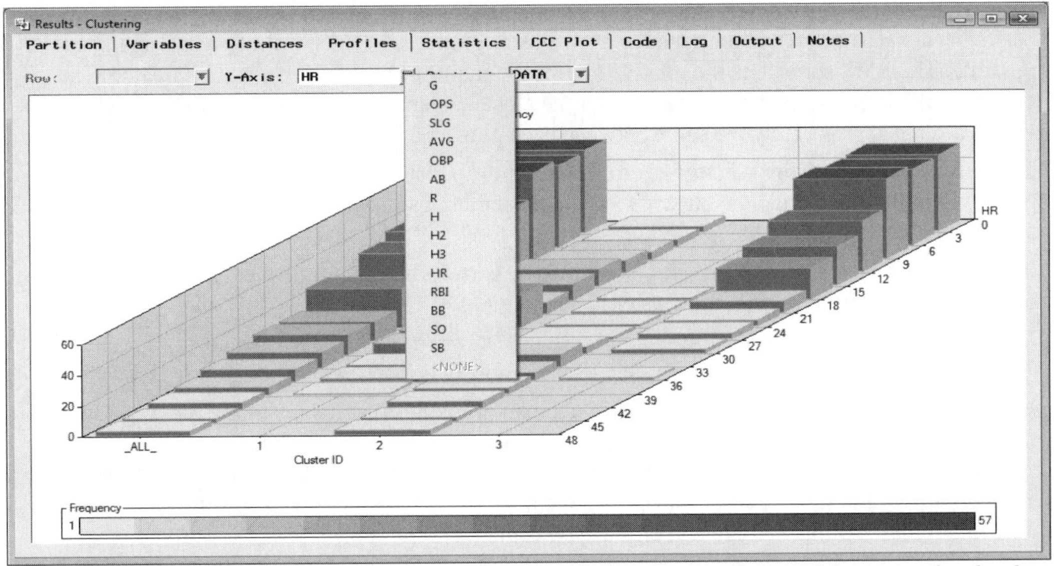

*The **Profiles** tab displays the profile plot used to identify the characteristics of the interval-valued inputs.*

The 3-D frequency bar chart displays the range of values of the interval-valued input variable against the various cluster groupings that are created. The heights of the bar charts represent the frequency counts that the range of values fall into for the corresponding cluster grouping. Besides viewing the various characteristics, the frequency bars are useful in viewing if there are any extreme values within the various clusters. From the **Y-Axis** pull-down menu, select from the list of available input variables in the data set to select a separate interval-valued input variable to be displayed on the vertical Y-axis of the 3-D bar chart. Select the **View Info** tool icon and drag the pointer to the various frequency bars to view the text box that displays the value, cluster id, and frequency of the input variable. The plot indicates that the third cluster with the weakest hitters in the game hit the fewest home runs and the second cluster with the best hitters in the game hit the most home runs.

Statistics tab

The **Statistics** tab is designed to display the clustering criterion statistics of the various clusters that are created. The table listing that is displayed within the tab is the **Seed Statistics** data set that can be viewed from the **Statistics Data Sets** subtab. The following table will allow you to view the discrepancies in the average values among the various input variables that are assigned to the different groups. For example, the third cluster grouping has a significantly lower batting average and average number of home runs in comparison to the other two cluster grouping of baseball hitters, and so on.

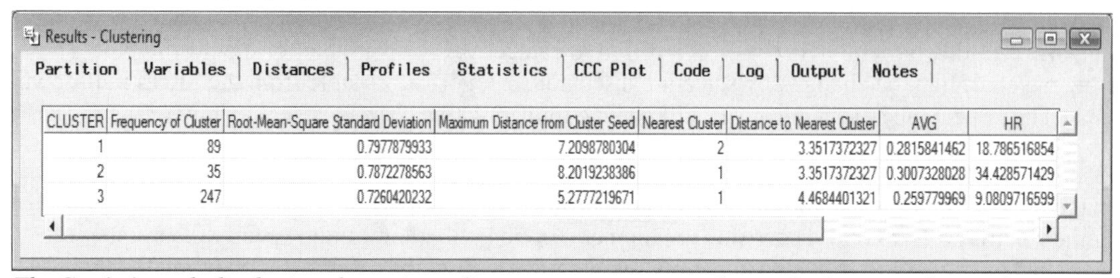

CLUSTER	Frequency of Cluster	Root-Mean-Square Standard Deviation	Maximum Distance from Cluster Seed	Nearest Cluster	Distance to Nearest Cluster	AVG	HR
1	89	0.7977879933	7.2098780304	2	3.3517372327	0.2815841462	18.786516854
2	35	0.7872278563	8.2019238386	1	3.3517372327	0.3007328028	34.428571429
3	247	0.7260420232	5.2777219671	1	4.4684401321	0.259779969	9.0809716599

*The **Statistics** tab displaying the various distance measurements for each cluster in the analysis.*

The following are the various clustering statistics that are displayed within the tab:

- **CLUSTER:** The segment or cluster identifier.
- **Frequency of Cluster:** The number of observations in each cluster. Clusters with a small number of observations will usually indicate that the data has some extreme values. Small frequency counts might indicate the most appropriate cluster to remove from the analysis in order to improve cluster separation and provide a better representation of the various cluster groupings.
- **Root Mean Square Standard Deviation:** The standard deviation or the root mean square standard deviation within each cluster that is the square root of the summarized squared distance between each data point and the seed for each cluster group, which we want to be as small as possible. The RMS statistic is the relative size of the radius of the clusters that are created. The statistic will allow you to determine whether the clusters should be partitioned when the RMS values are not appropriately the same between each cluster.
- **Maximum Distance from Cluster Seed:** The Euclidean distance between the most extreme data point within the cluster and the cluster seed. In cluster analysis, we want this statistic to be as small as possible.
- **Nearest Cluster:** The cluster that lies closest to the corresponding cluster that is determined by the distance between the two separate cluster means.
- **Distance to Nearest Cluster:** The squared distance between the two separate cluster groupings. The difference between the two separate cluster means, which we want to be as large as possible.
- **Mean of the Input Variables:** The computed mean of the input variables in the training data sets for each cluster. The various columns will allow you to interpret the characteristics of each cluster from the average values of the listed input variables. The following illustration indicates that the second cluster of baseball players have the most hits, doubles, home runs, and runs scored, along with the highest batting average and on-base percentage.

Select the **View > Statistics Plot** main menu options to display the following 3-D frequency bar chart. The bar chart displays the cluster frequencies defined by the range of values of the various interval-valued input variables in the analysis across each cluster.

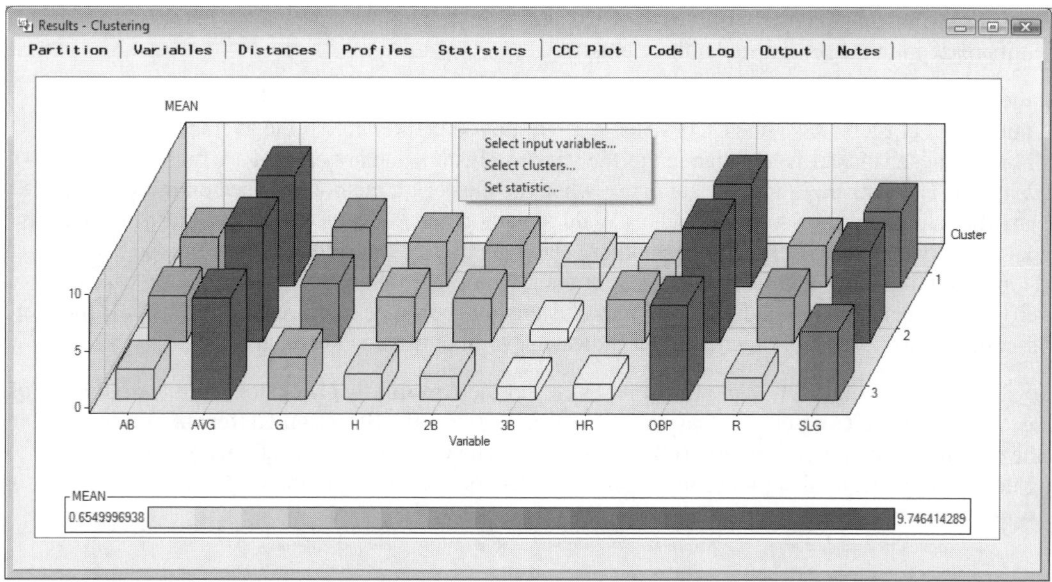

The ***Statistics Plot*** *displays the average values of the interval-valued inputs for each cluster to identify certain characteristics between the separate clusters from the active training data set.*

At times, the plot might be convoluted, with too many input variables or clusters in the analysis. Therefore, you may right-click the mouse and select the various pop-up menu options in order to make the interpretation of the analysis easier to visualize, as illustrated in the previous diagram.

- **Select input variables:** The **Select Input Variable** window will appear in order for you to select any number of the listed input variables of the statistic plot. By default, all input variables are selected from the frequency bar chart.
- **Select clusters:** The **Select Clusters** window will appear with a text box listing of the various clusters that are created. Press the Ctrl button to select any number of clusters to be displayed within the frequency bar chart. By default, all the clusters are automatically selected.
- **Set statistic:** By default, the heights of the bar charts are based on the frequency counts for each cluster. However, you may select from the following statistics:
  - **Mean:** The heights of the bar charts are determined by the average value of the various input variables for each cluster that is created.
  - **Std. Deviation:** The heights of the bar charts are determined by the standard deviation of the various input variables for each cluster that is created.
  - **Cluster seed:** The heights of the bar charts are determined by the cluster seed, which is essentially the cluster mean for each input variable in the analysis.
  - **Frequency** (default)**:** The heights of the bar charts are determined by the frequency counts of the various input variables for each cluster that is created.

CCC Plot tab

For hierarchical clustering, the **CCC Plot** tab is designed to display the plot of the cubic clustering criterion against the various clusters. The tab will be available for viewing when hierarchical clustering is performed. The plot is based on the specified cubic clustering criterion cutoff value used in determining the optimum number of clusters in the analysis. By default, a random sample of 2,000 observations is selected to construct the plot. However, you may change the default sample size that is automatically selected from the **Preliminary Training and Profiles** subtab. A solid vertical line displayed within the plot indicates the number of clusters that the **Clustering** node has automatically selected according to the first sudden peak in the CCC statistic. The CCC statistic is the ratio between the observed r-square statistic and the expected r-square statistic obtained by applying a variance-stabilizing transformation to the data in order to stabilize the variability across the number of observations, variables, and clusters. The expected r-square statistic is based on the **Clustering cubic criterion cutoff** value specified from the **Selection Criterion** window. However, the CCC statistic is computed under the assumption that the clusters that are created follow a uniform distribution of hypercubes of the same size. The following diagram displays the CCC plot from the baseball data set, where there seems to be a sudden increase in the CCC statistic at three clusters.

It is recommended that the best way to use the CCC statistic is to plot about one-tenth of the observations across the number of clusters. Also, the CCC statistic might not be that reliable if the average number of observations within each cluster is less than ten or the various clusters that are created are highly elongated or irregularly shaped. The statistic is appropriate to use when the clustering method is based on minimizing the trace of the within cluster sum-of-squares, such as Wald's method and *k*-means clustering. Furthermore, it is assumed that the variables are unrelated to each other. The general guidelines for determining the correct number of clusters will display a CCC statistic greater than two or three. Sudden peaks in the CCC plot between zero to two clusters might indicate potential clusters but should be interpreted cautiously. If the plot displays large negative CCC values, then it might indicate several outliers in the data.

In order to view the following **CCC Plot** tab, you must select the **Automatic** check box from the **Number of Clusters** section within the **Cluster** tab. This will allow you to select the **Selection Criterion** button in order to access the **Clustering cubic criterion cutoff value** option. The CCC plot is only shown when the **Clustering** node automatically determines the number of clusters to create from the CCC statistic.

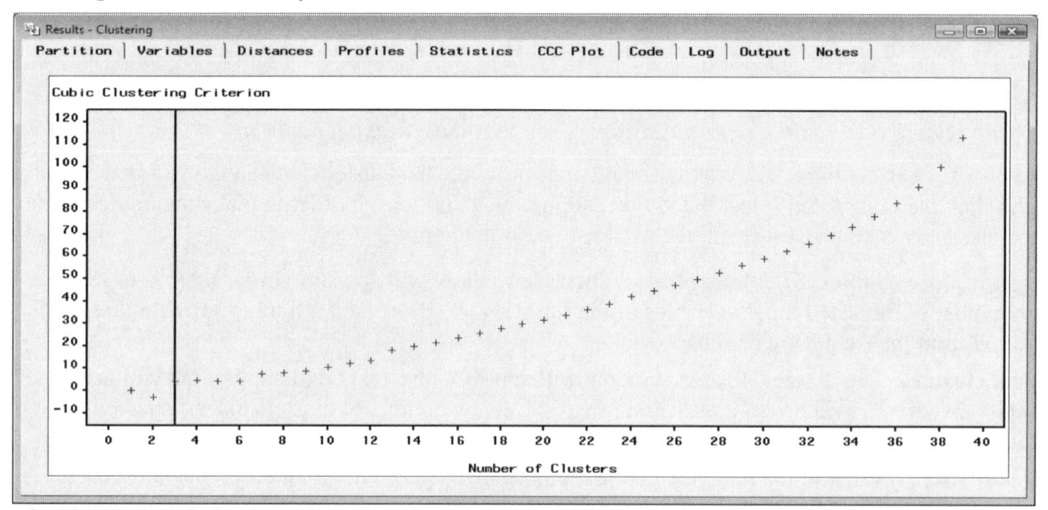

*The **CCC Plot** tab displays the cubic clustering criterion statistic across the various clusters.*

Code tab

The **Code** tab is designed to display the internal SEMMA training code that created the various clustering results by running the node. The training code displays the various clustering options that are specified from within the node to generate the procedure output listing that is displayed in the following **Output** tab. By default, the **Training** radio button is selected which displays the following PROC FASTCLUS procedure since the **Least Squares (Fast)** is the default cluster criterion that is automatically applied.

Selecting the **Scoring** radio button will display the scoring code that can be included in any SAS session in order to score new observations. Cluster analysis is considered to be a descriptive technique. However, since this unsupervised training technique calculates cluster means, the clustering assignments can be applied to new observations based on the minimum squared distance between each input variable and its own cluster mean. The scored code will first determine if any one of the input variables have missing values and set the segment identifier variable to missing. If the input variables do not have any missing values, the procedure will then standardize the input variables in the analysis, compute the squared distance in determining the most appropriate cluster assignment to each observation, and create the segment identifier in distinguishing between the corresponding clustering assignments based on a new set of values from the input variables in the analysis. In addition, the scoring code imputes missing values by the cluster seeds and is the default imputation method. The following scoring code is a partial listing of the scoring code that can be used in assigning the clustering assignments by fitting the model to a new set of observations from the major league baseball data set.

Results - Clustering

Partition | Variables | Distances | Profiles | Statistics | CCC Plot | Code | Log | Output | Notes

(•) Training ( ) Scoring

```
00001 title;
00002 options nodate;
00003 proc fastclus data=_CLUSTMP maxc=3
00004 outseed=EMPROJ.CLS99VPN(label="Clustered Seeds for EMDATA.VIEW_69F")
00005 outstat=EMPROJ.CLSHSMP9(label="Clustered Statistics for EMDATA.VIEW_69F")
00006 cluster=_SEGMNT_ radius=0 replace=Full maxiter=1 conv=0.0001 std=std
00007 impute=NS
00008 ;
00009 var
00010 G
00011 SLG
00012 AVG
00013 OBP
00014 AB
00015 R
00016 H
00017 H2
00018 H3
00019 HR
00020 RBI
00021 BB
00022 SO
00023 SB
00024 ;
00025 id TEAM;
00026 run;
00027 quit;
00028 run;
00029 quit;
00030 *** END OF FILE ***
```

*The **Code** tab displays the training code from the PROC FASTCLUS procedure.*

Results - Clustering

Partition | Variables | Distances | Profiles | Statistics | CCC Plot | Code | Log | Output | Notes

( ) Training (•) Scoring

```
00001 format _SEGMNT_ CLS_RDL_. distance best12.;
00002 ************************************;
00003 *** begin scoring code for clustering;
00004 ************************************;
00005 label _SEGMNT_ = "Cluster ID"
00006       Distance = 'Distance to Cluster Seed';
00007 drop _nonmiss; _nonmiss = n(
00008    G , SLG , AVG , OBP , AB , R ,
00009    H , H2 , H3 , HR , RBI , BB ,
00010    SO , SB );
00011 if _nonmiss = 0 then do;
00012    _SEGMNT_ = .; distance = .;
00013 end;
00014 else do;
00015    array _CLScads[3] _temporary_;
00016    drop _clus;
00017    do _clus = 1 to 3; _CLScads[_clus] = 0; end;
00018    if n(G) then do;
00019       _clusvar = G * 0.0311470971 ; drop _clusvar ;
00020       _CLScads[1] + (_clusvar - 4.7103994566 )**2;
00021       _CLScads[2] + (_clusvar - 4.7014812706 )**2;
00022       _CLScads[3] + (_clusvar - 3.4106588739 )**2;
00023    end;
 ...
00096    if n(SB) then do;
00097       _clusvar = SB * 0.1106057256 ; drop _clusvar ;
00098       _CLScads[1] + (_clusvar - 2.0632221885 )**2;
00099       _CLScads[2] + (_clusvar - 0.7619505539 )**2;
00100       _CLScads[3] + (_clusvar - 0.480638834 )**2;
00101    end;
00102    _SEGMNT_ = 1; distance = _CLScads[1];
00103    do _clus = 2 to 3;
00104       if _CLScads[_clus] < distance then do;
00105          _SEGMNT_ = _clus; distance = _CLScads[_clus];
00106       end;
00107    end;
00108    distance = sqrt(distance*14/_nonmiss);
00109 end;
00110 *** impute missing values;
00111 label _Impute_ = 'Number of Imputed Values';
00112 _impute_ = 14 - _nonmiss;
00113 if _impute_=0 then goto _CLSclex;
00114 if _nonmiss = 0 then do; *** for all missing, impute mean;
00115    G = 3.6554602578 * 32.105720677 ;
00116    SLG = 5.8279406903 * 0.0745384522 ;
00117    AVG = 8.7139091662 * 0.0308557377 ;
00118    OBP = 8.1586232644 * 0.0408292079 ;
00119    AB = 2.6349566256 * 150.98884465 ;
00120    R = 1.9934955588 * 28.909383575 ;
00121    H = 2.2902843028 * 47.671196479 ;
00122    H2 = 2.0774398001 * 10.610717424 ;
00123    H3 = 0.8881490059 * 2.5280476632 ;
00124    HR = 1.3265434011 * 10.403383012 ;
00125    RBI = 1.927025546 * 28.643504812 ;
00126    BB = 1.5631528225 * 25.516886051 ;
00127    SO = 2.1831015842 * 32.734899913 ;
00128    SB = 0.7161049941 * 9.0411232767 ;
00129 end; *** end all missing;
00130 else do; *** impute=nearseed;
00131 array _CLScasd [3,14] _temporary_ (
00132    /* cluster 1 */ 4.7103994566 6.2871772981 9.3868479839 8.6231703606 3.9082102735
00133 3.3147342222 3.5935261743 3.0919391212 2.1603779632 1.9612703032
00134 2.7600565234 2.2232638072 2.9126758933 2.0632221885
00135    /* cluster 2 */ 4.7014812706 8.0358716903 10.114824325 10.234865632 3.5712719272
00136 3.7819323629 3.521809841 3.534162536 0.8131000003 3.5672156903
00137 3.8810427494 3.8601888884 3.1855230367 0.7619505539
00138    /* cluster 3 */ 3.4106588739 5.6165683927 8.5138782979 7.9542086079 2.3590002631
00139 1.6582919941 1.9914934822 1.8150644714 0.6728499239 1.0828958925
00140 1.6662616906 1.3117494325 1.9971166998 0.480638834
00141    ); retain _CLScasd;
00142    if nmiss(G) then G = _CLScasd[_SEGMNT_,1 ]* 32.105720677 ;
00143    if nmiss(SLG) then SLG = _CLScasd[_SEGMNT_,2 ]* 0.0745384522 ;
00144    if nmiss(AVG) then AVG = _CLScasd[_SEGMNT_,3 ]* 0.0308557377 ;
00145    if nmiss(OBP) then OBP = _CLScasd[_SEGMNT_,4 ]* 0.0408292079 ;
00146    if nmiss(AB) then AB = _CLScasd[_SEGMNT_,5 ]* 150.98884465 ;
00147    if nmiss(R) then R = _CLScasd[_SEGMNT_,6 ]* 28.909383575 ;
00148    if nmiss(H) then H = _CLScasd[_SEGMNT_,7 ]* 47.671196479 ;
00149    if nmiss(H2) then H2 = _CLScasd[_SEGMNT_,8 ]* 10.610717424 ;
00150    if nmiss(H3) then H3 = _CLScasd[_SEGMNT_,9 ]* 2.5280476632 ;
00151    if nmiss(HR) then HR = _CLScasd[_SEGMNT_,10 ]* 10.403383012 ;
00152    if nmiss(RBI) then RBI = _CLScasd[_SEGMNT_,11 ]* 28.643504812 ;
00153    if nmiss(BB) then BB = _CLScasd[_SEGMNT_,12 ]* 25.516886051 ;
00154    if nmiss(SO) then SO = _CLScasd[_SEGMNT_,13 ]* 32.734899913 ;
00155    if nmiss(SB) then SB = _CLScasd[_SEGMNT_,14 ]* 9.0411232767 ;
00156 end; *** end nearseed;
00157 _CLSclex:;
00158 ************************************;
00159 *** end scoring code for clustering;
00160 ************************************;
00161 *** END OF FILE ***
```

*The **Code** tab displays the scoring code used to create the various clusters.*

Log tab

The **Log** tab is designed to display the log listing from the internal data mining procedure code that is used in generating the clustering assignments and corresponding results that are displayed within the node. The **Log** tab will first display the temporary data set that is created by performing a random sample from the training data set. The PROC FASTCLUS procedure is then used in creating the initial cluster seeds and the temporary clusters from the training data set that is randomly sampled. The log listing from the PROC STANDARD procedure is displayed since the input variables are standardized from the active training data set. The scored output data sets are created by calculating the clustering assignments and squared distances from the previous listed scoring code.

Output tab

For both hierarchical and nonhierarchical clustering, the **Output** tab is designed to display the procedure output listing from both the underlying SAS/STAT PROC FASTCLUS and CLUSTER procedures.

The PROC FASTCLUS procedure is used in generating the various clustering seed assignments along with the PROC CLUSTER procedure that is used in creating the final clustering results. In other words, the PROC FASTCLUS is the underlying procedure that is used to determine the mutually exclusive clusters that are initially created. However, the PROC CLUSTER performs the default automatic selection of the number of clusters. For nonhierarchical clustering, the procedure output listing is generated from the PROC FASTCLUS procedure. For hierarchical clustering, the PROC FASTCLUS procedure will display the initial seed assignments to each input variable in the analysis. The PROC FASTCLUS procedure will also provide you with the cluster means of each input variable in the analysis. The procedure listing will allow you to profile and characterize the various clusters that are created. This will be followed by various clustering statistics such as the frequency counts and standard deviations of each cluster in determining the stability of the clustering assignments. The PROC CLUSTER procedure output listing will then be displayed that will generate various goodness-of-fit statistics in determining the number of clusters to select.

The output listing displays the following cluster statistics that will allow you to determine the number of clusters to select.

The SAS/STAT PROC FASTCLUS procedure output listing used in determining the initial clusters:

- The initial cluster seed for each input variable after one pass through the training data set
- Cluster Summary for each cluster
  - Frequency count that is the number of observations assigned to each cluster to determine the reliability of the various clustering assignments, that is, small frequency counts within each cluster are undesirable.
  - Root Mean Squared Standard Deviation within each cluster, which is an alternative to the r-square statistic in evaluating the degree of homogeneity between the clusters in choosing the best partitioning of the data. Generally, the best clusters to select will result in the lowest variance. However, the drawback to this statistic is that clusters with a single observation will always be preferred with a variance of zero.
  - Maximum Distance from Seed to Observation that measures the internal cohesion within each cluster, which is the maximum distance from the cluster seed and any observation assigned to the cluster that we want to be as small as possible.
  - Radius Exceed – the relative size of the radius of the clusters that are created.
  - Nearest Cluster to the input variable.
  - Distance Between Cluster Centroids of the current cluster and the nearest cluster, which we want to be as large as possible. This measurement is preferable for all types of clustering techniques.
- Statistics for Variables: The following assessment statistics are useful in identifying the number of clusters and are very effective only for interval-valued input variables in the analysis.
  - Total Standard Deviation for each input variable in the analysis.
  - Pooled Standard Deviation for each input variable in the analysis.

- RSQ Statistic for each input variable in the analysis, which is the ratio of the between-cluster and within-cluster variance. The statistic explains the proportion of variability accounted for by the clusters that are created. A r-square statistic close to one indicates that the observations that are within each cluster are very similar, that is, low $\sum\|x_i - \overline{x}_g\|$), and the clusters that are well separated from each other, that is, high $\sum\|\overline{x}_g - \overline{x}\|$.

  $R^2 = 1 - (\sum_{i=1}^{g}\|x_i - \overline{x}_k\| / \sum_{i=1}^{n}\|x_i - \overline{x}\|)$ for all g clusters based on the Euclidean distance $\|\cdot\|$

  where $\overline{x}$ is the sample mean vector and $\overline{x}_k$ is the mean vector for the cluster k.
- $R^2 / (1 - R^2)$ statistic – the ratio of the between and within cluster variance.
- Pseudo F-Statistic in estimating the number of clusters that is computed as follows:

$$PSF_{v(g-1),v(n-g)} = \frac{\sum_{i=1}^{n}\|x_i - \overline{x}\|^2/(g-1)}{\sum_{i=1}^{g}\|x_i - \overline{x}_k\|^2/(n-g)} = \frac{R^2/(g-1)}{(1-R^2)/(n-g)}$$

  where v = number of variables, g = number of clusters and n = number of observations.

  The pseudo F statistic measures the significance between the group separation of the clusters and is the ratio of the between-clusters sum-of-squares and the within-cluster sum-of-squares. However, this statistic is not distributed as an F random variable. The PSF statistic is interpreted similarly to the CCC statistic by looking for peaks in selecting the best number of clusters, with a trade-off between simplicity and accuracy in the interpretation of the results.
- Overall R^2 Statistic that measures the overall separation of the clusters.
- Cubic Clustering Criterion statistic that determines the final cluster groupings, with a CCC statistic greater than 2 generally indicating the number of clusters to select.

- Cluster Means for each input variable in the analysis, which will allow you to distinguish between the various characteristics by the average values of the input variables in the analysis across the separate cluster groupings.
- Cluster Standard Deviations for each input variable in the analysis.

For hierarchical clustering, the SAS/STAT CLUSTER procedure output listing displays the following clustering statistics in determining the number of clusters:

- Wald's Minimum Variance Cluster Analysis
- Eigenvalues of the Covariance Matrix
  - Eigenvalues that are multiplied by each input variable in the analysis, in which the linear combination of the input variables and eigenvalues is designed to explain a majority of the variability in the data. The table listing will display the number of eigenvalues associated with the number of input variables in the analysis.
  - Difference in each successive eigenvalue.
  - Proportion of the cumulative eigenvalues that measures the proportional variability explained by each component. It is the proportion of each corresponding eigenvalue over the total number of eigenvalues from the variance–covariance matrix.
  - Cumulative proportion of each eigenvalue that measures the cumulative variability in the data.
  - RMS Total-Sample Standard Deviation that is the root mean square distance between each observation in the cluster. It is the squared distance between each observation and the assigned cluster across the separate input variables in the analysis.
  - RMS Distance Between Observations.
- Cluster History ranked by the R^2 Statistic
  - Number of clusters that are created.
  - Clusters joined, that is, the two separate clusters merged to create the unique cluster.
  - Frequency count of the number of observations in the joined clusters that are created.

- Semi-partial R^2 statistic – the amount of decrease in the between-cluster sum-of-squares divided by the corrected total sum-of-squares. That is, this statistic indicates the decrease in the proportion of variance by merging the two separate clusters.
- R^2 statistic that measures the amount of variability in the data from the various clusters.
- Pseudo F-statistic that might be helpful in determining the number of clusters in the data set. It is distributed as an F random variable with $v(g-1)$ and $v(N-g)$ degrees of freedom where v = number of input variables in the coordinate data and N = number of observations.
- Pseudo T^2 statistic that compares the means between two separate multivariate populations or clusters. The PST2 statistic can be used to determine if two separate clusters should be combined. Larger PST2 values indicate that the two cluster means are different and should not be combined. Conversely, small PST2 values will indicate that the two clusters can be safely combined. The statistic is distributed as an F random variable with v and $v(n_k+n_L-2)$ degrees of freedom, where v=number of input variables, n_k = number of observations in cluster k, and n_l = number of observations in cluster l.

The PST2 statistic for joining clusters C_k and C_l into cluster C_m is calculated as follows:

$$\text{PST2} = \frac{w_m - w_k - w_l}{(w_k - w_l) / (n_m + n_l - 2)} \quad \text{where } w_k = \sum_{i=l}^{k} \| x_i - \bar{x}_k \|$$

The following listing displays the development of the clustering process with the various clustering statistics that are useful in estimating the number of clusters created at each step by hierarchical clustering. Each row of the table represents one partition that is performed. The respective cluster is identified by CL*n*, where *n* is the number of the cluster that is merged or joined. The following is the table listing of the procedure output listing from the Wald's method using the r-square statistic, that is, RSQ, and the semi-partial r-square statistic, that is, SPRSQ, in determining the optimum number of clusters. The FREQ column indicates the number of observations in the corresponding cluster. The Tie column will list any ties that might occur in the minimum distance between the two separate clusters joined. The semi-partial r-square statistic indicates the amount of decrease in the variability in the data from the two separate clusters that are merged together. The following listing suggests that two or three clusters should be selected with a significant increase in the r-square statistic and a significant decrease in the semi-partial r-square statistic. A general rule in interpreting the pseudo F statistic, that is, PSF, is done by scanning down the column until you observe a value significantly different than the previous value. Conversely, the best number of clusters to select from the pseudo t^2 statistic, that is, PST2, is done by scanning down the column until you observe a value significantly different than the previous value, then moving back one row in the column. The pseudo F statistic indicates that the best possible number of clusters to select from seems to be two or three clusters, with the pseudo t^2 statistic indicating that three or four clusters should be selected.

Cluster History

NCL	--Clusters	Joined---	FREQ	SPRSQ	RSQ	PSF	PST2	Tie
39	OB19	OB26	3	0.0001	1.00	49E3	201	
38	OB3	OB38	47	0.0001	1.00	29E3	640	
37	OB21	OB39	2	0.0002	1.00	2E4	.	
36	OB10	OB29	2	0.0002	.999	15E3	.	
35	OB8	OB14	12	0.0002	.999	12E3	546	
34	OB5	OB25	3	0.0002	.999	9973	849	
33	OB16	OB27	16	0.0002	.999	8626	594	
32	OB1	CL34	4	0.0002	.999	7514	2.4	
31	OB6	OB34	32	0.0003	.998	6603	1405	
30	OB12	OB28	26	0.0003	.998	5816	927	
29	OB4	OB37	18	0.0003	.998	5206	1182	
28	OB11	CL37	4	0.0004	.997	4633	5.1	
27	OB2	OB9	6	0.0004	.997	4160	1511	
26	OB33	OB35	11	0.0005	.996	3771	1506	
25	CL27	CL39	9	0.0005	.996	3443	7.0	
24	OB18	OB40	35	0.0006	.995	3176	2200	
23	OB17	OB31	12	0.0008	.994	2846	3045	
22	OB22	OB24	22	0.0008	.994	2605	2484	
21	CL36	CL28	6	0.0009	.993	2394	5.1	
20	CL30	OB13	30	0.0009	.992	2240	82.4	
19	OB30	CL26	33	0.0009	.991	2126	59.8	
18	CL35	OB20	18	0.0011	.990	2012	89.8	
17	OB7	CL33	33	0.0014	.988	1889	201	
16	CL38	CL31	79	0.0015	.987	1783	290	
15	CL25	CL18	27	0.0017	.985	1697	17.7	
14	CL32	CL23	16	0.0018	.983	1632	20.4	
13	CL22	OB23	58	0.0021	.981	1568	143	
12	OB15	CL19	46	0.0022	.979	1529	68.8	
11	CL14	CL20	46	0.0027	.976	1489	28.1	
10	CL16	OB36	101	0.0038	.973	1422	195	
9	CL15	CL21	33	0.0039	.969	1397	21.6	
8	CL29	CL17	51	0.0044	.964	1396	114	
7	CL24	OB32	36	0.0047	.959	1435	285	
6	CL11	CL8	97	0.0082	.951	1424	58.7	
5	CL7	CL13	94	0.0097	.942	1473	108	
4	CL6	CL9	130	0.0141	.927	1563	58.0	
3	CL12	CL5	140	0.0395	.888	1457	252	
2	CL10	CL3	241	0.2126	.675	768	759	
1	CL4	CL2	371	0.6753	.000	.	768	

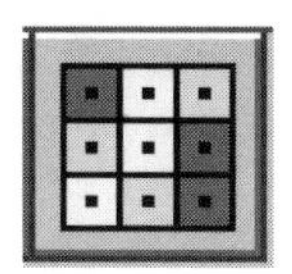

3.6 SOM/Kohonen Node

General Layout of the Enterprise Miner SOM/Kohonen Node

- **Data tab**
- **Variables tab**
- **Clusters tab**
- **General tab**
- **Advanced tab**
- **Seeds tab**
- **Missing Values tab**
- **Output tab**
- **Notes tab**

The purpose of the **SOM/Kohonen** node in Enterprise Miner is to perform various Kohonen network training techniques that are unsupervised training techniques. The three Kohonen network training unsupervised learning techniques that are applied within the **SOM/Kohonen** node are Kohonen vector quantization (VQ), Kohonen self-organizing maps (SOMs), or batch SOMs with Nadaraya–Watson and local-linear smoothing. The Kohonen VQ technique is a nonhierarchical clustering method as opposed to the SOMs techniques that are primarily dimension-reduction methods. However, SAS recommends that the **Clustering** node should be first considered for cluster analysis as opposed to the following Kohonen SOM data mining techniques.

The three unsupervised learning techniques applied within the **SOM/Kohonen** node have the added flexibility of handling the input variables of any level of measurement. However, if there are several categorically-valued input variables in the network training design, then the node will take a longer time to compile. Also, the input variables are typically standardized in order to assure convergence during network training. The reason is because the clustering assignments from the input variables are based on the squared distance in the cluster seeds from the output layer.

An Overview to Kohonen SOM Design

The objective of the Kohonen training techniques is to identify clusters in the data. The Kohonen SOM network techniques consist of two separate layers: an input layer and an output layer. In other words, where each input unit is mapped to each output unit in which each output unit represents its own unique cluster. Kohonen SOM training is similar to neural network training. The similarity between both network training designs is that both have an input layer and output layer. However, the Kohonen design consists of strictly two separate layers, that is, the input layer and the output layer with entirely different topologies. The topological mapping in the Kohonen SOM design is such that the output units are structured into squared grids, similar to squares on a checkerboard, in order to reduce the dimensionality of the data. Each squared grid represents a cluster. Therefore, a smaller number of squared grids will result in fewer clusters, which will make it easier to interpret. The size of the two-dimensional squared grid is determined by the number of rows and columns. Selecting the right number of rows and columns for the squared grid requires trial and error. Typically, large maps are usually preferred, assuming that there are a significant number of observations assigned to each cluster grouping. Similar to the neural network designs, each unit in the output layer is fully connected to each unit in the input layer, with the vector of weights connecting the input layer units to the output layer units. However in the Kohonen techniques, the output layer consists of several units, where every output layer unit is connected to every unit in the input layer. At the beginning, the weights are assigned at some small random values and the output layer in the SOM design is arranged in a two-dimensional squared grid, like a checkerboard, in order to reduce the dimensionality of the data. The objective to the SOM algorithm is that it tries to find clusters so that any two clusters that are close to each other in the grid space will have seeds close to each other in the input space. In Kohonen SOM training, each observation flows through the network from the input layer to the output layer. As each observation is fed into SOM training for cluster membership, the output units compete with one another based on the input unit assigned to the numerous output layer units, where the rules of the game are winner takes all. For each training case, the winning output unit is the unit with the smallest squared distance between it and the input unit. The reward is that the weight that is associated with the winning unit is adjusted to strengthen the connection between it and the input layer units. In other words, the squared distance is based on the difference between the vector of input values in the multidimensional input space and the vector of weights in the two-dimensional grid. Again, the objective of SOM training is superimposing the two-dimensional grid on the p-dimensional input space of p inputs such that every training case is close to some seed, with the adjustment to the grid performed as little as possible.

Also in SOM training, in order to preserve the topological ordering of the Kohonen mapping an added neighborhood concept is incorporated into the iterative clustering membership algorithm. That is, as each observation is entered for cluster membership, the neighboring units that are adjacent to the winning output unit are adjusted closer to the winning unit in order to increase the likelihood of the input units being assigned to it and the surrounding neighboring units. At the beginning, the size of each grid is the same. However, once Kohonen SOM training develops the square grids tend to form into nonrectangular-shaped grids based on the amount of movement of the units assigned to each grid due to the winner-take-all philosophy, as illustrated in the following diagram. The adjustment that is applied is defined by the specified tuning parameter called the *neighborhood size*. The tuning parameter controls both the size of the neighborhood and the amount of adjustment to the winning units and all units surrounding the winning unit. The adjustment that is performed is proportional to the Euclidean distance between the winning output unit and the input unit. Also, other neighboring units are also moved closer to the input unit. The amount of adjustment is determined by the distance between the neighboring unit and the input unit. The closer the neighboring unit is to the input unit, the more the adjustment. In other words, the closer the neighboring unit is to the input unit, then the more the neighboring unit moves. In order to achieve convergence, the neighborhood size is adjusted at each step. At the beginning, the neighborhood size tuning parameter applies a large adjustment to the various neighborhoods. However, as training develops the size of the neighborhoods and the amount of adjustment will gradually decrease. The reason that the adjustments are applied to the output units is that Kohonen SOM network training would tend to identify as many clusters as there are output layer units in the design, thereby defeating the purpose of the entire analysis. This iterative process is repeated any number of times until the neighborhood size becomes smaller and smaller with each subsequent iteration.

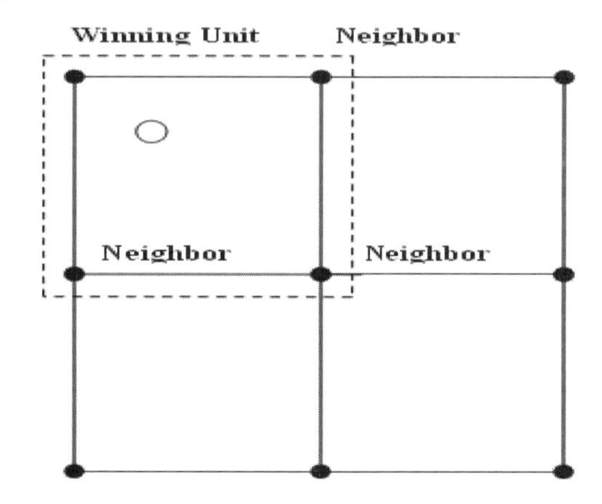

The initial squared grid of the Kohonen Self-Organizing Map design.

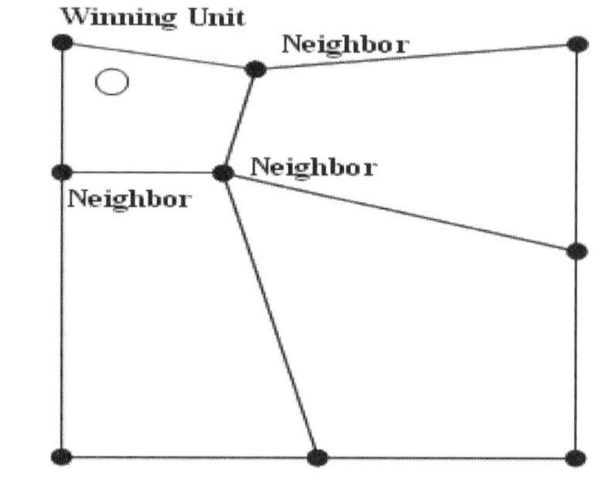

The first adjustment to the winning unit and all neighboring units surrounding the winning unit.

Once the iterative training has been completed, the **SOM/Kohonen** node will allow you to observe the number of winning units, that is, the frequency counts within each cluster based on the number of rows and columns of the Kohonen SOM topological map. The map will allow you to determine the meaning of each cluster or the relationships between the various clusters that are created from the Kohonen training design.

The final configuration to the Kohonen SOM network design is such that the output units with no winning units or a small number of winning units are removed from the design. The reason for discarding these output units is to speed up network training in order to reduce processing time, with all other output units retained in the design that contributed to creating the final clusters. Therefore, network training depends on the size of the network mapping. Also, adjacent output units will be similar to each other, whereas output units that are further away from each other will be quite different.

Kohonen Vector Quantization Method

The *Kohonen vector quantization method* is closely related to the partitive clustering algorithms, like the *k*-means clustering method. However, the *k*-means clustering method is recommended over the Kohonen VQ method. One reason for this is due to the various tuning constants in the Kohonen VQ design that require a lot of trial and error in determining the most appropriate values. In Kohonen VQ networks, each output unit represents a cluster and the center of each cluster is called the cluster seed. The difference in the unsupervised learning technique is that this network training technique adjusts the winning cluster seed within the corresponding cluster closer to the observation, as opposed to the traditional clustering technique that adjusts every cluster seed once an observation is assigned to its respective temporary cluster grouping. In other words, the cluster seeds of every cluster that is created are adjusted at the same time. A specified learning rate determines the rate in which the cluster seeds are updated. The seed of the winning cluster is updated by moving the seed in some proportion based on the squared distance between it and the input unit. At each step of Kohonen training, the winning cluster seed is updated at the *s* step as follows:

$$C_n^{s+1} = C_n^s \cdot (1 - L^s) + X_i \cdot L^s \quad \text{for some learning rate L at the } s \text{ step}$$

where C_k is the seed at cluster k and X_i is the input vector at the ith observation, where the index *n* of the winning cluster is such that $n = \|C_k - X_i\|$, that is, the squared distance between the cluster seed at the kth cluster and the X_i input vector at the ith observation.

MacQueen's k-means algorithm is another clustering method. However, MacQueen's *k*-means is slightly different from the Kohonen VQ network training. The MacQueen's *k*-means algorithm defines the learning rate based on the reciprocal of the number of observations that have been assigned to the winning cluster. The learning rate of MacQueen's *k*-means algorithm is defined so that the learning rate is defined as $L^s = 1/(N_n+1)$. From the **Clustering** node, select the **Use incremental training for one pass** check box from the **Initial** subtab within the **Seeds** tab that will perform MacQueen's *k*-means algorithm.

As the sample size increases, the gradual reduction of the learning rate will result in the seed of each cluster approaching the mean of all observations assigned within each cluster. It will also assure convergence in the iterative algorithm based on the minimum squared distance between the cluster seeds and the input units. However, setting the learning rate at a fixed rate will result in the failure of Kohonen training to converge.

Kohonen Self-Organizing Maps

Kohonen self-organizing maps are designed around a topological mapping from the input space to the clusters formed in the grid space. In the Kohonen SOM design, the various clusters are usually organized into a two-dimensional squared grid-like design, but sometimes one-dimensional and rarely three- or more dimensional. The SOM algorithm determines the clusters that are created based on any two adjacent clusters in the grid space that have their cluster seeds closest to each other in the input space. In other words, the SOM algorithm tries to embed the two-dimensional grid in the *p*-dimensional input space so that every input unit is as close as possible to some seed, while at the same time preserving the grid space in the output layer as much as possible. The two-dimensional grid can be determined by the number of rows and columns in the topological map that can be specified from the **General** tab. The one restriction is that the number of input variables in the training data set must be greater than the dimensionality of the SOM map. Kohonen SOMs are recommended for highly nonlinear training data or highly nonlinear cluster boundaries that are created.

The SOM algorithm and the previous mentioned Kohonen vector quantization algorithm are very similar. However in the SOM algorithm, a kernal function K is applied in updating the seeds or weights based on the winning cluster. At each step, the kernal function identifies the cluster seeds in the output layer that are as close as possible in the grid space, based on the squared distance between the cluster seeds in the grid space. The cluster seeds are updated during network training as follows:

$$C_n^{s+1} = C_n^s \cdot (1 - K^s(j,n) \cdot L^s) + X_i \cdot K^s(j,n) \cdot L^s$$

where $K^s(j,j) = 1$ and $K^s(j,n)$ is usually a non-increasing function of the distance between the seeds *j* and *n* in the grid space and $K^s(j,n) = 0$ for seeds that are far apart from each other in the two-dimensional grid space. The kernal function $K(j,n)$ then adjusts the size of the surrounding neighborhoods, that is, the area around each cluster in the grid, that are adjacent to the winning unit that is defined as follows:

$$K(j,n) = \{1 - [|\text{Row}(j) - \text{Row}(n)|^p + |\text{Col}(j) - \text{Col}(n)|^p]^{2/p} / Size^2\}^{2k}$$

where Row(j) is the row number for the j cluster, Col(j) is the column number for the j cluster, *Size* is the neighborhood size, k is the kernal shape, and p is the kernal metric. The various terms that are listed can be specified from the **Advanced** tab within the **Neighborhood Options** subtab.

In order to avoid undesirable results and bad local minima, it is usually a good idea to initially specify a large neighborhood and let the **SOM/Kohonen** node gradually decrease the neighborhood size during network training. If the kernal function at the s step is zero, that is, $K^s(j,n) = 0$, then the SOM update formula reduces to the previous formula used in updating the cluster seeds in the Kohonen vector quantization algorithm. When the neighborhood size is zero, then the SOM algorithm reduces into the Kohonen vector quantization algorithm. Therefore, it is important that the neighborhood size does not approach zero during training in order to preserve the topological mapping in the two-dimensional squared grid. The neighborhood size is the most important tuning parameter in the SOM training process. It is recommended in order to achieve an adequate topological map, that the neighborhood size should be some integer greater than one.

Kohonen SOM training is performed as follows:

1. Specify the number of rows and columns to the two-dimensional grid.
2. The observation is designated as the winning output unit. The winning output unit is moved closer to the input unit. The amount of movement in the two-dimensional grid is proportional to the Euclidean distance between the winning output unit and the input unit.
3. All other neighboring units in the area around the winning output unit in the two-dimensional grid are also moved closer to the winning output unit. The amount of movement of the neighboring unit is determined by the distance from the winning output unit. The closer the neighboring unit is to the winning unit, the more the neighboring unit moves.
4. Repeat steps 2 and 3 until convergence with the neighborhood size gets smaller at each iteration.

The default method is *batch SOM training* in the **SOM/Kohonen** node. In batch SOM training, a smoothing effect is applied to the cluster seeds. That is, the algorithm is very similar to both kernal estimation and k-means clustering. The reason that the SOM technique is similar to kernal estimation is that smoothing is performed on the cluster seeds. However, the difference is that smoothing is performed in the neighborhoods of the grid space as opposed to the input space. Also, the batch SOM algorithm is similar to the batch k-means clustering algorithm, where an extra smoothing procedure is performed to define the area around each winning unit in the grid. At the first step, the seeds are initialized by some random weights. The next step is assigning each observation to the nearest cluster seed. And finally, by applying nonparametric regression by using a kernal smoothing function $K^s(j, n)$ to the grid points and the number of observations within each cluster, the winning cluster seed is adjusted from the Kohonen SOM updated formula.

There are two separate nonparametric regression methods that can be specified within the node from the two Kohonen SOM training techniques. They are the *Nadaraya–Watson* and *local-linear smoothing methods*. The smoothing functions are designed to define the area around each cluster seed in the grid. The purpose of the local-linear smoothing method is to eliminate the border effect in which the seeds located near the border of the two-dimensional grid are compressed in the input space. The border effect is especially problematic when performing a high degree of smoothing in the Kohonen SOM method since the seeds will tend to move toward the center of the input space. This same SOM bordering effect phenomenon is similar to the boundary effect in kernal regression, where the estimated classification function tends to flatten or straighten out near the boundaries of the multidimensional input space. Therefore, local-linear smoothing is a remedy to the bordering effect in the batch SOM design. These nonparametric smoothing methods are specifically designed to achieve better convergence and improve the accuracy of the Kohonen SOM algorithm based on the borderline clustering assignments. The advantage of Nadaraya–Watson SOMs is that they are more robust to bad configurations as opposed to local-linear SOMs. If the Nadaraya–Watson kernal regression method is applied, then the batch SOM algorithm will generate the same results as the Kohonen SOM algorithm, assuming that the batch SOM algorithm does not get stuck in a bad local minimum. The main difference between the methods is that the batch SOM algorithm often converges without the existence of a learning rate to the design. However, the advantage of the Kohonen SOM technique, as opposed to the batch SOM technique, is that the Kohonen SOM method is less likely to get stuck in a bad local minimum for highly nonlinear training data.

Many of the properties of nonparametric regression are also applied in the batch SOM algorithm, that is, independence, constant variance, and normality in the data, in order to achieve adequate clustering results and good generalization. The shape of the kernal function is not a critical concern in both nonparametric regression and the Kohonen SOM method. However, the amount of smoothing that is used in nonparametric regression is critical. In other words, the choice of the final neighborhood size used in the Kohonen SOM method is important. This is compounded by the fact that there is no standard method that is currently known in determining the final neighborhood size that should be applied. Therefore, determining a good neighborhood size is usually very tedious and requires a lot of trial and error.

In the **SOM/Kohonen** node, incremental training is performed on the three Kohonen training algorithms so that each observation is successively passed through the data set any number of times in determining the optimum clustering assignments. From the **Advanced** tab, you may select the number of times each data point is repeatedly passed though the training data set from the **Maximum number of steps** option.

Kohonen self-organizing maps are discussed in greater detail in Sarle (2002), Hastie, Tibshirani, and Friedman (2001), Berry and Linoff (2004), and Giudici (2003).

Data tab

The **Data** tab is designed for you to select the training, validation, test, and scoring data set for the data mining analysis. If there are several data sets that are available, then Kohonen network training is performed for each one of the data sets once the node is executed. By default, the training data set is automatically selected. However, the subsequent cluster analysis can be performed for each one of the partitioned data sets in order to verify the accuracy of the cluster assignments. Furthermore, the scored data set can be selected in creating different clustering assignments by fitting the model to a new set of values of the input variables. The tab will also allow you to specify the sample size of the training data set for preliminary training. The **Data** tab consists of the following two subtabs:

- **Inputs subtab**
- **Cluster Profiles subtab**

Inputs subtab

The **Inputs** subtab is designed similarly to many of the other **Data** tabs in the other nodes. Therefore, the subtab will not be illustrated. The **Inputs** subtab will allow you to select the active training data set for the analysis. From the subtab, you may view the file administrative information or browse the table listing of the selected data set. The purpose of the partitioned data sets in Kohonen network training is defined as follows:

- **Training:** The training data set is automatically selected and used as the active data set to perform the following SOM/Kohonen analysis.
- **Validation:** The purpose of the validation data set is to compare or cross-validate the accuracy of the clustering assignments during network training.
- **Test:** The purpose of the test data set is to obtain an unbiased evaluation of the clustering assignments based on data that is entirely separate from the sample that created the clusters. At times, a single validation data set can generate imprecise results. Therefore, a test data set might be used in obtaining an unbiased comparison of the accuracy in the clustering assignments.
- **Score:** The purpose of the score data set is to create entirely different clustering assignments from a new set of observations of the input variables.

The subtab will also allow you to view the **Imports Map** window, where you may assign an entirely different partitioned data set to the analysis from the corresponding data mining data sets created within the process flow diagram that are connected to the node. By default, the training data set is selected. Therefore, the **Imports Map** window will enable you to select an entirely different training data set that exists within the currently opened process flow that has been created from the corresponding node that is connected to the **SOM/Kohonen** node.

Cluster Profiles subtab

The **Cluster Profiles** subtab is designed for you to specify the sample size to the active training data set that is used for preliminary training. In Kohonen network training, the purpose of preliminary training is to create the subsequent profile plots that are used to characterize the cluster groupings. By default, a random sample of 2,000 observations is selected from the active training data set to perform preliminary training. From the **Sample size** entry field, enter the appropriate number of observations to randomly sample the training data set.

*The **Cluster Profiles** subtab is used to set the number of training cases for preliminary training.*

Variables tab

The **Variables** tab will be displayed once you open the **SOM/Kohonen** node. The tab is designed for you to view the input variables in the analysis along with the various attributes that are automatically assigned to each variable from the metadata sample. Similar to the **Variables** tab in the **Clustering** node, the tab is also designed for you to standardize the numeric input variables. By default, standardization is not performed on the input variables. However, as in *k*-means clustering, if the input variables have a wide range of values or are measured in entirely different units, that it will have a profound effect in the final clusters that are created. Again, standardization is applied in order to interpret the input variables to a common unit with the same range of values. Therefore, standardization is highly recommended. For categorically-valued input variables, the node automatically constructs *k* dummy variables for each *k* class levels.

Similar to the **Clustering** node, the **Variables** tab will allow you to remove certain input variables from the analysis. By default, all input variables are automatically included in the analysis with a variable attribute status set to **use**. However, simply select any number of the listed input variables to remove from the analysis by setting the variable attribute status to the variable within the **Status** cell to **don't use**. Note that even though the input variables are removed from the analysis, these same input variables will still remain in the active training data set. Furthermore, all other columns are grayed-out, therefore, they cannot be changed.

Cluster tab

The **Cluster** tab is similar to the tab in the **Clustering** node, therefore, the tab will not be displayed. The tab is designed to set the variable attributes to the segment or cluster identifier variable. The segment identifier is automatically created in the output data set. The purpose of the variable identifier is to identify the cluster in which the observation has been assigned to the output data set.

General tab

The **General** tab is designed to specify the three types of Kohonen training methods, the size of the topological map from Kohonen SOM training, or the number of clusters from Kohonen VQ cluster training.

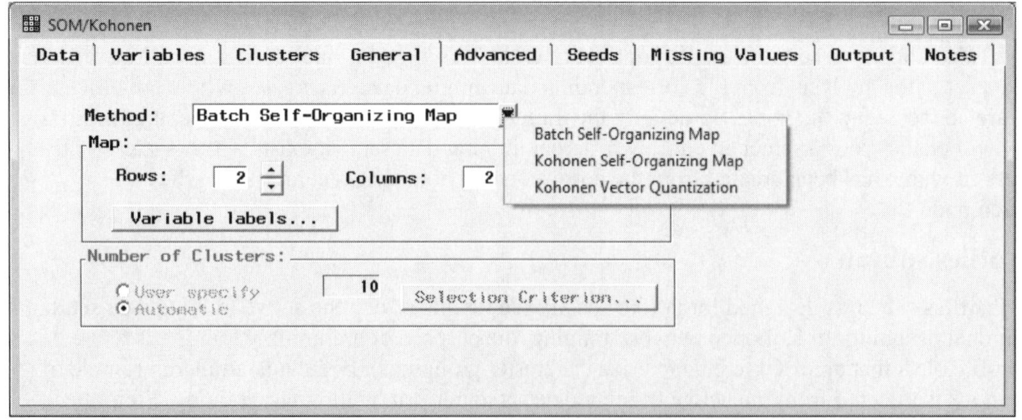

*The **General** tab is used to specify the Kohonen training method and the number of grids or clusters.*

From the **Method** display field, you may specify the three different Kohonen network training methods as follows:

- **Batch Self-Organizing Map** (default)**:** By default, batch Kohonen SOM training is selected. In batch Kohonen SOMs, it is important to specify the correct topological map of both the number of rows and columns for the SOM map and the final neighborhood size. Larger maps are usually the best choice as long as each cluster contains a sufficient number of observations. However, the drawback is that it will increase processing time based on the increased size of the SOM topological map. The number of rows and columns may be determined from preliminary training. Alternatively, the initial number of rows and columns can be determined from the number of clusters created by performing clustering analysis from the previous **Clustering** node. For example, assuming that ten clusters were created from the **Clustering** node, then you may specify five separate rows and two separate columns for the topological map. The number of rows and columns of the map can be specified from the **Map** section. It is recommended that the size of the neighborhood should change proportionally with the number of rows and columns in the SOM map. For example, if you double the number of rows and columns in the map, then you should also double the size of the neighborhood accordingly.

  In addition, the node will allow you to apply the smoothing techniques to the neighborhoods in the grid space of the winning unit by specifying either one of the Kohonen SOMs. The nonparametric regression is based on a kernel function with the grid points as the input values, cluster means as target values, and number of cases in each cluster as the case weights. The batch SOM replaces seeds with outputs of the nonparametric regression function evaluated at its grid point.
- **Kohonen Self-Organizing Map:** This option selects Kohonen SOM training. Specifying both the learning rate and the number of rows and columns for the Kohonen map are the most important option settings in the design. It is important that most of the clusters be adequately represented. Again, the number of rows and columns should change proportionally with the neighborhood size. For example, if you double the number of rows and columns, then you should also double the neighborhood size. In addition, most of the clusters should consist of at least five to ten observations within each cluster. Choosing the learning rate and map size usually requires spending a lot of time trying to determine the most appropriate values. It is recommended that the learning rate should be initially set at a high value, say .9. However, if you perform preliminary analysis, then you may select a much lower initial learning rate. In addition, the following smoothing options are available to adjust the neighborhoods of the winning unit to the squared grid.
- **Kohonen Vector Quantization:** This option selects Kohonen VQ network training. The most important parameter specifications for this option are the number of clusters and the learning rate, which usually requires a lot of trial and error. The learning rate may be specified as previously discussed. It is recommended that a good initial learning rate should be set at .5.

By specifying either **Batch Self-Organizing Map** or **Kohonen Self-Organizing Map** network methods, the **Maps** section becomes available with the **Number of Clusters** section dimmed and unavailable. From the **Maps** section, you may then set the size of the grid space of the topological map, that is, the number of rows and columns of the two-dimensional grid of the topological map. Selecting the **Kohonen Vector Quantization Network** method, the **Number of Clusters** section will become available with the **Maps** section dimmed and unavailable. The **Number of Clusters** section will allow you to specify the number of initial clusters to create in Kohonen VQ network training. The number of initial clusters might be determined by performing clustering analysis from the **Clustering** node.

Specifying the Number of Rows and Columns for the Kohonen Map

Once you have selected either one of the SOM techniques from the **Method** display field, you may then specify the number of rows and columns to determine the two-dimensional topological map from the **Map** section within the **General** tab. By default, Enterprise Miner sets the map size or the dimension of the squared grid to four rows and six columns. You may specify the number of rows and columns from the **Rows** and **Columns** entry fields within the **Map** section. The one drawback in specifying the correct number or rows and columns is that it requires a lot of trial and error. That is, specifying the size of the map too small will result in the map inaccurately reflecting the nonlinearity in the data. However, setting the map size too large will result in excessive processing time, with empty clusters resulting in misleading results. Large maps are usually the best choice, assuming that there are a significant number of observations for the data mining analysis. Since the amount of smoothing depends on the ratio between the neighborhood size and the map size, it is important that you change the map size proportionally with the neighborhood size.

The reason that the number of rows and columns of the Kohonen map are set to two for both the number of rows and columns in the previous diagram and the subsequent analysis is to retain the consistency in the clustering assignments from the previous *k*–means cluster analysis in order to approximate the number of clusters specified from the previous **Clustering** node that was set to three separate clusters.

Select the **Variable labels** button that will open the **Variable Labels** window to reassign the variable name and label to the row, column, and SOM cluster identifier variable that is written to the output data set.

Specifying the Number of Clusters

The **Kohonen Vector Quantization Network** option will allow you to specify the number of clusters that are created once you execute the node. By default, ten clusters are automatically created during training. From the **Number of Clusters** section, the **Automatic** option is the default. Selecting the **Selection Criterion…** button will allow you to open the **Selection Criterion** window in order to specify the clustering method, the clustering cubic criterion cutoff value that determines the number of clusters, and the minimum and maximum number of clusters. Alternatively, select the **User specify** option that will allow you to specify the number of clusters that the **SOM/Kohonen** node will automatically create.

The following are the various options available within the **Selection Criterion** window:

- **Clustering method:** This option will allow you to select from the three separate clustering methods, that is, the average linkage method, centroid linkage method, and Ward's method for Kohonen Vector Quantization network training. The default is Ward's clustering method.
- **Clustering cubic criterion cutoff** (default is 3)**:** The clustering cubic criterion cutoff is the ratio between the expected and the observed r-square statistic using a variance stabilizing transformation. From the entry field, enter the expected r-square value. The observed r-square statistic is calculated by the ratio of the overall variability between each cluster and the overall variability within each cluster. Large r-square values are most desirable, indicating large differences between the clusters associated with small differences within each cluster, that is, internal cohesion and external separation.
- **Minimum number of clusters** (default is 2)**:** This option will allow you to select the minimum number of clusters that Kohonen VQ network training will create.
- **Maximum number of clusters** (default is 40)**:** This option will allow you to select the maximum number of clusters to create in Kohonen VQ network training. Since Kohonen VQ network training is a nonhierarchical clustering technique, it is important that the correct number of initial clusters be specified. Therefore, traditional clustering analysis might be considered in determining the optimum number of clusters to create, which is usually determined by trial and error. Also, preliminary cluster analysis might be performed in determining the appropriate number of clusters and detecting outlying observations in the data.

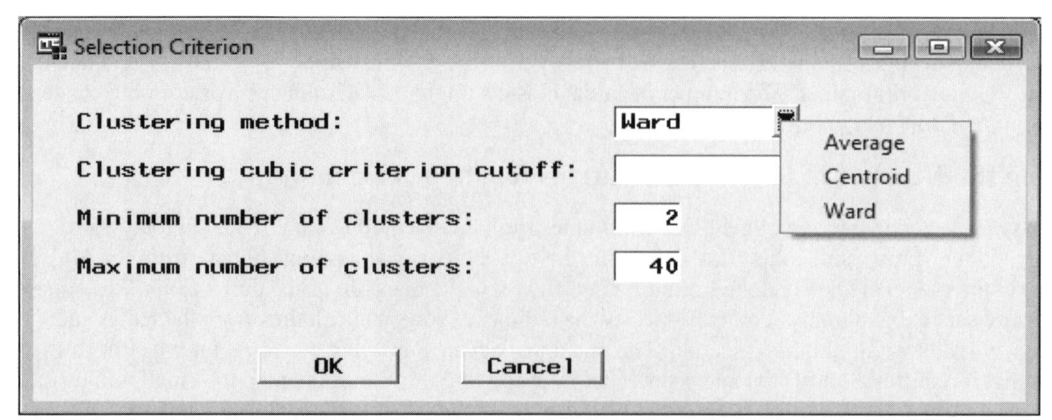

*The **Selection Criterion** window is used to specify the clustering criteria for Kohonen VQ network training.*

Advanced tab

The **Advanced** tab is designed to specify the various option settings for both SOM training techniques and the various parameters to control the creation of the SOMs and Kohonen networks. The tab consists of the following three subtabs:

- **Batch SOM Training subtab**
- **Neighborhood Options subtab**
- **Kohonen Training subtab**

Batch SOM Training subtab

By selecting the **Batch Self-Organizing Maps** or **Kohonen Self-Organizing Maps** methods from the previous **General** tab, then the **Batch SOM Training** subtab will become available. By default, the **Defaults** check box is selected, therefore, clear the **Default** check box to specify the nonparametric regression functions for the SOM network training. The Nadaraya–Watson and local-linear smoothing functions are used to adjust the neighborhood or the area around the winning unit in the two-dimensional grid. By default, the **Methods** section with the two smoothing techniques is dimmed and unavailable for selection. Clear the **Default** button, which will result in the **Methods** section becoming available for you to select either one or both the **Local-linear** and **Nadaraya–Watson** smoothing methods. The reason that Kohonen SOMs is the default is because it is recommended to first perform Kohonen SOMs, Nadaraya–Watson and local-linear training in that order. If the local-linear and Nadaraya–Watson smoothing methods are selected, then each smoothing technique is performed in succession. Nadaraya–Watson smoothing is first performed until the number of iterations or the convergence criteria are met. Local-linear smoothing is then performed until the number of iterations or the convergence criteria are satisfied in determining the appropriate clustering seed assignments for the training data set. The convergence criterion is determined by the relative change in the cluster seeds in successive iterations.

The **Defaults** check box is automatically selected to prevent the kernel function from being used in network training. When you clear the check box, the SOM training and the various methods will become available, as shown in the following diagram. In Enterprise Miner, you cannot set the **SOM Training** option to **No** if you have chosen either one of the SOM training methods in the previous **General** tab. Clear the smoothing method options to prevent the smoothing methods from being used during Kohonen SOM network training. By default, the maximum of number of iterations is set to 10 with a convergence criterion set at .0001 for the two smoothing methods. Select the **Options...** button to adjust both the maximum number of iterations and the convergence criterion. The **SOM Training Options** window will not be illustrated.

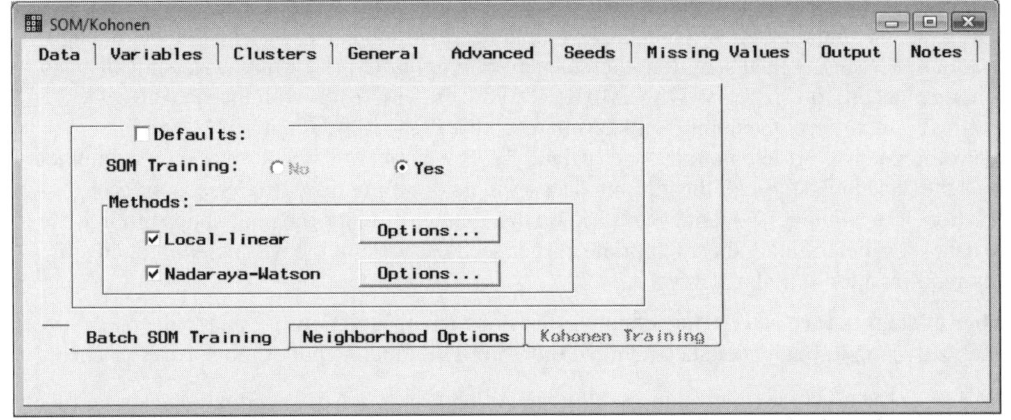

*The **Batch SOM Training** subtab is used to specify the two smoothing methods and option settings.*

Neighborhood Options subtab

By selecting the **Batch Self-Organizing Maps** or **Kohonen Self-Organizing Maps** methods from the previous **General** tab, then the **Neighborhood Options** subtab will become available. By default, the **Default** check box is selected indicating to you that the default settings are automatically applied to both SOM methods. Clear the **Default** check box to specify the following neighborhood option settings.

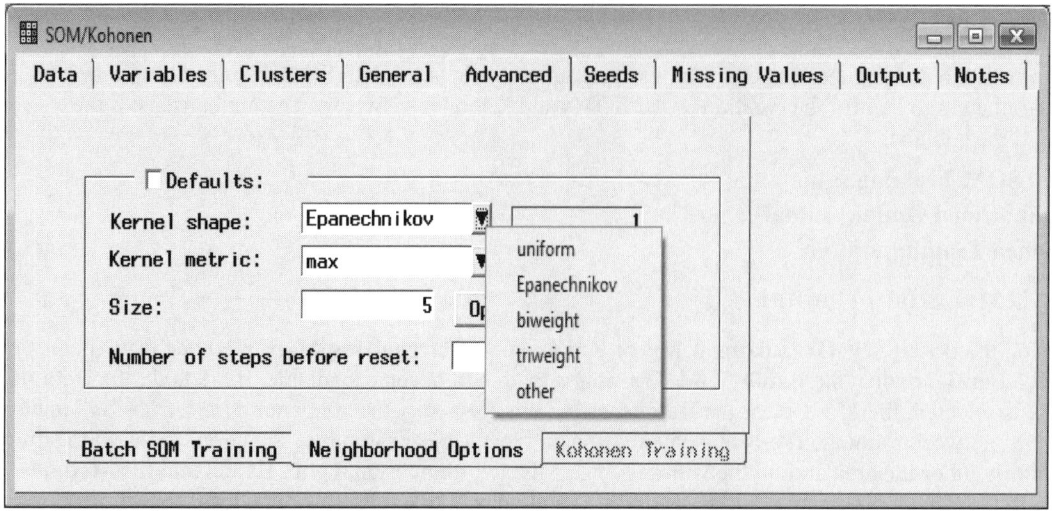

*The **Neighborhood Options** subtab is used to specify the various option settings for the kernal function.*

The following are the various options that you may select from to configure the kernal function in adjusting the neighborhood of the winning unit in the two-dimensional grid for Kohonen SOM training within the **Neighborhood Options** subtab:

- **Kernal shape:** Specify the shape of the neighborhood. The default kernal shape is Epanechnikov (1). Otherwise, you may select uniform (0), biweight (2), triweight (3), and other (nonnegative integer value *k*). However, specifying the exact shape assigned to the neighborhoods in the output layer is not that critical in the SOM training techniques.
- **Kernal metric:** The default is max (0). However, you may either select neighborhood metrics of cityblock (1), Euclidean (2), and other (nonnegative integer value *p*). This option determines the amount of adjustment to the seed of the winning cluster in the two-dimensional grid space and all neighborhoods adjacent to it that are determined by the distance function.
- **Size:** This option sets the initial size of the neighborhood, which must be some integer greater than zero. By default, the initial neighborhood size is set to half the size of the SOM grid. For batch Kohonen SOMs, the neighborhood size is gradually reduced to zero for the first ten iterations. For Kohonen SOMs, the neighborhood size is gradually reduced to zero for the first 1,000 training steps. Select the **Options...** button that will open the **Neighborhood Options** window to set the initial neighborhood size, the final neighborhood size, and the number of steps or iterations to reach the final size from incremental training. Enterprise Miner automatically sets the initial neighborhood size to MAX(5, MAX(ROWS,COLUMNS)/2) with the final size automatically set to zero, the number of steps to reach the final size (default 1000), and the number of iterations to reach the final size (default 3). The number of steps is based on the number of times the node loops through the training data set in determining the cluster seeds for each observation. The number of iterations is based on the number of times the node loops through the entire data set in determining the appropriate cluster seeds. Choosing the best neighborhood size usually requires a lot of trial and error.
- **Number of steps before reset:** The default is 100. For each observation, the data will automatically loop through the data a hundred times until the final neighborhood size is reached.

As explained earlier, based on the various option settings specified from the **Advanced** tab, the kernal function $K(j, n)$ adjusts the size of the surrounding neighborhoods adjacent to the winning unit that is defined as follows:

$$K(j, n) = \{1 - [|\text{Row}(j) - \text{Row}(n)|^p + |\text{Col}(j) - \text{Col}(n)|^p]^{2/p} / Size^2\}^{2k}$$

where Row(j) is the row number and Col(j) is the column number for the j cluster that can be specified from the **General** tab. From the **Neighborhood Options** subtab, the *Size* is the neighborhood size that can be specified from the **Size** display field, k is the kernal shape that can be set from the **Kernal shape** field, and p is the kernal metric specified from the **Kernal metric** field.

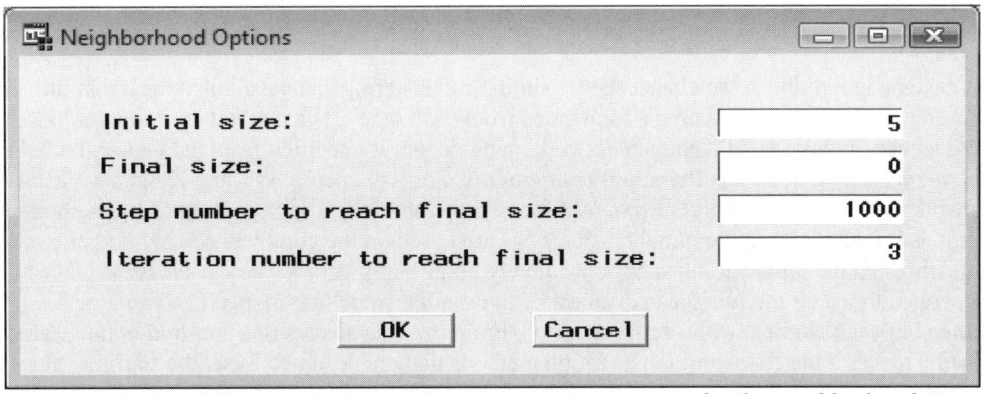

*The **Neighborhood Options** displaying the various option settings for the neighborhood size.*

Kohonen Training subtab

By selecting the **Kohonen Self-Organizing Maps** or **Kohonen Vector Quantization** methods from the previous **General** tab, the **Kohonen Training subtab** will become available. The subtab will allow you to specify both the learning rate and the termination criteria for either the Kohonen SOM or Kohonen VQ training techniques. The iterative network training routine stops when any one of the following termination criterion values are met.

Clearing the **Defaults** check box will allow you to specify the various termination criterion settings for Kohonen network training as follows:

- **Learning rate:** The learning rate must be a number between zero and one. By default, the learning rate is set to .9 for Kohonen Self-Organizing Map. For Kohonen Vector Quantization, the learning rate is automatically set at .5. In batch Kohonen SOM training, the learning rate is not needed. In the first 1,000 steps, that is, the number of times network training repeatedly passes through the data for each observation, the learning rate is reduced by an increment of .02 during incremental network training. The reason for the reduction in the learning rate is to assure convergence during network training. The **Options** button will allow you to specify the initial learning rate, the final learning rate, and the number of steps to reach the final learning rate. Specifying the same initial and final learning rate will prevent the reduction of the learning rate during network training, which is not recommended.
- **Maximum number of steps:** The default is set to 500 times the number of clusters. That is, the maximum number of times network training repeatedly passes through the entire data set for each observation in updating the cluster seeds.
- **Maximum number of iterations:** The default is set to 100. That is, the number of times that network training repeatedly passes through the entire data set in updating the cluster seeds.
- **Convergence Criterion:** The convergence criterion is automatically set to .0001.

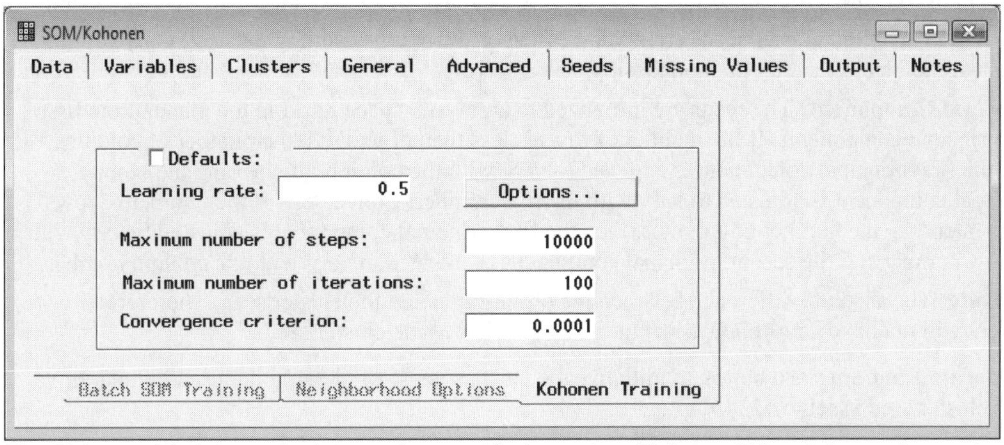

*The **Kohonen Training** subtab is used to specify the various Kohonen VQ training option settings.*

Seeds tab

The **Seeds** tab is designed to initialize the cluster seeds. Similar to clustering analysis, it is critical that the starting values for the initial cluster seeds be well separated from each other. The method used in initializing the cluster seeds depends on the type of Kohonen network methods that is specified from the **General** tab. For batch self-organizing maps, the default is **Principal components**. For Kohonen SOMs and Kohonen Vector Quantization network methods, the default is **First**. With the exception to the principal component method, the initialization methods for determining the initial cluster seeds are based on the complete cases. In other words, complete cases in which all the input variables have absolutely no missing values. The cluster seeds are required to be separated based on the Euclidean distance by at least the value that is specified from the **Minimum distance between cluster seeds (radius)** option. From the **Initial selection method** option, select the drop-down arrow to select the following initialization methods that are designed to set the starting values to the initial cluster seeds.

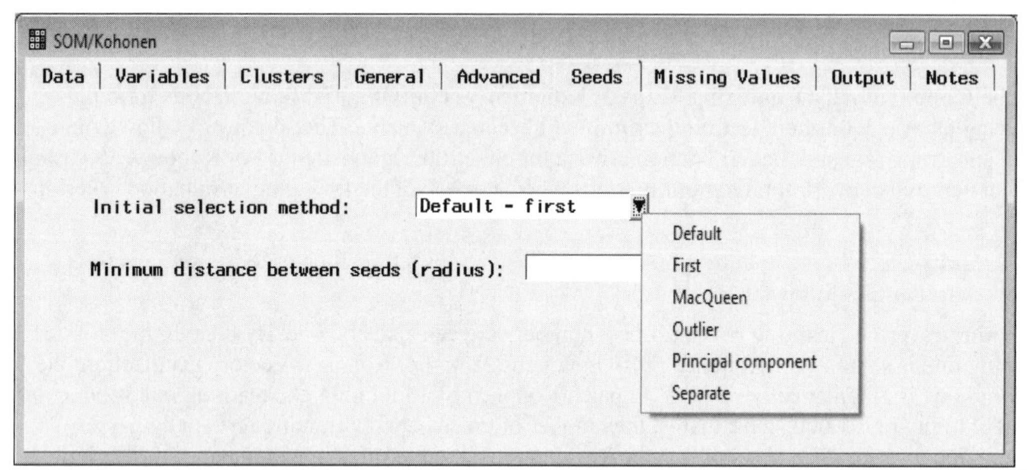

*The **Seeds** tab is used to specify the initial cluster seeds for the Kohonen network training methods.*

- **First:** The initial cluster seeds are created from the first k records, where all the values of the input variables are nonmissing.
- **MacQueen:** Initialization of the cluster seeds is computed by the k-means algorithm as follows:
  1. Initialize the cluster seeds with a seed assigned to each cluster.
  2. Repeat the following two steps until convergence.
     a. Read the data and assign each observation from the training data set to the nearest seed based on the Euclidean squared distance.
     b. Replace each seed with the average value of the observation that is assigned to the corresponding cluster.
- **Outlier:** Selects the initial seeds that are very well separated using the full-replacement algorithm.
- **Random:** The initial cluster seeds are randomly selected from the training data set, where all the values of the input variables are nonmissing.
- **Principal Component:** The seeds are initialized to an evenly spaced grid in the plane of the first two principal components. If the number of rows is less than or equal to the number of columns, then the first principal component is adjusted to vary with the column number and the second principal component is adjusted to vary with the row number. Conversely, if the number of rows are greater than the number of columns, then the first principal component is adjusted to vary with the row number and the second principal component is adjusted to vary with the column number.
- **Separate:** The absolute difference between the observation and the cluster seed. Therefore, outliers will not have a profound effect in the calculation of the cluster seeds.

Select the **Generate Random Seed** button to initialize the cluster seeds. By default, the pseudo random number of the cluster seed is set to 12345.

Missing Values tab

The **Missing Values** tab controls the way in which observations with missing values are treated during cluster initialization and written to the output scored data set. Missing values in any one of the input variables are automatically removed from the cluster seed assignments. The node is designed to either remove or impute the missing observations from network training. Imputation of the missing values will occur after the various cluster groupings are created from network training. By default, imputing missing values is not performed during network training. From the **Imputation method** field, press the drop-down arrow to select the following imputation methods:

- **None** (default)**:** Imputation is not performed for missing values during network training.
- **Seed of Nearest Cluster:** Replaces missing observations with the cluster seed that is closest to it.

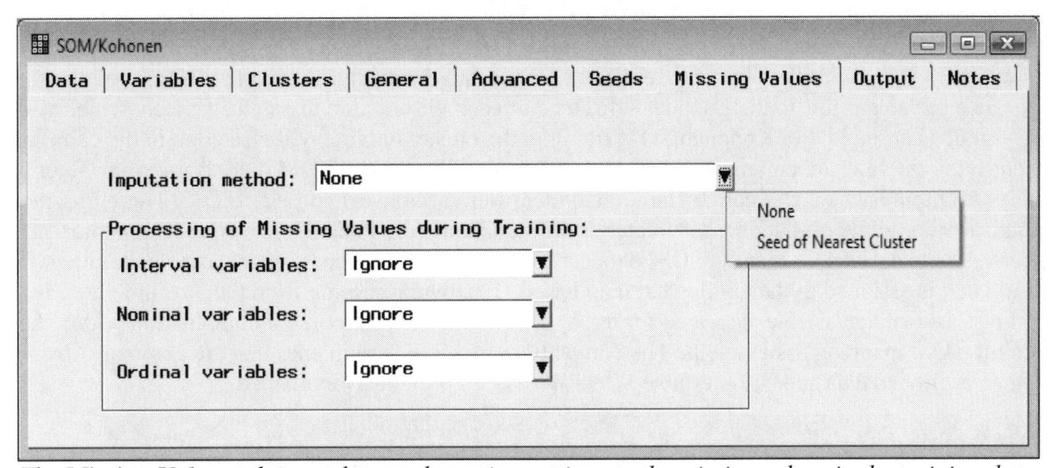

The ***Missing Values*** *tab is used to set the option settings to the missing values in the training data set.*

From the **Processing of Missing Values during Training** section, you may specify the replacement values based on the level of measurement of the input variables with missing observations. The following are the various options that are available in replacing missing values during the clustering initialization process.

For interval-valued input variables:

- **Ignore** (default)**:** Ignore the observations with missing values in any one of the interval-valued input variables during cluster initialization of the seed assignments. For observations with ignored missing values, the distance from the cluster seed is calculated from the summed variances ratio that is the total variance of all the nonmissing variables in the training data set divided by the total variance of all the nonrejected variables.
- **Mean:** Replace the input variables with missing values by their own average values during cluster initialization.
- **Midrange:** Replace input variables with missing values by the average of the minimum and maximum value, that is, the ratio between the sum of the minimum and maximum value over 2.
- **Omit:** Removes all the observations from the cluster analysis with missing values in any one of the input variables. That is, the observation will not be assigned to a cluster grouping.

For nominal or ordinal-valued input variables:

- **Category:** Missing values are considered a valid class level during the cluster initialization. This option is available when the **Imputation** method is set to **None**.
- **Ignore** (default)**:** See above.
- **Mean:** Replace input variables with missing cases by their average value during the clustering initialization process. This option is only available for nominal-valued input variables.
- **Mode:** Replace missing values with the most frequent class level.
- **Omit:** Remove observations from the cluster analysis with missing values in any one of the input variables. That is, the observation will not be assigned to a cluster grouping.

Output tab

The **Output** tab is designed to list both the output scored data set based on the partitioned data sets along with the data set containing the various clustering statistics and clustering seed statistics that are automatically created from the node. The appearance and layout design of the following **Output** tab is identical to the **Output** tab from the **Clustering** node that has the following two subtabs.

- **Clustered Data subtab**
- **Statistics Data Sets subtab**

Clustered Data Subtab

The **Clustered Data** subtab lists the data libraries and the associated output data sets from the partitioned data sets. The corresponding output data set will allow to you to view the various clustering assignments that were created for each training case that is identified by the segment identifier variable. The output data sets consist of the input variables along with the segment identifier variable and the distance variable for each training case. The distance variable refers to the Euclidean distance of each observation to its cluster seed, which you want to be as small as possible. For Kohonen SOM training, the cluster data set will also contain the coordinate variables identifying the row and column assignments of the two-dimensional SOM map. If missing values are imputed, then the output data set will contain an impute identifier variable called _IMPUTE_. Therefore, the data structure is similar to the output data set created from the **Clustering** node. However, the one major difference between the data sets is that the SOM data set contains a mapping coordinate variable that identifies the observation that is assigned to the topological map based on network training from the SOM method. In Enterprise Miner, a separate variable is created for the row, column, and row:column combination within the data set from the SOM mapping assignments. The conventional SAS data set names that are automatically assigned to the partitioned data sets in Enterprise Miner have been previously explained.

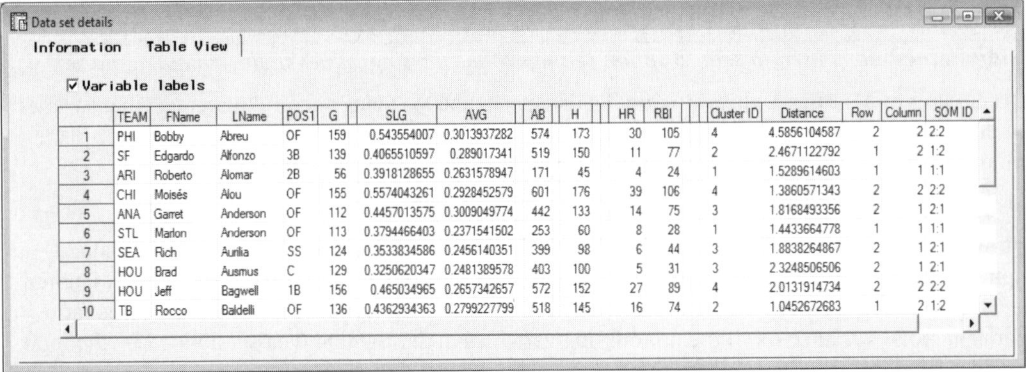

Data set details

Information | Table View

☑ Variable labels

	TEAM	FName	LName	POS1	G	SLG	AVG	AB	H	HR	RBI	Cluster ID	Distance	Row	Column	SOM ID
1	PHI	Bobby	Abreu	OF	159	0.543554007	0.3013937282	574	173	30	105	4	4.5856104587	2	2	2:2
2	SF	Edgardo	Alfonzo	3B	139	0.4065510597	0.289017341	519	150	11	77	2	2.4671122792	1	2	1:2
3	ARI	Roberto	Alomar	2B	56	0.3918128655	0.2631578947	171	45	4	24	1	1.5289614603	1	1	1:1
4	CHI	Moisés	Alou	OF	155	0.5574043261	0.2928452579	601	176	39	106	4	1.3860571343	2	2	2:2
5	ANA	Garret	Anderson	OF	112	0.4457013575	0.3009049774	442	133	14	75	3	1.8168493356	2	1	2:1
6	STL	Marlon	Anderson	OF	113	0.3794466403	0.2371541502	253	60	8	28	1	1.4433664778	1	1	1:1
7	SEA	Rich	Aurilia	SS	124	0.3533834586	0.2456140351	399	98	6	44	3	1.8838264867	2	1	2:1
8	HOU	Brad	Ausmus	C	129	0.3250620347	0.2481389578	403	100	5	31	3	2.3248506506	2	1	2:1
9	HOU	Jeff	Bagwell	1B	156	0.465034965	0.2657342657	572	152	27	89	4	2.0131914734	2	2	2:2
10	TB	Rocco	Baldelli	OF	136	0.4362934363	0.2799227799	518	145	16	74	2	1.0452672683	1	2	1:2

*The **Table View** tab is used to view the mapping assignments from Kohonen SOM training.*

Statistics Data Sets Subtab

The **Statistics Data Sets** subtab lists the data sets that are automatically created when running the node. The data set consists of both the various clustering statistics and the clustering seed statistics.

For Kohonen VQ network training, the clustering statistics are identical to the listing generated from the **Clustering** node along with various descriptive statistics such as the mean, median, minimum, and maximum for each cluster that is created across each input variable in the analysis and the overall statistics by every input variable that is combined. For Kohonen SOM training, the data set consists of the various clustering statistics such as the segment identifier variable, distance variable, and coordinate variables identifying the row and column assignments of the two-dimensional SOM map. The distance variable refers to the Euclidean distance of each observation to its cluster seed.

The clustering seed output listing is identical to the **Clustering** node. The data set will list the number of observations within each cluster, root mean square error, maximum distance from the cluster seed, nearest cluster, distance to nearest cluster, and the cluster seeds for each input variable in the selected active data set. The data set that contains the clustering seeds can be passed along to the PROC FASTCLUS procedure in order to initialize the seeds of other data sets.

Viewing the SOM/Kohonen Node Results

General Layout of the Results Browser within the SOM/Kohonen Node

- **Map tab** *(Kohonen/Batch SOM method only)*
- **Partition tab** *(Kohonen vector quantization method only)*
- **Variables tab**
- **Distances tab**
- **Profiles tab**
- **Statistics tab**
- **CCC Plot tab** *(Kohonen vector quantization method only)*
- **Output tab**
- **Log tab**
- **Code tab**
- **Notes tab**

As before in clustering analysis, the objective of the following analysis is grouping the major league hitters in the game during the 2004 baseball season from the various hitting categories.

Map tab

If the Kohonen SOM methods have been specified from the **General** tab, then the **Map** tab will automatically appear as you first open the **Results Browser** to view the compiled results. The **Map** tab displays two separate plots. The tab is only available for viewing if you select either one of the two SOM clustering methods that are designed to reduce the dimensionality of the data. In the following example, the topological grid map is defined by two separate rows and columns. The charts and plots provide a graphical representation of both the size of each cluster and the relationship between the clusters.

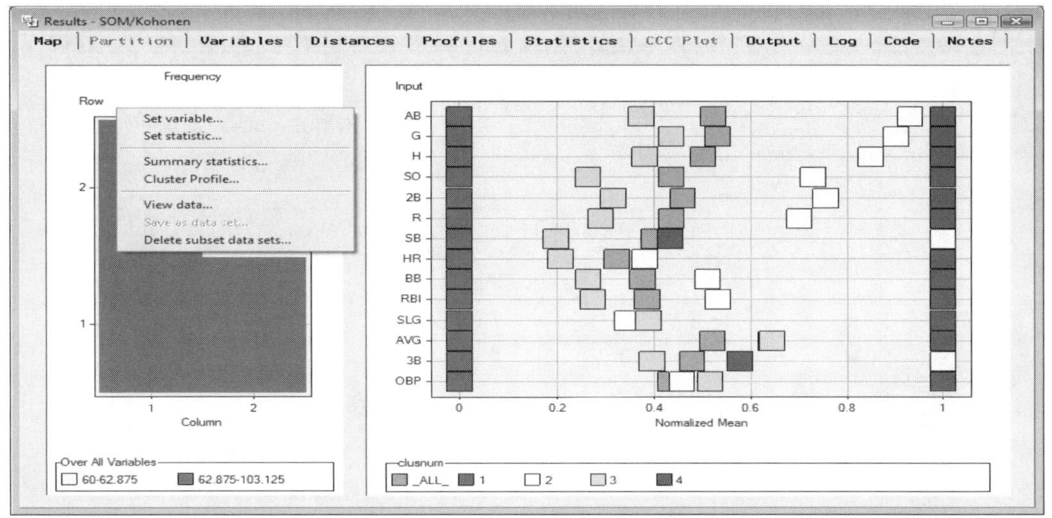

The ***Map*** *tab that displays the frequency plot and the input means plot of the Kohonen SOM results.*

The plot that is positioned to the left of the tab displays the topological mapping of the input space for the clusters. The stacked frequency bar displays the number of cases assigned to the winning clusters defined by the rows and columns from the two-dimensional grid space specified within the **Advanced** tab. Ideally, we would expect large frequency counts for each winning cluster. The number of columns are positioned on the horizontal axis and the number of rows are positioned on the vertical axis. The plot will allow you to determine the dimensionally of the topological map based on the number of rows and columns to the SOM map. The coordinates for the first cluster are identified by the first row and column. The second cluster corresponds to the first row and second column, and so on. The color coding of the winning clusters in the SOM map is such that light colors correspond to low frequency counts and darker color corresponds to larger frequency counts between the final clusters. The legend positioned at the bottom of the chart displays the color coding associated with the minimum and maximum frequency values. Select the **Scroll data** icon to scroll the color coding legend through every possible legend entry. To display the various clustering statistics of each cluster from the bar chart, select the **Select Points** icon from the toolbox and select the corresponding cluster listed within the frequency bar chart, then right-click the mouse to select the **Summary statistics...** pop-up menu option. This is illustrated in the previous diagram.

The adjacent plot positioned to the right of the tab displays the overall mean across all the input variables in the active training data set. The plot will allow you to compare the overall mean of all of the clusters that are formed from network training in comparison to the mean of the selected cluster for each input variable in the analysis. The node automatically standardizes the input variables by dividing the mean by the maximum value of each input variable under analysis that will fall within the range of zero and one. The reason for this is due to the fact that some input variables in the analysis might have entirely different variability with comparison to other input variables. From the input means plot, input variables that display a wide discrepancy in the cluster means will indicate to you that these same input variables best characterize the selected cluster. The plot is similar to the input means plot within the **Clustering** node. Therefore, the various options that are available within the tab are discussed in greater detail in the **Clustering** node.

From the Kohonen SOM results, map (1,1) consist of some of the weakest hitters in the major leagues, the map (1,2) consists of base stealers and triple hitters, map (2,1) consists of the above average hitters in the major leagues in comparison to the overall batting average, and, finally, map (2,2) consists of the greatest hitters in the major leagues with their group averages in all hitting categories greater than the overall average.

Partition tab

The **Partition** tab will automatically appear as you first open the **Result Browser**, assuming that you have specified the Kohonen vector quantization clustering method. The **Partition** tab is similar to the tab within the **Clustering** node. Again, the tab is available for viewing only if you have selected the Kohonen VQ method from the **General** tab. The tab is designed to display a graphical representation of the various statistical summaries of the clusters that are created from Kohonen VQ network training.

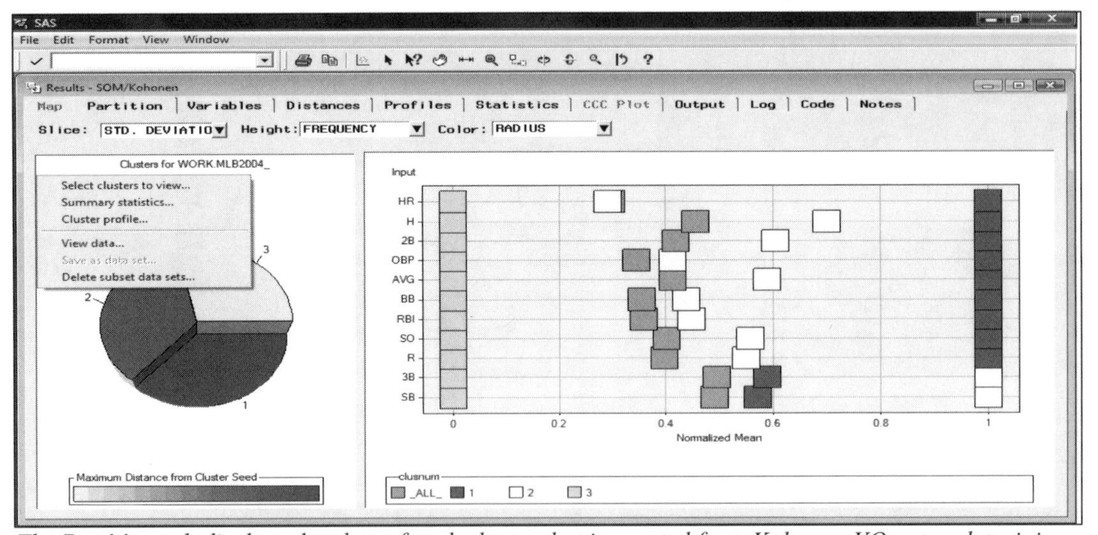

*The **Partition** tab displays the plots of each cluster that is created from Kohonen VQ network training.*

In Kohonen vector quantization network training, three separate clusters were specified. The various clusters that are created are very similar, with approximately the same variability within each group that is displayed by the equal proportion in each slice of the pie chart. Also, the second and third clusters had an approximately equal number of cases assigned to each cluster of about 150 baseball hitters in the game, which is quite different than the number of players assigned to the separate cluster groupings from the previous *k*-means cluster procedure. From the Kohonen vector quantization network training results, the third cluster grouping contains some of the weakest hitters and the first cluster grouping has some of the greatest hitters in the major leagues during the 2004 baseball season. However, the clustering assignments from Kohonen vector quantization network training are inconsistent with the cluster groupings that were created from the traditional clustering technique.

Variables tab

The **Variables** tab displays the input variables in the active training data set. The variables are listed in the order of importance based on the importance statistic. The importance statistic is calculated in decision tree analysis to measure the strength of the input variable in partitioning the data. Importance statistics that are close to one will indicate to you which input variables make the greatest contribution in creating the cluster groupings. In the following listing, the number of hits and the number of runs batted in are the most important input variables in determining the Kohonen VQ cluster groupings. None of the input variables that are important variables in the previous clustering analysis are consistent with the current clustering technique. Refer back to the **Clustering** node to determine the way in which the importance statistic is calculated.

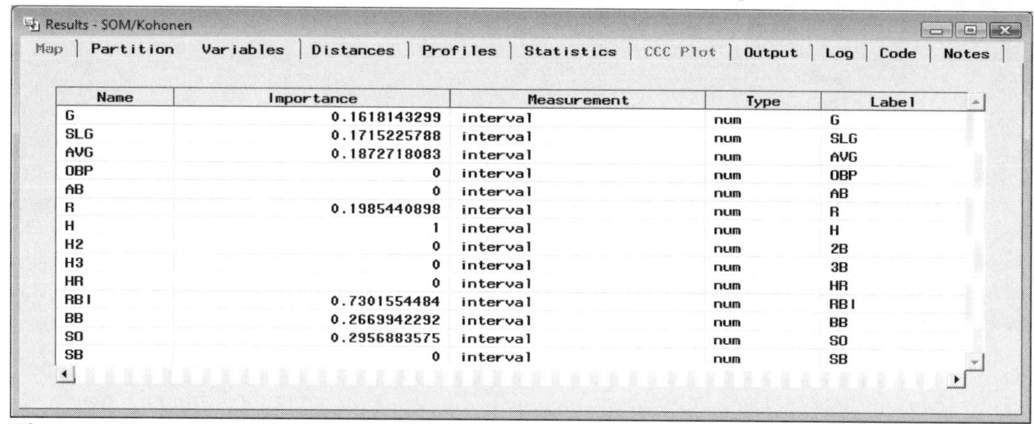

Results - SOM/Kohonen

Map | Partition | Variables | Distances | Profiles | Statistics | CCC Plot | Output | Log | Code | Notes

Name	Importance	Measurement	Type	Label
G	0.1618143299	interval	num	G
SLG	0.1715225788	interval	num	SLG
AVG	0.1872718083	interval	num	AVG
OBP	0	interval	num	OBP
AB	0	interval	num	AB
R	0.1985440898	interval	num	R
H	1	interval	num	H
H2	0	interval	num	2B
H3	0	interval	num	3B
HR	0	interval	num	HR
RBI	0.7301554484	interval	num	RBI
BB	0.2669942292	interval	num	BB
SO	0.2956883575	interval	num	SO
SB	0	interval	num	SB

*The **Variables** tab is used to observe the most important input variables in creating the cluster grids.*

Distance tab

The **Distance** tab is designed to display a bubble plot and a table listing of the various clusters created from network training. The bubble plot will display the amount of variability within each cluster that is based on the size of the bubbles and the variability between each cluster based on the distance between each bubble. Ideally, we are expecting small bubbles for these clusters that are well separated from each other. The asterisks represent the cluster centers and the circles indicate the cluster radii. The cluster radius represents the distance between the cluster center and the most distant case within the corresponding cluster.

The tab will also allow you to view the table listing of the distance matrix based on the distances between each cluster created from Kohonen vector quantization network training.

Profiles tab

For Kohonen vector quantization network training, the **Profiles** tab is designed to display the graphical representation of the clustering assignments between all categorically-valued input variables and interval-valued input variables from the training data set. The purpose of the tab is to determine the various characteristics to the clusters that are created. The following profiles plots and corresponding clusters that are displayed within the **Profiles** tab are based on the number of observations that are randomly sampled from the training data set that can be specified from the **Cluster Profiles** subtab. The **Profiles** tab displays several three-dimensional pie charts. The number of pie charts that are displayed depends on the number of clusters that are created and the number of class levels of the categorically-valued input variable. Selecting the **View > Categorical Variables** main menu options will allow you to view the three-dimensional pie charts, where each pie chart is displayed by the clusters created from Kohonen VQ network training against the various class levels positioned on the opposite axis. The height of the pie charts is determined by some interval-valued variable. Each slice of the pie chart is color coded to distinguish between the separate class levels of the selected categorical input variable from the output data set. Although the profile plot will not be display, the third cluster has the weakest hitters in baseball who hit the fewest number of home runs and the first cluster has the best hitters in the game who hit the largest number of home runs.

Selecting the **View > Interval Variables** main menu options will display the following three-dimensional frequency bar chart. The axes to the bar chart will display the various clusters created from preliminary training across every value of the interval-valued variable selected from the output scored data set.

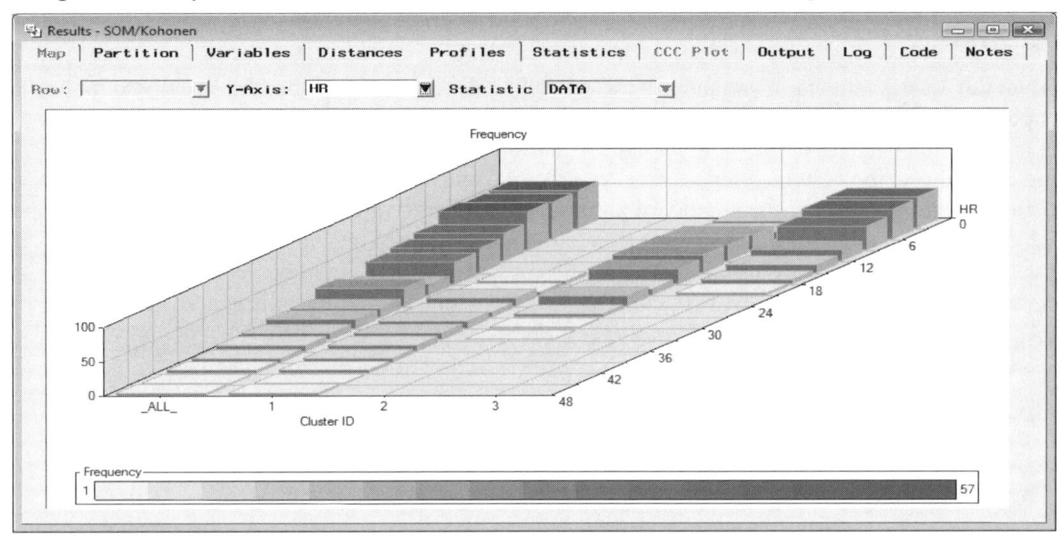

*The **Profiles** tab displays the frequency counts for each cluster across the range of values of each input variable to determine the various characteristics of each cluster from Kohonen VQ network training.*

Statistics tab

The **Statistics** tab is designed to display a table view of the various statistics for each cluster created from network training. The same data set can be displayed from the **Statistics Data Sets** subtab from the **Seed Statistics** data set. The following statistics are calculated from the active training data set. Some of the clustering statistics that are displayed are the following:

- **_SEGMNT_ Segment Identifier:** The cluster identifier variable.
- **Clustering Criterion:** The cluster criterion statistic.
- **Maximum Relative Change in Cluster Seeds:** The relative change in the cluster seeds.
- **Improvement in Clustering Criterion:** Improvement in the clustering criterion statistic.
- **_FREQ_ Frequency Count:** The number of cases assigned to each cluster.
- **_RMSSTD_ Root Mean Square Standard Deviation:** The root mean square distance between each observation in the cluster. That is, the squared distance between each observation and the assigned cluster (the corresponding cluster seed) across the separate input variables in the active training data set.
- **_RADIUS_ Maximum Distance from the Cluster Seed:** The maximum distance from the cluster seed and any observation assigned to the corresponding cluster.
- **_NEAR_ Nearest Cluster:** The cluster closest to the current cluster based on the distance between the means of each cluster.
- **Distance to Nearest Cluster:** The distance between the mean of the current cluster and the mean of the closest cluster.
- **Row:** Row identifier of the current cluster from Kohonen SOM training.
- **Column:** Column identifier of the current cluster from Kohonen SOM training.
- **SOM ID:** Row:Column identifier of the current cluster from Kohonen SOM training.
- **Mean of Each Input Variable:** Average value of each one of the input variables in the active training data set with regard to the currently listed cluster.

Results - SOM/Kohonen

Map | Partition | Variables | Distances | Profiles | Statistics | CCC Plot | Output | Log | Code | Notes

SEGMNT	Clustering Criterion	Maximum Relative	Frequency	Root-Mean-Square Standard Deviation	Maximum Distance from Cluster Seed	SOM ID	HR
1	0.6532094895	0.0480004882	104	0.5125159511	3.2324730264	1:1	5.4711538462
3	0.6532094895	0.0480004882	106	0.5933648305	4.5103542633	2:1	10.773584906
2	0.6532094895	0.0480004882	101	0.7614938265	8.1470384246	1:2	15.237623762
4	0.6532094895	0.0480004882	60	0.7838441785	9.7807815765	2:2	31.166666667

*The **Statistics** tab that displays the clustering statistics and assignments in Kohonen SOM training.*

CCC Plot tab

For the Kohonen vector quantization clustering method, the **CCC Plot** tab displays the plot of the cubic clustering criterion statistic across the possible clusters that are created. The CCC statistic is designed to determine the best number of clusters defined by the ratio between the expected and the observed r-square statistic. The tab will be available for viewing when the Kohonen VQ clustering method is specified. The plot is similar to the previously displayed CCC plot. Therefore, the tab will not be displayed.

Output tab

The **Output** tab displays the output scored data set that is automatically created when you execute the **SOM/Kohonen** node. For network training, the procedure output listing is generated from the PROC DMVQ data mining procedure that is running behind the scenes. The output listing displays the overall descriptive statistics across each input variable and the same descriptive statistics for each cluster that is created. However, for Kohonen VQ training, the procedure output listing will display the various clustering statistics that are generated from the SAS/STAT PROC CLUSTER and FASTCLUS procedures.

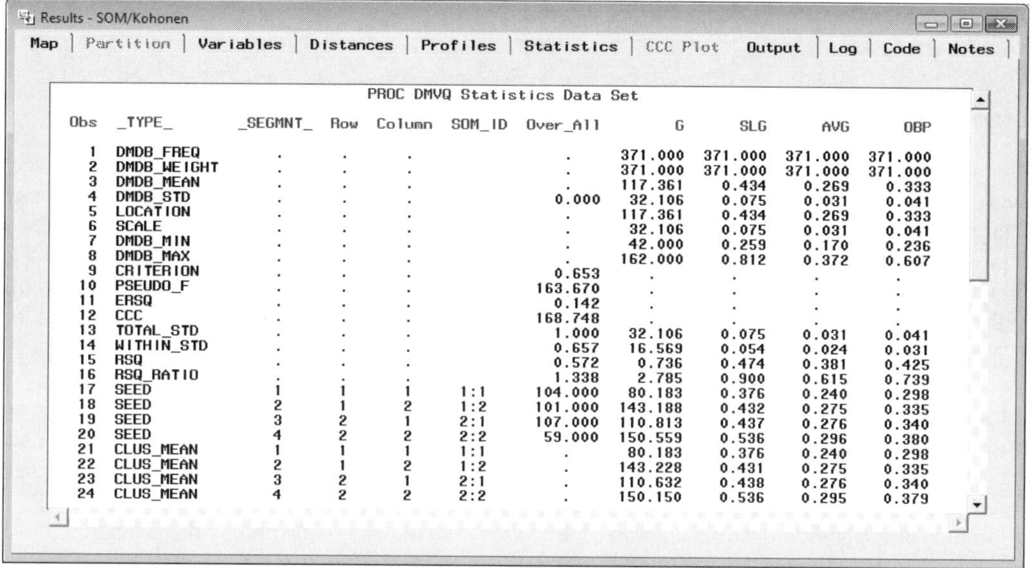

Results - SOM/Kohonen

Map | Partition | Variables | Distances | Profiles | Statistics | CCC Plot | Output | Log | Code | Notes

PROC DMVQ Statistics Data Set

Obs	_TYPE_	_SEGMNT_	Row	Column	SOM_ID	Over_All	G	SLG	AVG	OBP
1	DMDB_FREQ	.	.	.		.	371.000	371.000	371.000	371.000
2	DMDB_WEIGHT	.	.	.		.	371.000	371.000	371.000	371.000
3	DMDB_MEAN	.	.	.		.	117.361	0.434	0.269	0.333
4	DMDB_STD	.	.	.		0.000	32.106	0.075	0.031	0.041
5	LOCATION	.	.	.		.	117.361	0.434	0.269	0.333
6	SCALE	.	.	.		.	32.106	0.075	0.031	0.041
7	DMDB_MIN	.	.	.		.	42.000	0.259	0.170	0.236
8	DMDB_MAX	.	.	.		.	162.000	0.812	0.372	0.607
9	CRITERION	.	.	.		0.653	.	.	.	.
10	PSEUDO_F	.	.	.		163.670	.	.	.	.
11	ERSQ	.	.	.		0.142	.	.	.	.
12	CCC	.	.	.		168.748	.	.	.	.
13	TOTAL_STD	.	.	.		1.000	32.106	0.075	0.031	0.041
14	WITHIN_STD	.	.	.		0.657	16.569	0.054	0.024	0.031
15	RSQ	.	.	.		0.572	0.736	0.474	0.381	0.425
16	RSQ_RATIO	.	.	.		1.338	2.785	0.900	0.615	0.739
17	SEED	1	1	1	1:1	104.000	80.183	0.376	0.240	0.298
18	SEED	2	1	2	1:2	101.000	143.188	0.432	0.275	0.335
19	SEED	3	2	1	2:1	107.000	110.813	0.437	0.276	0.340
20	SEED	4	2	2	2:2	59.000	150.559	0.536	0.296	0.380
21	CLUS_MEAN	1	1	1	1:1	.	80.183	0.376	0.240	0.298
22	CLUS_MEAN	2	1	2	1:2	.	143.228	0.431	0.275	0.335
23	CLUS_MEAN	3	2	1	2:1	.	110.632	0.438	0.276	0.340
24	CLUS_MEAN	4	2	2	2:2	.	150.150	0.536	0.295	0.379

*The **Output** tab displays the procedure output listing from Kohonen SOM training that computes the various clustering statistics along with the standard descriptive statistics for each input variable by each cluster that is created.*

Log tab

The **Log** tab displays the standard log listing when executing the **SOM/Kohonen** node. The log file will allow you to diagnose any problems that might occur when executing the node.

Code tab

The **Code** tab displays the SAS training code that created the Kohonen network training results by executing the node. The tab will allow you to display either the training code or the score code. By default, the training code is automatically displayed. The training code will list the PROC DMVQ data mining procedure and the corresponding option statements. This is illustrated in the following diagram. The various option settings that are specified within the **SOM/Kohonen** node will instruct SAS to automatically generate the corresponding scored data sets and statistical output listings from the following procedure code.

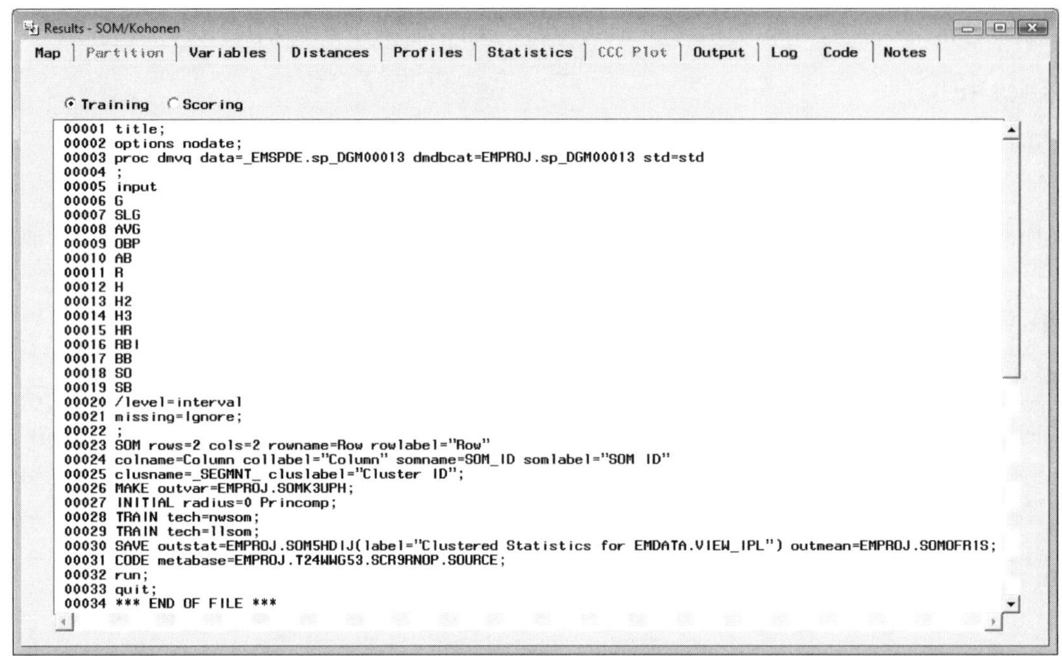

*The **Code** tab displays the internal procedure training code that generated the Kohonen SOM results.*

Selecting the **Scoring** button will instruct the node to display the following scoring code. The purpose of the scoring code is that it can be applied to a new set of values from the input variables in the data set to generate entirely different clustering assignments in Kohonen SOM training. The scoring code will allow you to view the internal programming code that is performed within Enterprise Miner to create the various cluster groupings in either Kohonen SOM or Kohonen VQ network training. The following scoring code is a partial listing from the Kohonen SOM training. The first step is standardizing each interval-valued input variable in the analysis. The next step is to compute the squared distance for each cluster that is created from the squared distance of each input variable in the model by its own cluster seed. An adjustment is performed on the accumulated squared distance that is changed proportionally to the specified number of rows and columns in the SOM map and the neighborhood size. From the adjusted squared distance, a segment identifier is created to identify which cluster the observation is assigned to, which is determined by the smallest accumulated squared distance of each input variable in the Kohonen SOM network model. The final step is to assign the observations to the corresponding row and column of the SOM map that are identified by the segment identifier variable. The scoring code from Kohonen VQ network training is essentially the same as the *k*-means clustering technique with the exception that the distance function is adjusted by the learning rate. Therefore, the scoring code from Kohonen VQ network training will not be illustrated.

Results - SOM/Kohonen

Map | Partition | Variables | Distances | Profiles | Statistics | CCC Plot | Output | Log | Code | Notes

Training   Scoring

```
00001 format _SEGMNT_ CLS_D2S_.;
00002 ****************************************;
00003 *** Begin Scoring Code from PROC DMVQ ***;
00004 ****************************************;
00005
00006
00007 *** Begin Class Look-up, Standardization, Replacement ;
00008 drop _dm_bad; _dm_bad = 0;
00009
00010 *** Standardize G ;
00011 drop T_G ;
00012 if missing( G ) then T_G = .;
00013 else T_G = (G - 117.361185983827) * 0.0311470971186;
00014
00015 *** Standardize SLG ;
00016 drop T_SLG ;
00017 if missing( SLG ) then T_SLG = .;
00018 else T_SLG = (SLG - 0.43440567864593) * 13.4158943513423;
• • •
00083 *** Omitted Cases;
00084 if _dm_bad then do;
00085    _SEGMNT_ = .; Distance = .;
00086    goto _vlex ;
00087 end; *** omitted;
00088
00089 *** Compute Distances and Cluster Membership;
00090 label _SEGMNT_ = 'Cluster ID' ;
00091 label Distance = ' ' ;
00092 array _vads [4] _temporary_;
00093 drop _vqclus _vqmvar _vqnvar;
00094 _vqmvar = 0;
00095 do _vqclus = 1 to 4; _vads [_vqclus] = 0; end;
00096 if not missing( T_G ) then do;
00097    _vads [1] + ( T_G - -1.15800215325418 )**2;
00098    _vads [2] + ( T_G - 0.80443398507119 )**2;
00099    _vads [3] + ( T_G - -0.2039543649397 )**2;
00100    _vads [4] + ( T_G - 1.03402556770836 )**2;
00101 end;
00102 else _vqmvar + 1;
00103 if not missing( T_SLG ) then do;
00104    _vads [1] + ( T_SLG - -0.78459890949339 )**2;
00105    _vads [2] + ( T_SLG - -0.0283034698872 )**2;
00106    _vads [3] + ( T_SLG - 0.03787994314276 )**2;
00107    _vads [4] + ( T_SLG - 1.36277598524818 )**2;
00108 end;
• • •
00194 _vqnvar = 14 - _vqmvar;
00195 if _vqnvar <= 2.2282620193436E-11 then do;
00196    _SEGMNT_ = .; Distance = .;
00197 end;
00198 else do;
00199    _SEGMNT_ = 1; Distance = _vads [1];
00200    _vqfzdst = Distance * 0.99999999999988; drop _vqfzdst;
00201    do _vqclus = 2 to 4;
00202       if _vads [_vqclus] < _vqfzdst then do;
00203          _SEGMNT_ = _vqclus; Distance = _vads [_vqclus];
00204          _vqfzdst = Distance * 0.99999999999988;
00205       end;
00206    end;
00207    Distance = sqrt(Distance * (14 / _vqnvar));
00208 end;
00209
00210 *** SOM Row and Column;
00211 label Row = 'Row' ;
00212 label Column = 'Column' ;
00213 label SOM_ID = 'SOM ID' ;
00214 length SOM_ID $8;
00215 if _SEGMNT_ > 0 then do;
00216    Row = 1 + floor(( _SEGMNT_ - 1) / 2);
00217    Column = 1 + mod( _SEGMNT_ - 1, 2 );
00218    SOM_ID = put( Row ,8. );
00219    SOM_ID = left( SOM_ID );
00220    _vqlen = 1 + length( SOM_ID ); drop _vqlen;
00221    substr( SOM_ID , _vqlen , 1 ) = ':';
00222    length _dm8 $8; _dm8 = put( Column , 8. );
00223    _dm8 = left(_dm8); drop _dm8;
00224    substr( SOM_ID , _vqlen+1 ) = _dm8;
00225 end;
00226 else do; Row = .; Column = .; SOM_ID = ' '; end;
00227 _vlex :;
00228
```

*The **Code** tab displays the internal scoring code from Kohonen SOM network training to generate entirely different clustering assignments for a new set of input values.*

3.7 Time Series Node

General Layout of the Enterprise Miner Time Series Node

- **Data tab**
- **Variables tab**
- **Output tab**
- **Notes tab**

The purpose of the **Time Series** node in Enterprise Miner is to collapse and condense time-stamp transitional data sets and accumulate the data into an equally spaced interval of time. Time-stamp transitional data is often recorded in an unequal period of time. Therefore, the time-stamp transitional data needs to be accumulated and condensed in order to form an equally spaced interval of time. Time-stamp transitional data is similar to the data structure that is required in sequential analysis such as bank transactions or purchasing transactions over an unequal period of time. The **Time Series** node will allow you to impute missing observations, define the time interval variable within the data set by consolidating the input data set into specified time intervals, and transpose the file layout of the active input data set to create the output seasonal time series data set. The advantage in collapsing the enormous data sets into a time series structure is that it might allow you to observe trends and seasonal variability in the data over time that might not be discovered otherwise. The node is designed to accumulate and collapse the values of the interval-valued target variable into equally spaced time intervals in the time-stamped transitional data by a wide assortment of descriptive statistics. By default, the values of the interval-valued target variable are summarized within each time interval.

The purpose of time series modeling is to predict the seasonal variability of the target variable over time. In time series modeling, the past values of the target variable are used as the input variables for the model. Time series forecasting is usually performed over a short period of time. The reason is because the correlation between the target variable and its past values tend to weaken over time. Although there are many different forms of time series models to consider, there are three major classes of time series modeling: the autoregressive AR models, integrated autoregressive moving average ARIMA models, and moving average MA models. The autoregressive ARMA models consist of two separate components: the autoregressive AR component and the moving average MA component. The models are usually referred to as ARMA(p, q) with p representing the autoregressive component and q representing the moving average component. The integrated autoregressive moving average ARIMA(p, d, q) model is a generalization of the ARMA model, where d represents the differencing component. The purpose of differencing will be discussed shortly. Moving average MA models basically calculate the average value of the consecutive past values of the target variable over time. The moving average estimate is then used as the target variable for the model.

One of the purposes of the node is to prepare the data in order to perform subsequent time series modeling or repeated measures modeling based on cross-sectional analysis. However, the **Time Series** node is specifically designed to create the output data set that can be passed along to the SAS/ETS time series procedures, that is, PROC ARIMA, FORECAST, AUTOREG, STATESPACE, X11, and SPECTRA procedures, to perform time series modeling. By default, the **Time Series** node will automatically create a data set with twelve accumulated data points under the assumption that the time series data is defined in months. The main goal of time series modeling is to identify the underlying nature or the hidden pattern represented by the sequence of data points over time and to generate accurate forecasting estimates. Time series modeling predicts future values of the target variable by its own previous values. The main goal to repeated measures modeling is to determine the relationship between the input variables and the target variable from repeated measurements that are recorded within each subject, that is, **crossid** variable, that can be viewed over time.

The difference between the data structure in time series modeling and repeated measures modeling is that the longitudinal data set used in repeated measures modeling for cross-sectional analysis will usually consist of several short series of time points that are recorded within each subject that is identified by the **crossid** variable, in comparison to time series data sets that consists of a continuous series of time points identified by the **timeid** variable. The **Time Series** node requires that the input training data set consist of both a time identifier variable and an interval-valued target variable. The active training data set must have a time stamp variable with a variable role of **timeid** that identifies the numeric interval or the interval of time along with an interval-valued **target** variable in order to accumulate its values. Again, the node requires a time interval that is used in defining the way in which the numeric-valued target values are estimated or accumulated. If the time interval or **timeid** variable does not exist in the data set, then the node will fail to run. In addition, any number of cross-sectional or grouping variables can be specified with a model role set to **crossid**. The data layout consists of a time interval, that is, **timeid**, within each group, that is, **crossid**, that is analogous to the file layout in sequence discovery analysis or viewing the relationship between the target variable or any number of input variables over time within each subject that is assigned a **crossid** variable role in the repeated measures model. The subsequent time trend plots that are automatically created within the node will allow you to view the change in the values of the target variable over time for each subject, that is, the **crossid** variable, in the repeated measures model.

Imputation of the Missing Values

Time series modeling should only be applied to data structures of equally spaced time series data over time, where the data does not contain any missing values. In time series analysis, it is important to retain the chronological order of the data structure. Therefore, the node is designed for you to impute or fill in the problematic and undesirable missing values in the time series data.

In time series analysis, it is important to retain the chronological order of the equally spaced time intervals intact. Therefore, various imputation or interpolation methods are applied to replace each missing value by some estimate, then perform the analysis as if there were no missing data. This can be performed from the **Missing Value Assignment** option within the **Options** tab. Also, in order to keep the time interval in the data intact, the node is designed for you to define the interval of time or the seasonality in the data based on the **timeid** variable in the active training data set from the **Time Interval** tab, that is, defining each successive observation based on the specified interval of time. By default, the node assumes that the interval of time in the time series data is defined in months. The node will also allow you to retain the chronological order of the time series by collapsing the target values with various statistics such as the minimum, maximum, median, or average value in the time interval to name a few from the **Accumulation Method** option within the **Options** tab.

Repeated Measures Modeling Assumptions

In repeated measures modeling, it is assumed that the values in the longitudinal data set that are measured between each subject or **crossid** variable are independent. However, it is also assumed that the values recorded within each subject are correlated with one another, that is, measurements recorded close together are more highly correlated than measurements recorded further apart, and that the variability often changes over time. These potential patterns in the correlation and the variability recorded over time within each subject may at times contribute to a very complicated covariance structure. Therefore, the corresponding covariance structure must be taken into account to generate valid statistical inferences with an accurate assessment of the variability in the parameter estimates from the repeated measures model. In model designing, like repeated measures modeling, the first step is to determine the relationship between the target variable and the input variables over time. The following trend plots that are generated from the transposed output data set will allow you to identify general trends and the functional form of each subject over time. In addition, the plot will allow you to identify the variability between the subjects over time.

Time Series Modeling Assumptions

In time series analysis, it is assumed that the chronological observations or error terms are related to one another over time. Chronological observations are defined as a sequence of data points that occur in a non-random order over an equally spaced interval of time. The forecast modeling assumptions are such that the error terms are independent, identically distributed (iid), normal random variables with a mean of zero and a constant variance about the mean of zero. In time series modeling, when the residuals in the time series model have a constant mean of zero with an unsystematic uniformly random variability over time, the result is called *white noise*. However in time series modeling, the independence assumption is usually violated since the data structure is an equally spaced interval of time with various adjacent data points being influenced by one another over time. For example, in forecasting stock market prices over time, the stock market prices for one day might be influenced by the stock market prices of the previous day.

Trend and Seasonality Time Series Patterns

In time series analysis, there are two general patterns to be identified: the trend and seasonality in the time series data. *Trend* is a constant, slow-growing change in data over time. *Seasonality* is a pattern in the data that repeats itself over time. In other words, trend is a constant systematic pattern that changes over time and does not repeat itself. There are two different types of seasonality in time series called additive seasonality and multiplicative seasonality. *Multiplicative seasonality* is the most common in seasonal time series data. Multiplicative seasonality occurs in seasonal time series data when the magnitude in the fluctuating behavior of the seasonal pattern over time increases with the trend in the time series data, indicating that the mean of the target values is correlated with its variance over time. Both the trend and the seasonal variability are increasing in a fanned-out pattern over time. *Additive seasonality* occurs in seasonal time series data when both the trend and the seasonality stay the same over time.

Stationarity in the Time Series Data

In time series modeling, it is critical that the data be stationary. When performing time series modeling, stationarity is one of the first assumptions to check in the data. *Stationarity* in time series data is when the target variable does not display any trend or seasonality over time. In other words, the time series is stationary when the target values display a uniform nonsystematic fluctuating behavior with any increasing or decreasing trend in the data over time. This suggests that once the data set is processed in the **Time Series** node, then further transformations might be required to achieve stationarity in the data that is required in time series modeling. Stationarity can be checked by simply creating a scatter plot, that is, a time trend plot of the target values over time. From the **Detailed Results** window within the **Results Browser,** the node is designed to create a time trend plot that will allow you to observe both the trend and seasonality in the time series data set to check for stationarity in the time series data. Also, it is extremely important to check for outliers and missing data points in the same time series plot. If the time series data displays a nonstationary pattern over time, then the forecasting model will generate unstable parameter estimates. Unstable estimates in time series modeling means that the parameter estimates in the model are highly correlated with each other. That is, the parameter estimates are stable if their values in the model are within a certain range. For example, an autoregressive time series model with one autoregressive parameter estimate must fall between the interval of [–1, 1], otherwise, past values would tend to accumulate over time due to nonstationarity in the time series data. Nonstationarity might be due to outliers, random walk, drift, trend or changing variance. Again, in time series modeling, when the residuals in the time series model have a constant mean of zero with an unsystematic uniformly random variability over time, the result is called white noise. The following are the three most popular ways for achieving stationarity in the time series data.

- **Transformation:** Typically, the logarithm transformation or a Box-Cox transformation is applied to the target variable in the model to stabilize the target variance over time. However, these transformations, like the logarithm transformation, are not designed as remedies to eliminate the trend in the data.
- **Detrending:** To eliminate the increasing or decreasing trend or seasonality in the time series data, detrending is performed by taking the difference in the target values and the fitted values from a simple linear regression model with a time identifier variable that is the only input variable for the model, that is, where $x_t = \beta_0 + \beta_1 t + w_t$, and detrending the time series data over time by taking the difference of $y_t - x_t$ as the target variable for the time series model.
- **Differencing:** An alternative to detrending that is often used in time series modeling is by taking the difference of successive target observations, for example, $y_t - y_{t-1}$, that is used as the target variable to the time series model in order to eliminate both the trend and the seasonal variability in the data over time. For example, calculating the sth seasonal difference of $y_t - y_{t-s}$ will eliminate the sth seasonality in the data. At times, it might be necessary to compute differences of differences until stationarity is achieved. It is important to point out that some time series data might require little or no differencing, while at the same time over differencing some time series data can result in unstable parameter estimates. Detrending and differencing are commonly used if you are interested in modeling changes in the target values over time, for example, price changes, temperature changes, and so on.

Time series modeling is discussed in greater detail in Shumway and Stoffer (2002), Brocklebank and Dickey (2003), SAS/ETS Software Time Series Modeling (2000), Forecasting Examples for Business and Economics Using the SAS System (1996), Yaffe (2000), Bowerman and O'Connell (1993), and Longitudinal Data Analysis with Discrete and Continuous Responses Course Notes, SAS Institute (2005).

Data tab

The **Data** tab is designed to list the active input data set. If several data sets are available for processing, then each one of the partitioned data sets may be created in preparation for time series modeling outside the node. By default, the active training data set is consolidated for time series preparation. However, creating a validation data set from the **Data Partition** node can be applied in cross-validating the forecasting estimates from the subsequent time series model and to verify the stability in the time series estimates. Select the **Properties...** button that will allow you to view the file administrative information and the table view of the active input data set. The tab will also allow you to select an entirely different training data set by selecting the **Select...** button since the training data set is automatically selected from the tab. The **Imports Map** window will appear for you to select from the list of training data sets that are available within the currently opened process flow diagram in assigning the active training data set to the time series analysis.

Variables tab

Opening the **Time Series** node will result in the **Variables** tab automatically appearing. The **Variables** tab will display the variables in the active training data set. The tab will allow you to remove modeling terms from the time series model within the **Status** column. Notice that the time series data set must contain a time identifier variable in the data set with a **timeid** model role to identify the time points in the active training data set.

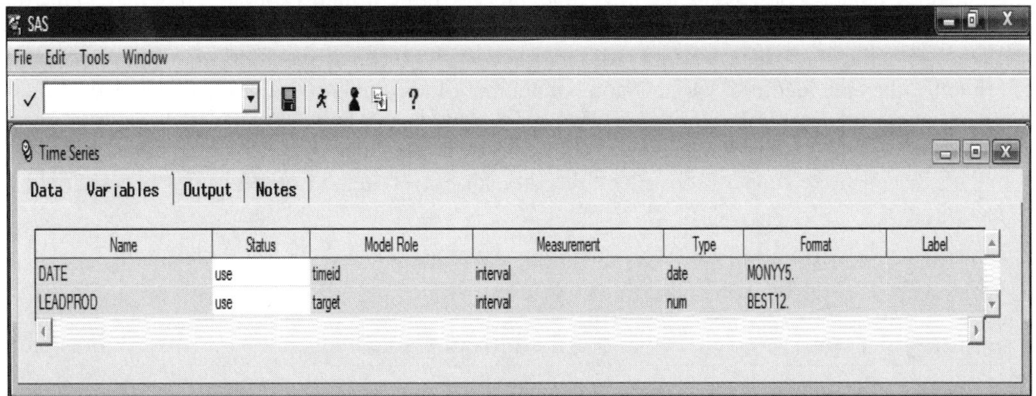

*The **Variables** tab that lists the target variable and the time id variable in the time series data set.*

Output tab

The **Output** tab is designed to display the output seasonal data set that is automatically created by executing the node. The output seasonal data set will consist of a single row of accumulated data points. By default, twelve separate columns are displayed for the twelve accumulated data points that are calculated from the active training data set. The length of the interval in which the data points are accumulated over time is defined by the number specified from the **Length of Cycle** option within the **Time Series – Settings** window. The node will also create a trend data set and a seasonal data set. The trend data set will consists of a time identifier variable and the various descriptive statistics such as the mean, median, sum, minimum, maximum and standard deviation value for each time interval from the active training data set. The seasonal data set will consist of the same summary statistics based on the accumulated time series interval listing of the various descriptive statistics. The output data set can then be passed to the PROC ARIMA time series procedure to perform subsequent time series modeling.

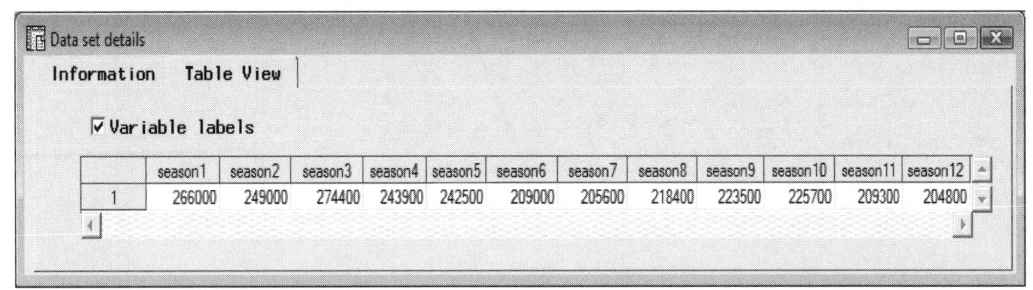

*The **Table View** tab displays the active training data set of the accumulated data points over time.*

Time Series Node Settings

From the tools bar menu, select the **Settings** toolbar icon or the **Tools > Settings** main menu options to specify the various configuration settings for the time series model. The **Time Series – Settings** window will appear that consists of the following three tabs:

- **Interval Time tab**
- **Options tab**
- **Output Data Set tab**

Time Interval tab

The **Time Interval** tab will automatically appear as you view the **Time Series - Settings** window. The **Time Interval** tab is designed to define the interval of time based on the **timeid** model role variable. This variable is critical in order for the node to execute. The requirement of the node is that there must exist a formatted date time identifier variable assigned within the input data set in order for the node to execute. By default, both sections are dimmed and unavailable with the **Automatic** button selected and the **Apply time range** check box unchecked. In Enterprise Miner, the node will automatically estimate the accumulated data points with the node assuming that the model is based on a monthly time interval of the **timeid** variable. The range in the time interval, that is, start time and end time, is automatically set from the minimum and maximum dates in the time identifier variable. Selecting the **User specified** radio button will allow you to specify the time interval or the increment in time for the time identifier variable and the number of accumulated values calculated from the node. The number of accumulated values or the number of seasonal values that are calculated from the active training data set is based on the value displayed in the **Length of Cycle** entry field. Once the interval type is selected from the subsequent **Settings** window, then the value displayed in the **Length of Cycle** entry field will be updated accordingly.

The following displays the relationship between the various interval types that are available and the associated length of the cycle that is displayed in the **Length of Cycle** entry field.

Interval Type	Length of Cycle
Year	1
Qtr	4
Month	12
Week	52
Day	7
Hour	24
Minute	60

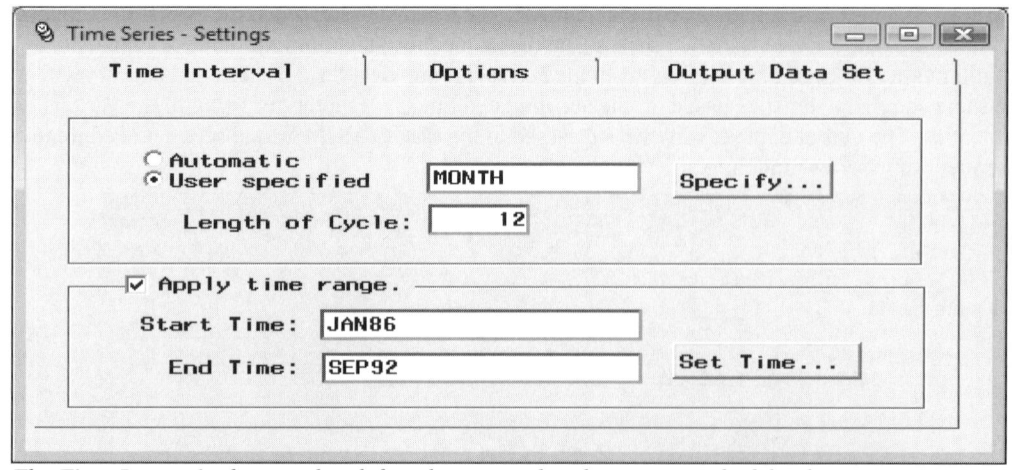

*The **Time Interval** tab is used to define the accumulated time interval of the data points.*

Time Interval Specification

Press the **Specify...** button to open the following **Interval Specification** window. The **Interval Specification** window will allow you to specify the interval of time for the time interval variable that is required in the **Time Series** node. The various interval types that can be specified are YEAR, SEMIYEAR, QTR, MONTH, SEMI-MONTH, TENDAY, WEEK, WEKEDAY, DAY HOUR, MINUTE, and SECOND. Again, the length of the cycle depends on the type of interval that is specified. The relationship between both values is displayed in the previous table.

The **Multiplier** field will allow you to change the number of interval types that are used to define the time interval. The default value is one. For example, let us assume that you have selected an interval type of WEEK from the **Type** display field. However, suppose that you want the time interval or increment to be every two weeks. Simply select the drop-down arrow next to the **Multiplier** field to change the value to 2, that is, creating a two-week interval of time. This will result in the **Length of Cycle** option in the **Settings** window changing from 52 to 26 cycles in time. From the **Shift** field, you may specify the start of the cycle on a unit of the time interval other than the first unit. For example, assume that the first time point begins in JAN07, then changing the **Shift** value to 7 will result in the time interval starting in JUL07 instead of JAN07.

By selecting the WEEKDAY interval type, then the **Weekend** field will become available. By default, Saturday and Sunday are specified as the weekend days. You can select any number of days to represent the weekend. Press the Ctrl key to select any number of the listed days. Select the **Date values include time of day** check box to include the time of day in your date values. That is, assuming that the time identifier variable is assigned a DATETIME format, then the check box is available for the time interval types of HOUR, MINUTE, and SECOND.

The **Description** field will display a short description of the selected interval specification. Press the **OK** button to return back to the **Time Series – Setting** window.

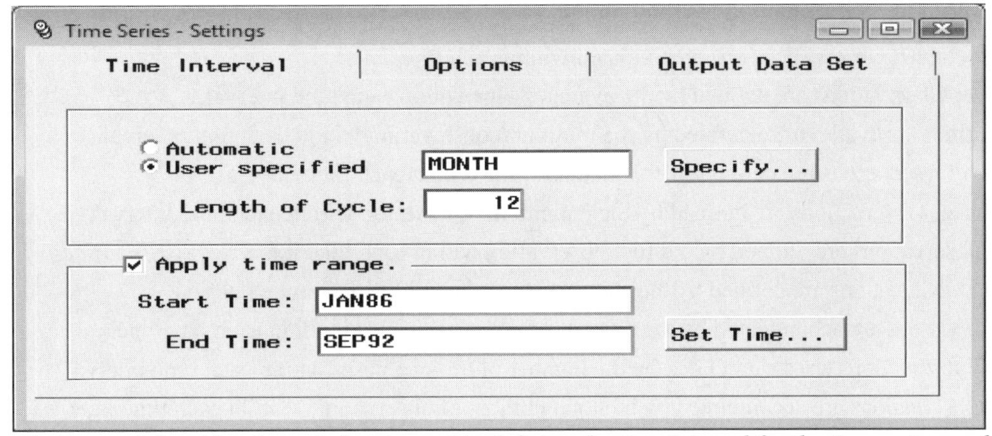

The ***Interval Specification*** *window is used in defining the time interval for the time series model.*

Apply Time Range

The **Apply time range** section will allow you to specify the range in time for the time identifier variable in the time series data set. By default, the time interval is defined by the minimum and maximum dates from the active input data set. For multiple cross-sectional (**crossed**) variables with overlapping time intervals, this option will allow you to set the same time interval for each cross-sectional variable. This will result in the same number of observations being assigned to each cross-sectional variable, where missing values are added to the data set that extend above or below the time interval of each cross-sectional variable in the training data set.

Selecting the **Apply time range** check box will allow you to enter the minimum and maximum dates in the appropriate entry fields. Otherwise, press the **Set Time...** button to open the following **Select values** window. The **Select values** window will display the univariate frequency bar chart. The sliders to the bar chart that are located at each end of the chart will allow you to set the time interval for the time identifier variable in the time series data set. This option is designed for you to reduce the number of observations to accumulate for each data point. Select the boundary boxes and move the line to set the appropriate minimum or maximum dates for the time series data set.

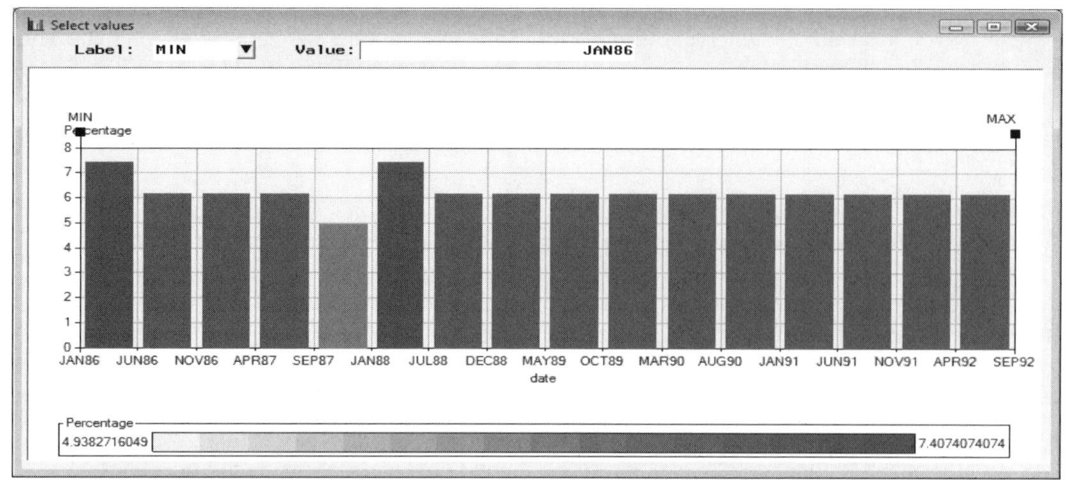

*The **Select values** window displaying the frequency bar chart for placing a bound on the time interval.*

Options tab

The **Options** tab is designed to specify how the time series values are defined, how the node handles missing values, and to control the listed output to the **Output** tab within the **Results Browser**.

The **Accumulation Method** option will allow you to define the data points in the output data set and the way in which the target values are accumulated based on the time interval variable. The following are the various options to select from the **Accumulation Method** option:

- **Total** (default)**:** Observations are accumulated within each time interval.
- **Average:** Observations are defined by the average value within each time interval.
- **Minimum:** Observations are defined by the minimum observation within each time interval.
- **Median:** Observations are defined by the median value within each time interval.
- **Maximum:** Observations are defined by the maximum observation within each time interval.
- **First:** Observations are defined by the first observation within each time interval.
- **Stddev:** Observations are defined by the standard deviation within each time interval.
- **N:** Observations are accumulated by the number of nonmissing cases within each time interval.
- **Nmiss:** Observations are accumulated by the number of missing cases within each time interval.
- **Nobs:** Observations are accumulated by the total number of observations within each time interval.

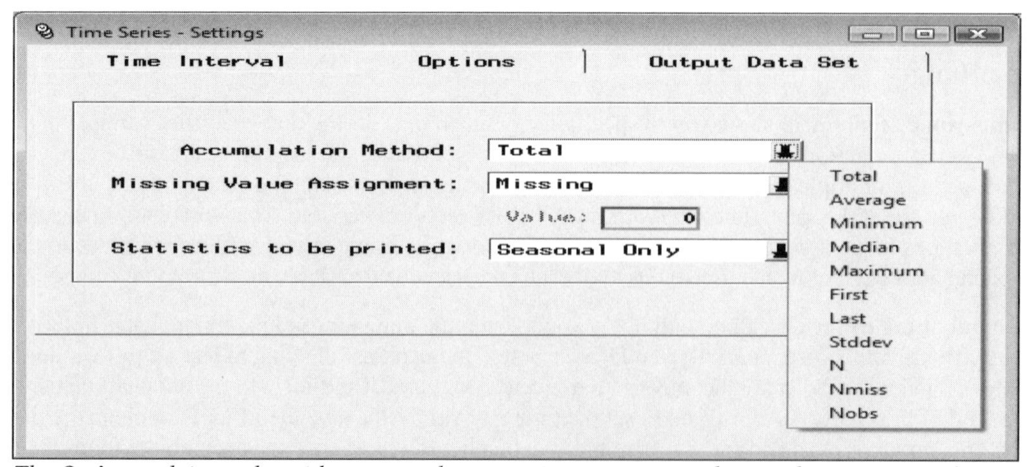

*The **Options** tab is used to either accumulate or estimate missing values in the time series data set.*

The **Missing Value Assignment** option will allow you to impute or estimate the missing values of the target variable. As previously mentioned, the time series model should only be fit to data that is equally spaced over time without any missing values. Therefore, the following options will allow you to fill in the missing values.

- **Missing** (default)**:** Missing values in the target variable stay the same.
- **Average:** Missing values are estimated by the average of all nonmissing target values.
- **Minimum:** Missing values are estimated by the minimum of the nonmissing target values.
- **Median:** Missing values are estimated by the median of the nonmissing target values.
- **Maximum:** Missing values are estimated by the maximum of the nonmissing target values.
- **First:** Missing values are estimated by the first nonmissing observation.
- **Last:** Missing values are estimated by the last nonmissing observation.
- **Previous:** Missing values are estimated by the previous nonmissing observation, that is, the previous nonmissing observation that is adjacent to the missing value.
- **Next:** Missing values are estimated by the next adjacent nonmissing observation, that is, the next nonmissing observation that was recorded after the missing value.
- **Constant:** Missing values are estimated by a specified constant value. Selecting this option will allow you to enter the value from the **Value** entry field.

The **Statistics to be printed** option will allow you to control the listed output that is displayed in the **Output** tab from the **Results Browser**. The following are the various listed options to select.

- **Seasonal Only** (default)**:** Displays the various descriptive statistics such as the standard descriptive statistics, that is, the mean median, sum, standard deviation, total number of observations, and total number of missing observations based on the accumulated data points of the target variable over time.
- **Trend Only:** Displays the various descriptive statistics based on each observation in the output data set.
- **Seasonal and Trend Only:** Displays both the seasonal and trend time series results.

Output Data Set tab

The **Output Data Set** tab will allow you to transpose the output seasonal data set. Again, the output data set that is automatically created consists of each record from the active training data set that is consolidated by each time period of the time identifier variable. However, transposing may be performed by assuming that the time series data structure originates from a sequential series of time with the various time intervals that are defined within each cross-sectional grouping. The output data set is formulated in a time series format with a one-record layout by each corresponding cross-sectional variable in the data set.

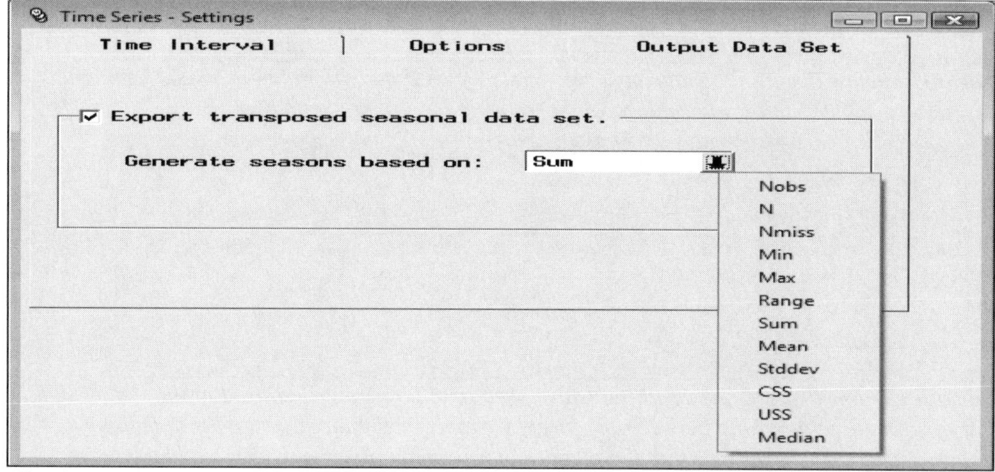

*The **Output Data Set** tab is used to redefine and transpose the seasonal output data set.*

The **Generate season based on** option will allow you to change the values of the variables that are used in generating the seasonal output data set. The number of records that are created for the transposed data set is determined by the value entered in the **Length of Cycle** option within the **Time Interval** tab.

The following are the various options used in defining the seasonality of the time series model:

- **Nobs:** The seasonal values are defined by all nonmissing observations.
- **N:** The seasonal values are defined by the total number of observations.
- **Nmiss:** The seasonal values are defined by the number of nonmissing observations.
- **Min:** The seasonal values are defined by the minimum value.
- **Max:** The seasonal values are defined by the maximum value.
- **Range:** The seasonal values are defined by the range of the time series interval.
- **Sum** (default)**:** The seasonal values are defined by the sum of the values.
- **Mean:** The seasonal values are defined by the average values.
- **Stddev:** The seasonal values are defined by the standard deviation of the values.
- **CSS:** The seasonal values are defined by the corrected sum-of-squares.
- **USS:** The seasonal values are defined by the uncorrected sum-of-squares.
- **Median:** The seasonal values are defined by the median of the values.

The following data set is taken from one of the course note manuals SAS publishes, called "Longitudinal Data Analysis with Discrete and Continuous Responses" to illustrate the following example from repeated measures modeling. The first diagram displays the initial time series, that is, repeated measures, data set. The second diagram displays the corresponding transposed data set. The time series data set is transposed in such a way that the rows of the original data set are now the columns of the transposed data set based on the **crossid** variable. The target variable is the response variable to predict in the repeated measures model. The following diagram displays the cross-sectional data set with the time identifier variable, that is, time, nested within each cross-sectional variable, that is, subject. Notice that the cross-sectional data structure is designed so that the values of the **crossid** grouping variable are unique within each **timeid** value.

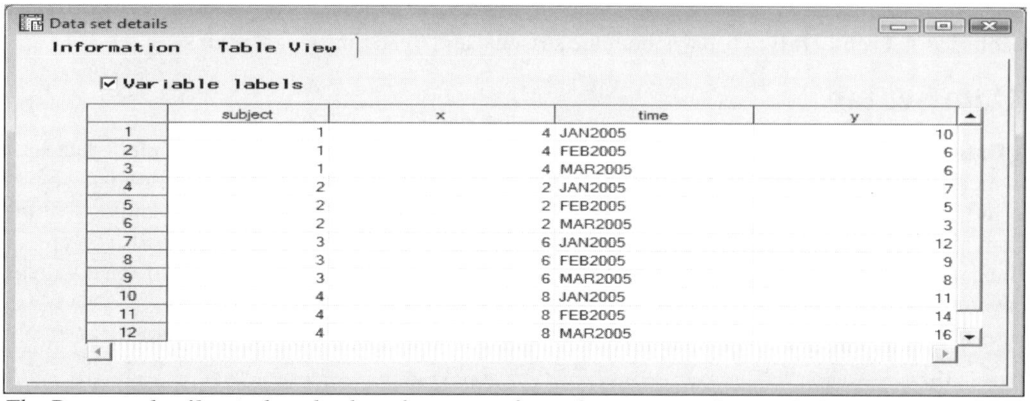

Data set details

Information Table View

☑ Variable labels

	subject	x	time	y
1	1	4	JAN2005	10
2	1	4	FEB2005	6
3	1	4	MAR2005	6
4	2	2	JAN2005	7
5	2	2	FEB2005	5
6	2	2	MAR2005	3
7	3	6	JAN2005	12
8	3	6	FEB2005	9
9	3	6	MAR2005	8
10	4	8	JAN2005	11
11	4	8	FEB2005	14
12	4	8	MAR2005	16

*The **Data set details** window displays the untransformed time series data with the two cross-sectional input variables, that is, subject and x, in preparation for repeated measures modeling.*

Data set details

Information Table View

☑ Variable labels

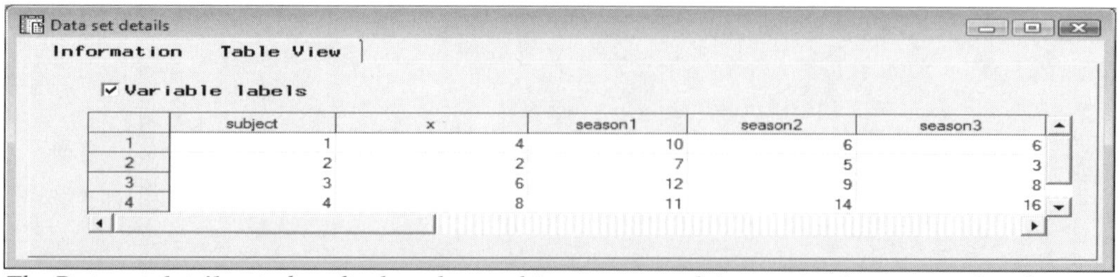

	subject	x	season1	season2	season3
1	1	4	10	6	6
2	2	2	7	5	3
3	3	6	12	9	8
4	4	8	11	14	16

*The **Data set details** window displays the resulting transposed time series data set used in the time series plot to view the relationship between the target variable or the input variables in the repeated measures model over time. Note that the number of repeated values is determined by the number of observations within each **crossid** variable.*

Viewing the Time Series Node Results

General Layout of the Results Browser within the Time Series Node

- **Model tab**
- **Code tab**
- **Log tab**
- **Output tab**
- **Notes tab**

The following data set is listed in a book published by SAS called "Forecasting Examples for Business and Economics Using the SAS System". The data set consists of monthly totals from a five-year period of U.S. lead production measured in tons, from January 1986 to September 1992, with 81 observations. The main idea of the time series modeling design is to apply the time series model to the data set in forecasting monthly U.S. lead production over time.

Model tab

The **Model** tab will allow you to view the file administrative information, the seasonal output data set, and the configuration settings for the time series model. The tab consists of the following four subtabs.

- **General subtab**
- **Variables subtab**
- **Settings subtab**
- **Status subtab**

General subtab

The **General** subtab displays the data description, creation date, and last modified date, and the target variable to the seasonal output data set.

Variables subtab

The **Variables** subtab displays a tabular view of the variables in the output time series data set. The data set consists of the target variable, time identifier variable, and the seasonal variables for each time series value that is accumulated. Note that all the columns are grayed-out and, therefore, cannot be changed.

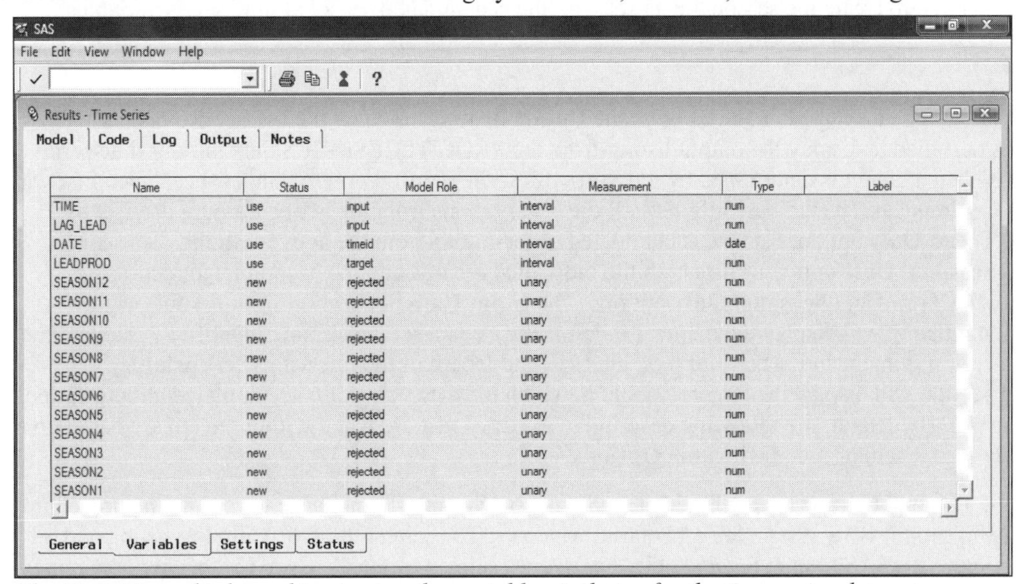

*The **Variables** subtab used in viewing the variable attributes for the time series data set.*

Settings subtab

The **Settings** subtab displays the configuration settings specified from the **Time Series** node.

Status subtab

The **Status** subtab displays the processing times by executing the **Time Series** node.

Code tab

The **Code** tab is designed to display the internal SEMMA training code that was executed behind the scenes that creates the following time series results and output data sets. The data mining time series results and output data sets are generated from the PROC TIMESERIES procedure. The following diagram displays the option settings used in the PROC TIMESERIES procedure from the various configuration settings that are specified within the **Time Series** node. The output data sets that are created from the procedure generate the various descriptive statistics from the training data set and the accumulated 12-month seasonal data set.

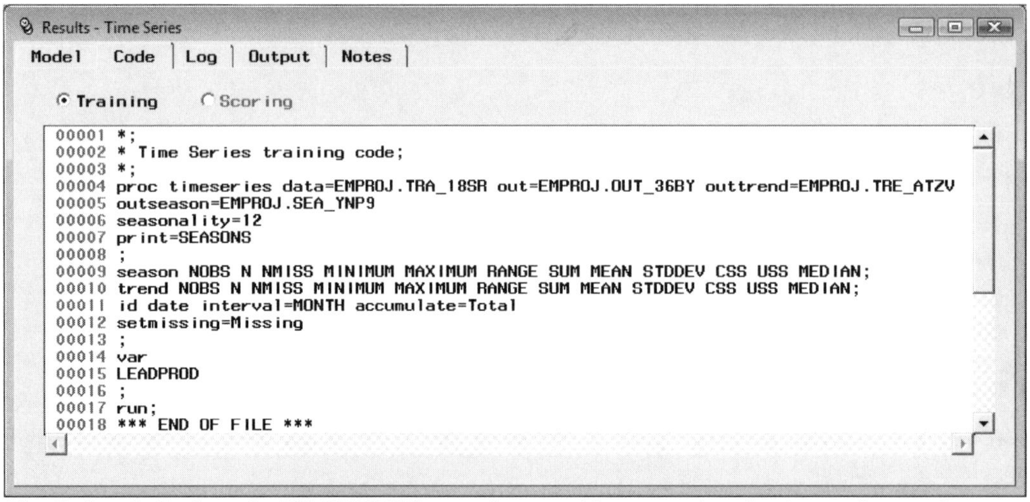

Results - Time Series

Model | Code | Log | Output | Notes

Training | Scoring

```
00001 *;
00002 * Time Series training code;
00003 *;
00004 proc timeseries data=EMPROJ.TRA_18SR out=EMPROJ.OUT_36BY outtrend=EMPROJ.TRE_ATZV
00005 outseason=EMPROJ.SEA_YNP9
00006 seasonality=12
00007 print=SEASONS
00008 ;
00009 season NOBS N NMISS MINIMUM MAXIMUM RANGE SUM MEAN STDDEV CSS USS MEDIAN;
00010 trend NOBS N NMISS MINIMUM MAXIMUM RANGE SUM MEAN STDDEV CSS USS MEDIAN;
00011 id date interval=MONTH accumulate=Total
00012 setmissing=Missing
00013 ;
00014 var
00015 LEADPROD
00016 ;
00017 run;
00018 *** END OF FILE ***
```

*The **Code** tab displays the TIMESERIES procedure used in generating the time series results.*

Log tab

The **Log** tab is designed to display the log listing of the previous training code along with an additional data step that is generated to create the seasonal data set with the accumulated seasonal time series variables.

Output tab

The **Output** tab will be displayed once you open the **Result Browser** to view the time series results. The **Output** tab is designed to display the output listing from the PROC TIMESERIES data mining time series procedure. The listed output is determined by the **Statistics to be printed** option that has been specified within the previous **Options** tab. The tab basically lists all the observations from the active training data set by selecting the **Trend Only** option, lists the accumulated observations from the active training data set by selecting the **Seasonal Only** option or lists both the individual records and the accumulated records by selecting both the **Trend** and **Seasonal Only** options. The **Input Data Set** section from the following procedure output listing will display the name of the time series data set, label, time identifier variable, time identifier interval, minimum and maximum time interval, and length of the interval. The **Variable Information** section will display the target variable in which the data points are accumulated at both endpoints of the time period. By default, the following descriptive statistics that are listed within the tab are based on the **Seasonal only** option setting specified within **Options** tab.

Note: Both the **Log** and **Output** tabs will be not available for viewing if you chose to reroute the procedure log and output listings to the SAS system log and output windows. To redirect the procedure log listing and output procedure listing to the SAS system log and output windows, select **Options** > **User Preferences** main menu options from the Enterprise Miner **Diagram Workspace**, then select the appropriate check boxes.

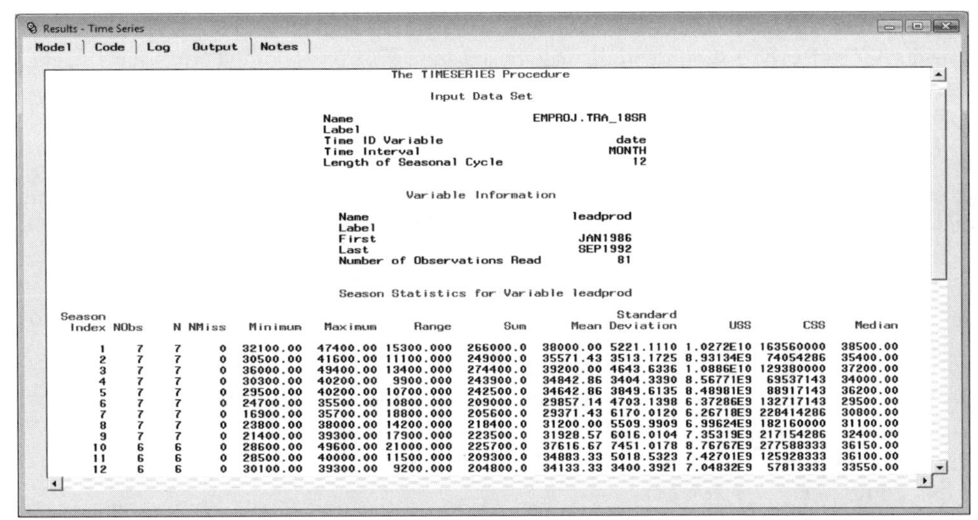

Results - Time Series

Model | Code | Log | Output | Notes

The TIMESERIES Procedure

Input Data Set

Name	EMPROJ.TRA_18SR
Label	
Time ID Variable	date
Time Interval	MONTH
Length of Seasonal Cycle	12

Variable Information

Name	leadprod
Label	
First	JAN1986
Last	SEP1992
Number of Observations Read	81

Season Statistics for Variable leadprod

Season Index	NObs	N	NMiss	Minimum	Maximum	Range	Sum	Mean	Standard Deviation	USS	CSS	Median
1	7	7	0	32100.00	47400.00	15300.000	266000.0	38000.00	5221.1110	1.0272E10	163560000	38500.00
2	7	7	0	30500.00	41600.00	11100.000	249000.0	35571.43	3513.1725	8.93134E9	74054286	35400.00
3	7	7	0	36000.00	49400.00	13400.000	274400.0	39200.00	4643.6336	1.0886E10	129380000	37200.00
4	7	7	0	30300.00	40200.00	9900.000	243900.0	34842.86	3404.3390	8.56771E9	69537143	34000.00
5	7	7	0	29500.00	40200.00	10700.000	242500.0	34642.86	3849.6135	8.48981E9	88917143	36200.00
6	7	7	0	24700.00	35500.00	10800.000	209000.0	29857.14	4703.1398	6.37286E9	132717143	29500.00
7	7	7	0	16900.00	35700.00	18800.000	205600.0	29371.43	6170.0120	6.26718E9	228414286	30800.00
8	7	7	0	23800.00	38000.00	14200.000	218400.0	31200.00	5509.9909	6.99624E9	182160000	31100.00
9	7	7	0	21400.00	39300.00	17900.000	223500.0	31928.57	6016.0104	7.35319E9	217154286	32400.00
10	6	6	0	28600.00	49600.00	21000.000	225700.0	37616.67	7451.0178	8.76767E9	277588333	36150.00
11	6	6	0	28500.00	40000.00	11500.000	209300.0	34883.33	5018.5323	7.42701E9	125928333	36100.00
12	6	6	0	30100.00	39300.00	9200.000	204800.0	34133.33	3400.3921	7.04832E9	57813333	33550.00

*The **Output** tab that displays the PROC TIMESERIES output listing of the descriptive statistics.*

Detailed Results

The **Detailed Results** window is designed to display the time series plot of the accumulated values over time.

Simply select the **Detailed Results** toolbar icon or select the **Tools > Detailed Results** main menu option that will result in the following **Detailed Results** window appearing to view the accumulated values over time.

The following time series plot is based on the transposed output data set that has been previously displayed. The time series plot is designed to display the accumulated data points by the cross-sectional variable in the input data set that will allow you to view the range in the values of the target variable over time within each subject, identified by the **crossid** variable. The time identifier variable is located on the horizontal axis and plotted against the accumulated data points of the target variable that is located on the vertical axis by each class level of the **crossid** variable in the repeated measures model. The plot indicates that there is a slight interaction in the recorded observations between each subject that is due to subject 4 over time.

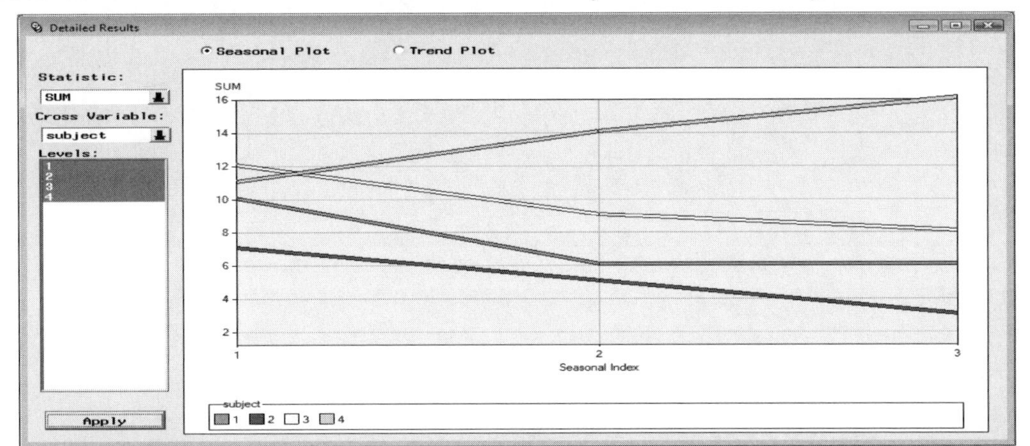

*The **Detailed Results** window to view the accumulated data points over time to either view the trend of the data point over time within each group variable for time series modeling or the change in the target values over time within each subject in the repeated measures model.*

By default, the seasonal time series plot is displayed. The seasonal plot displays the accumulated data points over time, that is, collapsed within each specified time interval. The trend plot will display the accumulated data points over the active training data set. Select the **Trend Plot** radio button to display the time series plot. The time series plot is based on the accumulated sum of each time interval. However, from the **Statistics** pull-down menu, you may display the time series plots by either the average values or the median values for each time interval. Summarizing the data points by the median value might be advantageous when there exists outliers in the time series data. By default, the time series plot displays the 12-month accumulated data points.

However, selecting the **Cross Variable** pull-down menu will allow you to select the cross-sectional variables in the time series data set. The **Levels** list box will display the separate class levels of the cross-sectional variable that can be selected from the **Cross Variable** pull-down menu. Select the desired class levels from the **Levels** list box, then press the **Apply** button to display the accumulated data points by each class level, as illustrated in the previous time series plot.

Since the time series data will usually consist of several data points. One big handicap of the trend plot is that it will not allow you to view every data point. In other words, the trend plot is not designed to display the accumulated data points over time as one overall snapshot. From the tab, select the **Scroll Data** toolbar icon that will result in the open hand appearing that will allow you to shift the axis of the graph from left to right or right to left across the entire time interval in the time series plot. Select the **Viewport** toolbar icon and drag the mouse over a particular time interval of interest to display the time series graph within the selected time interval. The **Reset** toolbar icon will allow you to reset the changes that have been performed on the time series plot.

The following seasonal and trend plots will allow you to verify an extremely important time series property called stationarity. Stationarity in the time series data is one of the first assumptions to check for in time series modeling. In time series analysis, it is important to first visually check for outliers, missing data points, and stationarity from the following trend plot before fitting the time series model. The time series data is considered stationary when the target values display a uniform fluctuating behavior with any increasing or decreasing trend over time. The following time series plot displays nonstationarity in the data over time in the first few time series data points. Therefore, an appropriate transformation might be suggested. However, differencing is typically applied to remedy nonstationarity in the time series data.

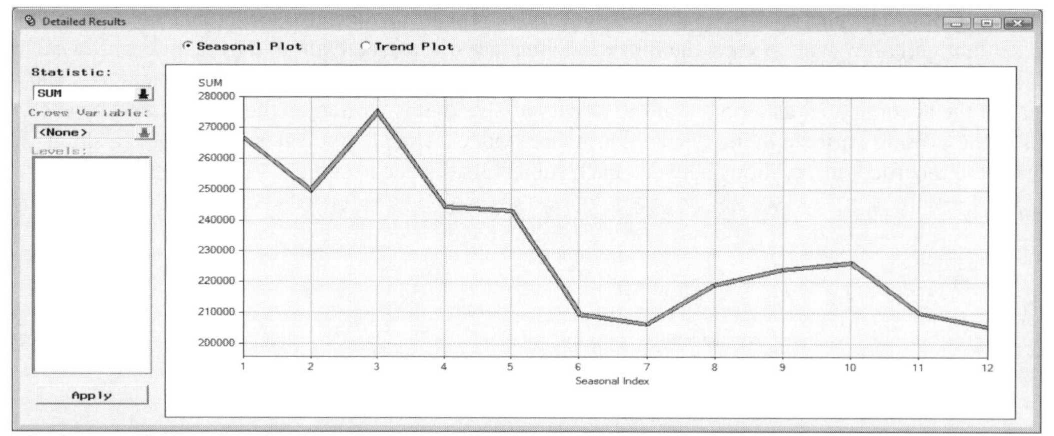

*The **Seasonal Plot** that displays the accumulated data points within each specified time period.*

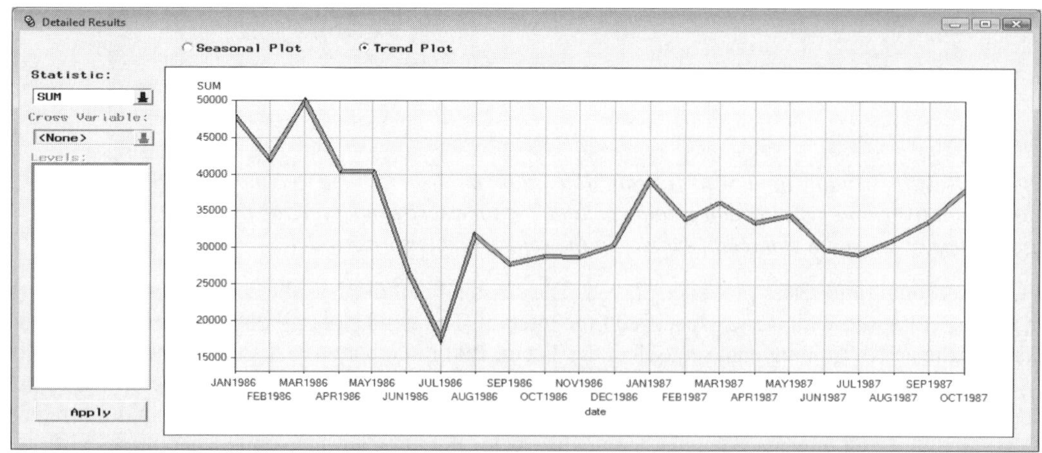

*The **Tread Plot** that displays the initial training data set to verify stationarity over time.*

3.8 Interactive Grouping Node

General Layout of the Enterprise Miner Interactive Grouping Node

- **Data tab**
- **Variables tab**
- **Output tab**
- **Notes tab**

The purpose of the **Interactive Grouping** node in Enterprise Miner is to automatically create group variables for the data mining analysis. The node is designed to transform integer-valued variables into categorically-valued variables or redefine the input variables into entirely different groups. Once the node is executed, Enterprise Miner will automatically create temporary data sets based on the final grouping results from each input variable in the analysis. These same data sets and the listed input variables may then be used as input variables in subsequent classification modeling designs such as fitting scorecard models. One of the purposes of the interactive grouping node is that, at times, grouping variables might lead to better classification modeling performance. In decision tree modeling, grouping the input variables might be considered in order to accelerate the exhaustive split search procedure and improve the stability and the interpretation of the decision tree. In addition, one remedy for removing extreme values from these same interval-valued variables in the classification model is to transform them into categorical variables to achieve normality.

The one requirement of the node is that a unique binary-valued target variable must exist in order to perform interactive grouping that is usually defined from the **Input Data Source** node. The variable groupings are created by the various option settings that are specified within the node. However, interactive grouping can also be performed by manually redefining the groupings that have been created from automatic grouping based on the various option settings in order to optimize binning and grouping.

The various groupings that are automatically created are determined by certain option settings. The interactive grouping settings window will allow you to specify certain requirements in automatically grouping the variables into separate classes, such as an adjustment factor for the WOE measurement, minimum group size, automatic grouping, the amount of precision in assigning the interval variables into groups, and preventing automatic grouping from being performed. The interactive grouping results will provide you with a table view of the summarized information. This will allow you to manually modify the variable groupings, that is, modify the variable groupings based on the range of the interval-valued input variables or certain discrete values based on the categorically-valued input variables. An output data set is created that can be used in subsequent nodes for modeling purposes or group processing analysis.

The groupings are defined by certain cutoff scores that are typically used in scorecard analysis, called the *WOE weight of evidence statistic* based on a binary-valued target variable. The WOE statistic measures the relative risk of input grouping variable that is the logarithm difference of the response rate calculated by the difference between the relative frequency of the target nonevent and the target event in the respective variable group that is calculated as follows:

$$\text{WOE}_{attribute} = \ln\frac{P_{nonevent\ attribute}}{P_{event\ attribute}} \quad \text{where } P_{event\ attribute} = \frac{n_{event\ attribute}}{N_{event}} \quad \text{and } P_{nonevent\ attribute} = \frac{n_{nonevent\ attribute}}{N_{nonevent}}$$

where $n_{event\ attribute}$ and $n_{nonevent\ attribute}$ are the total number of events and nonevents of the attribute and N_{event} and $N_{nonevent}$ are the total number of events and nonevents in the data.

The **Information Value** and the **Gini** statistic are used to measure the predictive power of the input variable to distinguish between the separate classes of the target event and the target nonevent. The *information value statistic* calculates the weighted difference between the proportion of the target nonevents and the proportion of the target events with respect to each class level, and is created as follows:

$$\text{Information Value} = \Sigma\,(P_{nonevent\ attribute} - P_{event\ attribute}) \cdot \text{WOE}_{attribute}$$

Therefore, an information value statistic that is greater than zero will indicate that the proportion of the target nonevent is greater than the target event. Conversely, an information value statistic less than zero will indicate that the proportion of the target event is greater than the target nonevent.

The *Gini statistic* is a measure of variability for categorical data. A Gini value of zero will indicate to you that the grouping input variable consists of all the target events. Conversely, a Gini value of one will indicate to you that the target events and nonevents are evenly distributed within the grouping input variable. The Gini statistic is calculated by first sorting the proportion of target events in descending order by each input variable, where the first group has the highest proportion of events. The next step is calculating the number of target events, $n_{event\ attribute}$, and the number of target nonevents, $n_{nonevent\ attribute}$, for each sorted group. Therefore, the Gini statistic is calculated in the ith group as follows:

$$\text{Gini} = 1 - \frac{2 \cdot \sum_{i=2}^{m} \left(n_{event\ attribute} + \sum_{i=1}^{i-1} n_{nonevent\ attribute}\right) + \sum_{i=1}^{m} \left(n_{event\ attribute} \cdot n_{event\ attribute}\right)}{N_{event} \cdot N_{nonevent}} \quad \text{for } 1,\ldots,m \text{ groups}$$

The **Interactive Grouping** node is designed to select the input grouping variables that best characterizes the separate target groups based on the information value or the Gini score statistic, then grouping the selected input variables based on the WOE statistic to capture the attributes representing risk rating trends from the two separate class levels of the binary-valued target variable. The WOE statistic is very similar to the log odds-ratio statistic in logistic regression modeling. Therefore, the various modeling assumptions must be satisfied in the input variables in the analysis for the procedure to provide good interactive grouping results.

Credit Scoring System

The purpose of a credit scoring system is designed to measure such things as determining the creditability of various customers applying for a certain loan. It is basically a scoring system that is used as a decision mechanism in evaluating the reliability of good or bad customers. Therefore, a certain cutoff point or threshold value can be selected in determining the creditability of a certain customer as a good or bad customer. The following table displays a certain credit scoring system from two separate characteristics or attributes, that is, the amount of the loan and number of customers defaulting on the same loan, which is used in determining the creditability of a certain customer based on a credit score to measure the amount of risk for a certain loan. For example in the following table, suppose that the amount of the loan that is lent to a customer is more than $20,000. This same customer defaulted on a certain loan two separate times. This will result in a credit score of –40 or (10 + (– 50)), that is, the individual scores are summarized by each of the attributes, which results in one overall score.

Amount of the Loan	**Score**
Less Than $10,000	50
Between $10,000 and $20,000	20
More Than $20,000	10
Number of Times Defaulting on a Loan	**Score**
More Than 2	-100
2	-50
1	-25
0	25

Therefore, the Information Value or Gini statistic can be used to measure the predictive power of each input variable in the model, that is, the amount of the loan and the number of times defaulting on a loan, in identifying customers defaulting on their loan. The various WOE statistics can be used as input variables to subsequent classification modeling for such things as estimating the probability of the loan not being repaid. Therefore, this will allow the creditor to evaluate certain threshold values from the range of estimated probabilities in establishing a certain cutoff probability in distinguishing good creditors from bad creditors.

The following home equity loan HMEQ data set was selected for interactive grouping. The binary-valued variable BAD (defaulting on the home loan) was selected as the binary-valued target variable in distinguishing between the various groups that were automatically created.

Data tab

The **Data** tab is designed for you to the select the active training data set. The tab displays the name and description of the previously created training data set that is connected to the node. Select the **Properties...** button to either view the file administrative information of the active training data set or browse the table view listing of the active training data set. Press the **Select...** button to open the **Imports Map** window that will allow you to select from a list of the currently available training data sets that are connected to the **Interactive Grouping** node. For instance, the process flow diagram might have created several previously created data sets that have the same model role from two separate training data sets that are connected to the node. Also, the scored data set may be selected to create entirely different groupings by fitting the model to a new set of values with the same data structure as the input data set. The tab has an appearance similar to the other tabs. Therefore, the tab will not be displayed.

Variables tab

The **Variables** tab will appear as you open the **Interactive Grouping** node. The **Variables** tab is designed to display a tabular view of the variables in the active training data set. The tab displays the name, status, model role, level of measurement, type format, and label for each variable in the active training data set. The tab will allow you to prevent certain variables from being automatically grouped into separate classes from the **Status** column. By default, all variables are included in the analysis with a variable status that is automatically set to **use**. However, you can remove certain variables from the analysis by simply selecting the variable rows and scrolling over to the **Status** column to set the variable status to **don't use**. The tab will allow you to view the distribution of a selected variable from the metadata sample by selecting the **View Distribution <variable name>** option. If the input variables display an approximately normal distribution, then it will result in the various groups that are created to be approximately evenly distributed. However, if the distribution of the selected variable is not normally distributed, then certain transformations might be recommended since outlying data points will result in the interactive grouping results creating their own separate groupings.

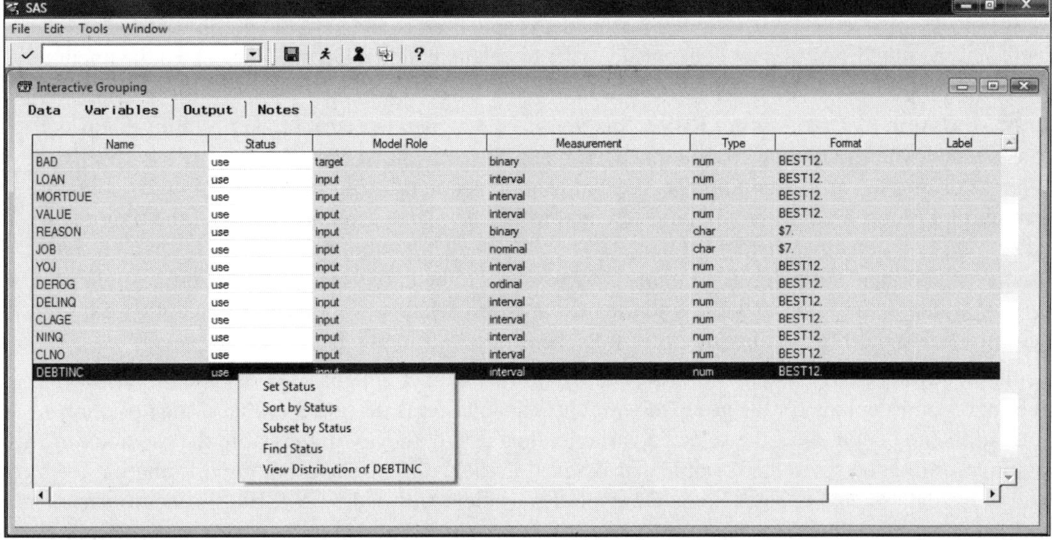

*The **Variables** tab is used to remove certain input variables from the analysis and view the distribution of each variable in the active training data set.*

Output tab

The **Output** tab is designed for you to view the scored data set with the interactive grouping results such as the assigned class levels and the corresponding WOE statistics by each input variable in the active training data set. The tab will allow you to browse the selected data set that will be passed along to the subsequent nodes within the process flow diagram. The **Output** tab will be unavailable for viewing along with the corresponding output data sets until the node is executed. Select the **Properties...** button to either view the file information or view the table listing of the selected output data set. The output data set created from interactive grouping can be passed along to the subsequent classification modeling nodes or group processing analysis.

Interactive Grouping Settings

The node is designed to specify the way in which automatic grouping is performed by the following option settings that can be specified from the **Interactive Grouping Settings** window in defining the way in which the variables are grouped into separate classes. Select the **Tools** > **Settings** main menu option or select the **Settings** toolbar icon that will open the following **IGN Settings** window. The window will allow you to specify the various settings for interactive grouping.

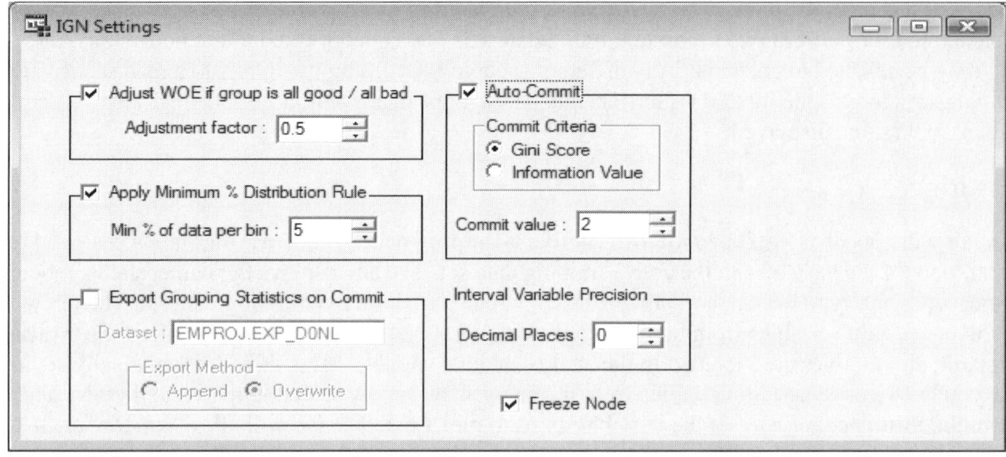

*The **IGN Settings** window is used to specify the various option settings for interactive grouping.*

- **Adjust WOE if group is all good / all bad:** This option applies an adjustment to the WOE statistic for each group in which all observations have the same target value. By default, the check box is selected, indicating that the adjustment factor is automatically applied. By default, the adjustment factor is automatically set at .5. From the spin box, you may change the adjustment factor from as low as zero or up to ten. The adjustment factor that is specified is added to the number of target events and the number of target nonevents in order to calculate the WOE statistic. The adjustment to the target event is calculated as follows: $P_{event\ attribute} = (n_{event\ attribute} + adjustment\ factor)\ /\ N_{event}$
- **Apply Minimum % Distribution Rule:** This option is designed to restrict the minimum group size that is created when automatic grouping is performed. Enter a value or use the spin box to specify the minimum proportion of the training data set that is allocated in creating a specific group. By default, the minimum group size of a group that is created from interactive grouping is 5% of the training data set. This suggests that there must be at least 5% of the data within the group to be created. For nominal-valued input variables, if certain a category fails to meet the specified total proportion, then it will be collapsed into the category with the next smallest relative proportion. If the collapsed categories fail to meet the minimum group size, then further grouping is performed.
- **Export Grouping Statistics on Commit:** This option will either append or overwrite the saved information of groupings for the grouped variable. The data set is designed so that if the grouping information can be found for the selected variable, then it will allow you to restore the previous grouping information set to the variable that is saved in the data set. The grouping information is saved in the data set that is listed in the **Dataset** display field within the EMPROJ project folder.
- **Auto-Commit:** Selecting this check box will finalize the automatic grouping results. Selecting this option will result in the automatic grouping procedure choosing the most appropriate groups to create according to the values of the variable. The automatic grouping procedure categorizes the variables from the **Gini Score** or the **Information Value** statistic that measures the predictive power of each variable. The results from automatic grouping are finalized by the **Gini Score** or the **Information Value** statistic greater than the **Commit value** specified from the entry field.
- **Interval Variable Precision:** This option will allow you to define the precision of the values of the interval-valued variables that are grouped. The default value is zero. Therefore, this will result in the interval-valued variable being grouped by integer values. In Enterprise Miner, the maximum amount of precision in categorizing the interval-valued variable by its own values is up to 10 decimal places.
- **Freeze Node:** This option prevents automatic grouping from being performed.

Performing Interactive Grouping

General Layout for Interactive Grouping

- **Chart tab**
- **Statistics tab**
- **Coarse Detail tab**
- **Fine Detail tab**
- **Output Groupings tab**

The purpose of performing interactive grouping is to interactively create the various groupings for each input variable through the use of the following tabs. From the process flow diagram, select the **Interactive Grouping** node, then right-click the mouse and select the **Interactive** pop-up menu item. The following results that are generated from interactive grouping are defined by the various options specified from the previous **Settings** window. The following windows are designed in such a way that they depend on the level of measurement of the grouped variables either being binary, interval, nominal or ordinal. For example, the following windows will allow you to view the grouping performance by every interval-valued input variable in the active training data set with the window name of **Interactive Grouping - Interval Variable**. The windows are designed to view the grouping results by selecting one input variable at a time. The following tabs are interconnected to each other in such a way such that updates performed within the **Chart** tab will be reflected in the other tabs.

Chart tab

Grouping the Interval-Valued Input Variables

The **Charts** tab is designed to display various charts of the selected input variable. For interval-valued input variables, the upper right-hand corner of the tab will display the standard frequency bar chart, allowing you to view the range of values that are automatically grouped into separate intervals that are identified by the vertical reference lines superimposed on to the chart. From the chart, twelve separate reference lines are displayed, indicating to you that thirteen separate groups were created. The vertical lines can be modified interactively in redefining the separate intervals of the transformed categorical variable. Simply reposition the vertical line by selecting the slider bars and dragging the reference line to a different location along the horizontal axis. The stack bar chart that is positioned in the lower left-hand corner of the tab displays the proportion of the target values for each group that is created from the selected interval-valued input variable. The line plot positioned to the right of the stacked bar chart displays the WOE statistic or the response rate for each group. By default, the WOE statistic is automatically displayed. However, select the **View > Statistics Plot > Response rate** main menu options or right-click the mouse to select the **Response Rate** pull-down menu option within the plot that will then display the line plot of the event rate. The line plot displays the event rate, that is, the proportional rate, between the two separate target groups across each group of the selected interval-valued input variable.

From the frequency bar chart, the value of the grouped loans displays a skewed distribution due to some of the larger loans. In addition, the WOE line plot indicates that the overall proportion of these same customers who have received large loans tend to default on their loan with the WOE statistic below zero. The stacked bar chart indicated that there is an abnormally high proportion of customers defaulting on a loan in the first group with the amount of the loan of up to $6,050. The WOE statistic is well below zero in the first group, indicating to you that the proportions of clients defaulting on their loan (13.2%) exceeds the proportion of clients paying their loan (3.7%). Further analysis might be required to determine the various characteristics as to the reason why these same clients fail to pay their small home loans. From the stacked bar chart, the number of good customers far exceeds the number of bad customers in nearly every group except the first group. In addition, the eighth group displays the largest bar, indicating to you that this group that was created has the largest number of customers who were granted a loan. Conversely, the ninth and the eleventh groups are the most reliable customers paying their home loan since both these groups display a large proportion of customers repaying their home loan as determined from the line plot of the WOE statistic. WOE values greater than zero will indicate the more reliable customers who paid off their home loan. This same conclusion can be determined from the stacked frequency bars based on the proportionally between the binary target levels.

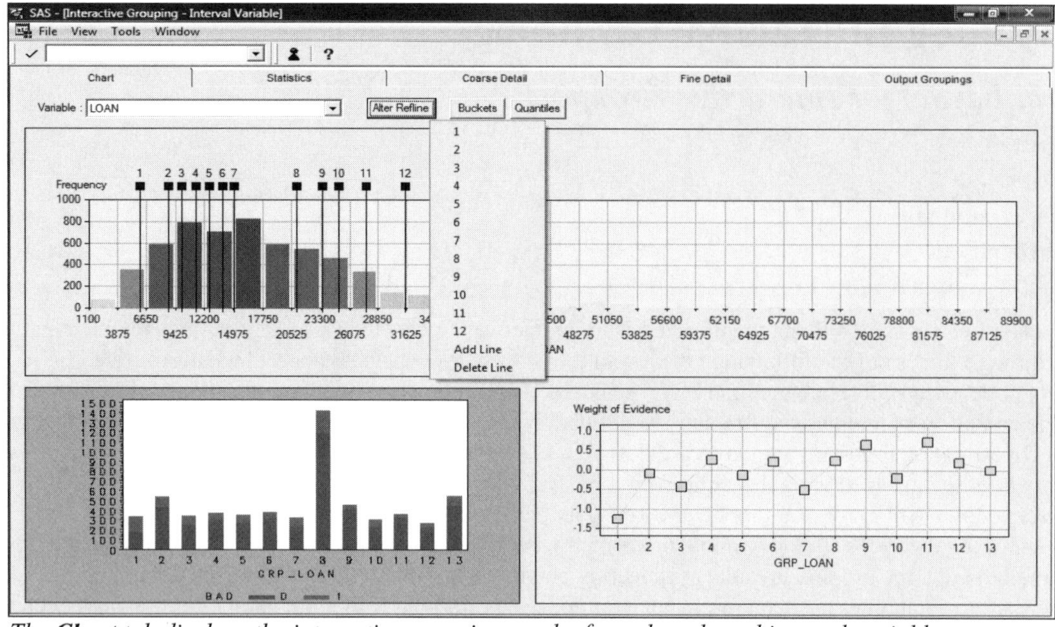

*The **Chart** tab displays the interactive grouping results from the selected interval variable.*

The **Alter Refline** button is designed to redefine the value set to the reference line from the frequency bar chart. Simply select the **Alter Refline** button or the **Tools > Grouping > Alter refine** main menu options that will allow you to redefine the value set to the vertical reference line of the chart. Selecting anyone of the listed group ids will result in the **Update Reference Line** dialog window appearing, which will allow you to redefine the upper bound to the selected interval. However, the upper bound that is entered from the **New Value** entry field must be no greater than the upper bound of the neighboring interval. To add an additional reference line to the chart, select the **Alter Refline** button and the **Add Line** pull-down menu item that will then create an additional class level for the selected variable. Conversely, to delete a reference line from the chart in order to remove a class level from the selected variable, select the **Alter Refline** button and the **Delete Line** pull-down menu item.

The **Buckets** or the **Quantiles** buttons are designed to redefine the class levels of the selected variable. Selecting the **Quantiles** button or selecting the **Tools > Grouping > Quantile binning** main menu options will allow you to interactively redefine the class levels of the selected variable into ten uniform groups: 10th, 20th up to the 100th percentile. Selecting the **Buckets** option or selecting the **Tools > Grouping > Bucket binning** main menu options will allow you to interactively redefine the class levels of the selected variable based on the difference between the minimum and maximum values. The difference between the separate grouping techniques is that quantiles grouping defines each group by the same number of observations as opposed to the bucket groupings that defines each group usually by an unequal number of observations within each group or bucket. The **Number of bins** window will appear, enabling you to specify the number of buckets or quantiles to create as shown below:

Number of bins

Number of bins : .

OK Cancel

*The **Number of bins** window is used to specify the number of bins for the variable.*

Grouping the Categorically-Valued Input Variables

For categorically-valued variables, the upper portion of the tab will display a frequency bar chart that will allow you to view the number of observations that fall into the separate groups that are automatically created. The **Allocate to Bin** button will allow you to redefine the groups that are created from the selected categorical variable. At times, combining groups might be advantageous if there are an unequal number of observations within groups. Select any one of the displayed bars to combine the values of the selected class level into the corresponding class level selected from the **Allocate to Bin** option settings. To create new interactive groupings, press the **Allocate to Bin** button and select the group to which you want to move the values of the selected class level of the corresponding frequency bar. Alternatively, select any number of bars by simultaneously pressing the Ctrl button, then selecting the **New Bin** option setting that will automatically combine all the selected categories into one unique category.

From the following results, the most reliable customers paying off their loan were customers with an unknown job occupation, that is, the missing class level, who had a low event rate of 8.24% along with a WOE statistic well above zero, that is, 1.02. Conversely, both sales representatives and self-employed customers, who were automatically collapsed into one group (group 1) were the least reliable customers, with a WOE statistic well below zero, i.e. –.626 with an event rate of 31.79%, that is, 96 out of 302 total customers defaulted on their home loan.

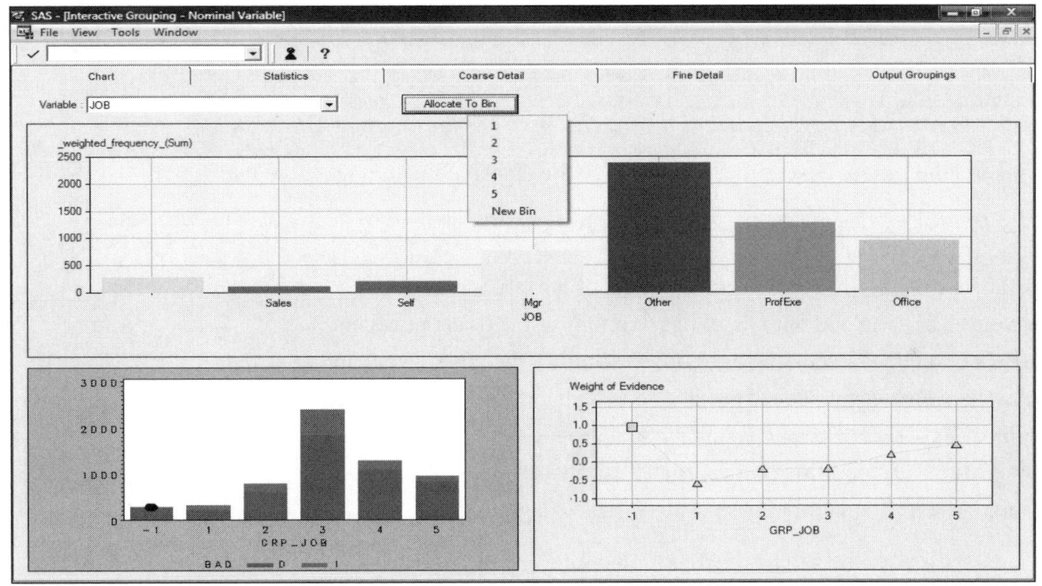

*The **Chart** tab that displays the interactive grouping charts from the selected categorical variable.*

Statistics tab

The **Statistics** tab is designed to display a table view of various descriptive statistics of each class level of the selected variable. The tab displays the summarized number of occurrences of the target event and the target nonevent along with the WOE and the Gini score statistic for each group that is created. Conversely, the summarized statistics positioned to the right of the table display the overall predictive power of the selected grouping input variable for all the groups that are automatically created. The listed chi-square statistic tests the equality in the marginal proportions between the various class levels of the selected grouping variable and the two separate class levels of the binary-valued target variable. The following statistics are listed in the EXPORT GROUPING STATISTICS data set for each input variable that was a part of interactive grouping.

Pressing the **Commit Grouping** button will result in finalizing the interactive grouping results for the selected variable that will be written to the corresponding output data set.

The following table listing indicates that the first group consists of all customers, with the amount of their home loan up to $6,050, has the highest probability in defaulting on their home loan since this group displays the largest negative weight in evidence statistic that is well below zero. This indicates that the proportion of clients defaulting on their home loan significantly outweighs clients paying their home loan. Conversely, both the ninth and the eleventh groups are the most reliable customers with the smallest proportion of the target event, that is, defaulting on their home loan (10.99 and 10.28, respectively). The overall information value statistic of .2485 indicates that customers who pay their home loan outweigh customers that do not pay their loan with regard to the amount of the loan.

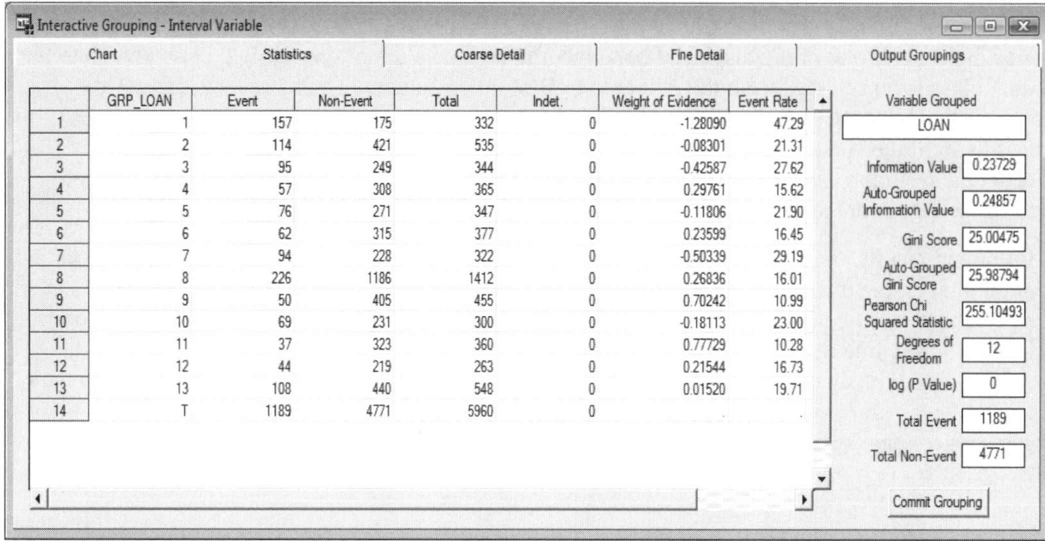

	GRP_LOAN	Event	Non-Event	Total	Indet.	Weight of Evidence	Event Rate
1	1	157	175	332	0	-1.28090	47.29
2	2	114	421	535	0	-0.08301	21.31
3	3	95	249	344	0	-0.42587	27.62
4	4	57	308	365	0	0.29761	15.62
5	5	76	271	347	0	-0.11806	21.90
6	6	62	315	377	0	0.23599	16.45
7	7	94	228	322	0	-0.50339	29.19
8	8	226	1186	1412	0	0.26836	16.01
9	9	50	405	455	0	0.70242	10.99
10	10	69	231	300	0	-0.18113	23.00
11	11	37	323	360	0	0.77729	10.28
12	12	44	219	263	0	0.21544	16.73
13	13	108	440	548	0	0.01520	19.71
14	T	1189	4771	5960	0	.	.

*The **Statistics** tab that displays the various descriptive statistics for the selected variable.*

The following are the various descriptive statistics that are displayed within the table listing:

- **GRP_<*Input Variable Name*>:** The group identifier variable used to distinguish between the separate groups that are created.
- **Event:** The weighted number of observations of the target event.
- **Nonevent:** The weighted number of observations of the target nonevent.
- **Total:** The weighted total number of observations of the target event and nonevent.
- **Indet:** The weighted number of observations with a missing target values in a group.
- **Weight of Evidence:** The WOE statistic of a group.
- **Event Rate:** The proportion of the target event in a group.
- **Missing Values:** The number of missing values in a group.

The following list displays the summarized grouping information that measures the predictive power of the selected variable. The following statistics are located on the right-hand portion of the tab:

- **Variable Grouped:** The listing of the selected variable.
- **Information Value:** The information value statistic of the selected variable.
- **Auto-Grouped Information Value:** The information value statistic of the selected variable based on automatic grouping that was performed.
- **Gini Score:** The Gini Score statistic of the selected variable.
- **Auto-Grouped Gini Score:** The Gini score statistic of the selected variable based on automatic grouping that was performed.
- **Pearson Chi-Squared Statistic:** This statistic is defined by the $k \times 2$ contingency table that tests for equal proportions between the two separate groups that are created by the binary-valued target variable.
- **Degrees of Freedom:** The degrees of freedom of the Pearson chi-square statistic that is essentially the number of groups that are created minus one.
- **log(P-Value):** The log of the Pearson chi-square statistic.
- **Total Events:** The total number of observations of the target event.
- **Total Nonevents:** The total number of observations of the target nonevent.

Course Detail tab

The **Course Detail** tab displays a table view of the frequency counts of the target event and the target nonevent along with the WOE log-odds statistic and the response rate across the separate intervals or class levels that have been interactively created from the selected input variable. For nominal-valued input variables, the tab will display the various groups that have been collapsed, which failed to meet the minimum group size specified from the previous **IGN Settings** window. The tab will allow you to interactively create new groupings by manually creating additional groups or combining separate groups that are created. For interval-valued or ordinal-valued variables, only neighboring class levels can be merged. The tab has an appearance similar to the previous **Statistics** tab, with the exception of an additional **Group Label** column that will allow you to view the range of values that are defined within each group of the selected variable.

From the table, select the group to be merged, then right-click the mouse and select the **Merge** option setting to merge the two separate groups that are next to each other. This is illustrated in the following diagram. Conversely, to create two separate groups by splitting the interval into two different groups, select the listed row that you want to divide into two separate groups, then right-click the mouse and select the **Split** option setting. The **Split Interval Groupings** window will appear. From the **Split Interval Groupings** window, enter the value to split the range of values of the selected interval from the **Split Value** entry field. The value entered in the entry field must fall between the listed lower bound and upper bound. Otherwise, a dialog box will appear to inform you that the value entered must lie within the range of the listed lower and upper bounds. The following table listing is written to an output data set called COARSESTATSTAB that is stored in the temporary SAS work directory.

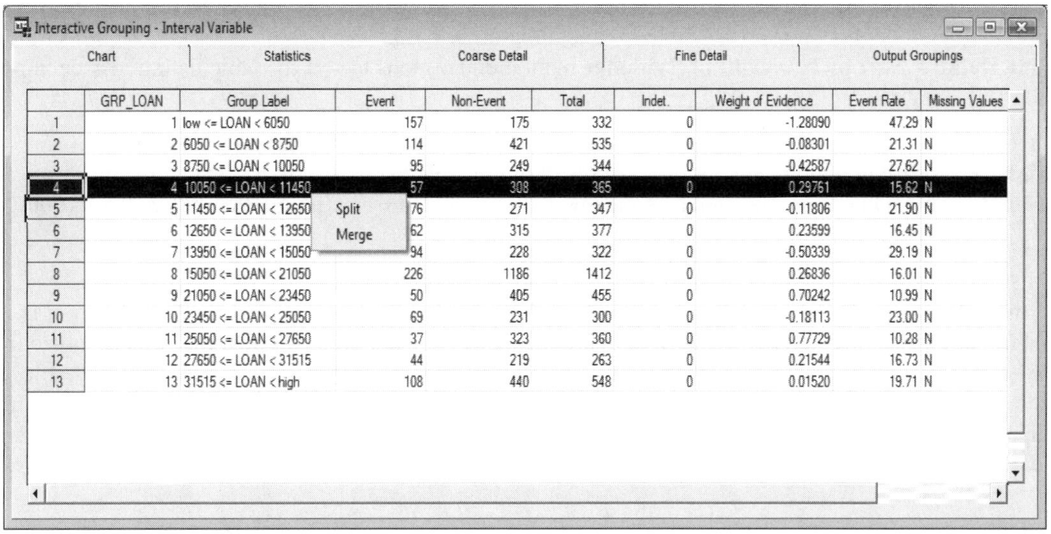

Interactive Grouping - Interval Variable

Chart | Statistics | Coarse Detail | Fine Detail | Output Groupings

	GRP_LOAN	Group Label	Event	Non-Event	Total	Indet.	Weight of Evidence	Event Rate	Missing Values
1	1	low <= LOAN < 6050	157	175	332	0	-1.28090	47.29	N
2	2	6050 <= LOAN < 8750	114	421	535	0	-0.08301	21.31	N
3	3	8750 <= LOAN < 10050	95	249	344	0	-0.42587	27.62	N
4	4	10050 <= LOAN < 11450	57	308	365	0	0.29761	15.62	N
5	5	11450 <= LOAN < 12650	76	271	347	0	-0.11806	21.90	N
6	6	12650 <= LOAN < 13950	62	315	377	0	0.23599	16.45	N
7	7	13950 <= LOAN < 15050	94	228	322	0	-0.50339	29.19	N
8	8	15050 <= LOAN < 21050	226	1186	1412	0	0.26836	16.01	N
9	9	21050 <= LOAN < 23450	50	405	455	0	0.70242	10.99	N
10	10	23450 <= LOAN < 25050	69	231	300	0	-0.18113	23.00	N
11	11	25050 <= LOAN < 27650	37	323	360	0	0.77729	10.28	N
12	12	27650 <= LOAN < 31515	44	219	263	0	0.21544	16.73	N
13	13	31515 <= LOAN < high	108	440	548	0	0.01520	19.71	N

Split
Merge

*The **Course Detail** tab is used to collapse or split the range of values of the interval-valued variable.*

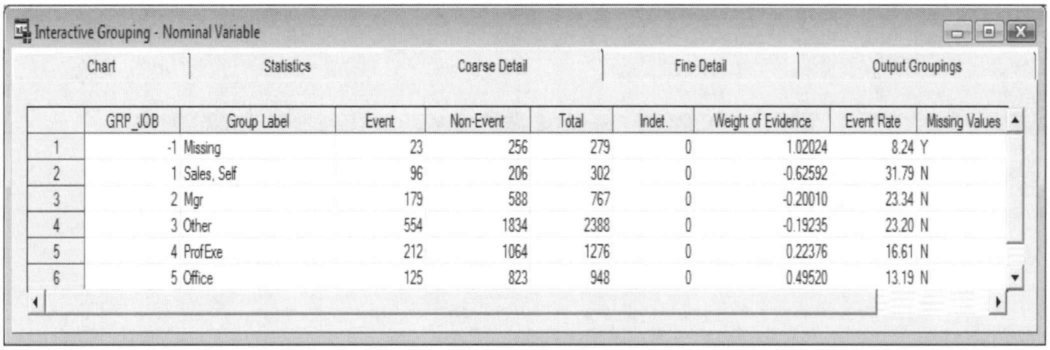

Interactive Grouping - Nominal Variable

Chart | Statistics | Coarse Detail | Fine Detail | Output Groupings

	GRP_JOB	Group Label	Event	Non-Event	Total	Indet.	Weight of Evidence	Event Rate	Missing Values
1	-1	Missing	23	256	279	0	1.02024	8.24	Y
2	1	Sales, Self	96	206	302	0	-0.62592	31.79	N
3	2	Mgr	179	588	767	0	-0.20010	23.34	N
4	3	Other	554	1834	2388	0	-0.19235	23.20	N
5	4	ProfExe	212	1064	1276	0	0.22376	16.61	N
6	5	Office	125	823	948	0	0.49520	13.19	N

*The **Course Detail** tab with the collapsed class levels from the categorically-valued input variables.*

Fine Detail tab

The **Fine Detail** tab is designed to display a table view of each value within each group that is created by automatic grouping. The tab has the similar appearance to the previous **Statistics** tab, with exception to an additional column that displays a detailed listing of each value of the selected variable. The tab displays the number of occurrences of the target event and the target nonevent with the WOE statistic and the proportion of events for each value of the selected variable. The table listing will provide you with an even more detailed calculation of the response rate in the binary-valued target variable across each distinct value of the selected input variable. In addition, the tab will allow you to interactively create new cutoff points and create entirely different groupings by manually moving values from one group to other. Select the table row, then right-click the value in the table to move the value from one group to another by selecting the interval or group to which you want to move the value. This is illustrated in the following diagram. For interval-valued or ordinal-valued input variables, the tab is designed so that the value that is selected can only be moved between neighboring groups, and all values between the selected value and the neighboring group will also be moved to the new group. For example, the amount of the loan in the first group range from $1,100 to $6,000, therefore, redefining the amount of the loan at $2,000 that is currently assigned to the first group into the second group will result in the rest the other values that are greater than $2,000 that are currently assigned in the first group to the second group. This will redefine the amount of the loan in the second group from $2,000 to $6,000. Alternatively, by highlighting several rows or values and selecting the **New Bin** option setting will result in an entirely new interval or class level being created next to the currently selected interval or class level.

By default, the table listing is sorted in ascending order by the grouping variable, that is, the **GRP_<variable name>** column. However for nominal-valued grouping variables, select the **View > Fine Detail > Sort by > Event Rate** main menu options to sort the table in ascending order by the event rate. Alternatively, select the **View > Fine Detail > Sort by > Weight of Evidence** main menu options to sort the table listing in ascending order by the WOE statistic.

By default, the tab will automatically display 100 rows. However, selecting the **View > Fine Detail > Observations > Unlimited** main menu options will result in the table listing every discrete value of the selected input variable. Otherwise, the menu settings will allow you to display up to 100, 200, 500, or 1000 rows.

The table listing is written to a temporary data set in the SAS work directory called FINESTATSTAB. Although, the data set is collapsed by both class levels of the binary-valued target variable, the file layout is somewhat similar to the scored data set. The SAS data set consists of each input variable that was involved in interactive grouping, the group identifier variable, and the WOE statistic, along with all other input variables in the analysis data set. Both the group identifier variable and the WOE values can be used as input variables in subsequent classification modeling designs, where the subsequent classification model can then be designed to establish a certain threshold probability in identifying good and bad creditors from the estimated probabilities.

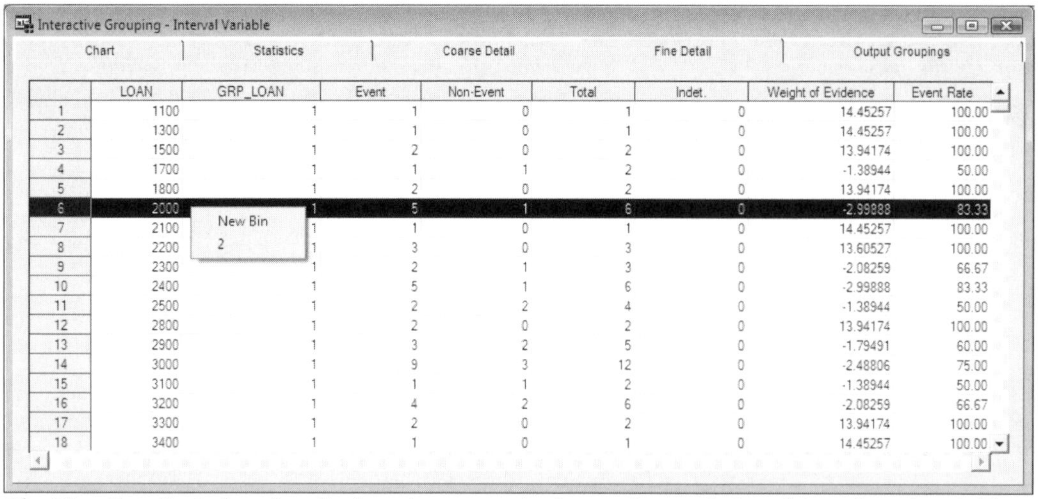

Interactive Grouping - Interval Variable

Chart | Statistics | Coarse Detail | Fine Detail | Output Groupings

	LOAN	GRP_LOAN	Event	Non-Event	Total	Indet.	Weight of Evidence	Event Rate
1	1100	1	1	0	1	0	14.45257	100.00
2	1300	1	1	0	1	0	14.45257	100.00
3	1500	1	2	0	2	0	13.94174	100.00
4	1700	1	1	1	2	0	-1.38944	50.00
5	1800	1	2	0	2	0	13.94174	100.00
6	2000	1	5	1	6	0	-2.99888	83.33
7	2100	1	1	0	1	0	14.45257	100.00
8	2200	1	3	0	3	0	13.60527	100.00
9	2300	1	2	1	3	0	-2.08259	66.67
10	2400	1	5	1	6	0	-2.99888	83.33
11	2500	1	2	2	4	0	-1.38944	50.00
12	2800	1	2	0	2	0	13.94174	100.00
13	2900	1	3	2	5	0	-1.79491	60.00
14	3000	1	9	3	12	0	-2.48806	75.00
15	3100	1	1	1	2	0	-1.38944	50.00
16	3200	1	4	2	6	0	-2.08259	66.67
17	3300	1	2	0	2	0	13.94174	100.00
18	3400	1	1	0	1	0	14.45257	100.00

New Bin
2

*The **Fine Detail** window that displays the summarized grouping statistics of the selected variable.*

Output Groupings tab

The **Output Groupings** tab is designed to display a table view of the committed grouping results from the output data set. That table listing displays the summarized grouping results that can be passed on to subsequent nodes within the process flow for further analysis. For each input variable, the table listing displays both the committed and automatically generated Information value and Gini statistic. The committed values are determined by the various adjustments performed within the interactive grouping session for each input variable in the analysis. The various summarized grouping results are stored in a temporary data set called OUTPUTGROUPINGS that is written to the SAS work directory.

Interactive Grouping - Interval Variable

Chart | Statistics | Coarse Detail | Fine Detail | Output Groupings

	Variable	Keep	Commit Type	Auto-Grouped Infoval	Committed Infoval	Auto-Grouped Gini	Committed Gini
1	CLAGE	YES	AUTO	0.25386	0.25386	27.79579	27.79579
2	CLNO	YES	AUTO	0.1069	0.1069	18.30475	18.30475
3	DEBTINC	YES	AUTO	1.92403	1.92403	66.7268	66.7268
4	DELINQ	YES	AUTO	0.56532	0.56532	33.04442	33.04442
5	DEROG	YES	AUTO	0.27659	0.27659	16.63788	16.63788
6	JOB	YES	AUTO	0.12303	0.12303	17.60817	17.60817
7	LOAN	YES	USER	0.24857	0.23729	25.98794	25.00475
8	MORTDUE	YES	AUTO	0.1054	0.1054	17.88647	17.88647
9	NINQ	NO		0.1732	.	20.13102	.
10	REASON	YES	AUTO	0.00862	0.00862	4.31088	4.31088
11	VALUE	YES	AUTO	0.53494	0.53494	29.73449	29.73449
12	YOJ	YES	AUTO	0.08538	0.08538	15.96654	15.96654

*The **Output Groupings** tab that displays the committed grouping results for each variable.*

The following is a short description of the listed columns that are displayed within the **Output Groupings** tab:

- **Variable:** Name of the grouping variable.
- **Keep:** Indicates the variable status of the listed variable from the active training data set with a variable status of YES, indicating to you that the associated output grouping results are outputted from the **Interactive Grouping** node.
- **Commit Type:** Indicates if automatic grouping was performed for the listed variable or not. AUTO means that automatic grouping was performed without interactive modification of the listed grouping variable. USER means that the groups were manually created for the grouping variable. A blank value means that the variable was not a part of the grouping process, with a variable status of NO.
- **Auto-Grouped Infoval:** Information value statistic that is calculated from automatic grouping performed for the grouping variable.
- **Committed Infoval:** The final information value statistic that is outputted and passed into the subsequent nodes in the process flow diagram. Both the auto-grouped infoval information value statistic and the committed infoval information value statistic will be identical if interactive modification in redefining the various groups was not performed.
- **Auto-Grouped Gini:** Gini score statistic that is calculated from automatic grouping performed for the grouping variable.
- **Committed Gini:** The final Gini score statistic that is outputted and passed into the subsequent nodes in the process flow diagram. Both the auto-grouped Gini statistic and the committed Gini statistic will be identical if you have not redefined the various groups that were automatically created from interactive grouping.

Viewing the Interactive Grouping Node Results

General Layout of the Results Browser within the Interactive Grouping Node

- **Model tab**
- **Code tab**
- **Log tab**
- **Output tab**
- **Notes tab**

The following tabs display the file administrative information, the committed grouping variables, the internal SAS code and various interactive grouping settings based on the scored data set that is created from interactive grouping analysis.

Model tab

The **Model** tab is designed to display the file information of the scored data set, a table view of the variables in the scored data set, the current settings specified for interactive grouping, and the status of the **Interactive Grouping** node. The tab consists of the following four subtabs.

- **General subtab**
- **Variables subtab**
- **Settings subtab**
- **Status subtab**

General subtab

The **General** subtab from within the **Model** tab displays the file administrative information for the scored data set, such as the date of creation and the last modification date and the binary-valued target variable that is selected for interactive grouping analysis. The subtab is similar to many of the other subtabs within the various modeling nodes.

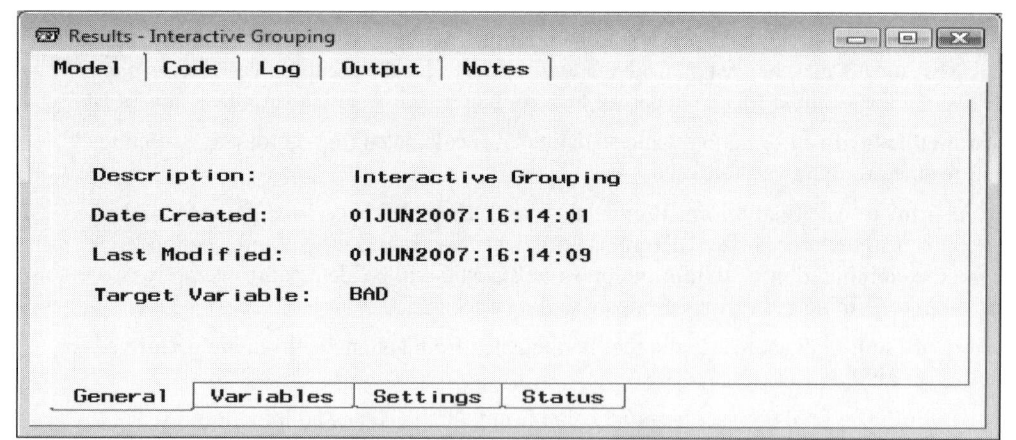

*The **General** subtab displays the file information of the scored data set from interactive grouping.*

Variables subtab

The **Variables** subtab displays a table view of the variables from the output scored data set. The table displays the output variables automatically created from interactive grouping such as the grouping label, WOE statistic and the group identification number for each committed variable in the analysis. The table will allow you to redefine both the model role and the level of measurement of the selected variable. This will allow you to control the selected variables and corresponding WOE statistics that are written to the output data sets and included in the subsequent classification modeling procedures.

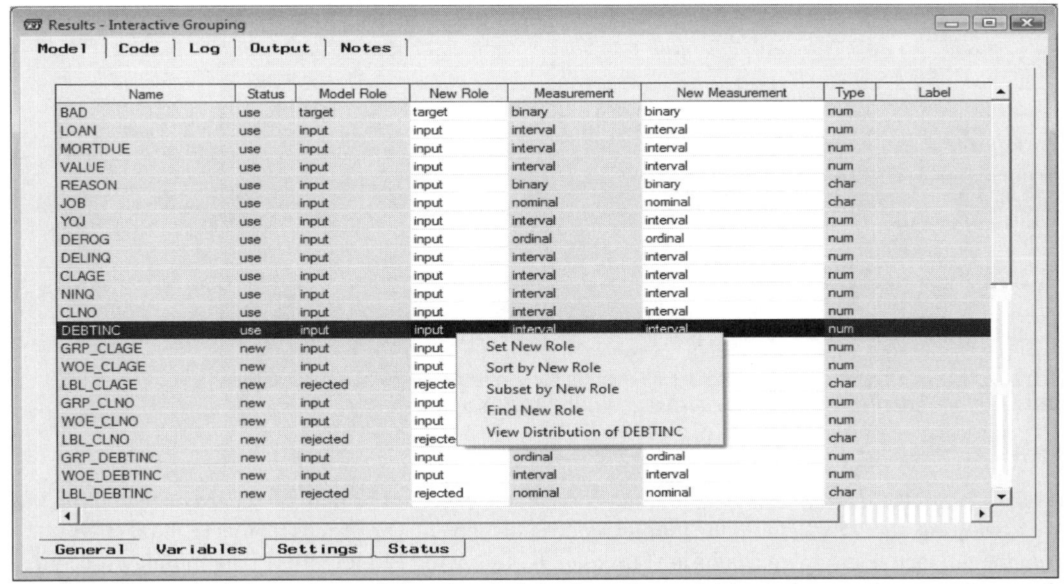

*The **Variables** subtab is used to redefine the model role to the variables in interactive grouping analysis.*

Settings subtab

The **Settings** subtab displays an output listing of the various options specified from the **Interactive Grouping Settings** window that are applied in interactive grouping.

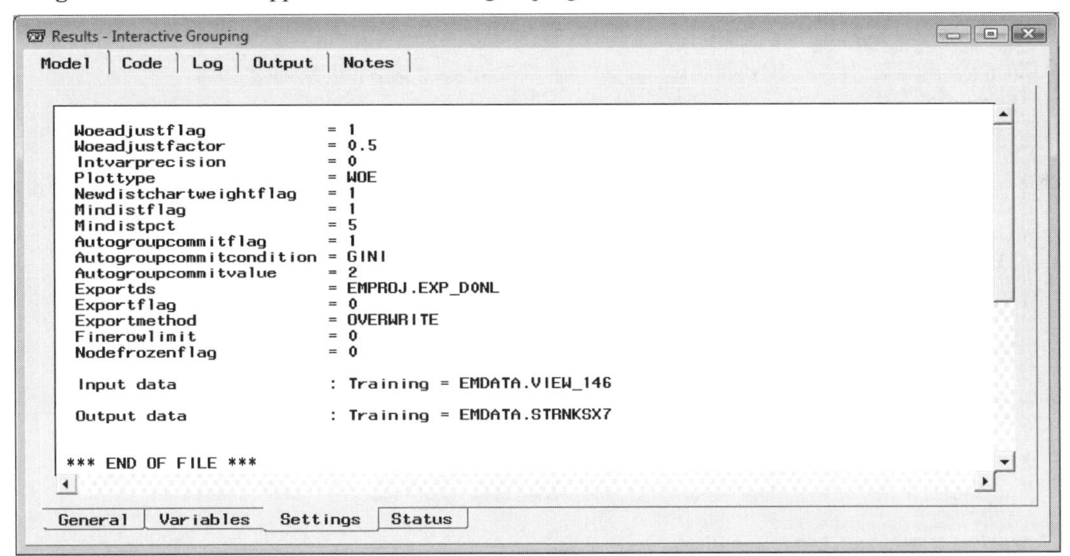

*The **Settings** subtab displays the various option settings for the current interactive grouping analysis.*

Status subtab

The **Status** subtab has been added to Enterprise Miner v4.3. The subtab displays the status of the node such as the creation date of the training and scored data sets, training time and scoring time from interactive grouping that was performed, and the processing time in both interactive training and scoring the interactive grouping results.

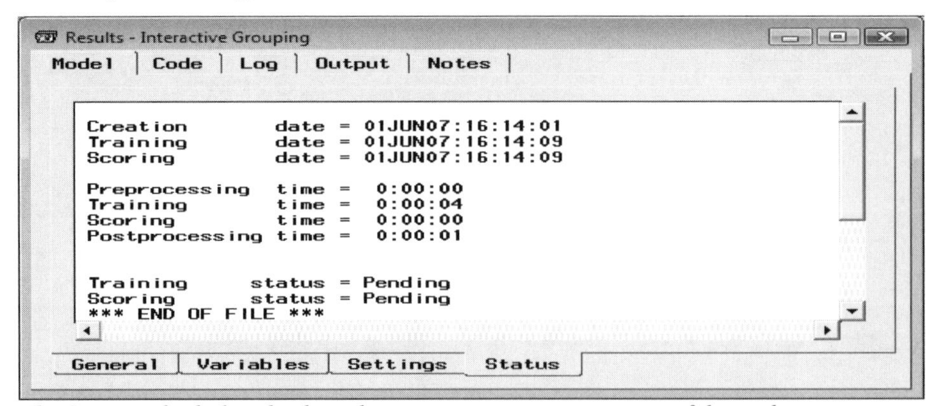

*The **Status** subtab that displays the interactive grouping status of the node.*

Code tab

The **Code** tab displays the internal SEMMA training code that is executed behind the scenes once you run the **Interactive Grouping** node. By default, the training code is displayed. The groups that were interactively created from the interactive grouping results are based on the following PROC SPLIT data mining procedure. The subsequent scoring code displays a partial listing of the series of *if-then* statements that are used in creating the optimal class levels for each committed input variable in the analysis based on the associated WOE statistic that is calculated for each group.

```
Results - Interactive Grouping
Model   Code | Log | Output | Notes |

(•) Training     ( ) Scoring

00001 proc split data=EMDATA.VIEW_146 criterion=chisq leafsize=298 outtree=outtree excludemiss;
00002 input CLAGE/level=interval;
00003 target BAD/level=binary;
00004 run;
00005 data grouped(keep = CLAGE BAD GRP_CLAGE dmign_resp dmign_nresp dmign_indet _weighted_frequency_
00006 );
00007 set varSummary;
00008 dmign_resp = 0;
00009 dmign_nresp = 0;
00010 dmign_indet = 0;
00011 if BAD = 1 then dmign_resp = _weighted_frequency_;
00012 else if BAD = 0 then dmign_nresp = _weighted_frequency_;
00013 else dmign_indet = _weighted_frequency_;
00014 if CLAGE = . then GRP_CLAGE = -1;
00015 else
00016 if CLAGE LT 71 then GRP_CLAGE = 1;
00017 else
00018 if CLAGE LT 92 then GRP_CLAGE = 2;
00019 else
00020 if CLAGE LT 112 then GRP_CLAGE = 3;
00021 else
00022 if CLAGE LT 121 then GRP_CLAGE = 4;
00023 else
00024 if CLAGE LT 130 then GRP_CLAGE = 5;
00025 else
00026 if CLAGE LT 150 then GRP_CLAGE = 6;
00027 else
00028 if CLAGE LT 173 then GRP_CLAGE = 7;
00029 else
00030 if CLAGE LT 189 then GRP_CLAGE = 8;
00031 else
00032 if CLAGE LT 201 then GRP_CLAGE = 9;
00033 else
00034 if CLAGE LT 214 then GRP_CLAGE = 10;
 • • •
00828 data infoval(keep=infoval) gini(keep=dmign_resp dmign_nresp grouprespráte) stats(keep=GRP_YOJ dm
00829 );
00830 if _n_=1 then set summ(keep=dmign_resp dmign_nresp rename=(dmign_resp=totresp dmign_nresp=totnre
00831 set summ(where=(_type_ = 1) keep=GRP_YOJ dmign_resp dmign_nresp dmign_indet _type_
00832 ) end=lastobs;
00833 respper = dmign_resp / totresp;
00834 nrespper = dmign_nresp / totnresp;
00835 if ((dmign_resp = 0) OR (dmign_nresp = 0)) then wtev = log(((dmign_nresp + 0.5) * totresp) / ((d
00836 else
00837 wtev = log(nrespper / respper);
00838 infoval + ((nrespper - respper) * wtev);
00839 grouprespráte = dmign_resp / (dmign_resp + dmign_nresp) * 100;
00840 output gini stats;
00841 if lastobs then output infoval;
00842 format wtev 8.5;
00843 format grouprespráte 6.2;
00844 run;
00845 proc sort data=gini nothreads;
00846 by descending grouprespráte;
00847 run;
00848 data gini(keep=gini);
00849 set gini(keep=dmign_resp dmign_nresp) end=lastobs;
00850 between + 2 * cumnresp * dmign_resp;
00851 within + dmign_nresp * dmign_resp;
00852 cumresp + dmign_resp;
00853 cumnresp + dmign_nresp;
00854 if lastobs then do;
00855 gini = 100 * (1 - (within + between) / (cumnresp * cumresp));
00856 output;
00857 end;
00858 run;
00859 *** END OF FILE ***
```

*The **Code** tab displays the PROC SPLIT training code that creates the various groups.*

```
Results - Interactive Grouping
Model  Code | Log | Output | Notes |

 Training   Scoring

00001 length LBL_CLAGE $ 19;
00002 if CLAGE = . then do;
00003 GRP_CLAGE = -1;
00004 WOE_CLAGE = -0.30807282659172;
00005 LBL_CLAGE = "Missing";
00006 end;
00007 else
00008 if CLAGE LT 71 then do;
00009 GRP_CLAGE = 1;
00010 WOE_CLAGE = -0.94475748756387;
00011 LBL_CLAGE = "low <= CLAGE < 71";
00012 end;
00013 else
00014 if CLAGE LT 92 then do;
00015 GRP_CLAGE = 2;
00016 WOE_CLAGE = -0.47942218996476;
00017 LBL_CLAGE = "71 <= CLAGE < 92";
00018 end;
00019 else
00020 if CLAGE LT 112 then do;
00021 GRP_CLAGE = 3;
00022 WOE_CLAGE = -0.25900458854553;
00023 LBL_CLAGE = "92 <= CLAGE < 112";
00024 end;
00025 else
00026 if CLAGE LT 121 then do;
00027 GRP_CLAGE = 4;
00028 WOE_CLAGE = -0.16396429927842;
00029 LBL_CLAGE = "112 <= CLAGE < 121";
00030 end;
00031 else
•••
00698 if YOJ LT 21 then do;
00699 GRP_YOJ = 8;
00700 WOE_YOJ = -0.12720159637541;
00701 LBL_YOJ = "17 <= YOJ < 21";
00702 end;
00703 else
00704 do;
00705 GRP_YOJ = 9;
00706 WOE_YOJ = 0.5503255722079;
00707 LBL_YOJ = "21 <= YOJ < high";
00708 end;
00709 *** END OF FILE ***
```

The interactive grouping score code that can be used in creating new groups from the HMEQ data set.

Log tab

The **Log** tab will allow you to diagnose any compiling errors that might have occurred in creating the various groups by running the previous listed data mining code in the background once the **Interactive Grouping** node is executed. The **Log** tab will display numerous compiling results and will not be illustrated.

Output tab

The **Output** tab will be displayed once you select the **Results Browser** option. The tab displays a table listing of the finalized summarized grouping statistics for each committed input variable in the analysis that is displayed in the **Output Groupings** tab. These same committed grouping results can be viewed from the temporary data set called WORK.OUTPUTGROUPINGS.

Results - Interactive Grouping

Model | Code | Log Output | Notes |

Interactive Grouping Output Summary

Variable	Keep	Commit Type	Auto-Grouped Infoval	Committed Infoval	Auto-Grouped Gini	Committed Gini
CLAGE	YES	AUTO	0.25386	0.25386	27.7958	27.7958
CLNO	YES	AUTO	0.10690	0.10690	18.3048	18.3048
DEBTINC	YES	AUTO	1.92403	1.92403	66.7268	66.7268
DELINQ	YES	AUTO	0.56532	0.56532	33.0444	33.0444
DEROG	YES	AUTO	0.27659	0.27659	16.6379	16.6379
JOB	YES	AUTO	0.12303	0.12303	17.6082	17.6082
LOAN	YES	USER	0.24857	0.23729	25.9879	25.0048
MORTDUE	YES	AUTO	0.10540	0.10540	17.8865	17.8865
NINQ	NO		0.17320	.	20.1310	.
REASON	YES	AUTO	0.00862	0.00862	4.3109	4.3109
VALUE	YES	AUTO	0.53494	0.53494	29.7345	29.7345
YOJ	YES	AUTO	0.08538	0.08538	15.9665	15.9665

*** END OF FILE ***

*The **Output** tab that displays the finalized committed grouping results for each input variable.*

Chapter 4

Model Nodes

Chapter Table of Contents

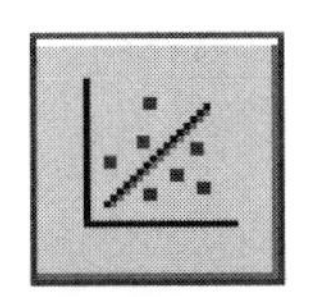

4.1 Regression Node

General Layout of the Enterprise Miner Regression Node

- **Data tab**
- **Variables tab**
- **Model Options tab**
- **Selection Method tab**
- **Initialization tab**
- **Advanced tab**
- **Output tab**
- **Notes tab**

The purpose of the **Regression** node in Enterprise Miner is to perform either linear regression modeling or logistic regression modeling. *Multiple linear regression modeling* or *least-squares modeling* is designed to predict an interval-valued target variable. *Logistic regression modeling* is designed to predict a categorically-valued target variable. Both regression models may consist of any number of continuous or categorical input variables in the model. The general idea of regression modeling is to explain or predict the variability in the target variable based on the input variables in the model, that is, to determine the functional relationship that is usually a linear relationship between the various input variables in the model and the target variable that you want to predict. One of the objectives of regression modeling is to determine which input variables best predict the target values. In other words, the objective of regression analysis is to determine the optimal balance of choosing the model with the smallest error and the fewest number of parameters.

The **Regression** node is one of the few modeling nodes that has a variable selection routine within the node along with various modeling selection criteria. The node will allow you to perform forward, backward, or stepwise regression. After the variable selection routine terminates, the model that optimizes the modeling selection criteria from the validation data set is selected as the final model. The node also has the added flexibility that will allow you to select the most important input variables that will be automatically included in the final model despite the input variables that been selected from the modeling selection routine. By default, the modeling selection procedure is not applied. The tab will automatically construct the regression model from all the input variables in the data with a variable status of **use**. However, failure to reduce the number of input variables will result in overfitting and bad generalization, increased computational time, and computational difficulties in obtaining good parameter estimates. It is generally recommended to create many different models from the three separate modeling selection routines, then use prior knowledge in addition to the various modeling assessment statistics in selecting the final model. The **Regression** node supports binary, interval, ordinal, and nominal target variables. In addition, the node will also support interval or nominal-valued input variables. In other words, the node treats ordinal-valued input variables as nominal-valued input variables by ignoring the ordering levels of the categorically-valued input variable. The node is specifically designed to model a single target variable of the regression model. Therefore, once you first open the **Regression** node, a **Target Selector** dialog box will appear that will force you to select the most appropriate target variable for the regression model, assuming that multiple target variables have been specified from the previous nodes.

Some of the options that are available within the node are removing the intercept term from the model, including higher-order input variables or interaction terms in the regression model, specifying dummy indicator coding for the regression model for nominal-valued or ordinal-valued input variables, performing the modeling selection routines such as forward, backward, or stepwise regression, and specifying the order in which the input variables are entered into the regression model. For logistic regression modeling, the node has various convergence options from the nonlinear link function incorporated into the modeling design. An output scored data set can be created that consists of the predicted values and the residual values along with the rest of the variables in the data. From the **Result Browser**, the node generates various scatter plots that will allow you to validate the functional form and the various modeling assumptions that must be satisfied in the least-squares model. Similar to the other modeling nodes, the node will allow you to view the **Model Manager** that is a directory table facility of the various models that have been created within the **Regression** node.

Interpretation of the Parameter Estimates

Multiple Regression Modeling

In multiple linear regression, the regression equation is written in the following form:

$$\hat{Y} = \alpha + \beta_1 X_1 + \beta_2 X_2 + \varepsilon$$

the target variable Y is a linear combination of the input variables X_1 and X_2 for some intercept α and unknown parameter estimates β_1 and β_2 with an error term ε where ε iid Normal(0, σ^2), that is, the random error is assumed to be independent and normally distributed with a constant variance σ^2 that is centered about zero. If these modeling assumptions are satisfied and the least-squares estimates are unbiased estimates of the true parameter estimates along with having a minimum variance, then the regression parameters are often called the *best linear unbiased estimates* (BLUE), where best is in reference to the minimum variance property.

In multiple regression modeling, the slope estimates or parameter estimates will explain the linear relationship between the corresponding input variable and the target variable, and the intercept term is simply the sample mean of the target variable that you want to predict. The parameter estimates measure both the rate of change and the correlation between both variables after controlling for all other input variables in the model. A parameter estimate of zero will indicate that there is no increase or decrease in the target values as the values of the input variable increases. Therefore, the input variable is usually not very useful for the model. Parameter estimates greater than one will indicate that the rate of change of the target variable is increasing at a faster rate than the input variable after setting all other inputs to zero. Conversely, parameter estimates less than one will indicate that the rate of change of the input variable is increasing at a faster rate than the target variable, that is, assuming all other input variables are the same. The parameter estimate will also indicate the correlation between both variables. Parameter estimates greater than zero will indicate that the values of the input variable increases linearly along with the target values after controlling for all other input variables in the model. Parameter estimates less than zero will indicate that as the values of the input variable increase, then the values of the target variable decrease linearly, or vice versa. Also, a correlation of ± 1 will indicate a perfect positive or negative correlation with regard to the linear relationship between both variables in the model. It is unwise to interpret the parameter estimates as an indication of the importance of the input variables to the model. This is because some of the input variables might contain a wide range of values and there might be multicollinearity in the model, where some of the input variables might be highly correlated with the other input variables in the regression model, affecting the size and magnitude of the parameter estimates.

Measuring the Accuracy of the Least-Squares Model

The basic measurement in evaluating the accuracy of the least-squares regression model is based on the sum-of-squares difference between the target values and the predicted values and is called the *sums of squared error* or the *unexplained variability*, $\sum e_i^2 = \sum(y_i - \hat{y}_i)^2$. The sum-of-squares error is the statistic that you want to minimize in least-squares modeling by fitting the interval-valued target variable. In linear regression modeling, this is determined by the line that minimizes the sum-of-squares vertical distance from the data points to the regression line that is illustrated in the following diagram. The difference between the regression line and the overall target mean is the amount of variability explained by the least-squares model, called the *explained variability*, $\sum(\hat{y}_i - \bar{y})^2$. The difference between the target values and the overall target mean that is the sum of the explained variability and the unexplained variability is called the *total variability*, $\sum(y_i - \bar{y})^2$. The F-test statistic is used to measure the significance of the fit that is the ratio of the explained variability over the unexplained variability. When the F-test statistic is relatively large, it will indicate to you that the linear regression model is generally a good fit to the data.

The *R^2 statistic*, also called the *coefficient of determination*, is often used to assess the accuracy in the regression model that measures the total variability of the target variable explained by the model. The r-square statistic generally indicates how much better the model is in estimating the target values than simply always guessing the mean. The r-square value measures the percentage of the total variability in the target variable explained by the linear use of the input variables in the regression model. Generally, higher r-square values will indicate that the model is a good fit to the data. The purpose of multiple linear regression is to determine the highest correlation between the target variable and the linear combination of input variables in the model. The pictorial interpretation of the coefficient of determination is illustrated in the following diagram.

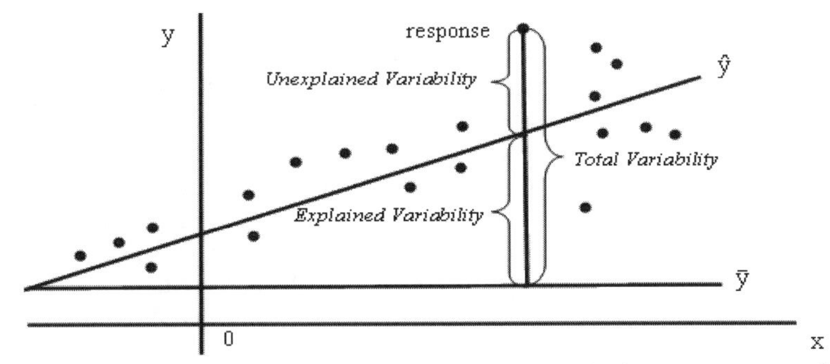

Pictorial interpretation of the r-square statistic in the least-squares model.

$$R^2 = \frac{\text{Explained Variability}}{\text{Total Variability}} = \frac{\text{Total Variability} - \text{Unexplained Variability}}{\text{Total Variability}} = 1 - \frac{\text{Unexplained Variability}}{\text{Total Variability}}$$

Logistic Regression Modeling

If the level of measurement of the target variable is categorical, then the statistical model that fits the linear regression model is called *logistic regression.* An interval-valued target variable is impractical in logistic regression since its values usually do not fall within the desired probability range between the interval of zero and one. Logistic regression modeling is essentially predicting the response variable by redefining the variable through the use of an indicator variable. The reason traditional linear regression should not be applied to the categorical target variable that you want to predict is that at times the predicted values used to estimate the probability of the target event will not result in the desired interval in predicting the probabilities that must be between the interval of zero and one. Like multiple regression modeling, logistic regression models can have a mixture of interval-valued and categorically-valued input variables that can be included into the model. In addition, similar to traditional regression modeling, the input variables for the logistic model can be selected by applying the various modeling selection routines. The general logistic mathematical form is considered to be nonlinear regression modeling for the various input variables in the statistical model. The reason is because a nonlinear transformation is applied to the target variable in the classification model. This nonlinear transformation is called a link function, which is applied in order to generate the probability of the expected target between the desired interval between zero and one. In addition, the probabilities are transformed since the relationship between the probabilities and the input variables are nonlinear. In logistic regression modeling, once the link function is applied, it is assumed that the transformed probabilities have a linear relationship with each one of the input variables in the model. For binary-valued input variables, the assumption can be ignored since a straight line is connected to two separate points.

The Link Function

In logistic regression modeling, there is an added transformation to be considered. It is called the *link function.* Since the relationship between the estimated probabilities and the input variable is nonlinear, then a link function is applied to transform the probabilities in achieving linearity to the logistic regression model. The link function is applied in order to transform the fitted values of the target variable that are generated from the model into a desirable range between zero and one.

In order to obtain the best predictive performance, it is important that the model be correctly specified. Correct specification of the model involves selecting the correct set of input variables in the model and the appropriate error distribution that matches the range of values of the target variable. However, in logistic regression modeling, it is also extremely important to select the correct link function that conforms to the range of values of the target variable. Furthermore, once the link transformation is applied, then it is assumed that the probabilities of the target variable will result in a linear relationship with the input variable that does not interact with the other input variables in the logistic regression model. In the node, the following logit link function is automatically applied to the logistic regression model that is written in the following form:

$\text{logit}[p(x_i)] = \ln[p(x_i) / (1 - p(x_i))] = \alpha + \beta X_i$ where $\text{logit}[p(x_i)]$ is called the *logit link function* to x_i

The probability of the target event in the logit link probability model is defined as follows:

$p(x_i) = \exp(\alpha + \beta_i x_i) / [1 + \exp(\alpha + \beta_i x_i)]$ is called the *logit link probability model*

$p(x_i) = \exp(\alpha + \beta_i x_i) / [1 + \Sigma\exp(\alpha + \beta_i x_i)]$ is called the *generalized link probability model*

where there are $k - 1$ intercept terms in the logistic regression model for the k target categories.

In logistic regression modeling, the functional form of the estimated probabilities of the target event across the range of values of the input variable has a sigmoidal S-shaped logistic function, where the relationships in the range of probabilities are fairly linear in the middle and fairly nonlinear at each tail. When the parameter estimate of the input variable is greater than zero, ($\beta_i > 0$), this will indicate that the estimated probabilities of the target event are increasing as the values of the input variable increase. Conversely, when $\beta_i < 0$, then this will indicate that the estimated probabilities of the target event are decreasing as the values of the input variable increase. In addition, as the absolute value of β_i increases, then the curve will have a steeper rate of change, and when the parameter estimate $\beta_i = 0$, then the curve will resemble a straight diagonal line.

For an ordinal-valued target variable, a cumulative (logit) link function is applied. The cumulative logit link function performs an ordered logistic regression that depends on the ordering levels of the target variable. The cumulative logit link function that is applied depends on the "*proportional odds assumption*" of the logistic model. The reason is because the cumulative probabilities of the ordered target categories assumes a common slope with different intercepts for each input variable, where there is no overlap between the cumulative probabilities of the target categories across the values of each one of the input variables in the model.

In logistic regression modeling, the odds ratio can be calculated by taking the exponential of the parameter estimates, that is, $\exp(\beta_i)$. The odds ratio is the proportion of the target level between both levels of the binary-valued input variable. The odds ratio measures the strength of the association between the categorically-valued input variable and target variable. Parameter estimates greater than one will indicate that the event level of the categorically-valued input variable is more likely to have the target event as opposed to the reference level of the categorically-valued input variable. Conversely, the value of the parameter estimate less than one will indicate that the event level of the categorically-valued input variable is less likely to have the target event, and a parameter estimate of one will indicate even odds. By default, in logistic regression modeling, the last class level of the nominal-valued input variable plays the role of the denominator for the odds ratio statistic. Also, taking the exponential of the parameter estimate of the interval-valued input variable, then subtracting one and multiplying by 100, that is, $100 \cdot [\exp(\beta_i) - 1]\%$, is the percent change in the estimated odds with each unit increase in the input variable with respect to the target event.

The probit and complementary log-log link functions are the other two link functions that are available within the **Regression** node as follows.

Probit Link Function

At times, the binary response variable might be the result of an underlying variable that is normally distributed with an unlimited range of values. The response variable can be expressed as a standard linear regression model in which the response variable is normally distributed with a wide range of values of its associated probability of the target event $p(x_i)$ that is distributed within the normal curve. Therefore, the probit function is applied that transforms each side of the linear regression model in order to conform its values to the normal probabilities. The probit function is the inverse of the standard normal cumulative distribution, called the z-value to the probability of the target event $p(x_i)$ that is defined as follows:

$\text{probit}[p(x_i)] = \Phi^{-1}[p(x_i)] = \alpha + \beta x_i$

where $p(x_i) = \Phi[\alpha + \beta x_i]$ is the probability of the target event in the *probit link probability model.*

A *cumulative probit link function* is applied to ordinal-valued targets. The results from both the logit and probit probability models will usually draw the same conclusion, with the exception of extremely large sample sizes. In addition, both the logit and probit link functions are usually applied in modeling binomial probabilities.

Complementary Log-Log Link Function

Complementary log-log probability model is also known as the proportional hazards model and is used in survival analysis in predicting target event occurring in some fixed interval of time. This link function can be applied when it is assumed that the probability of the event occurs within some interval of time. However, the one assumption of the proportional hazards model is that the model assumes that the range of probability values between the separate categorical input variables are proportional within each category. The complementary log-log function is the inverse of the extreme value distribution that applies the log transformation to the negative of the transformation of $[1 - p(x_i)]$ as follows:

$$\text{c-log-log}[p(x_i)] = \log[-\log(1 - p(x_i))] = \alpha + \beta x_i$$

where $p(x_i) = 1 - \exp[-\exp(\alpha + \beta x_i)]$ is the probability of the event in the *complementary log-log model.*

Calculation of the Logistic Regression Parameter Estimates

In logistic regression modeling, since a link transformation is applied to the linear combination of inputs, this will result in the normal equations not having a closed form solution. Therefore, the parameter estimates need to be iteratively calculated by a maximum likelihood method rather than least squares that is analogous to fitting iteratively reweighed least-squares. The weighting of the parameter estimates is performed in this iterative procedure, where the weights are the reciprocal of the variance of the binary-valued target variable, that is, $p \cdot (1 - p)$, for some estimated probability p at each target class level. The iterative process is repeatedly performed until the maximum difference in the parameter estimates between successive iterations reaches the convergence criterion. For binary-valued target variables, the maximum likelihood estimates of the parameter estimates are usually determined by maximizing the Bernoulli likelihood function from the appropriate link function. The PROC IML code that computes the parameter estimates and log-likelihood statistic by fitting the binary-valued target variable to the logistic regression model can be viewed from my website.

The Modeling Assumptions

Linearity Assumption: Initially, it is assumed that the functional form between all the input variables and the target variable that you want to predict is linear. The general idea is fitting a straight line through all the data points. This can be verified by constructing a bivariate scatter plot of each input variable and the target variable that you want to predict. If both the input variable and the target variable are strongly related to each other, then the functional distribution between each variable will either be a straight line or a clear curve. Otherwise, if both variables are not related, then the functional distribution between each variable will form a cloud of points. In general, traditional linear regression modeling will provide a better fit than the other modeling procedures when there is a linear relationship between the target variable and the input variables in the model.

In linear regression modeling, the error terms or residual values are calculated by taking the difference between the actual target values and the predicted target values. Therefore, the smaller the residual value, the better the prediction is. Again, this can be done by fitting a regression line through the data points. Theoretically, the best regression line that can be determined is by minimizing the squared distance between each data point and the regression line that you want to fit.

It is important in linear regression modeling that the residual values have a uniform random variability with an approximately normal distribution about its mean of zero. Therefore, it is important that the following modeling assumptions be satisfied. Otherwise, transformations to the input variables and the target variable might be considered. Failure to address the various modeling assumptions will result in bad predictions and poor generalization. The following assumptions are very subjective, with the exception of the independence assumption. Therefore, gross violations of the following modeling assumptions should be dealt with accordingly.

Constant Variance Assumption: A bivariate scatter plot of the error values across the fitted values will indicate increasing, decreasing, or constant variability in the residuals. It is not only important that the residual values display a uniform pattern across the fitted values or the various input variables in the model, but, it is also extremely important that the residuals be distributed randomly over the fitted values with no systematic pattern or trend about zero. One way to determine if the variability has been stabilized in the model is to compute the Spearman's rank correlation statistic between the absolute value of the residuals and the predicted values. If this statistic is close to zero, then it is an indication that the residual values do not have an increasing or decreasing trend about their mean of zero. Constant variability in the error values will indicate both the non-existence of outliers and normality in the data. Randomness in the residual values will indicate independence and an unbiased sample in the data. The reason for correcting for constant variability is that nonconstant variance will lead to biased estimates and misleading results. In logistic regression modeling, the variability assumption can be ignored since the binary target variable is concentrated between two separate values, which will result in its variability approaching zero.

Normality Assumption: The normality assumption is the least important modeling assumption that can be a very subjective decision. The first step in regression modeling is to check for reasonable normality in the input variables and the target variable in the model. The reason that the normality assumption is so important is because the various test statistics that are applied depend on the normality assumption. Although the various statistical tests, such as the F-test statistic, are very robust to the violations of the statistical assumptions, it is always a good idea before making any final conclusions to review the distribution of the variables in the regression model. Normality can usually be satisfied with a large sample size. Although, there is no general rule as to an optimum sample size to fit the model, the sample size should be as large as possible in order to improve the power of the test and better meet the distributional assumptions of the various statistical tests. The normality assumption can be verified by constructing histograms or, even better, normal probability plots of the residuals in order to view the distribution of the error values. In logistic regression modeling, the normality assumption can be ignored, since the binary-valued target variable is composed of two separate values that are obviously not normal.

Independence Assumption: This is the basic statistical assumption in a large majority of statistical analysis. The independence assumption is the most important statistical assumptions and is the hardest to determine. The common remedy for the independence assumption is performing an effective random sample of the data that you want to fit. At times, the data that is being modeled might be recorded in chronological order or some sequence of time; then the independence assumption in the data must be confirmed. In predictive modeling, the error terms are correlated if the value of the error term depends upon the other error values. The phenomenon often occurs in time series modeling and repeated measures modeling. Plotting the residuals over time can verify the independence assumption by observing an unsystematic random uniform pattern of the residual values over time or the order in which the observations were recorded. For positive autocorrelation, the values in the error terms tend to be followed by values with the same sign over time. For negative autocorrelation, the values in the error terms tend to alternate between positive and negative values over time. The Durbin–Watson statistic is used to measure the significance in the correlation between the residual values and its lagged values over time. Failure to recognize independence in the data can produce bias standard errors that will result in inaccurate confidence intervals and test statistics of the parameter estimates. Again, the statistical inferences are questionable and meaningless with a sample that is not independent, called a biased sample.

Logistic Regression Modeling Assumptions: The statistical assumptions that must be satisfied in traditional regression modeling still hold true in logistic regression modeling. Since the level of measurement of the target variable is either nominal or ordinal, the distribution of the target values is concentrated at a single point that is obviously not normal, with the variance approaching zero. Therefore, both the normality and constant variance assumptions are violated. Since the variability in the target variable is not constant across the values of the input variables, the least-squares estimators for solving the parameter estimates are not used in logistic regression. The least-squares parameter estimates are not efficient, that is, other estimation methods generate smaller standard errors due to nonconstant variability in the target variable. Therefore, the conventional maximum likelihood method is used to produce the best set of linear unbiased estimates by calculating BLUE estimators that are consistent, asymptotically efficient, and asymptotically normal.

The modeling assessment statistics and the various diagnostic statistics used to evaluate the predictability of the regression models are also used in logistic regression modeling. In addition, the multicollinearity issue of the input variables related to each other must not be overlooked in logistic regression modeling. In logistic regression modeling, the common problems to look for are extreme values in the data, interaction, nonlinearity, overdispersion, confounding, multicollinearity in the input variables, and the iterative procedure failing to converge, usually due to quasi-complete separation with zero cell counts.

It is recommended that the first step in constructing the logistic regression model be to first perform preliminary data analysis by constructing various logit plots to visualize the linearity assumption in the log of the ratio between the target event and nonevent (or the log of the odds ratio) and the input variables in the model and constructing contingency tables, and to carefully look for cells with no observations, which leads to quasi-complete separation. *Quasi-complete separation* occurs when the levels of the categorically-valued input variable perfectly predicts the categorically-valued target variable with a zero cell count occurring. Quasi-complete separation will result in undefined parameter estimates of the input variables in the classification model. A relatively large number of input variables in the logistic regression model will increase the likelihood of complete separation or quasi-complete separation. The first step in logistic regression modeling is to perform exploratory data analysis. Exploratory data analysis, such as constructing contingency tables to identify any numerical problems due to a small sample size, creating separate stratified groups in detecting confounding and interaction between the target levels and each input variable, and creating logit plots to identify any nonlinear relationships between the logit of the log odds ratio and the input variables.

When the target event occurs over time, size, population, and so on, it is more informative to model the rate or proportion instead of the actual counts. The general idea is to standardize the rate of occurrences by the length of the time interval, size or population in order to allow the outcomes to be comparable. For example, you would need to adjust for the number of subjects at risk with a certain disease by each country if it is assumed that the rate of incidence of the disease in each country is very different. In logistic regression modeling, when the target variable consists of rates and proportions, it is important to check that the various modeling assumptions are satisfied since the logistic regression model is fitting an interval-valued target variable with its range of values that vary between zero and one.

The Relationship Between Input Variables

Interaction can be detected by plotting the relationship between the input variable and the target variable in the model with the line plot displaying an overlap between each value of the input variable that is plotted across the range of values of another input variable in the model, or vice versa. If there is no interaction, then each line will be relatively parallel to each other. In statistical modeling, line plots can be used in detecting both confounding and interaction effects in the model by observing the relationship between the target variable and the primary input variable along with the existence of an additional variable in the model. *Confounding* is present in the model when the primary input variable might not explain the variability in the model that well without the existence of an additional confounding input variable in the model. As an example, suppose you are interested in predicting the price of a book based on the number of pages. Confounding would be present in the multiple linear regression model when there might be a weak negative linear relationship between both variables in the model. However, including the confounding input variable of the type of book, such as either being a hard cover or soft cover book, into the model changes the relationship into a strong positive linear relationship between the price of the book and the number of pages by each type of book. This is illustrated in the following diagram. For interval-valued variables, confounders can be identified by fitting the regression model with and without the possible confounder input variable and observing any dramatic changes in the parameter estimates of the other input variable in the model. Confounding can also be detected by constructing line plots. For categorical input variables with a small number of class levels, a stratified table may be constructed to help identify confounders. For the primary binary-valued input variable and binary-valued target variable, the marginal table might display absolutely no association at all between both variables, that is, an odds-ratio statistic of 1, without the existence of an additional confounding categorical variable, which results in significant association between both categorical variables from the partial tables that are created by each class level of the additional confounding categorical variable. This phenomenon is called *Simpson's paradox.*

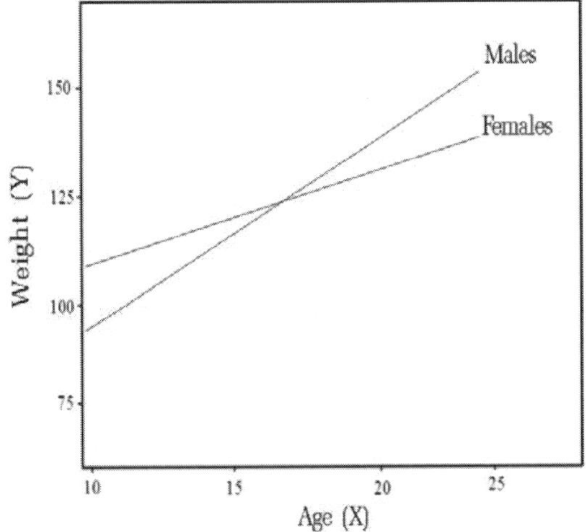

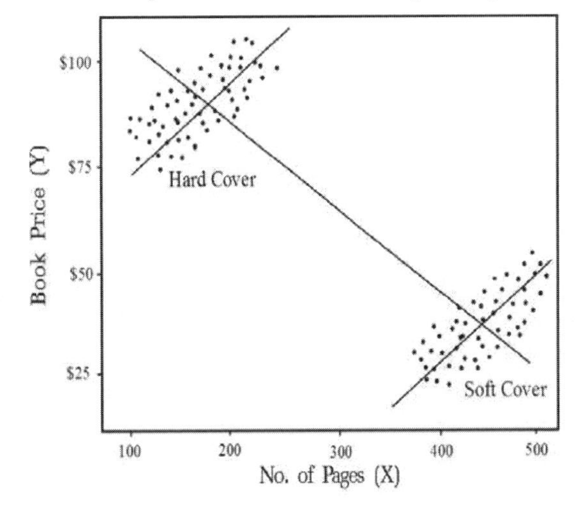

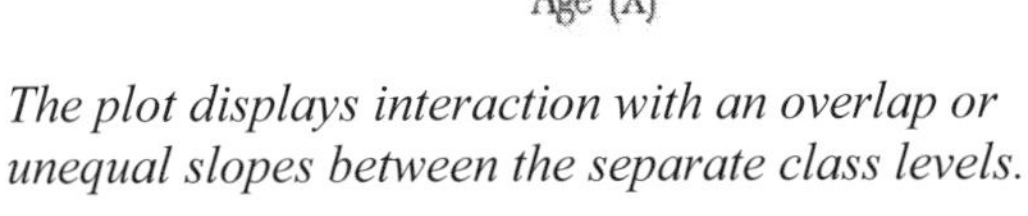
The plot displays interaction with an overlap or unequal slopes between the separate class levels.

The plot displays confounding between two separate input variables and the target variable in the model.

Extreme Values

It is important that the data points that are being fitted conform to the same trend or pattern. Data points that are well separated from the other data points are called extreme values or outlying data points. Extreme values or outlying data points will have a profound effect on the modeling fit that will lead to biased parameter estimates. To achieve the best predictions, these same extreme data points that do not follow the same trend in the rest of the data need to be removed from the analysis. Conversely, these same extreme values might reveal important information about the underlying data that might require further investigation. However, it might indicate that critical input variables, interaction or nonlinear effects have been omitted from the regression model, suggesting that the outlying data point might well be the most important data point in the analysis.

Correlation Analysis

In model building, a good starting point in determining the importance of the input variables to the model is performing correlation analysis, which involves calculating the individual correlations between the input variable and the target variable while controlling for all other input variables in the multiple linear regression model. Again, the correlation statistic indicates the increasing or decreasing linear relationship between the input variable and the target variable that you want to predict. Basically, the input variables with the highest correlations are retained in the multiple linear regression model. However, the shortcoming is that the procedure ignores the partial correlations between the other input variables in the multiple linear regression model. This might result in certain input variables being mistakenly added or removed from the multiple linear regression model. In addition, the presence of interaction in the model can result in misleading univariate partial correlation statistics. For this reason, the three standard modeling selection procedures are preferred, which will be introduced shortly.

Multicollinearity

It is not uncommon to have different input variables highly related to each other in the regression model. Multicollinearity in the input variables makes it difficult or even impossible to interpret the relationship between the input variables and the target variable you want to predict. Multicollinearity can be detected when the overall F-test statistic is significant along with no highly significant modeling terms in the model. Multicollinearity can also be detected when there is a high partial correlation between the two separate input variables along with insignificant *p*-values from the t-test statistics of both the related input variables. The remedy is to fit two separate regression models. That is, fit a regression model to each one of the correlated input variables that are related to each other and select the best model based on the largest r-square statistic. Multicollinearity will lead to ill-conditioning in the design matrix, which will result in a cross-product matrix that cannot be inverted. The reason is because one or more columns of the design matrix can be expressed as a linear combination of the other columns. There are three common remedies to multicollinearity. The first remedy is to fit the regression model with a smaller number of input variables. This will decrease the chance that two separate input variables will be related to each other. The second remedy is to simply add more data points and, hopefully, the collinearity between the input variables in the regression model will disappear. The third common remedy is performing principal components analysis that is designed to reduce the number of input variables for the model, where the principal components that are selected are uncorrelated to each other.

Diagnostic Analysis

The objective of predictive modeling is to determine the best set of parameter estimates that results in the smallest error with the fewest number of input variables possible. In linear regression modeling, the error function that you want at a minimum is the sum-of-squares error. For logistic regression modeling, the error function that you want at a minimum is the difference in the likelihood functions, the likelihood function of the model you are trying to fit and the reduced model with the corresponding terms removed from the model one at a time. Therefore, assessing the significance of the input variables that are added or removed from the model one at a time can be determined from the difference in the likelihood functions. The error distribution should account for the relationship between the fitted values and the variability in the target variable. This can be determined by plotting the residual values from the model across the fitted values by plotting the residuals against the fitted values to validate the constant variance assumption in the error terms. In regression modeling, it is assumed that the residual values are normally distributed about zero. In logistic regression modeling, the incorrect link function, violation of the linearity assumption of the log transformed target variable and the input variables, outliers or influential data points and violation of the independence assumption will contribute to a poor fitting model. In addition, selecting the correct specification of the range of the target values and the correct error distribution will increase the probability of selecting good input variables for the regression model. By selecting good input variables for the statistical model with the correct functional form with regard to the target variable that you want to predict will result in the correct degree of flexibility that can be incorporated into the regression model by adapting to the variability in the target values that will result in the best predictions.

Regression modeling is discussed in greater detail in Kutner, Nachtsheim, Neter, and Li (2004), Bowerman and O'Connell (1990), Draper and Smith (1981), Freund and Littell (2000), and Trumbo (2002). Logistic regression modeling is discussed in greater detail in Agresti (2002), Stokes, Davis, and Koch (2000), Allison (1999), and Categorical Data Analysis Using Logistic Regression Course Notes, SAS Institute (2001).

Data tab

The **Data** tab will allow you to view the selected partitioned data sets or the scored data set. The tab will also allow you to select an entirely different data set with a similar role from a predecessor node that is connected to the node. The tab will display both the data mining data set name located in the EMDATA directory and a short description of the selected data set. By default, the training data set is automatically selected. The **Role** option will allow you to select the desired partitioned data set.

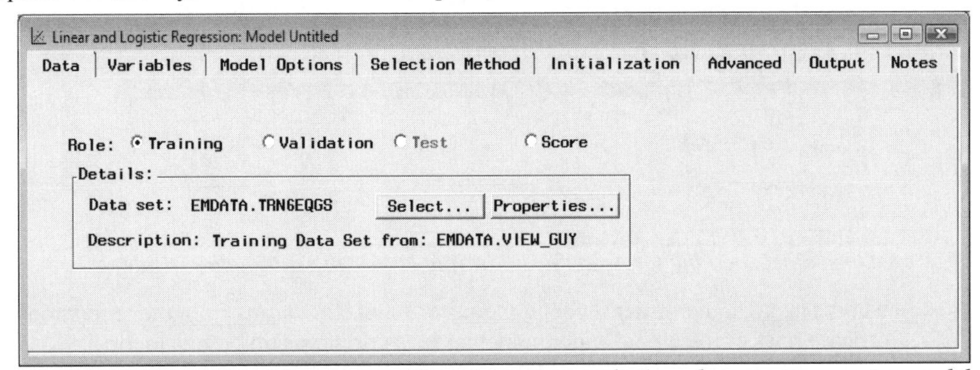

*The **Data** tab is used to view the partitioned data sets to fit the subsequent regression model.*

Select the **Properties...** button that will display the **Data set details** window to view both the file information such as the creation date, last modified date, and number of rows and columns of the selected data set. In addition, you may view a table listing of the selected data set.

Press the button **Select...** button that will open the **Imports Map** window. By default, the training data set is automatically selected. The purpose of the window is that it will allow you to redefine an entirely different training data set to fit. A directory tree will be displayed that will list each node that was previously connected to the **Regression** node. Select any one of the directory tree check boxes that is automatically created for each node that was connected to the **Regression** node. The tree will expand in a hierarchical fashion with a listing of the various training data sets that are created within the selected node. Select any one of the data mining data sets that are displayed. This will result in an entirely different training data set to be fitted. For instance, the exported training data set that was created from the **SAS Code** node was selected from the **Imports Map** window in order to fit the multiple linear regression model to a unique set of records by performing bootstrap sampling without replacement from the **Group Processing** node. The **Role** display that is positioned just below the directory tree will list the role type that is selected. Since the training data set is automatically selected from the **Data** tab, TRAIN will be displayed in the **Role** field. The **Selected Data** and **Description** fields will display both the name and a short description of the selected training data set. Unless the sample size of the data set to fit is extremely small, it is not advisable to remove either the validation or test data sets from the modeling assessment process by selecting the **Clear** button from within the window. Select the **OK** button to reassign the selected training data set to the current **Regression** node. The **Imports Map** window was described in the previous **Sampling** node chapter, therefore, the window will not be displayed.

Variables tab

The **Variables** tab will appear once you open the node. The tab displays a table view of the variables in the active training data set. The tab will give you one last chance to remove the input variables before fitting the regression model. By default, all the variables are automatically set to a variable status of **use**, with the exception of the variables set to **rejected** from the **Input Data Set** node or unary variables that contain one value or date fields, which are automatically excluded from the analysis with a variable status of **don't use**. To set the variable status to a selected input variable, select the desired variable row and drag the mouse over the **Status** column, right-click the mouse, and select the **Set Status** pop-up menu item. From the **Set Status** pop-up menu item, select the variable status of **don't use** to remove the selected variables from the modeling fit. Press the Ctrl key or the Shift key to select several rows to add or remove any number of input variables from the regression model. Note that all other columns are grayed-out and cannot be changed.

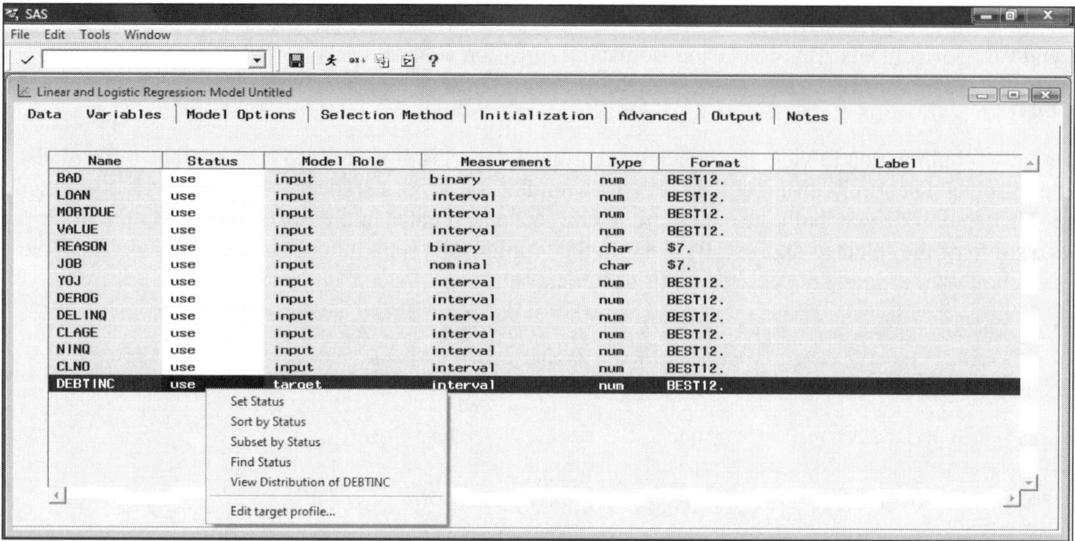

*The **Variables** tab is used to set the modeling terms in the model and access the **Interaction Builder.***

The tab will also allow you to delete all higher-order modeling terms or interaction terms from the regression model. Select the desired variable rows of the higher-order modeling terms or interaction terms in the regression model, then right-click the mouse and select the **Delete interaction** pop-up menu item.

The tab is also designed to view the distribution of each variable that is selected. It is important that each input variable and the target variable in the regression model be approximately normally distributed. The reason is because extreme values have a profound effect on the modeling fit that will result in imprecision in the parameter estimates and fitted values. Select the desired variable row, then right-click the mouse and select the **View Distribution of <variable>** pop-up menu item that will display the standard frequency bar chart in order to view the distribution of the selected variable. The metadata sample is used to create the histogram. In predictive modeling, the normality assumption is very subjective. Therefore, if the variable displays a highly skewed distribution, then a transformation to the variable is strongly recommended.

The target profile of the target variable can be viewed from the tab. Simply highlight the target variable row, then right-click the mouse and select the **Edit target profile...** pop-up menu item. This will allow you to perform updates to the target profile within the **Regression** node that will then be passed along to the **Assessment** node for modeling assessment.

Interaction Builder

By default, the regression model will initially fit a model without any interaction terms or higher-order terms. Therefore, the purpose of the **Interaction Builder** is to increase the complexity in the modeling design by adding higher-order modeling terms or interaction terms to the model. Although increasing the complexity of the modeling design might lead to overfitting and poor generalization. Selecting the appropriate complexity of the modeling design might result in a better fit and more accurate predictions.

Higher-order terms such as polynomial terms are recommended when there exists a curvature functional form between the input variables and the target variable that you want to predict. Therefore, use of polynomial terms is advised for increasing the accuracy of the prediction estimates. However, fitting the data perfectly might result in poor generalization, resulting in absurd estimates by fitting the model to new data. It is important to point out that in nonlinear modeling, as well as all the other forecasting modeling designs, it is extremely risky to make predictions beyond the range of the actual data points since there is absolutely no guarantee that the functional form will continue to follow the same distributional form beyond the range of the actual data points. This condition especially holds true for nonlinear models like polynomial models as opposed to linear regression models.

At times, use of interaction terms in the regression model is advisable if there exists a relationship between the predicted values and any one of the input variables in the model that depends on the values of another input variable in the model. Interaction between two separate input variables can be viewed by plotting a different trend, slope, or rate of change in the target values across the values of one of the input variables in the model as the values of the other input variable is set to separate values, or vice versa. The interaction plot has been previously displayed. An easier approach to detecting interaction in the input variables is by simply including interaction terms to the model and evaluating the significance of the p-value from the t-test statistic. If the interaction effect is significant to the model, then it is advised that the interaction term and corresponding main effects should be retained in the model; this is called *model hierarchy*. As an initial step in model building, correlation analysis is commonly applied, where each input variable is individually selected from a predetermined cutoff value. However, the shortcoming to this univariate input screening technique is that it ignores the partial associations and the presence of interaction of the input variables in the model.

In backward elimination, the first step is fitting the full model with all main effects and interaction terms and then eliminating significant effects one at a time. Initially, the least significant highest-order interaction term with the largest p-value should be removed before removing the lower-order interaction terms. The next step is eliminating all nonsignificant main effects that are not involved in any significant interaction terms. Therefore, the final regression model should result in all significant interaction terms, main effects that are a part of the interaction terms, and all significant main effects. In forward elimination, a reasonable approach to testing all interaction effects is by fitting each main effect with all possible interaction effects one at a time. That is, fitting the first model with all main effects along with all possible two-way interaction effects based on the first main effect. The next step is fitting the second model with all main effects along with all possible two-way interaction effects based on the second main effect, and so on. The biggest shortcoming to introducing interaction effects to the model is that it makes the relationship between the target variable and the input variables in the model much harder to interpret. Generally, the interaction effects to include in the model should be based on the final modeling selection results along with prior knowledge or general intuition.

The **Interaction Builder** can only be accessed from the **Variables** tab within the **Regression** node. To activate the **Interaction Builder**, simply select the **Interaction Builder** toolbar icon or select the **Tools > Interaction Builder** main menu options from the **Variables** tab. The **Interactive Builder** window will appear that will display two separate list boxes. The **Input Variables** list box that is positioned to the left of the tab displays all the input variables in the regression model. The **Terms in Model** list box positioned to the right of the tab displays the modeling terms that will be added to the regression model. Simply highlight the various input variables in the regression model from the **Input Variables** list box to either add higher-order terms or interaction effects to the regression model.

Adding Interaction or Polynomial Terms to the Regression Model

To add higher-order terms of second degree or higher, simply select the desired input variable from the **Input Variables** list box and press the **Polynomial** button. Press the **Degrees** scroll button to set the number of degrees to the selected input variable. The limitation to adding higher-order terms is that you may add input variables in the regression model of up to 5 degrees. Press the **Cross** button to add interaction terms to the regression model by simply selecting two separate input variables from the **Input Variables** list box. Press the **Expand** button by selecting two or more input variables from the **Input Variables** list box that will result in adding all possible two-way, three-way, and so on, and interaction terms to the model. Press the **Remove** button to remove terms from the model. Select the interaction terms that are listed in the **Terms in Model** list box, then select the **Remove** button. Press the **Standard Model** button to select no interaction terms to the regression model, which is the default; all possible modeling terms up to two-way interactions; or adding all possible modeling terms up to three-way interactions to the regression model. This is illustrated in the following diagram. Press the **OK** button to close the **Interaction Builder** window and return to the **Linear and Logistic Regression** window.

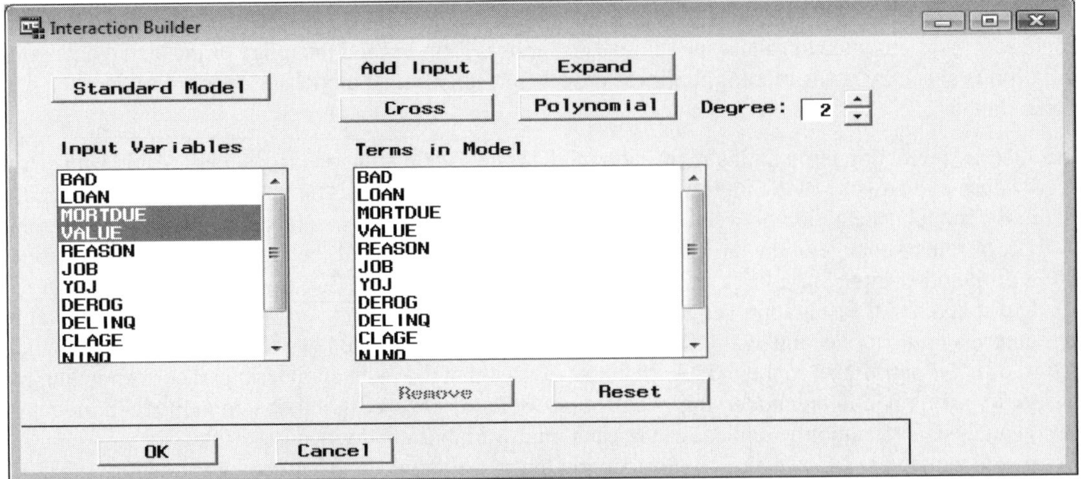

*The **Interaction Builder** window is used to set higher-order or interaction terms for the regression model.*

Model Options tab

The purpose of the **Model Options** tab is designed to display the variable name, measurement level, and event level of the target variable. In addition, the node will allow you to select either linear or logistic regression modeling, the link function, the input coding schema, or remove the intercept term from the least-squares model. The **Model Options** tab consists of the following two subtabs.

- **Target Definition subtab**
- **Regression subtab**

Target Definition subtab

The **Target Definition** subtab displays the variable name and level of measurement of the target variable. For a categorical target variable, the tab will also display the target event that the logistic regression model is trying to predict. For logistic regression modeling, the tab is designed to display extremely important information about the target event that the logistic regression model is trying to predict by fitting the categorically-valued target variable. The **Event** field will inform you as to which value of the response variable is being modeled. It is important to understand the binary-valued target level that the logistic regression model is trying to predict in order to make the interpretation of the lift values from the various lift charts much easier to understand.

Regression subtab

The **Regression** subtab displays the type of regression model, link function, input coding schema, and the removal of the intercept term from the regression model. The input coding schemas are used to determine which effects are most interesting to test. The subtab will indicate to you the type of regression that is performed, which is determined by the level of measurement of the target variable to predict. For logistic regression, the tab will allow you to select different link functions to transform the target values that are generated from the model into probabilities. Regression modeling is performed if there are categorical and continuous-valued input variables in the model with an interval-valued target variable to predict. Regression modeling with a categorical and continuous-valued input variables and an interval-valued target variable is called *analysis of covariance. Generalized linear modeling* is performed if there are only categorically-valued input variables in the predictive model. The purpose of analysis of covariance is to test for significant differences between the various class levels of the corresponding categorically-valued input variable. In multiple regression modeling, the first step in analysis of covariance that is extremely important is to determine if there is no interaction between the separate class levels of the categorically-valued input variable and the interval-valued input variable in the model, that is, in the simplest scenario. This can be done by including the interaction effect between the categorical and continuous input variable in the model. The next step is fitting the multiple regression model to the categorically-valued input variable in order to determine significant differences between the various class levels. In other words, if interaction is present in the model, then it must not be ignored. Otherwise, remove the interaction term from the predictive model. If interaction is not present in the predictive model, then the next step is to determine whether the intercepts are the same.

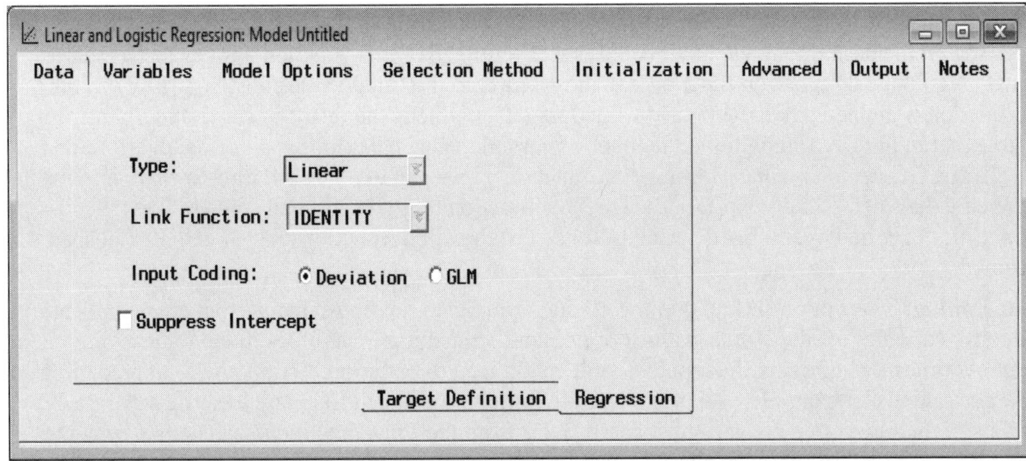

*The **Model Options** tab for linear regression modeling to fit the interval-valued target variable.*

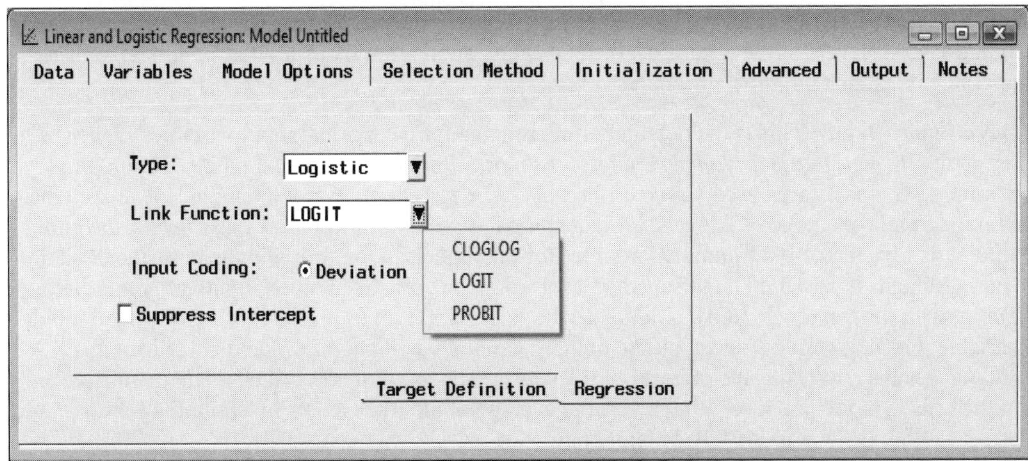

*The **Model Options** tab for logistic regression modeling to fit the categorically-valued target variable.*

The following are the various options available from the **Regression** subtab:

- **Type:** This option displays the type of regression model. The node will automatically determine the type of regression model to fit based on the level of measurement of the target variable that you want to predict. For an interval-valued target variable, the **Type** field will display **Linear** indicating to you that the node is automatically fitting a linear regression model. This cannot be changed. For a categorically-valued target variable, the **Type** field will display **Logistic**, indicating to you that the node will automatically fit a logistic regression model based on either a binary or ordinal-valued target variable to predict. However, you can select a linear regression model that is essentially no transformation applied to the predicted probabilities of the categorically-valued target variable. The advantages to fitting the binary-valued target variable to the least-squares model are documented in the previous **Variable Selection** node. However, the drawback to fitting the categorically-valued target variable to the least-squares model is that at times this method might inadvertently generate undesirable predicted values greater than one or less than zero. For a nominal-valued target variable, the **Type** field will display **Multinomial** to fit the generalized logistic regression model.
- **Link Function:** This option applies to logistic regression modeling only. The link function to apply should match the fitted values and the range of values of the target variable. For linear regression modeling by fitting the interval-valued target variable, this option is grayed-out and unavailable. The link function that is applied is the IDENTITY link function or basically no transformation of the predicted values. The purpose of the link function is to transform the predicted values from the logistic regression model into a desirable range between zero and one. For logistic regression modeling, the LOGIT link function is automatically applied. The logit link function performs an exponential transformation of the predicted values of exp(E(Y) / {1 + exp[E(Y)]}). Therefore,

regardless of the values of the expected target, the transformation will always produce predicted values within the desirable range of zero and one. The PROBIT link function applies the standard normal cumulative distribution function, i.e. the z-score, to the fitted values of $Z_{\text{p-value}}[E(Y)]$. This link function is applied when the binary-valued target variable is the result of some underlying variable that might have an unlimited number of possible values that follow a normal distribution. The CLOGLOG complementary log-log link function applies an exponential transformation to the predicted values of $1 - \exp(-\exp[E(Y)])$. This link function might be applied if the range of probabilities occurred over a fixed period of time. The three link functions have been previously explained.

- **Input Coding:** This option sets up dummy coding or indicator coding schemas that create indicator or dummy variables for the ordinal-valued or nominal-valued input variables in the regression model. At times, categorizing the input variable might lead to better prediction power in predicting the target values. For example, you might want to form various age groups by creating separate intervals of their age or create various income levels from their personal income. The two separate coding schemas that can be selected will result in different prediction estimates and *p*-values. The reason is because one tests the effect in comparison to its overall average and the other tests the difference between the individual effects. Each row of the design matrix corresponds to the unique combination class levels of the input variable and each column to the design matrix corresponds to each parameter estimate in the regression model.

  - **Deviation** (default)**:** This is the default coding schema for categorical input variables in the regression model. In the *deviance coding schema* or differential effects parameterization, the parameter estimates are based on the difference in the effect between the reference level and the average of all the categorical levels. For categorical input variables with *k* class levels, the node automatically creates $k - 1$ dummy variables for the model. In the following matrix, the deviation coding schema is based on three separate levels of a categorically-valued input variable called Temp, with the last level HIGH assigned as the reference level to the categorically-valued input variable. For the reference level, all the dummy variables will have a value of –1. Since this coding schema constrains the effects of all levels to sum to zero, you can estimate the difference in the effects of the last level HIGH with the average of all other levels by computing the average effect of both LOW and MEDIUM temperature levels.

Temp	A_1	A_2
Low	1	0
Medium	0	1
High	–1	–1

  - **GLM:** In the *GLM coding schema*, the parameter estimates are the incremental effect between each level of the categorically-valued input variable and the reference level of the input variable. In GLM coding or incremental effects parameterization, the last level is automatically set as the reference level when the levels are sorted in ascending numeric or alphabetical order. Representing a grouping variable completely requires one less indicator variable than there are class levels in the categorically-valued input variable. The following input variable, TEMP, has three levels that will generate two separate dummy variables to the model, where the first dummy variable A_1 is 1 if low, 0 otherwise. Conversely, the second dummy variable A_2 is 1 if medium, 0 otherwise. For example, the effect in the LOW temperature level is estimated as the difference between the LOW and HIGH temperature levels.

Temp	A_1	A_2
Low	1	0
Medium	0	1
High	0	0

- **Suppress Intercept:** This option will remove the intercept term from the least-squares model. By default, the option is not selected, indicating to you that the intercept term is automatically included in the least-squares model. Therefore, select the check box to remove the intercept term from the least-squares model when it is assumed that the regression line passes through the origin.

Selection Method tab

The purpose of the **Selection Method** tab is to allow you to specify the type of variable selection routine to apply, the order in which the input variables will be entered into the regression model and the criteria values used during the modeling selection routine. The tab consists of the following two subtabs.

- **General subtab**
- **Criteria subtab**

General subtab

The **General** subtab will allow you to perform backward, forward, or stepwise regression modeling. By default, the modeling selection procedure is not applied. Stepwise regression is a process by which the procedure determines the best set of input variables in the regression model that best predicts the target variable based on a pool of all possible combination of input variables. The following variable selection routines can be applied to both linear regression modeling and logistic regression modeling. The purpose of the modeling selection routine is to revert to a simpler model that is more likely to be numerically stable and easier to generalize and interpret. Since the role of the validation data set is to fine-tune the regression model, then the best linear combination of input variables that are selected to the model from the variable selection routine is based on fitting the validation data set. However, if the split sample procedure is not performed, then the final regression model that is selected from the node is based on fitting the training data set. It is important to point out that when you are performing the modeling selection routine, you do not remove input variables from the regression model that best predict the target variable. Therefore, it is important that the input variables are not highly correlated with each other. The reason is because the variable selection routine might exclude these highly related input variables from the model altogether. That is, the drawback to the following modeling selection procedure is that it is not designed to be used on data sets with dozens or even hundreds of potential input variables and their interactions. Again, although all the modeling terms that are removed from the final regression model and set to a role model of **rejected**, these same variables that are removed from the model are automatically passed on to the subsequent nodes within the process flow. Selecting the different selection routines or different criteria in adding or removing variables from the model can result in many different models. Therefore, it is not a bad idea to apply the numerous modeling selection routines with many different criteria, which will result in many different models to select from in order to select the best fitting model.

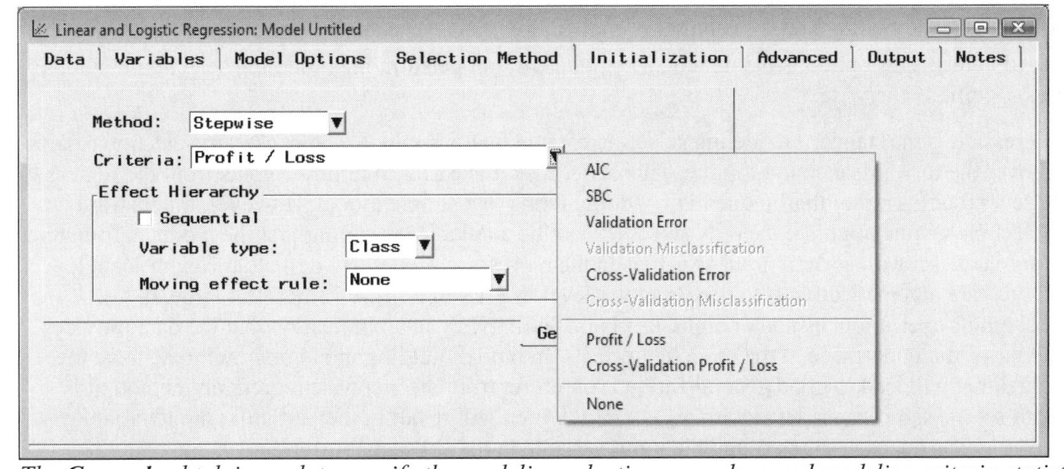

*The **General** subtab is used to specify the modeling selection procedure and modeling criteria statistic.*

The following are the three modeling selection routines to select from the **Method** display field:

- **Backward:** This option performs the backward modeling selection routine. Initially, all the modeling terms are included in the regression model. The backward procedure systematically removes the modeling term that is least significant in comparison to the other modeling terms one at a time. This process continues until there are no more modeling terms with a p-value from the F-statistic that is greater than the α_{stay}, that is, the alpha value specified in the **Stay** display field from within the **Significance Level** alpha value, or when the **Stop** value from within the **Number of variables** section in the regression model is met. Once the input variable is removed from the model, it is never added to the model. It is important to mention that this method is not recommended for binary-valued or ordinal-valued target variables when there are many potential input variables in the regression model or several categorically-valued input variables with many class levels. This is due to the iteratively procedure that is performed in estimating the parameter estimates and standard errors that can be computationally demanding with several different classification models to fit.

- **Forward:** This option performs the forward modeling selection routine. This technique may be useful in determining significant interaction effects. The reason is that the routine will allow you to force all main effects into the model, then sequentially add each interaction term entering into the model one at a time. However, the entry significance level, that is, the **Entry** display field, should be set relatively low, that is, .01 or less, which will result in only very significant interaction effects entering into the regression model. The variable selection procedure begins by fitting the intercept-only model, then the best one-variable model, then the best two-variable model among those that contain the first selected variable, and so on. This method systematically includes modeling terms into the regression model one at a time that are significantly related to the target variable until none of the remaining input variable terms with a *p*-value from the F-statistic is significantly greater than the α_{entry}, that is, the alpha value specified in the **Entry** display field within the **Significance Level** section, or when the **Stop** value from the **Number of variables** section in the regression model is met. In general, once the input variable is entered into the model, it is never removed. This process is repeated until none of the remaining input variables meet the α_{entry} alpha value.
- **Stepwise:** This option performs the stepwise regression modeling selection routine. The procedure begins with no input variables in the regression model, then starts sequentially adding or removing the modeling terms from the regression model based on the significance levels from the **Criteria** subtab. The procedure stops when the α_{stay} **Stay** alpha-level value criterion or the **Stepwise stopping criterion** is met that limits the number of steps performed in the iterative variable selection process.

  The basic stepwise regression steps are such that if the *p*-value associated with the modeling term is significantly below the **Entry Significance Level** α_{entry} alpha-level value, then the modeling term is retained in the regression model. Conversely, the modeling terms that are already in the regression model that are not significantly related to the target variable are removed from the regression model with a *p*-value greater than the **Stay Significance Level** α_{stay} alpha-level value. Although, this method requires more processing time in selecting the best set of modeling terms with comparison to the previous two selection methods, the advantage is that the method evaluates many more potential models before the final model for each set of modeling terms is selected.
- **None** (default)**:** This option is the default. The node will not perform the modeling selection methods unless specified otherwise.

Stepwise regression is the standard modeling selection routine that is applied. The backward selection routine is preferred over the forward selection routine. This is because it is easier to remove inputs from the regression model with several terms rather than sequentially adding terms to a simple model. However, the forward modeling selection routine might be the only selection routine available, assuming that the model is fitting an extremely large data set with several input variables in the regression model. By default, the significant levels of α_{stay}, and α_{entry} are automatically set to the .05 alpha-level in SAS Enterprise Miner. For large data sets, the alpha-level should be set at much smaller alpha-levels that will give you an indication that the data provides strong evidence of the importance in the modeling effects. In model building, it is important to achieve the simplest model that will lead to good generalization. Therefore, from the stepwise regression method, it is advised not to set the significance level with $\alpha_{entry} > \alpha_{stay}$, which will result in the particular input variable added to the model as opposed to removing the input variable in the subsequent steps.

It is recommended to perform the modeling selection routine several times using many different option settings with the idea of evaluating several potential models to determine which model best fits the data. The next step is then to select the input variables that result in the best model based on prior knowledge, previous studies, or general intuition. In addition, it is important that an entirely different data set be applied to the modeling selection routine in order to get a more realistic picture in evaluating the performance of the modeling fit, that is, using one portion of the data to select the best set of input variables and the other portion of the data to evaluate the accuracy of the selected model. This is the general procedure that is performed within the node. In conclusion, after the final models have been selected, it is important that the models be evaluated by the various modeling assessment statistics, that they satisfy the various modeling assumptions, and that you check for multicollinearity in the input variables and influential data points in the data.

The following are the various modeling assessment statistics used in selecting the best model that can be selected from the **Criteria** display field. Once the modeling selection routine terminates, then the candidate model that optimizes the modeling selection criteria from the validation data set is selected as the final model.

- **AIC:** The best model is selected with the lowest Akaike's Information Criteria statistic from the validation data set.
- **SBC:** The best model is selected with the lowest Schwarz's Bayesian Criteria statistic from the validation data set. The SBC statistic severely penalizes models with several modeling terms in the statistical model in comparison to the AIC modeling assessment statistic. Therefore, the AIC goodness-of-fit statistic tends to select the more complex model in comparison to the SBC assessment statistic, which tends to select the simplest model that is most desirable.
- **Validation Error:** The best model is selected with the lowest validation error. For linear regression modeling, the smallest error is based on the sum-of-squares error. For logistic regression modeling, the smallest error is determined by the negative loglikelihood statistic.
- **Validation Misclassification:** The best model is selected with the lowest misclassification rate from the validation data set by fitting the logistic regression model.
- **Cross-Validation Error:** The best model is selected with the lowest validation error.
- **Cross-Validation Misclassification:** The best model is selected with the lowest cross-validation misclassification error rate from the training data set by fitting the categorical-valued target variable.
- **Profit/Loss** (default)**:** This is the default modeling assessment statistic. The best model is selected with the largest average profit or the smallest average loss from the validation data set, that is, assuming that the decision matrix is specified from the target profile. If the validation data set does not exist, then the training data set is used. If the decision matrix is not specified from the target profile, then the best model is selected with the lowest training error.
- **Cross-Validation Profit/Loss:** The best model is selected with the largest average profit or the smallest average loss from the validation data set. Note that none of these cross-validation statistics can be applied to ordinal-valued target variables.
- **None:** This is the default method that is used when you have not defined a decision matrix, that is, with two or more decisions, from the target profile. By selecting this option, the last model selected from the variable selection routine is chosen as the final model.

The **Effect Hierarchy** display field allows you to set requirements for the modeling terms entered or removed from the regression model assuming that interaction effects exist in the model called model hierarchy. For example, specifying a hierarchical effect requires that both main effects A and B must be in the regression model in order to include the two-way interaction term of A*B in the model. Similarly, the main effects A and B cannot be removed from the regression model, assuming that the two-way interaction effect A*B is already present in the regression model. The **Effect Hierarchy** option enables you to control the way in which the modeling terms are entered or removed from the regression model during the variable selection procedure.

- **Sequential:** Forces you to sequentially add input variables into the regression model based on the specified order from the **Model Ordering** option until the termination criterion values are met with a p-value greater than the **Entry Significance Level** from the **Criteria** subtab. This option is useful when fitting polynomial models by including linear, quadratic, and cubic terms in that order.
- **Variable type:** Determines if categorically-valued input variables (**Class**) or both the interval-valued or categorically-valued input variables (**All**) are subject to hierarchy.
- **Moving effect rule:** Determines whether hierarchy is maintained and whether single or multiple effects are entered or removed from the model in one step.

  Single: A single modeling term is entered or removed from the regression model that is subjected to hierarchy.

  Multiple: Multiple modeling terms are entered or removed from the regression model that are subjected to hierarchy. This option will accelerate the variable selection process.

  None (default): Modeling terms entered or removed from the regression model are not subject to hierarchy, that is, any single effect can be added or removed from the model at any given step of the modeling selection process.

Criteria subtab

The **Criteria** subtab will allow you to specify various criteria for the modeling selection routine. The subtab is divided into two separate sections. At times, common knowledge or general know-how is needed to determine if certain input variables should be automatically entered into the regression model that best predicts the target variable. Therefore, the **Number of variables** section that is located to the left of the tab will allow you to specify the number of input variables that must exist in the regression model no matter which input variables are included in the final model from the modeling selection routine. It is important to understand that the input variables that are displayed at the top of the list from the **Model Ordering** option are either automatically added to the final model or automatically added to the initial regression model as the procedure begins. The **Significance Levels** section allows you to specify the significance levels for each modeling term with the *p*-values and the associated t-test statistics from the modeling terms meeting these significance levels either being added or removed from the regression model during the iterative modeling selection procedure.

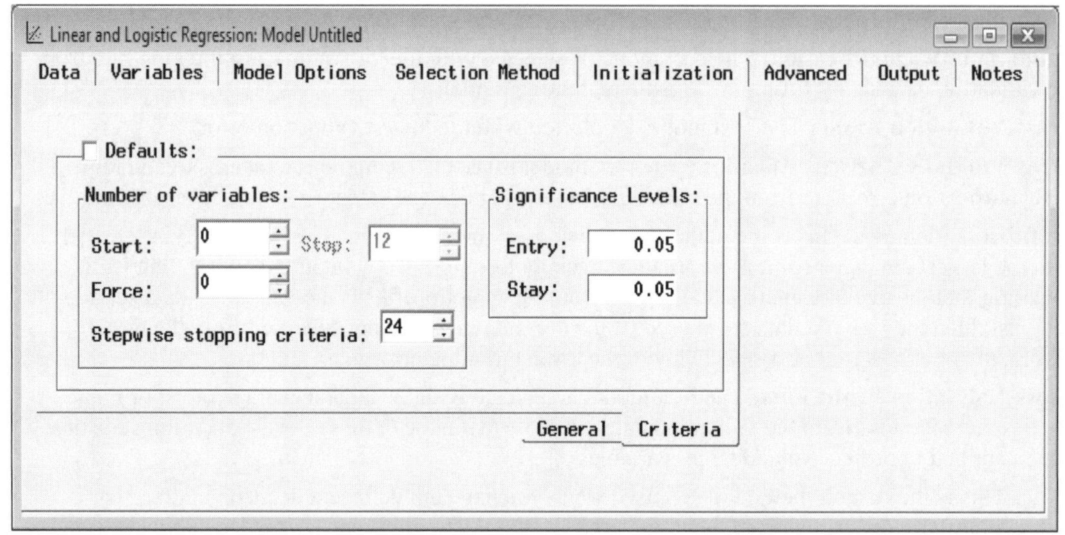

*The **Criteria** subtab is used to set the various criteria settings for the modeling selection procedure.*

The **Number of variable** section will allow you to control the most important modeling terms that should be automatically included in the regression model when either the iterative modeling selection procedure starts or finishes. Again, the modeling terms entered or removed from the regression model can be specified from the **Model Ordering** option.

- **Start:** The **Start** display field will allow you to specify the number of modeling terms to automatically include in the regression model before the iterative variable selection procedure begins. For both **Forward** and **Stepwise** variable selection methods, the **Start** value is automatically set to zero. For the **Backward** method, the value is set to the number of input variables. That is, the first *n* effects that are listed at the beginning of the **Modeling Order** list box are automatically selected for the model when the variable selection procedure begins.
- **Force:** The **Force** display field will allow you to specify the number of modeling terms that must exist in the regression model even if these same modeling terms are not significant. These same modeling terms can be specified from the **Model Ordering** option that appears at the top of the list. For example, entering three in the **Force** display field will result in the first three modeling terms that must exist in the regression model after the modeling selection routine terminates. In other words, the first three modeling terms that are listed at the top of the list from the **Model Ordering** list box will be automatically included into the final model.
- **Stop:** The **Stop** display field will allow you to specify the number of modeling terms to include in the regression model when the iterative modeling selection procedure terminates. For the **Forward** variable selection procedure, the default **Stop** value is the maximum number of modeling terms in the final model. For the **Backward** method, the **Stop** value is automatically set to the minimum number of modeling terms in the final model.

The **Stepwise stopping criteria** display field will allow you to control the maximum number of steps in the **Stepwise** selection procedure. By default, the number that is displayed is twice the number of modeling terms in the regression model.

The **Significance Levels** section will allow you to set the alpha-level to either the modeling terms entered or removed from the regression model. The **Entry** display field is the alpha-level for the modeling terms entering into the regression model. In the variable selection routine, a modeling term with a *p*-value greater than the **Entry** alpha-level value will prevent the modeling terms from entering into the regression model. Conversely, modeling terms with a significant *p*-value greater than the **Stay** significant level will result in the modeling term being removed from the regression model.

Model Ordering Option

The **Model Ordering** option will allow you to control the order in which the modeling terms are entered into the regression model. The option is also designed for you to specify which input variables are extremely important for the regression model. Again, there might be times that certain input variables are known to be extremely important for the regression model. Therefore, this option will allow you to specify the most important input variables to the final model. The most important modeling terms will be positioned at or near the top of the list. By default in Enterprise Miner, all modeling terms are entered into the regression model in alphabetical order followed by the corresponding interaction terms, then the polynomial terms. The reason why it is so important that the main effects are first entered in the regression model is to obtain the simplest model with the fewest number of parameters in predicting the target values. This will result in a larger number of degrees of freedom in the statistical model that will lead to more powerful tests in testing the accuracy of the regression model. In backward modeling selection routine, the input variables that are forced in the model will be automatically included in the model once the iterative modeling selection routine begins. In both forward and stepwise regression, the input variables that are forced in the model will be automatically included in the final regression model once the modeling selection routine terminates. The **Model Ordering** window is illustrated in the following diagram.

Select the **Tools > Model Ordering...** main menu options that will result in the following **Model Ordering** window appearing. Once the **Model Ordering** window is initially displayed, then the order in which the input variables are listed in the list box will display no coherent order as to the importance of the input variables in the regression model. Therefore, highlight the modeling terms that are listed from the list box and select the up or down arrow buttons to change the order in which the modeling terms are entered into the regression model. The modeling terms that are positioned at or near the top of the list are the most important modeling terms for the regression model, therefore, these modeling terms are first entered into the regression model. Conversely, the modeling terms that are listed at or near the bottom of the list are the last few terms entered into the regression model. Press the **Reset** button to undo any changes that have been made. This will reset the order in which the listed variable first appeared when you opened the window. Press the **OK** button and the **Linear and Logistic Regression Model** window will reappear.

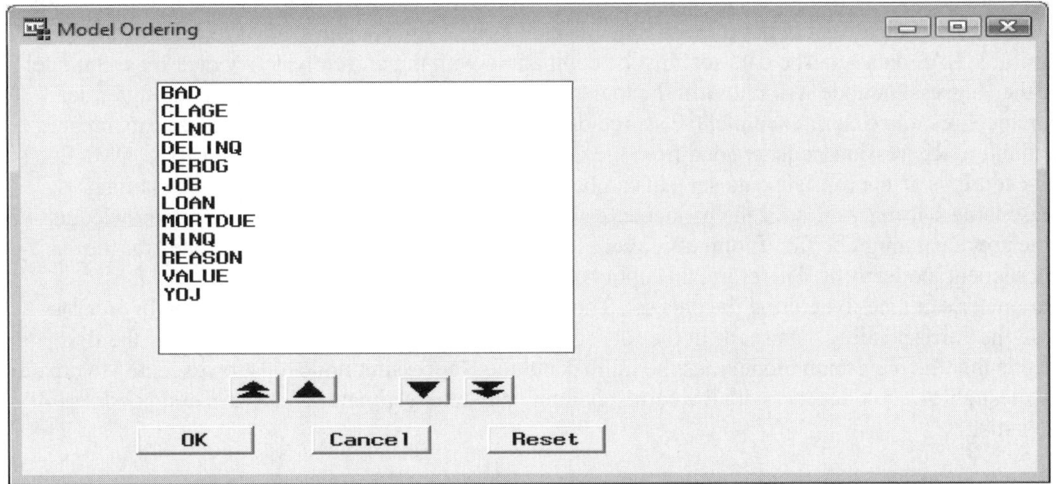

The ***Model Ordering*** *window is used to set the order of the effects entering into the regression model.*

Initialization tab

Both the **Initialization** tab and the **Advanced** tab are specifically designed for logistic regression modeling. The reason is because a link function is applied that transforms the expected target into a desirable range between zero and one. However, this transformation results in a nonlinear system of equations. In other words, the partial derivatives from the nonlinear error function with respect to the parameter estimates results in no unique solution. Therefore, the parameter estimates must be solved iteratively by some optimization technique. The **Initialization** tab will allow you to specify the initial starting values of the parameter estimates for the subsequent logistic regression modeling fit.

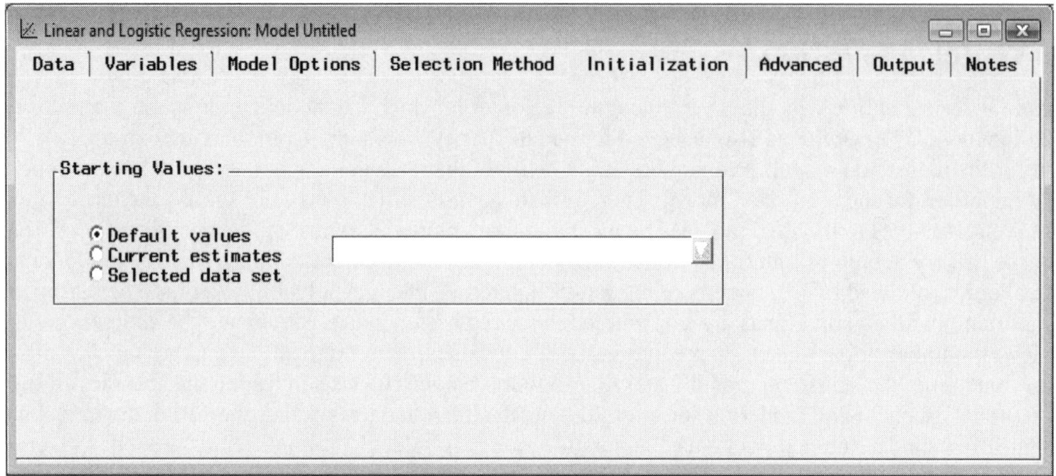

*The **Initialization** tab is used to set the starting values of the parameter estimates in the logistic model.*

The following options are available within the **Starting Values** section:

- **Default values:** The default starting values for the intercept and slope estimates from the logistic regression model are automatically set to zero.
- **Current estimates:** The parameter estimates from the initial fit are used as the current starting estimates in refitting the logistic regression model. One reason for selecting this option is to assure you convergence that the correct logistic regression parameter estimates have been achieved. This can be determined by observing small changes in the parameter estimates between the successive fits. It should be noted that selecting this option, the **Regression** node will always rerun even though it has converged.
- **Selected data set:** This option will allow you to specify a data set that contains the parameter estimates that are used as the initial parameter estimates of the subsequent fit to the logistic regression model. Selecting the **Selected data set** radio button will active the corresponding list box. Click the drop-down arrow button within the display field to browse the system folders for the desirable SAS data set. The data set must be compatible with the currently active data set and model or the **Regression** node will fail to fit the logistic regression model with the corresponding initial parameter estimates. For example, the scored data set, with the parameter estimates that are created within the **Regression** node, created from the OUTEST option statement from the PROC DMREG procedure, is an appropriate data set that can be specified from this option. At times, obtaining reasonable starting values for the parameter estimates might be determined by general knowledge. One approach might be first fitting a simpler model to obtain reasonable starting values for the subsequent model to fit. Therefore, this option will allow you to set your own values for the parameter estimates by editing the data set. The next step would be to select this option in order to read the corresponding data set to fit the subsequent logistic regression model. However, the first step is that the regression model must be trained and the **Regression** node must be executed to create the desirable output data set with the corresponding parameter estimates and associated goodness-of-fit statistics.

Advanced tab

The **Advanced** tab is specifically designed for logistic regression modeling. The tab will allow you to specify the optimization or minimization technique along with the various convergence criteria values. In Enterprise Miner, convergence is assumed to an iterative procedure when any one of the listed convergence criterion values are met. By default, the default optimization techniques are applied when training the logistic regression model. The tab will also allow you to specify the maximum number of iterations, maximum function calls and the maximum amount of computational time before terminating the procedure, with an added option to minimize resources to avoid memory limitations with several categorical input variables in the logistic regression model.

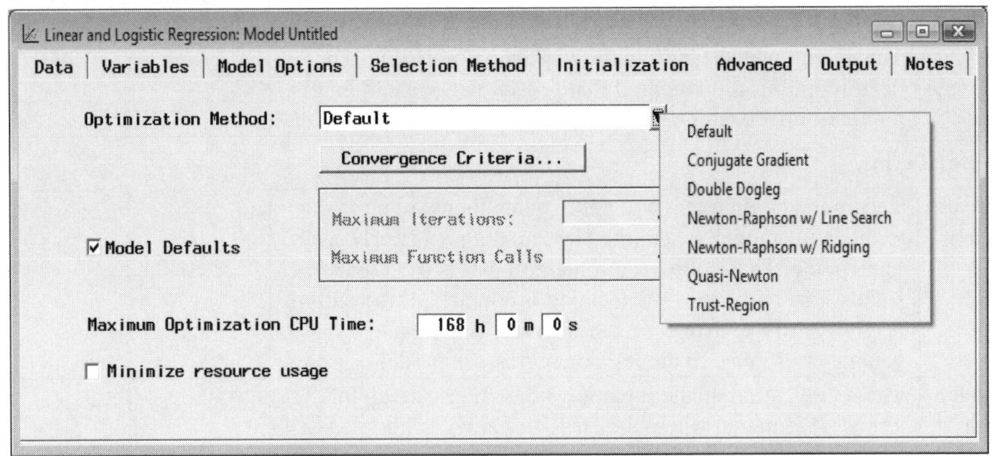

*The **Advanced** tab is used to specify the optimization technique for the logistic regression model.*

The following are the various options available for selection within the **Advanced** tab.

- **Optimization Method:** This option will allow you to specify the minimization technique that is used in finding the best set of parameter estimates for the model. Since there is no superior optimization technique that is considered the standard optimization technique to apply, the optimization technique that is automatically applied depends on the number of parameters in the logistic regression model.
- **Model Defaults:** By default, the **Model Defaults** check box is selected. Clearing the **Models Default** check box will allow you to enter the maximum number of iterations to the iterative process and the maximum number of function calls that are evaluated at each iteration. Both the **Maximum Iterations** and **Maximum Function Calls** options depend on the optimization method that is applied. The number of *function calls* is the number of internal calculations performed on the error function, gradient function, and Hessian matrix. Normally, several function calls are needed to achieve convergence in the iterative procedure. By default, the maximum number of iterations is missing, indicating to you that the value will be set at run time that is determined by the amount of data to fit and the complexity of the nonlinear logistic regression design.
- **Maximum Optimization CPU Time:** This option will allow you to set a time limit in training the logistic regression model. By default, the maximum amount of computational time is automatically set to 168 hours or 7 days in the iterative process for calculating the parameter estimates in the logistic regression model.
- **Minimize resource usage:** This option is useful when the logistic regression model consists of a categorically-valued target variable with many class levels, where it is not uncommon to have a large number of parameters in the logistic regression model. By default, this option is unchecked. However, selecting this option will result in the optimization technique defaulting to the conjugate gradient method, where the Hessian matrix is not needed. Therefore, this will result in standard errors of the parameter estimates that will not be computed in the logistic regression model. The standard errors from the logistic regression model are based on the diagonal elements of the negative of the inverted Hessian, which can be computationally demanding with several input variables in the model. Therefore, Wald's statistic will not be computed. Wald's statistic is used to measure the significance of the input variables in the logistic regression model. The statistic is calculated by

taking the parameter estimates and dividing by their standard errors, which follow a chi-square distribution, assuming that you have a large sample size. When this option is selected, it will prevent the various modeling selection routines from being performed.

The following table displays the various optimization methods that can be selected from the **Optimization Method** option and the default maximum number of iterations and function calls for each method along with the default optimization technique that is applied based on the number of modeling terms in the logistic regression model.

Optimization Method	Description	Max. # of Iterations	Max. # of Function Calls
Default	The default method for logistic regression modeling. The default optimization method that is used will depend on the number of parameter terms in the logistic regression model.		
Small Number of Terms			
Newton–Raphson w/ Ridging (default)	The optimization method to apply when there are up to 40 terms in the model, where the Hessian matrix is fairly easy to compute. This is the default method that is used for small-sized problems. The Hessian is a matrix of the partial derivative of the error function with respect to each parameter estimate in the logistic regression model.	50	125
Newton–Raphson w/ Line Search	The optimization method to apply when there are up to 40 terms in the model where the Hessian matrix is fairly easy to compute.	50	125
Trust-Region	Similar to Newton–Raphson with ridging but more stable that is based on restricted steps during updates.	50	125
Medium Number of Terms			
Quasi-Newton (default)	The optimization method to apply when the number of parameters in the logistic model is greater than 40 and up to 400. This is the default method used for medium-sized problems where the objective error function or the gradient function is much faster to compute than the Hessian matrix. The *gradient error function* is the vector of partial derivatives of the objective error function with respect to the parameters that are evaluated for each term in the logistic model. The gradient points in either the steepest uphill or downhill search direction.	200	500
Double Dogleg	Requires more iterations than the previous quasi-Newton method.	200	500
Large Number of Terms			
Conjugate Gradient (default)	The optimization method to apply when the number of parameters in the logistic model is greater than 400. The method has slow convergence where the objective error function or the gradient function is much faster to compute than the Hessian matrix. This method does not require the Hessian matrix due to the large number of parameters in the logistic regression model.	400	1000

By selecting the **Convergence Criteria...** button, the following **Convergence Criteria** window will appear in order to set the various convergence criteria values for the iterative grid search procedure.

Convergence Criteria window

The purpose of the *convergence criterion* or the *termination criterion* is to set a threshold value in terminating the iterative grid search procedure. The iterative gradient-descent grid-search is a convergence technique used in logistic modeling and the back-propagation algorithm, which is also applied in neural network modeling. The iterative *gradient-descent search* is an optimization technique that is applied to minimize the vector of error values at each iteration by starting with an arbitrary initial vector of parameter estimates and repeatedly minimizing error values in small steps. At each iteration, the parameter estimates are modified with the new values replaced by the past values that produce a smaller squared error along the error surface in searching through a space of possible parameter estimates. The gradient search is a "hill climbing" technique where the gradient is the partial derivatives of the error function with respect to the parameter estimates that is designed to determine the direction that produces the steepest increase in the error surface. Therefore, the negative of the gradient determines the direction of the steepest decrease in the error function. The process continues until hopefully the smallest error is reached. The drawback to the minimization techniques is that if there are numerous local minimums in the bumpy multidimensional error surface, then there is no guarantee that the iterative procedure will find the desirable minimum. This is especially true if the initial parameter estimates are not even near the correct parameter estimates.

The node uses several different convergence criteria. The reason is because there is no universal convergence criterion that is used to stop the iterative procedure. The default is the relative gradient convergence criterion. In Enterprise Miner, when any one of the following convergence criterion values are met, then convergence is assumed and the iteration procedure stops. When the error function falls below any one of the convergence criterion values, then convergence is assumed and the iterative grid search procedure stops. In Enterprise Miner, the optimization procedure looks at one or more of the various convergence criteria to determine if convergence is met. However, it is important to point out that setting convergence criterion value that is too strict will force SAS Enterprise Miner to produce even more accurate parameter estimates for the logistic regression model. This might lead to Enterprise Miner failing to report convergence more often than not. In addition, it is important to remember that even if one of the convergence criterions has been met, however, there is no guarantee that the optimal set of parameter estimates has been found.

The optimization technique and the various convergence criteria relate to logistic regression modeling. These optimization procedures do not apply to linear regression modeling in predicting an interval-valued target variable. Although the appearance of the **Convergence Criteria** window and the **NLP** tab in the **Neural Network** node look very similar. It would be incorrect to assume that their functionality is the same. In the following **Convergence Criteria** window for logistic regression modeling, convergence is assumed when any one of the listed convergence criterion values are met in any one iteration, as opposed to the neural network model where convergence is assumed when any one of the convergence criterion values are met in successive iterations at a specified number of times, which can be more than once.

Convergence can be verified from the **Result Browser**. From the **Result Browser**, simply view the message within the **Optimization Results** section in the **Output** tab from the PROC DMREG procedure output listing, which will inform you that one of the convergence criterion values have been met.

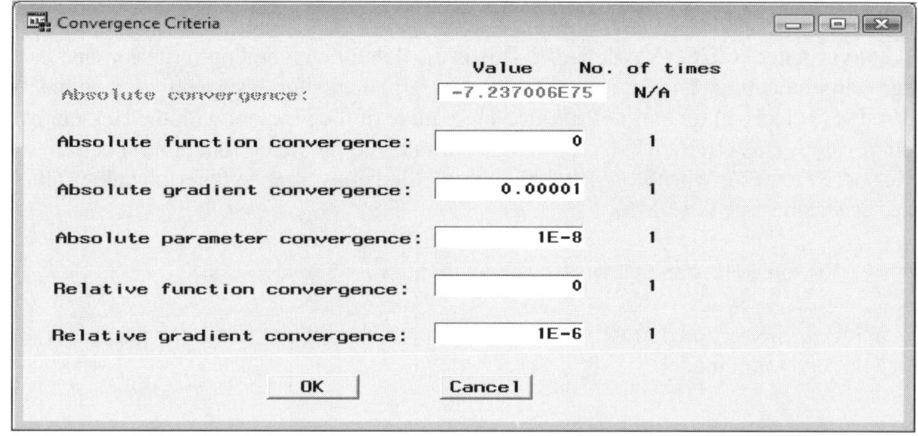

*The **Convergence Criteria** window is used to set the convergence criteria values.*

The abbreviations represent the option settings for the convergence criteria value based on the NLOPTIONS option statement within the PROC DMREG data mining procedure. In Enterprise Miner, the procedure is used to generate the subsequent regression results. The following are the various convergence criteria values to select from the **Convergence Criteria** window:

- **Absolute convergence (ABSCONV):** Specifies the absolute convergence criterion for the error function. Requires a small change in the error function in successive iterations. The default convergence criterion value is extremely small.

  $| f(x^k) | \le c$ for some error function f at the kth iteration.

- **Absolute function convergence (ABSFCONV):** Instead of looking at just the absolute value of the error function, look at the absolute difference in the error function value between successive iterations, where convergence is considered to occur when it falls below the specified convergence criterion value. The default convergence criterion value is zero.

  $|f(x^{k-1}) - f(x^k)| \le c$ for some error function f at the kth iteration.

- **Absolute gradient convergence (ABSGCONV):** Looks at all of the values of the gradient error function, that is, the first derivative of the error function with respect to the parameter estimate in the logistic regression model, from the current iteration, takes the maximum of those values, and considers convergence to occur when the maximum gradient objective function value falls below the specified convergence criterion value. The default convergence criterion value is .00001.

  $\max_j | g_j^{(k)} | \le c$ for some gradient function g at the kth iteration.

- **Absolute parameter convergence (ABSXCONV):** Looks at the difference in the parameter estimates between successive iterations, takes the maximum of those values, and considers convergence to occur when the maximum difference in the parameter estimate falls below the specified convergence criterion value. The default convergence criterion value is 1.0E-8.

  $| \beta^{k-1} - \beta^k | \le c$ for some parameter estimate β at the kth iteration.

  Note: The PROC IML code that calculates the parameter estimates and the likelihood ratio goodness-of-fit statistic by fitting the binary-valued target variable, BAD, from the HMEQ data set can be viewed from my website. An iterative process is repeated until the maximum difference in the parameter estimates between successive iteration is less than 1.0E-8.

- **Relative function convergence (FCONV):** Rather than looking at the absolute difference in the objective function values between successive iterations, it looks at a ratio of the absolute difference in the error function between successive iterations and the absolute value of the current error function value. Convergence is considered to occur when it falls below the specified convergence criterion value. This is rarely a good tool to use for convergence identification because it is most likely only going to occur at a local solution when the gradient values are still large. The default convergence criterion value is zero.

  $\frac{|f(x^{k-1}) - f(x^k)|}{MAX(|f(x^{k-1})|, c)} \le c$ for some error function f at the kth iteration.

- **Relative gradient convergence (GCONV)** (default)**:** This is the default convergence criterion that is used in logistic regression modeling. That is, if none of the convergence criteria is specified, then the default is GCONV=1E-6. It looks at the ratio of a scaled magnitude of the gradient with the Hessian over the objective function value. This criterion is looking at the normalized predicted function reduction. Convergence is considered to occur when it falls below the specified convergence criterion value. The default convergence criterion value is 1.0E-6.

  $\frac{(g^k)' [H^k]^{-1} (g^k)}{MAX(|f(x^k)|, c)} \le c$ for some Hessian matrix H at the kth iteration.

  Note: The Hessian matrix is the second derivative of the error function with respect to the parameter estimates in the logistic regression model.

Output tab

The purpose of the **Output** tab is designed to score the partitioned data sets. A scored data set is a SAS data set with the computed predicted values and the associated residual values from the regression model along with all other variables in the data set. The node also creates an additional output data set with the parameter estimates from the regression model along with its associated t-test statistic that are calculated from the training data set along with the various assessment statistics or goodness-of-fit statistics from each partitioned data set. This same parameter estimates data set can then be used in computing different predicted values given a new set of values for the input variables in the regression model. The **Regression** node must first be executed in order to view the corresponding output data sets. The **Output** tab consists of the following three subtabs.

- **Scored Data Sets subtab**
- **Parameter Estimates subtab**
- **Printed Output subtab**

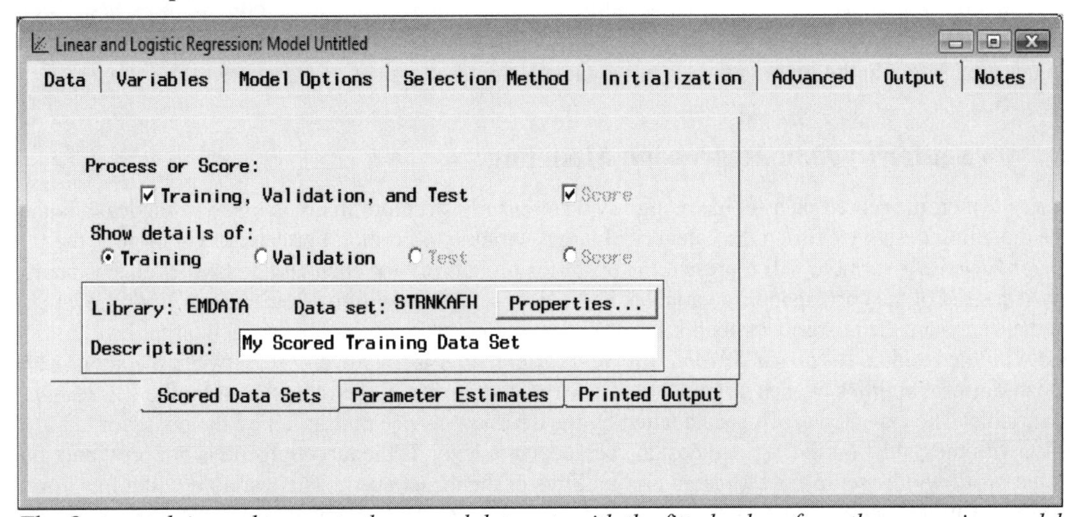

*The **Output** tab is used to create the scored data sets with the fitted values from the regression model.*

Scored Data Sets subtab

The **Scored Data Sets** tab will allow you to create the scored data set and view the scored data set from the selected partitioned data set. By default, the **Process or Score: Training, Validation, and Test** check box is automatically unchecked. Select the **Process or Score: Training, Validation, and Test** check box to create a scored data set for each partitioned data set once the **Regression** node is executed and the model is trained. The scored data set can be passed on to the **Assessment** node for modeling assessment. Select the appropriate radio button from the **Show detail of:** section to view the file administrative information and table listing of the corresponding partitioned data set. The output data set will display the fitted values from the statistical model along with all other variables in the partitioned data set. By selecting the corresponding radio button, a scored data set will be created for each one of the partitioned data sets. Creating a scored data set will also result in separate macro variables that will be created within the **SAS Code** node for each scored data set. The output scored data set will be overwritten and updated accordingly when changes have been applied to the currently fitted model from the **Regression** node.

From the **Show details of** section, select the appropriate radio button, then select the **Properties...** button to view both the file administrative information and a table view of the scored data set from the selected partitioned data set. The tab will display file administrative information such as the creation date, modification data, number of columns or variables, and number of rows or observations of the selected scored data set.

The section that is listed at the bottom of the subtab will display the EMDATA library that is automatically assigned to the scored data set along with the data set name assigned to the scored data set. The **Description** entry field will allow you to enter a short description of the scored data set.

Scored Data Set for Linear Regression Modeling

The file structure of the scored data set for multiple linear regression modeling will consist of the variables in the data set along with the corresponding P_*target name* and R_ *target name* variables. The P_ *target name* variable represents the fitted values from the predictive model for each observation. The R_ *target name* variable represents the residuals of the predictive model that is the difference between the actual target values and the fitted values. In addition, a missing value indicator called _WARN_ is automatically created to identify all records with missing values in any one of the input variables in the data set. When comparing models, it is essential that each model under comparison have the same number of observations. If observations are missing in one modeling design that are not in another, then you will get invalid, "apples-and-oranges" modeling comparisons. Therefore, the modeling nodes are designed to generate predictions, even in situations where there are missing values in any one of the input variables. Unlike the standard modeling procedures in SAS, the Enterprise Miner modeling nodes will automatically compute the predicted values even when any one of the input values are missing. For categorically-valued input variables, the missing values are treated as an additional category. For interval-valued target variables, missing values in any one of the input variables in the least-squares model, the fitted values are imputed by the average values of the target variable. Otherwise, when all the values are present in the input variables in the model, the fitted values are estimated from the least-squares model.

Scored Data Set for Logistic Regression Modeling

In Enterprise Miner, the scored data set has a slightly different file structure in comparison to the least- squares predictive modeling design by fitting the categorical target variable to predict. For categorical targets, the P_ *target variable <level>* variable will represent the posterior probability for each target class level. The scored data set will consist of the corresponding variables in the data set along with the P_*target variable <level>* variables that represent the posterior probabilities or the estimated proportions for the each target level associated with the residual R_ *target variable <level>* variable that is the difference between the actual values and the posterior probabilities at each target level. The expected profit will be represented by the EP_*target variable* variable. The expected profit is calculated by the decision entries multiplied by the posterior probability, with the reduction in the fixed cost at each decision level. If the target profile is not provided, then the expected profit will be set to the estimated probabilities of the target event. This variable will allow you to list all the observations that exceed a predetermined profit threshold value. Typically, this threshold value is set at zero. The I_ *target name* variable is the classification identifier variable that identifies which class level the observation is classified that is determined by the largest estimated probability. For binary-valued target variables, the observation is identified as the target event if the posterior probability of the target event is greater than the estimated probability of the target nonevent or greater than .5. For nominal-valued or ordinal-valued target variables, the observation is classified as the appropriate target category based on the largest estimated probability. Also using the Enterprise Miner modeling nodes and data mining analysis, the predicted values are computed even when the observations in any one of the input variables are missing. The estimated probabilities P_*target variable <level>* value will be automatically set to the prior probabilities for each class level of the categorically-valued target variable. If the prior probabilities are not specified, then the proportion of each target class level among all nonmissing target values is used.

Parameter Estimates subtab

The **Parameter Estimates** subtab will display the data mining data set that consists of the parameter estimates along with the goodness-of-fit statistics. In other words, the data set consists of two separate records. The first record lists the parameter estimates and the various assessment statistics from the partitioned data sets. The second record displays the associated t-test statistic for each modeling term in the regression model. The listed statistics are calculated from the training data set. The **Regression** node must be executed and the model must be trained in order to view the data set. In other words, the parameter estimates data set is automatically created once the node is executed. This same data set can be specified from the **Selected data set** option within the **Initialization** tab to refit the logistic regression model with the listed parameter estimates that will be set to the initial parameter estimates in the subsequent modeling fit. The first step would be opening this subtab in order to view the data set name that is automatically assigned to this scored data set, then updating the parameter estimates through data step programming. The next step is selecting the **Selected data set** option in order to select the appropriate data set that is automatically created in the EMPROJ folder of the currently opened diagram. As an alternative, this same scored data set can be accessed from the **SAS Code** node, thereby modifying the parameter estimates within the corresponding node. The **Description** entry field will allow you to enter a short description to the parameter estimates data set. Select the **Properties...** button that will open both the file administrative information of the data set and a table view of the data set with the parameter estimates for the current regression model.

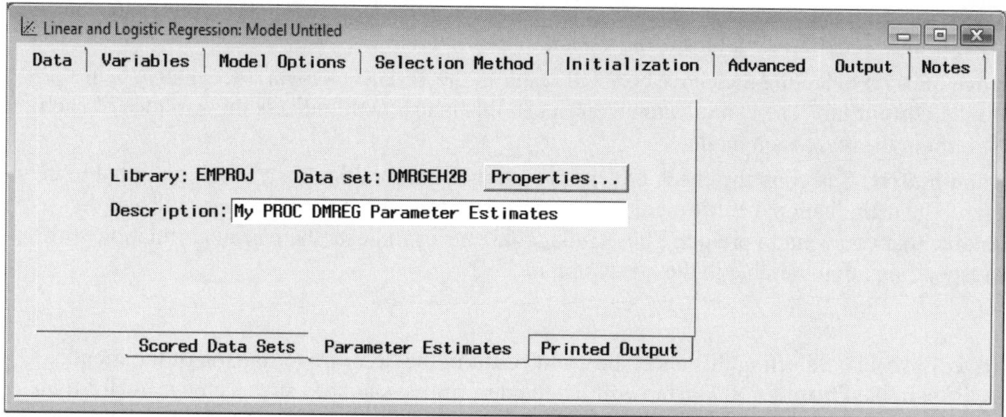

*The **Parameter Estimates** subtab that displays the data set with the parameter estimates.*

Note: For linear regression modeling, the parameter estimates measures the linear relationship between each input variable in the model and the target variable to predict. For logistic regression modeling, the parameter estimates measures the rate of change in the logit scale, that is, the log of the odd ratio, with respect to a one unit increase in the input variable with all other input variables held constant. Therefore, the coefficients will provide the odds ratio statistic by taking the exponential of the parameter estimates.

Printed Output subtab

The **Printed Output** subtab is designed to control the amount of listed output that is generated by executing the **Regression** node. By default, the following options are unselected. However, selecting the various options will result in the listed output being displayed in the **Output** tab that is created from the PROC DMREG regression procedure running behind the scenes once the node is executed and the regression model is trained.

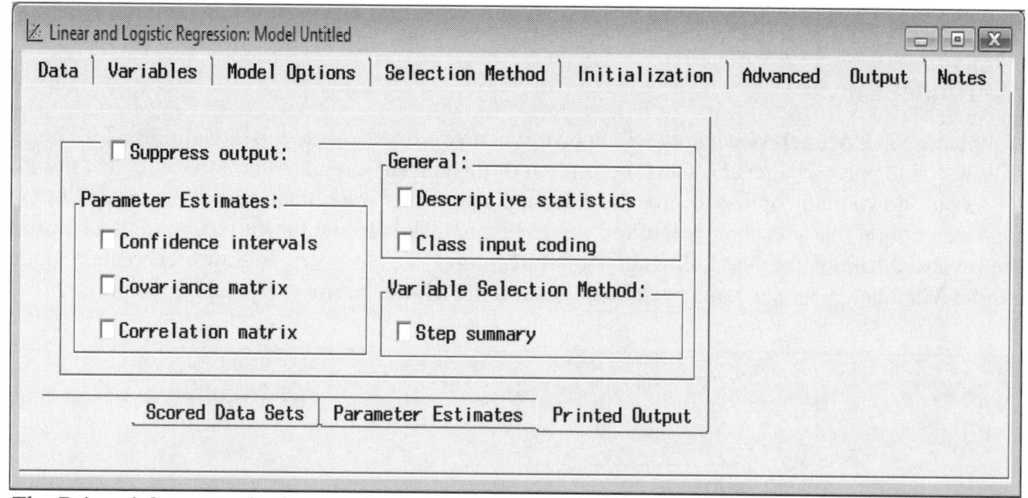

*The **Printed Output** subtab is used to control the listed output to the **Output** tab in the **Result Browser**.*

The following are the various options to select from the **Printed Output** subtab to control the listed results that will be displayed within the **Result Browser**.

- **Suppress Output:** By default, the check box is unselected. Selecting the check box will prevent the procedure output listing from being displayed within the **Output** tab. From the **Result Browser**, selecting this option will result in the **Output** tab being dimmed and unavailable for viewing.

Parameter Estimates

- **Confidence intervals:** Selecting the check box will result in the confidence interval of the parameter estimates listed in the **Output** tab. All parameter estimates that contain the value of zero within their own confidence interval will indicate that the modeling term is a bad predictor for the regression model. Conversely, all parameter estimates beyond their own confidence interval will indicate the significance of the modeling term for the regression model.
- **Covariance matrix:** Selecting the check box will result in the variance–covariance matrix being printed to the **Output** tab. The main diagonal entries in this matrix will indicate the variance of each modeling term in the regression model.
- **Correlation matrix:** Selecting the check box will result in the correlation matrix being printed to the **Output** tab. The main diagonal entries indicate the correlation between the input variable and the target variable that you want to predict. The off-diagonal entries indicate the partial correlation with each corresponding input variable in the regression model.

General

- **Descriptive Statistics:** Selecting this check box will result in the descriptive statistic of the input variables listed in the **Output** tab. The tab will list the descriptive statistics such as the standard five-point summary statistics of the mean, median, standard deviation, minimum, and maximum. For a categorically-valued target variable, a two-way frequency table is also displayed for each categorically-valued input variable in the regression model. This will allow you to determine how well each categorically-valued input variable in the logistic regression model predicts a particular target level.
- **Class input coding:** Selecting this check box will result in the design matrix being displayed in the **Output** tab, which could result in a tremendous amount of printed output if there are several categorically-valued input variables in the regression model with many class levels.

Variable Selection Method

- **Step summary:** Selecting this check box will result in the listing of each step performed in the modeling selection procedure. The listing will display the iterative variable selection routine in which the modeling terms are entered or removed from the regression model at each step. However, this could amount to a tremendous amount of listed output, assuming that there are several modeling terms in the regression model.

Creating the Model Entry

Initially, the following **Save Model As** window will appear when you first execute the modeling node. The window will allow you to enter a name and short description of the current model. Alternatively, select the **File > Save Model As** pull-down menu option to save the current changes that were made within the node. Both the model name and description that you enter will allow you to distinguish between the various models that can be selected and reviewed from within the following **Model Manager**. To close the window and create a model entry in the **Model Manager**, a model name must be entered in the **Model Name** entry field, then press the **OK** button.

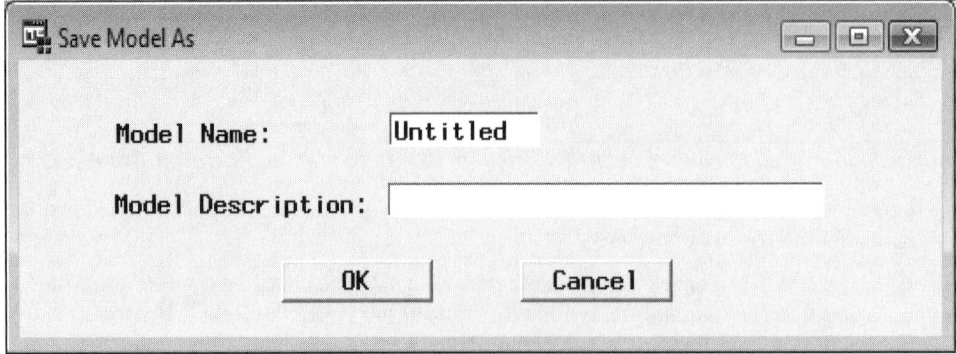

*The **Save Model As** window is used to enter the name and description of the model.*

Viewing the Regression Node Results

General Layout of the Results Browser within the Regression Node

- **Model tab**
- **Estimates tab**
- **Plot tab**
- **Statistics tab**
- **Code tab**
- **Log tab**
- **Output tab**
- **Notes tab**

It is important to understand that the following parameter estimates of the regression model are calculated from the training data set. Therefore, assuming that the split sample procedure is applied, then the purpose of the validation data set is to produce unbiased estimates, since the same training data set that is used to fit the model will result in biased estimates. The best linear combination of parameter estimates is based on the smallest assessment statistic from the validation data set, assuming that the modeling selection procedure is applied. Also, the scored data set from the validation data set is automatically selected for modeling assessment. In other words, the various performance charts that are displayed in the **Model Manager** or the **Assessment** node are automatically created from the validation data set for modeling assessment.

Also, it is important to know that once the **Regression** node is executed, then the PROC DMREG data mining regression procedure will be running behind the scenes. Therefore, a large portion of the output listing that is displayed within the **Result Browser** is generated from the PROC DMREG procedure output listing. However, it is important to remember that not all the options used in the corresponding data mining procedures are incorporated in the listed results within the node, and vice versa.

From the following least-squares modeling results, the **Filter Outliers** node was applied to the process flow diagram to remove outliers in the target variable, DEBTINC. In addition, extreme values in the input variables, MORTDUE, VALUE, and CLAGE were removed from the model with values that were well separated from the rest of the other data points. The **Transform Variables** node was applied to transform the DEROG and DELINQ interval-valued variables into indicator variables in order to increase the predictive power of the least-squares model. And, finally, the stepwise regression was performed where the validation error was selected as the modeling criteria statistic to select the best set of input variables to predict the interval-valued target variable, DEBTINC. The process flow diagram can be viewed from the previous **Filter Outliers** node.

Model tab

The **Model** tab will display the various configuration settings that are applied to the node. However, the various configuration settings that are displayed cannot be changed. The appearance of the various subtabs is quite similar to the previous tabs within the **Regression** node. The difference is that the various configuration settings that are displayed in the following subtabs cannot be changed. The tab consists of the following four subtabs.

- **General subtab**
- **Regression subtab**
- **Selection Method subtab**
- **Advanced subtab**

General subtab

The **General** subtab is designed to display the name, description, creation date, and last modified date of the scored data set along with the name of the target variable to predict. Press the **Profile Information...** button that will display the target profile of the target variable. The **Assessment Information** tab will display the objective function and the target profile settings for the regression model. For a categorical target variable, the tab will display the objective function, decision matrix, and a separate subtab that displays the assessment objective and the various assessment costs at each decision level. For an interval-valued target variable, the tab will display the objective function, profit function, and a separate subtab that displays the assessment costs assigned to each decision level.

*The **General** subtab displays various file information for the scored data set by executing the node.*

Regression subtab

The **Regression** subtab is designed to display the various configuration settings that are specified for the current model from the **Regression** subtab within the **Model Options** tab. Again, the appearance of the subtab is very similar to the previous **Model** tab. However, the various option settings cannot be changed. Configuration settings include the type of regression modeling performed, the link function, the input coding schema for the categorically-valued input variables in the regression model, and the removal of the intercept term from the regression model.

Selection Method subtab

The **Selection Method** subtab is designed to display the various configuration settings for the modeling selection procedure that was applied. Many of the option settings that are displayed on the left side of the subtab can be specified from the **General** subtab within the **Selection Method** tab. And the option settings displayed on the right-hand portion of the **Selection Method** subtab can be specified from the **Criteria** subtab. Again, notice that the validation error was selected as the modeling criteria statistic in order to achieve an unbiased evaluation or an honest assessment in the accuracy of the model from the validation MSE in comparison to the MSE statistic from the training data set.

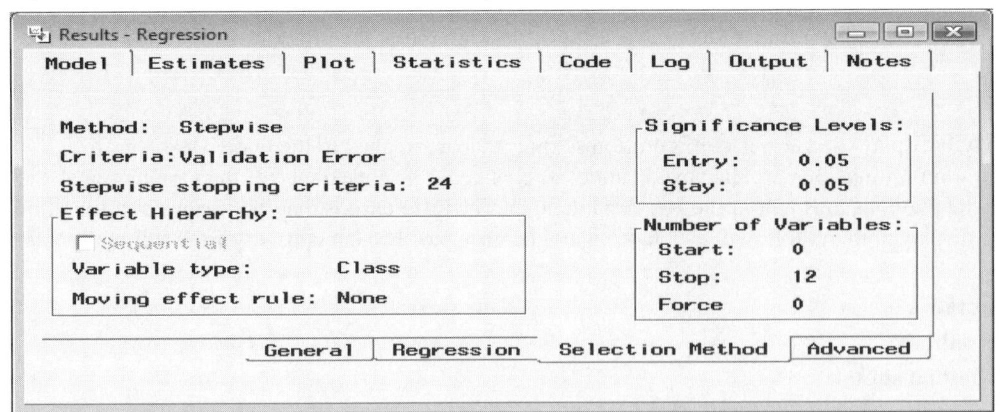

*The **Selection Method** subtab that displays the various modeling selection criteria.*

Advanced subtab

The **Advanced** subtab is designed to display the various configuration settings for nonlinear logistic regression modeling. The tab displays various configuration settings such as the optimization method that was applied, the maximum number of iterations, function calls, and the maximum amount of computational time spent in fitting the nonlinear logistic regression model. The various option settings can be specified from the previous **Advanced** tab.

Estimates tab

The **Estimates** tab is automatically displayed once you first open the **Result Browser**. The tab is designed to display a bar chart of the parameter estimates and their associated *p*-values. The tab will also display a table view of each modeling term in the regression model along with the parameter estimates and their associated *p*-values. The parameter estimates that are displayed within the tab are calculated from the training data set.

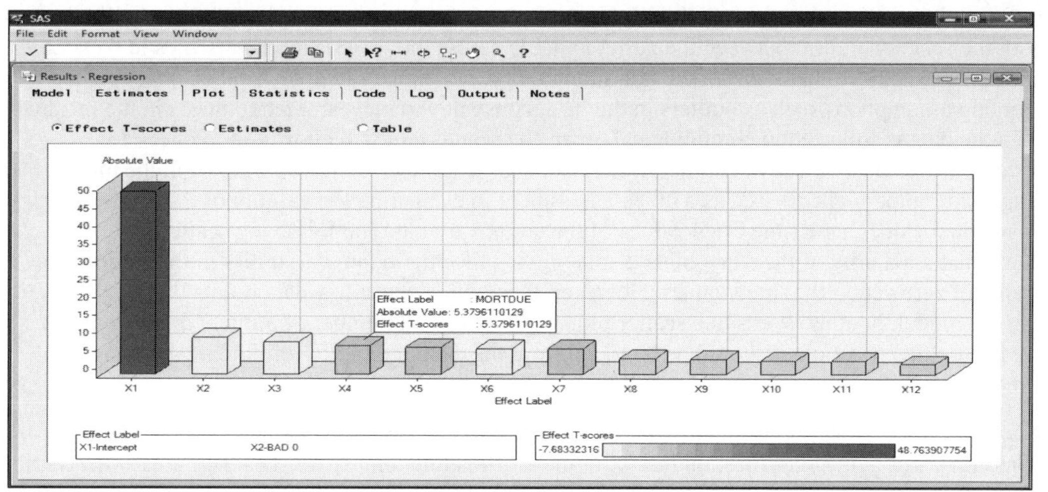

*The **Estimates** tab displays bar charts of the t-test statistic and the corresponding parameter estimates.*

By default, the tab will display a bar chart of the t-test statistics with the **Effect T-scores** radio button automatically selected. The bars of the t-test statistics of the associated parameter estimates will be displayed in decreasing order in absolute value, with the largest bars displayed to the left of the chart. Therefore, all bars greater than 1.96 in absolute value will indicate the importance of the modeling term to the regression model based on two-sided testing, assuming that the sample size of the training data set has more than 30 observations with a .05 alpha-level. Selecting the **Estimates** radio box will result in the bar chart displaying the regression coefficients or the parameter estimates for each modeling term in the regression model. In other words, not only are the main effect displayed from the tab, but also all possible higher-order terms and all possible interaction effects in the regression model. From the sign and magnitude of the parameter estimates, it will allow you to interpret the linear relationship between each input variable in the model and the target variable to predict. In addition, if the input variables are standardized to a common unit with the same range of values, it would be possible to select the best set of input variables to the regression model by observing the size of the parameter estimates in absolute value.

In Enterprise Miner, there exists a color coding schema to the bars. For instance, all bars colored in red will indicate a relatively large positive value, yellow bars will indicate a relatively large negative value, and orange bars will indicate values close to zero.

By default, selecting any one of the bars will result in the value of either the parameter estimate or the associated t-test statistics being displayed in the lower left-hand corner of the tab. A small black dot will appear that is positioned above the color bar legend that is displayed at the bottom of the chart. It will indicate the value of the bar. This can be performed by selecting the **Select Points** toolbar icon from the tools bar menu. Select the **View Info** toolbar icon, then select any one of the bars that are displayed and hold down the left-click button. This will result in a text box appearing that will display the modeling effect, parameter estimate in absolute value, and its associated t-test statistic. This is illustrated in the previous diagram. If there are a large number of modeling terms in the regression model, then select the **Move Data** toolbar icon from the tools bar menu to see any additional modeling terms that have been truncated off the bottom vertical axis of the graph by dragging the mouse along the vertical axis of the graph with the open hand.

Selecting the **Table** radio button will allow you to view the table listing of each modeling term in the regression model along with the parameter estimate and the associated t-test statistic.

Plot tab

For an interval-valued target variable, the **Plot** tab will automatically display the predicted plot or a scatter plot of the fitted values across the target values from the training data set. However, select the **Y-axis** pull-down menu to select any other input variable or the residual values for the model, which will automatically appear on the Y-axis or the vertical axis of the bivariate plot. Conversely, select the **X-axis** pull-down menu to select a different variable that will be displayed on the X-axis or the horizontal axis of the plot. The plot will allow you to check both the functional form between the target variable across the various input variables in the regression model or verify the various modeling assumptions that must be satisfied. That is, checking the various modeling assumptions such as outliers in the data, constant variance, and randomness in the residual values. When the data is collected in chronological order, the scatter plot will allow you to check for independence in the residual values over time. If there are a large number of input variables in the model, then the following scatter plot might be used as a quick assessment in evaluating the validity of the model by plotting the residual values across the fitted values of the target variable. The following scatter plot of the residuals show that a majority of the data points displaying a uniform random variability in the error terms about the value of zero across the fitted values. However, there still exists a few data points that might be deleted from the model. In other words, the scatter plot displays a small number of outliers in the data that either needs further investigation or should be removed from the data in order to refit the least-squares regression model.

For categorically-valued target variables, the **Plot** tab will display the frequency counts of the actual and predicted class levels from the classification matrix or the cross-validation matrix. That is, the tab will display frequency bars of the frequency counts between each category of the actual target levels and the predicted target levels from the training data set. All bars displayed along the main diagonal of the chart will indicate to you the accuracy of the classification performance of the logistic regression model.

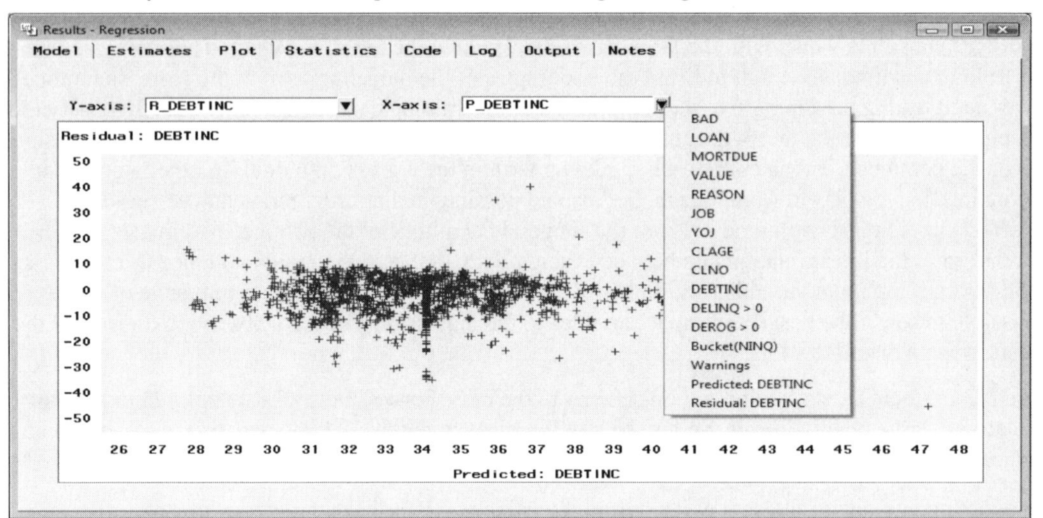

*The **Plot** tab that displays the residual plot of the residual values across the fitted values in order to check for uniform random variability in the residual terms and outliers in the least-squares model.*

Statistics tab

The **Statistics** tab displays the various goodness-of-fit statistics or the modeling assessment statistics for each partitioned data set that was created. The listed statistics will allow you to evaluate the accuracy of the modeling fit. The various goodness-of-fit statistics are calculated from the training and validation data sets. From the table listing, the modeling assessment statistics are slightly different between the partitioned data sets. This suggests that there is some overfitting to the multiple linear regression model. Therefore, you might consider reverting to a simpler model in order to avoid overfitting in the model, achieve stability in the estimates, and increase generalization to the multiple linear regression model.

Results - Regression

Model | Estimates | Plot | Statistics | Code | Log | Output | Notes

Fit Statistic	Label	Training	Validation	Test
AIC	Akaike's Information Criterion	9396.8469292	.	.
ASE	Average Squared Error	53.33540179	50.246288201	.
AVERR	Average Error Function	53.33540179	50.246288201	.
DFE	Degrees of Freedom for Error	2345	.	.
DFM	Model Degrees of Freedom	12	.	.
DFT	Total Degrees of Freedom	2357	.	.
DIV	Divisor for ASE	2357	2297	.
ERR	Error Function	125711.54202	115415.724	.
FPE	Final Prediction Error	53.881265177	.	.
MAX	Maximum Absolute Error	43.83607895	37.082121603	.
MSE	Mean Square Error	53.608333484	50.246288201	.
NOBS	Sum of Frequencies	2357	2297	.
_NW_	Number of Estimate Weights	12	.	.
RASE	Root Average Sum of Squares	7.3031090496	7.0884616245	.
RFPE	Root Final Prediction Error	7.3403859011	.	.
RMSE	Root Mean Squared Error	7.3217711985	7.0884616245	.
SBC	Schwarz's Bayesian Criterion	9466.028668	.	.
SSE	Sum of Squared Errors	125711.54202	115415.724	.
SUMW	Sum of Case Weights Times Freq	2357	2297	.

*The **Statistics** tab that lists the various modeling assessment statistics for each partitioned data set from the multiple linear regression model.*

For predicting interval-valued target variables, the following is a brief explanation to the various goodness-of-fit statistics that are listed in the **Statistics** tab.

- **Akaike's Information Criterion:** the main objective of model building is achieving the smallest error with the fewest parameters. Both the AIC and SBC modeling assessment statistics are designed to measure the difference between the actual target values and the fitted values while at the same time penalizing the regression model with several parameters in the model. The objective is selecting the best model with the smallest AIC and SBC statistic.

  $$\text{AIC} = n \cdot \ln(\frac{\text{SSE}}{n}) + 2p$$

  where n = the number of nonmissing target values and p = the number of fitted parameters

- **Average Squared Error:** This statistic is similar to the mean square error that is the residual values from the predictive model squared, divided by the number of nonmissing target values as follows:

  $$\text{ASE} = \frac{\sum_{i=1}^{n}(y_i - \hat{y}_i)^2}{n} = \frac{\text{SSE}}{n}$$

  where n = number of nonmissing cases of the target variable.

- **Average Error Function:** The average value of the error function that is the ratio between the error function and the number of nonmissing target values.

- **Degrees of Freedom for Error:** The degrees of freedom of the error term to the model.

  $$\text{df}_{\text{error}} = (n - 1) - (p - 1) = (n - p)$$

- **Model Degrees of Freedom:** The degrees of freedom of the regression model.

  $$\text{df}_{\text{model}} = \text{df}_{\text{total}} - \text{df}_{\text{error}} = (n - 1) - (n - p) = p - 1$$

- **Total Degrees of Freedom:** The number of nonmissing cases in the target variable.

  $\text{df}_{\text{total}} = n - 1$ where n = number of nonmissing cases in the target variable

- **Divisor of ASE:** The divisor for the average squared error statistic that is the number of nonmissing target values.

- **Error Function:** The final value of the objective function that is based on the error function. For least-squares regression, the final value of the objective function is the squared error for each observation, that is, $SSE = \sum_{i=1}^{n} (y_i - \hat{y}_i)^2$ assuming that there is no profit or loss matrix that has been previously defined. Otherwise, the error function is based on the maximum average profit or minimum average loss. For generalized linear models or other methods based on minimizing the deviance, the error value is the deviance. For other types of maximum likelihood estimation, the error value is the negative log likelihood. In other words, the error value is whatever the statistical model is trying to minimize.
- **Final Prediction Error:** Akaike's *final prediction error* is called the FPE. The FPE statistic is essentially the mean square error statistic multiplied by an adjustment factor. The limitation of the FPE statistic is that it is not a good modeling assessment statistic for adequately penalizing models with several parameters in the statistical model.

  $FPE = \frac{(n+p) \cdot SSE}{n \cdot (n-p)} = \frac{n+p}{n} \cdot MSE$ therefore, FPE is larger than MSE by a factor of $\frac{n+p}{n}$
- **Maximum Absolute Error:** The maximum value of the difference between the target values and the fitted values in absolute value. This robust statistic is useful if it is assumed that the data you want to fit contains several outliers. The statistic is calculated as follows:

  $$\text{Maximum Absolute Error} = \max |y_i - \hat{y}_i|$$
- **Mean Square Error:** The mean square error is the ratio between the residual values squared and the degrees of freedom in the regression model that is calculated as follows:

  $$MSE = \sum_{i=1}^{n} (y_i - \hat{y}_i)^2 / df_{error} = SSE / df_{error}$$
- **Sum of Frequencies:** The total number of observations in the partitioned data sets. The frequency variable is a nonnegative integer, where the total number of observations from the each partitioned data set is replicated as many times as the value of the frequency variable. The frequency variable may be used to perform weighted regression modeling. The sum of the frequencies is defined by the **freq** model role variable in the partitioned data sets that represents the number of occurrences within each record. Otherwise, this statistic is the sample size of the partitioned data set.
- **Number of Estimate Weights:** The number of modeling terms in the regression model.
- **Root Average Sum of Squares:** Square root of the sum of squares error, $\sqrt{SSE}$.
- **Root Final Prediction Error:** Square root of the final prediction error, $\sqrt{FPE}$.
- **Root Mean Squared Error:** Square root of the mean square error, $\sqrt{MSE}$.
- **Schwartz's Bayesian Criterion:** This statistic is quite similar to the previous AIC statistic. The SBC statistic is more severe, with several parameters in the model as opposed to the AIC statistic. Therefore, the SBC is considered one of the best modeling assessment statistics. Note that the best model to select from will be indicated by the smallest AIC or SBC goodness-of-fit statistic.

  $$SBC = n \cdot \ln(\frac{SSE}{n}) + p \cdot \ln(n)$$
- **Sum of Squared Error:** The sum-of-squared errors is the difference between the actual target values and the fitted values squared that is calculated as follows:

  $$SSE = \sum_{i=1}^{n} (y_i - \hat{y}_i)^2$$
- **Sum of Case Weights Times Freq:** The number of nonmissing cases of the target variable that you want to predict.
- **Total Profit for <target variable>:** The total profit amount based on each decision entry multiplied by the frequency count of each case (usually set to one), and the posterior probability p that is subtracted from the corresponding fixed cost of revenue at each target interval.
- **Average Profit for <target variable>:** The average profit that is the ratio between the total profit and the number of nonmissing target values.

Results - Regression

Model | Estimates | Plot | Statistics | Code | Log | Output | Notes

Fit Statistic	Label	Training	Validation	Test
AIC	Akaike's Information Criterion	2385.1377565	.	.
ASE	Average Squared Error	0.1229173562	0.1257887281	.
AVERR	Average Error Function	0.3951573417	0.40472829	.
DFE	Degrees of Freedom for Error	2965	.	.
DFM	Model Degrees of Freedom	15	.	.
DFT	Total Degrees of Freedom	2980	.	.
DIV	Divisor for ASE	5960	5960	.
ERR	Error Function	2355.1377565	2412.1806087	.
FPE	Final Prediction Error	0.1241610394	.	.
MAX	Maximum Absolute Error	0.9786446588	0.9724111843	.
MSE	Mean Square Error	0.1235391978	0.1257887281	.
NOBS	Sum of Frequencies	2980	2980	.
_NW_	Number of Estimate Weights	15	.	.
RASE	Root Average Sum of Squares	0.3505957162	0.3546670665	.
RFPE	Root Final Prediction Error	0.3523649237	.	.
RMSE	Root Mean Squared Error	0.3514814331	0.3546670665	.
SBC	Schwarz's Bayesian Criterion	2475.1329352	.	.
SSE	Sum of Squared Errors	732.58744309	749.70081918	.
SUMW	Sum of Case Weights Times Freq	5960	5960	.
MISC	Misclassification Rate	0.1637583893	0.1734899329	.
PROF	Total Profit for BAD	591	598	.
APROF	Average Profit for BAD	0.1983221477	0.2006711409	.

The ***Statistics*** *tab lists the various modeling assessment statistics for each partitioned data set from the logistic regression model that is one of the models under assessment in the modeling comparison procedure from the* ***Assessment*** *node. Overfitting in the classification model seems to be of no concern with a small difference in the misclassification rates between both partitioned data sets.*

For predicting categorically-valued target variables, the following is a brief explanation of the various goodness-of-fit statistics that are listed in the **Statistics** tab.

- **AIC:** For maximum likelihood training, the AIC statistic is calculated as follows:

  $\text{AIC} = -2 \cdot \text{Log}(L) + 2p$ for some likelihood ratio statistic $-2 \cdot \text{Log}(L)$

- **Average Squared Error:** For classification problems, the squared error between the identifier that determines if the observation was correctly classified as the target event and the estimated probability is calculated as follows:

  $$\text{ASE} = \frac{\sum_{i=1}^{n} (y - \hat{p}_i)^2}{n}$$ where y = 1 if the observation is correctly classified, y = 0 otherwise

  n = number of nonmissing cases of the target variable,
  $\hat{p}_i$ = estimated probability of the target event at the ith observation.

- **Average Error Function:** For the logistic regression model, the average error function is the ratio between the error function of the negative likelihood ratio statistic, $-2 \cdot \text{Log}(L)$, and the ASE divisor as follows:

  $$\text{Average Error Function} = \frac{-2 \cdot \text{Log}(L)}{\text{ASE divisor}}$$ for some likelihood ratio statistic $-2 \cdot \text{Log}(L)$

- **Degrees of Freedom for Error:** (see previous definition).
- **Model Degrees of Freedom:** (see previous definition).
- **Total Degrees of Freedom:** The total sample size of the active training data set.
- **Divisor of ASE:** The divisor of the average squared error statistic that is calculated by multiplying the number of target class levels and the number of nonmissing target values to predict.
- **Error Function:** For classification problems, the error function is the negative of the likelihood ratio statistic, $-2 \cdot \text{Log}(L)$, where likelihood ratio statistic is calculated by the maximum likelihood method that is analogous to iteratively reweighted least-squares regression.
- **Final Prediction Error:** (see previous definition).

- **Maximum Absolute Error:** The largest error value that is based on the difference in the identifier that determines if the observation was correctly classified in the correct target class level and its estimated probability of the target event in absolute value as follows:

  Maximum Absolute Error = $\max | y - \hat{p}_i |$

  where y = 1 if the target event is correctly classified, y = 0 otherwise.

- **Mean Square Error:** The accumulated squared error of the identifier that determines if the observation was correctly classified in the correct target class level and its associated estimated probabilities, then divided by the error degrees of freedom as follows:

  $$\text{MSE} = \sum_{i=1}^{n} (y - \hat{p}_i)^2 \,/\, \text{df}_{\text{error}}$$

  where y = 1 if the target event is correctly classified, otherwise y = 0.

- **Sum of Frequencies:** The sum of the frequency weights from the active training data set. The frequency weight represents the frequency of occurrence for the corresponding observation. That is, each observation is treated as if it has occurred n times, where n is the value of the *frequency variable* for each observation as follows:

  $$\text{Sum of Frequencies} = \sum_{i=1}^{n} F_i$$

  where F_i is the frequency at the ith observation.

- **Number of Estimate Weights:** (see previous definition).

- **Root Average Sum of Squares:** Square root of the average squared error, $\sqrt{\text{ASE}}$.

- **Root Final Prediction Error:** Square root of the final prediction error, $\sqrt{\text{FPE}}$.

- **Root Mean Squared Error:** Square root of the mean square error, $\sqrt{\text{MSE}}$.

- **Schwartz's Bayesian Criterion:** For maximum likelihood training, the SBC statistic is as follows:

  $\text{SBC} = -2 \cdot \text{Log}(L) + p \cdot \ln(n)$, where L is the likelihood ratio statistic

- **Sum of Squared Error:** For classification problems, the accumulated squared error of the identifier that determines if the observation was correctly classified to the correct target class level and its associated estimated probabilities by each class level of the target variable as follows:

  $$\text{SSE} = \sum_{i=1}^{n} [(y - \hat{p}_{1i})^2 + (y - \hat{p}_{0i})^2]$$, based on a binary-valued target variable

  where y = 1 if the ith observation is correctly classified, otherwise y = 0,
  for some posterior probability $\hat{p}_{1i}$ of the target event and $\hat{p}_{0i}$ of the target nonevent.

- **Sum of Case Weights Times Freq:** The number of nonmissing observations of the target variable from each partitioned data set.

- **Misclassification Rate:** The percentage of cases that have been incorrectly classified. By default, the observation is assigned to the appropriate target class level from the largest estimated probability. The misclassification rate is computed as follows:

  Misclassification Rate = number of misclassified cases / total number of cases classified

- **Total Profit for <target variable>:** The total profit amount is calculated by multiplying each decision entry and both the frequency of each case (usually set to one) and the posterior probability p that is then subtracted by the corresponding fixed cost of revenue at each target class level. By default, the total profit equals the number of cases of the target event.

- **Average Profit for <target variable>:** The average profit is calculated by taking the ratio between the total profit and the number of nonmissing cases of the target variable as follows:

  $$\text{Average Profit} = \frac{\text{Total Profit}}{n}$$

Code tab

The **Code** tab is designed to display both the internal training code and the scoring code. The training code is automatically displayed once you open the tab. The training code is the PROC DMREG data mining procedure code that is automatically executed once you run the **Regression** node and train the regression model. The training code will allow you to view the various option settings that are automatically applied by the various option settings that are specified from the **Regression** node. For instance, the fitted values that are created in the scored data set that is listed within the SCORE option statement might be plotted across the range of values of the interval-valued input variables in order to view the accuracy of the statistical model.

Selecting the **Scoring** radio button will then display the internal score code. The scoring code will allow you to view the internal SEMMA programming code that is performed to calculate the fitted values in the statistical model. The scoring code can be used in calculating new prediction estimates by specifying entirely different values for the input variables in the regression model. The following scoring code is a partial listing of the scoring code that is listed within the node that is based on all interval-valued input variables in the HMEQ data set that were selected from the modeling selection routine. The scored code will first identify any missing values in each one of the input variables in the multiple linear regression model. If there are any missing values in any one of the input variables, then the target variable is estimated by its own average value. Otherwise, the target variable is estimated by the linear combination of least-squares estimates. The scored code calculates the linear combination of least-squares estimates by accumulating the least-squares estimates by multiplying each input variable in the model by its corresponding slope estimate. The intercept term is then added to the model to calculate the predicted values. The final step is calculating the residual values that is the difference between the target values and fitted values.

From the main menu, you must select the **Edit > Copy** menu options to copy the listed code or the highlighted text into the system clipboard.

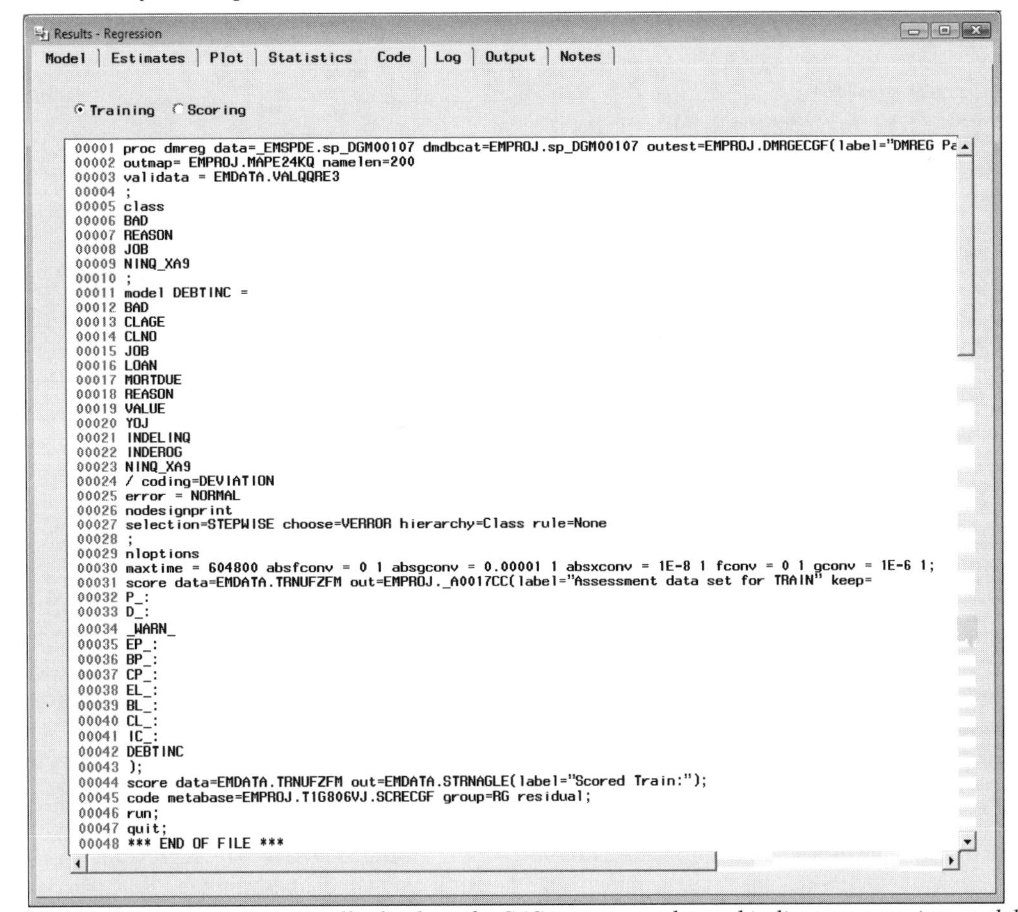

*The **Code** tab that automatically displays the SAS training code used in linear regression modeling.*

Results - Regression

Model | Estimates | Plot | Statistics | Code | Log | Output | Notes

○ Training ◉ Scoring

```
00001 *************************************;
00002 *** begin scoring code for regression;
00003 *************************************;
00004
00005 length _WARN_ $4;
00006 label _WARN_ = 'Warnings' ;
00007
00008 drop _Y;
00009 _Y = DEBTINC ;
00010
00011 drop _DM_BAD;
00012 _DM_BAD=0;
00013
00014 *** Check CLNO for missing values ;
00015 if missing( CLNO ) then do;
00016    substr(_warn_,1,1) = 'M';
00017    _DM_BAD = 1;
00018 end;
•••
00164 *** If missing inputs, use averages;
00165 if _DM_BAD > 0 then do;
00166    _LP0 =      34.0613868387657;
00167    goto RGDR1;
00168 end;
00169
00170 *** Compute Linear Predictor;
00171 drop _TEMP;
00172 drop _LP0;
00173 _LP0 = 0;
00174
00175 ***  Effect: BAD ;
00176 _TEMP = 1;
00177 _LP0 = _LP0 + (    -2.2573441613809) * _TEMP * _0_0;
00178
00179 ***  Effect: CLNO ;
00180 _TEMP = CLNO ;
00181 _LP0 = _LP0 + (     0.08198916403656 * _TEMP);
00182
00183 ***  Effect: JOB ;
00184 _TEMP = 1;
00185 _LP0 = _LP0 + (     0.56550046819045) * _TEMP * _2_0;
00186 _LP0 = _LP0 + (     0.76187078856867) * _TEMP * _2_1;
00187 _LP0 = _LP0 + (      0.6150907950906) * _TEMP * _2_2;
00188 _LP0 = _LP0 + (    -1.83222682834733) * _TEMP * _2_3;
00189 _LP0 = _LP0 + (     1.35152814753051) * _TEMP * _2_4;
00190
00191 ***  Effect: LOAN ;
00192 _TEMP = LOAN ;
00193 _LP0 = _LP0 + (     0.00007333541095 * _TEMP);
00194
00195 ***  Effect: MORTDUE ;
00196 _TEMP = MORTDUE ;
00197 _LP0 = _LP0 + (     0.00002332054529 * _TEMP);
00198
00199 ***  Effect: NINQ_XA9 ;
00200 _TEMP = 1;
00201 _LP0 = _LP0 + (    -1.39668930643646) * _TEMP * _3_0;
00202 _LP0 = _LP0 + (     0.07115563394829) * _TEMP * _3_1;
00203 *--- Intercept ---*;
00204 _LP0 = _LP0 + (     31.2767362170143);
00205
00206 RGDR1:
00207
00208 *** Predicted Value, Error, and Residual;
00209 label P_DEBTINC = 'Predicted: DEBTINC' ;
00210 P_DEBTINC = _LP0;
00211
00212 drop _R;
00213 if _Y = . then do;
00214    R_DEBTINC = .;
00215 end;
00216 else do;
00217    _R = _Y - _LP0;
00218     label R_DEBTINC = 'Residual: DEBTINC' ;
00219    R_DEBTINC = _R;
00220 end;
00221
00222 *************************************;
00223 ***** end scoring code for regression;
00224 *************************************;
00225 *** END OF FILE ***
```

*The **Code** tab that displays the SAS scoring code that can be used in predicting new observations from the multiple linear regression model.*

Log tab

The **Log** tab displays the log listing from the training code that is automatically executed once the **Regression** node is run. The tab will allow you to view any warning or error messages that might have occurred by executing the **Regression** node.

Output tab

The **Output** tab displays the procedure output listing from the PROC DMREG data mining regression procedure. The procedure output listing is very similar to the output listing generated from the PROC REG regression procedure.

For multiple regression modeling, the DMREG procedure output listing displays the **Analysis of Variance** table with the corresponding F-test statistic. The F-test statistic measures the overall significance of the modeling fit. The statistic is the ratio between the mean square of the regression model and the mean square error used to determine if the regression model fits better than the baseline mean model. The **Model Fit Statistic** table displays the various modeling assessment statistics. The **Type 3 Analysis of Effects** table will display the partial F-test statistic that is the ratio between the variability of each input variable in the model and the unexplained variability in the regression model, that is, the mean square error. The statistic will allow you to determine whether the input variable explains a significant amount of variability in the target variable. The **Analysis of Maximum Likelihood Estimates** table will display the parameter estimates, standard error, t-test statistic and the associated *p*-value for each input variable in the model to determine the significance of each modeling term in the regression model. Examining the *p*-values of the parameters from the table, a majority of the modeling terms in the regression model are very significant in explaining the variability in the target values which are the best set of input variables that were selected from the stepwise regression procedure to predict the interval-valued variable, DEBTINC. However, notice that the class level of ProfExe in the categorically-valued variable JOB is the only class level that is significantly different from zero at the .05 alpha-level. Therefore, one strategy is creating a binary-valued variable by collapsing all the insignificant class levels into one group with ProfExe defined as the other class level and refitting the multiple linear regression model.

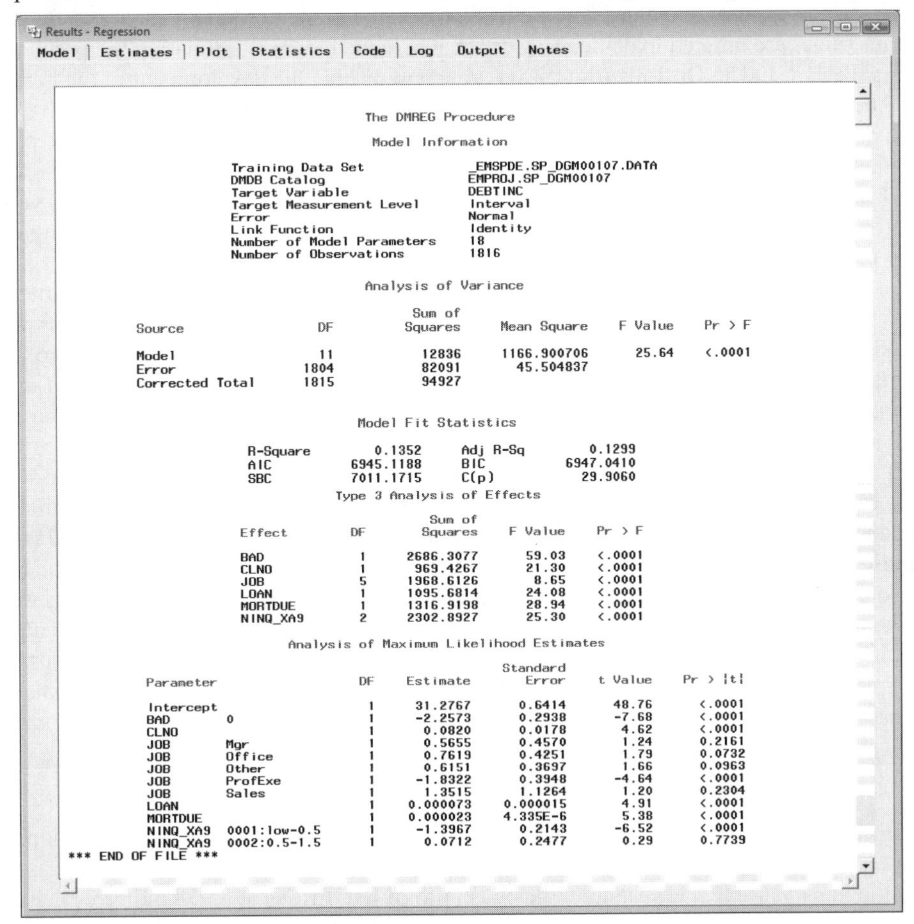

Results - Regression

Model | Estimates | Plot | Statistics | Code | Log | Output | Notes

The DMREG Procedure

Model Information

Training Data Set	_EMSPDE.SP_DGM00107.DATA
DMDB Catalog	EMPROJ.SP_DGM00107
Target Variable	DEBTINC
Target Measurement Level	Interval
Error	Normal
Link Function	Identity
Number of Model Parameters	18
Number of Observations	1816

Analysis of Variance

Source	DF	Sum of Squares	Mean Square	F Value	Pr > F
Model	11	12836	1166.900706	25.64	<.0001
Error	1804	82091	45.504837		
Corrected Total	1815	94927			

Model Fit Statistics

R-Square	0.1352	Adj R-Sq	0.1299
AIC	6945.1188	BIC	6947.0410
SBC	7011.1715	C(p)	29.9060

Type 3 Analysis of Effects

Effect	DF	Sum of Squares	F Value	Pr > F
BAD	1	2686.3077	59.03	<.0001
CLNO	1	969.4267	21.30	<.0001
JOB	5	1968.6126	8.65	<.0001
LOAN	1	1095.6814	24.08	<.0001
MORTDUE	1	1316.9198	28.94	<.0001
NINQ_XA9	2	2302.8927	25.30	<.0001

Analysis of Maximum Likelihood Estimates

Parameter		DF	Estimate	Standard Error	t Value	Pr > |t|
Intercept		1	31.2767	0.6414	48.76	<.0001
BAD	0	1	-2.2573	0.2938	-7.68	<.0001
CLNO		1	0.0820	0.0178	4.62	<.0001
JOB	Mgr	1	0.5655	0.4570	1.24	0.2161
JOB	Office	1	0.7619	0.4251	1.79	0.0732
JOB	Other	1	0.6151	0.3697	1.66	0.0963
JOB	ProfExe	1	-1.8322	0.3948	-4.64	<.0001
JOB	Sales	1	1.3515	1.1264	1.20	0.2304
LOAN		1	0.000073	0.000015	4.91	<.0001
MORTDUE		1	0.000023	4.335E-6	5.38	<.0001
NINQ_XA9	0001:low-0.5	1	-1.3967	0.2143	-6.52	<.0001
NINQ_XA9	0002:0.5-1.5	1	0.0712	0.2477	0.29	0.7739

*** END OF FILE ***

The following is a short explanation to the various goodness-of-fit statistics that can be used in comparing one model to another listed in the **Model Fit Statistics** table.

- **R-Square:** *R-square* statistic measures the proportion of the variability observed in the data that is explained by the regression model. R-square values close to one will indicate a very good fit and r-square values close to zero will indicate an extremely poor fit. From the table, the regression model explains 14.32% of the total variability in the target variable.
- **AIC:** (see previous definition).
- **SBC:** (see previous definition).

- **Adj R-Sq:** (*Adjusted R-Square* statistic) This statistic is based on the r-square statistic that takes into account the number of modeling terms in the regression model in which the statistic is adjusted by the degrees of freedom of the corresponding regression model. The statistic does not automatically increase with an increasing number of input variables added to the model as opposed to the r-square statistic. The adjusted r-square statistic is calculated as follows:

$$\text{ADJRSQ} = 1 - [(\text{SS}_{\text{error}} / (n - p)) / (\text{SS}_{\text{total}} / (n - 1))] = 1 - [(n - 1) / (n - p)] \cdot (1 - \text{R}^2)$$

- **BIC:** (*Bayesian information criterion*). This statistic is very similar to the previous SBC statistic.
- **C(p):** (*Mallow's Cp* statistic). This statistic determines the best-fitting model as follows:

$$\text{C}_P = p + \frac{(\text{MSE}_p - \text{MSE}_{full})(n - p)}{\text{MSE}_{full}}$$

where MSE_p is the mean square error with p parameters. The main idea of this statistic is to select the first model where $\text{C}_P \leq p$ has the fewest number of parameters p in the model.

For logistic regression modeling, the **Model Information** table displays information such as the data set name, target variable, level of measurement and number of levels of the target variable, error function and link function, and number of parameters and observations. In our example, the logit link function was used. The **Target Profile** table displays the target values that are listed by their assigned ordered values and frequencies. By default, the target values are ordered in descending order, where the model is estimating the probability of bad creditors, that is, BAD = 1. The **Optimization Start** table will display the development of the iterative grid search procedure. The table listing will display the change in the parameter estimates and the negative log-likelihood function at each iteration. The table listing will allow you to observe small changes in the parameter estimates and the likelihood function between successive iterations, indicating convergence in the iterative grid search procedure. The table displays a small number of iterations that were performed, indicating to you that convergence has probably been achieved. The **Optimization Results** table displays an extremely important message indicating to you that convergence has been met. The default is the gradient convergence criterion with a default value of 1E-8. The **Likelihood Ratio Test for Global Null Hypothesis: BETA=0** table will indicate to you how well the overall model fits the data, with a null hypothesis that all the parameters in the model are zero. A significant p-value will indicate that at least one parameter in the model is nonzero. Both likelihood ratio statistics can be used as modeling assessment statistics in determining the best set of input variables in the logistic regression model. In other words, remove any number of input variables from the model and then calculate the difference in the likelihood ratio statistics to determine the significance of the input variable(s) in the logistic regression model. The **Type III Analysis of Effect** table is displayed when there is at least one categorically-valued input variable in the logistic regression model. The table will display the overall significance of each categorically-valued input variable in the model. The **Analysis of Maximum Likelihood Estimates** table displays the modeling terms and the associated parameter estimates, standard error, Wald's chi-square statistic and p-value and its odd ratio statistic. The parameter estimate measures the rate of change in the logit, that is, the log of the odd ratio, corresponding to a one-unit change in the input variable. Note that large parameter estimates and standard errors for the corresponding input variables in the model will be an indication of sparseness in the data. In calculating the fitted probabilities, then estimating probability that a bad client has a derogatory report (with all other input variables in the model set to zero) is calculated as follows:

$$\hat{p} = \exp(\alpha + \beta \cdot \text{X}_{\text{inderog}}) / [1 + \exp(\alpha + \beta \cdot \text{X}_{\text{inderog}})] = 1 / \{1 + [\exp(9.1545 + (1.3655 \cdot 1))]^{-1}\} = .999$$

The **Odds Ratio Statistic** table displays the odds ratio statistic. For example, the categorical-value input variable, JOB, displays the odds of each class level in comparison to its fourth class level that is the assigned reference level for the input variable. For example, the odds ratio of INDEROG at 3.918 indicates that it is almost 4 times more likely that a person with a derogatory report will not pay their home loan as opposed to a person without a derogatory report. If classifying the observations is the main objective of the analysis, then the two-way frequency table will allow you to interpret the accuracy of the classification model by evaluating the misclassification rate. The table displays the observed target levels as the row variable and the predicted target levels as the column variable. By default, each case is assigned to the target class level based on the largest estimated probability. The following assessment statistics are calculated from the training data set. The results are based on the modeling comparison procedure that is performed from the **Assessment** node.

Note: It is important to be aware that the PROC DMREG procedure will not generate a warning message to inform you of complete separation or quasi-complete separation in the logistic regression model.

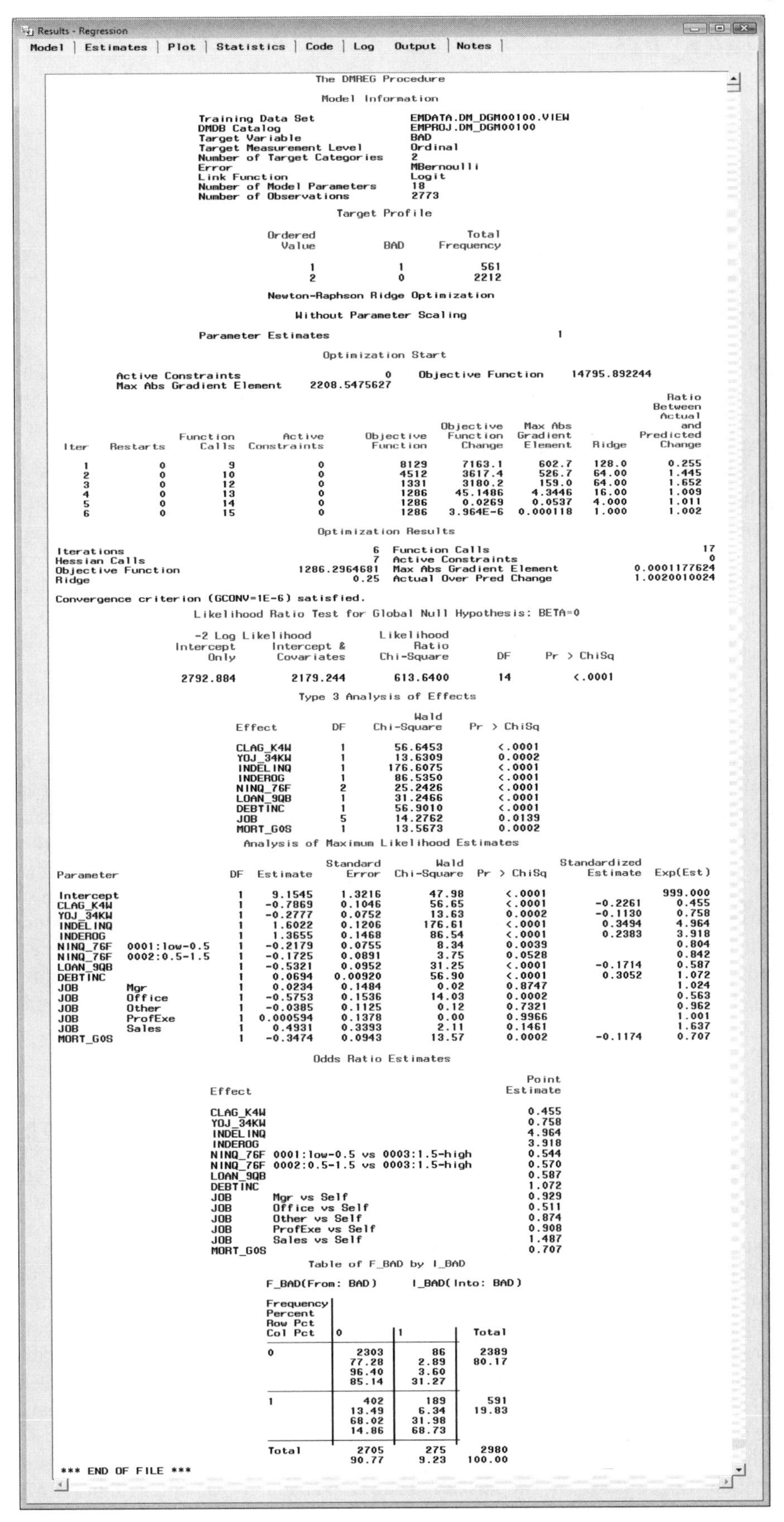

Results - Regression

Model | Estimates | Plot | Statistics | Code | Log | Output | Notes

The DMREG Procedure

Model Information

Training Data Set	EMDATA.DM_DGM00100.VIEW
DMDB Catalog	EMPROJ.DM_DGM00100
Target Variable	BAD
Target Measurement Level	Ordinal
Number of Target Categories	2
Error	MBernoulli
Link Function	Logit
Number of Model Parameters	18
Number of Observations	2773

Target Profile

Ordered Value	BAD	Total Frequency
1	1	561
2	0	2212

Newton-Raphson Ridge Optimization

Without Parameter Scaling

Parameter Estimates 1

Optimization Start

Active Constraints	0	Objective Function	14795.892244
Max Abs Gradient Element	2208.5475627		

Iter	Restarts	Function Calls	Active Constraints	Objective Function	Objective Function Change	Max Abs Gradient Element	Ridge	Ratio Between Actual and Predicted Change
1	0	9	0	8129	7163.1	602.7	128.0	0.255
2	0	10	0	4512	3617.4	526.7	64.00	1.445
3	0	12	0	1331	3180.2	159.0	64.00	1.652
4	0	13	0	1286	45.1486	4.3446	16.00	1.009
5	0	14	0	1286	0.0269	0.0537	4.000	1.011
6	0	15	0	1286	3.964E-6	0.000118	1.000	1.002

Optimization Results

Iterations	6	Function Calls	17
Hessian Calls	7	Active Constraints	0
Objective Function	1286.2964681	Max Abs Gradient Element	0.0001177624
Ridge	0.25	Actual Over Pred Change	1.0020010024

Convergence criterion (GCONV=1E-6) satisfied.

Likelihood Ratio Test for Global Null Hypothesis: BETA=0

-2 Log Likelihood Intercept Only	-2 Log Likelihood Intercept & Covariates	Likelihood Ratio Chi-Square	DF	Pr > ChiSq
2792.884	2179.244	613.6400	14	<.0001

Type 3 Analysis of Effects

Effect	DF	Wald Chi-Square	Pr > ChiSq
CLAG_K4W	1	56.6453	<.0001
YOJ_34KW	1	13.6309	0.0002
INDELINQ	1	176.6075	<.0001
INDEROG	1	86.5350	<.0001
NINQ_76F	2	25.2426	<.0001
LOAN_9QB	1	31.2466	<.0001
DEBTINC	1	56.9010	<.0001
JOB	5	14.2762	0.0139
MORT_G0S	1	13.5673	0.0002

Analysis of Maximum Likelihood Estimates

Parameter		DF	Estimate	Standard Error	Wald Chi-Square	Pr > ChiSq	Standardized Estimate	Exp(Est)
Intercept		1	9.1545	1.3216	47.98	<.0001		999.000
CLAG_K4W		1	-0.7869	0.1046	56.65	<.0001	-0.2261	0.455
YOJ_34KW		1	-0.2777	0.0752	13.63	0.0002	-0.1130	0.758
INDELINQ		1	1.6022	0.1206	176.61	<.0001	0.3494	4.964
INDEROG		1	1.3655	0.1468	86.54	<.0001	0.2383	3.918
NINQ_76F	0001:low-0.5	1	-0.2179	0.0755	8.34	0.0039		0.804
NINQ_76F	0002:0.5-1.5	1	-0.1725	0.0891	3.75	0.0528		0.842
LOAN_9QB		1	-0.5321	0.0952	31.25	<.0001	-0.1714	0.587
DEBTINC		1	0.0694	0.00920	56.90	<.0001	0.3052	1.072
JOB	Mgr	1	0.0234	0.1484	0.02	0.8747		1.024
JOB	Office	1	-0.5753	0.1536	14.03	0.0002		0.563
JOB	Other	1	-0.0385	0.1125	0.12	0.7321		0.962
JOB	ProfExe	1	0.000594	0.1378	0.00	0.9966		1.001
JOB	Sales	1	0.4931	0.3393	2.11	0.1461		1.637
MORT_G0S		1	-0.3474	0.0943	13.57	0.0002	-0.1174	0.707

Odds Ratio Estimates

Effect	Point Estimate
CLAG_K4W	0.455
YOJ_34KW	0.758
INDELINQ	4.964
INDEROG	3.918
NINQ_76F 0001:low-0.5 vs 0003:1.5-high	0.544
NINQ_76F 0002:0.5-1.5 vs 0003:1.5-high	0.570
LOAN_9QB	0.587
DEBTINC	1.072
JOB Mgr vs Self	0.929
JOB Office vs Self	0.511
JOB Other vs Self	0.874
JOB ProfExe vs Self	0.908
JOB Sales vs Self	1.487
MORT_G0S	0.707

Table of F_BAD by I_BAD

F_BAD(From: BAD) I_BAD(Into: BAD)

Frequency / Percent / Row Pct / Col Pct	0	1	Total
0	2303 77.28 96.40 85.14	86 2.89 3.60 31.27	2389 80.17
1	402 13.49 68.02 14.86	189 6.34 31.98 68.73	591 19.83
Total	2705 90.77	275 9.23	2980 100.00

*** END OF FILE ***

4.2 Model Manager

General Layout of the Enterprise Miner Model Manager

- **Models tab**
- **Options tab**
- **Reports tab**
- **Output tab**

The **Model Manager** is a directory table facility that will allow you to store, retrieve, update, and access models that are created from the currently opened modeling node. The **Model Manager** is designed to evaluate the selected active predictive or classification model that is constructed within the modeling node. The **Model Manager** will also allow you to select the partitioned data set with its associated scored data set that will be used in creating the various performance charts for the currently selected model. Performance charts such as lift charts, threshold charts, and ROC charts will be discussed in greater detail in the **Assessment** node. The **Model Manager** is available within all of the modeling nodes such as the **Regression, Tree, Neural Network, User-Defined, Ensemble,** and **Two-Stage Model** modeling nodes. The interface and the functionality of the **Model Manager** are identical within each one of the modeling nodes. The appearance of the **Models** tab within the **Model Manager** window is similar to the **Assessment** tab within the **Assessment** node. From each modeling node, the **Models** tab will list every model that was created within the currently opened modeling node with a name and a short description of each model that is adjacent to several columns that display the various modeling assessment statistics for each partitioned data set used in evaluating the accuracy of the listed models. This is illustrated in the following diagram. To create a separate entry in the **Model Manager** with the corresponding settings, simply select the **File > Save Model As** pull-down menu options from within the corresponding modeling node.

The difference between the **Model Manager** and the **Assessment** node is the way in which both nodes manage the target profile. Within the **Model Manager**, you can define a target profile of the selected model as opposed to the **Assessment** node that uses the previously defined target profiles from the corresponding models under assessment. To save the target profile permanently, you should redefine the target profile from the **Model Manager** of the modeling node or from the **Input Data Source** node. However, the big advantage of the **Assessment** node is that the node will allow you to assess several models at one time, as opposed to the **Model Manager**, which is restricted to comparing one and only one model at a time. The advantage of the **Model Manager** is that it will allow you to evaluate the modeling performance based on updates applied to the target profile from the selected model within the **Model Manager** window. Conversely, from the **Assessment** node, modifications to the target profile can be performed to evaluate the impact of different priors or different decision consequences have on the modeling performance that must be previously specified from the corresponding modeling node in order to generate the assessments.

To open the **Model Manager** directory table facility, select any one of the modeling nodes from the **Diagram Workspace**, then right-click the mouse to select the **Model Manager** pop-up menu option. Alternatively, from any one of the modeling nodes, select the **Model Manager** menu item from the **Tools** main menu option, that is, select the **Tools > Model Manager** main menu option that will open the **Model Manager** window. Again, the appearance of the **Model Manager** window is similar to the **Assessment** node.

The **Model Manager** will allow you to display and create the available performance charts, that is, the diagnostic charts, lift charts, threshold charts, and ROC charts, based on the selected partitioned data set of the predictive model under assessment in the **Assessment** node. The various performance charts that can be created will be discussed in greater detail in the **Assessment** node. From the **Models** tab, highlight any one of the row model entries from the list of all existing predictive models that are displayed, and both the **Options** and **Reports** tabs will then become available for viewing. The option settings of both of these tabs can only be defined from the **Model Manager**. The **Reports** tab is designed to control the various performance charts that will be created from the currently selected model. The **Option** tab is designed to create the scored data sets in addition to controlling the number of records that are generated for each data set that is selected for assessment in the **Assessment** node. The **Output** tab is always grayed-out and unavailable for viewing within the **Model**

Manager. The tab is designed to pass the scored data sets from the **Assessment** node on to the subsequent nodes within the process flow diagram.

Models tab

The purpose of the **Models** tab is to select the active model from the list of all the available models that are created within the modeling node. The tab will display various goodness-of-fit statistics that will allow you to determine the best fitting models that were generated from the corresponding modeling node. By selecting the listed model, the node will allow you to view the appropriate performance chart and access the following **Options** and **Reports** tabs. The **Models** tab will automatically appear once you open the **Model Manager**. From the **Models** tab, the most recently created model is active that is always the last model listed. Therefore, any changes made within the modeling node are applied to the active model when running the node. To add a new model entry to the **Model Manager,** repeat the basic steps in saving the corresponding model option settings by selecting the **Save Model As** menu option setting within the corresponding modeling node. Both the **Name** and **Description** cells are available for you to enter an appropriate name and short description to distinguish between the various listed models that are created within the modeling node. This feature is especially useful if you are fitting several models from the same modeling node. To delete the active model, select the desired row model entry, then select the **Edit > Delete** main menu option. To set the active model, open the modeling node and make the appropriate changes to the option settings to the model, then save the new model by selecting the **File > Save Model As** main menu option. To make a previous model active in the **Model Manager**, you must delete all models created after the listed model. To add a new model entry to the **Model Manager,** repeat the previously mentioned steps in saving the model option settings. From the **Models** tab, you will be able to save, print or delete the modeling assessment statistics, view the various modeling assessment statistics and modeling results, edit the target profile, and view the various performance charts from the selected partitioned data set. To view any one of the performance charts, select the desired model row entries, then select the **Tools** main menu option to select any one of the performance charts that are listed.

Alternatively, select the **Draw Lift Chart** toolbar icon or the **Draw Diagnostic Chart** toolbar icon from the toolbar menu. It is important to understand that the modeling node must be executed for the various modeling assessment statistics to be listed within the tab based on the scored data set that is created from the corresponding modeling node. By default, both the **Options** and **Reports** tabs are dimmed and unavailable for viewing. Therefore from the **Models** tab, you must select any one of the currently listed models to open either tab.

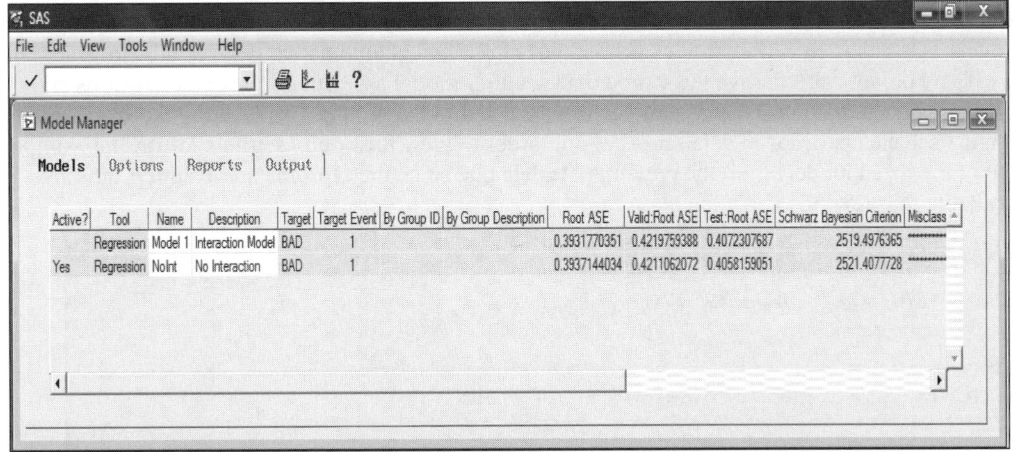

*The **Models** tab that lists the various models created within the **Regression** modeling node.*

Options tab

The **Options** tab will allow you to select the training, validation or test data sets that will be used to create the various performance charts, with the added option of saving the scored data set after modeling assessment. By default, the validation data set is automatically selected. The assessment plots and the associated scored data set from the validation data set will be automatically passed along to the **Assessment** node. However, assuming that each partitioned data set is created, usually from the **Data Partition** node, then the tab will allow you to select all three partitioned data sets along with their associated assessment plots and scored data sets to be passed along for modeling assessment within the **Assessment** node.

From the **Assess** section, select the partitioned data set and its associated scored data set for assessment. By default, the performance charts are based on the scored validation data set, assuming that the input data set has been partitioned. Therefore, the assessment plots and the associated scored validation data set will be automatically passed along to the **Assessment** node for modeling assessment. The option is designed to save the generated modeling estimates from the corresponding model to the selected data set. This will allow you to view the stability of the modeling estimates from the same statistical model that can be determined by viewing small deviations in the fitted values from the training and validation data sets. For predictive modeling, the stability in the estimates can be determined by plotting the residual values across the fitted values between both partitioned data sets. For classification modeling, the stability in the estimates can be determined by plotting the estimated proportions across the grouped percentile estimates.

From the **Score Data Set** section, the tab will allow you to control the sample size of the scored data set for each selected partitioned data set to reduce processing time during interactive assessment. *Interactive assessment* is when you modify the target profile or the horizontal scale of the lift charts during assessment within the **Assessment** node. By default, the **Save after model assessment** check box is automatically selected indicating to you that the scored data set will be saved after assessment. Otherwise, clear the check box if you do not want to save the scored data set after model assessment. However, selecting the check box will result in the **Sample** radio button automatically being selected within the **Score Data Set** section. This will indicate to you that the node will perform a random sample of 5,000 observations from the selected partitioned data set for assessment. In other words, the sample is used only for interactive assessment. It is important to understand that when you first create the lift charts, then all the observations from the scored data set are used to construct the initial lift charts. That is, the random sample is used for interactive assessment only. Select the **Entire Sample** radio button that will select all the observations from the selected partitioned data set. This option might be selected if your data consists of rare target class levels in order to make sure that the data is large enough for interactive assessment so that the classification model accurately classifies these same infrequent target class levels. Also, the **Save after model assessment** check box is automatically selected. However, clear the check box if you do not want to save the scored data set after model assessment. Both the **Options** and **Reports** tabs are designed to control the type of performance charts that will be displayed in the **Assessment** node based on the selected partitioned data set. Again, in order to view the option settings for this tab, simply select any one of the listed models from the previous **Models** tab, since the **Options** and **Reports** tabs are initially unavailable for viewing.

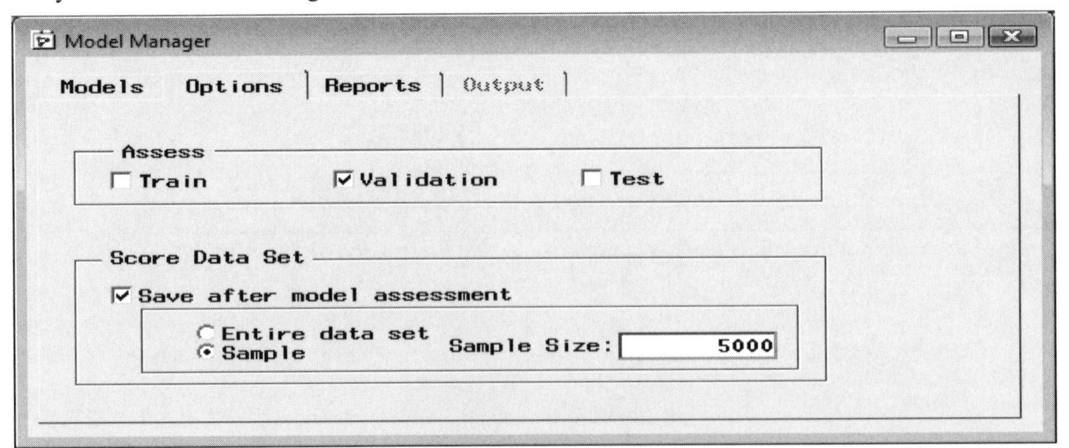

*The **Options** tab is used to specify the partitioned data set that is used to create the performance charts.*

Reports tab

The **Reports** tab is designed for you to control the various performance charts that will be available for assessment within the **Assessment** node for each model that is selected. By default, all of the performance charts and diagnostic charts are automatically selected and, therefore, passed along to the subsequent **Assessment** node. All the performance charts and diagnostic charts will be available for viewing within the **Assessment** node that is determined by the level of measurement of the target variable that you want to predict. However, the tab will allow you to select the various assessment charts that will be available for viewing within the **Assessment** node for each model that is selected. Although, this tab will determine which graphs will be viewed within the subsequent **Assessment** node, all the performance charts and diagnostic charts are still available for viewing within the **Model Manager** based on the level of measurement of the target variable in the model. For instance, the lift charts are generated by fitting the categorically-valued target variable or plotting the target profile information from the stratified interval-valued target variable. The diagnostic charts generate bar charts of the predicted values and residuals from the model fitting the interval-valued target variable. And the response threshold, ROC, and threshold-based charts plot the predictive power of the classification model by fitting the binary-valued target variable. From the **Models** tab, in order to view the various option settings for this tab, select the currently listed model.

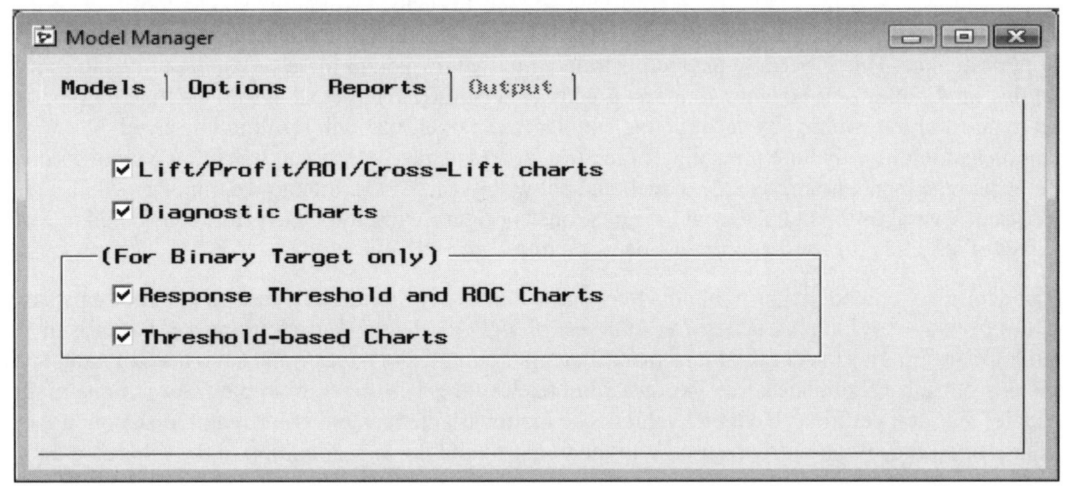

*The **Reports** tab is used to control the type of performance charts that are passed along for assessment.*

Output tab

The **Output** tab is always grayed-out and unavailable for viewing within the **Model Manager**. The tab can only be viewed within the following **Assessment** node. The tab is designed to pass the scored data sets from the **Assessment** node on to the subsequent nodes within the process flow diagram. The tab will be displayed in the following **Assessment** node.

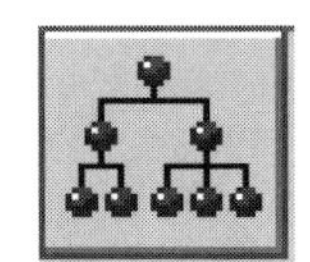

4.3 Tree Node

General Layout of the Enterprise Miner Tree Node

- **Data tab**
- **Variables tab**
- **Basic tab**
- **Advanced tab**
- **Score tab**
- **Notes tab**

The purpose of the **Tree** node in Enterprise Miner is to perform decision tree modeling. The **Tree** node will allow you to automatically construct the tree or interactively build your own tree. That is, the **Tree** node will automatically determine the best decision tree from the selected criterions by choosing the most appropriate input variables and splits. However, in the interactive mode the **Tree** node will allow you to construct your own custom-designed decision tree. Interactive training is designed to increase the predictive or classification modeling performance. This is because interactive training will allow you to force certain input variables into the model that might otherwise be ignored to best predict or separate the target values. Interactive training will allow you to perform tree pruning by constructing simpler tree models that will result in improved generalization and improve the interpretability of the tree. Also, interactive training will allow you to specify the way in which the input variables in the model are split within each node. Interactive training will also allow you to construct several different trees from the numerous configuration settings, which is considered one of the best ways of selecting the most appropriate decision tree.

The goal of decision tree modeling is to build a tree that will allow you to identify various target groups based on the values from a set of input variables. *Decision tree modeling* is based on performing a series of if-then decision rules that forms a series of partitions that sequentially divide the target values into a small number of homogenous groups that formulate a tree-like structure. Each split is performed from the values of one of the input variables that best partitions the target values. The input variables can either be continuous, nominal or ordinal. The *if* condition of the rule corresponds to the tree path and the *then* condition of the rule represents the connection to the leaf node. The modeling algorithm performs a recursive procedure by which the target variable is divided into various groups based on a decision rule that is designed to maximize the impurity measurement as each one of the successive groups are formed. Division rules are applied that determine the way in which the input variables in the model are split. For categorical target variables, the model is called a *classification tree*, in which the leaves which contain the target proportions can be interpreted as predicted probabilities or predicted proportions of the categorically-valued target variable. For interval-valued target variables, the model is called a *regression tree*, where the leaves which contain the target means can be interpreted as predicted values of the target variable.

The node requires only one target variable and any number of input variables for the tree model. Again, decision trees are designed to perform both classification and predictive modeling by fitting either the categorical or interval-valued target variable. For categorically-valued target variables, the data is repeatedly partitioned by its corresponding class levels. For interval-valued target variables, the data is repeatedly partitioned by its average value. Each rule is sequentially assigned to an observation based on a value of an individual input variable in the tree branching process. The root of the tree consists of the entire data set where the strongest splits occur, with the final tree branching process stopping at the leaves of the tree. The decision tree begins at the root node and is read from the top-down. At each node of the tree, an *if-then* rule is applied to the values of one of the input variables in the node. For interval-valued inputs, the splits are a mutually exclusive range of input values. For nominal-valued input variables, the splits are mutually exclusive subsets of various categories. The tree splitting procedure repeats itself until the partitioning of the input values reaches the leaves of the tree. The leaves of the tree indicate that there are no input variables that are found to significantly decrease the splitting criterion statistic any further. At each stage of the tree splitting process, the input variable is selected from the pool of all possible input variables that best splits the target values, then searches for the best value of the input variable in which to split the target variable into homogenous groups.

However, keep in mind that certain input variables can repeatedly appear in any number of splits throughout the decision tree. For interval-valued input variables, either binary splits or multiway splits can be performed, where the values of the input variables are partitioned from a simple inequality or a predetermined bound or interval. Again, the **Tree** node automatically performs binary splits. The reason that binary splits are automatically performed in Enterprise Miner is because binary splits simplify the exhaustive search of choosing the best splits from a multitude of possible splits. However, the advantage of multiway splits is that they will often result in more interpretable decision trees since the splitting variable tends to be used fewer times in the recursive tree branching process.

The Splitting Criterion

A splitting criterion is applied in the recursive tree branching process to select the best splits. The *splitting criterion* measures the reduction in the distribution of the target variable between the subsequent node and the root node. The criterion for a split may be based on the statistical significance test, that is, the F-test or chi-square test or a reduction in the target variance, that is, entropy or Gini impurity measure. At each step, after the best set of splits is determined, then the splitting criterion is used to determine the best split. This recursive partitioning is repeatedly performed in each subsequent node or child node as if it were the root node of the decision tree. As the iterative process develops, the data becomes more fragmented through the decision tree. The splitting criterion is considered a two-stage process. The first step is to determine the best split for each input variable. The second step is to choose the best split among a multitude of possible splits from many different input variables. Both these steps might require adjustments to the tree splitting criterion statistics to ensure an unbiased comparison. A perfect split will result in the values being assigned to each target category or target range without any error. For instance, assuming that the input value k performs the best split with respect to the input variable x, or $x < k$, that will result in exactly identifying both binary-valued target groups in each node. As an example, assume that the parent node has half red balls and half black balls, then a perfect split will have one node containing all red balls, with the other node containing all black balls. In other words, the best splits are determined by separating the target values into unique groups where one class level predominates in each node. Stopping rules are also applied that determine the depth of the decision tree in order to terminate the recursive partitioning process. The various stopping criteria that can be specified from the **Basic** tab are designed to avoid overfitting to the tree and improve the stability and the interpretation of the decision tree. This can be performed by simply increasing the number of records required for a split search, the number of records allowed in a node, and reducing the depth of the tree. The importance of the stopping rules is that constructing the decision tree that is too small will result in unreliable estimates and constructing the tree too large will result in unstable estimates due to a small number of observations within each leaf.

The following are the basic steps of the recursive split search algorithm in selecting the best split at each node.

1. Start with an L-way split of the input variable in the decision tree model.
2. Collapse the two levels that are closest to each other based on the splitting criterion statistic.
3. Repeat Step 2 with the remaining $L - 1$ consolidated levels.
4. This will generate a number of different splits with different sizes. Choose the best split.
5. Repeat this process for every input variable in the tree model and select the best split of the node.

Assessing the Fit

The main objective of decision tree modeling is to determine the appropriate-sized tree that is sufficiently complex to account for the relationship between the input variables and the target variable and yet keep the tree as simple as possible. In decision tree modeling, the main goal is to achieve the greatest predictive and classification accuracy, and yet at the same time make the interpretation of the results as easy as possible, that is, growing the decision tree to the right size, which is usually determined by either prior knowledge, previous analyses, or general intuition. The best tree is selected based on the modeling assessment statistic from the validation data set. In Enterprise Miner, the appropriate modeling assessment statistic to use is based on the level of measurement of the target variable. Otherwise, the best tree is determined by the maximum expected profit or minimum loss, assuming that a decision matrix is incorporated into the decision tree design. The purpose of decision trees is to either classify observations by fitting the categorically-valued target variable, calculate predictions by fitting the interval-valued target variable, or predict decisions from the profit or loss decision matrix. Finally, as a general strategy in constructing a well-designed decision tree, it is recommended to construct several trees with many different configurations, then select the best decision tree based on the various modeling assessment statistics from the validation data set.

Decision Tree Diagram

The following diagram displays the decision tree used to predict the categorically-valued target response in clients defaulting on a loan, BAD, from the home equity loan data set. The reason that the model is called a decision tree is because the diagram of the recursive partitioning process can be viewed as a tree-like structure, where the two separate class levels of the binary-valued target variable are repeatedly partitioned throughout the decision tree. The numerical labels positioned above each node indicate at which point the **Tree** node found the best splits from the interval, nominal, or ordinal input variables. The decision tree consists of four separate leaves. The interpretation of the tree diagram indicates that 952 of the clients or 20% defaulted on the loan from the training data set. The ratio between debt and income, DEBTINC, was determined to be the best splitting variable for the tree model in which the first rule is applied. Therefore, clients with the debt to income ratio under the value of $45.18 will default on the loan at a little over 7% of the time. Conversely, those with a debt-to-income ratio of over $45.18 will not pay the loan nearly 64% of the time from the training data set. The other various splits of the decision tree can be interpreted accordingly. The number of delinquent lines, DELINQ, resulted in the next best splitting variable followed by age of the oldest delinquent trade line, CLAGE. The various association rules can be interrupted by following the path of the decision tree from the root to each one of the leaves. The leaves of the tree contain the final fitted probabilities of 7.0%, 64.9%, 30.1%, and 83.4% from the series of rules that are applied to the training data set. Within each node of the decision tree, the node statistics from the training data set are displayed in the left-hand column and the node statistics from the validation data set are displayed in the right-hand column. Reliability and stability in the estimates can be determined by observing similarities in the node statistics across the number of leaves between both the training and validation data sets. Notice that the lower bound of the binary split is always allocated to the left-handed split. Conversely, the upper bound of the binary split is allocated to the right-hand side. In addition, the right leaf nodes will almost always contain fewer observations than the left leaf nodes. In addition, it should be noted that the recursive splits that were performed display very consistent target proportions between the training and validation data sets, indicating stability in the tree.

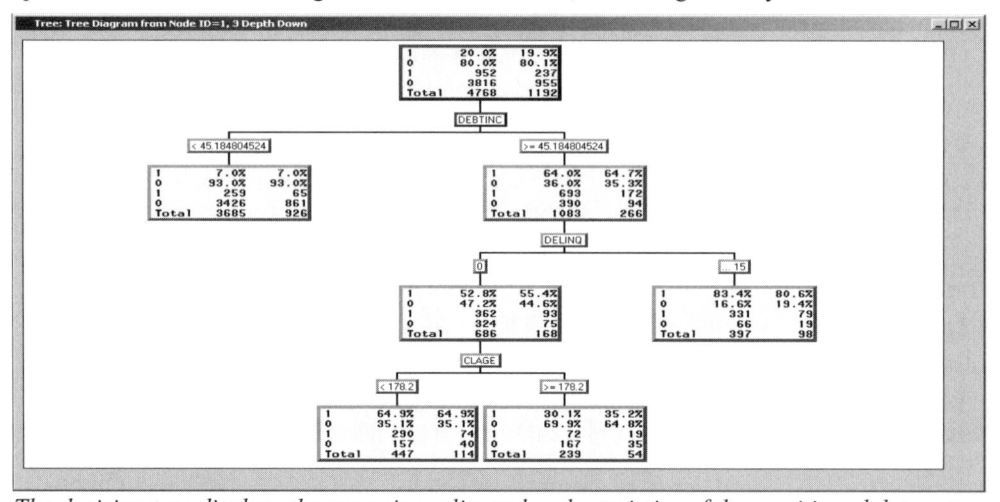

The decision tree displays the recursive splits and node statistics of the partitioned data sets.

Comparing Decision Trees to Some of the Other Statistical Techniques

The most important purpose in constructing predictive models is generating accurate predictions. However, it is also extremely important to understand the factors that are involved in explaining the target variable. Evaluating the predicted values or posterior probabilities of the tree would be done in the final leaves of the tree. For interval-valued targets, the fitted values are the target mean values at each leaf. For categorically-valued targets, the estimated probabilities are the target proportions at each leaf. The advantage of decision tree modeling as opposed to the other modeling techniques is that the interpretability of the predictive modeling results is simply a process of assessing a series of *if-then* decision rules that are used to construct the entire tree diagram, that is, from the root to each leaf of the decision tree. In decision tree modeling, various groups in the leaves may be formed based on the tree model identifying these same groups all meeting the rules that define these groups where some rules might perform better than others, as opposed to linear, logistic or discriminate analysis that calculate a classification boundary that is either a line or a curve to distinguish between the various target groups. However, the distinct advantage to classification tree modeling is that some of the target levels might form distinct groupings from the unique series of *if-then* rules, thereby making it impossible for the classification line to adequately capture these separate nonoverlapping rectangular response regions. As each split is performed, it will result in a perpendicular line or boundary that is drawn at the point in which the range of values of the input variable splits the target values. In two-dimensional scenarios, the rectangular regions are displayed in the following **Memory-Based Reasoning** node that is based on the Kd-tree search technique to determine the number of nearest neighbors in each rectangular region. However, the drawback to decision tree modeling is its limited ability to produce smooth prediction curves since the fitted tree model forms a multivariate step function.

Since the decision tree splits are determined by the input variables placed into separate intervals, decision trees are robust to outliers in the range of values in the input variables. Decision tree modeling is considered nonparametric prediction modeling since there are no assumptions being made about the functional form between the input variables and the target variable in the model. The reason is that the construction of the decision tree relies on the frequencies of the values in the training data set. In regression tree modeling, the constant variability assumption in the target values must not be overlooked in order to provide stability to the predictive model. In decision tree modeling, an asymmetrical distribution in the target values may result in isolated groups with a small number of observations formed from the outlying data points. Therefore, transformations might be performed in order to increase the accuracy of the regression tree. The big drawback to decision tree modeling is the instability in the construction of the tree, that is, even small changes in the training data set can result in entirely different tree structures. However, the overall performance of the decision tree remains stable, with the same accuracy. Even a small change in the data will introduce a different split being selected, which will result in different subsets in the subsequent child nodes in which the changes will continue to filter down the tree. For categorically-valued target variables, it is recommended to limit the number of class levels to ensure that the child nodes are as large as possible in order to achieve stability in the tree and improve generalization performance.

The Process in Constructing the Decision Tree

Initially, the **Tree** node first sorts the data. For interval and ordinal-valued input variables, the node sorts the input values with missing input values placed at the end. For nominal-valued inputs and interval-valued targets, the node sorts the class levels of the input variable by the average target values within each class level. For nominal-valued inputs and binary-valued targets, the node sorts the input class levels by the proportion of one of the target values. In all other situations, sorting is not performed. For a binary or interval-valued target, the best splits of the input variable are always found by searching every possible binary split while preserving the sorted data. If binary splits are specified from the **Maximum number of branches from a node** option, then the exhaustive search is finished. However, for all other situations, the data is first consolidated and all possible splits are evaluated or a heuristic search is applied. The best splits are then determined by consolidating the group of input values that seems likely to be assigned to the same branch. Therefore, the splitting search algorithm treats these consolidated data points as the same input value. The reason this process is performed is to accelerate the split search process with a smaller number of possible splitting values of the input variable to evaluate and consider. After consolidating the input values, the heuristic search is applied if the number of possible splits is greater than the value entered in the **Maximum tries in an exhaustive split search option** within the **Advanced** tab. A *heuristic search* attempts to calculate the best split of the input variable. For nominal-valued input variables, the heuristic algorithm is designed to perform a sequence of binary splits. Initially, the algorithm first groups the different tree branches, then reassigns and collapses several observations into different branches until binary splits are reached or any more target groups can be formed. The tree split that produces the smallest p-value is selected, that is, that best separates the target groups, based on all splits that were evaluated from the merge-and-shuffle heuristic procedure.

The **Tree** node performs various shortcuts in the split search strategy in order to limit the number of possible variable splits to evaluate during the construction of the tree. First of all, the node will automatically allocate 5,000 cases within each node if the sample size within the node is greater than 5,000. For categorically-valued target variables, the range of target values is consolidated into evenly distributed groups. However, the number of cases required for creating a particular node can be increased from the **Advanced** tab. Second, binary splits are automatically performed. However, if multiway splits are specified, then the initial phase consists of consolidating the range of input values into separate groups. Finally, all possible splits of the consolidated groups are evaluated unless the number of cases exceeds 5,000 observations, in which case an agglomerative algorithm is applied, where the **Maximum tries in an exhaustive split search** can be increased from the **Advanced** tab. The agglomerative algorithm evaluates every possible split of the input variables at every node in the tree.

The **Tree** node is designed to determine the best splits based on the *p*-values from the chi-square test statistic by the class levels of the categorically-valued target variable and the *p*-values from the F-test statistic by the range of values of the interval-valued target variable. For categorically-valued target variables, the chi-square statistic is based on the two-way contingency table where the columns represent the class levels of the target variable and the rows that correspond to the splits performed to the range of values or class levels of the input variable, or vice versa. For interval-valued target variables, the F-test statistic is analogous to the one-way ANOVA design where the variability in the target variable is divided into three separate parts. The between sum-of-squares measures the squared difference between the node means and the overall mean. The within sum-of-squares measures the variability within the node. Large F-test statistics will indicate differences between the node means. Although, the F-test statistic is rather robust to the normality assumption, however the nonconstant variance can not be overlooked. The reason is because the F-test statistic will indicate the significance of the split more often than not when there are nodes with a small number of cases that have a wide range of values. Therefore, various variance stabilizing transformations of the target values might be considered to improve the model.

The *p*-values can be adjusted from the **Tree** node to take into account for the successive testing that is applied in comparing the splits on the input variables, the number of input variables involved in the tree design, and the depth of the decision tree. For categorically-valued input variables, the chi-square statistic tends to increase as the number of splits or class levels of the input variable increases. Therefore, an adjustment is performed to the *p*-values in an attempt to correct for this bias. Assuming that the chi-square or the F-test statistics are specified, the *p*-values are adjusted even further before the tree split occurs in determining the best splits from the Kass CHAID adjustment. The node is also designed to limit the number of observations involved in the split of each node. An additional adjustment of the *p*-value can be performed from the **Tree** node that depends on the depth of the splits as the tree grows. The purpose of the depth adjustment is to severely limit the growth by retaining the highly significant splits to the tree from the recursive tree splitting algorithm. The tree split will stop if the adjusted *p*-value is greater than or equal to the significance level that is specified from the **Advanced** tab. The tree splitting criterion statistics, such as the F-test, chi-square test, reduction in variance, entropy, or Gini impurity measure statistic is used in evaluating the succession of tree branching processes. These tree splitting criterion statistics will generally achieve similar results. However, the various assessment statistics will tend to differ by increasing the number of observations in the training data set.

In predictive modeling and specifying too many input variables in the model will usually fit the noise level in the training data extremely well. Similarly in tree designs, too large of a tree might develop due to adapting to the random variation in the data, which will generally lead to poor generalization by fitting the decision tree to a new set of input values. Poor generalization will also occur by disabling the chi-square or the F-test statistic tree splitting criterion *p*-values, the number of observations required for a split search, and the maximum depth of the tree, which will result in the tree tending to grow to a point until each leaf contains a single target value. Many of the following **Tree** node options are designed to determine the growth of the tree, the termination of the tree branching process to each node, and the nodes to prune off or remove from the entire tree. These options are extremely important in the construction of decision trees that are designed to deliver improved classification or predictive modeling performance. In addition, the node will automatically perform a variable selection routine in order to select the best set of input variables in the construction of the decision tree. The **Tree** node is especially good at ignoring irrelevant inputs to the tree and using a small number of input variables to construct the decision trees given that the input data set contains several input variables.

Tree Pruning

The goal of decision tree modeling is to achieve accurate predictions and classifications. However, it is also extremely important to strive for the most parsimonious grouping with the fewest number input variables, the fewest splits and the fewest leaves. Therefore, it is important to stop the growth of the tree in a reasonable dimension to avoid overfitting and achieve good generalization, while at the same time constructing a decision tree that is easy to interpret. In developing a well-designed decision tree, the size of the tree might be the most important factor as opposed to creating effective splits. This process is called *tree pruning*. Pruning the tree will consist of a fewer number of rules, making it easier to interpret the results and achieve stability in the decision tree. Increasing the stability of the tree, such as fitting the tree model to an entirely different data set, that is, the validation data set, will result in the same set of input splitting variables used in the initial decision tree model, which will result in comparable estimates. There is no standard tree-growing or tree-pruning technique that is widely used for variable selection. However, a common tree-pruning strategy is to select the more complex tree as opposed to a simpler tree. The reason is because a smaller tree with a small number of branches may tend to underfit the data, which will result in poor generalization. The number of leaves, the number of splits, and the depth of the tree determine the complexity of the tree. Even in decision tree modeling, there is this bias-variance trade-off in achieving a well-fitted tree with low bias and low variance. Performing excessive pruning will result in an insufficient number of variables being selected. Therefore, it is often best to select a bushier tree among all other trees that are constructed. There are two separate tree-pruning techniques for selecting the right-sized tree, called pre-pruning and post-pruning. The **Tree** node performs both pre-pruning and post-pruning techniques in selecting the right-sized tree from the validation data set.

Pre-Pruning: The first tree splitting strategy uses the forward stopping rules that are designed to stunt the growth of the tree called *pre-pruning*. A common pre-pruning strategy is stopping the growth of the tree if the node is pure, that is, assigning the splits without error. The node consists of various stopping criterion options that will allow you to control the growth of the tree based on the pre-pruning idea. The various stopping rules that are used to terminate the construction of the tree are the specified number of leaves, the number of records required for a split search, the depth of the tree, and the various statistical tests that are based on distributional assumptions of the variables. Decision tree modeling might be considered in selecting the best set of input variables for the classification or predictive modeling design based on the input variables that perform the strongest splits with the largest logworth chi-square statistics in the tree model. Commonly, the pruning techniques that are used in removing extra leaves in determining the right-sized tree is by fitting the validation data set. The advantage of pre-pruning is that it is less computationally demanding than post-pruning. However, the drawback is that it runs the risk of missing future splits that occur below the weaker splits.

Post-Pruning: A second tree pruning strategy is creating a sequence of trees of increasing complexity in selecting the best tree, called *post-pruning*. The process of post-pruning is growing large trees and pruning back the tree branches. It is highly recommended to build several different trees using various configuration settings in selecting the best decision tree. There are two requirements for post-pruning. First, in order to make an honest assessment of the accuracy of the model and reduce the bias in the estimates, it is recommended to split the data into two separate data sets. However, data splitting is inefficient with small sample sizes since removing data from the training data set will degrade the fit. Therefore, techniques such as jackknifing are recommended to average the results over the *k* separate models. Second, an appropriate modeling assessment criterion should be applied, such as the misclassification rate from the validation data set. Post-pruning can be applied by performing interactive training within the **Tree** node.

Preprocessing the Input Data Set

One advantage of decision tree modeling is that each input variable may be evaluated separately as opposed to evaluating the modeling performance by every input variable that is combined in predicting the target variable. The tree diagram is useful in assessing which input variables are important based on the variables with the strongest splits located at or near the top of the tree, while at the same time interpreting the way in which the variables interact with each other. However, like all the other traditional modeling designs, in order to achieve good predictive or classification modeling performance, it is extremely important that the best set of input variables be selected that best predict the target variable. Therefore, one of the advantages of the **Tree** node is that it is one of the few modeling nodes in Enterprise Miner that has a built-in variable selection routine that is automatically performed once the **Tree** node is executed. This will result in selecting the most important input variables for the decision model that will result in a small number of splits to consider. As in all predictive modeling designs, it is important to preprocess the data when performing decision tree analysis by identifying the extreme values or outliers in the target variable. The reason is because these extreme values will result in the tree forming an unbalanced tree in which the outliers will tend to distort the structure of the tree, which is highly undesirable. Also, it is important that there are not too many class levels to the categorically-valued target variable. The reason is because a smaller number of class levels will improve the stability of the tree and improve the classification performance of the model. In addition, tree models are known to perform rather poorly with numerous interval-valued input variables in the model since the split search algorithm is designed for categorical input variables. A preliminary stage in grouping the interval input variables or the categorical input variables with more than two levels might be considered in order to accelerate the split search routine and improving stability in the decision tree. The reason is that there are a fewer number of candidate splits that need to be evaluated since an exhaustive split search routine is initially applied that evaluates all possible splits of every input variable at each node or branch of the decision tree. The goal is grouping the values of the input variables that seem likely to be assigned to the same branch in selecting the best split. This recursive grouping process is performed for every input variable in the model by continually collapsing the class levels from the splitting criterion until the best split is selected that best separates the target values. The CHAID algorithm will apply a backward elimination step to reduce the number of possible splits to consider. Reducing the number of input variables is also important for the tree branching process and will result in faster processing time in the split search routine since there are fewer input variables in the exhaustive tree splitting algorithm.

Once the preliminary stages of processing the data have been performed, then it is important to select the most appropriate tree modeling algorithm. The node is designed to perform either the CART (Classification And Regression Tree), CHAID (Chi-squared Automatic Interaction Detection), C4.5 and C5.0 algorithms. The CART and CHAID methods are the most popular algorithms. The CART method is the most widely used decision tree modeling algorithm. By default, the **Tree** node automatically performs the CART algorithm with binary splits to the input variables in the decision tree model. For interval-valued input variables, the splitting value to each tree branch is determined by the value that performs the best split. For categorically-valued input variables, the splitting values that are selected within each branch are determined by the two separate class levels that best separate the target groups.

Partitioning the Input Data Set

The **Tree** node and decision tree modeling is actually designed for large data sets. The reason is because a validation data set is recommended in the modeling design in order to evaluate the accuracy of the tree and to prevent overfitting in the classification or predictive modeling design. Overfitting results in poor generalization. In other words, fitting a tree model with too many leaves will result in poor classification or predictive modeling estimates when fitting the tree model to new data. Overfitting might be attributed to a falsified split that produces an accumulation of erroneous results in the data or an accumulation of small errors from many different splits. The various stopping and pruning options are designed to prevent overfitting in the decision tree model.

One way of detecting the overfitting phenomenon is by fitting the current tree model to an entirely different data set that was not involved in the construction of the tree. Overfitting can be detected when the current tree fits well to the training data set as opposed to fitting the same model to an entirely different data set that will result in a large discrepancy in the node statistics between the two data sets. The **Tree** node is designed to use the validation data set in order to revert to a simpler tree that is designed to stabilize the tree and increase the predictability of the tree model. The training data set is used to build the tree and the validation data set is used to measure the accuracy of the tree model. The **Data Partition** node is designed to divide the input data set into separate data sets. The type of modeling assessment statistic that should be applied depends on the level of measurement of the target variable. For an interval-valued target, the sum-of-squares assessment statistic is used. For a categorically-valued target, the misclassification rate is automatically applied. The best tree that is selected is determined by the modeling assessment statistic from the validation data set, assuming that the validation data set exists, which is highly recommended. However, if a decision matrix is applied to the design, then the cost of misclassification would then be the appropriate modeling assessment criterion in which the posterior probabilities are multiplied by the specified fixed costs at each target level. These misclassification costs along with the prior probabilities may be incorporated in the splitting criterion when constructing the tree.

The Standard Decision Tree Algorithms

The CHAID and CART methods are the two most widely known algorithms that are used in tree modeling.

CHAID Method

The CHAID algorithm is restricted to categorically-valued input variables where ordinal-valued input variables cannot be applied. The algorithm performs multiway splits using pre-pruning techniques. The first step is that the interval-valued input variables must be partitioned into separate categories with an approximately equal number of observations within each group before the tree is constructed. Then the algorithm performs a series of recursive binary splits by generating the greatest difference between the two groups. The best splitting variables for the node are determined by either applying the chi-square test based on a categorically-valued target variable, or the F-test statistic based on an interval-valued target variable. In the first phase, if the respective tests are not statistically significant based on the binary splits, then the process repeats itself by merging the separate categories to find the next pair of categories or merged categories. If the test is significant for the corresponding binary splits, then the CHAID algorithm uses a Bonferroni-adjusted p-value based on the number of categorical values in the input variable with the idea in adjusting the bias by choosing input variables for the tree with several class levels. The next step is to select the best splitting variable with the smallest Bonferroni-adjusted p-value. If no further branching is performed on the corresponding input variable, then the respective node will become a leaf. This process will continue until no further splits are performed.

Again, the split searching routine for this algorithm will initially construct the tree with each branch representing each value of the input variables in the tree model. Branches are then recursively merged together and then resplit based on the smallest chi-square p-values by fitting the categorically-valued target variable. A binary split is then performed with the input values consolidated into two separate groups, where the best splits are selected with the smallest chi-square p-values in comparison to all other possible splits. The strength of each split is evaluated by forming two-way frequency tables based on the number of cases for each tree branch from the consolidated group of target values. If the significance level from the chi-square test is below the significance level, that is, specified from the **Advanced** tab, then the binary branches are merged and the process is repeated. Otherwise, binary splits are formed by the three separate ranges of input values. The binary split is performed where the best splits are selected with the smallest chi-square p-values that performs the best split with the smallest chi-square p-values in comparison to all other possible splits.

This process of merging the branches and resplitting stops when an adequate p-value is reached, that is, the stopping rule is applied where there are no more splits or remerging of branches; this delivers significant differences in the input categories determined by a specified threshold value. The best splitting value of the input variable is determined from the last split performed on the input variable, as opposed to the other tree algorithms that initially form a fully grown tree that usually overfits the data, then pruning the tree. Instead, the CHAID method is designed to stop the branching process before overfitting occurs. After the initial splits are performed, then the p-values are adjusted in order to determine the best splitting input variables for the decision tree design. The construction of the tree stops when the adjusted p-values of the splitting variables for all unsplit nodes are above the significance level that is specified from the **Advanced** tab.

To perform the CHAID algorithm, the following steps must be applied within the **Tree** node as determined by the level of measurement of the target variable in the tree model.

For a nominal-valued target variable:

- From the **Basic** tab within the **Splitting criterion** section, set the splitting criterion statistic to the **Chi-square** test.
- In the **Advanced** tab, set the modeling assessment statistic to **Total leaf impurity (Gini Index)** from the **Model assessment measure** display field to avoid pruning being automatically performed.

For an interval-valued target variable:

- For an interval-valued target variable, set the splitting criterion statistic to the **F-test** statistic within the **Splitting criterion** section of the **Basic** tab.
- To avoid automatic pruning, set the **Subtree** method to **The most leaves** from the **Advanced** tab.

For either a nominal-valued or interval-valued target variable:

- For an interval-valued target variable, set the splitting criterion statistic to the **F-test** statistic within the **Splitting criterion** section of the **Basic** tab. For nominal-valued targets, set the splitting criterion statistic to the **Chi-square** option setting within the **Basic** tab
- Set the **Maximum number of branches from a node** entry field to the maximum number of class levels of the input variables in the model from within the **Basic** tab. As opposed to the following CART method, the tree does not need to be a sequence of binary splits for each of the tree nodes.
- From the **Basic** tab, set the **Surrogate rules saved in each node** entry field to zero.
- From the **Advanced** tab, set the **Maximum tries in an exhaustive split search** entry field to zero to force a heuristic search.
- From the **Advanced** tab, set the **Observations sufficient for split search** entry field to the number of records in the data set, with a limit of 32,000 observations.
- From the **Advanced** tab, select the **Apply Kass after choosing number of branches** check box within the **P-value adjustment** section.

CART Method

Assuming that a profit or loss matrix is not applied to the tree model, then the following CART method is very similar to the algorithm applied with the **Tree** node. The CART method determines the best possible binary splits rather than using the various post-pruning techniques. However, the drawback to this approach is that the data might naturally split into more than two separate groups. That is, binary splits are performed based on nominal or interval-valued input variables in predicting the nominal, ordinal or interval-valued target variable. An exhaustive search is performed and the best split is determined by maximizing the splitting measure. For a categorically-valued target variable, the assessment statistic is the reduction in the Gini index. For an interval-valued target variable, the assessment statistic is the reduction in the average sum-of-squares error. The drawback to this method is that the search can be lengthy if there are several input variables with many different class levels. However, since an exhaustive search is incorporated, it guarantees finding splits that produce the best classification.

The algorithm first overfits the data by constructing a fully grown tree. From the currently constructed tree, a subtree is found, with subtrees forming within other subtrees, which form a sequence of subtrees. Each subtree is determined by the smallest assessment statistic among all subtrees with the same number of leaves. For categorically-valued targets, the best subtrees are determined by the proportion of cases correctly classified. In order to accurately assess the classification performance of the tree model, it is highly recommended that a validation data set be applied, unless the training data set is too small. Therefore, the final tree is selected from the best candidate of subtrees based on the best assessment statistic from the validation data set. The main objective of this method is finding the best subtree from a fully grown tree by minimizing the *adjusted error rate* based on the number of leaves as follows:

$AE(\mathrm{T}) = E(\mathrm{T}) + \alpha \cdot N(\mathrm{T})$ for some tree T

where $E(\mathrm{T})$: is the impurity error function, and $N(\mathrm{T})$: is the number of leaves for some tuning constant $\alpha \geq 0$. The final subtree depends on the complexity parameter α. When $\alpha = 0$, this will result in a fully grown tree with small values of α resulting in larger trees and larger values of α, resulting in smaller trees. Gradually increasing the value produces subtrees that predict the target values better than the fully grown tree. The impurity error function may be specified from the **Splitting criterion** section. The **Tree** node is designed so that the structure of the tree is based on selecting the optimal number of leaves $N(\mathrm{T})$ for minimizing some impurity function $E(\mathrm{T})$.

A linear combination of the input variables is also used in determining the observation assigned to the one of the two tree branches from the linear combination of input splitting variables above or below some constant. The method performs a heuristic search to find the best binary splits but may not find the best linear combination of input splitting variables. Observations are excluded from the analysis if the splitting variable or the set of linear combination of input variables have any missing values. The **Tree** node is similar to the CART method with the exception of the linear combination of input splitting variables. Also for ordinal-valued target variables, the CART algorithm and the **Tree** node are different.

The following options must be applied in order to apply the CART method within the **Tree** node.

For a nominal-valued or interval-valued target variables:

- For a nominal-valued target variable, set the splitting criterion statistic to the **Gini reduction** statistic in the **Basic** tab, that is, neither the misclassification cost or misclassification rate is used. For a interval-valued target variable, set the splitting criterion statistic to the **Variance reduction** statistic in the **Basic** tab.
- From the **Basic** tab, set the **Maximum number of branches from a node** display field to two since the CART method automatically performs a binary split for each node based on the input variable with the best splitting criterion statistic. The CART method is designed to perform the best splits first.
- From the **Basic** tab, clear the **Treating missing as an acceptable value** check box to remove missing values in the input variables from the tree model.
- From the **Advanced** tab, set the **Subtree** display field to the **Best assessment value**, assuming that a validation data set is applied to the design. For a nominal-valued target, select the modeling assessment statistic to the **Total leaf impurity (Gini index)** option.

- From the **Advanced** tab, set the **Maximum tries in an exhaustive split search** entry field to 5000 or more, which will force the **Tree** node to search for every possible split of the input variable to determine the optimal binary split within each node. In other words, the method performs an exhaustive search in determining the best possible binary split.
- Apply the validation data set in evaluating the accuracy of the tree model that is used in conjunction with the **Best assessment value** subtree method.

Decision Tree Interactive Training

Interactive training is designed for you to construct your own custom-designed decision tree. If you already know a certain set of input variables that best predict or classify the target variable, then interactive training will allow you to include these extremely important input variables in the construction of the decision tree and the corresponding decision tree model. Interactive training will allow you to control the way in which the input variables are entered into the tree design as you interactively build the decision tree. For instance, you might want to build the decision tree by the frequency counts rather than the various assessment statistics specified from the **Advanced** tab. The other reason for using interactive training is to prune existing trees that have already been constructed and design the way in which the input variables are split within each tree branch. Interactive training will allow you to construct several decision trees to determine the best tree to fit based on the modeling assessment statistic from the validation data set. Initially, the best decision tree model is automatically selected from the validation data set once you first begin interactive training. In addition, interactive training enables you to compare the assessment statistics of the training and validation data sets, which will allow you to select the node to perform tree pruning. The purpose of interactive training is that it might increase the accuracy or improve the interpretability of the decision tree.

The *SAS Enterprise Miner Tree Desktop Application* introduced in Enterprise Miner v4.3 will allow you to either construct an entirely new decision tree or interactively modify an existing decision tree. The first step is activating the EMTree9.exe executable file that is located within the folder that the SAS application has installed the file on your machine in order to open the SAS Enterprise Miner Tree Desktop Application and train a new decision tree model. It should be noted that interactive training will not be performed until the model has been trained and the **Tree** node has been executed. The SAS Enterprise Miner Tree Desktop application and the various windows and option settings are discussed in greater detail from my website.

Missing Values

The **Tree** node automatically removes observations from the tree branching process with missing values in the target variable. Typically, missing values from the input variable can be estimated or imputed by the average values of the nonmissing cases. However, decision tree modeling does not use any of the various imputation techniques. The **Tree** node does not need the **Replacement** node in estimating the missing values. The advantage of decision tree modeling is the way in which missing values are handled. Since the tree branching process is based on all nonmissing cases of the input variable, handling missing values correctly will increase the classification or predictability of the decision tree model. Another advantage in the association between the missing values with the target values is that these same missing values can contribute to the predictive ability of the split. The imputation methods are applied prior to constructing the decision tree. For interval-valued input variables, the missing values might be treated as a special unknown nonmissing value. For categorically-valued input variables, the missing values may be identified as a separate class level. However, the **Tree** node will provide you with several different ways for handling missing observations in the input splitting variable based on the use of surrogate variables for the tree branching process. The node performs the various surrogate rules, applying each surrogate rule sequentially until a certain surrogate rule is met. However, if none of the surrogate rules are met, then the observation that is assigned to the branch is identified as a missing value. A *surrogate split* is basically a backup to the main splitting rule, that is, a surrogate split is partitioning performed using an entirely different input variable that will generate a similar split in comparison to the primary splitting variable. For example in predicting income, high-income people would be reluctant to respond to questions about their personal income. In other words, the predictability of the split is based on the relationship between the missing values and the target response. However, if there exist several missing values, then these same observations with missing values will be assigned to a single branch, creating nonhomogenous tree splits that are to be avoided in decision tree modeling. The advantage of imputing missing values during the recursive tree branching routine is that the worth of the split is computed with the same number of observations for each input variable in addition to avoiding nodes with a small number of observations with a wide range of values.

The surrogate splitting rule is applied to handle missing values. Again, the best surrogate split results in a partition of the target values using a different input variable that best replicates the selected split. The surrogate splitting rules are applied in sequence until one of the surrogate rules is applied to the missing observation in the primary input variable. The first surrogate rule is applied when the tree comes across missing values in the input variable. The input variable that is selected as the best surrogate variable for the branch is determined by the accuracy in the proportion of cases assigned to the corresponding branch in comparison to the primary input variable. If the best surrogate variable is missing as well, then the second-best surrogate variable is applied, and so on. The number of surrogate variables that are applied as a possible backup splitting variable due to missing values in the primary variable is determined by the value entered in the **Surrogate rules saved in each node** entry field from the **Basic** tab. The observation is identified as missing if the primary splitting variable (main rule) and all subsequent surrogate variables (surrogate rules) are missing. For example, assume that the tree assigns a leaf node based on the geographical state, where the state is unknown. Therefore, the main splitting rule depends on the state, whereas the surrogate rule relies on the geographical region. If the region is known and the state is unknown, then the surrogate rule is applied. Typically, the surrogate rule is related to the main rule. For example, income level and level of education may be related if they have the same predictive power. Generally, most people tend to respond to questions about their education level as opposed to their personal income level.

Calculation of the Decision Estimates

Typically, a separate decision is made for each case within the various modeling nodes, such as the **Neural Network** and **Regression** nodes, assuming that the decision matrices are applied to the modeling process. However in the **Tree** node, the same decision is made for all cases that are assigned to the same leaf. Since separate costs can be specified for each case in the training data set, the average cost of each leaf is used as opposed to the individual costs for each case in the calculation of the expected profit. From the **Tree** node, the calculated profit can be computed as the difference between revenue minus the average fixed cost within the same leaf, which will result in a single decision being made within each leaf.

Decision tree modeling is discussed in greater detail in Berry and Linoff (2004), Hastie, Tibshirani, and Friedman (2001), Giudici (2003), Hand, Mannila, and Smyth (2001), Mitchell (1997), and Decision Tree Modeling Course Notes, SAS Institute (2001).

Data tab

The **Data** tab is designed for you to view the previously created data sets that are connected to the **Tree** node. By default, the node automatically selects the training data set as the active input data set. However, it is highly recommended to also supply a validation data set that would be used in measuring the accuracy and stability of the decision tree modeling estimates and fine-tune the construction of the decision tree. The reason is that the training data set will tend to produce poor predictions with new data. The scored data set may be selected in predicting the target values, which can thought of as fitting the decision tree model to a new set of data.

Variables tab

The **Variables** tab is automatically displayed once you open the node. The tab displays a table view of the various properties that are assigned to the listed variables from the metadata sample. Variables that are assigned a status of **use** are automatically included in the tree modeling process. However, the tab will allow you to remove input variables that do not seem important in predicting the target variable by simply setting the status of the variables to **don't use** from the **Status** column.

The tab will allow you to edit, browse, or delete the target profile of the target variable. The target profile is designed to increase the modeling performance of the design by specifying appropriate prior probabilities to the class levels of the categorically-valued target or specifying the objective function that is either maximizing profit or minimizing loss instead of minimizing the sum-of-squares error or the misclassification rate by fitting the interval-valued or categorically-valued target variable in the model. SAS warns that the number of decision levels must match the number of class levels of the ordinal-valued target variable to perform decision tree modeling. Otherwise, a warning message will appear that will inform you that the tree modeling fit failed. However, this is not a problem for either the **Neural Network** or the **Regression** modeling nodes.

Basic tab

The **Basic** tab is designed to specify the splitting rule criterion. By default, binary splits are performed. The tab will also allow you to specify various stopping criteria by controlling both the growth and size of the decision tree. The **Splitting criterion** section contains the modeling assessment statistics used in determining the tree splitting process that depends on the measurement level of the target variable. As the tree is constructed, the number of splitting rules are evaluated for each leaf node. The tab is designed so that two separate **Splitting criterion** sections will appear, depending on the level of measurement of the target variable that you want to predict. For classification trees, the three splitting criteria that are used are the Gini index, entropy, and the chi-square test. For regression trees, the two splitting criteria that are used are the variance reduction and the F-test statistic. After the candidate splits are determined, then one of the following splitting criteria is used in determining the best splits. The following statistics are designed to measure the strength of the tree splitting rule in determining the best splits as the decision tree is growing.

For nominal or binary-valued target variables, the following three goodness-of-fit statistics will appear in the **Splitting criterion** section. Note that the following three methods will usually give similar results. However, as the sample size increases, the various assessment statistics tend to disagree more often.

- **Chi-square test** (default)**:** The Pearson chi-square statistic measures the significance of the proportion between the target variable and the branching node with a default **Significance level** of .20. The chi-square statistic tests if the column distribution, that is, the target categories, is the same in each row, that is, the child node or the range of values of the input variable that are collapsed into separate groups. The chi-square statistic tests if the distribution in the categorically-valued target variable is the same across the branches. The chi-square test is the splitting criterion that is used in the CHAID algorithm. When using the chi-square splitting criterion, the best splits are selected by the largest chi-square statistic. In other words, the best splits in grouping the data are determined by the smallest *p*-values. The logworth statistic is computed as follows:

  $\text{LOGWORTH} = -\log_{10}(p\text{–value})$, where the statistic will increase as the *p*-values decrease.

- **Entropy reduction:** This statistic is a measure of variability in categorical data. If the target values are the same within the node, then the entropy statistic is zero. If the two target groups are evenly distributed, then the entropy statistic is one. Therefore, the best split is selected by the smallest entropy statistic. The formula for the impurity reduction statistic is defined as follows:

  $$\text{Entropy statistic} = E_E(m) = -1 \cdot \sum_{\text{Leaves}} \left(p_{\text{leaves}} \cdot \left(\sum_{i}^{\text{classes}} p_i \cdot \log_2 \cdot p_i \right)\right)$$

  where p_i is the proportion of each class level in the corresponding node *m*. If prior probabilities are incorporated in the decision tree model, then the entropy statistic is adjusted accordingly.

  As an example, assume that 9 bad creditors and 11 good creditors are assigned to the root node with 3 bad creditors, 7 good creditors assigned to the left node, and 6 bad creditors and 4 good creditors assigned to the right node. The entropy reduction statistic is calculated as follows.

  $$\text{Entropy statistic} = -1 \cdot [(10/20) \cdot ((3/10) \cdot \log(3/10) + (7/10) \cdot \log(7/10)) + (10/20) \cdot ((6/10) \cdot \log(6/10) + (4/10) \cdot \log(6/10))] = .764$$

- **Gini reduction (Total Leaf Impurity):** This statistic measures the probability that two things chosen at random in a population will be the same, that is, it is a measure of purity. If the target values are the same within the node, then the Gini statistic is one. If the two target groups are evenly distributed, then the Gini statistic is .5. Therefore, the best split is selected by the largest Gini reduction statistic. The formula for the Gini index is very similar to the chi-square test for proportions that are based on the square of the proportions in each node, calculated as follows:

  $$\text{Gini index} = E_G(m) = \sum_{\text{Leaves}} p_{\text{leaves}} \cdot \left(\sum_{i}^{\text{classes}} p_i^2 \right)$$

  where $E_G(m)$ is the Gini index at node *m* and p_i is the proportion of each target class in the ith node. If prior probabilities are incorporated in the decision tree model, then the impurity reduction statistic is adjusted accordingly. The Gini reduction criterion statistic is calculated as follows:

  $$\text{Gini index} = (10/20) \cdot [(3/10)^2 + (7/10)^2] + (10/20) \cdot [(6/10)^2 + (4/10)^2] = .55$$

For ordinal-valued target variables, the default is the **Entropy reduction** criterion statistic. Otherwise, you may select the **Gini reduction** criterion statistic.

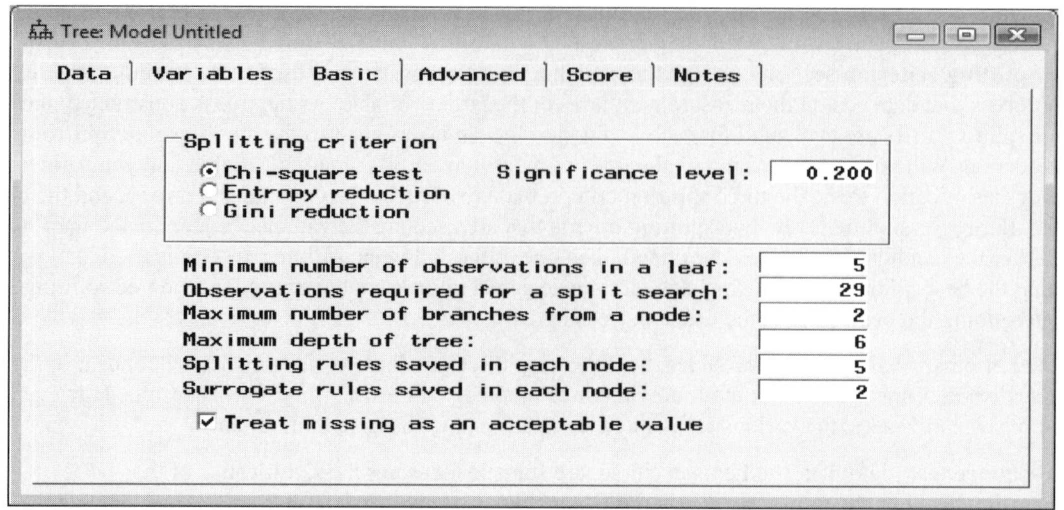

*The **Basic** tab and the splitting criteria of the tree are obtained by fitting the categorical target variable.*

For interval-valued target variables, the following criterion statistics will be listed within the tab:

- **F test** (default)**:** The F-test statistic is analogous to the chi-square test. The F test statistic is used in testing the variability of the node with a default **Significance level** of .20. The statistic is based on the ratio of the between sum-of-squares and the within sum-of-squares after the split. The between sum-of-squares measures the difference in the means of the target values between each node and the overall target mean at the root node. The within sum-of-squares measures the variability within each corresponding node. The F test statistic is the splitting criterion that is used in the CHAID algorithm. Similar to the chi-square statistic, the split that delivers the maximum logworth splitting criterion statistic is selected. The logworth statistic is calculated as follows:

  $$\text{LOGWORTH} = -\log_{10}(p\text{–value})$$

- **Variance reduction:** The reduction is simply the sample variance that is the squared difference between the observation and the overall target mean of the corresponding node as follows:

  $$\text{Variance reduction} = \frac{1}{n_m}\sum_{i}^{\text{classes}} (y_{im} - \bar{y}_m)^2 \text{ where } \bar{y}_m \text{ is the average target variable at node } m$$

  Note that the variance reduction is similar to the previous F-test statistic except that the split-induced variance is not divided by the residual variance after the split.

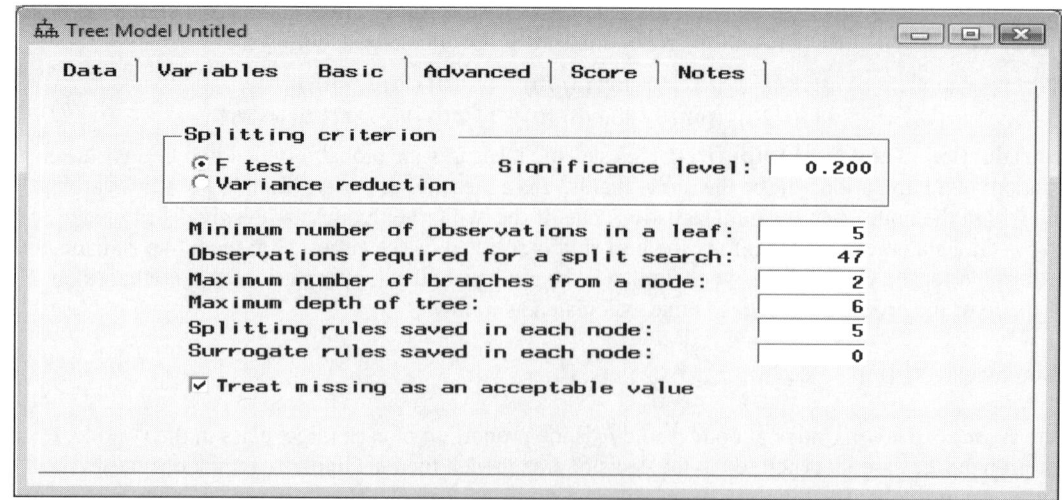

*The **Basic** tab displaying the splitting criterion by fitting an interval-valued target variable.*

The following few options are designed to limit the growth of the tree, that is, prepruning the tree, and will allow you to control the recursive process and stop the branching process by preventing the node from splitting any further. The following stopping criterion options are designed to avoid overfitting in the model and improve the stability and the interpretation of the decision tree. The growth and size of the tree can be stopped by specifying a positive integer value for any one of the following options:

- **Minimum number of observations in a leaf:** The minimum number of cases is one. This option limits the number of observations assigned within each leaf with the splitting process stopping when the leaves contain no more than the specified number of cases. The stability in the decision tree can be achieved by increasing the number of observations falling within each leaf. The **Tree** node requires the following restriction to the split search:

  Observations required for a split search $\geq 2 \cdot$ **Minimum number of observations in a leaf**

- **Observations required for a split search:** This option will allow you to specify the minimum number of observations required to perform a split, preventing the tree branching from occurring within a node due to a small number of observations. Again, the stability in the decision tree can be achieved by increasing the number of cases required for a split search. The default value is the total number of observations from the training data set that is determined by the following three steps:

  1. **Observations required for a split search** = INT[MAX(number of observations / 100, 2)]
  2. If the **Observations required for a split search** > 32766, then **Observations required for a split search** = 32766
  3. If the **Observations required for a split search** < $2 \cdot$ **Minimum number of observations in a leaf**, then the **Observations required for a split search** = $2 \cdot$ **Minimum number of observations in a leaf**

- **Maximum number of branches from a node:** This option controls the number of branches from each branching node. The default is two. The node will allow you to construct either binary trees or extremely bushy trees. By default, the **Tree** node automatically performs binary splits on the values of the input variables, preventing the tree from fragmenting the data too quickly, which would result in insufficient data filtering through the tree.
- **Maximum depth of tree:** This option determines the maximum number of successive tree branches that forms the depth of the decision tree. The default is no more than six splits. By performing the variable selection routine within the node, then it is recommended to increase the depth of the tree, since severe pruning will result in fewer input variables being selected. However, limiting the growth of the tree will result in an easier interpretation of the final results.
- **Splitting rules saved in each node:** This option will allow you to control the number of tree splitting rules displayed in the **Competing Splits** window from the **Results Browser**. The default is five splitting rules that will be saved.
- **Surrogate rules saved in each node:** This option will allow you to specify the maximum number of surrogate tree splitting rules to create. The default is zero surrogate rules. Once the best (primary) split and splitting value is found, then the **Tree** node will automatically create a list of corresponding surrogate variables and split values that best replicate the primary split. The number of surrogate variables to fall back on when the primary input variable has missing values depends upon the number entered in the entry field.
- **Treat missing as an acceptable value:** This option will allow you to specify missing values as a valid class level. By default, the node uses the missing values as an input in the construction of the decision tree and the tree splitting rule procedure. The reason is because the missing class levels might be associated to the variability of the target values, which will increase the predictability of the decision tree model.

Advanced tab

The **Advanced** tab is designed to specify from a wide variety of configuration settings in either assessing the tree or reducing the number of possible variable splits to evaluate during the construction of the decision tree. The tab will allow you to specify the modeling assessment statistic, and the subtree method, set the split search criteria, adjust the p-value that determines the tree splitting process, and allow you to incorporate the decision matrix or prior probabilities for the split search by fitting the categorically-valued target variable.

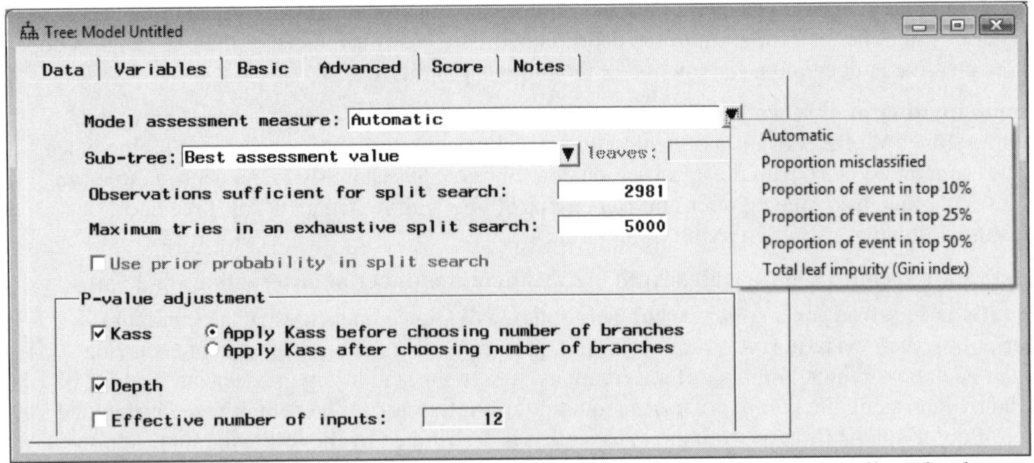

*The **Advanced** tab displaying the settings to the decision tree fitting the categorically-valued target.*

Modeling Assessment Statistics

The following are the various modeling assessment statistics to select from within the tab. The modeling assessment statistic is usually based on the validation data set in evaluating the overall accuracy of the prediction estimates from the decision tree model. The statistic depends on the measurement level of the target variable and whether you specified a decision matrix for the design. For a categorically-valued target, the modeling assessment statistic is the maximum proportion of correctly classified cases. For interval-valued target variables, the sum-of-squares error is the assessment statistic that you want at a minimum.
For interval-valued targets with no profit or loss matrix, the following assessment statistics will be displayed to evaluate the overall accuracy of the regression tree:

- **Average square error** (default)**:** The best tree is selected with the smallest average square error. The ASE statistic is the average sum-of-square error that is based on the difference between the target values and its predicted values.
- **Average in top 10%, 25% and 50%:** The best tree is selected with the smallest average square error in the top n% of the cases. These assessment statistics might be selected if the goal to the analysis is to obtain the best assessment values from the first n% of the observations.

For interval-valued targets with the existence of a decision matrix, the following assessment statistics will be displayed to measure the overall assessment of the regression tree.

- **Automatic** (default)**:** This is the default. The largest expected profit or smallest expected loss is automatically selected. Otherwise, the average squared error is selected.
- **Average square error:** The best tree is selected with the smallest average square error.
- **Average profit/loss:** The best tree is selected with the largest profit or smallest loss. However, the average square error is used if there are less than two decision levels in the profit or loss matrix.
- **Average profit/loss in top 10%, 25% and 50%:** The best tree is selected with the largest average expected profit or smallest average expected loss in the top n% of the cases with a profit or loss matrix. These assessment statistics might be selected if the goal of the analysis is to obtain the best lift values from the first n% of the observations. However, the average predicted values of the target in the top n% cases is used to select the best tree if there are less than two decision levels in the profit or loss matrix.

For categorically-valued targets with no profit or loss matrix, you may select from the following modeling assessment statistics to measure the overall assessment of the classification tree.

- **Automatic** (default)**:** This is the default. For categorically-valued target variables, the misclassification rate is automatically selected, assuming that the target profile has not been specified.

- **Proportion misclassified:** The best tree is selected with the lowest misclassification rate or the proportion of cases that have incorrectly classified the target class levels.
- **Ordinal proportion correctly classified** (ordinal-valued targets only)**:** The best tree is determined by the ordinal distances where the target levels are assigned ranks and the classification rate is weighted for ordinal distances between the separate ordered class levels.
- **Proportion of event in 10%, 25% and 50%:** The best tree is selected with the largest average expected profit or smallest average expected loss in the top *n*% of the cases with the existence of a profit or loss matrix. The node selects the best tree with the highest proportion in the target event level for the top *n*% of the cases if the decision matrix consists of a single decision level.
- **Total leaf impurity (Gini index):** The best tree is selected by the smallest Gini index statistic.

For categorically-valued targets with the existence of a profit or loss matrix, the following modeling assessment statistics are displayed to measure the overall assessment of the classification tree.

- **Automatic** (default)**:** The modeling assessment statistic is automatically selected. For categorically-valued targets, the largest expected profit or the smallest expected loss is automatically selected.
- **Proportion misclassified:** The best tree is selected with the lowest misclassification rate.
- **Average profit/loss:** The best tree is selected with the largest average expected profit or smallest average expected loss. The accuracy rate is used if there are less than two decision levels in the profit or loss matrix. However, the node selects the best tree with the lowest misclassification rate if there are less than two decision levels to the profit or loss matrix.
- **Proportion of profit, or loss in 10%, 25% and 50%:** The best tree is selected with the largest average expected profit or smallest average expected loss in the top *n*% of the cases with a profit or loss matrix. The node selects the best tree with the highest proportion in the target event level for the top *n*% of the cases if the decision matrix consists of a single decision level.
- **Total leaf impurity (Gini index):** The best tree is selected by the greatest reduction in the leaf impurity or smallest Gini index statistic.

Subtree options

The **Subtree** option is designed to control the way in which the subtree is selected from the fully grown decision tree. The node is designed to report all possible subtrees that are created within the fully constructed tree. The results from the node reports the number of leaves that are created from the best subtrees. Once the tree branching process stops, the **Tree** node will automatically construct a sequence of subtrees. Each subtree is based on the number of leaves that are created from the fully constructed tree. Therefore, the following three subtree methods may be specified in determining the best possible subtree.

The following options in determining the best subtree are based on the validation data set:

- **Best assessment value** (default)**:** The best subtree is selected based on the modeling assessment statistics that were previously specified from the validation data set.
- **The most leaves:** Selects the entire decision tree. The node uses the largest subtree after the nodes have been pruned that does not increase the assessment statistics.
- **At most indicated number of leaves:** Uses the subtree with at most *n* leaves. The number of leaves may be specified from the **leaves** entry field. The default is one.

Splitting Criterions to the Tree Branching Process

The following options will allow you to specify additional tree splitting criterions to the decision tree:

- **Observations sufficient for split search:** This option will allow you to specify the maximum number of observations involved in the tree branching process. By default, the maximum number of observations is set at 5000 or the total sample size of the active training data set. The reason is because the **Tree** node will then try to determine every possible split. This option is designed to reduce both memory and CPU time by specifying a smaller number of observations involved in the exhaustive search procedure and the corresponding tree branching routine.

  The difference between this option and the **Observations required for a split search** option from the previous **Basic** tab is that this option is designed to specify an upper limit to the number of observations to the tree split as opposed to the **Basic** tab option that is designed to restrict the tree branching procedure from occurring based on the value that is specified.

- **Maximum tries in an exhaustive split search:** This option is designed to specify the highest number of candidate splits that you want to find in the exhaustive search procedure. If additional splits have to be considered, then a hierarchical search is used instead. In order to determine the most optimum tree split, the **Tree** node might evaluate all possible values in determining the best possible split of the input variable at every node of the decision tree. The exhaustive search procedure tries to seek the best splitting rule that might involve making some adjustments to the splitting criterions in order to maximize the splitting criterion. At times, this procedure might be exhaustive, create extremely large trees, and might be extremely time consuming. Therefore, this option is designed for you to specify the maximum number of possible splits to search for within the leaf node. The default is 5,000 tries.
- **Use profit/loss matrix during split search:** This option is only displayed for a categorically-valued target variable in the model. The check box must be selected in order to make sure that the decision matrix is incorporated in the split search routine. Therefore, the best tree will be selected by the largest expected profit or the smallest expected loss. For ordinal-valued targets, the decision matrix must be applied, with the construction of the tree that is based on the ordering arrangement of the ordinal class levels of the target variable. By default, the profit or loss matrix is not applied to the nominal-valued target variable.
- **Use prior probability split search:** This option is only available when fitting a categorically-valued target variable to the model, the tree splitting will be determined by the posterior probabilities that are adjusted by the predetermined prior probabilities. Adjustments to the target proportions by the prior probabilities will result in the modification of the worth of the splitting rule.

P-Value Adjustments

The node is designed to make adjustments to the *p*-values by the tree splitting criterion F-test statistic or the chi-square statistic that are specified from the **Basic** tab. However, further adjustments of the *p*-values can be performed to take into account in the number of input variables or the depth of the node in the decision tree.

The following are the *p*-value adjustment methods that can be specified from the **P-value adjustment** section:

- **Kass:** This option is automatically selected with the *p*-values for the F-test and chi-square statistic automatically adjusted. The *p*-value is adjusted by multiplying the *p*-value by a Bonferroni factor that depends on the number of branches, the target values, and, at times, the number of distinct values in the input variables to the decision tree model. The chi-square statistic tends to increase as the number of splits increase. This will result in larger decision trees. Therefore, the following options are designed to adjust the *p*-values to account for this bias. Basically, the Kass adjustment subtracts $\log_{10}(m)$ from the logworth chi-square statistic, where m is the number of splits.
    - **Apply Kass before choosing number of branches** (default)**:** By default, the node selects the best splits based on the adjusted *p*-value by applying the Bonferroni factor before the split is performed for each input variable. Therefore, the *p*-values for determining the best splits on the same input variable are compared by the adjustment applied to the *p*-values.
    - **Apply Kass after choosing number of branches:** The Kass's CHAID algorithm was actually designed to apply adjustments to the *p*-value after the tree splitting process has occurred for each input variable. Therefore, the *p*-values for determining the best splits on the same input variable are compared without adjustments.
- **Depth:** By default, this option is checked. The drawback to the CHAID algorithm is that multiplicity adjustments are performed within each node, however, adjustments are not performed from the number of leaves in the tree model. Since the tree branching process involves a number of individual significance tests within each node, the CHAID algorithm might result in a number of corresponding false rejections of the null hypothesis, resulting in an extremely large tree, which is very undesirable. Therefore, this option controls the type I error rate (α-level) in testing all the nodes that are currently eligible for a split by reducing the error rate as the tree grows. Each individual test is rejected if the *p*-value is less than the adjusted alpha-level. The adjusted alpha-level is calculated as follows:

  $\frac{\alpha}{2^d}$ where d is the depth adjustment to the binary split for some significance level α

This option applies a Bonferroni factor to the final p-value based on the number of leaves to correct for the excessive number of rejections. The reason for adjusting the p-value is to simultaneous partition all the previous splits that are incorporated in creating the current partition. The depth option is designed to severely limit the size of the tree growth.

- **Effective number of inputs:** By default, this option is not selected. This option adjusts the p-values by the number of input variables in the decision tree model. The problem with several input variables in the tree branching process is that it will increase the occurrence that some input variables might be selected in the tree branching process rather than other input variables that best explains the target values. By default, the number displayed in the entry field is determined by the number of input variables automatically set to a status of **use** by the **Tree** node. Adjustments are made to the splitting criterion p-values by multiplying these p-values by the number specified in the entry field as follows:

  Adjustment factor = $\max[(m / M(root) \cdot M(\tau), 1]$ where m = **Effective number of inputs**, $M(root)$ = number of input variables, and $M(\tau)$ = number of input variables involved in the recursive splitting routine in which the **Tree** node searches for a splitting rule in a specific node.

Score tab

The **Score** tab is designed to create an output scored data set from the decision tree results. The scored file will create the standard output data set. For interval-valued target variables, the scored file will consist of the predicted values along with all other variables in the data set. Conversely, for categorically-valued target variables, the output data set will consist of the class probabilities and the classification identifiers. Since the **Tree** node performs an internal modeling selection routine, the output data set might be applied to the other modeling nodes since the data set automatically includes all input variables that best describes the target variable that you want to predict. The **Score** tab consists of the following two subtabs.

- **Data subtab**
- **Variables subtab**

Data subtab

The **Data** subtab is designed to display the scored data sets and view the file information and table view of the output data set. The subtab will allow you to view the scored data sets. The subtab is similar to the **Output** tab in the other modeling nodes. Therefore, the tab will not be displayed. The file structure of the scored data set from the decision tree model is different in comparison to the previous regression model. Each observation in the scored data set is listed by the way in which the observation fell through the tree splits that are identified by both the node and the leaf where the observation resides. This is illustrated in the following table listing.

To view the file information or the table view of the scored partitioned data sets, select the appropriate radio button, then select the **Properties...** button. The training data set is automatically selected. The **Tree** node must first be executed in order to view the corresponding output data sets.

Data set details

Information Table View

☑ Variable labels

	Node	Leaf	Predicted: BAD=1	Predicted: BAD=0	Into: BAD	Unnormalized Into: BAD	From: BAD	Residual: BAD=1	Residual: BAD=0	Warnings	BAD
1	10	6	0.7134328358	0.2865671642	1	1	1	0.2865671642	-0.286567164		1
2	10	6	0.7134328358	0.2865671642	1	1	1	0.2865671642	-0.286567164		1
3	10	6	0.7134328358	0.2865671642	1	1	0	-0.713432836	0.7134328358		0
4	10	6	0.7134328358	0.2865671642	1	1	1	0.2865671642	-0.286567164		1
5	10	6	0.7134328358	0.2865671642	1	1	1	0.2865671642	-0.286567164		1
6	10	6	0.7134328358	0.2865671642	1	1	1	0.2865671642	-0.286567164		1
7	10	6	0.7134328358	0.2865671642	1	1	1	0.2865671642	-0.286567164		1
8	10	6	0.7134328358	0.2865671642	1	1	0	-0.713432836	0.7134328358		0
9	10	6	0.7134328358	0.2865671642	1	1	1	0.2865671642	-0.286567164		1
10	7	10	0.9680851064	0.0319148936	1	1	1	0.0319148936	-0.031914894		1
11	5	5	0.85	0.15	1	1	1	0.15	-0.15		1
12	10	6	0.7134328358	0.2865671642	1	1	1	0.2865671642	-0.286567164		1
13	14	1	0.0431654676	0.9568345324	0	0	0	-0.043165468	0.0431654676		0
14	10	6	0.7134328358	0.2865671642	1	1	1	0.2865671642	-0.286567164		1
15	10	6	0.7134328358	0.2865671642	1	1	1	0.2865671642	-0.286567164		1

The scored output data set with each observation that is identified by each node and leaf in the tree with the target proportions at each leaf that are the estimated proportions to the decision tree model.

Variables subtab

The **Variables** subtab will allow you to select the variables to be written to the scored data set that can then be passed along to the subsequent modeling nodes.

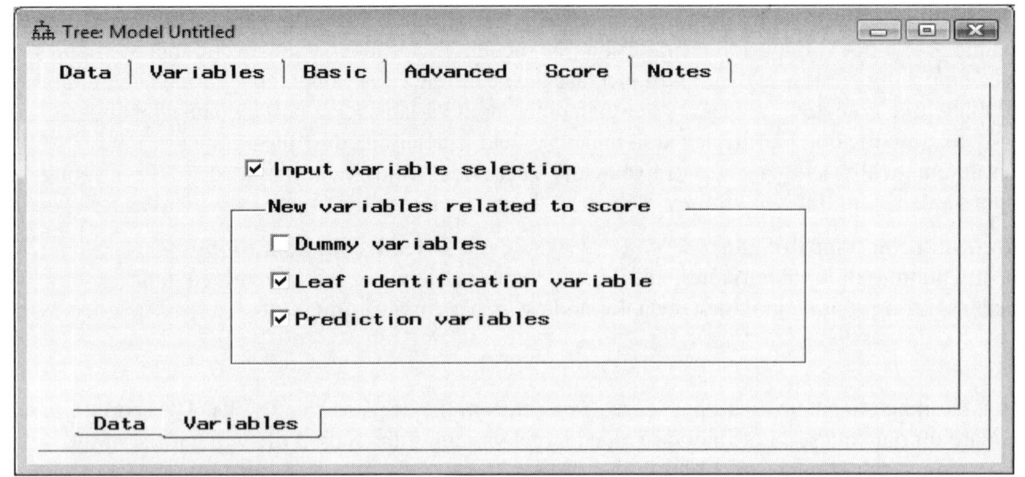

*The **Variables** subset controlling the variables listed in the output scored data set.*

- **Input variable selection:** This option is automatically checked to select the most important input variables for the tree branching process. This option is designed to perform a variable selection routine that is designed to select the most important input variables that best predict the target variable. The input variables are automatically included in the tree branching process with an importance value of greater than or equal to .05. Otherwise, the input variables are removed from the model and their model roles automatically set to **rejected**. Alternatively, the **Variable Selection** node could be used in determining the most important input variables that best describe the target variable. The drawback to the input selection routine in the **Tree** node is that it only selects the input variables that appear in the tree. Therefore, the variable selection routine in the **Tree** node does not take into consideration in the other input variables that do not appear in the decision tree, as opposed to the variable selection routine in the **Variable Selection** node that initially performs correlation analysis in measuring the linear relationship between each one of the input variables and the target variable, then performs forward stepwise regression in the subsequent step for all input variables that are not dropped from the initial correlation analysis procedure. In addition, even though the input variables are selected from the variable selection routine within the **Tree** node, the variable selection routine does not necessarily indicate a strong linear relationship with the target variable. Therefore, the variable selection routine within the **Tree** node might be more appropriate for nonlinear modeling designs like neural network modeling.

The following are the various options that you may select from the **New variables related to score** section within the **Score** tab. The following options will determine the type of variables that are created within the scored output data set:

- **Dummy variables:** By default, this option is not selected. Selecting this option will automatically create indicator variables for the output data set by each leaf in the current decision tree model. The option is designed to identify the interactions or partial associations between the input variables that can be used as inputs in the subsequent modeling nodes. The drawback in creating these indicator variables is that they will contribute to an undesirable increase in the dimensionality of the model.
- **Leaf identification variable:** This option is automatically selected. It creates a numeric id variable that assigns a number to the leaf. For example, these identification numbers might be applied to create separate predictive models or identify the final leaves of the tree that consist of the posterior probabilities to be used in the subsequent modeling nodes. In addition, the leaf identification data set can be passed along to the **Group Processing** node in order to perform group processing analysis.
- **Prediction variables:** If this option is checked, it automatically writes the predicted values of the target variable to the scored data set.

Viewing the Tree Node Results

General Layout of the Results Browser within the Tree Node

- **Model tab**
- **All tab**
- **Summary tab**
- **Tree Ring tab**
- **Table tab**
- **Plot tab**
- **Score tab**
- **Log tab**
- **Notes tab**

The following results are from the modeling comparison procedure that is displayed in the **Assessment** node in which the decision tree model fits the binary-valued target variable, BAD, from the HMEQ data set.

Model tab

The **Model** tab is designed to view the file information of the node, and the various configuration settings specified from the node, and view the imported tree data set. The tab consists of the following four subtabs.

- **General subtab**
- **Basic subtab**
- **Advanced subtab**
- **Import Tree subtab**

General subtab

The **General** subtab is designed to display the name, description, creation date, and last modified date of the scored data set that is automatically created from the node and the target variable for the decision tree model. Select the **Profile Information** button to view the target profile information such as the prior probabilities and the decision values to the profit and loss matrix in the modeling design.

Basic subtab

The **Basic** subtab is designed to display the configuration settings of the current decision tree model that were previously specified from the **Basic** tab. The displayed options can only be modified within the **Tree** node from the **Basic** tab.

Advanced subtab

The **Advanced** subtab is designed to display the configuration settings for the current decision tree model that were previously specified from the **Advanced** tab.

Import Tree subtab

The **Import Tree** subtab is designed to display the name of the imported tree data set. The **Tree** node is designed to create two separate output data sets: the scored data sets and the following tree definition data set. The subtab will be dimmed and unavailable for viewing if you have not imported a tree from either the **Variable Selection** node, **Tree** node, or **SAS Code** node. The **Tree** node will allow you to import a tree by fitting the binary-valued target variable to the model.

An imported tree data set can be created by performing the following steps. From the **Variable Selection** node, select the **Chi-square** selection criterion within the **Target Association** tab by fitting the categorically-valued target variable to the model. From the **Diagram Workspace**, connect the **Variable Selection** node to the **Tree** node. Next, open the **Tree** node and select the **Tools > Import** main menu options. Finally, execute the node to view the imported data set within the subtab. The imported data set lists the decision tree results that consist of the numerous assessment statistics from each node within the current tree. The imported data set is listed in the following diagram.

From the **Import Tree** subtab, select the **Properties...** button to view the file information and a table view of the imported tree data set as follows:

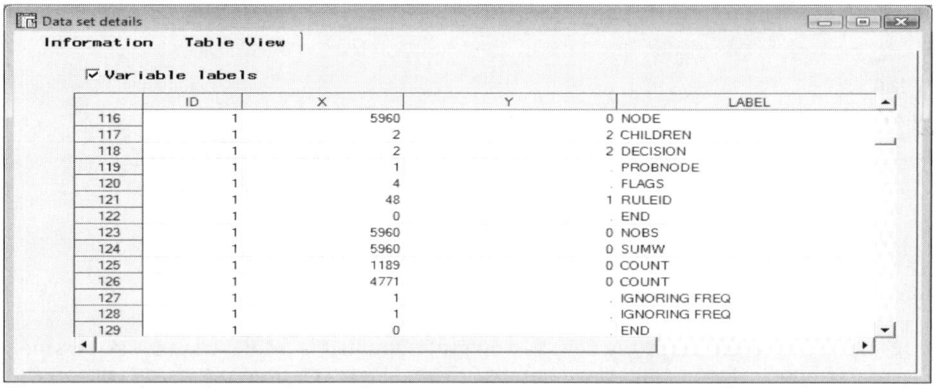

Data set details

Information Table View

☑ Variable labels

	ID	X	Y	LABEL
116	1	5960	0	NODE
117	1	2	2	CHILDREN
118	1	2	2	DECISION
119	1	1	.	PROBNODE
120	1	4	.	FLAGS
121	1	48	1	RULEID
122	1	0	.	END
123	1	5960	0	NOBS
124	1	5960	0	SUMW
125	1	1189	0	COUNT
126	1	4771	0	COUNT
127	1	1	.	IGNORING FREQ
128	1	1	.	IGNORING FREQ
129	1	0	.	END

The imported tree data set listing the various assessment statistics within each tree node.

All tab

The **All** tab will automatically appear once you open the **Results Browser**. The **All** tab will allow you to update the tree and the corresponding scored data set based on the selected assessment statistic and number of leaves that are specified. The following results are generated from the home equity loan data set in predicting clients defaulting on their home loan.

By default, the node will automatically display the best subtree with the fewest number of leaves that was determined by the assessment statistics by fitting the validation data set. The tab is designed to view the following four tabs within the **Results Browser** that are consolidated into one tab. The summary and assessment listing is displayed alongside two separate decision tree plots. The summary table and the assessment table will appear on the left-hand side of the tab and the tree ring and the assessment plot of the decision tree will be displayed on the right-hand side of the tab. Although these same listings and plots will be displayed in the following four tabs, the tab is designed for you to view all four tabs at one time, which will display the assessment statistic that is selected in choosing the corresponding decision tree. The assessment plot in the lower right-hand corner of the tab will allow you to view the development of the assessment statistic with a vertical reference line that will indicate to you the current subtree that was selected from the modeling assessment statistic. The assessment table that is displayed alongside the plot will display a table listing of the corresponding assessment plot with a particular row highlighted, indicating to you the best subtree that is currently selected. The current decision tree that is selected is defined by the best modeling assessment statistic from the validation data set. From the **Results Browser**, you may select an entirely different tree by either selecting a different row from the assessment table or selecting a different leaf from the assessment plot. By selecting the **View > Assessment**... main menu options, you may view an entirely different modeling assessment statistic that will be displayed for the current decision tree.

The various tabs are linked to one another so that by selecting an entirely different subtree from the assessment table or plot will result in the other tabs being updated accordingly. In other words, a different configuration of rings will be displayed from the **Tree Ring** tab and an entirely different table listing of the summarized statistics will be displayed within the summary table. Once you close the **Result Browser**, the **Tree** node will automatically create the scored data set in the background with the current subtree that was selected that can be passed along to the subsequent process flow. The tab will also allow you to prune the tree and view the corresponding assessment statistics. At times, pruning the decision tree will improve both the stability of the tree and the predictive or classification performance of the tree model by removing some of the weaker branches. This can be performed by paying particularly close attention to the weaker tree branches by viewing the listed leaves with high misclassification rates, large average sum-of-squares error, or small frequency counts. The difficulty in tree pruning is to determine how far back to remove the undesirable branches in order to avoid overfitting to the model. This can be performed by repeatedly pruning the tree to identify various subtrees. From each tab, you may view the currently constructed decision tree that is created by selecting the

View > Tree main menu options. The decision tree will allow you to view the evolution of the recursive tree branching process and the fitted values that are generated from the model.

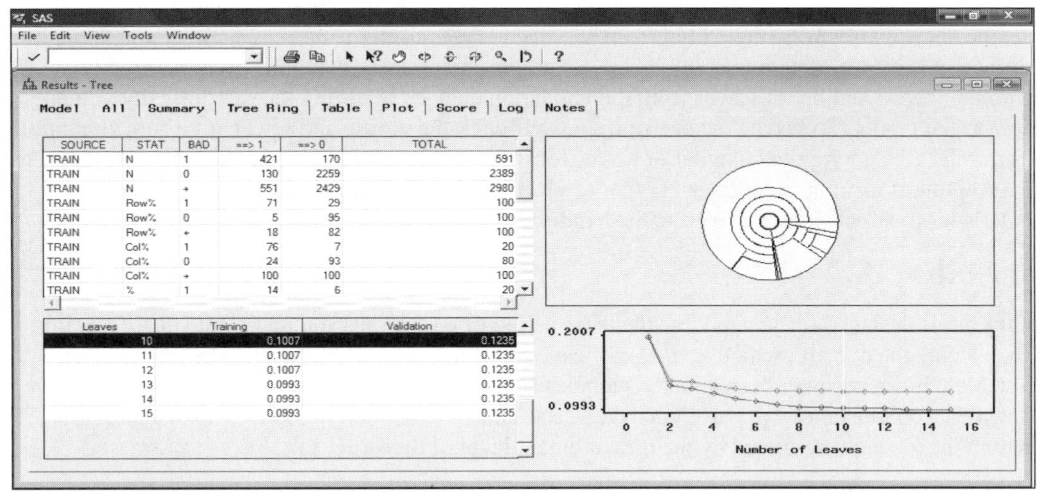

*The **All** tab that displays the development of the decision tree with the current assessment statistics.*

Summary table

The **Summary table** that is positioned in the upper left-hand corner of the tab will display the classification matrix, the row, column, and total percentages of each class level of the categorically-valued target variable from each partitioned data set. For interval-valued target variables, the summary table will display the number of observations, average target values, average squared error, and r-square statistic. These statistics will change as you select the corresponding row or subtree from the assessment table located at the bottom of the tab.

Tree Ring

The **Tree Ring** is designed for you to view the numerous splits that were performed. The rings will allow you to distinguish between the separate assessment values that are calculated within each tree leaf. The tab is designed for use as a navigational tool that will allow you to quickly scan through the various assessment statistics based on each split that was performed within the tree from the training data set. The tree ring will allow you to observe the complexity of the entire tree, the balance in the tree from the range of the target values that are assigned to each split, and the overall efficiency of the tree in creating the homogenous target groups. The center of the ring represents the root of the tree, which is the entire data. Each successive ring represents the recursive splits that are performed on the decision tree. The different color hues correspond to the different assessment statistics. For interval-valued target variables, the different color hues from the tree ring correspond to the assessment values. The tree ring displaying similar colors will indicate identical assessment values. The different sizes of the tree rings represent the different proportions of the target values assigned within the leaves of the decision tree from the training data set. The larger rings display the splits that you would like to identify. Conversely, small rings will indicate instability in the decision tree.

Assessment table

The **Assessment table** is designed for you to view the table listing of the selected assessment statistic from each partitioned data set by each leaf that is created. The corresponding values are plotted in the adjacent assessment line plot. By default, for categorically-valued target variables, the table will display the proportion of cases incorrectly classified to each leaf without the existence of the decision matrix. For interval-valued target variables, the table will display the average squared error of each leaf that is created in the decision tree. The row that is automatically highlighted within the table indicates the number of leaves that was selected from the validation data set. The table will allow you to select an entirely different tree by highlighting any one of the listed rows of the table.

Assessment Plot

The **Assessment Plot** is designed for you to view the development in the construction of the decision tree. The plot will allow you to visualize the stability of the tree, or if there is overfitting in the model, by observing a wide discrepancy between the line plots. The plot displays the assessment statistic that is located on the vertical axis across the separate number of leaves that are created from the recursive tree branching process. The vertical reference line of the plot will indicate to you the current tree that is selected with the best assessment statistic and the fewest number of leaves that are created for the decision tree from the validation data set. In our example, the decision tree was originally grown to fifteen leaves from the training data set and then pruned back to ten separate leaves, as determined by the lowest proportion of cases that have been incorrectly classified with the fewest number of leaves from the validation data set by fitting the categorically-valued target variable to the model. To update the decision tree, left-click the mouse and select anywhere along the plotted lines to select the appropriate number of leaves for the tree. In addition, the tab will allow you to select the **View > Assessment** main menu options in order to view a different subtree. The various modeling assessment statistics to select from are listed in the **Modeling Assessment Statistics** section.

Summary tab

The **Summary** tab is designed for you to view the table listing of the various summary statistics. The table will allow you to evaluate the overall predictive or classification performance of the model. The layout of the summarized table will first display the summary statistics from the training data set that are followed by the summary statistics from the other two partitioned data sets. The listed modeling assessment statistics that are displayed within the tab are determined by the measurement level of the target variable as follows:

Selecting **View > Confusion Matrix** main menu options will display the following table listing:

For categorically-valued target variables, the table will display the classification matrix that will allow you to determine how well the decision tree model predicts the target categories. The *<target name>* column corresponds to the actual target class levels with the table columns representing the predicted target levels. The **ROW%**, **COL%** and % represent the row, column, and total percentages of the classification matrix. The first classification matrix that is displayed is based on the training data set, **TRAIN**, that is followed by summary statistics from the validation data set, **VALID**. The table listing displays fairly similar cell proportions in the target class levels between both data sets, indicating stability in the decision tree.

(For interval-valued target variables, this option is dimmed and unavailable.)

Selecting **View > Tree Statistic** main menu options will display the following table listing:

(For categorically-valued targets, this option is dimmed and unavailable.)

For interval-valued target variables, the table will display the various assessment statistics, that is, the number of observations, average target value, average squared error, and r-square statistic for each partitioned data set. The summarized statistics from the training data set are displayed, followed by modeling assessment summary statistics from the validation data set.

Selecting **View > Leaf Statistics** main menu options will display the following table listing:

For categorically-valued target variables, the summary table will display the number of observations and the frequency percentages for each target class level within each leaf from the training and validation data sets. The **LEAF ID** column displays the identification number for the leaf. The **N** column displays the number of observations in each leaf from the training data set. The **N*PRIORS** column displays the number of observations in each leaf from the training data set based on the prior probabilities that are used to adjust the modeling assessment statistic. The **VN** column displays the number of observations in each leaf from the validation data set. The **VN*PRIORS** column displays the number of observations in each leaf from the validation data set based on the prior probabilities adjustments at each target level. **1%** and **0%** columns display the percentage of the target event and target nonevent level from the training data set. The **V1%** and **V0%** columns display the percentage of the target event and nonevent level from the validation data set.

For interval-valued target variables, the summary table will display the number of observations, squared error and average value of the target variable within each leaf from the training and validation data sets. The **LEAF ID** column displays the identification number to the leaf. The **N** column displays the number of observations in each leaf from the training data set. The **VN** column displays the number of observations in each leaf from the validation data set. The **ROOT ASE** column displays the square root of the average squared error modeling assessment statistic from the training data set. The **V ROOT ASE** column displays the square root of the average squared error modeling assessment statistic from the validation data set that is used to select the best decision tree. The **AVERAGE** column displays the average target values for each leaf node from the training data set and the **V AVERAGE** column displays the average target values for each leaf node from the validation data set.

Results - Tree

Model | All Summary | Tree Ring | Table | Plot | Score | Log | Notes

SOURCE	STAT	BAD	==> 1	==> 0	TOTAL
TRAIN	N	1	421	170	591
TRAIN	N	0	130	2259	2389
TRAIN	N	+	551	2429	2980
TRAIN	Row%	1	71	29	100
TRAIN	Row%	0	5	95	100
TRAIN	Row%	+	18	82	100
TRAIN	Col%	1	76	7	20
TRAIN	Col%	0	24	93	80
TRAIN	Col%	+	100	100	100
TRAIN	%	1	14	6	20
TRAIN	%	0	4	76	80
TRAIN	%	+	18	82	100
VALID	N	1	390	208	598
VALID	N	0	160	2222	2382
VALID	N	+	550	2430	2980
VALID	Row%	1	65	35	100
VALID	Row%	0	7	93	100
VALID	Row%	+	18	82	100
VALID	Col%	1	71	9	20
VALID	Col%	0	29	91	80
VALID	Col%	+	100	100	100
VALID	%	1	13	7	20
VALID	%	0	5	75	80
VALID	%	+	18	82	100

*The **Summary** tab displays a table listing of the summary statistics from the partitioned data sets.*

Tree Ring tab

The **Tree Ring** tab is designed for you to view the tree ring plot. The tree ring plot will allow you to observe the numerous splits that are performed from the current tree, the proportionality between the numerous splits, and the accuracy of each split. The plot is designed to distinguish between the various modeling assessment values or the range of target values that are assigned to each leaf of the tree from the training data set. At times, different colors will be displayed that are designed to make it easier for you to observe the different assessment values or range of target values that are assigned to each tree leaf.

The center of the ring represents the entire data set, that is, the root to the decision tree. The ring surrounding the center ring represents the initial split with the strongest split. Each successive ring represents the recursive splits that are performed to the decision tree. The size of the rings is proportional to the number of observations in each tree leaf from the training data set. In other words, the larger rings will indicate greater reliability of the split. The ring will be displayed in different shades of red and yellow. Nodes that contain all ones or zeros are colored in red and nodes that contain a mixture of ones and zeros will be colored in yellow. Also, nodes that consist of approximately the same assessment values or target values will display similar colors, as illustrated in the following diagram. In other words, the reason why the tree ring is colored yellow is that the color coding to the tree ring has assigned the minimum range of the misclassification rate between the interval of 0 to .5. In our example, the misclassification rates range between the interval of .09 to .21.

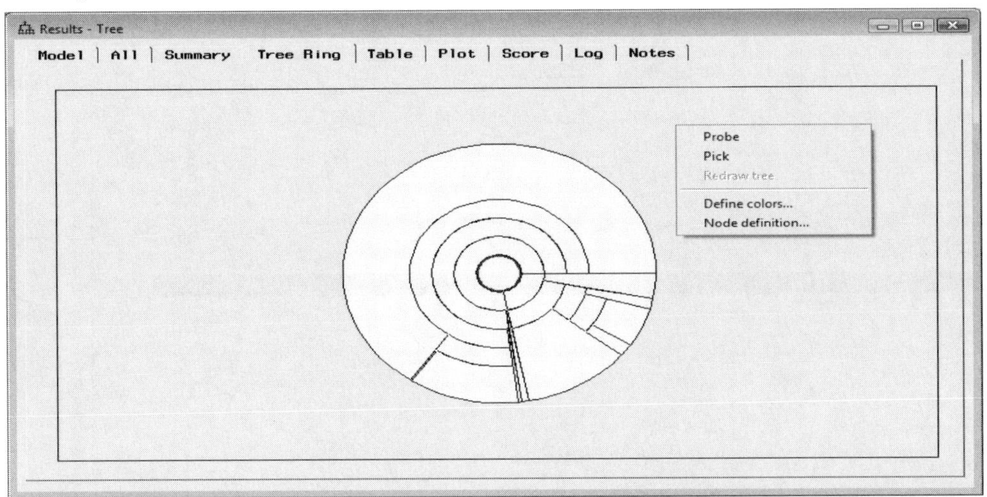

*The **Tree Ring** tab used to view the complexity, balance, and discriminatory power of the tree.*

Pop-up Menu Items of the Tree Ring

The **Tree Ring** tab provides you with a navigational tool to quickly view the various statistics that are created to each node from the recursive tree branching process. The tab will allow you to view a text box that will display the splitting variable, the cutoff value, or the assessment statistics used for each split of each tree ring of the tree. The **View > Probe Tree Ring Splits** main menu option will display the various tree splits that are the defaults. Alternatively, you may select the **View > Probe Tree Ring Statistics** main menu option that will display the assessment statistics of the various tree rings that are created from the corresponding tree splits. Right-click the mouse and select the **Probe** pop-up menu item, then drag the mouse over the various rings while holding down the left button of the mouse, which will display a text box listing of either the leaf statistics or the tree statistics of each node in the decision tree. Select the **Probe Tree Ring Splits** menu option, and the text box will appear that will display the node id, splitting variable name, and cutoff value for the split. Selecting the **Probe Tree Ring Statistics** menu option will result in the text box displaying the node id, name of the input variable, number of observations in the node, and frequency percentages for each class level of the categorically-valued target variable. For interval-valued target variables, the text box will display the node id, number of observations in the node, the average target value, standard deviation, and the input variable that was involved in the split.

Select the **Tools > Define Colors** main menu option to assign different color combinations to the tree ring. The **Color Palette** window will appear that is designed for you to customize the colors between the leaves displayed in the plot based on the target values, assessment values and the input variables. For categorically-valued target variables, the default coloring criterion is defined by the assessment values. For interval-valued target variables, the default coloring criterion is defined by the target values.

Select the **View > Node definition** main menu option that will result in the **Node definition** window displaying the node statistics of the selected leaf in the tree ring. The window will be displayed shortly.

Table tab

The **Table** tab is designed for you to view the various modeling assessment statistics of each leaf from the fully constructed decision tree. The modeling assessment statistics are displayed for each partitioned data set. The type of assessment statistic that is displayed depends on the measurement level of the target variable. The rows of the table will identify the various leaves for the current decision tree. For categorically-valued target variables, each row displays the misclassification rate or the proportion of cases that have incorrectly classified the target class levels. For an interval-valued target variable, the average squared error is displayed within each leaf of the tree. By default, the **Tree** node highlights the row with the best assessment statistic from the validation data set with the fewest number of leaves. The tab is designed in such a way that selecting any row will automatically update the current decision tree with the selected number of leaves.

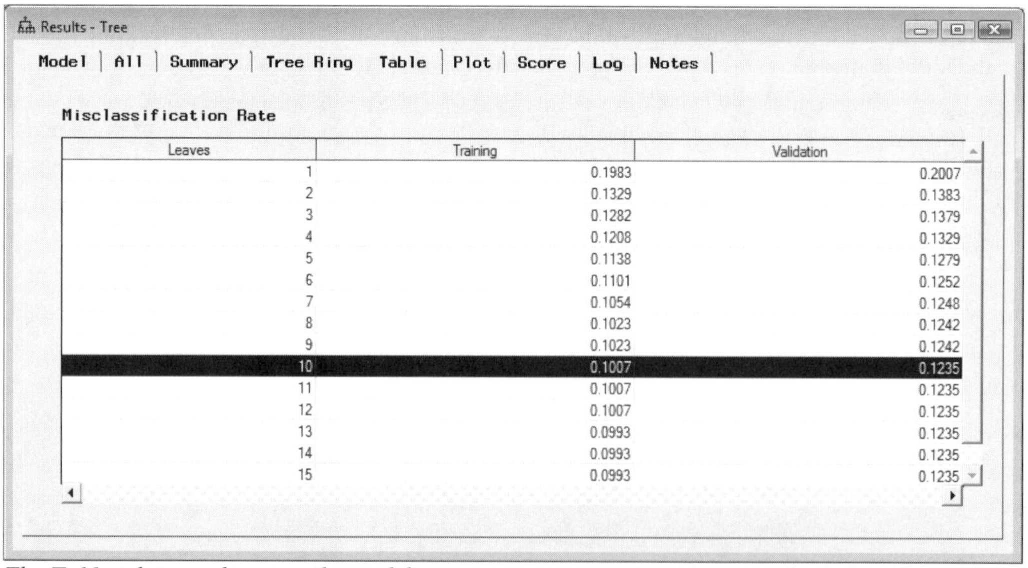

Results - Tree

Model | All | Summary | Tree Ring | Table | Plot | Score | Log | Notes

Misclassification Rate

Leaves	Training	Validation
1	0.1983	0.2007
2	0.1329	0.1383
3	0.1282	0.1379
4	0.1208	0.1329
5	0.1138	0.1279
6	0.1101	0.1252
7	0.1054	0.1248
8	0.1023	0.1242
9	0.1023	0.1242
10	0.1007	0.1235
11	0.1007	0.1235
12	0.1007	0.1235
13	0.0993	0.1235
14	0.0993	0.1235
15	0.0993	0.1235

*The **Table** tab is used to view the modeling assessment statistic within each leaf of the decision tree.*

Plot tab

The **Plot** tab is designed for you to display the assessment plot of the recursive tree branching process. The line plot will display the assessment statistic that is selected across the number of leaves that are created from the recursive tree branching process. The assessment statistic that is selected will be located on the vertical axis and the number of leaves to the tree will be located on the horizontal axis. The plot will allow you to select any number of leaves that are listed across the horizontal axis, which will update the current decision tree. The vertical reference line indicates the shortest tree with the smallest number of leaves that was created based on the best modeling assessment statistic in comparison to all other subtrees under assessment. The plot will allow you to observe the stability of the tree that will result in an increased generalization to the model with both lines displaying a similar trend over the number of leaves. Conversely, if there is a large discrepancy between both lines, then it will indicate to you that there is overfitting in the decision tree model. This will suggest to you that you might want to select a smaller tree before both lines tend to differentiate by selecting a smaller number of splits to the model. Generally, the decision tree model will fit the training data set better than the validation data set. The reason is because the splitting rules are constructed to fit the training data set. A discrepancy between both lines will tend to grow, resulting in an inadequate split in the recursive splitting procedure filtering through the tree, or a small number of observations that might be filtering down the tree. The magnitude in the discrepancy between both data sets that is displayed depends on the assessment statistic that is selected. For example, the same decision tree might overfit the data based on the misclassification rate while at the same time displaying no discrepancy at all based on the various lift statistics.

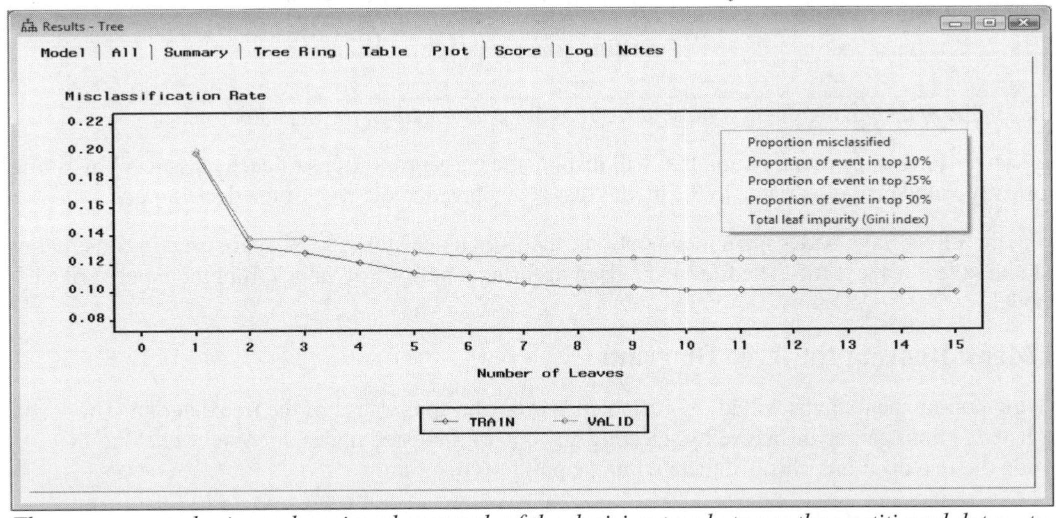

The assessment plot is used to view the growth of the decision tree between the partitioned data sets.

Tree Diagram

The diagram of the decision tree can be displayed by selecting the **View > Tree** main menu options. The tree diagram displays the root of the tree, the internal nodes that contain the splitting values, and the leaf nodes of the decision tree. It is important to understand that the **Tree** node will automatically display the following decision tree that is three levels deep. However, from the tree diagram, you may view deeper levels of the decision tree. The final predicted values or class probabilities can be viewed from the diagram, which might be used for further assessment. The tree displays the recursive process involved in constructing the tree by partitioning the input variables into separate subgroups from previous tree splits based on the specified assessment values. For interval-valued input variables, numerical splitting values of the input variable are displayed above each node indicating to you that the **Tree** node found the most significant split. For categorically-valued input variables, the class levels of the input variable are displayed for each split. The input variable is listed beneath each node and centered between each tree split, indicating to you which input variable performed the current split. Simply view the recursive tree branching routine from top to bottom in order to interpret the successive if-then rules that are used in calculating the final class probabilities or predicted values that are located in the leaves of the tree. For categorical valued target variables, each node will automatically display both the proportion and the frequency count for each class level of the target variable and the total frequency count from the partitioned data sets. For interval-valued target variables, the default statistics that are listed within each node of the tree are the frequency count and the average value of the target variable from the tree split.

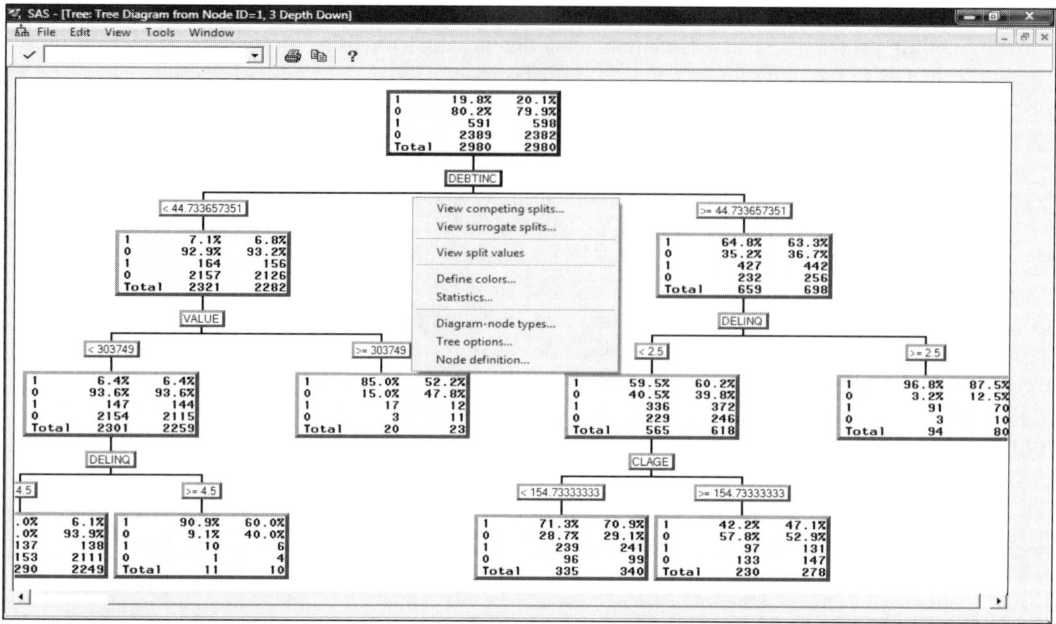

The ***Tree Diagram*** *displaying the first three splits that are performed on the decision tree.*

Select the **View > Path** main menu option that will display the class proportions of each class level by fitting the categorically-valued target variable to the model that is displayed at the root of the decision tree.

By selecting the **File > Save Rules** main menu options, the if-then decision rules that are used in constructing the tree can be saved to a text file. The file can be used in fitting a new set of values from the input variables in the tree model.

Pop-up Menu Items of the Tree Diagram

The following pop-up menu items will allow you to customize the appearance of the tree diagram. The following listed options can be displayed by selecting any one of the listed nodes, then right-clicking the mouse within the tree diagram. This is illustrated in the previous diagram.

- **View competing splits:** This option will open the **Competing Splits** window. The window is designed to display the next-best possible splitting variables and the corresponding logworth statistic that are listed in descending order. The purpose of this option is to show that the input variables displayed at or near the top of the list should be considered as the next-best splitting variables to consider for the currently selected node. In other words, the input variable that is listed at the top of the list is the next splitting variable that will be used in the subsequent split. The first **Variable** column will display the various input variables involved in the tree design. The second **Logworth** column lists the logworth value that measures how well the input variable divides the target values into each group. A large logworth statistic will indicate that the input variable is a good predictor for the tree model. The third **Groups** column will display the number of splits or groups that are formed by each input variable. By default, the **Groups** column will display a value of 2 for each input variable that is listed within the window since the node automatically performs binary splits for the input values in the decision tree. Select any one of the listed input variables, then press the **Browse Rule** button to view the **Splitting Rule** window that will list the upper and lower splitting criterion values. By default, the **Competing Splits** window will display the separate rules that can be specified from the **Splitting rules saved in each node** option in the **Basic** tab. However, the **OK** button is grayed-out in automatic mode. This is because you must be in interactive mode to modify the existing splits and the corresponding splitting values.

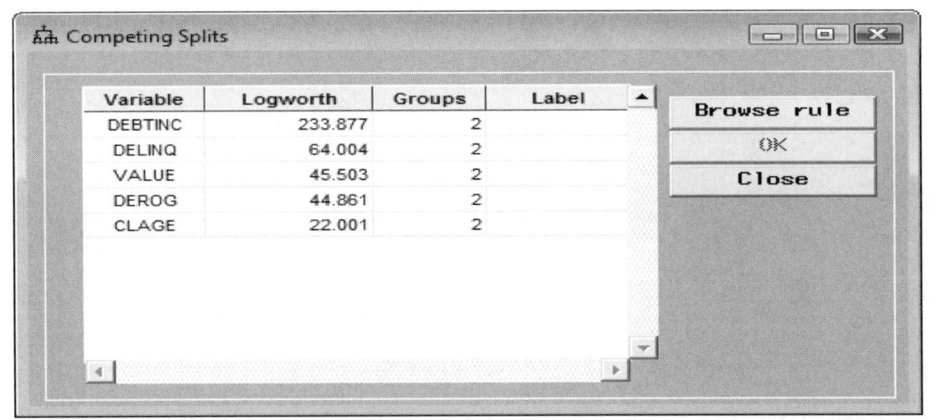

*The **Competing Splits** window displays the next-best splitting variables from the root node.*

- **View surrogate splits:** This option displays the surrogate splits in the **Competing Splits** window. The number of surrogate variables that are used can be specified from the **Basic** tab. When there are missing values in the primary splitting variable, then the surrogate variable is used. Press the **Browse Rule** button to view the **Splitting Rule** window that will list the upper and lower splitting criterion value of the selected surrogate variable. You must select a node before you can view the surrogate splits. However, if there are no surrogate rules saved that can be specified from the **Surrogate rules saved in each node** option within the **Basic** tab, then no window will be displayed.
- **View split values:** This option displays the split values, i.e. the range of values that created the node.
- **Define colors:** This option displays the **Color Palette** window that defines the colors to both the tree diagram and tree ring chart.
- **Statistics:** This option displays the **Statistics Selection** window in order to select the various statistics to display within each one of the nodes, such as each target level, frequency percent and count, predicted values and accuracy rate, training or validation data set statistics, or node id and label. Select the statistic that you would like to display or remove within the node by right-clicking the mouse within the **Select** cell of the table, then select the pop-up menu items of **YES** or **NO** as follows:

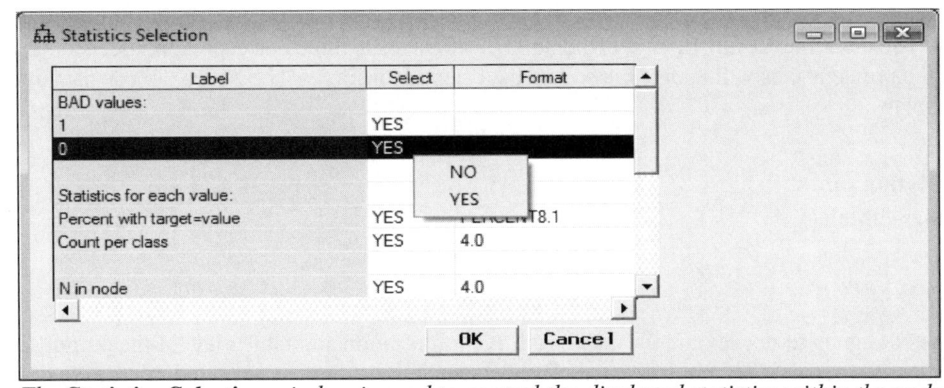

*The **Statistics Selection** window is used to control the displayed statistics within the nodes.*

- **Diagram-node types:** This option opens the **Diagram Node Types** window that will allow you to either display the (1) **Node statistics**, (2) **Node statistics and variables**, (3) **Node statistics, variables and values**, (4) **Leaves and variables** or (5) **Leaves, variables, and variable values**. By default, the tree is displayed by the **Node statistics, variables and values**.
- **Tree options:** This option will display the **Set Tree Depth Options** window to control the number of levels of the decision tree that are displayed. The **Tree depth up** and **Tree depth down** entry fields will either increase or decrease the depth of the tree diagram. By default, the **Tree depth down** entry field is set to three and **Tree depth up** entry field is set to zero. This will result in the tree diagram automatically displaying the decision tree in three separate levels, that is, displaying three separate successive splits from the root of the tree.

Set Tree Depth Options

Tree depth down: 3

Tree depth up: 0

OK Cancel

*The **Set Tree Depth Option** window is used to control the depth of the decision tree.*

- **Node definition:** This option will display the **Node definition** window that will list the node statistics of the currently selected node. The window will display the if-then splitting rule that was performed on the currently selected node. For interval-valued targets, the window will display the frequency count, average value, and standard deviation of the target variable. For categorically-valued targets, the window will display both the frequency count and percentages at each target class level as follows:

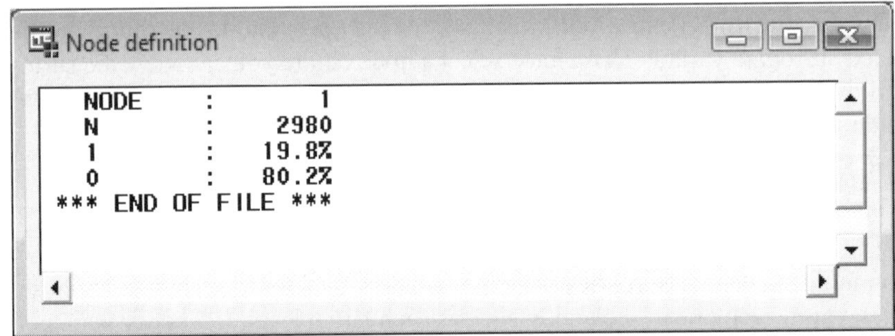

*The **Node definition** window that displays the node statistics from the root node.*

Score tab

The **Score** tab is designed to display the output scored data set that is created from the currently selected decision tree. The scored data sets can be created in the tab by selecting the **Training, Validation, and Test** check box. When you close the **Result Browser**, the number of leaves and the corresponding decision tree that is selected from the previous tabs will update the scored data sets accordingly. The **Score** tab consists of the following four subtabs.

- **Data subtab**
- **Variable Selection subtab**
- **New Variables subtab**
- **Code subtab**

Data subtab

The **Data** subtab is designed to display the file administrative information and table view of the output scored data sets for each partitioned data set that is created. The **Tree** node must be first executed in order to display the table view of the scored data set.

Variable Selection subtab

The **Variable Selection** subtab is designed to display the input variables that have either been retained in or removed from the decision tree model by the variable selection routine that is automatically performed within the **Tree** node. The variables that are listed in the output data set can then be used as input variables in the other modeling nodes. The subtab displays the input variable names, importance values, model roles, variable labels, and the number of times that the input variable was used in the splitting rules. The following subtab lists the order of importance of the splitting variables based on the importance statistics. Intuitively, the input variables that are used in the decision tree will have different levels of importance. Therefore, input variables with the largest importance statistics are positioned at or near the root of the tree that best divides the target groups or has been involved in several different splits. By default, the variables are sorted in descending order by their value of importance. For an interval-valued target, the importance of a split is the reduction in the sum of squared errors between the node and the immediate branches. For categorically-valued targets, the importance is the reduction in the Gini index.

The total variable importance of an input is equal to the sum over nodes of agreement multiplied by the reduction in impurity. The agreement is equal to 1 if the input variable is used to split the node; equal to the agreement measure of surrogate, if the input variable is used as a surrogate split; or equal to 0 if the input variable is neither a primary or surrogate split. For interval-valued target variables, impurity is equal to the sum of squared errors. For categorically-valued target variables, impurity is equal to the Gini index. The formulation to the importance is to ensure that the input variable that is used to split the root node will result in being the most important, and the accumulated reduction in impurity in the target values between the node and the immediate branches with regard to the input variable will get less importance in assigning observations to the leaves deeper down the decision tree. Therefore, the importance of the input variable to the decision tree model can be determined by the number of splits and the number of cases involved.

The values in the **Importance** column lists the percent relative importance of the input variable that is equal to 100•total importance divided by the maximum total importance. Therefore, the formulation of the variable importance in measuring the importance of each variable is to first measure the importance of the model in predicting the individual. In other words, the importance of an individual is equal to the absolute value of the difference between the predicted value of the individual with and without the model. The next step is to divide the individual importance among the variables in the model that are used in predicting the individual, and then average the variable importance over all individuals. The advantage of the importance statistic is that it involves all input variables that may not be involved in the best split at the node, but the second or third best split. Since the calculation of the statistic includes the surrogate variables, it will result in a more accurate assessment of the input variables in the decision tree. However, the drawback is that the surrogate input variables are correlated with the primary splitting input variables that will introduce redundancy in the variable selection process.

Input variables with a value of importance of greater than or equal to .05 are automatically assigned a role model of **input** and are, therefore, included in the output scored data set. These same input variables that best predict the target variable might be used as input variables to the other modeling nodes. The subtab will allow you to add or delete certain input variables from the output data set. Select the variable, then right-click the mouse from the **Role** column to select the **Set Role** pop-up menu item. Select the variable role of **input** or **rejected** to add or remove the selected input variables from the output scored data set.

By default, the **Exported role as indicated in this table** check box is automatically selected. This will instruct the node to export the variable roles to the subsequent nodes with the assigned roles that are displayed in the subtab. However, deselecting the check box will result in the **Tree** node assigning the input variables roles in the output data set and the subsequent nodes the same variable roles as they came into the **Tree** node.

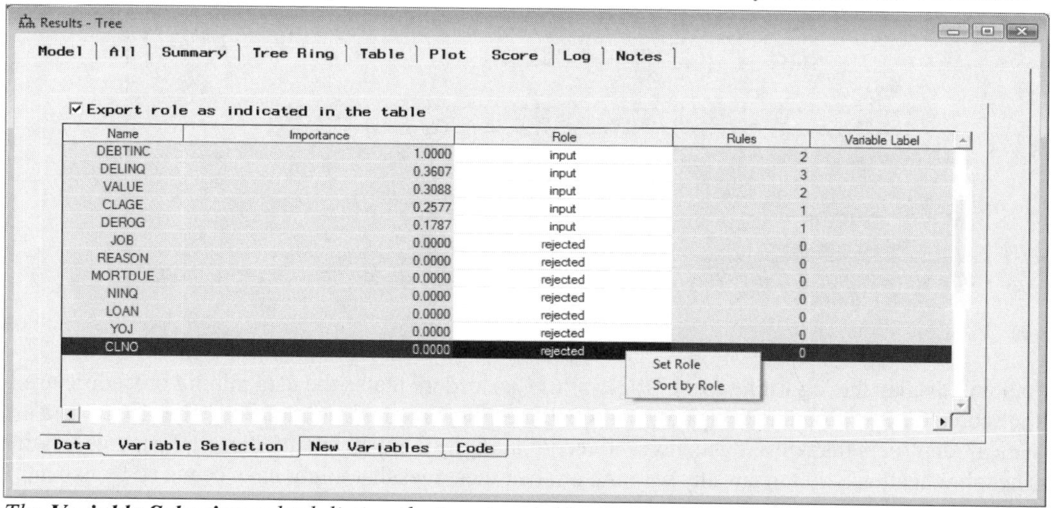

Results - Tree

Model | All | Summary | Tree Ring | Table | Plot | Score | Log | Notes

☑ Export role as indicated in the table

Name	Importance	Role	Rules	Variable Label
DEBTINC	1.0000	input	2	
DELINQ	0.3607	input	3	
VALUE	0.3088	input	2	
CLAGE	0.2577	input	1	
DEROG	0.1787	input	1	
JOB	0.0000	rejected	0	
REASON	0.0000	rejected	0	
MORTDUE	0.0000	rejected	0	
NINQ	0.0000	rejected	0	
LOAN	0.0000	rejected	0	
YOJ	0.0000	rejected	0	
CLNO	0.0000	rejected	0	

Set Role
Sort by Role

Data | Variable Selection | New Variables | Code

*The **Variable Selection** subtab listing the input variables that perform the best splits for the tree model.*

New Variables subtab

The **New Variables** subtab is designed to control the type of variables that are created within the output scored data set. By default, the leaf identification variable and the predicted values of the target variable are automatically created in the output scored data set with the **Leaf identification variables** and **Prediction variables** check boxes automatically selected. Indicator variables or dummy variables of the various tree splits can be written to the output data set by selecting the **Dummy variables** check box. By default, this option is not selected since the number of indicator variables that are created can be enormous depending on the growth of the decision tree.

Code subtab

The **Code** subtab is designed to display the internal SAS scoring code that can be used to generate entirely different decision tree estimates. The SAS scoring code displays the leaf nodes that are created from the range of values of the input variables in the model to predict the binary-valued target variable. Simply copy the SAS program code and the series of if-then partitioning rules of the input variables with the listed target proportions within each leaf node into a separate SAS program to calculate entirely different classification estimates from a new set of input values for the decision tree model. The following is a partial listing of the score code from the HMEQ data set in predicting the binary-valued target variable, BAD.

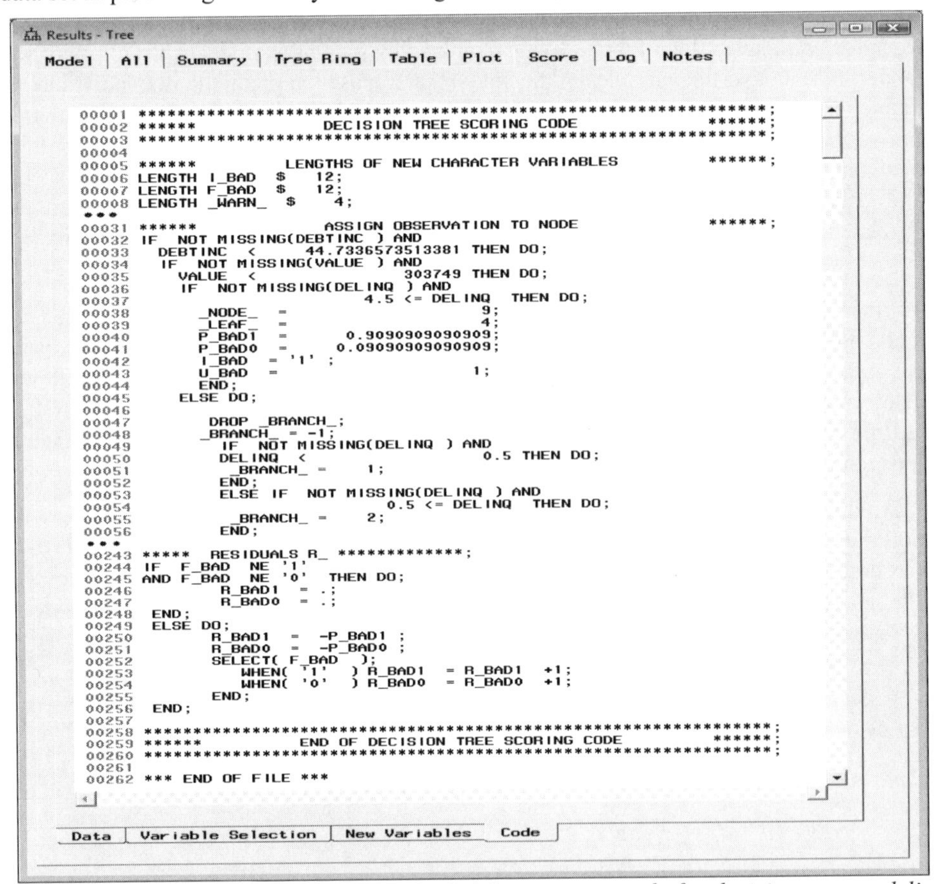

*The **Code** subtab that displays the internal if-then scoring code for decision tree modeling.*

Log tab

The **Log** tab will display the log listing from PROC SPLIT procedure that is the data mining procedure used in creating the scored data set from the decision tree model once the **Tree** node is executed. The scored data set will list the leaf identifiers that are assigned to each record along with all other variables in the active training data set. The tab will allow you to view any warning or error messages that might have occurred by executing the **Tree** node.

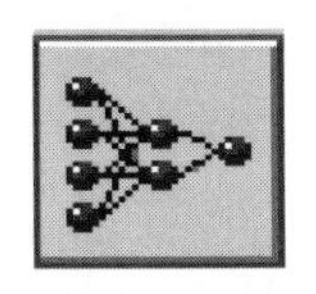

4.4 Neural Network Node

General Layout of the Enterprise Miner Neural Network Node

- **Data tab**
- **Variables tab**
- **General tab**
- **Basic tab**
- **Advanced tab**
- **Output tab**
- **Notes tab**

The purpose of the **Neural Network** node in Enterprise Miner is to model the various neural network architectures such as MLP and RBF designs. The node is generally designed to perform nonlinear modeling in the process flow diagram. By default, the node constructs multilayered, feed-forward neural network models. Neural network modeling performs both predictive and classification modeling based on the level of measurement of the target variable with the added capability of modeling a wide variety of extremely complex nonlinear functions and highly nonlinear classification boundaries.

The node will automatically display an optimization plot by executing the node when training the model that will allow you to view the modeling selection criterion at each iteration of the hold out method in evaluating both the progress of the grid search routine and the iteration where the final neural network model was selected and the weight estimates are determined. The node will allow you to specify various option settings for the layered nodes within the current neural network design such as the activation functions, combination functions or error functions that are mathematically compatible with the current network design. In the neural network design, many of the configuration settings are related to each other. For example, the activation function that is selected depends on the distribution of the target values. In addition, the error function that is selected should be consistent with the objective function that depends on the distribution of the target values. However, at times network training will not be performed due to some combination of the option settings that are incompatible with each other in the current neural network design. Therefore, the node will allow you to reset the option settings that have been changed back to their original configuration settings that SAS has carefully selected in order to perform neural network training.

The node will allow you to select from a wide variety of transfer or activation functions, combination functions, error functions, several optimization techniques, to perform preliminary training runs, add several hidden layers and hidden layer units, direct connections, perform weight decay regularization, apply various standardization methods to the input and target units, specify various initialization settings for the starting values of the hidden layer and output layer weights and biases, use the current weight estimates and refit the neural network model in order to accelerate network training and perform interactive training that is essentially sequential neural network modeling by reconfiguring and refitting the neural network model any number of times, all within the same interactive network interface. The node will allow you to select the best set of weight estimates that minimizes the objective error function that is determined by the modeling selection criterion. Therefore, the node will allow you to select the most appropriate error function that is based on the distribution between the residual values and the fitted values that will result in better prediction estimates while at the same time satisfying the various modeling assumptions. Also, the node will allow you to perform multivariate regression modeling by fitting several target variables to the model.

The **Neural Network** node is designed to create a separate node for the neural network workspace for each input layer and output layer based on the level of measurement of the input and target values. In addition, a separate node is also created in the network diagram for each additional hidden layer to the network design. The nodes are grouped together and each input layer node will have the same method of standardization. The hidden layer nodes are designed so that the nodes will have the same number of hidden units, and the same combination function and activation function. Each output layer node is designed to have the same standardization method and the same combination function, activation function, and error function.

In linear regression modeling, it is assumed that the target variable is linearly related to the input variables in the model as opposed to neural network modeling, which does not require any distributional assumptions as to the functional relationship between the target variable and the numerous input variables in the model. Typically, neural network modeling assumes a nonlinear relationship between the target variable and the input variables in the model that is based on the architecture and the activation function that is applied to the neural network model. Neural network modeling is analogous to nonlinear modeling, which is a general model fitting procedure used in estimating any kind of relationship between the target variable and the various input variables in the model. Generally, when the relationship between the target variable and the input variable is not linear, then nonlinear modeling such as neural network modeling should be considered. Neural network modeling is similar to regression modeling where unknown parameter estimates are estimated by the data. Therefore, network training is performed that fits the data in estimating the weight estimates. The *weights* represent the regression coefficients for the neural network model. It is important to select the most significant input layer and output layer weight estimates for the model. This is similar to traditional regression modeling where you are looking for a high correlation or a significant effect on the input variables in predicting the target variable.

The neural networks design consists of units or neutrons and connections between the units. There are three type of units. The *input units* represents the vector of input variables, where each input unit has its own weight in the input layer. The input units are standardized, assuming that they have an interval level of measurement. Standardizing the input variables is very effective when the variables are measured in different units or have a wide range of values. The input units are used to predict the values of the target variable. The *hidden units* perform an internal nonlinear transformation and the *output units* generate the predicted values, then compute the error that is the difference between the predicted values and the values of the output units in the target layer. Connections are used to pass information from units to other units between the layers as follows:

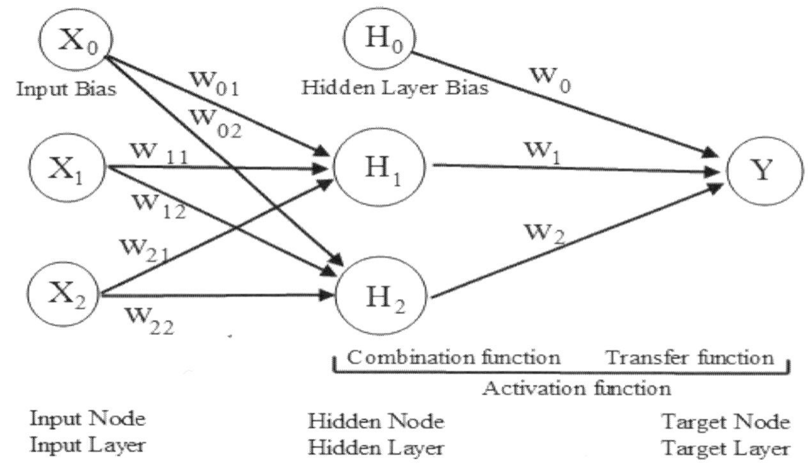

A feed-forward neural network MLP architecture with two input variables (X_1 and X_2), two hidden layer units (H_1 and H_2) and a single target variable (Y) to predict.

Neural Network Model: $Y = w_0 + w_1 \cdot H_1 + w_2 \cdot H_2 + e$

where $H_1 = \tanh(w_{01} + w_{11} \cdot X_1 + w_{21} \cdot X_2)$ and $H_2 = \tanh(w_{02} + w_{12} \cdot X_1 + w_{22} \cdot X_2)$

The *input layer* consists of units for each input variable. The *output layer* of units are called the target variables. The *hidden layer* of the neural network design contains hidden units or neurons. The hidden layer determines the multiple layer neural network that applies a nonlinear transformation to the linear combination of input layer and hidden layer units to generate the expected target values or the predicted values. In a single-layered network, each output receives inputs from all the input units, sums the units, and mathematically transforms the sum, typically using a nonlinear transfer function, to calculate the predicted values. Neural network modeling is designed to predict the values of the target variable, called the expected target values from other unknown input variables. The target variable may be continuous in order to perform linear regression modeling or categorical in order to perform classification modeling such as logistic regression modeling or discriminant analysis. Conversely, the level of measurement of the input variables can also be interval, ordinal or nominal. However, numerical codes must be assigned to the categorical values of both the target and input variables in the predictive or classification model.

A neural network may contain many units, with the units combined and connected between the hidden layers. The input layer is the first layer and the output layer is the last layer. The intermediate layers are called the hidden layers. Connecting two separate layers in a neural network design is performed such that every unit in the first layer is connected to every unit in the second layer. But at times units in the first layer can bypass one or more hidden layers and directly connect to the output layer, called a *skip layer* connection. A skip layer connection is essentially traditional linear regression modeling in a neural network design. Each input unit is connected to each unit in the hidden layer and each hidden unit is connected to each output unit. A neural network consists of a vector of units or neutrons. The units pass information to other units called *connections*, with the connections orchestrating the process of the neural network computations. The role of the *combination function* in neural networks is to combine or summarize the values received from the preceding nodes from both the hidden and output layers into one single value in which the activation transformation is then applied in producing the expected target response. The type of combination function to apply depends on the level of measurement of the target variable. A simple neural network design will have a feed-forward structure. That is, the connections will not loop and are restricted in connecting forward into the network from the input units to the hidden units and eventually reaching the output units. This is called a feed-forward network.

The hidden units combine the input values with a transfer or activation function that is applied. These hidden layers might have many different activation functions. The activation function applied to the output layer depends on the type of target variable and the values from the hidden units that are combined at the output units with additional activation functions applied that might be different. Again, all units in a given layer have the same attributes. For instance, all the units in the input layer share the same level of measurement and the same method of standardization. The hidden layer units share the same combination function and activation function. And all the units in the output layer share the same level of measurement, combination function, activation function, and error function. An *activation function* is a mathematical transformation of the summarized weighted input units to create the output from the target layer. In other words, the activation function is a two-step process. The first step applies a combination function that accumulates all the input units into a single value. The combination function is usually a general-linear combination function that is a weighted sum, where each unit is multiplied by its associated weight and added together into a single value. The second step of the activation function is called the *transfer function*, which usually applies a nonlinear transformation to these summarized units to generate an output unit. The sigmoidal transfer function applied in the target layer is analogous to the link function used in logistic regression modeling. The transfer function that is introduced into the neural network model provides incredible flexibility in the model to account for either the linear or nonlinear relationship between the target variable and the various input variables in the model. The type of transfer function applied in the hidden layer depends on the type of combination function that is applied. The choice in the correct transfer function to apply is not as important as the selection of the appropriate type of activation function to apply.

By default, a *MLP multilayer perceptron* is a feed-forward neural network architecture that uses various linear combination functions and nonlinear sigmoidal activation functions. A MLP architecture is a nonlinear regression model. But, there are other combination functions and activation functions to apply that connect the layers in many different ways. The hidden layer might have many hidden units with different activation functions. But, only one type of activation function is allowed for each hidden layer. A second hidden layer could be added to the design. But, since a MLP architecture with one hidden layer can virtually predict any model to any degree of accuracy, assuming that you have an adequate number of hidden layer units, a sufficient amount of data, and a reasonable amount of computational time, additional hidden layers are usually not required. But adding more hidden units to the neural network architecture gives the model the flexibility of fitting extremely complex nonlinear functions. This also holds true in classification modeling in approximating any nonlinear decision boundary with great precision. Adding additional hidden units to a feed-forward neural network design is similar to adding additional polynomial terms to a polynomial model. This process is called generalization. Generalization is a process of choosing the appropriate complexity for the model in generating accurate prediction estimates based on data that is entirely separate from the data that was used in fitting the model. For instance, adding additional higher-order modeling terms to the nonlinear polynomial regression model will always lead to a perfect fit. However, a perfect fit may also lead to the danger of extrapolation and generating absurd predictions from the polynomial regression model having an entirely different functional form by fitting the regression model beyond the range of the actual data points. This same analogy holds true in network modeling where adding too many hidden layer units might lead to the global minimum and poor generalization error, that is, an undesirable test error, based on data that is entirely separate from network training. Selecting the correct number of hidden units is important and usually requires training the network several times and evaluating the generalization error. The reason is because selecting not enough hidden units will result in underfitting and selecting too many hidden units will result in overfitting. However, both underfitting and overfitting lead to poor generalization. There are three conditions for good generalization. First, the input variables must accurately describe or predict the target values. Second, the function that you want to fit must be approximately smooth, with small changes in the weight estimates resulting in small changes in the target values. And last, good generalization requires a sufficiently large training data set to fit.

Shortcomings of Neural Networks

The big drawback to neural network modeling is that it is very difficult to measure the importance of the input variables in the model. Unlike traditional regression parameter estimates, the weight estimates do not tell you the effect, magnitude or the rate of change in the relationship between the target variable and the input variables. The reason is because the input variables in the model are associated with both the hidden layer and the output layer weight estimates. Therefore, the main objective of neural network modeling is making predictions without the capability of interpreting the relationship between the target variable and the input variables in the model. Furthermore, currently there do not seem to be available certain diagnostic statistics such as testing for lack of fit of the neural network model, and identifying influential and outlier data points, and tests to determine the significance of the neural network weight estimates. In addition, similar to nonlinear regression, there is no guarantee that the iteration algorithm that is applied will converge to the global minimum in determining the best linear combination of weight estimates in the neural network model.

Network Training

A term often used in the following section is network training. *Network training* is fitting the neural network model any number of times, usually between 10 to 50 times, in estimating the weights and biases. Network training begins by initializing the neural network model and setting all the weight estimates to zero, then fitting the neural network model to the target mean and performing small steps, assuming the weight estimates are small, in the first few iterations around the target mean. This is illustrated in the following diagram. The main idea of network training is generating a different set of randomly generated starting values of the initial weight estimates that are set at some small numbers, then selecting the best neural network model based on the smallest modeling assessment statistics, usually from the validation data set. It is vital that considerable thought should be considered in determining the initial target weight estimates to ensure an adequate modeling performance. Setting the initial weight estimates too small will result in the neural network model failing to converge to the correct solution. Conversely, setting the initial weight estimates too large will result in the weight estimates diverging away from the correct values, which will result in a bad fit to the data. In other words, specifying the correct initial weight estimates is critical in accelerating and assuring convergence to the desirable minimum in the objective function. In network training, it is recommended to try a different set of random numbers that are used as the starting values for the weight estimates, then assessing the different fits to the data based on the lowest goodness-of-fit statistics. However, a better idea is performing ensemble modeling by generating a different set of random numbers for each separate model and creating different initial weight estimates, then averaging the predicted values. That is, the main objective in performing ensemble modeling is to generate more accurate and more stable estimates in comparison to a single network model.

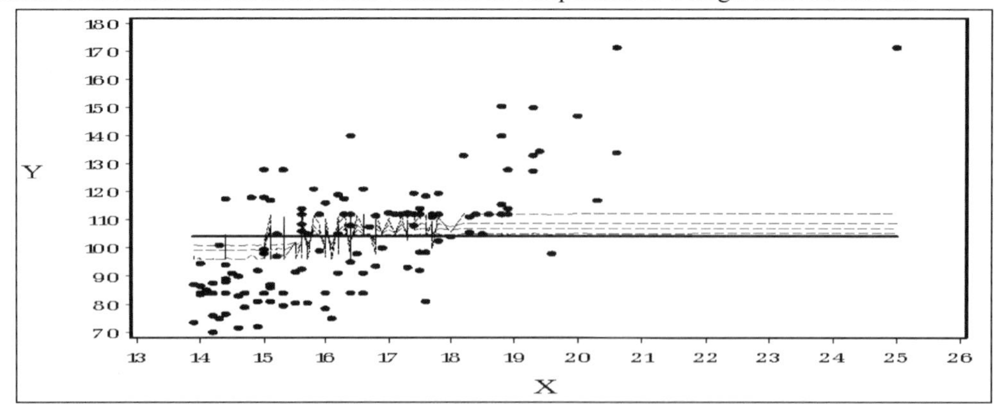

The bold line is the initial network model that is set to the target mean. The dashed line identifies the first few iterations to neural network training.

The Backpropagation Algorithm

In neural network modeling, once the number of hidden layers and hidden layer units are selected, then the initial neural network weight estimates need to be set in order to minimize the error function of the neural network model. The network training process begins by applying an iterative algorithm, either using a gradient-descent algorithm or a back-propagation convergence technique, with the training data set searching through the space of all possible weight estimates to fit in an attempt to minimize the squared error from a second data set, called the validation data set. The MLP neural network architecture typically incorporates the back-propagation technique that uses the gradient-descent algorithm. Back-propagation is a process that computes the derivatives of the sum-of-squares error function with respect to the weight estimates with the gradient-descent method used to determine a smaller error as the number of iterations increases. There are two distinct steps to the back-propagation algorithm as follows.

1. The first step is calculating the activation transformations of both the hidden layer and the output layer units with respect to the summarized weighted input variables in the neural network model, called *forward propagation*. It comes up with the predicted values and checks to see how well it did with the target values by calculating the error values.
2. The second step is evaluating the derivatives based on the error function with respect to the network weights. The reason that this algorithm is called *back-propagation* is because the error terms are passed backwards through the network and the corresponding hidden layers with the weight estimates adjusted to minimize the error.

This process continues until a perfect fit is achieved, when the vector of error values are all zero, or when any one of the convergence criterion values are met.

Radial Basis Function Architecture

The other nonlinear neural network architecture that can be performed within the **Neural Network** node is called the *Radial Basis Function* (RBF) architecture. The RBF design computes a nonlinear transformation, typically a Gaussian transformation, of the input vectors using a radial basis function or a radial combination function that is applied to the squared distance between the input values and their centers. The RBF configuration is a two-step process. The first step applies a basis function that is usually a nonlinear transformation, that is, a Gaussian transfer function, of the input layer to the hidden layer, which consists of Euclidian distances between the input values and their centers. However, at times a linear transformation or even a cubic basis function may be applied. The second step incorporates a linear transformation or an identity output activation function from the hidden layer into the output layer. Since these hidden layer connections are linear, convergence is extremely fast and at times does not depend on the various optimization techniques for determining the neural network model. However, other nonlinear output activation functions may be applied, although the shortcoming is that various optimization methods are required. Instead of using the input values from the source data, the centers are arbitrarily chosen and the distance from the input values to these centers represents the first layer of the weight estimates. Therefore, it is critical in a RBF design that the input variables be standardized so that the input variables span the same range of values. The center values are transformed in the second layer that produces the fitted values. Therefore, it is very important that correct values for the centers be specified for the RBF network. And it is also very important to select an accurate value for the variance when a Gaussian basis activation function is applied based on the input data set. However, the centers and the variances are usually determined by certain cluster groupings of the data. In the RBF designs, the centers, variances, and the hidden layer weights are the parameter estimates of the neural network design that can be adjusted in determining the minimum of the error function.

The Hold Out Method

The neural network modeling process usually divides the input data set into two separate files to generate good generalization. The input data set is divided into a training data set, validation data set, and even a third data set called the test data set. The role of the training data set is to calculate the neural network weight estimates and the role of the validation data set is to calculate the minimum error during the iterative process. The role of the test data set is to reduce the bias and generate unbiased estimates for predicting future outcomes. The test data set is entirely separate from the data that generated the estimates. The test data set is used at the end of the iterative process for evaluating the performance of the model from an independent sample drawn. The neural network performs the iterative process first using the training data set to calculate the weight estimates, then applies these same parameter estimates in the validation data to calculate the error values. The process then goes back and forth repeatedly, substituting parameter estimates from the training data set into the validation data set to find the smallest possible average error with respect to the validation data set. Therefore, the algorithm monitors the error with respect to the validation data set, while using the training data set to drive the gradient-descent grid-search routine. The final weight estimates are determined at the optimal point in the iterative process when the validation data set reaches the smallest average error. For interval-valued target variables, the error function you want to minimize is the sum-of-squares error. For categorically-valued target variables, the error function you want to minimize is the misclassification rate.

The Neural Network Layers

Input Layer: The input layer is the first layer of the neural network design. The purpose of the input layer is to introduce the input variables into the neural network model. At times, the neural network models are known to consist of hundreds or even thousands of input variables of the neural network model that depends on the number of observations in the training data set. In Enterprise Miner, the **Neural Network** node will automatically create one input layer node for interval, nominal and ordinal variables. That is, all the units in the input layer will automatically share the same level of measurement and the same method of standardization.

For nominal-valued inputs with k levels, $k-1$ input units will automatically be created. Therefore, at times the number of input units in the neural network design may be greater than the number of input variables. Hence, it is recommended to avoid including categorical input variables in the model with many more weights to optimize. By default, the interval-valued input variables in the input layer are standardized by a standard normal distribution, with the target variables in the hidden-to-target layer left unstandardized in the MLP or RBF design. However, standardization of the target variable will assure convergence. The drawback is that the target values will usually need to be remapped to their original values. The reason is because the transformed target values are not as informative as the original values. The method of standardization can be specified for each input unit. However, standardization can only be applied to interval-valued variables. The purpose of standardizing the input units in the neural network model is to produce reasonable values of the input-to-hidden layer weights close to zero to assure convergence, to reasonably compare the size of the weight estimates among the various input variables in the neural network model, and to select a meaningful range of the initial input layered weight estimates that will not result in large changes in the weight estimates between the separate iterations, which will result in good generalization. Standardizing the input variables will allow you to determine the appropriate number of input variables of the neural network model by evaluating the size or magnitude in absolute value of the standardized weight estimates that can be used as a crude assessment of the importance of the input variables to the model. Standardization is very effective when the input variables in the model have a wide range of values or are measured in different units in order to interpret the variables to a common scale or unit with the same range of values. Again, standardization is applied because the weight estimates cannot be compared to one another under the assumption that each input variable is measured entirely different. Other reasons for standardizing the input variables in the neural network model are to reduce the risk in the iterative process, resulting in a bad local minimum in the error function, accelerate network training, and alleviate the "*curse of dimensionality*" problem and ill-conditioning in the network model.

It is important to understand that as each input variable is entered into the nonlinear neural network model, then it will result in an additional dimension to the space in which the range of the data resides. Therefore, it is important that input variables be eliminated from the modeling design that provides very little information in describing or predicting the target values that will lead to an increase in the modeling performance and reducing the "*curse of dimensionality*" dilemma. The complexity of the neural network model, the number of internal calculations (function calls) and computational time increase when additional parameters are added to the neural network model such as adding additional input units, hidden layers and hidden layer units for the network design. This is especially true with extremely large training data sets in which the nonlinear model is trying to fit since the iterative grid search procedure must make several passes through the entire data set in order to determine the next best set of weight estimates at each iteration. In addition, increasing the number of input variables for the model will reduce the chance of determining the correct starting values for the weight estimates. Also, increasing the number of input variables for the model will also reduce the chance of the iterative grid search procedure terminating too early. In other words, increasing the number of input variables in the model will result in the optimization procedure inheriting the problem of deciding when the iterative grid search procedure should stop in determining the best linear combination of weight estimates. The reason is because the minimization technique will have problems in finding the minimum in the error function due to an increase in dimensionality in the neural network model. This is overcome by constructing neural network models with fewer input variables in the nonlinear model, which will result in the reduction in the dimensionality of the error space for finding the optimum minimum in the sum-of-squares error function.

It will usually take common knowledge to know which input variables to include in the neural network model that best explains the target values. Specifying the correct specification of the target range and error distribution will increase the chance of selecting good inputs, which should result in the best prediction estimates. Therefore, a simple rule of thumb is that if you have some general idea that the input variable will not be very useful in predicting the target variable, then do not include it in the neural network model. Reducing the number of input units in the design will also reduce the number of hidden layer units connected to them. This will result in better generalization since the network model will train more efficiently. However, if it is important to include several input variables in the neural network model, then it is critical that a MLP design be applied as opposed to some of the other RBF designs. The reason is because the MLP design is better at ignoring irrelevant input variables of the network model in comparison to some of the other RBF designs.

Hidden Layer: In a layered neural network design, the hidden layer is an intermediate layer between the input and output layer, where one or more activation functions can be applied. The hidden layer is also called the inner hidden layer or the input-to-hidden layer. In neural networks, the hidden layers may apply additional transformations to the general–linear combination of input weight vectors. Generally, each input unit is connected to each unit in the hidden layer and each hidden unit is connected to each output layer unit. An activation function is applied that is usually nonlinear and which calculates the units of the output by multiplying the values of each input unit and the hidden layer weight that are added up, then usually transformed using a sigmoidal activation function. The sigmoidal activation function is applied to the neural network design so that the output from the hidden units are close to zero with reasonable values of the hidden layer weights that are centered about zero. This will result in small changes in the weight estimates during network training that will lead to good generalization to the network model. However, since a nonlinear function is applied to the hidden layer, various optimization techniques must be applied in determining the weight estimates to the design. More hidden neutrons or hidden units added to the neural network architecture gives the model the ability to fit extremely complex nonlinear smoothing functions such as polynomial functions that have a high degree of oscillating behavior. The simplest neural network model consists of two hidden units since a network model with one hidden unit is simply a regression model. It is not uncommon to have several hidden layers, but it is usually not recommended. Each hidden layer is connected to the following hidden layer until the last hidden layer is connected to the output layer. A single hidden layer is restricted to one type of activation function. Therefore, several hidden layers with different activation functions in each hidden layer can be applied, as illustrated in following diagram.

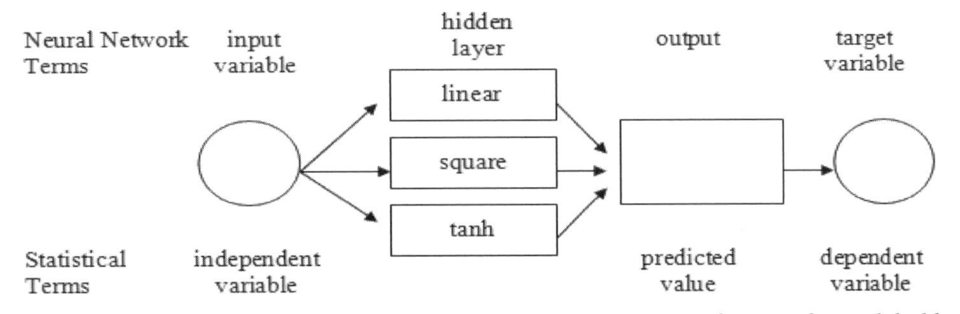

The MLP design with three hidden layers and a separate activation function for each hidden layer unit.

Applying a linear activation function to the hidden layer essentially creates a neural network model with no hidden units, called a single-layer perceptron or a general linear model. Usually, a single hidden layer is applied to several input variables in the neural network model, but adding more hidden units to the model will increase the complexity of the network design, which will result in a model capable of approximating any relatively smooth nonlinear function with any degree of accuracy. One of the biggest decisions in neural network modeling is the number of units in the hidden layer. Selecting the correct number of hidden units is an important aspect of producing good generalization performance, which is the main goal of both network training and predictive modeling. The best number of hidden units to apply to the neural network design depends on the number of input variables in the network model, the number of observations in the training data set and the noise level in the underlying distribution of the target values in the training data set. Since you do not know the complexity of the underlying training data set or the random error of the target values, it is recommended to fit the neural network model several times with a different number of hidden units and then analyze the various modeling assessment statistics and stop the iterative process when the goodness-of-fit statistic begins to increase. The simplest approach is to initially fit the network model with no hidden layer units and sequentially fit the model by adding an additional hidden unit each time. However, there is no general rule of thumb as to the exact number of hidden layer units to apply to the neural network model.

The complexity to neural network modeling is that if the neural network model has an insufficient number of hidden units, then it will lead to underfitting. Underfitting is basically a poor fit to the data. Conversely, specifying too many hidden units will result in overfitting. Overfitting is creating a model with too many parameters that does not generalize well. That is, increasing the complexity of the model will result in absurd predictions by fitting the model to new data. Generally, it is better to have too many units in the hidden layer than too few. In order for the neural network model to fit the nonlinear pattern in the underlying data or create highly nonlinear decision boundaries for classification modeling, a sufficient number of hidden layer units are needed. Bypassing the hidden layer and making a direct connection from the input layer to the output layer is called a skip layer design. A skip layer design is essentially least-squares regression modeling in a neural network design. Conversely, increasing the number of hidden layers in the neural network model will increase the accuracy of the estimates, but it might result in overfitting to the data and lead to ill-conditioning in the Hessian matrix. Therefore, one hidden layer is usually sufficient.

A second hidden layer can be applied to the neural network design that consists of a linear combination of weight estimates and associated input variables that are nonlinearly transformed. Since a MLP model with a single hidden layer is called a *universal approximator*, adding any additional hidden layers to the network model is usually not needed in estimating continuous nonlinear smoothing functions except in extremely rare circumstances when fitting extremely complicated target functions or reducing the number of weights in the neural network design. A universal approximation means that given a sufficient number of hidden layer units, the neural network is designed to approximate any functional form to any degree of accuracy. However, this level of accuracy in a MLP design may not be achieved since the network weights and biases need to be estimated from the data, which, at times, might require an enormous number of units in the hidden layer. In addition, there is no guarantee that there even exists a function that can be approximated from the target values based on the input variable that are selected.

The following table displays the relationship between the number of hidden layers that should be applied to the network design as determined by the type of activation function and combination function that is used in the neural network model that depends on the level of measurement of the target variable.

Type of Combination Function	# of Hidden Layers	Default Activation Function
Additive	Any	Identity
Linear	Any	Tanh
Radial	One	Exp
	Two or more	Softmax

Output Layer: The *output layer* is the final layer of the neural network design that represents the target variable of the model. The output layer is also called the outer hidden layer or the hidden-to-target layer. Usually, the output layer is fully connected to the hidden layer. The transfer function, combination function, and error function that is applied to the output layer depend on the level of measurement of the target variable. All the units in a given output layer have the same level of measurement, combination function, activation function, and error function. For interval-valued target variables, a unique output unit is created in the network design. For categorical target variables with k class levels, $k - 1$ output units are created in the network design. The predicted values from the neural network model are calculated from the output layer. In the back-propagation technique, the error values that are the difference between the target values and predicted values are calculated in the output layer and then passed backwards (back-propagated) through the network to calculate the error values. Also, for back-propagation, the derivative of the error function with respect to the weight estimates is also calculated from this layer and then passed backwards through the network design in order to adjust the weight estimates and minimize the error in network training. For multivariate models, the **Neural Network** node automatically creates one output layer unit for each categorically-valued target variable and one output layer unit for each interval-valued target variable in the model. The output layer may have more than one output layer unit in the neural network modeling design with no hidden layers, analogous to multivariate regression modeling with several target variables to predict. However, one restriction to the neural network design is that the separate target layered units can not be connected to each other.

Preliminary Training

In neural network modeling, there is no known standard method for computing the initial weight estimates. Therefore, preliminary training is performed before network training, which is designed to determine the most appropriate starting values to be used as the initial weight estimates for the subsequent network training run that is critical to the iterative convergence procedure. Furthermore, preliminary training is used to accelerate convergence in the iteration process, with the idea of avoiding bad local minimums in the error function. Preliminary training performs a succession of runs from the training data set in fitting the neural network model that is based on the number of preliminary training runs that are specified. In other words, the preliminary training is applied to the training data set that performs a sequence of independent preliminary training runs. The preliminary method initializes the starting weight estimates by some small random numbers for each preliminary training run. The reason that preliminary training is so critical is that many of the optimization methods that are applied in finding the best set of weight estimates with the smallest validation error depend on the initial weight estimates being a close appropriation to the final weight estimates, which will result in the acceleration of the iterative grid search procedure and avoid bad local minimums. Therefore, the preliminary training run is performed before network training by specifying a random number generator to generate a new set of random numbers assigned to the weights that are set to some small values for each separate preliminary training run that is performed. The preliminary training run usually performs a small number of iterations since the procedure must go through the entire data set for every combination of possible initial weight estimates. In preliminary training, usually twenty iterations are sufficient. However, if computational time and memory resources permit, then 1,000 preliminary runs may provide even better starting values for the network weight estimates. The final preliminary weight estimates from the training data set are determined by the particular preliminary training run that resulted in the smallest training error among all preliminary training runs that were performed. These final preliminary weight estimates are then used as the starting values for the subsequent network training run. By default, preliminary training is not performed.

Regularization Techniques

Early Stopping: Both early stopping and weight decay regularization techniques are designed to improve generalization. However, this regularization technique is easier to apply in comparison to the following weight decay. Early stopping is designed to improve generalization in controlling network training by terminating network training once the validation error begins to increase in order to prevent overfitting to the network model. Early stopping requires an enormous number of hidden layer units in order to avoid bad local minimums. In order to perform early stopping correctly, it is recommended to partition the input data set with a small percentage of the observations assigned to the validation data set and use one of the slow-converging training techniques that will result in slow, but efficient steps during the iterative grid search procedure.

Weight Decay: The weight decay is a nonnegative number that is used during network training in order to adjust the updated weight estimates by restricting the growth of the weight estimates at each iteration, which will result in improved generalization. A larger weight decay value will result in a greater restriction on the total weight growth during network training. This will result in a smoother fit to the bumpy nonlinear error function. The weight decay value is designed to alleviate overfitting by producing a much smoother nonlinear function or decision boundary while at the same time keeping the generalization error constant no matter how many weight estimates are applied to the neural network model, assuming that the correct weight decay value has been specified. However, the drawback to this technique is that estimating the correct weight decay value is usually extremely difficult to achieve and might require several network training runs. Achieving the correct weight decay value requires several network training runs that need to be performed by retaining the same network configuration and specifying a different weight decay value in each network training run.

The *regularization function* is the sum-of-squares of the weight estimates excluding the output bias term, multiplied by the weight decay parameter γ, defined as follows:

Penalty function: $P = \gamma \cdot \sum w_i^2$ for some weight estimate w_i

The main idea in network training is minimizing the objective function. The objective function is the sum of the total error function L and the regularization or penalty function P, divided by the total frequency f_i for the ith case that is defined by the following function:

$$\text{Objective function} = \frac{L+P}{\sum f_i}$$

Typically, weight decay regularization is very difficult to apply since the weight decay γ must be accurately specified, which might require a very long time to find the correct value. In SAS, the weight decay parameter is initially set to zero. Therefore, the regularization function can usually be ignored, resulting in the objective function equaling the total or average error function.

The Error Function

The flexibility in the **Neural Network** node means that you may specify the appropriate error function that conforms to the distribution between the fitted values and the residual values, which satisfies the various modeling assumptions. In statistical modeling, transformations are typically applied when the distribution of the residuals violate the modeling assumptions in order for the fitted values to conform to the range of values of the target variable in the predictive model. This is the same approach that is used in logistic regression modeling by specifying an appropriate link function. The drawback in transforming the fitted values is that complications might occur when trying to retransform the target values back to their original form. Therefore, as an alternative, you might consider selecting an entirely different error function to the model. In predictive modeling, it is extremely important that the correct error function conforms to the distribution of the target variable being modeled, which will result in the best predictions while at the same time correcting the various assumptions that must be satisfied in the error distribution. Assumptions of the error distribution, such as correcting for heteroscedasticity (nonconstant variance), skewness, and the residuals should be consistent with the numeric range of target values that the model is fitting. The most common error distribution is the normal distribution. However, at times the distribution of the target values might be all positive values, whereas the normal distribution might conform to both positive and negative values. In other words, if the residual plots display an increasing variability in the residual values, then better modeling results might be achieved with the Poisson, gamma, or lognormal error distribution. The error function that should be selected depends on the relationship between the predicted values and the variance of the target variable, that is, the relationship between the distribution of the fitted values and the residual values of the model. The error distribution should be consistent with the numeric range of the target values and at times might have an approximately skewed distribution. That is, not all error distributions are normal. The type of error function to apply should match the distribution between the residual values and the fitted values in the network model that can be viewed from the residual plot. If the values of the residuals display a fanned-out pattern across the fitted values, then a transformation to the target variable might be considered such as the log transformation or an entirely different error function should be applied in achieving uniformity in the residuals of the model. This suggests that a multitude of nonlinear regression models can be constructed within the node by excluding the hidden layer from the neural network design. The node supports a wide variety of link functions or activation functions and an assortment of error functions as well as robust estimation capabilities. For interval-valued target variables, the **Regression** node applies the identity link function and the normal error function. For categorically-valued target variables, the **Regression** node applies the logit link function and the binomial distribution error function. Both the activation function and error function can be specified from the **Advanced user interface** within the **Target** tab from the output layer unit. The error function that is selected should match the objective function that you want to minimize. For interval-valued targets, the default error function is either the **Normal** or the **Huber** error function. For categorically-valued targets, the default error function is either the **Multiple Bernoulli** or the **Huber** error function. For multiple targets, for which any one of the target variables has an error function with an **M Estimation** objective function, the **Huber** error function is the default.

By default, the **Neural Network** node tries to determine the most appropriate objective function from the specified error function, assuming that you have not selected an objective function from the **Objective Function** display field within the **Optimization** tab. In neural network modeling, the objective function depends on the type of error function to apply. The error function to select is determined by the level of measurement of the target variable, the predicted values, and the objective function that is applied. Again, the error distribution that is selected should match the variance, skewness, and range of the target values. If the error function does not exist, then you may want to transform the target variable so that the distribution of the residual values conforms to the error function that you have selected for the neural network model.

The Relationship Between the Link Function and Error Function

The following table displays the relationship between the output activation function and the error function that can be specified within the **Target** tab from the output layer unit in the **Neural Network** node. The selection of the link function should match the fitted values and the range of values of the target variable. The selection of the output activation function and the error function depends on the level of measurement of the target variable. Usually, the best error function to apply depends on the distribution between the residual values and the fitted values from the model.

Target Measurement	Link Function	Output Activation Function	Error Function
Counts / Amounts	Identity	Identity	Normal
	Identity	Identity	Huber
	Log	Exponential	Gamma
	Log	Exponential	Poisson
Names / Grades	Logit	Logistic	Bernoulli
	Cumulative Logit	Logistic	Multiple Bernoulli
	Generalized Logit	Softmax	Multiple Bernoulli
Proportions or Rates	Logit	Logistic	Entropy
	Generalized Logit	Softmax	Multiple Entropy

The Basic User Interface versus the Advanced User Interface

In the **Neural Network** node, there are two separate user interfaces for specifying the various configuration settings of the neural network model. From the **Basic** tab, either the basic user interface or the advanced user interface can be specified. The basic user interface is the default. The various basic configuration settings that can be specified are the modeling selection criterion; the neural network architecture; specifying the number of hidden layer units, direct connections, preliminary runs, training techniques or the optimization algorithms; accumulating the network training runs and specifying the maximum amount of computational time; creating output scored data sets; or applying the various pruning input techniques. The **Advanced user interface** will allow you to gain more control in configuring the neural network design by increasing the complexity of the neural network model, which will hopefully generate better prediction estimates. For instance, the **Neural Network** node will allow you to select various advanced user interface option settings such as specifying the various standardization methods for either the input variables or the target variable, specifying the initialization settings for the starting values of the input-to-hidden and hidden-to-target layer weight estimates and biases, specifying the error function for the hidden-to-target layer, preliminary training, early stopping, weight decay, specifying the type of transfer or activation function and combination function of either the input-to-hidden layer or the hidden-to-target layer, several optimization techniques, convergence criterion settings, and adding any number of hidden layers. Interactive training may be activated by selecting the **Advanced user interface** check box. *Interactive training* is basically sequentially training the neural network model by reconfiguring and refitting the model any number of times, all within the same interactive network interface.

The advantage of selecting the **Advanced user interface** is that it has many more configuration settings for the network design in comparison to using the **Basic user interface**. The **Advanced** tab will allow you to specify the various advanced user interface configurations for the neural network model. If you are relatively new to neural networks, then you may want to first experiment with the basic user interface option settings. However, it is encouraged and highly recommended to select from the many more advanced user interface options in order to increase the complexity and precision of the neural network modeling design. The **Neural Network** node will allow you to change the configuration of the neural network model by adding additional hidden layers to the neural network model in order to fit extremely complex nonlinear functions, specifying an entirely different hidden layer activation function that conforms to the distribution of the target values in order to increase the accuracy of the neural network estimates, selecting an entirely different error function that is determined by the distribution between the residual values and the fitted values to achieve better prediction estimates while at the same time satisfying the various modeling assumptions, adjusting the starting values to the hidden-to-target layer weights to help alleviate the convergence algorithm in finding the best validation error by specifying an entirely different random seed value, refitting the network model any number of times, and creating separate input or target layered nodes for the neural network design.

Neural network modeling is discussed in greater detail in Bishop (2002), Ripley (2002), Sarle (1997), Mitchell (1997), Matignon (2005) and Neural Network Modeling Course Notes, SAS Institute (2000).

Data tab

The **Data** tab will allow you to specify the partitioned data sets to fit the subsequent neural network model. In addition, the tab will allow you to specify the number of records allocated to preliminary training. The tab contains the following two subtabs.

- **General subtab**
- **Options subtab**

General subtab

The **General** subtab is similar to the standard **Data** tab in the previous nodes. The **Data set** and **Description** fields will display the partitioned data set name and description that is usually created from the **Data Partition** node. Select the **Properties...** button, then the subtab will allow you to view both the file administrative information and the table view of the selected partitioned data sets. Otherwise, press the **Select...** button that will open the **Imports Map** window. The window will allow you to select an entirely different data set to model by browsing the tree directory listing of all the available data sets, usually created from the **Input Data Source** node, that are connected to the node. Since the training data set is automatically selected, then the tree directory listing within the **Imports Map** window will display all the available training data sets that are connected to the node. Press the **OK** button to close the **Imports Map** window and the **Neural Network** window will then reappear.

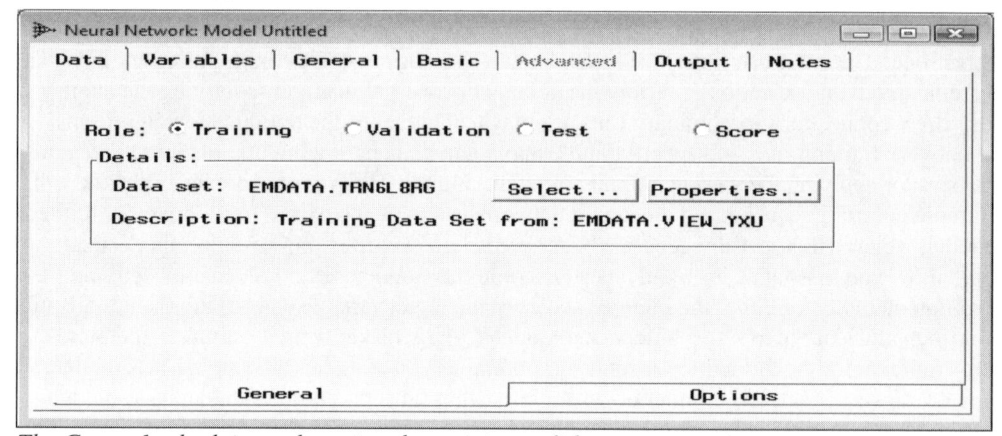

*The **General** subtab is used to view the training, validation, or test data sets.*

From the **Neural Network** window, press the **Properties...** button and the **Data set details** window will appear to view the data set information or view the table listing of the selected data set. Select the **Information** tab to view the file administration information of the SAS data set such as the name, label, type, creation date, last modified date, number of columns or fields, number of rows or observations, and the number of deleted rows of the selected data set. Select the **Table View** tab to display the table listing of the partitioned data mining DMDB data set. As a reminder, the table view that displays the records from the scored data set must be closed in order to perform updates to the source data set that is associated with the selected partitioned data set. Press the **OK** button to return to the **Neural Network** window.

Options subtab

The **Options** subtab will allow you to specify whether or not to sample the training data set for preliminary training and set the maximum number or rows allocated to interactive training. The training data set is used to perform preliminary training that is designed to determine the best set of initial weight estimates before network training begins. From the **Neural Network** node, interactive training will allow you to construct neural network models any numbers of times all within the same interactive network interface.

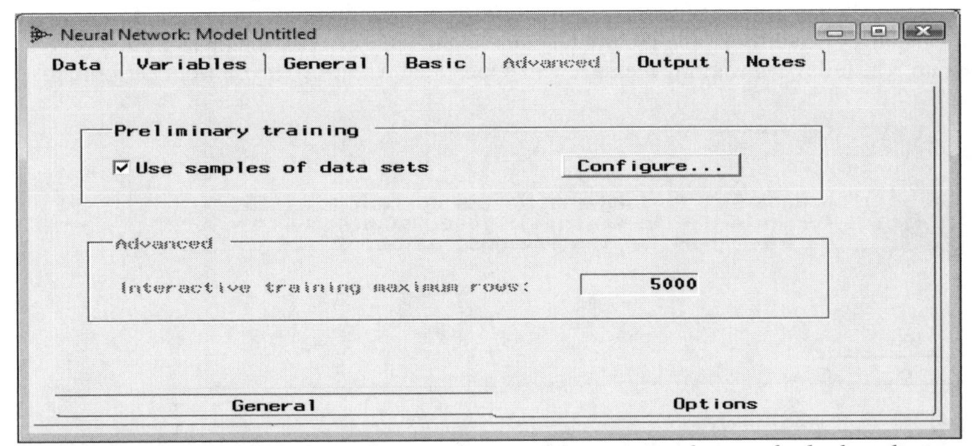

*The **Options** tab is used to create the sample size of the training data set for both preliminary training and interactive training.*

The **Options** subtab will allow you to set the number of observations that are randomly selected from the training data set that is used for preliminary training. By default, 2,000 observations are randomly selected from the training data set to perform preliminary training. However, by selecting the **Use samples of data sets** check box and pressing the **Configure...** button, the following **Sampling options** window will appear. From the **Sampling options** window, you may be able to enter the number of records that are randomly sampled from the training data set that are then used to fit the neural network model for preliminary training.

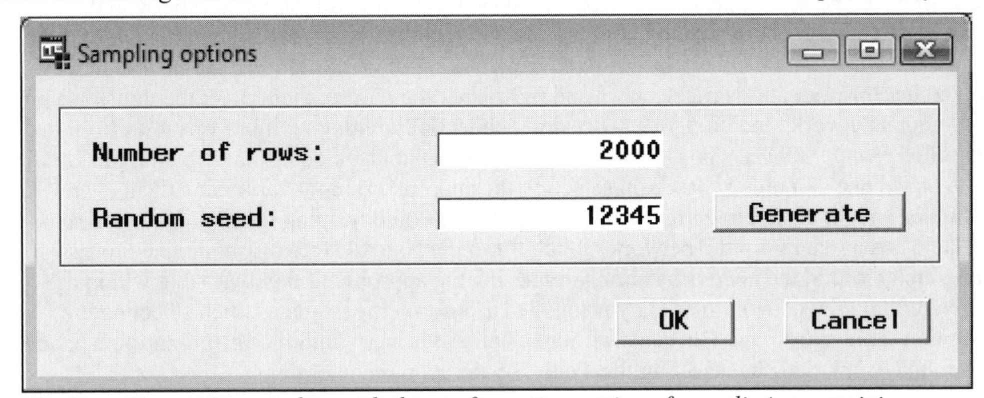

*The **Sampling option** window with the configuration settings for preliminary training.*

Number of Rows: Sets the number of records that are randomly selected from the training data set to fit the neural network model for preliminary training.

Random Seed: The **Random seed** entry field will allow you to set the seed value that is used in randomly selecting observations from the training data set. Specifying the same positive integer seed value will result in the same observations being randomly selected from the training data set for preliminary training. Conversely, setting the seed value to zero will result in the system using the computer's clock at run time to initialize the seed stream. Therefore, setting the seed to zero will generate a different sample for preliminary training every time you fit the network model. The default is a random number seed of 12345.

Click the **Generate** button that will result in the **Random seed** entry field displaying an entirely different random seed number. This will result in an entirely different sample that is randomly selected from the training data set that is used in the preliminary training runs. Click the **OK** button to return to the **Neural Network** window and the **Options** subtab.

From the **Advanced** section, the **Interactive training maximum rows** entry field will allow you to specify the number of records for interactive training within the advanced user inference. The advanced user inference can be activated by selecting the **Advanced user interface** check box from the following **General** tab. This will also activate the **Advanced** section for editing from the **Options** subtab. By default, the **Interactive training maximum rows** entry field is unavailable for editing within the basic user interface. However, the restriction is that if you specify the maximum number of rows from the **Interactive training maximum rows** entry field that is less than the number of records within the training data set and then activate interactive training from the process flow diagram. The following warning message will appear that will prevent you from performing interactive training.

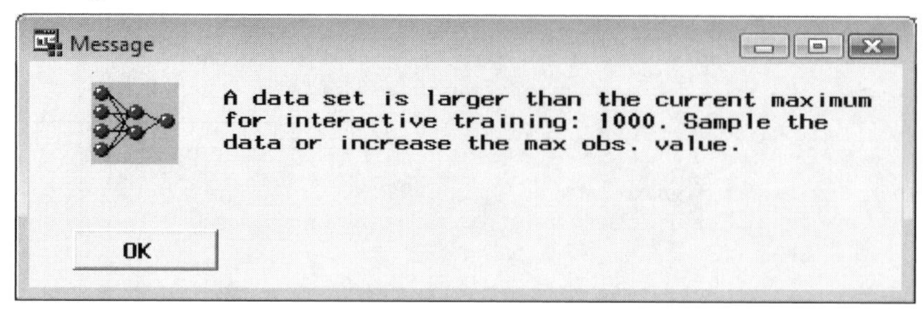

Variables tab

Similar to many of the other modeling nodes, the **Variables** tab will automatically appear when you first open the **Neural Network** node. The tab is designed to view the various properties of the listed variables such as the model roles, level of measurement, variable type, formats, and labels. The tab will allow you to add or remove variables from the neural network model before performing network training, view the frequency distribution of each selected variable and edit, browse, or delete the target profile of the target variable to predict. The **Tree** node was included in the process flow in selecting the best set of input variables to the neural network model.

From the **Status** column, the tab will allow you to add or remove certain input variables from the network model by simply setting the variable status of the listed variables. By default, all the listed input variables are automatically set to a variable status of **use**, indicating to you that the input variables are automatically included in the neural network model. Otherwise, to remove certain input variables from the neural network model, simply select the corresponding variable rows and right-click the mouse, then select the **don't use** pop-up menu option. In neural network modeling, this process of sequentially removing input variables from the network model is called *pruning inputs*. The criterion used in pruning inputs to determine which input variable to remove from the model one at a time is based on selecting the appropriate input variable with an overall average input-to-hidden weight closest to zero or some threshold value and refitting the neural network model after the input variable is removed from the network model. From the **Neural Network** node, the process of performing pruning inputs within the node is by simply removing the appropriate input variable within the **Status** column, then refitting the network model by using the current weight estimates, then selecting the **Starting Values** option settings from the **Current estimates** option setting within the **Initialization** tab, and rerunning the **Neural Network** node by selecting the **Tools > Run** main menu options.

SAS — Neural Network: Model Untitled

Data | Variables | General | Basic | Advanced | Output | Notes

Name	Status	Model Role	Measurement	Type	Format	Label
BAD	use	input	binary	num	BEST12.	
LOAN	use	input	interval	num	BEST12.	
MORTDUE	use	input	interval	num	BEST12.	
VALUE	use	input	interval	num	BEST12.	
REASON	don't use	rejected	binary	char	$7.	
JOB	use	input	nominal	char	$7.	
YOJ	don't use	rejected	interval	num	BEST12.	
DEROG	don't use	rejected	ordinal	num	BEST12.	
DELINQ	don't use	rejected	interval	num	BEST12.	
CLAGE	use	input	interval	num	BEST12.	
NINQ	use	input	interval	num	BEST12.	
CLNO	use	input	interval	num	BEST12.	
DEBTINC	use	target	interval	num	BEST12.	

Set Status
Sort by Status
Subset by Status
Find Status
View Distribution of DEBTINC
Edit target profile...

*The **Variables** tab is used to view the input variables selected to the network model from the **Tree** node.*

Viewing the Distribution of the Variable

Similar to the other modeling nodes, the **Variables** tab will allow you to view the frequency distribution of each one of the listed variables within the tab. From the **Model Role** column, simply select the appropriate variable row, then right-click the mouse to select the **View Distribution of <variable name>** pop-up menu item. This will result in the **Variable Histogram** window appearing that will display a histogram of the frequency distribution of the selected variable. In neural network modeling, it is assumed that all the variables in the neural network model are normally distributed. The reason is because extreme values may have a profound effect on the final estimates, which will result in slower convergence with no assurance that the iterative grid search procedure will find the optimum minimum of the error function that will result in the best fit. For the input variables, it is important that the input variables be normally distributed since the input variables are automatically standardized, which will result in better initial weight estimates assigned to each one of the input variables, leading to faster convergence and a better fitting neural network model.

Editing the Target Profile

Highlight the row of the target variable in the model, then scroll over to the **Model Role** column and right-click the mouse to select the **Edit target profile...** pop-up menu item in order to edit the target profile of the selected target variable. The corresponding updates performed within the target profile from the **Neural Network** node will then be passed along to the subsequent **Assessment** node in order to evaluate the various lift charts.

General tab

The **General** tab is designed to specify the modeling selection criterion in determining the final weight estimates, activate the advanced user interface environment, specify whether or not to accumulate neural network training results from each network training run and specify the way in which you want to monitor the training process during network training.

The advanced user inference may be specified within the **General** tab. From the advanced user inference, you can access the various neural network advanced option settings. By default, the basic user interface is automatically selected within the **Neural Network** node, but, selecting the **Advanced user interface** option will allow you to access the more advanced configuration settings for the neural network design.

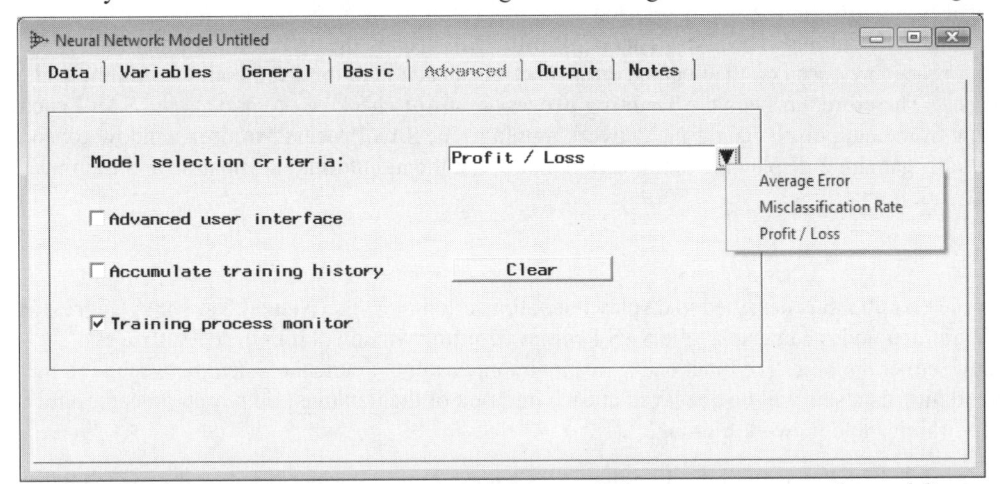

*The **General** tab is used to specify the advanced user interface environment or the modeling selection criteria for the neural network modeling design.*

Modeling Selection Criteria

The following are the various modeling selection criteria that can be used in selecting the best neural network model. From the **Model selection criteria** option, click the drop-down arrow button in the field display to select the appropriate criterion statistic that is used as the criterion value in selecting the final weight estimates for the neural network model. The final weight estimates are determined during network training based on the following modeling selection criteria that are selected.

The following are the three modeling selection criterions to select the best neural network model:

- **Average Error:** For interval-valued target variables, the smallest average squared error is the modeling selection criterion statistic that is used in selecting the best neural network model from the validation data set, assuming that the profit matrix has not been specified for the neural network modeling design. However, if the validation data set is unavailable, then the training data set is used.
- **Misclassification Rate:** For categorical target variables, the smallest misclassification rate is the modeling selection criterion statistic that is used in selecting the best neural network model from the validation data set without the existence of the profit matrix for the neural network modeling design.
- **Profit / Loss** (default): This is the default modeling selection criterion statistic that is used in selecting the best neural network model with the largest expected profit or the smallest expected loss from the validation data set. However, a profit or loss matrix must be defined within the target profile by fitting the categorically-valued or the stratified interval-valued target variable.

Accumulating Training History

By default, the **Accumulating training history** option is not selected. However, selecting the **Accumulating training history** check box will result in the node accumulating the various weight estimates and goodness-of-fit statistics by refitting neural network model any number of times. The history estimates just shows you how things change in the weight estimates and goodness-of-fit statistics as network training proceeds forward toward a desirable solution. Well-defined weight estimates do not change that much, so you may want to consider each estimate at each iteration to see where the weight estimates are unstable and undefined. The accumulated neural network estimates can either be viewed from the **Tables** tab by selecting the **History estimates** from the entry field or from the **Plot** tab that displays the optimization plot. Press the **Clear** button to clear the accumulated training history of all previous network training runs.

SAS Process Monitor

In Enterprise Miner 4.3, enhancements have been made to the **SAS Process Monitor** window. The purpose of the **SAS Process Monitor** is to graphically view the development of the error function at each iteration of both the training and validation data sets. From the training monitor, you may either stop or interrupt the iterative training process. From the **General** tab, you may elect to view the built-in process monitor or not. By default, the **Training process monitor** check box is selected. This will result in the **SAS Process Monitor** being displayed during network training. The **SAS Process Monitor** will display the optimization plot of the neural network hold out procedure based on the updated error function at each iteration from both the training and validation data sets. Therefore, uncheck the **Training process monitor** check box to prevent the **SAS Process Monitor** window from being displayed during network training. The **SAS Process Monitor** window consists of two separate tabs that either display the optimization plot or list the monitoring information and training history from network training.

Graph tab

The purpose of the **Graph** tab is designed to display the optimization plot. The **Neural Network**, **Regression** and the **SOM/Kohonen** nodes can display the **SAS Process Monitor** window. From the **SAS Process Monitor** window, either the objective function or the maximum absolute gradient element between both the training and validation data sets will be displayed at each iteration of the iterative grid search procedure for both preliminary training and network training.

Running the **Neural Network** node will result in the **SAS Process Monitor** window appearing, which will display the optimization plot. Selecting the **Stop Current** button will result in the interruption of the iterative process at the currently selected step during network training before any one of the termination criterion values are satisfied without error and using the current weight estimates to begin network training. Press the **Stop All** button to stop network training altogether and begin scoring. That is, scoring the neural network results such as the current weight estimates and the various goodness-of-fit statistics from the final neural network model. The **Close** button will close the **SAS Process Monitor** window; however, network training will continue. By default, the objective error function is displayed. Press the arrow button from the **Plot Group** field and select the option of **1** to view the plot of the maximum gradient element at each iteration between both the training and validation data sets. From the table listing, you are looking for small changes in the gradient values between successive iterations that will indicate convergence to the iterative grid search procedure.

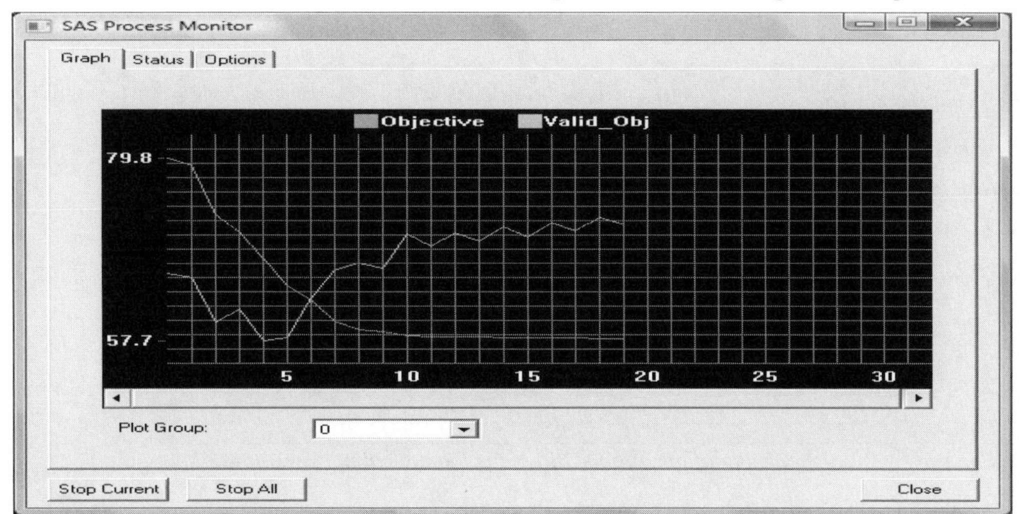

*The SAS **Process Monitor** window displaying the optimization window during network training by fitting the interval-valued variable, DEBTINC, from the HMEQ data set.*

Status tab

The **Status** tab is designed to display a table view of the process monitoring information from network training. The tab will display the process monitoring information such as the starting time, current time, and the duration time of network training with various network server connection information. The table listing displays the accumulating training history from network training. The table will display the objective error function and the maximum gradient element from both the training and validation data sets at each iteration of the iterative grid search procedure.

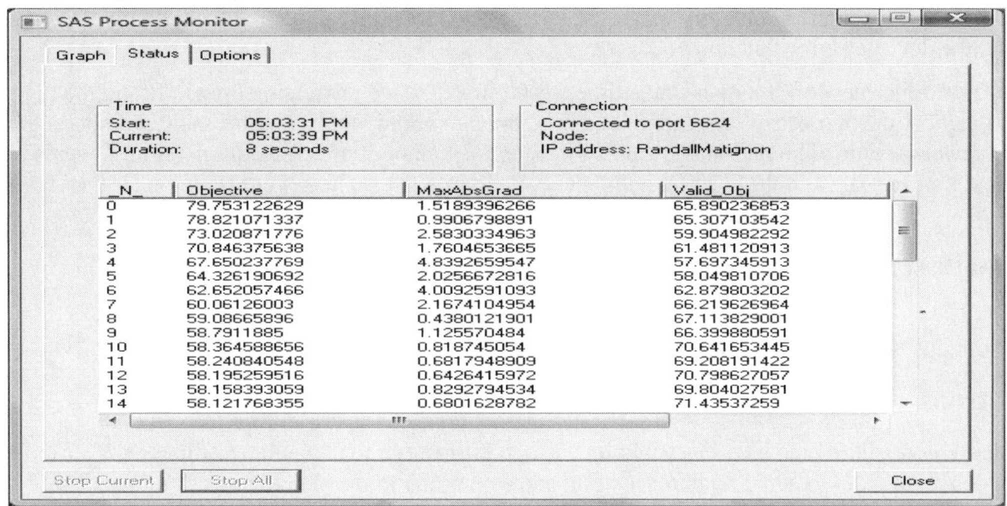

_N_	Objective	MaxAbsGrad	Valid_Obj
0	79.753122629	1.5189396266	65.890236853
1	78.821071337	0.9906798891	65.307103542
2	73.020871776	2.5830334963	59.904982292
3	70.846375638	1.7604653665	61.481120913
4	67.650237769	4.8392659547	57.697345913
5	64.326190692	2.0256672816	58.049810706
6	62.652057466	4.0092591093	62.879803202
7	60.06126003	2.1674104954	66.219626964
8	59.08665896	0.4380121901	67.113829001
9	58.7911885	1.125570484	66.399880591
10	58.364588656	0.818745054	70.641653445
11	58.240840548	0.6817948909	69.208191422
12	58.195259516	0.6426415972	70.798627057
13	58.158393059	0.8292794534	69.804027581
14	58.121768355	0.6801628782	71.43537259

*The **Status** tab displays the various monitoring information and history training for the network model.*

Basic tab

The **Basic** tab is designed for you to select the basic configuration settings to apply in network training. The tab will allow you to specify the basic neural network configuration settings such as the network architecture, preliminary training runs, the minimization technique, the number of hidden units, direct connections, and the maximum amount of computational time to network training. By default, the basic user interface is automatically selected once you first open the **Neural Network** node. By default, the **Advanced user interface** check box is automatically unchecked within the **General** tab, which will allow you to view the following **Basic** tab and the corresponding basic configuration settings for the neural network model.

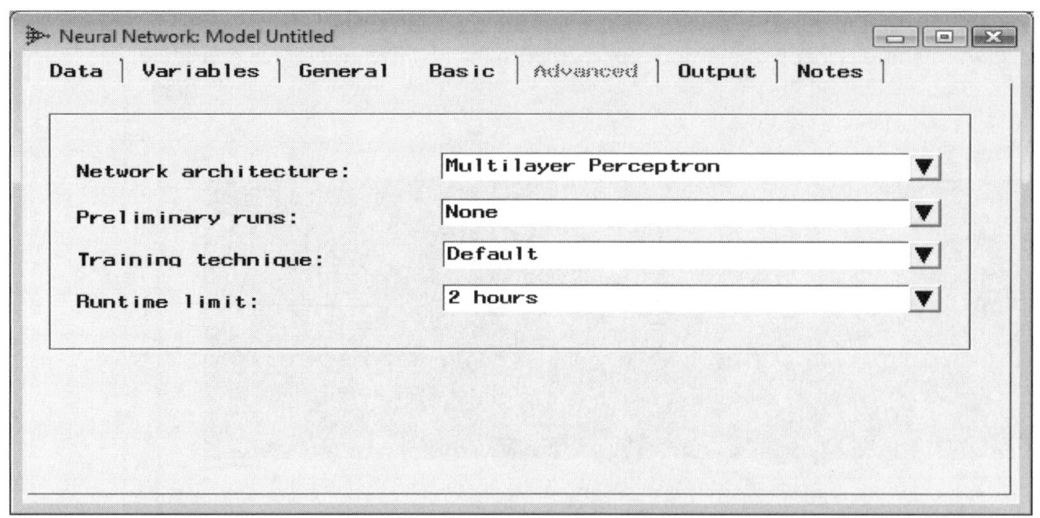

*The **Basic** tab is used to set the basic configuration settings for the neural network model.*

The following are the available options to select from the **Basic** tab:

- **Network Architecture:** Sets the type of neural network architecture. The default is the multilayer perceptron MLP architecture.
- **Preliminary Run:** Preliminary training runs are performed on the training data set in an attempt to find the best set of starting values for the weight estimates in the subsequent network training run that will result in accelerating convergence to the iteration process. The best set of weight estimates is determined by the smallest error among all preliminary training runs. By default, no preliminary training runs are performed.
- **Training Technique:** The default training technique that is used depends on the number of weights applied to the neural network model during execution. Levenberg – Marquardt is the standard convergence method that is used when the number of weights in the model is less than a hundred with the objective of minimizing the deviance error function.
- **Runtime limit:** Since network training can at times result in excessive processing time in fitting the highly nonlinear function, the purpose of this option is to limit the maximum length of processing time in order to terminate network training in a reasonable amount of time. The amount of computational time depends on the complexity of the data and the corresponding network architecture. The default is no longer than two hours.

Select the **Network Architecture** drop-down button and the following **Set Network Architecture** window will appear.

The following options within the **Set Network Architecture** window are as follows:

Hidden neutrons

The **Hidden neutrons** display field provides you with a rough estimate as to the number of hidden layer units to the network design in order to avoid both underfitting and overfitting to the neural network model. The default is three hidden units. The following are the various option settings used in specifying the number of hidden layer units to the neural network model based on the noise level in the active training data set:

- **High noise data:** 1 hidden layer unit to prevent overfitting. However, this is generally a generalized linear model.
- **Moderate noise data:** 2 hidden layer units. This is the simplest neural network model.
- **Low noise data:** 3 hidden layer units
- **Noiseless data:** 5 hidden layer units
- **Set number...:** Enter or set the number of hidden units to the hidden layer.

The interpretation of the noise level in the data can be defined from the following modeling assessment statistics by fitting the categorical or interval-valued target variable to the network model as follows:

- **High noise data:** Misclassification Rate: < 40% or R-square: .5
- **Moderate noise data:** Misclassification Rate: 20% or R-square: .8
- **Low noise data:** Misclassification Rate: 10% or R-square: .9
- **Noiseless data:** Misclassification Rate: 0% or R-square: 1.0

The first four hidden neutron menu items that are listed within the **Hidden neutrons** display field are determined by Enterprise Miner at run time based on the total number of input levels, total number of target levels and the total number of observations in the training data in addition to the noise level in the target values from the underlying training data set. *Noise* is the random error that is inherited in the underlying data set that you want to fit. Noise in the underlying data is not good since it reduces the accuracy of generalization even if you have an enormous amount of data to fit. Noise in the target values will also increase the risk of overfitting. Note that the reason why the **Neural Network** node sets a small number of hidden units for high noise data is to prevent overfitting during network training. In other words, the goal is finding the proper balance in specifying the appropriate number of hidden units that best explains the variability in the target values while at the same time avoiding overfitting to the neural network model.

Setting the number of hidden units to the neural network model is one of the most critical decisions to the network architecture. In addition, there is no exact science as to the correct number of hidden units to select. However, the number of hidden units to select depends on the number of observations in the training data set, the amount of noise or random error in the underlying training data set, the complexity of the neural network model, and the underlying function you are trying to fit. The problem is that you usually do not know the variability and the exact distribution of the underlying data set that you want to fit. Therefore, in solving this problem, network training might be performed by fitting the neural network model any number of times, that is, perform network training by selecting a different number of hidden units each time to find the correct number of hidden units to apply, which is determined by the smallest error from the validation data set. Again, the simplest modeling strategy is fitting the neural network model by starting with no hidden units and sequentially increasing the number of hidden units in the neural network model each time, then stopping when the assessment statistics begin to increase.

Relying on Enterprise Miner and the **Neural Network** node to provide you with the correct number of hidden units for neural network modeling design might be very risky since the node might result in an unexpectedly large number of hidden units that may lead to overfitting and bad generalization of the neural network model. Therefore, you might want to specify the number of hidden units for the architecture by selecting the **Set number...** menu item. The number of hidden neutrons must be greater than zero in the network architecture. The simplest neural network model is two hidden units, which is the default.

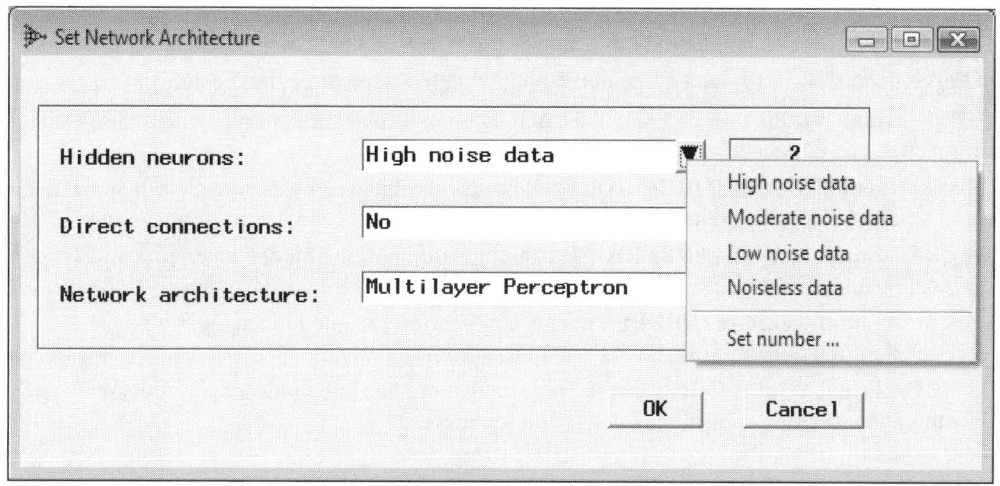

*The **Basic** tab is used to specify the number of hidden layer units for the neural network design.*

Direct connections

The default is no direct connections. In other words, each input unit is connected to each unit in the hidden layer and each hidden unit is connected to each output unit. Direct connections is a special type of network architecture in which additional connections are added to the neural network design where each input unit bypasses the input-to-hidden layer and connects directly to each output unit. Direct connections are also called a skip layer design. Creating a direct connection within the network architecture can be achieved by selecting the **Direct connections** display field to **Yes**. This will result in a direct connection that is applied to each input variable in the neural network model. In time series modeling, a direct connection or a skip layer design is typically applied to adjust to nonstationarity in the data over time.

Network architecture

The purpose of the **Network architecture** option is that it will allow you to specify the neural network architecture. The first four listed neural network architectures are available within the **Basic user interface**. The default is the multilayer perceptron MLP design. In the following listing, the abbreviations in brackets represent the option settings for the respective architecture based on the TECH option statement from the PROC NEURAL network procedure.

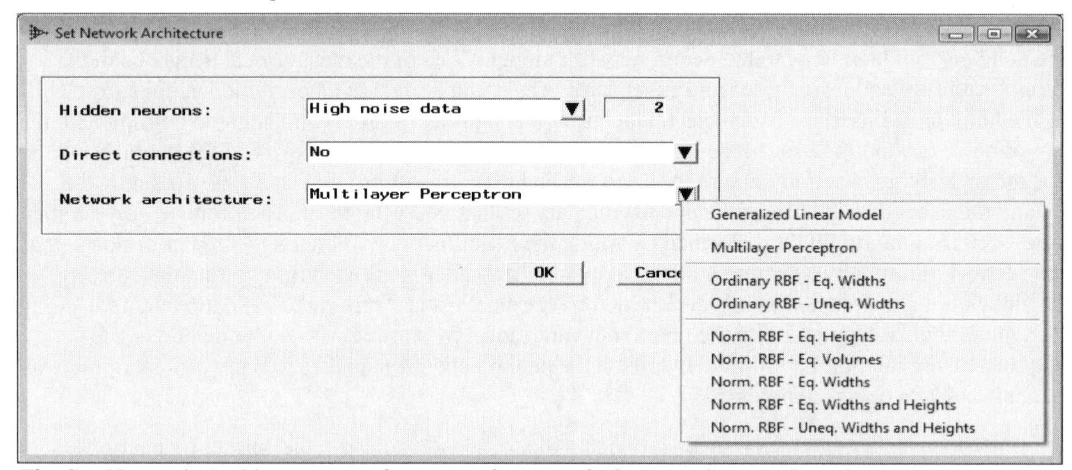

*The **Set Network Architecture** window is used to specify the neural network architecture.*

The following are the various network architectures to select from the **Network architectures** field:

- **Generalized Linear Interactive Model (GLIM):** This configuration is essentially a neural network skip layer design where there is no hidden layer and the input and target variables are not standardized. You may select this method in order to perform multivariate regression modeling within the process flow by fitting numerous target variables in the predictive model that you want to predict.
- **Multilayer Perceptron (MLP)** (default)**:** The default neural network architectural design.
- **Ordinary RBF – Equal Widths (ORBFEQ):** This architecture has one hidden layer with an EQRadial combination function and exponential activation function.
- **Ordinary RBF – Unequal Widths (ORBFUN):** This architecture has one hidden layer with an EHRadial combination function and exponential activation function.
- **Normalized RBF – Equal Heights (NRBFEH):** This architecture has one hidden layer with an EHRadial combination function and softmax activation function.
- **Normalized RBF – Equal Volumes (NRBFEV):** This architecture has one hidden layer with an EVRadial combination function and softmax activation function.
- **Normalized RBF – Equal Widths (NRBFEW):** This architecture has one hidden layer with an EWRadial combination function and softmax activation function.
- **Normalized RBF – Equal Widths and Heights (NBRFEQ):** This architecture has one hidden layer with an EQRadial combination function and softmax activation function.
- **Normalized RBF – Unequal Widths and Heights (NRBFUN):** This architecture has one hidden layer with an Xradial combination function and softmax activation function.

Training technique

The following are the various training techniques or optimization techniques to apply within the **Basic user interface** environment. The purpose of the various training techniques that are applied is finding the best linear combination of weight estimates that will result in the smallest generalization error. The reason that the following training techniques are applied is because the hidden layer units apply the activation function that is usually a nonlinear transformation to the linear combination of input units, which results in a nonlinear system of equations that must be solved iteratively. These minimization techniques are applied in searching for a global minimum to the multidimensional error surface. The training techniques are performed by determining every possible combination of the weight estimates at each iteration in which the weight estimates are updated to reduce the error function. The difference between many of the various training techniques is based on the distance and the direction of the updates performed on the weight estimates during the iterative grid search procedure. The various training techniques to apply are explained in greater detail in subsequent sections.

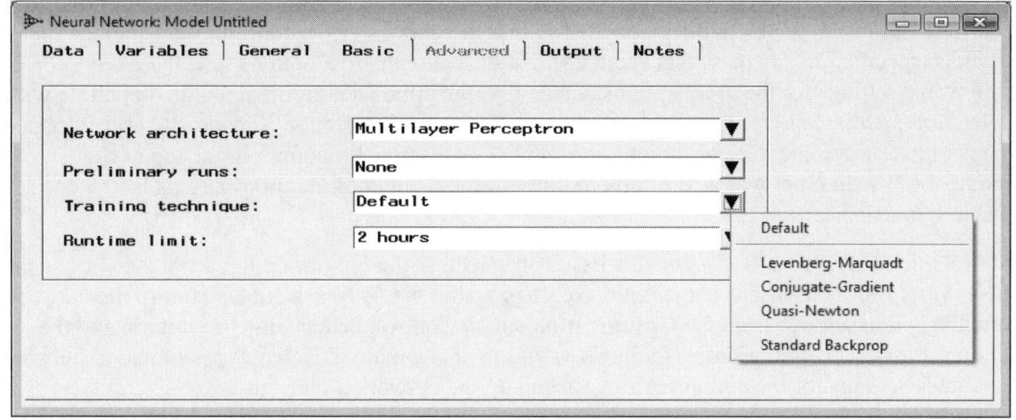

*The **Neural Network** window is used to specify the optimization method for network training.*

- **Default** (default)**:** The values and the number of network weights determine the optimization technique that is applied. This option is the default method that is used in network training.

Advanced tab

The **Advanced** tab is designed for you to gain access to many of the more advanced neural network configuration settings of the neural network design. From the **General** tab, select the **Advanced user interface** check box to activate the **Advanced** tab.

- **Network subtab**
- **Initialization subtab**
- **Optimization subtab**
- **Train subtab**
- **Prelim subtab**

Some of the Purposes of the Advanced tab

- **Create the network:** Allows you to view the neural network design while at the same time specifying several different option settings for each layered unit in the neural network design.
- **Diagram the network:** Create connections or additional hidden layers for neural network design.
- **Add hidden layers:** Add additional hidden layers or hidden layer units to the neural network design.
- **View and set connection properties:** Set the distribution, scale, and location parameters and the random seed number in generating a set of random numbers that are assigned to the starting values of the weight estimates. These random numbers are usually set to some small values that are centered about zero.
- **Preliminary training, regularization and optimization techniques:** Specify preliminary training, weight decay or early stopping regularization and several optimization techniques for the preliminary and network training runs.
- **View and set node properties:** Group variables into each separate node, create nodes for each input variable in the design, or completely remove the node from the neural network design.

Resetting the SAS Default Settings

From the **Advanced user interface**, the **Neural Network** node has an added option for resetting all the changes that have been made to the layered nodes back to their original SAS default settings. Selecting the **Network > Reset nodes** main menu option will result in Enterprise Miner resetting the various node properties that have been previously specified within the node back to their original configuration settings that SAS has carefully selected.

The **Neural Network** node consists of several option settings with regard to the neural network architecture where many of the options are associated to each other. In other words, network training will not be performed if there exists some combination of the selected option settings that are incompatible within the neural network architecture. Many of the default settings that SAS has carefully selected are displayed in green. These same default settings are highly recommended by SAS. Changing the default settings, will result in the specified option settings being displayed in black. However, by resetting one or more nodes, the default settings will then be restored. The **Neural Network** node is designed so that the options or the configuration settings for the layered nodes can be specified in any particular sequence, but be aware that the options selected often influence the allowable settings for the other options. Generally, the option settings that are incompatible with the previous selections are displayed in gray. In Enterprise Miner, as each selection is made, the user interface checks with the neural network engine. The option setting for the objective function is at the top of the hierarchy for compatibility. In other words, the compatibility of the other configuration settings is first checked against the objective function setting.

The process of designing a neural network model is based on selecting the option settings from the various tabs in the **Neural Network** node. For instance, the option settings within the **Network** subtab control the network architecture, and the option settings from the **Optimization** subtab control the convergence criteria and the objective function settings. The option settings will allow you to create many different types of neural network designs. The available settings for the option settings depend on the network architecture, the size of the training data set, and so on.

Press the **OK** button to reset the node properties to the default settings and the **Neural Network** window will reappear.

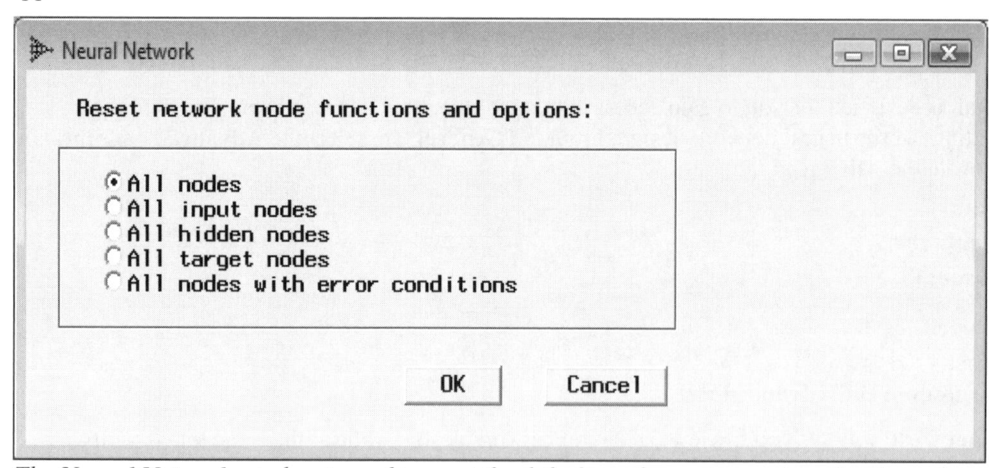

*The **Neural Network** window is used to reset the default configuration settings for the nodes.*

The following are the various options in resetting the configuration settings for the **Neural Network** node by selecting the **Reset nodes** main menu option:

- **All nodes:** Resets all the configuration settings for the **Neural Network** node or the node properties back to their original SAS default settings.
- **All input nodes:** Resets all the configuration settings that were applied to the input variable nodes back to their original default settings.
- **All hidden nodes:** Resets all the configuration settings that were applied to the hidden layer nodes back to their original SAS configuration settings.

- **All target nodes:** Resets all the configuration settings that were applied to the target layer nodes back to their original SAS configuration settings.
- **All nodes with error conditions:** Resets all the neural network nodes with error condition option settings back to their original SAS default settings. In Enterprise Miner, many of the option settings for the neural network nodes are associated with one another. The node is designed so that many of the option settings that are incompatible are colored in gray. Therefore, this option will restore the incompatible option settings back to their default settings in order to perform network training.

It is very important to understand that some of the configuration option settings within the following **Advanced user interface** are associated with one and other. For example, you may specify the number of preliminary training runs from the **Prelim** subtab. However, preliminary training will not be performed unless you specify preliminary training from the **Optimization Step** field display within the **Optimization** subtab. Conversely, you might specify the various configuration settings for the network training run from the **Train** subtab. However, the corresponding network training options will be ignored and network training will not be performed unless you specify network training from the **Optimization Step** option within the **Optimization** subtab.

Advanced User Interface

The advanced user interface environment will give you more flexibility in specifying many more configuration settings for the neural network model.

- **Network subtab**
- **Initialization subtab**
- **Optimization subtab**
- **Train subtab**
- **Prelim subtab**

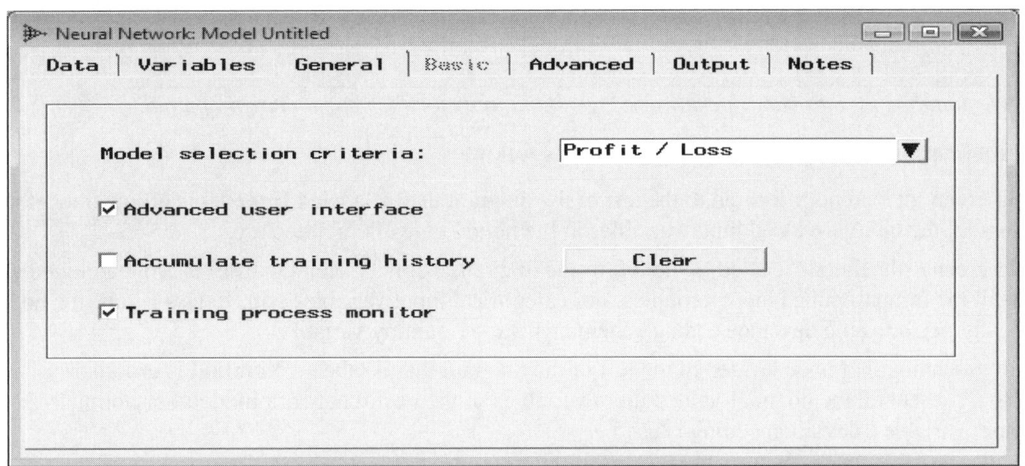

*Select the **Advanced user interface** check box to display the **Advanced** tab and access the advanced user interface option settings and perform interactive training.*

The following subtabs are available within the advanced user interface environment:

- **Network subtab:** Specify the type of neural network architecture in addition to the various option settings within each layered node of the neural network design.
- **Initialization subtab:** Specify the initialization settings of the starting values of the weight estimates in the neural network model.
- **Optimization subtab:** Specify the optimization settings such as the type of network training, preliminary training, objective function, weight decay regularization, and the convergence criteria.
- **Train subtab:** Specify the type of training technique used in finding the minimum error.
- **Prelim subtab:** Specify the configuration settings for the preliminary training runs such as the number of preliminary training runs and the type of optimization technique to apply.

Network subtab

The **Network** subtab displays the neural network workspace where you may view the neural network design and specify the various configuration settings for each one of the layered nodes within the neural network architecture.

The network diagram automatically places the input layered nodes to the left of the diagram, the hidden layer nodes are located in the middle, and the output layer nodes are positioned to the right of the neural network workspace. For the input and target layer nodes, separate nodes are grouped together by their corresponding level of measurement. In addition, it is important to point out that a separate weight estimate is created for each class level of the categorically-valued input variable or target variable in the neural network model.

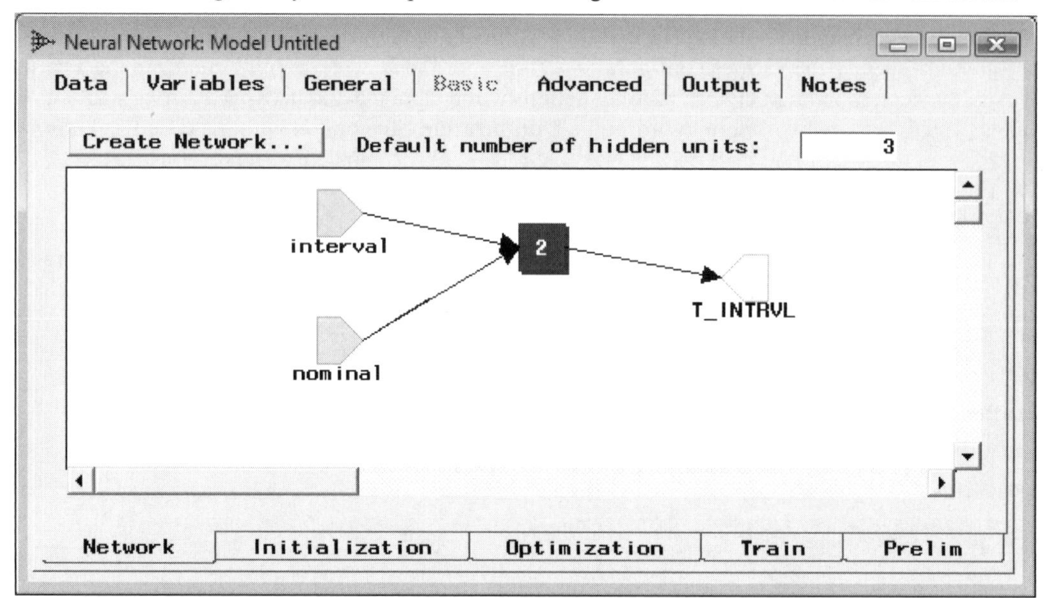

*The **Network** subtab is used to view the network layers and units for the neural network model.*

The default groupings of the diagram nodes are defined as follows:

- Each light green interval node located to the left of the diagram that is labeled **Interval** is grouped together to represent all the interval-valued input variables in the neural network architecture.
- Each light green ordinal node located to the left of the diagram that is labeled **Ordinal** is grouped together represent all the ordinal-valued input variables. For categorical input variables with k class levels, the node automatically performs the deviance coding schema with $k - 1$ dummy variables.
- Each light green nominal node located to the left of the diagram that is labeled **Nominal** is grouped together to represent all the nominal-valued input variables of the neural network model. For nominal-valued input variables, deviation coding is used.
- All interval-valued target variables are grouped together into one yellow interval-type target node that is located to the right of the network diagram.
- A separate nominal target node is created for each nominal-valued target variable. For categorically-valued target variables with k class levels, the node automatically generates k dummy variables using the GLM coding schema.
- A separate ordinal target node is created for each ordinal-valued target variable.

Set the number of hidden units from the **Default number of hidden units** entry field in order to specify the number of units that are automatically assigned to the hidden layers in the neural network architecture. The default is three hidden units.

- The light green icon or the icons located to the left of the diagram represents the input node. Double-click the icon or right-click the mouse and select the **Properties** pop-up menu item to configure the input node settings of the neural network model.

- The blue icon or the icon located in the middle of the diagram represents the hidden layer node. Double-click the icon or right-click the mouse and select the **Properties** pop-up menu item to configure the hidden layer option settings. The number that is displayed within the node indicates the number of hidden layer units. In our example, there are two separate hidden units that were selected in the neural network model.
- The yellow icon or the icon located to the right of the diagram represents the hidden-to-target layer node. Double-click the icon or right-click the mouse and select the **Properties** pop-up menu item to configure the hidden-to-output layer neural network settings. Note that a separate node will be created for fitting several target variables to the neural network model. In other words, a separate target layer node will be created by each measurement level of the target variable.

In the subsequent sections, I will first explain the other subtabs within the **Advanced** tab and the corresponding configuration settings that can be specified from the various subtabs, then come back to this same neural network diagram and the corresponding **Network** subtab that is designed to set the various option settings for the network layered nodes. Again, it is important to point out that the following tabs and the associated configuration settings are related to each other.

Initialization subtab

The **Initialization** subtab is designed to set the initialization options of the starting values that are used to estimate each one of the weight estimates in the neural network model. Specifying the correct initial weight estimates is extremely critical in many of the training techniques. This is especially true to many of the Newton–Raphson minimization techniques that apply the Hessian matrix. Specifying the correct initial weight estimates will result in faster convergence to the correct weight estimates resulting in the desirable minimum in the objective error function. The following options will allow you to specify whether to randomize the weight estimates and bias to the network model. The **Initialization** subtab will allow you to randomize the scale parameter estimates, target weights, and target bias weights by simply clicking the appropriate check box.

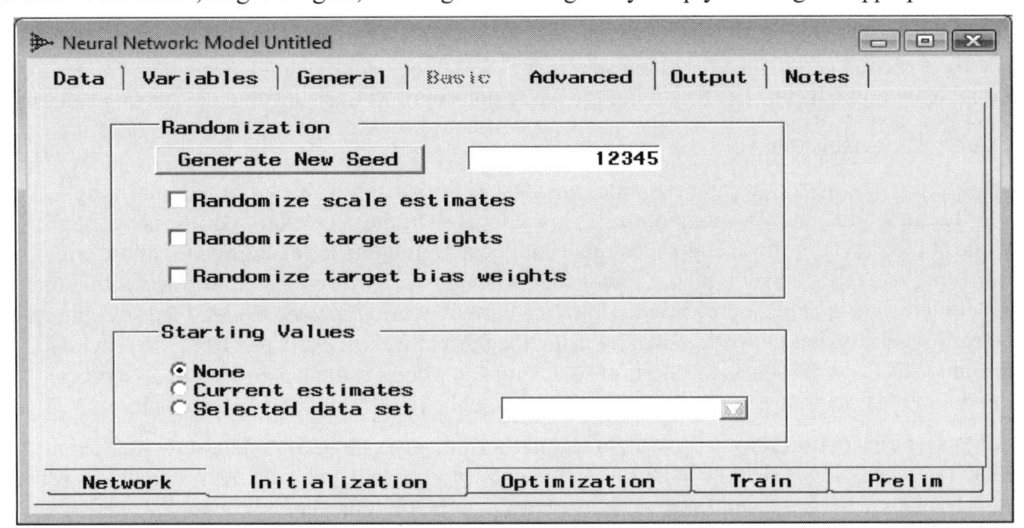

*The **Initialization** subtab is used to initialize the weight estimates for the neural network model.*

- From the **Randomization** section, click the **Generate New Seed** button to display a new random seed number that generates a new set of random numbers that are used as the starting values for either the initial scale parameter, target weight estimates or target bias weights. The default is a random seed of 12345. By default, these randomly generated numbers and the corresponding initial weight estimates are set to some small values that follow a normal distribution centered about zero. Again, the purpose of setting the weight estimates to some small values is to avoid large changes in the weight estimates between successive iterations, which will result in unstable estimates and bad generalization.

The following are the various option settings within the **Randomization** section used in randomizing the initial weight estimates for the neural network model:

- **Randomize scale estimates:** Sets the value to the scale parameter that controls the variability of the initial weights and biases. Each weight estimate is divided by its scale estimate that is determined by the distribution in which the weight estimates are randomly drawn. The importance of the scale estimate is that it is used to adjust the weight estimates. These adjustments to the weight estimates will prevent network training from computing extremely large weight estimates that might cause the output units to produce wild predictions far beyond the range of the target values. Select the check box to randomize the target scale estimates. By default, the scale parameter is initialized to the standard deviation of the target variable.
- **Randomize target weights:** Sets the starting values to the hidden-to-target weight estimates. Thereby, the random initial target weights are adjusted by the randomized scale estimate that is divided by the square root of the number of connections to the target unit, excluding the target bias and altitudes. The random numbers assigned to the initial weight estimates follow a normal distribution that is centered about zero. This will result in a small chance of producing large values. By default, the starting values of the hidden-to-target layer weight estimates are set to zero. Hence, at the first iteration the hidden-to-target layer weights are set to the mean from the output activation function. In network training, the neural network model is initialized by fitting a constant function through the data and then performing small steps, assuming that the weight estimates are small, in the first few iterations around the target mean. This is illustrated in the introductory section.
- **Randomize target bias weights:** Sets the starting values to the target layer bias weights that are adjusted by the randomized scale estimate, then divided by the square root of the number of connections to the target unit, excluding the target bias and altitudes. By default, the target layer bias weights are set to the average value of the target variable, then transformed by the activation function that is applied to the output layer.

The **Starting Values** section is designed to set the starting values of the neural network weight estimates for the subsequent network training run. The following are the various options in specifying the starting values:

- **None** (default)**:** The default is none.
- **Current estimates:** The current weight estimates are used as the starting values for the initial weight estimates in the subsequent network training run. This will result in the acceleration of the convergence process while at the same time avoiding undesirable bad local minimums in the highly nonlinear error function. This option is used in the input pruning technique. *Input pruning* can be performed within the node by deleting the undesirable input variable from the current neural network model, then selecting this option and refitting the neural network model by using the current weight estimates from the previous network training run that are used as the starting values for the weight estimates in refitting the neural network model in order to assure convergence and accelerate the iterative grid search procedure.
- **Selected data set:** This option selects the weight estimates from an existing SAS data set with a set of values replacing the missing weight estimates to assure convergence and speed up the optimization process and network training. Clicking the drop-down arrow button in the field display will give you access to the list box that will allow you to browse the available SAS data libraries for the appropriate SAS data set. The current weight estimates from the scored data set created within the **Neural Network** node in the EMPROJ library is an example of a SAS data set that is compatible with this option. By default, the node will automatically assign random numbers to all initial weight estimates with missing values. However, this option will allow you to assign your own weight estimates by editing the appropriate scored data set that will be used to assign the initial weight estimates in the subsequent neural network fit.

Note: In neural network modeling, it is recommended that the nonlinear model be fitted numerous times. This can be achieved by fitting the neural network model with different starting values or randomized values assigned to the weight estimates in the neural network model from the previously displayed options. The purpose of refitting the neural network model multiple times is to assure that the iterative grid search procedure has obtained the minimum error in the highly nonlinear error function that will result in an adequate modeling performance and good generalization.

Optimization subtab

The **Optimization** subtab will allow you to specify network training, preliminary training, type of objective function, regularization techniques, and the various convergence criteria values that are used in terminating the iterative grid search procedure.

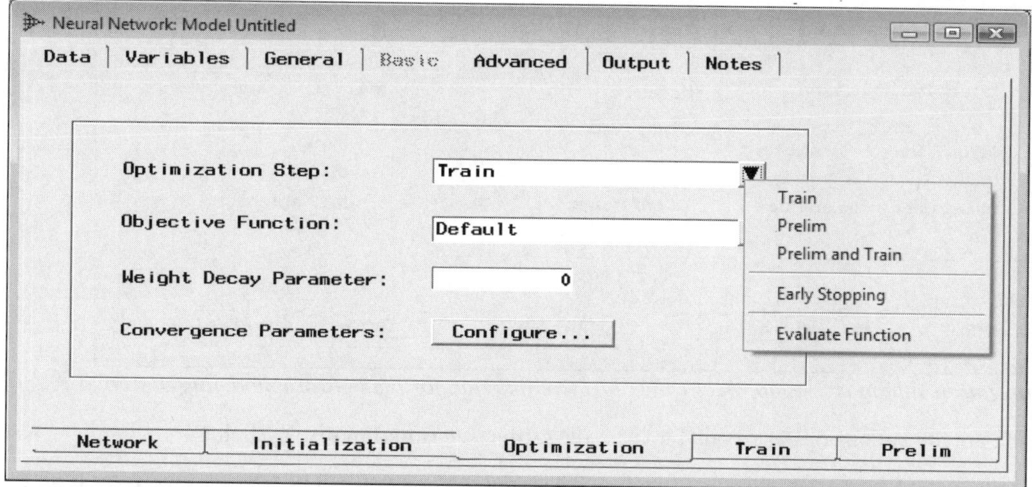

*The **Optimization** subtab is used to specify the optimization technique for the neural network model.*

The following are the various options available from the **Optimization Step** display field:

- **Train** (default)**:** Performs network training that is the default.
- **Prelim:** Performs preliminary training runs to the training data set, then stops.
- **Prelim and Train:** Performs preliminary training, then network training in succession. The initial weight estimates in the network training run are estimated by the best set of weight estimates that is determined by the smallest objective function value from a particular preliminary training run. Network training is then automatically performed in selecting the final neural network model with the smallest validation error.
- **Early Stopping:** This network training technique is designed to prevent overfitting by stopping network training at the first iteration when the optimization function you want to minimize begins to increase. There are two separate steps to the early stopping routine. The first step is to determine the weight estimates during the network training run that results in the best validation error. In the second step, a single network training run is performed using the same weight estimates as the starting values that merely evaluate the error function.
- **Evaluation Function:** This option prevents network training from iterating and takes the current weights and uses them to compute the objective function by performing a single network training run.

Objective Function

The main idea in neural network modeling is to minimize the objective function that will result in the best linear combination of network weight estimates. The type of objective function that you want to minimize during network training is one of the most important neural network configuration settings for the neural network design. Enterprise Miner is designed so that any changes made to the objective function may cause the other configuration settings to become incompatible. There are three different types of objective functions that depend on the type of error function that is applied, which you want to minimize during network training. Also, the weight decay parameter estimate may be specified from the **Optimization subtab** in order to overcome overfitting in the model and improve generalization.

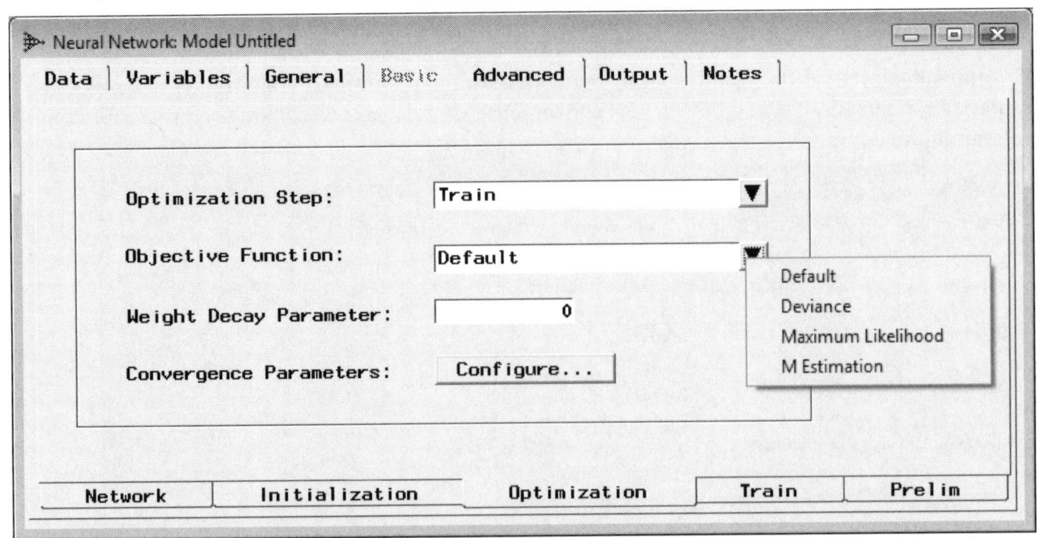

*The **Optimization** subtab is used to specify the objective function for the network modeling design.*

The following are the various options available from the **Objective function** display field:

- **Default** (default)**:** Chooses an objective function that is compatible with all of the specified error functions that is the default.
- **Deviance:** The deviance objective function can be used when the error distribution is a part of the exponential family. Otherwise, the maximum likelihood function is applied. For interval-valued targets, the deviance function is similar to the sum-of-squares function. For categorical targets, the deviance is the difference between the likelihood from the current network model that you want to fit and the likelihood from a saturated model in which there is one weight estimate for every observation. The other difference between these objective functions is that the **Neural Network** node does not estimate the scale parameter using the deviance function, as opposed to the likelihood function that estimates the scale parameter estimate from the training data. In other words, the deviance function automatically sets the scale estimate to one for both the normal and the gamma distributions. Conversely, the maximum likelihood function estimates the scale parameter estimate by the conventional maximum likelihood method.
- **Maximum Likelihood:** For other types of maximum likelihood estimation, this option minimizes the negative log likelihood function that estimates the parameter estimates based on the conventional maximum likelihood estimation technique. The maximum likelihood estimation technique is designed to find the weight estimates that are most likely to occur given the data. The weight estimates are derived by maximizing the likelihood function that expresses the probability of the observed data as a function of the unknown weight estimates in the neural network model. The likelihood function can be maximized by conventional approaches to the maximum likelihood estimation. In particular, calculating the logarithm of the likelihood function in order to linearize the function, then computing the derivative of each parameter estimate with all other input variables in the model held fixed and solving the system of linear equations by means of the various training techniques.
- **M Estimation:** This objective function is applied for robust regression estimation, assuming that the interval-valued target variable contains several outliers. The M-estimation statistic incorporates an absolute error by calculating the absolute value of the difference between the target values and the fitted values since outliers will inflate the squared error function.

- **Weight Decay Parameter:** The regularization weight decay parameter is part of the objective function. The weight decay parameter value must be a nonnegative number. The purpose of the weight decay parameter value is to reduce the growth of the weight estimates at each iteration, which is important in avoiding bad generalization. Adjusting the weight parameter accordingly will improve the modeling performance of the neural network design and overcome both overfitting in the model and ill-conditioning in the data. By default, the weight decay parameter estimate is automatically set to zero.

The objective function is defined as the sum of the total error function and a penalty term. The penalty term is the product of the weight decay estimate and the sum of all the network weights other than the output bias. However, the weight decay is usually set to zero since it is extremely difficult to apply. Therefore, the objective function is essentially the total error function that you want to minimize.

Convergence Parameters

Press the **Configure** button and the following **Convergence Criteria** window will appear. The tab is designed for you to set the convergence parameters. The **NLP** tab will be displayed when you first open the window. The **NLP** tab is designed to set the common nonlinear convergence criterion values and the **Adjustment** tab is designed to adjust or relax the criterion values.

NLP tab

The **NLP** (Non Linear Processes) tab is designed for you to set the various convergence criterion values that control the nonlinear optimization processes.

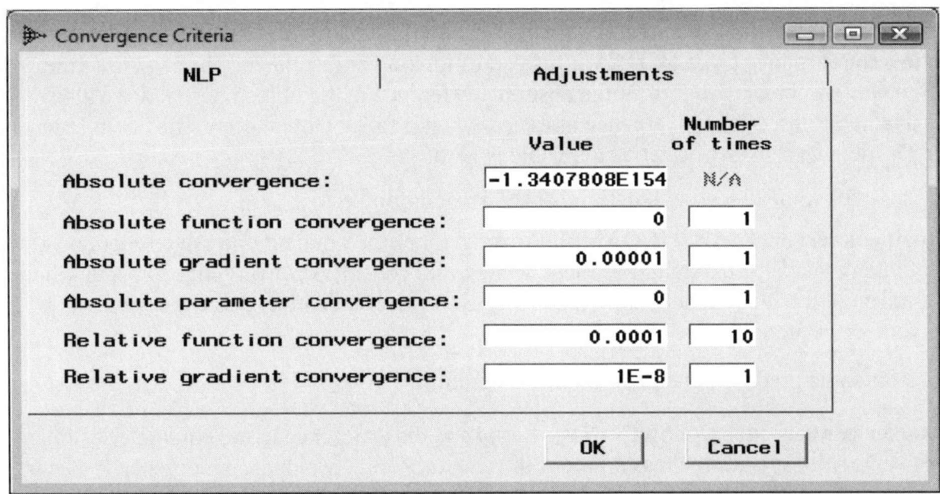

*The **NLP** tab is used to specify the convergence criteria values for the training technique.*

This tab is very similar to the **Convergence Criteria** window in the **Regression** node for setting the convergence configuration settings for logistic regression modeling. However, the tab has an additional option that will allow you to specify the number of successive iterations for which the associated convergence criteria must be satisfied before the iterative process can be terminated. The purpose of the tab is to assure you that the objective function has reached a desirable minimum through the use of the minimization technique. When the error function is less than any one of the convergence criterion values during network training, then convergence is assumed and the iterative algorithm stops. This will result in the best linear combination of weight estimates among all possible combinations of weight estimates. However, the complexity of nonlinear modeling, such as neural network modeling, is such that even though one of the convergence criteria values has been met, it does not necessarily guarantee you that the optimal solution has been found. For instance, if it is reported that convergence has been achieved by some certain convergence criterion, the gradient value for some of the other parameters in the neural network model may still be not be met. Therefore, it is highly recommended to perform network training by refitting the neural network model any number of times until the best solution is reached.

Although Enterprise Miner automatically sets the various convergence criteria values, the choice of the convergence criteria value is very important. The reason is because setting the criteria value too high will result in the algorithm performing an inadequate number of iterations that will result in the failure of the iterative grid search routine in finding the smallest validation error. Conversely, setting the criteria value too low might result in a number of useless iterations that will result in overfitting, instability and bad generalization to the neural network model.

In network training, involves an iterative process of repeatedly passing through the network to calculate the appropriate weights of each input variable in each one of the hidden layers, then calculating the predicted values and the associated error values from the target layer with the weight estimates, then adjusted accordingly back through the network layers. Therefore, convergence might mean that from one pass to another that the weights do not change very much or that they change less than the convergence criteria from one iteration to the next. Furthermore, it is also important to point out that not all of the convergence criteria are good for all of the optimization methods that are applied.

The abbreviations represent the option settings in specifying the convergence criteria value to apply in neural network training based on the NLOPTIONS option statement from the PROC NEURAL network procedure.

The following are the various convergence criteria values to select from the **NLP** tab:

- **Absolute convergence (ABSCONV):** Specifies the absolute convergence criterion for the objective function that requires a small change in the objective function value in successive iterations. The default convergence criterion value is extremely small.

  $| f(x^k) | \leq c$ for some objective function f at the kth iteration

- **Absolute function convergence (ABSFCONV):** Instead of looking at just the absolute value of the objective function, you might want to look at the absolute difference in the objective function value between successive iterations, with convergence assumed when it falls below the specified convergence criterion value. The default convergence criterion value is zero.

  $|f(x^{k-1}) - f(x^k) | \leq c$ for some objective function f at the kth iteration

- **Absolute gradient convergence (ABSGCONV):** It looks at all of the gradient objective function values for the current iteration, takes the maximum of those values, and considers convergence to occur when the maximum gradient objective function value falls below the specified convergence criterion value. The default convergence criterion value is .00001.

  $\max_j | g_j^{(k)} | \leq c$ for some gradient function g at the kth iteration

- **Absolute parameter convergence (ABSXCONV):** It looks at the difference in the parameter estimates between successive iterations, takes the maximum of those values and considers convergence to occur when the maximum difference in the parameter estimate falls below the specified convergence criterion value. The default convergence criterion value is zero.

  $| w^{k-1} - w^k | \leq c$ for some weight estimate w at the kth iteration

- **Relative function convergence (FCONV):** Rather than looking at the absolute difference in the objective function values between successive iterations, you might want to look at a ratio of the absolute difference in the objective function between successive iterations and the absolute value of the current objective function value. Convergence is considered to occur when the error function falls below the specified convergence criterion value. This is rarely a good tool to use for convergence identification because it is most likely only going to occur at a local solution when the gradient values are still large. The default convergence criterion value is .0001.

  $\frac{|f(x^{k-1}) - f(x^k)|}{\mathrm{MAX}(|f(x^{k-1})|, c)} \leq c$ for some objective function f at the kth iteration

- **Relative gradient convergence (GCONV):** This convergence criterion value looks at the ratio of a scaled magnitude of the gradient with the Hessian over the objective function value. This criteria looks at the reduction of the normalized predicted function. Convergence is considered to occur when it falls below the specified convergence criterion value. The default criterion value is extremely small.

  $\frac{(g^k)' [H^k]^{-1} (g^k)}{\mathrm{MAX}(|f(x^k)|, c)} \leq c$ for some Hessian matrix H and some gradient function g at the kth iteration

- **Number of times:** The tab has an additional constraint in terminating the iterative process by specifying the number of times the convergence criterion value is met in successive iterations in which the convergence criterion value is met, assuring you that the iterative minimization process has reached the desirable minimum in the objective function.

Adjustments tab

The **Adjustments** tab is designed for you to adjust the previously listed convergence criteria values for each one of the objective functions that is applied.

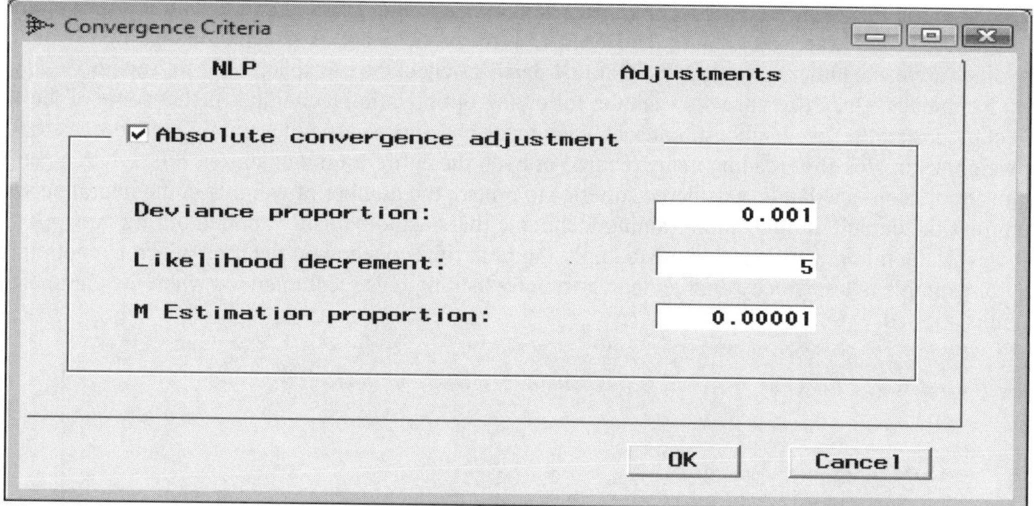

The ***Adjustments*** *tab is used to specify the adjustments to the convergence criteria values.*

If you have the feeling that the iterative procedure is terminating too early in reaching the desirable minimum of the objective function, then the various convergence criterion values may be adjusted and relaxed. For example, when you believe the gradient values from the objective function are still too high even though the objective error function met the absolute convergence criteria, then you might want to set the absolute convergence value to a smaller value. The reason for adjusting the convergence criterion values is to prevent early termination during network training in failing to meet the smallest validation error, which will result in an unacceptable minimum error and an undesirable linear combination of weight estimates. Adjustment to the convergence criterion values might be applied in order for the **Neural Network** node to perform more iterations to assure you that convergence has been achieved in the iterative grid search routine.

Select the **Absolute convergence adjustment** check box to set the following adjustments to the convergence criterion values that is associated with the previously specified objective function. After configuring the convergence criteria options, click the **OK** button to return to the **Optimization** subtab:

- **Deviance proportion:** The proportional change in the deviance statistic between successive iterations falls below the specified criterion rate. For interval-valued target variables, the deviance proportion is the proportional change between the sum-of-squares errors between successive iterations. The adjustment in the deviance proportion is applied when the deviance objective function can be used. The deviance objective function can be used when the distribution in the residual values from the neural network model is a part of the exponential family.
- **Likelihood decrement:** The difference in the log likelihood functions between the saturated model and the reduced model. The saturated model consists of one weight estimate for each observation in the training data set and the reduced model is the neural network model that you are trying to fit. The likelihood decrement is applied when the deviance objective function cannot be used. Therefore, the adjustment to the maximum likelihood objective function is applied.
- **M Estimation proportion:** The proportional change in the M Estimation statistic between successive iterations falls below the specified criterion rate. The M Estimation statistic is applied in robust estimation modeling when it is assumed that there might exist several outliers or extreme values in the target variable or if the distribution in which the sample was drawn is not precisely known.

Train subtab

The **Train** subtab will allow you to specify the type of optimization technique to apply during network training. In addition, the tab will allow you to set the learning rate and the various tuning constants that are required in some of the following training techniques. The purpose of the **Train** subtab is to optimize network training in your data or hardware. By default, the **Default Settings** check box is selected, indicating to you that the default values are used. Otherwise, clear the **Default Settings** check box to specify the following optimization techniques. One difference between the following optimization techniques is that some of the numerical techniques adjust the weight estimates by each record as opposed to other training techniques that adjusts the weight estimates after reading many records or even the entire training data set. Since there is no single optimization technique that is considered superior to others, the number of weights in the neural network model determines the default minimization training technique that is automatically applied during network training. However, the minimization technique to apply might at times depend on the type of error function that is selected. For example, a less greedy double dogleg training technique is recommended when the gamma error function is applied.

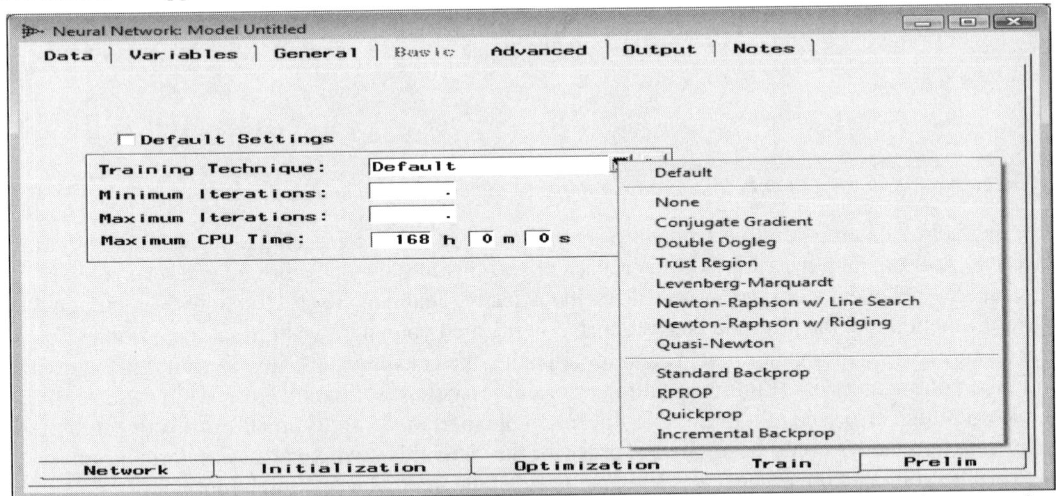

*The **Train** subtab is used to specify the optimization method and configuration settings for network training.*

The following are some of the optimization techniques to select from the **Training Technique** display field:

- **Default:** The default method that is used when the values and the number of weights of the neural network model determine the type of optimization technique that is applied.
- **None:** No optimization technique is applied.
- **Conjugate Gradient:** This is the default method that is used for large-sized problems. This technique is recommended for network architectures with several weight estimates in the neural network model of at least 500 weights. This method does not require the Hessian matrix that will result in slower convergence. This is the standard algorithm used in the early stopping regularization technique due to slow convergence in the iterative technique.
- **Double Dogleg:** This method performs a combination of the quasi-Newton and the trust-region training techniques and works best for medium-sized problems. This method is typically used in early stopping regularization since this technique performs slow but efficient steps.
- **Trust Region:** This method is usually applied to small-to-medium-sized problems. It updates the weight estimates within a restricted bound or region by performing a quadratic approximation about a restricted hyperelliptic trust region.
- **Levenberg–Marquardt:** This is the default method that is used for small-sized problems. This technique is recommended for smooth least-squares objective functions and neural network architectures with a small number of weights (no more than 100). The algorithm uses the Hessian matrix that will result in faster convergence. Since, it is generally recommended to limit the number of weights to prevent overfitting in the neural network model, this is typically the iterative algorithm that is usually applied. One reason this method is the default training technique that is used is because it works well even if the input

variables are highly correlated with each other when the algorithm adjusts the initial Hessian matrix if it is not positive definite in order to avoid bad local minimums in the error function.

- **Newton – Raphson w/ Line Search:** Performs the Newton – Raphson method assuming that the Hessian matrix is positive definite. Otherwise, a succession of line searches are performed using a quadratic approximation in finding the minimum to the error function. If the Hessian matrix is not positive definite, then several identity matrices are added to the Hessian matrix until it is positive definite.
- **Newton – Raphson w/ Ridging:** Performs the similar Newton step as the previous **Newton – Raphson w/ Line Search** technique. This technique uses a succession of the standard Newton steps when the Hessian matrix is positive definite, which will converge successfully at each step. However, if the Hessian matrix is computationally intensive, then the following quasi-Newton or the conjugate gradient methods are recommended.
- **Quasi-Newton:** This is the default method that is used for medium-sized problems. This technique is recommended for network architectures with a moderate number of weights in the neural network model of between 100 to 500. This minimization technique applies an identity Hessian matrix, then calculates an approximation to the Hessian matrix at each subsequent step.

The following "PROP" optimization techniques are based on a specified learning constant. However, SAS does not recommend using the following training techniques due to slow convergence, memory limitations and the various configuration settings that need to be accurately specified. In addition, the following training techniques require constant training to the neural network model that needs to be performed several times in order to assure convergence.

- **Standard Backprop:** This is the most popular training technique that is applied. However, this method is slow, unreliable, and requires several adjustments to the learning rate by sequentially fitting the network model any number of times, which can be extremely tedious and time-consuming. The *learning rate* controls the size of the steps that are performed in the iterative grid search procedure in locating the minimum error. The other tuning constant is called the momentum. The *momentum* forces to make the grid search keep going in the same direction instead of turning to a new steepest direction after every step. A higher momentum will make it less likely to getting stuck at a local minimum, but it will also make it more difficult for the grid search to stop once the minimum is reached. Selecting this minimization technique will result in the following **Backprop Options** window appearing for you to enter the desired tuning constants, that is, the learning rate and momentum, to the algorithm. The default is a learning rate set at .1 and a momentum of zero. This technique is generally not recommended since it depends on accurately specified tuning constants in order to assure convergence in the iterative grid search procedure.
- **RPROP:** This training technique uses different learning rates for each weight and is more stable than the other three "prop" techniques since the algorithm does not depend on an exact learning rate at the initial step. This method applies an acceleration and deceleration to the learning rate in order to control the rate of convergence at each step of the weight updates. The method usually does not require constant updates to the learning rate since the algorithm sets a predetermined bound about the learning rate to achieve convergence. This method can be used in the early stopping regularization technique due to slow convergence. The following **Rprop Options** window will appear when you select this technique so you can specify the minimum and maximum bound to the learning rate in assuring converge and the acceleration and deceleration values that adjust the learning rate to control the acceleration at each step of the weight updates in the iterative grid search procedure.
- **Quickprop:** Iteration is fast but requires more iterations than the conjugate gradient technique. Quickprop applies both a line search and the gradient descent technique. The basic algorithm of this routine uses a small number of weights to fit the multidimensional parabola, then approaches directly to the minimum of the parabola. The major drawback to this method is that convergence depends on accurately specified values for the tuning constants and weight decay parameter. However, this method is faster and more reliable than the standard back-prop algorithms, with fewer adjustments required. The following **Quickprop Options** window will appear by selecting this convergence technique. From the window, you can specify the learning constant, weight decay, maximum momentum, and the tilt. These parameters will usually need constant adjustments during network training by sequentially fitting the network model any number of times. The default settings is a learning rate of .1, a weight decay of .0001, a tilt of 0 that is used to modify the gradient, and a maximum momentum of 1.75 that is needed to avoid problems with nearly flat regions in the error function.

- **Incremental Backprop:** This method updates the weight estimates after each record with trial and error adjustments applied to the learning rate as opposed to the standard backprop training technique where the weight estimates are updated after reading many records or perhaps the entire training data set. However, this training technique is not a convergence algorithm and if the estimates are near a local minimum, then you should select the other training techniques to assure convergence.

From the **Train** subtab, the minimum and maximum number of iterations and the maximum amount of computational time for network training can be specified as follows:

- **Minimum Iterations:** Minimum number of iterations for the neural network training technique. The default is zero.
- **Maximum Iterations:** Maximum number of iterations for the neural network iterative procedure. The default is missing, indicating to you that maximum number of iterations is set at run time based on the amount of data in the training data set and the complexity of the neural network architecture.
- **Maximum CPU Time:** Maximum amount of computational time for the iterative process that is designed to avoid lengthy computational network training runs. The default is seven days.

Backprop Options window

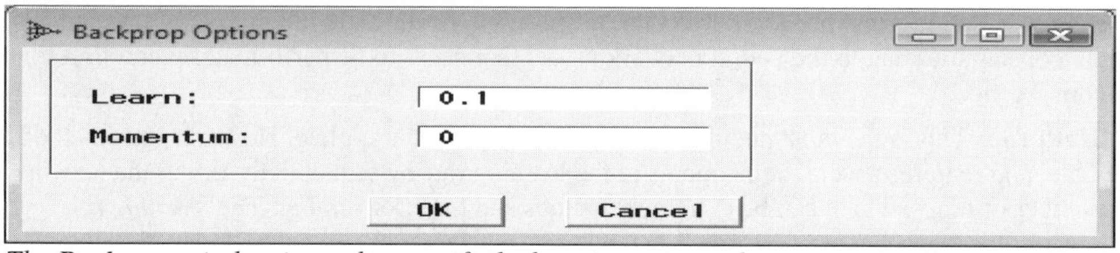

Backprop Options

Learn: 0.1
Momentum: 0

OK Cancel

*The **Backprop** window is used to specify the learning rate, and momentum tuning constants.*

Rprop Options window

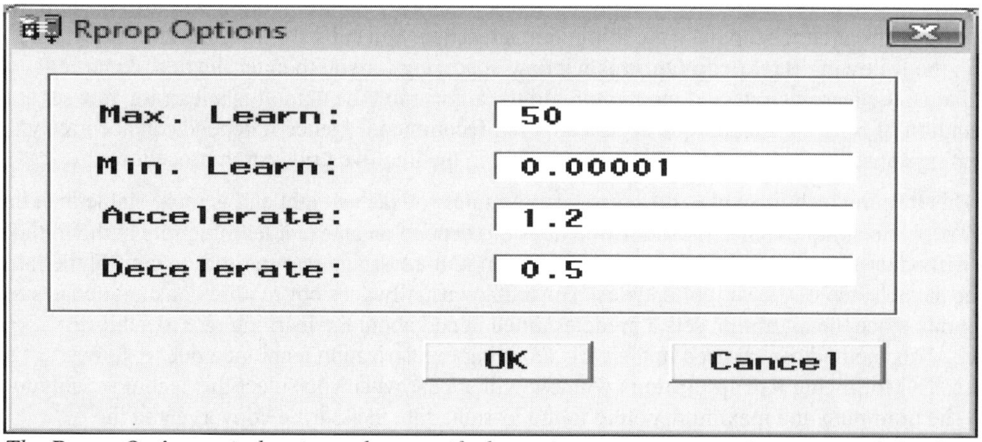

Rprop Options

Max. Learn: 50
Min. Learn: 0.00001
Accelerate: 1.2
Decelerate: 0.5

OK Cancel

*The **Rprop Options** window is used to specify the tuning constants, acceleration, and deceleration rates.*

Quickprop Options window

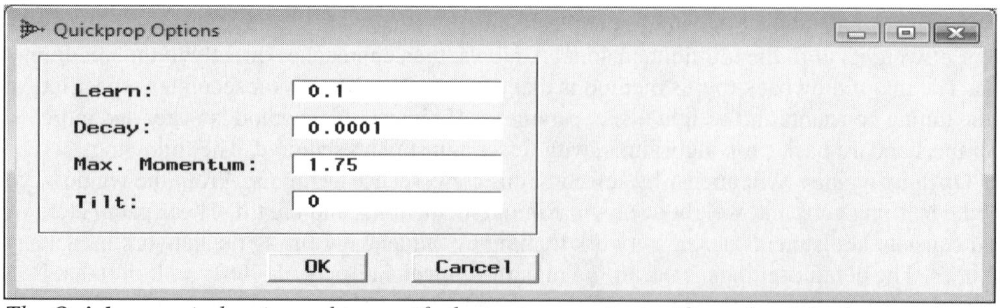

Quickprop Options

Learn: 0.1
Decay: 0.0001
Max. Momentum: 1.75
Tilt: 0

OK Cancel

*The **Quickprop** window is used to specify the tuning constants, tilt, and weight decay estimates.*

Prelim subtab

The **Preliminary** subtab controls the configuration settings for the preliminary training run. By default, preliminary training runs are not performed, unless otherwise specified from the **Optimization** subtab.

By default, the **Default Setting** check box is selected. However, clearing the **Default Setting** check box will allow you to specify the various configuration settings for preliminary training. The option settings are similar to the previous **Train** tab settings with regard to the preliminary training run from the training data set. From the tab, you may specify the optimization technique, minimum and maximum number of iterations, and the maximum amount of CPU time for the preliminary training run. This subtab has the added option for you to specify the number of preliminary training runs. However, it is important to point out again that preliminary training will not be performed unless you specify preliminary training runs from the **Optimization Step** field display within the **Optimization** subtab.

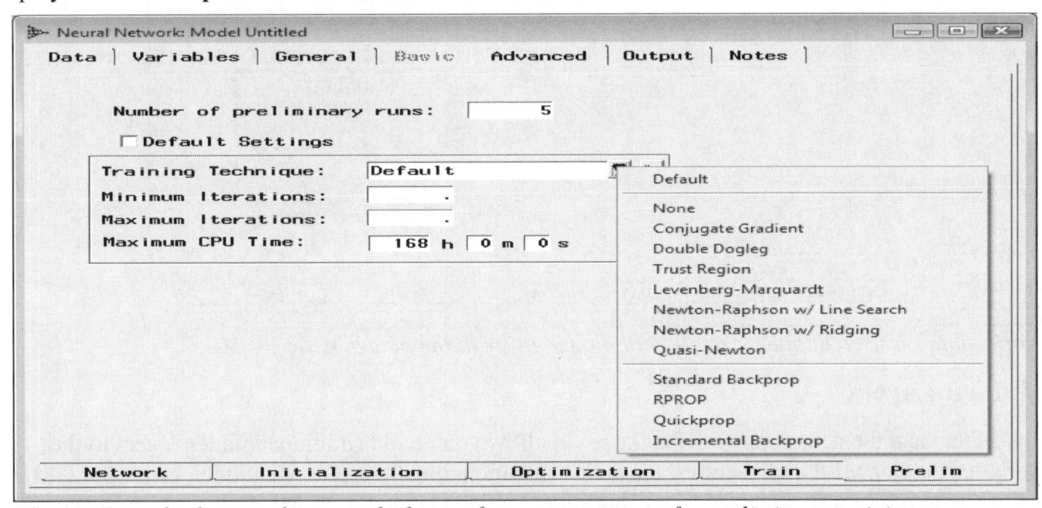

*The **Prelim** subtab is used to specify the configuration settings for preliminary training.*

Preliminary training is performed by fitting the neural network model to the training data set. The initial weight estimates that are used from the preliminary training run are based on the preliminary training run that produced the smallest modeling assessment statistic, based on the **Model selection criteria** option setting, from the training data set among the separate preliminary runs performed. That is, the preliminary training performs a number of independent preliminary training runs, specified from the **Number of preliminary runs** option setting, by fitting the training data set and selects the best set of weight estimates that produces the smallest training error. It's important that these starting values are an adequate approximation to the final parameter estimates to assure convergence to the correct parameter estimates in the iterative process and avoiding bad local minimums in the highly nonlinear error function. In other words, the importance of preliminary training is to achieve starting values that are a close approximation to the correct parameter estimates to accelerate convergence, decrease computational time in the iterative process and more importantly increase the prediction or classification performance. In Enterprise Miner, the default number of preliminary runs that are automatically performed depends on the complexity of the model, assuming that preliminary training is specified from the **Optimization** subtab. Increasing the number of preliminary iterations increases the chance that the correct starting values are applied to the weight estimates in the subsequent neural network training run, which is extremely important in many of the training techniques. Furthermore, preliminary training is highly recommended when fitting extremely nonlinear error function that you want to minimize.

Constructing a well-designed neural network model, involves fitting several different models to the same data. However, if the neural network model is still unsatisfactory, then preliminary runs are recommended in determining the most appropriate initial weight estimates for the neural network model. It should be noted that the default number of preliminary runs of ten were performed in the following modeling comparison procedure that is displayed in the **Assessment** node which resulted in a significant increase in the accuracy of the classification estimates from the neural network model.

Customizing the Neural Network Input, Hidden, and Output Layer Nodes

We now return to the **Network** subtab to explain the various configuration settings for the corresponding neural network layered nodes by fitting the interval-valued variable, DEBTINC, from the HMEQ data set.

Network subtab

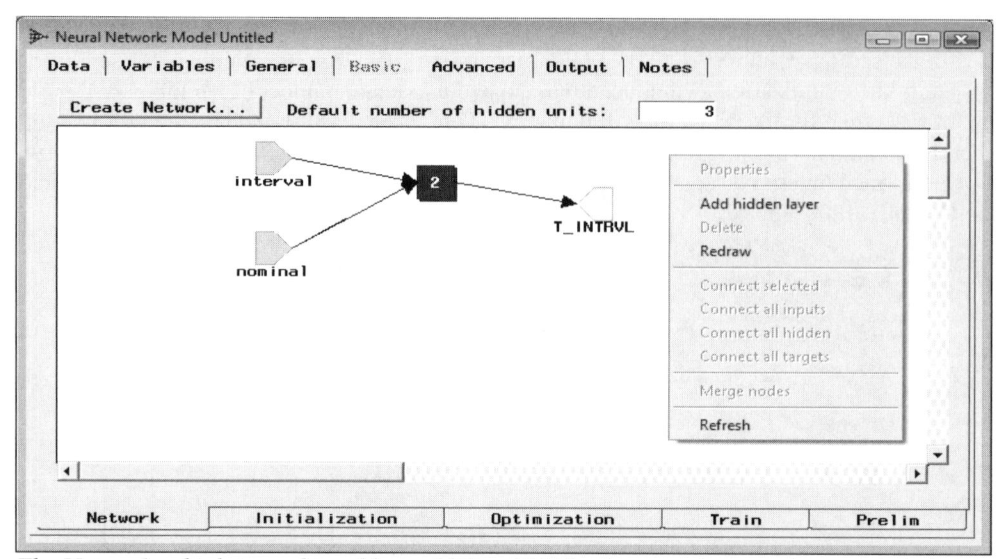

*The **Network** subtab is used to add an additional hidden layer to the neural network design.*

Adding Hidden Layers

The **Network** subtab and the neural network diagram will allow you to add additional hidden layers to the neural network design. For MLP architectures, the node will automatically create one hidden layer. For GLIM architectures, no hidden layers are created. To add additional hidden layers to the MLP network architecture, place the cursor anywhere in the open area of the neural network diagram workspace, then right-click the mouse and select the **Add hidden layer** pop-up menu item that is illustrated in the previous diagram. This will create an additional hidden layer node icon. The hidden layer node will automatically appear between both the input and output layer nodes within the network diagram. The next step is that you must provide connections to and from the additional hidden layer node. The process of connecting the input nodes to the currently added hidden layer node or the same hidden layer node to the output layer node within the neural network diagram is similar to connecting the Enterprise Miner nodes within the process flow diagram. From the **Default number of hidden units** field entry, you may enter the number of units that will automatically be assigned to the hidden layer node that is created within the network design each time. The default is three hidden units.

Two-Stage Modeling from the Network Diagram

Neural network modeling will allow you to fit several target variables. Therefore, two-stage modeling may be performed in predicting the expected values of a binary-valued and an interval-valued target variable in succession. This can be done by constructing two separate neural network designs by creating two separate hidden layer nodes, then connecting the hidden layer node from the binary-valued target model to the hidden layer node from the interval-valued target model. The first step is to select an existing input layer node within the network diagram. The next step is to open the node and select the appropriate input variables from the **Variables** tab that best predicts the target variable and then select the **Single new node** option setting. This will instruct Enterprise Miner to automatically create an entirely separate input layer node, with the selected input variables, that will appear in the network diagram. And, finally, add an additional hidden layer node to the network diagram by selecting the **Add hidden layer** pop-up menu to construct the additional neural network design.

It is also worth noting that a direct connection may be applied to the network design from the **Network** subtab by connecting the input layer node icon to the target layer node icon.

Select the **Create Network** button and the following **Set Network Architecture** window will appear.

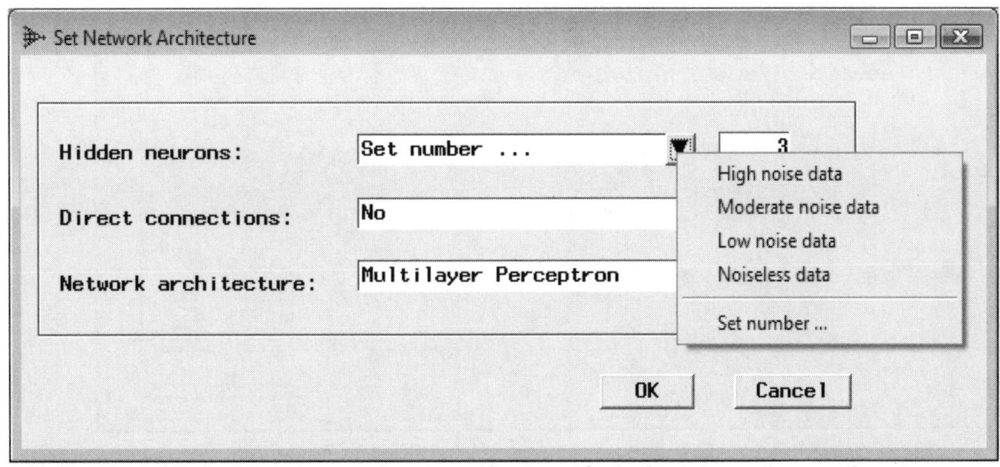

*The **Set Network Architecture** window is used to specify the basic neural network settings.*

The **Set Network Architecture** window is designed to set the number of hidden units, direct connections and the neural network architecture. These configuration settings are automatically applied and reconfigure the entire neural network design and any additional hidden layer nodes that are added to the existing neural network design. The default is three hidden layer units with no direct connections that are applied to the MLP neural network design. These same neural network option settings have been previously explained in the basic user interface.

From the network diagram, select the light green **Interval** icon or the icon located to the left of the network diagram, double click the mouse or right-click the mouse and select **Properties** from the pop-up menu items that will result in the **Node properties – INTERVAL** window appearing.

Input Layer tabs

The input layer tabs are designed to configure the interval-valued input variables for the neural network model. The tab will allow you to specify the various option settings of the input units, such as the type of standardization method to apply and select the desirable input units in the neural network model to transfer into a single input unit, multiple input units, existing input units, or remove the input units altogether from the neural network model.

The input layer of the neural network design can potentially amount to an enormous number of units. Therefore, the Enterprise Miner **Neural Network** node will automatically create a separate input layer node icon within the neural network diagram based on the level of measurement of the input variables. In other words, Enterprise Miner will automatically combine all input variables in the neural network model with the same level of measurement in the neural network diagram. This will allow you to better maintain the various configuration settings such as the standardization methods that should be applied to each one of the interval-valued input variables in the neural network model, or specify the combination of input variables that best predicts each target variable in the neural network model.

General Layout of the Input Layer Node

- **General tab**
- **Input tab**
- **Variables tab**

General tab

The **General** tab will automatically appear when you open the input layer node. The tab will display the type of node and objective function applied to the interval-valued input units.

Node properties - INTERVAL
General | Input | Hidden | Target | Initial | Variables
Node type : INPUT
Objective function: Default
OK

*The **General** tab is used to display the type of node and objective function for the input layer unit.*

Input tab

The purpose of the **Input** tab is to allow you to specify the method of standardization to apply to the interval-valued input nodes. By default, only the interval-valued input variables are standardized to accelerate the convergence process. The input variables are automatically standardized for all the modeling architectures with the exception to the GLIM architecture. Standardizing the input variables will allow you to evaluate the importance of the input variables in the neural network model that can be determined by the magnitude of the weight estimates, since the input variables are adjusted to a mean of zero and a sample variance of one. In other words, the input variables are adjusted to the same range of values.

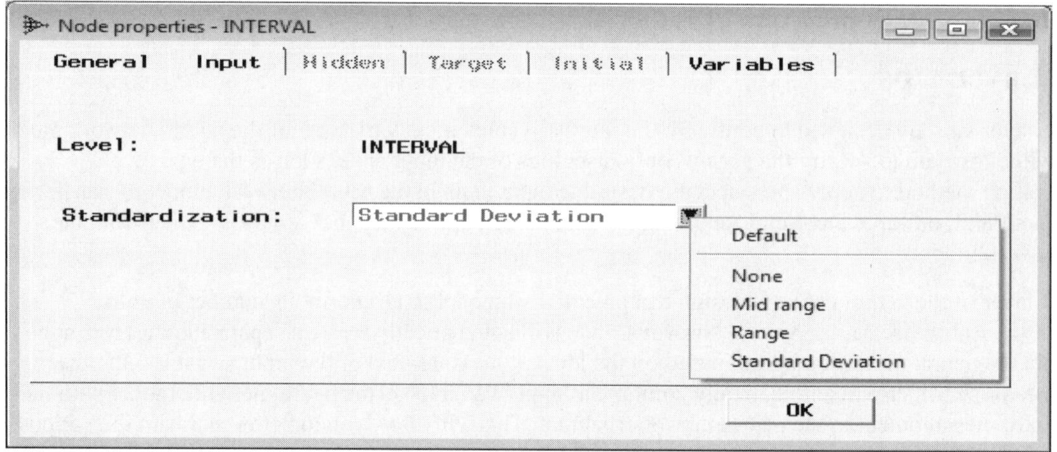

*The **Input** tab is used to specify the standardization method for the input variables in the network model.*

The following are the standardized methods to select from the **Standardized** display field:

- **None:** Selecting this option will prevent standardizing the input variables and is the default for categorically-valued input variables in the neural network model or GLIM architectures.
- **Midrange:** Standardizes the interval-valued input variables by subtracting the midrange value, then dividing by the midpoint range of the values. The midrange value is calculated by adding the minimum and maximum value and dividing by two, then dividing by half the range, resulting in the range of values with a minimum of –1 and a maximum of +1. This standardization technique is usually applied to the tanh, arctangent, Elliott, sine, and cosine activation functions.
- **Range:** Standardizes the interval-valued input variable by subtracting its minimum value, then dividing by the range that transform its values into a [0, 1] interval. This technique is usually applied with a nonnegative target variable with values consisting of rates or proportions resulting in a minimum of zero and a maximum of one. Even though this method is available within the **Input** tab, SAS does not recommend using this standardization method for the input variables in the neural network model.

- **Standard Deviation:** This is the default standardization method that can only be applied to the interval-valued input variables. The standard deviation method subtracts the mean and divides by the standard deviation, which will result in its values being converted into z-scores. That is, this standardization method will transform the input variable to a standard normal distribution with a mean of zero and a variance of one.

Variables tab

The **Variables** tab is designed to select the input variables in the neural network model to transfer into a single input unit, multiple input units, existing input units, or remove the input units from the neural network model. This tab is designed for you to create a separate node of the neural network workspace for each selected input variable that will allow you to specify entirely different configuration settings such as the standardization method for the separate input layer nodes. Furthermore, the node will allow you to reset the separate nodes back into a single node, that is, select the separate network nodes from the neural network workspace, then right-click the mouse and select the **Merge nodes** pop-up menu option.

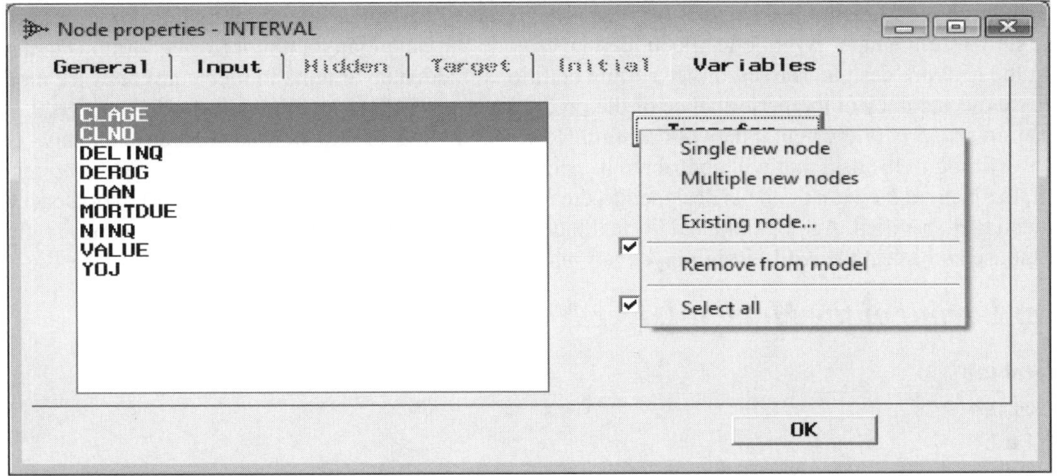

*The **Variables** tab is used to select the transfer settings for the selected input units in the network design.*

Select the **Transfer** button and the following options will be displayed:

- **Single new node:** This option creates a new single input node containing all the selected input variables by highlighting the listed input variables from the displayed list box. Since neural network modeling can predict several target variables, therefore this option might be selected in creating a separate input layer node based on several input variables that are selected from the list box that best predicts a separate target variable in the neural network model.
- **Multiple new node:** This option creates a new input node for each input variable that is selected. For example, there might be times when you might want to control the standardization method that is automatically applied to all of the interval-valued input variables grouped together in the neural network diagram workspace. Therefore, select this option in order to create a separate network node in the neural network workspace based on each input variable selected from the list of existing input variables in the neural network model.
- **Existing node:** This option will allow you to move the selected input variables into an existing input node that has the same level of measurement as the corresponding input variable. For example, you may want to move interval-valued input variables into different interval-valued input variable nodes. However, the restriction is that you cannot combine input variables with different levels of measurements.
- **Remove from the model:** This option removes the selected input nodes from the neural network model.
- **Select all:** This option selects all the input variables from the list of input variables in the network model.

Select the **Confirmation** check box and a dialog message box will appear to confirm the selection of the listed options. By default, the check box is automatically selected.

By selecting the **Auto-connect** check box, the number of nodes are changed and automatically connected in the network diagram. By default, the check box is automatically selected.

Select the **OK** button to return to the **Neural Network** window and the neural network workspace.

From the network diagram, select the blue numbered icon or the icon located in the middle of the network diagram that represents the hidden layer node, double click the mouse or right-click the mouse and select **Properties** from the pop-up menu items, which will result in the **Node properties – H1** window appearing.

Hidden Layer tabs

The hidden layer tabs are designed to set the various configuration settings for the hidden layer units. The purpose of this tab is to specify the number of hidden units, the type of activation function and combination function, remove the hidden layer bias term from the neural network model, and initialize the starting values of the hidden layer weight estimates and biases that are set at some small random values.

In Enterprise Miner, one hidden layer node is automatically created for each hidden layer. Typically, a single hidden layer node is sufficient for the neural network design due to the universal approximator property. However, similar to the input layer node, the hidden layer node can potentially amount to an enormous number of units in the network design. The reason for adding additional hidden layer units to the neural network model is to increase the accuracy of the performance of the predictive or classification model due to the universal approximation property of the neural network design. However, adding additional hidden layer units may result in overfitting in the data that will contribute to poor generalization in the model. In order to keep things organized, the **Neural Network** node is designed to create a separate node in the neural network workspace for each hidden layer specified. Again, to add additional hidden layers to the neural network design, simply right-click the mouse and select the **Add hidden layer** pop-up menu option from the neural network diagram.

General Layout of the Hidden Layer Node

- **General tab**
- **Hidden tab**
- **Initial tab**

General tab

The **General** tab will appear as you open the hidden layer node. The tab simply displays the type of node and the objective function applied to the hidden layer unit.

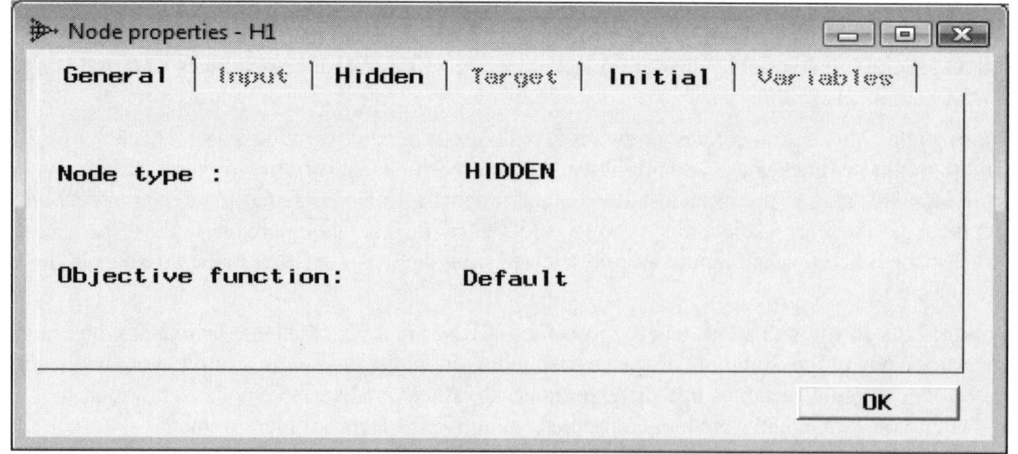

*The **General** tab is used to display the type of node and objective function for the hidden layer unit.*

Hidden tab

The **Hidden** tab is designed to specify the number of hidden layer units, the type of hidden layer activation function and combination function with the option of removing the input-to-hidden layer bias term from the neural network model.

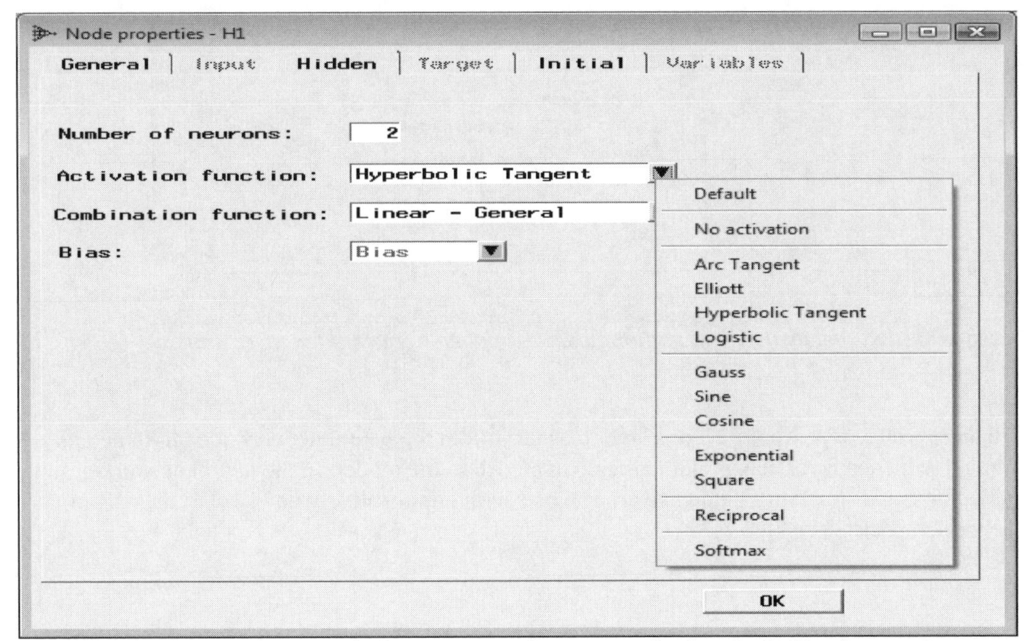

*The **Hidden** tab is used to specify the activation function applied to the hidden layer units.*

Number of neutrons

The entry field enables you to set the number of hidden units for the network. The default is three hidden units. Adding more hidden units to the neural network architecture increases the complexity of the neural network design, which might lead to overfitting. In the sequential network training techniques, such as the sequential network construction and cascade correlation designs, the number of hidden layer units must be set to one.

Activation function

The activation function is introduced into the model to account for the nonlinearity in the neural network model. The purpose of the input-to-hidden layer activation function is to transform the linear combination of weight estimates and input variables in the design. This option will allow you to specify the type of activation function to apply to the input-to-hidden layer. The type of activation or transfer function applied to the input-to-hidden layer depends on the type of combination function that is applied. The default is the hyperbolic tangent activation function that is applied for Linear–General and Linear–Equal Slopes combination functions. The hyperbolic tangent is applied so that the output from the hidden units are close to zero with reasonable values of the hidden layer weights that are centered about zero that will result in good generalization. At times, it might be advantageous to select an entirely different hidden layer activation function in order to increase the precision in the neural network prediction estimates since any activation function may be applied to the hidden layer. However, the type of hidden layer activation function to apply will usually depend on the distribution of the target variable.

Combination function

The combination function determines the way in which the linear combination of weight estimates and input variables are accumulated in the design. The option will allow you to specify the type of combination function to apply to the input-to-hidden layer. The default combination function that is applied depends on the level of measurement of the target variable. For interval and nominal targets, the default is the Linear–General combination function. For ordinal target values, the default is the Linear–Equal Slopes combination function.

*The **Hidden** tab is used to specify the combination function applied to the hidden layer units.*

Bias

This option will allow you to remove the hidden layer bias term from the neural network model. In neural network modeling, the hidden layer bias is automatically included to the model. In the neural network design, there is one hidden layer bias for every hidden layer. The bias term has absolutely no effect to the performance of the neural network model.

Initial tab

The purpose of the **Initial** tab is designed to initialize the starting values for the input-to-hidden layer weight estimates and bias term. By default, the initial input-to-hidden layer weights and biases follow a random normal distribution with a location or mean of zero and the following variance:

σ^2 = scale^2 / (# of connections connected to the hidden layer unit excluding the hidden layer bias and altitudes)

where the scale is set to one assuming that a linear combination function to the hidden layer units is applied. This will result in a small probability of generating large initial weight estimates for the hidden layer weights, that is, the random initial hidden layer weights are adjusted by the square root of the number of connections connected to the hidden unit, assuming a scale estimate of one.

From **Altitude** radio button, the node will allow you to specify the configuration settings to the altitude parameter, whether or not it is applied to the transfer function. The purpose of the altitude parameter is to measure the maximum height of the bell-shaped transfer function. However, this option be can ignored unless a bell-shaped input-to-hidden layer activation, such as the Gaussian activation function, is applied. The Gaussian transfer function is typically applied to the input-to-hidden layer in the RBF designs or fitting MLP neural network designs in order to fit bumpy nonlinear error surfaces.

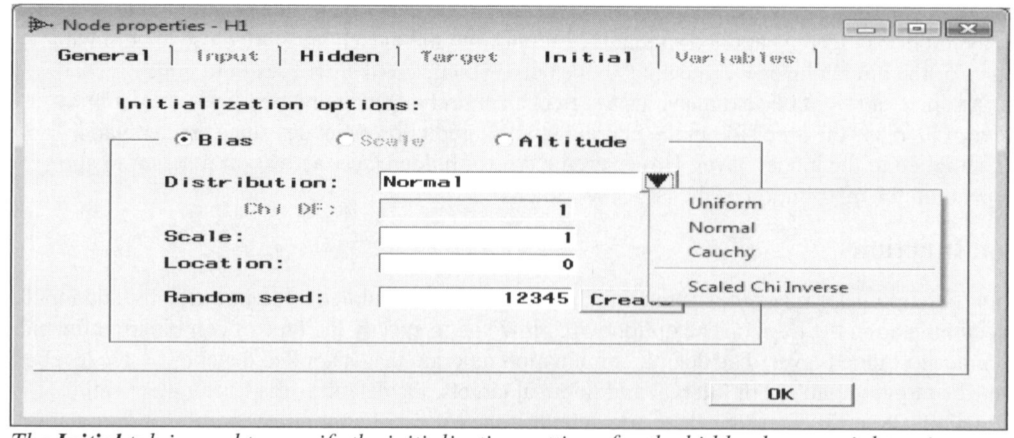

*The **Initial** tab is used to specify the initialization settings for the hidden layer weight estimates.*

Select the **Bias**, **Scale** or **Altitude** radio button to set the distribution, standardization and random starting values to the input-to-hidden layer bias, scale or altitude term separately in the neural network model.

- **Distribution**: Specify the **Uniform**, **Normal**, **Cauchy** or **Scaled Chi Inverse** distribution applied to the initial input-to-hidden layer weight estimates and bias.

  By selecting the **Scaled Chi Inverse** distribution, the **Chi DF** entry field will become available for you to supply the corresponding degrees of freedom for the distribution.
- **Scale**: Specify the scale parameter for the initial weight estimates. The scale parameter controls the variability in the initial weight estimates. The scale parameter is usually set to some small number between zero and one. The choice of the scale value with regard to the corresponding distribution can be important in order to avoid generating large initial weight estimates that will result in bad generalization. The default is a scale of one.
- **Location**: Specify the location parameter for the initial weight estimates. The distribution of the random numbers or the starting values of the input-to-hidden layer weight estimates and bias term are centered about the value of the location parameter. The default is zero.
- **Random seed**: Specify a random seed number to generate a set of random numbers that are used as the starting values for the weight estimates and bias terms that are set to some small randomly generated values. Press the **Create** button to create a different random seed in order to generate a new set of random numbers, thereby creating a new set of starting values for the weight estimates. The default is a random seed number of 12345.

Select the **OK** button to return to the **Neural Network** window and the neural network workspace.

From the network diagram, select the yellow icon or the icon located to the right of the diagram that represents the output layer and double click the mouse or right-click the mouse and select the **Properties** pop-up menu item and the following **Node properties–T_INTRVL** window will appear.

Output Layer tab

The output layer tab is designed to configure the hidden-to-target layer configuration settings such as the type of standardization, activation function, combination function, and error function. In addition, the tab will allow you to remove the bias term, initialize the starting values for the output layer weight estimates and biases that are set at some small random values, and select all the target units in the model to transfer or convert into a single target unit, an existing target unit or remove the target unit from the neural network design.

The neural network design can model several target variables in the same neural network model, analogous to multivariate regression modeling. Therefore, the Enterprise Miner **Neural Network** node is designed to create a separate node in the neural network workspace based on the level of measurement of the target variable that you want to predict. In other words, multivariate regression modeling may be performed within the **Neural Network** node without the existence of the hidden layer unit in the neural network design. However, the restriction is that determining the best fitted model based on the best expected profit or loss can not be performed when fitting multiple target variables from the **Neural Network** node.

From the node, two-stage modeling may also be performed in predicting the fitted values of a binary-valued target variable and an interval-valued target variable in succession. Two separate neural network designs can be constructed with both a binary-valued target variable and an interval-valued target variable to predict. From the network diagram, construct the two separate neural network designs by selecting the best set of input variables that best predicts the target variables and then connect the hidden layer node by fitting the binary-valued target variable to the target layer unit by fitting the interval-valued target variable. One of the restrictions in the network diagram is that the separate target layer nodes can not be connected to each other.

General Layout of the Output Layer Node

- **General tab**
- **Target tab**
- **Initial tab**
- **Variables tab**

General tab

The **General** tab will first appear when you open the target layer node. The tab displays the type of node and the type of objective function that is applied. The tab has the same appearance as the previously displayed tabs. Therefore, the tab will not be illustrated.

Target tab

The **Target** tab will allow you to specify the type of standardization, activation function, combination function, and error function based on the level of measurement of the target variable. Also, you may add or delete the bias term from the hidden-to-target layer.

In order to achieve the best neural network estimates with the various modeling assumptions satisfied, it is important that the specification of the range of the target values that the model is currently fitting be correct along with the correct specification of the error distribution. The error distribution should match the variability, skewness, and the range of values of the target variable. The correct error distribution that is selected depends on the relationship between the target variance and mean. Therefore, the correct error function to select should match the residual values and the fitted values in the neural network model that can be viewed from the residual plot. The **Target** tab will allow you to specify the type of error function from the **Error function** display field along with the type of output activation function to apply from the **Activation function** display field. Also, the type of error function that is applied should match the objective function that can be specified from the **Optimization** subtab. The type of error function and corresponding output activation function to apply depends on the level of measurement of the target variable that is displayed in the introductory section.

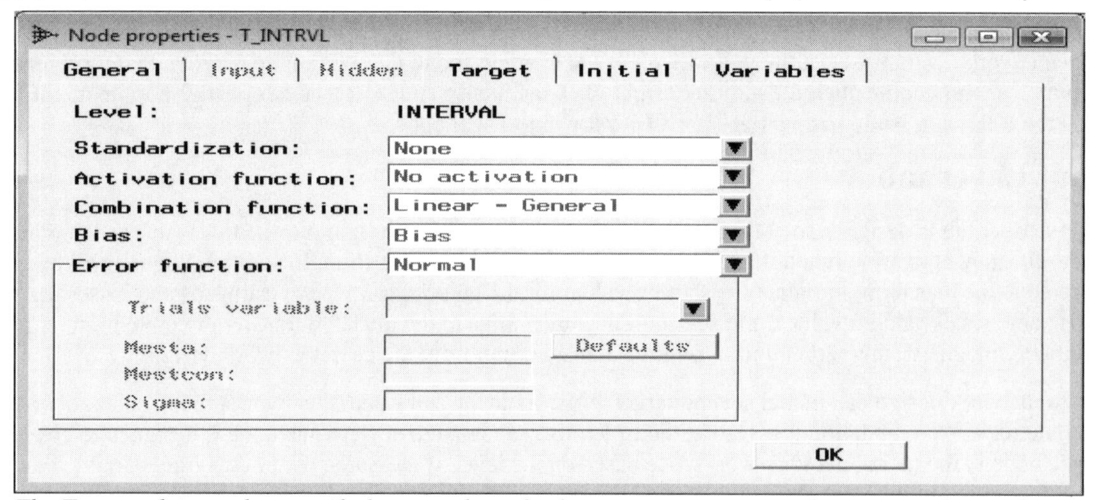

*The **Target** tab is used to specify the type of standardization, activation, combination, and error function.*

Level

The **Level** section displays the level of measurement of the target variable.

Standardization

This option will allow you to specify the type of standardization that is applied to the interval-valued target variable. By default, the target variable is not standardized. The reason is because standardizing the target variables makes the fitted values very difficult to interpret. Yet, standardizing the interval-valued target variable helps to assure convergence to the iterative process. Also, standardization of the input and the target variables helps alleviate ill-conditioning with several input variables or hidden layer units in the design. Again, only interval-valued target variables can be standardized. The purpose of standardization is to treat the target values equally. However, the drawback to standardizing the target variable is that its values must be remapped to their original scale. The reason that the target values need to be retransformed to their original values is because the standardized values are usually not as informative as the original values. The same standardization methods that are applied to the input variables that have been previously displayed within the **Input** tab can be applied to the target variable.

Activation function

This option will allow you to specify the activation function applied to the hidden-to-target layer. The hidden-to-target layer activation function depends on the type of combination function that is applied and the level of measurement of the target variable. The selection of the activation function must conform to the range of the target values, which corresponds to the distribution of the target variable even after standardization is applied. By default, there is no activation function applied for interval-valued targets; a softmax activation function is automatically applied for nominal or binary-valued targets and a logistic activation function is applied for ordinal-valued targets. The reason that a hidden-to-target layer activation function is not applied in the network design for an interval-valued target variable is to simplify the computational complexity of the network design. However, it is recommended, and at times more desirable, to consider other types of output activation functions that are suited to the range or the distribution of the target values in order to improve the prediction results and neural network generalization performance. Note that for nominal-valued targets having either a multiple Bernoulli, multiple entropy or multinomial distribution error function, the only output transfer function that can be applied is the softmax activation function. The softmax activation function is analogous to the multiple logistic function that automatically computes proportional target values. For ordinal-valued targets, you may use only monotonically increasing output transfer functions such as the identity, exponential, logistic, Elliott, tanh, and arctan functions.

The following are the various output activation functions that can be selected from the **Target** tab that are determined by the distribution of the target variable:

- **ARCtangent:** The arctangent output activation function that transforms the values of the target variable into a (–1, 1) interval, $\hat{y} = w_0 + w_1[(2/\pi)\tan^{-1}(x_1)] + w_2[(2/\pi)\tan^{-1}(x_2)]$
- **COSine:** The cosine output activation function that transforms the values of the target variable into a (0, 1) interval, $\hat{y} = w_0 + w_1[\cos(x_1)] + w_2[\cos(x_2)]$
- **ELLiott:** The Elliott output activation function that transforms the values of the target variable into a (–1, 1) interval, $\hat{y} = w_0 + w_1[x_1 / (1 + |x_1|)] + w_2[x_2 / (1 + |x_2|)]$
- **EXPonential:** The exponential output activation function that transforms the values of the target variable into a (0, ∞) interval, that is, $\hat{y} = w_0 + w_1[\exp(x_1)] + w_2[\exp(x_2)]$
- **GAUss:** The Gaussian output activation function that transforms the values of the target variable into an (0, 1) interval, $\hat{y} = w_0 + w_1[\exp(-x_1^2)] + w_2[\exp(-x_2^2)]$
- **IDEntity** (default)**:** The identity output activation function where no transformation is applied to the target variable that is the default output activation function, that is, $\hat{y} = w_0 + w_1(x_1) + w_2(x_2)$. Note that a MLP design with no hidden layers is analogous to a multiple linear regression model.
- **LOGistic:** The logistic output activation function that transforms the values of the target variable into an (0, 1) interval where a MLP design with no hidden layers is analogous to a standard logistic regression model, $\hat{y} = w_0 + w_1[1 / (1+\exp(-x_1))] + w_2[1 / (1+\exp(-x_2))]$
- **MLOGistic:** The multiple logistic output activation function that transforms the values of the target variable into a (0, 1) interval
- **RECiprocal:** The reciprocal output activation function that transforms the values of the target variable into a (0, ∞) interval, $\hat{y} = w_0 + w_1(1/x_1) + w_2(1/x_2)$
- **SINe:** The sine output activation function that transforms the values of the target variable into a (0, 1) interval, $\hat{y} = w_0 + w_1[\sin(x_1)] + w_2[\sin(x_2)]$
- **SOFtmax:** The softmax output activation function that transforms the values of the target variable into a (0, 1) interval, $\hat{y} = w_0 + w_1[\exp(x_1)/\sum\exp(x_i)] + w_2[\exp(x_2)/\sum\exp(x_i)]$
- **SQUare:** The square output activation function that transforms the values of the target variable into an interval between (0, ∞), $\hat{y} = w_0 + w_1(x_1^2) + w_2(x_2^2)$
- **TANh:** The hyperbolic tangent activation function that transforms the values of the target variable into a (–1, 1) interval, $\hat{y} = w_0 + w_1 \tanh(w_{01} + w_{11}x_1) + w_2 \tanh(w_{02} + w_{12}x_2)$
  where $\tanh(x) = [\exp(x) - \exp(-x)] / [\exp(x) + \exp(-x)]$ and w_0 is the bias weight estimate

Combination function

This option will allow you to specify the type of combination function that is applied to the hidden-to-target layer target. For interval or nominal-valued target variables, the default is the General–Linear combination function. For ordinal-valued target variables, the default is the Linear–Equal Slopes combination function. The combination function that is applied depends on the level of measurement of the target variable.

Bias

The **Bias** option setting will allow you to add or remove the hidden-to-target layer bias from the network model. By default, the hidden-to-target layer bias is automatically included into the neural network design where the bias is set to the inverse activation function of the target mean. The hidden-to-target layer bias term is comparable to the intercept term in linear regression modeling. The value of the output layer bias can be greater than the other weight estimates in the model given that the target values are not standardized. However, the prediction or the classification performance of the network model does not depend on the bias term.

Error function

In statistical modeling, you want the error function to be minimized. In other words, the node will allow you to specify other error distributions other than the normal or lognormal. The type of error function to apply depends on the level of measurement and the range of values of the target variable, the number of target variables in the output layer, and the objective function. For continuous targets, the default error function that is applied is the normal error function or the sum-of-squares error. For categorical targets, the default is the multiple Bernoulli error function. The type of error function to apply should match the distribution between the residual values and the fitted values in the neural network model, which can be viewed from the residual plot.

The following are the various error functions that can be specified from the **Error function** option setting:

- **BIWeight:** The Biweight M-estimator is a robust error function that can be applied to any kind of target variable. However, this distribution is often used for interval-valued target variables with several outliers.
- **BERnoulli:** The Bernoulli error function that is the default error function applied to binary-valued target variables that consists of values of either zero or one.
- **BINomial:** The Binomial error function is used for interval-valued target variables that are proportions between the interval of zero and one.
- **CAUchy:** The Cauchy distribution is a robust error function that can be used for any kind of target variables.
- **ENTropy:** The entropy error function is applied to binary-valued target variables. It is identical to the Bernoulli distribution which is also applied to interval-valued target variables between zero and one.
- **GAMma:** The gamma error function is usually applied to target variables that consist of count data with an approximately skewed distribution, where the target variance is approximately equal to the target mean squared. Unlike the Poisson distribution, its skewness is independent of the expected target values that will usually have large average values in the target variable. Note that both the gamma and lognormal distributions are appropriate for interval-valued targets where the residual values are increasing in a fanned-out pattern to the fitted values squared that can be viewed from the residual plot.
- **HUBer:** The Huber's M-estimator for robust regression modeling. This estimate is based on a robust error function for predictive modeling that is unaffected by outliers in the target variable. This method is used to reduce the magnitude of the extreme values in the target values. Robust regression is at least as efficient as least-square regression, assuming normality, and will usually generate better estimates when the distribution of the target variable is concentrated at the tails or contains several outliers or extreme values. However, the robust M-estimators perform best when the distribution of the target values has a symmetric distribution. This distribution should be first applied in obtaining initial weight estimates before applying the other two robust error functions, that is, the biweight or wave distributions. The reason is because the biweight or wave error functions have severe problems of getting stuck in bad local minimums.
- **LOGistic:** The logistic error function where the target values range between the interval of zero and one.
- **MBErnoulli:** The multiple Bernoulli error function is the default for nominal or ordinal-valued target variables. This distribution can be used for single nominal-valued or ordinal-valued target variables or two or more interval-valued target variables in the model.

- **MENtropy:** The multiple entropy error function that is applied to binary-valued targets for polychotomous outcomes of two or more target levels. It is similar to the multiple Bernoulli distribution that can also be used when the set of proportions of the target values sum to one.
- **MULtinomial:** The multinomial error function by fitting two or more interval-valued target variables with nonnegative values that generates proportional outcomes between zero and one.
- **NORmal:** The normal error function that is the default error function applied to interval-valued targets. The normal distribution may also be used when the continuous target values are proportions between the interval of zero and one. The lognormal distribution can be applied to the neural network model by simply computing the log of the target variable and specifying the normal error distribution with an identity link function. The lognormal distribution is a highly skewed distribution with large target mean values.
- **POIsson:** The Poisson error function is usually applied to target variables that consist of count data of rare events over time, where the variance of the target variable is approximately equal to its mean. The distribution is usually highly skewed and should have a fairly small target mean. The target mean should be below 10, preferably below 5, and ideally in the neighborhood of 1. Although, this distribution is typically used for positive integer data, it can also be applied to interval-valued target variables as well, especially when the residuals are increasing in a fanned-out pattern across the fitted values.
- **WAVe:** The Wave M-estimator for robust errors that can be used for any kind of target, but often used for interval-valued target variables with several outliers.

For interval-valued target variables, the following objective functions may be applied to the corresponding error function:

- **Maximum Likelihood:** Cauchy and logistic distributions.
- **Deviance** or **Maximum Likelihood:** Normal, Binomial, Poisson, gamma, Bernoulli, and cross-entropy distributions.
- **M Estimation:** Biweight, Huber and wave distributions.

For nominal or ordinal-valued target variables, the following objective functions may be applied to the corresponding error function:

- **Maximum Likelihood:** Cauchy and logistic distributions.
- **Deviance** or **Maximum Likelihood:** Normal, Poisson, gamma, Bernoulli, cross-entropy, multiple Bernoulli, and multiple entropy distributions.
- **M Estimation:** Biweight, Huber, and wave distributions.

In summary, the following restrictions are applied in specifying the appropriate error function based on the level of measurement of the target variable as follows:

- The multinomial, multiple Bernoulli, and multiple entropy error function cannot be applied for an interval-valued target variable.
- The Binomial and the multinomial error functions cannot be applied for a nominal-valued or ordinal-valued target variable.

Note: Changing some of the error functions from the **Error Function** display field will result in the computer beeping at you with a warning message appearing in the lower left-hand corner of the window alerting you of the appropriate deviance or maximum likelihood function that must be applied to the selected error function.

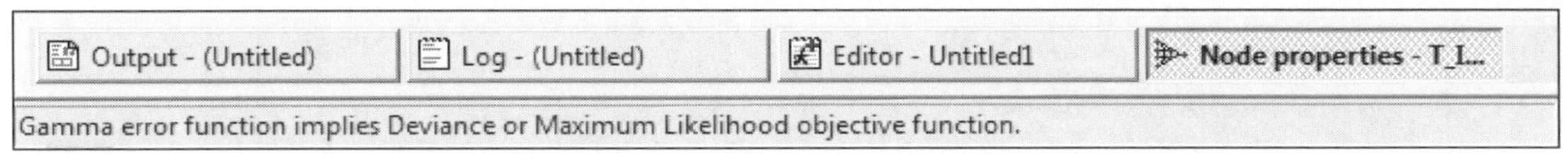

The error message that appears by selecting the Gamma error function from the ***Error Function*** *field.*

The **Trials variable** entry field is used to specify the trials variable for Binomial error functions.

Option Settings for Robust Error Functions

By default, the error function option settings are grayed-out except when you specify either the Biweight, Huber or Wave robust error functions that are used for robust estimation when it is assumed that the target values may contain several outliers. The robust error function to select from should be such that the estimates are robust to efficiency, which will assure you that the estimates are appropriate when repeatedly sampled from a distribution that is not exactly known, while at the same time resistant to a small number of outliers in the data. The **Mesta** specifies the scale constant for M-estimation. The **Mestcon** is the tuning constant for M-estimation. The **Sigma** parameter is the fixed value of the scale parameter with a default value of one.

Mestcon Default Settings **Biweight:** 9 **Huber:** 1.5 **Wave:** 2.1 π

Node properties - T_INTRVL
General | Input | Hidden | Target | Initial | Variables
Level: INTERVAL
Standardization: None
Activation function: No activation
Combination function: Linear - General
Bias: Bias
Error function: Normal
Trials variable:
Mesta:
Mestcon:
Sigma:
Defaults
Default
Cauchy
Logistic
Normal
Binomial
Poisson
Gamma
Bernoulli
Cross Entropy
Multinomial
Multiple Bernoulli
Multiple Entropy
Biweight
Huber
Wave
OK

*The **Target** tab is used to specify the type of error function for the target layer unit.*

Initial tab

The **Initial** tab and the **Variables** tab within the hidden-to-target layer are similar to the hidden layer configuration settings explained earlier. The purpose of the **Initial** tab is designed to initialize the starting values of the hidden-to-target layer weight estimates, bias and altitude. By default, the initial output layer weights are set to zero and the output layer bias is initialized by applying the inverse of the activation function to the mean of the target variable. However, the tab is designed to set the hidden-to-target layer weights and biases at some small random values that follow a normal distribution with a location or mean of zero and a variance of the following:

$\sigma^2 = \text{scale}^2$ / (# of connections to the output layer unit excluding the target layer bias and altitude).

The scale parameter is initialized to the standard deviation of the target variable with a default value of one. This will result in a small chance of generating large initial weight estimates, which is important in generating good generalization. Again, it is worth noting that the altitude may be ignored unless you are applying a Gaussian output activation function.

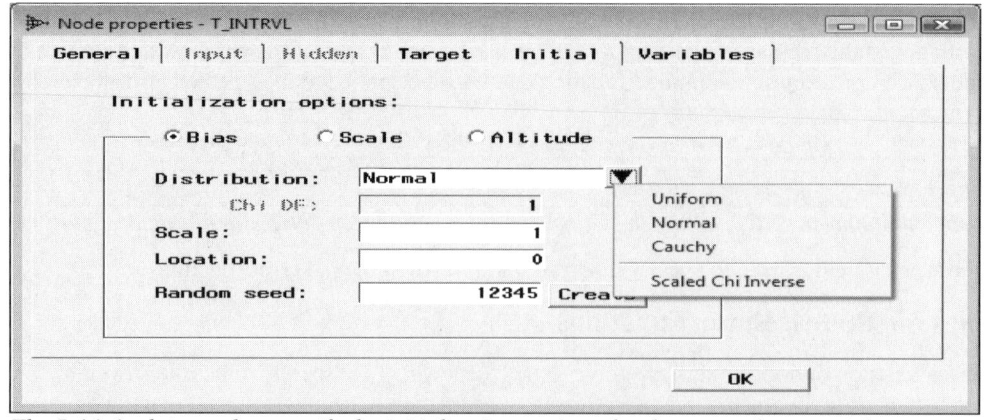

*The **Initial** tab is used to specify the initialization settings for the output layer weight estimates.*

Variables tab

The **Variables** tab is designed to select the target variables in the model to transfer into a single target unit, an existing target unit, or completely remove the target unit from the neural network design. The tab is similar to the input layered node with the difference obviously being in regard to the output layer nodes. Since neural network modeling can also perform multivariate modeling by fitting several target variables in the model, the advantage of the following option is that there could be times that you might want to specify entirely different neural network configuration settings for the separate target variables that you want to predict. However, it should be warned that performing neural network modeling with several target variables in the model is usually not recommended. The reason is because added complexity in the multiple response surfaces will result in the probability of the iterative grid search procedure failing to locate the optimum minimum in the numerous error functions.

Select the **Transfer** button and the following options will be displayed:

- **Single new node:** This option creates a new single target node by simply highlighting the target variable that is listed within the list box.
- **Multiple new node:** This option creates a new target node for each target variable selected from the list box.
- **Existing node:** This option will allow you to move the selected target variable to an existing target node that has the same level of measurement as the target variable.
- **Remove from the model:** This option removes the selected target node from the neural network model.
- **Select all:** This option selects all the target nodes in the neural network model.

Select the **Confirmation** check box and a dialog message box will appear to confirm the selection of the listed options.

Selecting the **Auto-connect** box will result in the number of nodes changing and the nodes being automatically connected in the neural network workspace.

Select the **OK** button to return to the **Neural Network** window and the neural network diagram.

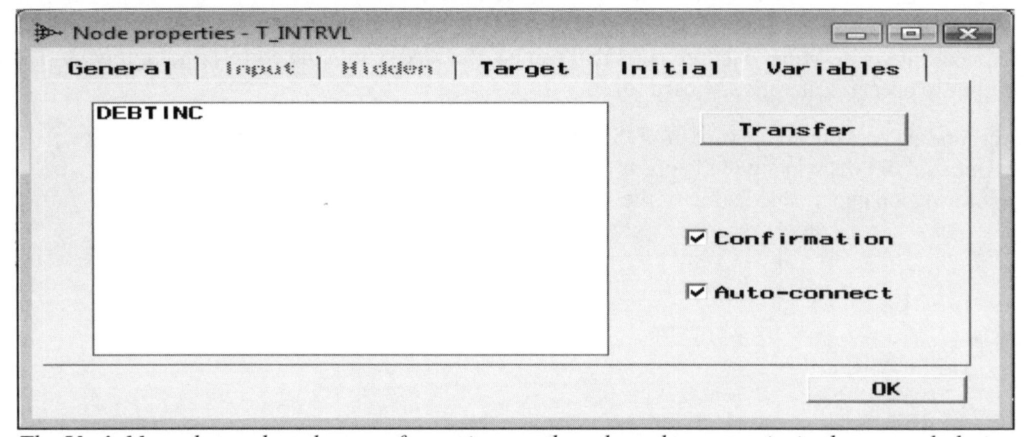

*The **Variables** tab to select the transfer settings to the selected target units in the network design.*

Output tab

- **Data subtab**
- **Variables subtab**

The **Output** tab displays the file administration information and a table listing of the scored training, validation and test data sets. A scored data set can be generated for each one of the partitioned data sets that are created. The scored data set is a permanent output data set that consists of the calculated predicted values generated from the neural network model along with all other variables in the corresponding data set. These same scored data sets are then passed along to the following nodes within the process flow diagram. For instance, the scored data set is passed on to the **Assessment** node for modeling assessment. The scored data sets will not be created until the network model has been trained and the **Neural Network** node has been executed.

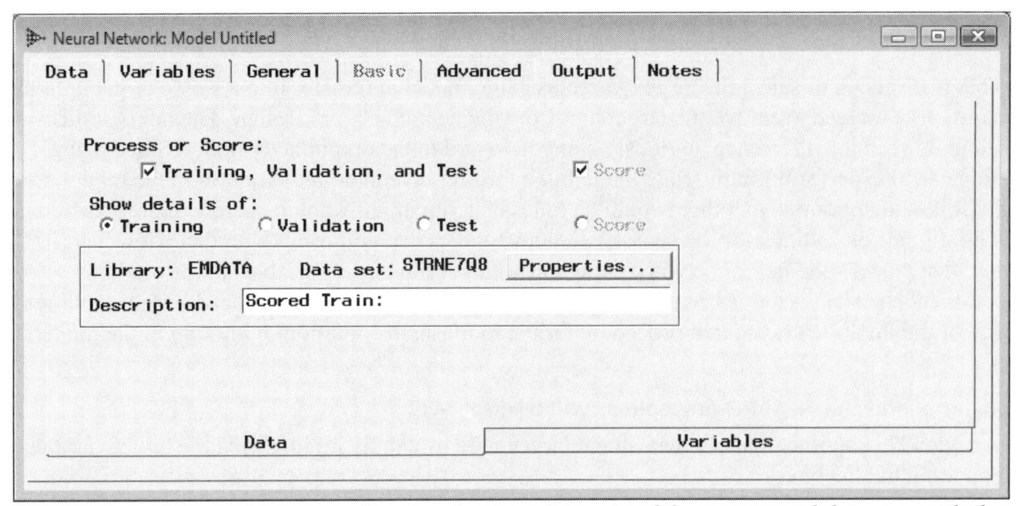

*The **Output** tab is used to create the scored output data sets of the partitioned data sets with the associated predicted values from the neural network model.*

Data subtab

Select the **Process or Score: Training, Validation, and Test** check box to create the scored data sets for each partitioned data set. By default, the **Process or Score: Training, Validation, and Test** check box is not selected in order to save disk space. In addition, Enterprise Miner will also create separate macro variables in the **SAS Code** node that will allow you to access the scored data sets with the associated predicted values, the final weight estimates, and the final model assessment statistics from the neural network model. The data sets can be accessed by the corresponding macro variable within the **SAS Code** node or the currently opened SAS session in order to perform modeling comparison analysis or calculate entirely different neural network prediction estimates from a different set of input values. Alternatively, the scored data set can be imported, saved, and then exported to an external SAS file from the **Score** node that can then be included in the currently opened SAS session. The output data set is automatically created within the EMDATA folder. The tab will display the assigned Enterprise Miner data set name that can be accessed within the corresponding EMDATA folder of the currently opened Enterprise Miner project.

Select the corresponding partitioned data set from the **Show details of** option, then select the **Properties...** button and the **Data set details** window will appear. The window will allow you to view both the file administrative information and a table listing of the selected scored data set with the fitted values.

Data set details

Information Table View

Variable labels

	Predicted: BAD=1	Predicted: BAD=0	Into: BAD	Unnormalized Into: BAD	From: BAD	Warnings	Decision: BAD	Expected Profit: BAD	Best Profit: BAD	Computed Profit: BAD	BAD
1	0.5549010421	0.4450989579	1	1	1		1	0.5549010421	1	1	1
2	0.3985364576	0.6014635424	0	0	1		1	0.3985364576	1	1	1
3	0.5570129486	0.4429870514	1	1	0		1	0.5570129486	0	0	0
4	0.9590499151	0.0409500849	1	1	1		1	0.9590499151	1	1	1
5	0.3985364576	0.6014635424	0	0	1		1	0.3985364576	1	1	1
6	0.6715719756	0.3284280244	1	1	1		1	0.6715719756	1	1	1
7	0.5718832975	0.4281167025	1	1	1		1	0.5718832975	1	1	1
8	0.4499194626	0.5500805374	0	0	0		1	0.4499194626	0	0	0
9	0.6706991383	0.3293008617	1	1	1		1	0.6706991383	1	1	1
10	0.993889713	0.006110287	1	1	1		1	0.993889713	1	1	1
11	0.039204631	0.960795369	0	0	1		1	0.039204631	1	1	1
12	0.6360458046	0.3639541954	1	1	1		1	0.6360458046	1	1	1
13	0.0582740609	0.9417259391	0	0	0		1	0.0582740609	0	0	0
14	0.6720704389	0.3279295611	1	1	1		1	0.6720704389	1	1	1
15	0.6104780798	0.3895219202	1	1	1		1	0.6104780798	1	1	1
16	0.0772476688	0.9227523312	0	0	1		1	0.0772476688	1	1	1
17	0.0515556742	0.9484443258	0	0	0		1	0.0515556742	0	0	0
18	0.4712660064	0.5287339936	0	0	1		1	0.4712660064	1	1	1
19	0.3427415432	0.6572584568	0	0	0		1	0.3427415432	0	0	0
20	0.9999756827	0.0000243173	1	1	1		1	0.9999756827	1	1	1

*The scored data set from the neural network model that was applied to the model comparison procedure in the **Assessment** node by fitting the categorical target variable, BAD, from the HMEQ data set.*

Variables subtab

Select **Standardized inputs**, **Hidden Units**, **Residuals**, and **Errors** check boxes to include the corresponding variables into the scored data set since these variables are automatically excluded from the scored data set. By default, the residuals are not automatically written to the scored data set. In addition, the standardized input variables and hidden layer units are not automatically written to the scored data set. Selecting the **Residuals** option will result in residual values that will be written to the scored data set and selecting the **Error** option will result in the final value of the objective function that will be written to the scored data set. The final value of the objective function is determined by the error function that is selected. For least-squares regression, the final value of the objective function is the squared error for each observation. For generalized linear models or other methods that are designed to minimize the deviance, the error value is the deviance. For other types of maximum likelihood estimation, the error value is the negative log likelihood. In other words, the error value is whatever the neural network model is trying to minimize.

Select the **Residuals** or **Error** check box in order to have both the residual values and the final value of the objective function for each observation included in the scored data sets. The **rejected** and **input** pull-down menu options are designed to set a model role for the standardized input variables or the hidden units to be carried along into the subsequent nodes. Selecting the **rejected** model role will remove the residual values and the final value of the objective function from the subsequent predictive model in the process flow. Conversely, selecting the **input** option will include the residual values and the final value of the objective function into the subsequent predictive model.

The difference in the file structure of the scored data set in comparison to the **Regression** node scored file is that an additional E_ *target name* variable is created that represents the value of the objective function for each observation. Since the interval-valued input variables are automatically standardized in the neural network design, a S_ *input name* variable is created in the scored data set for each standardized input variable in the neural network model. In addition, the scored data set will consist of the predicted values, P_ *target name,* for each interval-valued target variable in the neural network model that you want to predict.

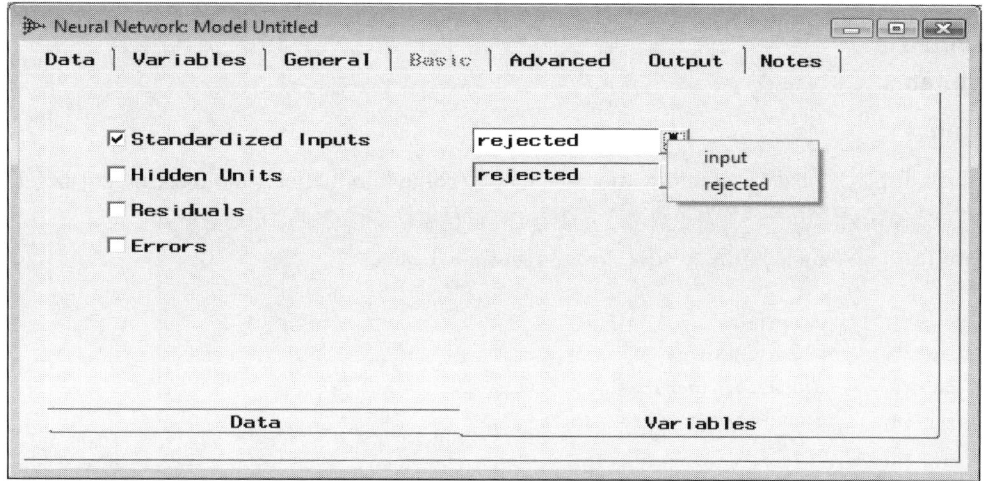

*The **Variables** subtab is used to specify the variables to be created in the scored SAS data sets and setting the model roles for the standardized input variable and hidden layer units.*

Notes tab

The **Notes** tab is designed to enter notes or record any other information in the Enterprise Miner **Neural Network** node.

Viewing the Neural Network Node Results

General Layout of the Results Browser within the Neural Network Node

- **Model tab**
- **Tables tab**
- **Weights tab**
- **Plot tab**
- **Code tab**
- **Output tab**
- **Notes tab**

It is important to understand that the results that are listed from the various Enterprise Miner nodes are generated from the associated data mining procedure. The following results from the **Neural Network** node are created by executing the node that results in the PROC NEURAL network procedure running behind the scenes. However, it is important to remember that not all the options used in the corresponding data mining procedure are incorporated in the listed results within the node, and vice versa.

The following results are from the neural network model that is one of the classification models under assessment in the following model comparison procedure displayed in the **Assessment** node by fitting the binary-valued variable, BAD.

Model tab

The **Model** tab displays the data set information, target profile information, and the current neural network configuration settings that have been previously specified within the node. The tab consists of the following two subtabs.

- **General subtab**
- **Network subtab**

General subtab

The **General** subtab displays the data set information and target profile information from the scored data set.

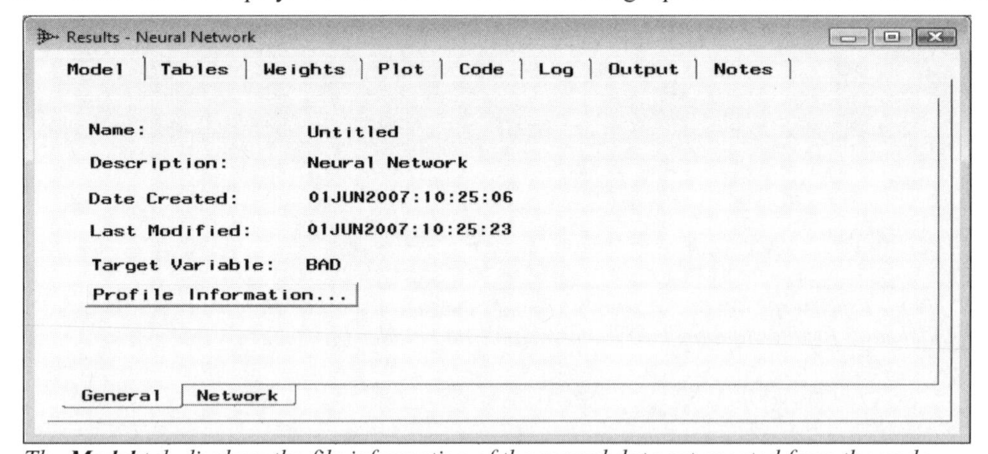

*The **Model** tab displays the file information of the scored data set created from the node.*

Network subtab

The **Network** subtab displays the neural network configuration settings that have been specified for the current neural network model. The tab will display a summary listing of the various configuration settings that have been applied to the input layer, hidden layer, output layer, preliminary training run, optimization technique, objective function, weight decay, and the various convergence criterion option settings. From the following listing, the CLAGE, DEBTINC, DELINQ, and DEROG input variables were selected from the modeling selection procedure that was performed within the **Tree** node. The modeling selection procedure was applied to avoid the "curse of dimensionality" problem by reducing the complexity of the error surface and increasing the chance for the neural network optimization procedure to locate the desirable minimum in the error function and obtain the best possible linear combination of the weight estimates.

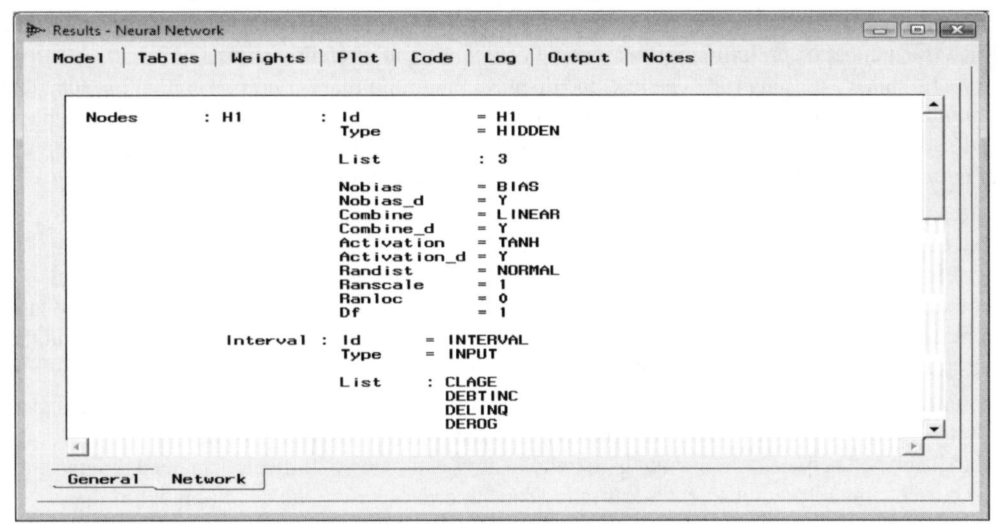

*The **Network** subtab displays a listing of the current neural network configuration settings.*

Tables tab

The **Tables** tab will first appear when you view the neural network results. The tab displays a table listing of the various modeling assessment statistics in alphabetical order for each partitioned data set that is created. The purpose of the tab is not only to evaluate the accuracy of the modeling fit, but it will also allow you to detect overfitting to the neural network model by observing a wide discrepancy in the listed assessment statistics between the partitioned data sets. However, overfitting to the neural network model seems to be of no concern with both partitioned data sets having very comparable estimates with three separate hidden layer units that was automatically selected by the **Neural Network** node in the nonlinear model.

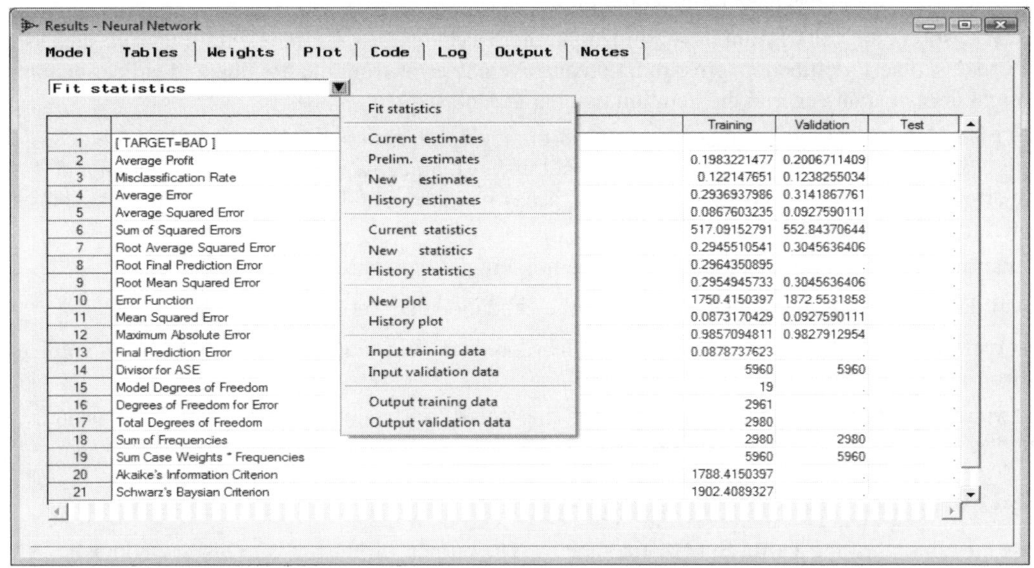

		Training	Validation	Test
1	[TARGET=BAD]	.	.	.
2	Average Profit	0.1983221477	0.2006711409	.
3	Misclassification Rate	0.122147651	0.1238255034	.
4	Average Error	0.2936937986	0.3141867761	.
5	Average Squared Error	0.0867603235	0.0927590111	.
6	Sum of Squared Errors	517.09152791	552.84370644	.
7	Root Average Squared Error	0.2945510541	0.3045636406	.
8	Root Final Prediction Error	0.2964350895	.	.
9	Root Mean Squared Error	0.2954945733	0.3045636406	.
10	Error Function	1750.4150397	1872.5531858	.
11	Mean Squared Error	0.0873170429	0.0927590111	.
12	Maximum Absolute Error	0.9857094811	0.9827912954	.
13	Final Prediction Error	0.0878737623	.	.
14	Divisor for ASE	5960	5960	.
15	Model Degrees of Freedom	19	.	.
16	Degrees of Freedom for Error	2961	.	.
17	Total Degrees of Freedom	2980	.	.
18	Sum of Frequencies	2980	2980	.
19	Sum Case Weights * Frequencies	5960	5960	.
20	Akaike's Information Criterion	1788.4150397	.	.
21	Schwarz's Baysian Criterion	1902.4089327	.	.

*The **Tables** tab displays a table listing of the goodness-of-fit statistics from the partitioned data sets.*

- **Fit Statistics:** Displays a table listing of the various goodness-of-fit statistics such as the average error, AIC, SBC, and mean square error from both the training and validation data sets based on the most recent network training run. Again, if the same fitted statistic displays a large discrepancy between the partitioned data sets, then it is a strong indication of overfitting in the neural network model.

- **Current estimates:** Displays a one-line listing of the current network estimates such as the optimization technique, weight decay parameter, objective function, average error that is by fitting the partitioned data sets, the iteration number, and the final weight estimates with the smallest average validation error from the training data set. These same weight estimates might be used in preliminary training.
- **Preliminary estimates:** Displays a one line listing of each preliminary training run that is performed. This option will list the number of preliminary runs, weight estimates, and modeling assessment statistics at each separate preliminary training run. The last line displays the preliminary training run that produced the smallest training error among all of the preliminary training runs that are performed. The weight estimates are used as the initial weight estimates for the subsequent network training run. A blank listing will appear if there were no preliminary training runs. By default, no preliminary training runs are performed.
- **New estimates:** Displays the new and history estimates that are related to interactive training. Initially, both the **New estimates** and **History estimates** table listing will be the same. By applying interactive training and refitting the model more than once, both the **New estimates** and **History estimates** table listing will be different. The **New estimates** represents the most recent neural network model fit. The table lists the neural network weight updates, optimization technique, weight decay, objective function, and the average error from the training and validation data sets and the iteration number at each step.
- **History estimates:** Displays the history estimates that are based on all of the network training runs that are performed in the interactive session. The variables listed are similar to the **New estimates** listing. However, the difference between the listings is that the history estimates will display an accumulated listing of the current and all previous network training runs in order to view the development of the objective function at each iteration of the iterative grid search procedure. This will allow you to determine if convergence has been achieved in the iterative grid search procedure by observing small changes in the gradient error function between successive iterations.
- **Current statistics:** Displays a one-line listing of the current modeling assessment statistics from the partitioned data sets.
- **New statistics:** Displays a table listing of the most recent network training run that displays the various modeling assessment statistics at each step of the iterative grid search procedure.
- **History statistics:** Displays a table listing of all the iterative sequential network training runs performed in interactive training that accumulates a listing of the various modeling assessment statistics at each step of the iterative grid search procedure. Both the **New statistics** and **History statistics** listings are identical if the neural network model has been trained only once.
- **New plot:** Displays the values from the optimization plot based on the most recent training run, such as the weight updates, objective function, error function and average error from the training and validation data sets, weight decay parameter, and the iteration number at each step.
- **History plot:** Displays a table listing of the values from the optimization plot based on all of the interactive sequential training runs such as the weight updates, objective function, error function and average error from the partitioned data sets, weight decay parameter, and the iteration number at each step of the iterative grid search procedure.
- **Input training data:** Displays a table listing of the training data set that is being fitted.
- **Input validation data:** Displays a table listing of the validation data set that is being fitted.
- **Output training data:** Displays a table listing of the scored training data set with the associated predicted values from the neural network model.
- **Output validation data:** Displays a table listing of the scored validation data set with the associated predicted values from the neural network model.

Weights tab

The **Weights** tab consists of the **Table** and **Graph** subtabs. The purpose of the tab is to display both a table listing and a graphical display of the final neural network weight estimates. The weight estimates are the inner hidden layer, outer hidden layer weights, and bias estimates of both layers from the neural network model.

The **Weights** tab consists of the following two tabs.

- **Table subtab**
- **Graph subtab**

Table subtab

The table listing of weight estimates will allow you to observe the magnitude of the weights in the neural network model. Since the input variables are automatically standardized, the magnitude in absolute value of the standardized weight estimates can be used as a rough guide in determining the importance of the input variables to the neural network model. The weights associate with the input variables that are close to zero will indicate small changes in the input variable which corresponds to small changes in the output.

The following weight estimates are based on the home equity loan data set for predicting the binary-valued variable, BAD. The listed input variables to the neural network model were selected from the variable selection routine that was performed from the **Tree** node. The weight estimate from the target layer bias term of 3.87 is analogous to a small intercept term in the least-squares regression model. Although, it is tempting to assume that the input variable DEBTINC has a large effect on the output variable, BAD, with the largest weight estimate of –70.49. However, the high input-to-hidden weight does not imply that the input variable has a large effect on the output since there also exists the association in the effect of the other hidden-to-target layered weight estimates. The outer layer weight estimates were –3.11, 3.63, and –7.59 since three separate hidden layered units were specified for the neural network model. Since the input variables have been standardized, this will allow you to compare the input-to-hidden layer weight estimates among the various input variables in the neural network model. Therefore, you may perform input pruning. The input variable of the number of delinquent trade lines, DELINQ, seems to be the most likely candidate for pruning since it has the smallest overall average, that is, in absolute value, in the three separate input-to-hidden layer weight estimates in the neural network model in comparison to the average weight estimates for the other input variables. Therefore, the next step would be to refit the neural network model and observe a lower misclassification rate to assure you that the input variable should be removed from the model. Otherwise, the input variable should be reinserted back into the model.

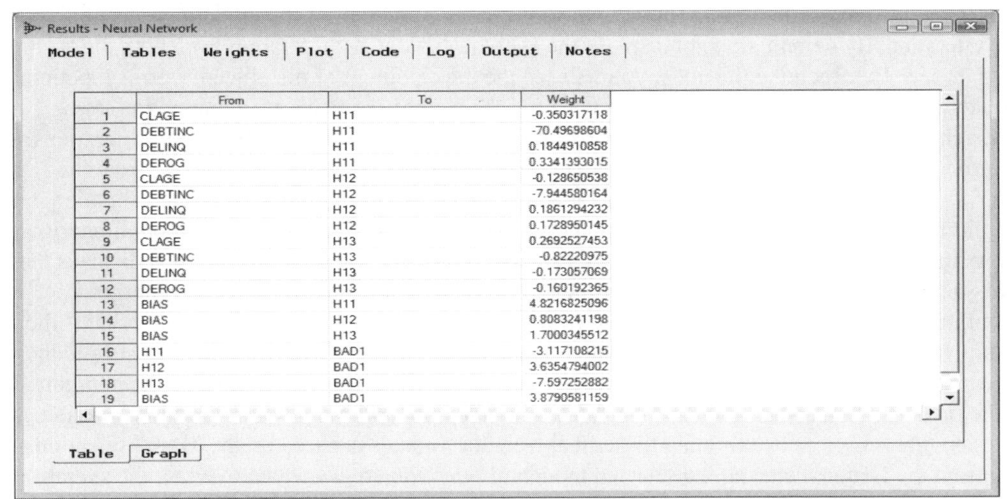

	From	To	Weight
1	CLAGE	H11	-0.350317118
2	DEBTINC	H11	-70.49698604
3	DELINQ	H11	0.1844910858
4	DEROG	H11	0.3341393015
5	CLAGE	H12	-0.128650538
6	DEBTINC	H12	-7.944580164
7	DELINQ	H12	0.1861294232
8	DEROG	H12	0.1728950145
9	CLAGE	H13	0.2692527453
10	DEBTINC	H13	-0.82220975
11	DELINQ	H13	-0.173057069
12	DEROG	H13	-0.160192365
13	BIAS	H11	4.8216825096
14	BIAS	H12	0.8083241198
15	BIAS	H13	1.7000345512
16	H11	BAD1	-3.117108215
17	H12	BAD1	3.6354794002
18	H13	BAD1	-7.597252882
19	BIAS	BAD1	3.8790581159

*The **Weights** tab displays a table view of the scored weight estimates for the neural network model based on the modeling comparison procedure from the **Assessment** node.*

Graph subtab

The listed neural network weights from the **Weight** tab are the final scored weight estimates that are indicated by the vertical line drawn from the optimization plot located in the **Plot** tab. The **Weights** tab can be used in pruning inputs by viewing the input-to-hidden weight estimates that are closest to zero or some threshold value, then removing these same input variables and refitting the neural network model. The following plot that is displayed in the **Graph** subtab is called a *Hinton diagram*. The vertical (FROM) axis represents the inner hidden layer weights and the horizontal (TO) axis represents the outer hidden layer weights. The magnitude of the weight estimate is displayed by the size of the box that is proportional to the absolute value of the weights. The color of the box represents the sign of the weights, with red boxes indicating positive weight estimates and blue boxes indicating negative weight estimates. You may select anywhere in the box and hold down the left-click mouse button to display the text box listing of the outer hidden layer variable label, inner hidden layer variable label, weight estimate, and the sign of the weight estimate. Select the **Scroll data** toolbar icon from the tools bar menu to see the additional weight estimates in the plot that have been truncated off the bottom vertical axis of the plot by dragging the mouse along the vertical axis of the graph with the open hand. The Hinton diagram is illustrated in the following diagram.

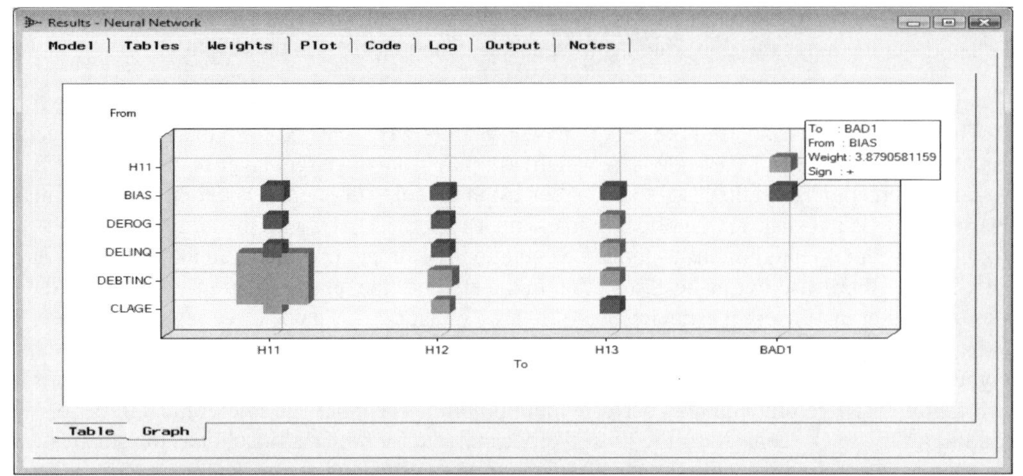

*The **Graph** subtab displays the Hinton diagram of the neural network weight estimates between the four separate hidden layer units in the neural network model.*

Plot tab

The **Plot** tab displays the optimization plot. The *optimization plot* is a line plot of the modeling selection criterion at each iteration. By default, the plot displays the average error at each iteration from both the training and validation data sets. By executing the node, you will see the development of the objective error function based on the hold out procedure that is performed in updating the weight estimates at each iteration from iterative grid search procedure. In our example, the network training performed up to thirty-seven iterations, however, the vertical line indicates the iteration in which the smallest average error was reached from the validation data set. The smallest average error was achieved at the eighteenth iteration from the validation data set. The reason for the erratic behavior at the beginning of the optimization plot is because ten preliminary network training runs were performed. For interval-valued target variables, the optimization plot displays the average sum-of-squares error from the training and validation data set. For categorically-valued target variables, the optimization plot displays the average error from the training and validation data set that is the rate of misclassification at each iteration. The blue line is the average error from the training data set and the red line is the average error from the validation data set. A vertical line is displayed to indicate the iteration step at which the final neural network estimates were determined by the optimum model selection criterion value. The criterion that is used in determining these final weight estimates is based on the **Model selection criteria** option from the **General** tab and the selected modeling selection criteria value of either the average error, misclassification rate, or maximum or minimum expected loss from the validation data set.

In viewing the optimization plot, the **Plot** tab is designed to update and score the weight estimates for the network architecture by right-clicking the mouse and selecting the **Set network at...** pop-up menu item that is illustrated in the following diagram. Selecting the **Set network at selected iteration** options will give you the ability to change the scored weight estimates by selecting a separate iteration step at any point along both lines in order to move the vertical line in updating the current weight estimates and modeling assessment statistics. Simply select any point along either one of the two lines and the vertical line will then move to the selected iteration that will then adjust the neural network weight estimates and associated goodness-of-fit statistics to the corresponding iteration. This technique is used in early stopping. The regularization technique determines the most appropriate weight estimates of the neural network model based on the smallest validation error, then refits the model with the current weight estimates until a desirable minimum of the validation error is reached. However, you must select the **Tools > Score** main menu option to score the current estimates that will result in writing the selected weight estimates and corresponding modeling assessment statistics to the scored data set based on the selected weight estimates. The scored data set can then be passed on to the **Model Manager** or the subsequent **Assessment** node for modeling assessment.

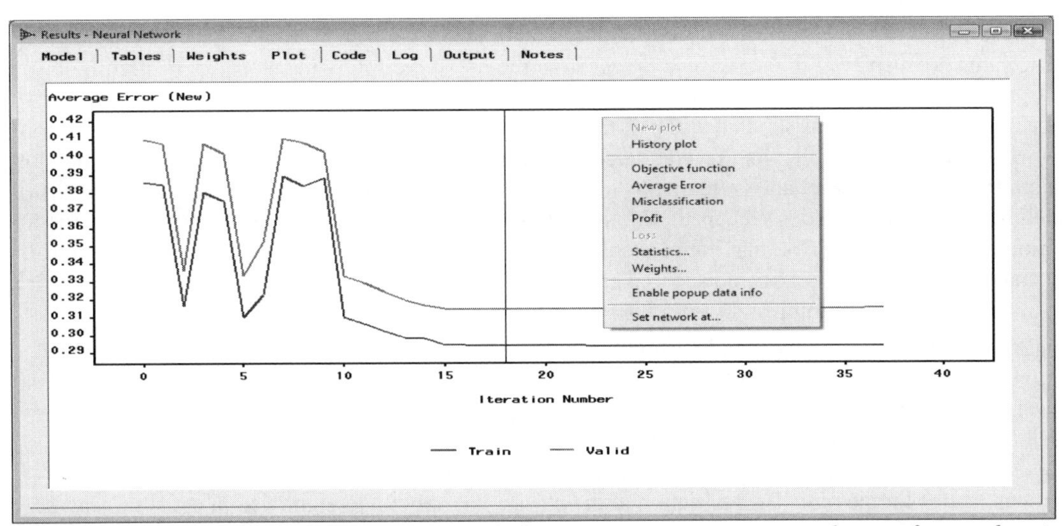

The ***Plot*** *tab displays the optimization plot with the assessment statistics at each step of network training.*

Right-click the mouse and select from the various modeling selection criteria that the optimization plot will display to view the development of the neural network hold out method:

- **New Plot** (default)**:** The optimization plot displays the most recent network training run that is the default.
- **History plot:** The optimization plot displays all of the accumulated neural network training runs that are performed in the interactive session.
- **Objective Function:** The optimization plot displays the objective function at each step of the neural network hold out procedure. For interval-valued targets, the objective function is the average squared error. For categorical targets, the objective function is the misclassification rate.
- **Average Error:** The optimization plot displays the average error at each step of the hold out method that is the default assessment statistic displayed in the plot.
- **Misclassification:** The optimization plot displays the misclassification rate at each iteration of the split sample procedure by fitting the categorically-valued target variable.
- **Profit:** The optimization plot displays the expected profit at each iteration, given that a profit matrix has been previously specified from the target profile. The profit plot will display more variability across each iteration as opposed to the average error plot.
- **Loss:** The optimization plot displays the expected loss at each iteration, given that a loss matrix has been previously specified from the target profile. The loss plot will display more variability across each iteration as opposed to the average error plot.

Right-click the mouse and select the **Statistics…** pull-down menu option to display the following **Select statistics** window in order to view the optimization plot from the neural network hold out method at each iteration based on the selected goodness-of-fit statistics that are listed in alphabetical order within each partitioned data set that is created.

Select statistics

Train: Akaike's Information Criterion.
Train: Average Squared Error.
Train: Error Function.
Train: Final Prediction Error.
Train: Maximum Absolute Error.
Train: Mean Squared Error.
Train: Objective Function (Error Part)
Train: Objective Function Divisor
Train: Root Average Squared Error.
Train: Root Final Prediction Error.
Train: Root Mean Squared Error.
Train: Schwarz's Bayesian Criterion.
Train: Sum of Squared Errors.
Valid: Average Squared Error.
Valid: Error Function.
Valid: Maximum Absolute Error.
Valid: Mean Squared Error.
Valid: Objective Function (Error Part)
Valid: Objective Function Divisor
Valid: Root Average Squared Error.
Valid: Root Mean Squared Error.
Valid: Sum of Squared Errors.
Weight Decay

OK Cancel

The ***Select statistics*** *window displays the list of goodness-of-fit statistics for the plot.*

- **Statistics:** This option will list the various goodness-of-fit statistics from both the training or validation data sets that are displayed in the list box. Select from the list of the following statistics by highlighting the row of the desired statistic and pressing the **OK** button. The optimization plot will then be displayed by the selected goodness-of-fit statistic that will plot the development of the selected assessment statistic during network training. All the following listed modeling assessment statistics may be displayed from the plot for both the training data set and validation data set with the exception of the weight decay value. That is, modeling assessment statistics such as the Akaike's information criterion, average square error, error function, final prediction error, maximum absolute error, mean squared error, root average square error, root final prediction error, root mean squared error, Schwartz Bayesian criterion, sums of squared error can be plotted between the training and validation data sets.
- **Weights:** Displays the selected weight estimates in the neural network model at each iteration of the iterative grid search procedure. Plotting the weight estimates will allow you to evaluate the magnitude of the weights changing at each iteration during network training. Small changes in the weight estimates will contribute to good generalization as the number of iterations increase, showing that convergence has been achieved in the iterative process. From the listed weight estimates, press the Ctrl key to select several weight estimates for the plot. The restriction to the plot is that only up to eight weight estimates can be viewed in the optimization plot at one time.
- **Enable popup data info:** Left-click the mouse at a particular point in the plot to display the iteration information such as the run number, the optimization step, iteration number, and the objective function values from both the training and validation data sets. By default, this option is disabled.
- **Set network at ...:** Displays a pair of choices to select from: the **Set network at selected iteration** and **Set network at minimum error function** pop-up menu options. This option automatically shifts the vertical line in the optimization plot to any point along either line. Select the **Set network at selected iteration** and left-click the mouse at any point in the optimization plot to set the iteration number in which the neural network weight estimates are automatically scored and updated. This option must be selected for the updated weight estimates to be listed in the adjacent **Weights** and **Tables** tabs in order to observe the updated goodness-of-fit statistics. The current weight estimates will be written to the output data set by selecting the **Tools > Score** main menu options. This will result in the neural network model being refitted with these same weight estimates and does not perform any further iterations. This method is commonly used in the early stopping regularization technique. The **Set network at minimum error function** pop-up menu option will reset the vertical line back to the initial point at which the weight estimates were originally selected with the minimum error from the validation data set.

Code tab

The **Code** tab is designed to display the internal Enterprise Miner SAS training code that was compiled to produce the neural network results and the final neural network weight estimates. By default, the **Training** radio button is automatically selected, which displays a listing of the internal Enterprise Miner SAS training code that created the neural network results and scored data sets. In other words, the listed SAS training code is compiled in the background when you execute the **Neural Network** node and train the neural network model that generates the various results and scored data sets that are listed within the node. By selecting the **Scoring** radio button, the tab will display the internal SAS scoring code. Select the **Tools > Score** menu options to execute the listed SAS scoring code that will score the scored data sets with the currently displayed weight estimates. The scored code will first determine if there are any missing values in any one of the input variables in the model, in which case the target variable will be estimated by its own average value. Otherwise, the score code standardizes each interval-valued input variable in the network model. The code then generates the linear combination of input layer weight estimates with the previously computed standardized input variables for each hidden layer unit. The input layer bias term is added to each hidden layer. The hidden layer weight estimates are applied to the linear combination of weight estimates and standardized input variables, and the activation function is applied to each hidden layer unit. The hidden layer units are multiplied by the hidden layer weight estimates that are added together along with the hidden layer bias term to generate the final neural network estimates. And finally, if there are any missing values in anyone of the input variables, then the predicted value is computed by the average value of the target variable.

```
Results - Neural Network
Model | Tables | Weights | Plot   Code | Log | Output | Notes |
(•) Training   ( ) Scoring
00001 filename remote socket "127.0.0.1:6652";
00002 *;
00003 * SAS NEURAL NETWORK - TRAINING FUNCTION;
00004 *;
00005 proc neural data=emdata.dm_DGM00047 dmdbcat=EMPROJ.dm_DGM00047
00006 network=EMPROJ.NNS_NFEC.NETWORK
00007 validdata=EMDATA.VALKK3MQ
00008 predata=EMDATA.NNDD8GVH
00009 ;
00010 REMOTE socket=remote;
00011 decision decisiondata= EMPROJ.BAD_ decvars=
00012 _DEC1
00013 ;
00014 *;
00015 prelim 10
00016 outest= EMPROJ.NNPUXMK0
00017 ;
00018 *;
00019 train
00020 outest= EMPROJ.NNE170AM estiter=1
00021 outfit= EMPROJ.NNFZ5P6I
00022 ;
00023 *;
00024 code noerror nores metabase=EMPROJ.NNS_NFEC.DATASTEP.SOURCE;
00025 *;
00026 RUN;
00027 *;
00028 * SAS NEURAL NETWORK - SCORING FUNCTION;
00029 *;
00030 proc neural data= emdata.dm_DGM00047 dmdbcat=EMPROJ.dm_DGM00047
00031 network=EMPROJ.NNS_NFEC.NETWORK;
00032 decision decisiondata= EMPROJ.BAD_ decvars=
00033 _DEC1
00034 ;
00035 *;
00036 initial inest=EMPROJ.NNX7IX5E BYLABEL;
00037 *;
00038 NLOPTIONS noprint;
00039 TRAIN tech=none;
00040 *;
00041 code noerror nores metabase=EMPROJ.NNS_NFEC.DATASTEP;
00042 score noerr nostd nohidden nores data=EMDATA.TRNY0GWY role=TRAIN outfit=_tfit
00043 out=EMDATA.STRNVFQN (label='Neural scores:TRAIN' type=TRAIN
00044 );
00045 score noerr nostd nohidden nores data=EMDATA.VALKK3MQ role=VALID outfit=_vfit
00046 out=EMDATA.SVALKUE3 (label='Neural scores:VALID' type=VALID
00047 );
00048 score noerr nostd nohidden nores data=EMDATA.VALKK3MQ role=VALID
00049 out=EMPROJ._A0013AX (label='Neural scores:VALID' type=VALID
00050 keep=
00051 P_:
00052 D_:
00053 _WARN_
00054 EP_:
00055 BP_:
00056 CP_:
00057 EL_:
00058 BL_:
00059 CL_:
00060 IC_:
00061 BAD
00062 );
00063 RUN;
00064 * Save fit statistics;
00065 data EMPROJ.NNYSKHBJ;
00066 merge _tfit _vfit;
00067 RUN;
00068 *** END OF FILE ***
```

*The **Code** tab displays the training code from the neural network procedure.*

```
Results - Neural Network
Model | Tables | Weights | Plot   Code | Log | Output | Notes |
( ) Training   (•) Scoring
00054 IF _DM_BAD EQ 0 THEN DO;
00055    S_CLAGE   =    -2.12683050688911 +    0.01177148130024 * CLAGE ;
00056    S_DEBTINC  =    -4.23565806211598 +    0.12460936685911 * DEBTINC ;
00057    S_DELINQ  =    -0.36236948891134 +    0.86388886156464 * DELINQ ;
00058    S_DEROG   =    -0.27736891913986 +    1.31617735515411 * DEROG ;
00059 END;
00060 ELSE DO;
00061    IF MISSING( CLAGE ) THEN S_CLAGE  = . ;
00062    ELSE S_CLAGE   =    -2.12683050688911 +    0.01177148130024 * CLAGE ;
00063    IF MISSING( DEBTINC ) THEN S_DEBTINC  = . ;
00064    ELSE S_DEBTINC   =    -4.23565806211598 +    0.12460936685911 * DEBTINC ;
00065    IF MISSING( DELINQ ) THEN S_DELINQ  = . ;
00066    ELSE S_DELINQ   =    -0.36236948891134 +    0.86388886156464 * DELINQ ;
00067    IF MISSING( DEROG ) THEN S_DEROG  = . ;
00068    ELSE S_DEROG   =    -0.27736891913986 +    1.31617735515411 * DEROG ;
00069 END;
00070 *** *************************;
00071 *** Writing the Node H1 ;
00072 *** *************************;
00073 IF _DM_BAD EQ 0 THEN DO;
00074    H11  =   -0.35031711792682 * S_CLAGE  +   -70.4969860436658 * S_DEBTINC
00075          +   0.18449108575908 * S_DELINQ  +    0.33413930152401 * S_DEROG
00076          ;
00077    H12  =   -0.12865053781085 * S_CLAGE  +   -7.94458016426562 * S_DEBTINC
00078          +   0.18612942323723 * S_DELINQ  +    0.17289501453247 * S_DEROG
00079          ;
00080    H13  =    0.26925274534102 * S_CLAGE  +   -0.82220975009789 * S_DEBTINC
00081          +  -0.17305706864271 * S_DELINQ  +   -0.16019236495051 * S_DEROG
00082          ;
00083    H11  =     4.8216825096159 + H11 ;
00084    H12  =    0.80832411979041 + H12 ;
00085    H13  =    1.70003455124041 + H13 ;
00086    H11  = TANH(H11 );
00087    H12  = TANH(H12 );
00088    H13  = TANH(H13 );
00089 END;
00090 ELSE DO;
00091    H11  = .;
00092    H12  = .;
00093    H13  = .;
00094 END;
00095 *** *************************;
00096 *** Writing the Node BAD ;
00097 *** *************************;
00098 IF _DM_BAD EQ 0 THEN DO;
00099    P_BAD1  =     -3.1171082152167 * H11  +    3.63547940017616 * H12
00100          +     -7.5972528821107 * H13 ;
00101    P_BAD1  =     3.87905811594848 + P_BAD1 ;
00102    P_BAD0  = 0;
00103    _MAX_ = MAX (P_BAD1 , P_BAD0 );
00104    _SUM_ = 0.;
00105    P_BAD1  = EXP(P_BAD1  - _MAX_);
00106    _SUM_ = _SUM_ + P_BAD1 ;
00107    P_BAD0  = EXP(P_BAD0  - _MAX_);
00108    _SUM_ = _SUM_ + P_BAD0 ;
00109    P_BAD1  = P_BAD1  / _SUM_;
00110    P_BAD0  = P_BAD0  / _SUM_;
00111 END;
00112 ELSE DO;
00113    P_BAD1  = .;
00114    P_BAD0  = .;
00115 END;
00116 IF _DM_BAD EQ 1 THEN DO;
00117    P_BAD1  =      0.198322147651;
00118    P_BAD0  =     0.80167785234899;
00119 END;
```

*The **Code** tab displays a partial listing of the neural network score code.*

Log tab

The **Log** tab will allow you to view the PROC NEURAL procedure log listing in order to look for any warning or error messages that might occur when running the **Neural Network** node. The note in the **Log** tab from the PROC NEURAL procedure will enable you to determine if one of the convergence criteria has been met and convergence has been achieved in the neural network model. In addition, the log listing will allow you to observe the various scoring data sets that have been created by executing the node.

Output tab

The **Output** tab is designed to display the PROC NEURAL procedure output listing. Initially, the procedure output listing displays the ten separate preliminary training runs that were performed. The **Optimization Start Parameter Estimates** table displays the initial weight estimates and associated gradient error function for each parameter estimate based on the weight estimates from the best preliminary training run. Since the network model is fitting an binary-valued target variable, therefore the procedure applies the multiple Bernoulli error function. By default, the initial weight estimates are automatically set to some small random number that follows a normal distribution centered at zero with a small variability. Therefore, the initial weight estimates will have a small chance of generating large values that might lead to bad generalization. Also, the target layer weight bias estimate is automatically set to the average value of the target variable. The subsequent table listing displays the default Levenberg – Marquardt optimization technique since there are less than 100 weight estimates in the model, along with the various convergence criterion statistics. Convergence is assumed in the nonlinear model when any one of these values are met. The procedure output listing displays the iteration history, such as the objective function at each step of the iterative grid search procedure based on the Levenberg – Marquardt optimization technique. If the final gradient values of all the weight estimates are close to zero or if there is a small change in the final objective function, then it will be a strong indication that convergence in the minimum final objective error value has been achieved. In addition, quick convergence in the iterative grid search procedure in finding the optimum weight estimates from the neural network model might also be a strong indication that convergence has been met. Conversely, slow convergence might indicate that the iterative search procedure had problems in seeking the best set of weight estimates in the model. From the **Optimization Results** table, an extremely important message will be displayed to inform you that at least one of the convergence criteria has been met. The **Optimization Results Parameter Estimates** table displays the final weight estimates and gradient values from network training. It is desirable to have the gradient values associated with each weight estimate close to zero, indicating a small change in the weight estimate between successive iterations. Again, since the input variables have been standardized, then the magnitude in absolute value of the standardized weight estimates can be used as a crude assessment in determining the best set of input variables for the neural network model. The following listed results are generated from the internal SEMMA neural network procedure training code that is compiled by running the **Neural Network** node and training the neural network model.

Results - Neural Network

Model | Tables | Weights | Plot | Code | Log | Output | Notes

The NEURAL Procedure

Preliminary Training Run	Starting Random Seed	Objective Function Value	Number of Iterations	Terminating Criteria
1	12345	0.385886387312	20	
2	494012387	0.384824281085	20	
3	936745137	0.316004483399	20	
4	337281485	0.38035594086	20	
5	42129862	0.37530125124	20	
6	462622703	0.310163559513	20	
7	1962974514	0.323002271495	20	
8	267208285	0.389189329009	20	
9	1715549360	0.383681213846	20	
10	422345732	0.388366778606	20	

Optimization Start
Parameter Estimates

N	Parameter	Estimate	Gradient Objective Function
1	CLAGE_H11	0.015804	0.016787
2	DEBTINC_H11	-6.975953	0.001784
3	DELINQ_H11	-0.068180	0.000895
4	DEROG_H11	0.110614	0.000290
5	CLAGE_H12	-0.096862	-0.010020
6	DEBTINC_H12	-5.588524	-0.000023766
7	DELINQ_H12	0.129575	-0.000529
8	DEROG_H12	0.213357	0.000740
9	CLAGE_H13	0.311790	0.001133
10	DEBTINC_H13	-1.125634	-0.000336
11	DELINQ_H13	-0.179570	0.000292
12	DEROG_H13	-0.188607	0.001164
13	BIAS_H11	0.192153	-0.000103
14	BIAS_H12	0.957224	-0.003360
15	BIAS_H13	1.598884	-0.001025
16	H11_BAD1	-3.499788	0.000988
17	H12_BAD1	4.095753	-0.000183
18	H13_BAD1	-4.447787	-0.000364
19	BIAS_BAD1	0.533077	-0.000805

Value of Objective Function = 0.3101635595

The NEURAL Procedure

Levenberg-Marquardt Optimization

Minimum Iterations	0
Maximum Iterations	100
Maximum Function Calls	2147483647
Maximum CPU Time	604800
ABSGCONV Gradient Criterion	0.00001
GCONV Gradient Criterion	1E-8
GCONV2 Gradient Criterion	0
ABSFCONV Function Criterion	0
FCONV Function Criterion	0.0001
FCONV2 Function Criterion	0
FSIZE Parameter	0
ABSXCONV Parameter Change Criterion	0
XCONV Parameter Change Criterion	0
XSIZE Parameter	0
ABSCONV Function Criterion	0.001838911
Trust Region Initial Radius Factor	1
Singularity Tolerance (SINGULAR)	1E-8

Levenberg-Marquardt Optimization

Scaling Update of More (1978)

Parameter Estimates 19

Optimization Start

Active Constraints	0	Objective Function	0.3101635595
Max Abs Gradient Element	0.0167866531	Radius	1

Iter	Restarts	Function Calls	Active Constraints	Objective Function	Objective Function Change	Max Abs Gradient Element	Lambda	Ratio Between Actual and Predicted Change
1	0	3	0	0.30617	0.00399	0.00454	0.121	0.884
2	0	4	0	0.30219	0.00398	0.0133	0.0249	0.905
3	0	5	0	0.29849	0.00370	0.00695	0.00218	1.047
4	0	6	0	0.29776	0.000731	0.0371	0.00104	0.208
5	0	7	0	0.29432	0.00344	0.0102	0.00049	0.724
6	0	9	0	0.29417	0.000153	0.0139	0.00182	0.172
7	0	10	0	0.29383	0.000337	0.00327	0.00077	0.438
8	0	12	0	0.29369	0.000134	0.00590	0.00968	0.294
9	0	13	0	0.29360	0.000090	0.00110	0.00223	0.445
10	0	14	0	0.29358	0.000029	0.00293	0.00318	0.179
11	0	15	0	0.29353	0.000043	0.00107	0.00207	0.289
12	0	16	0	0.29351	0.000023	0.00251	0.00292	0.146
13	0	17	0	0.29348	0.000030	0.00130	0.00187	0.200
14	0	18	0	0.29346	0.000020	0.00249	0.00285	0.114
15	0	19	0	0.29344	0.000019	0.00161	0.00178	0.122
16	0	20	0	0.29342	0.000019	0.00252	0.00291	0.0999
17	0	21	0	0.29336	0.000063	0.00121	0.0111	0.491
18	0	22	0	0.29334	0.000019	0.00127	0.00317	0.410
19	0	23	0	0.29333	0.000014	0.00101	0.00265	0.356
20	0	24	0	0.29331	0.000016	0.00104	0.00300	0.364
21	0	25	0	0.29330	0.000011	0.00107	0.00248	0.293
22	0	26	0	0.29328	0.000015	0.000945	0.00289	0.319
23	0	27	0	0.29327	8.351E-6	0.00111	0.00273	0.210
24	0	28	0	0.29326	0.000014	0.000898	0.00323	0.286
25	0	29	0	0.29325	6.98E-6	0.00114	0.00263	0.175
26	0	30	0	0.29324	0.000014	0.000872	0.00314	0.270
27	0	31	0	0.29323	5.664E-6	0.00116	0.00255	0.140

Optimization Results

Iterations	27	Function Calls	33
Jacobian Calls	29	Active Constraints	0
Objective Function	0.2932335758	Max Abs Gradient Element	0.0011621331
Lambda	0.0025542235	Actual Over Pred Change	0.1397009261
Radius	0.0774765595		

Convergence criterion (FCONV=0.0001) satisfied.

NOTE: At least one element of the gradient is greater than 1e-3.

Optimization Results
Parameter Estimates

N	Parameter	Estimate	Gradient Objective Function
1	CLAGE_H11	-0.459649	0.000056922
2	DEBTINC_H11	-92.229218	0.000037844
3	DELINQ_H11	0.225036	0.000840
4	DEROG_H11	0.336182	-0.000186
5	CLAGE_H12	-0.135026	0.000018237
6	DEBTINC_H12	-7.850722	-0.000009089
7	DELINQ_H12	0.199675	-0.000198
8	DEROG_H12	0.172511	0.000104
9	CLAGE_H13	0.263037	-0.000200
10	DEBTINC_H13	-0.802945	-0.000051264
11	DELINQ_H13	-0.166919	0.001162
12	DEROG_H13	-0.155336	-0.000562
13	BIAS_H11	6.750446	0.000313
14	BIAS_H12	0.851728	-0.000204
15	BIAS_H13	1.725110	0.000164
16	H11_BAD1	-2.966340	0.000062681
17	H12_BAD1	3.503802	-0.000038087
18	H13_BAD1	-8.297742	-0.000056704
19	BIAS_BAD1	4.559732	-0.000076344

Value of Objective Function = 0.2932335758

*** END OF FILE ***

*The **Output** tab displays the development of the objective function and network weights.*

Neural Network Interactive Training

Interactive training will allow you to sequentially construct a neural network model. The purpose of interactive training is trying to avoid bad local minimums and, hopefully, moving to a desirable minimum in the nonlinear error function during network training. Iterative training is designed to modify the neural network properties and configuration settings as you fit the network model any number of times, all within the same interactive training interface. In order to perform interactive training, you must first execute the **Neural Network** node. In addition, you must select the **Advanced user interface** option setting from the **General** tab. Once the node is executed, then right-click the mouse and select the **Interactive** pop-up menu item from the **Neural Network** node within the process flow diagram. This will result in the following **Interactive Training** window appearing.

The interactive training interface performs a wide variety of routines that are designed to improve the current neural network estimates. Interactive training performs various neural network routines such as performing preliminary training runs, network training runs, early stopping and weight decay regularization techniques, and architecture selection routines such as sequential network construction, cascade correlation, and input pruning techniques. From the interactive training environment, you have the ability to clear all previously listed output training and accumulated history runs, set values to the current weight estimates, reset all the current weights and modeling assessment statistics, reinitialize the weight estimates by randomizing the current weight estimates and associated modeling assessment statistics, or perturb the current weights by resetting the current weight estimates in making small changes to the values in each one of the weight estimates.

The **Interactive Training** environment is similar to the **Advanced user interface** environment within the **Neural Network** node and has the following tabs:

- **Data tab:** Displays the file administrative information and a table view of the partitioned data sets.
- **Variables tab:** Displays a listing of the variables in the neural network model so you can remove certain input variables from the model by performing input pruning techniques or edit the target profile by fitting the target variable in the neural network model.
- **Advanced tab:** Displays the configuration settings for the neural network design so you can change several of the option settings each time preliminary or network training is performed.
- **Plot tab:** Displays the optimization plot in order to update and score the current weight estimates and modeling assessment statistics each time the network is trained. This network training strategy is used in regularization techniques such as early stopping.
- **Output tab:** Displays the neural network procedure output listings to observe such things as the final weight estimates, verifying that convergence has been met with a reduction in both the objective function and gradient values from network training by observing small changes in the objective function or the gradient values between successive iterations based on the iterative grid search procedure.

Clearing the Output

From the **Interactive Training** window, select the **Edit > Clear Output** main menu items to clear the output of all previous neural network training runs. Selecting the **Edit > Clear interactive history** menu options will clear all previous accumulated training history runs.

Interactive Preliminary Training

From the **Interactive Training** window, select **Tools > Preliminary optimization...** main menu items or select the **Preliminary** toolbar icon located in the tools bar menu in order to perform preliminary runs from the training data set in determining the initial weight estimates to the neural network model. Again, preliminary training runs are recommended since there is no standard method that is used in neural network modeling to compute the initial weight estimates in a neural network design. The two main reasons for applying preliminary training are to accelerate convergence and avoid bad local minimums in the objective function during network training.

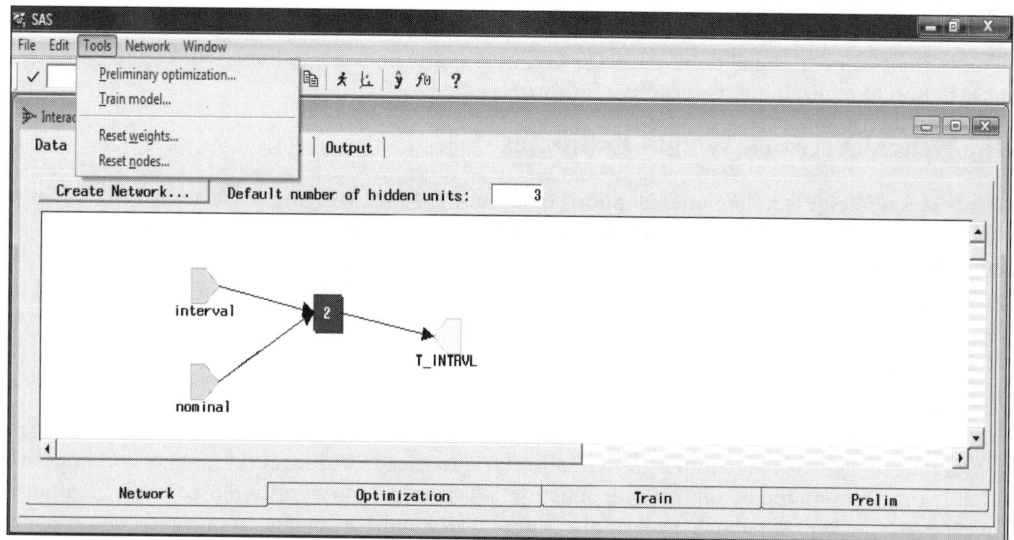

*The **Interactive Training** window displaying the menu options for network training.*

Before the preliminary iterative training run is performed, the following **Train network** window will appear in order to view the current settings of the subsequent preliminary training run. The window will display the number of training observations, the number of active weights in the network model, and the number of frozen weights. *Frozen weights* are weight estimates that do not change during the preliminary or network training runs. The preliminary iterative training run is designed for you to specify the number of preliminary runs, the maximum number of iterations, and the maximum amount of computational time. Clearing the **Show this prompt always** check box will prevent the **Train network** window from appearing again the next time you perform preliminary training. Press the **OK** button and the preliminary training procedure will begin. Otherwise, press the **Cancel** button to return to the **Interactive Training** window.

Train network window

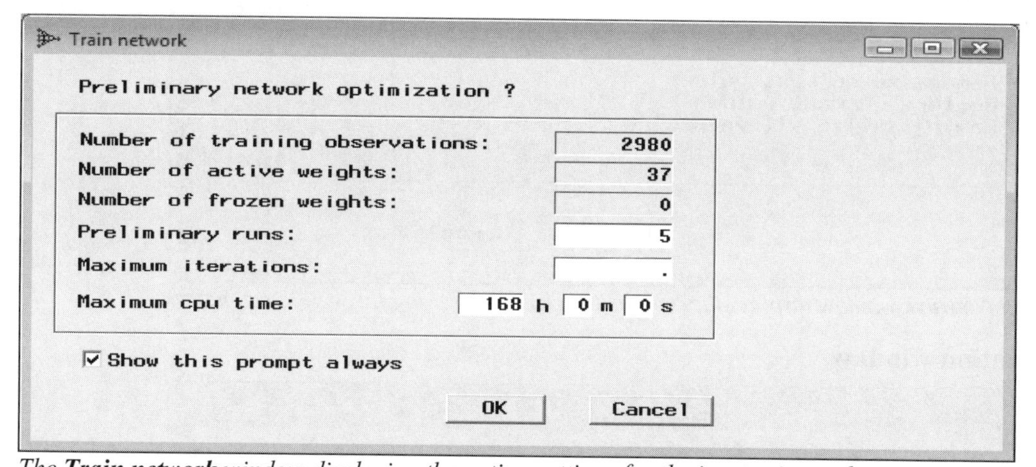

*The **Train network** window displaying the option settings for the interactive preliminary training run.*

Running Interactive Training

Select the **Run SAS Code** toolbar icon from the tools bar menu to train the network model several times based on the current neural network configuration settings.

Interactive Training

Select the **Tools > Train model...** main menu items and the **Train network** window will appear. The previously displayed **Train network** window will appear before the network training is performed. From the **Train network** window, you have the option of specifying the various configuration settings for the subsequent network training run such as the number of training runs, the maximum number of iterations, and the maximum amount of computational time. Press the **OK** button and network training will begin. Otherwise, press the **Cancel** button to return to the **Interactive Training** window.

Resetting the Neural Network Weight Estimates

Select the **Tools > Reset Weights...** main menu options or select the **Reset weights** toolbar icon that is located from the tools bar menu and the **Neural Network** window will appear with the following **Reset network weights** options. This will give you the option of randomizing the current neural network weight estimates, perturbing the current weight, or reinitializing the current weight from the previous network training run.

The first two **Randomize current values** and **Perturb current values** reset network weight options will display the same **Randomization** window for resetting the weight estimates and resetting the scale and location estimates to some small values that is illustrated in the following page. The reason for randomizing or perturbing the weight estimates is that, at times, it will allow you to escape bad local minimums and hopefully move to a desirable minimum in the nonlinear error function. During the randomization process, it is important that the weight estimates be kept small, particularly the input layer weight estimates, in order to compute the output units, which will prevent the network model from generating predicted values beyond the range of the actual target values. The options are designed to reset the current weight estimates and corresponding goodness-of-fit statistics. The window will allow you to select an entirely different distribution for the random values that will be assigned to the initial weight estimates–of either a uniform distribution, a normal distribution, a Cauchy distribution or a scaled chi-square distribution with specified degrees of freedom. The default is the normal distribution. The difference between the methods is that the *perturb weight* estimates are reset with a small shift in the existing weight values with comparison to *randomizing weights*, which reinitializes the weight estimates at some small random values. Press the **Create** button to set a different random seed number in computing different random values that are applied to the initial weight estimates.

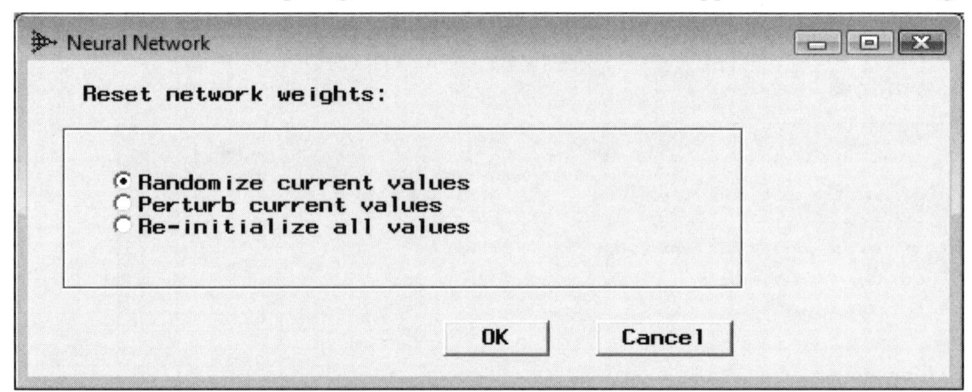

*The **Neural Network** window displays the options to select in resetting the weight estimates.*

Randomization window

Randomization
Distribution: Normal
Chi DF: 1
Scale: 1
Location: 0
Random seed: 1134979935 Create
Uniform
Normal
Cauchy
Scaled Chi Inverse
OK
Cancel

*The **Randomization** window displays the option settings in randomizing the weight estimates.*

By selecting the **Re-initialize all values** check box from the **Neural Network** window, the following **Initialization** window will appear. The **Initialization** window is designed for you to reinitialize the weight estimates to some small random values. Reinitializing the weight estimates such as the scale estimate, target weights, or target bias weights by randomizing the initial weight estimates by a specified random number generator to create a new set of random numbers of some small values that will be assigned to these same weight estimates. Select the **Create** button to generate a new global random seed value.

Initialization window

Initialization

Global seed: 12345 Create

Randomize scale estimates

Randomize target weights

Randomize target bias weights

OK Cancel

*The **Initialization** window that reinitializes the initial weight estimates for network training.*

Resetting the SAS Default Settings

The **Neural Network** node has an added option for you to automatically reset all changes that have been made to the layered nodes from interactive training back to the SAS default settings. By selecting **Network > Reset nodes** main menu options or selecting the **Reset node functions** tools bar icon that is located from the main menu will allow you to automatically reset either all the nodes, input nodes, hidden nodes, target nodes, or all nodes with error conditions back to the original **Neural Network** option settings that SAS has carefully selected. Again, some of the option settings to the neural network design are associated with each other. Furthermore, if some of these same option settings are incompatible with each other, then preliminary training or network training will not be performed. Therefore, the window will allow you to reset the nodes in order to achieve compatibility with the other configuration settings of the neural network design in order to perform network training. Press the **OK** button to restore the selected nodes to their default settings and the **Neural Network** window will then reappear.

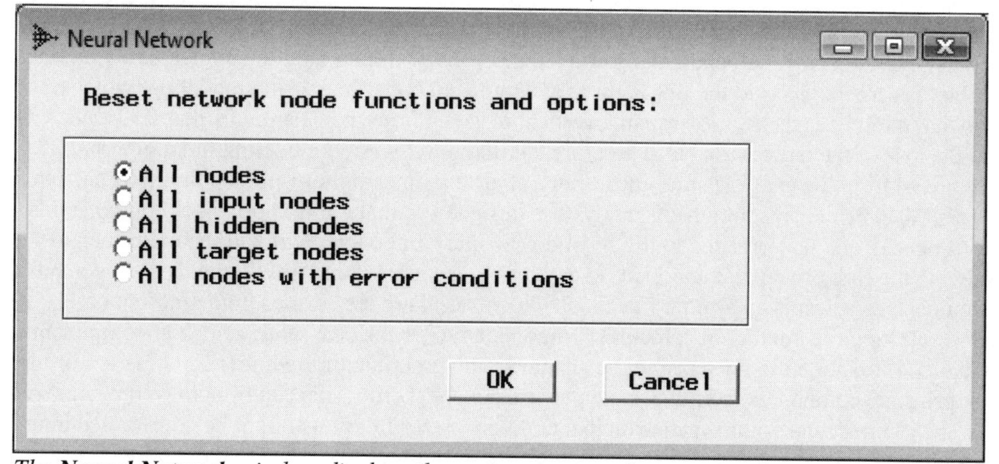

*The **Neural Network** window displays the various options for resetting the configuration settings.*

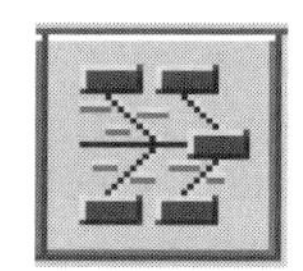

4.5 Princomp/Dmneural Node

General Layout of the Enterprise Miner Princomp/Dmneural Node

- **Data tab**
- **Variables tab**
- **Advanced tab**
- **PrinComp tab**
- **Output tab**
- **Notes tab**

The purpose of the **Princomp/Dmneural** node in Enterprise Miner is to perform both principal component analysis and dmneural network training. *Dmneural network training* is an additive nonlinear model that uses a set of principal components that are placed into separate buckets to predict either a binary-valued or interval-valued target variable. Although, the dmneural modeling design will extend to nominal and ordinal-valued target variables in future releases. Dmneural network training is designed to overcome some of the shortcomings of nonlinear modeling, such as neural network modeling. The various shortcomings will be discussed shortly. The purpose of *principal components* is both data reduction and interpretation of a linear combination of the input variables in the data that best explains the covariance or correlation structure. The analysis is designed to reduce the dimensionality of the data while at the same time preserving the structure of the data. The advantage of principal components is that you can create a smaller number of linear independent variables, or principal components, without losing too much variability in the original data set, where each principal component is a linear combination of the input variables in the model. In addition, each principal component is uncorrelated to each other.

Principal Components Analysis

Principal components modeling is based on constructing an independent linear combination of input variables in which the coefficients (*eigenvectors*) capture the maximum amount of variability in the data. Typically, the analysis creates as many principal components as there are input variables in the data set in order to explain all the variability in the data where each principal component is uncorrelated to each other. This solves one of two problems in the statistical model. First, the reduction in the number of input variables solves the dimensionality problem. Second, since the components are uncorrelated to each other, this will solve collinearity among the input variables, which will result in more stable statistical models. The goal of the analysis is first finding the best linear combination of input variables with the largest variance in the data, called the first principle component. The basic idea is to determine the smallest number of the principal components to account for the maximum amount of variability in the data. Principal components analysis is designed so that the first principle component consists of a line that is perpendicular to each data point by minimizing the total squared distance from each point that is perpendicular to the line. This is analogous to linear regression modeling, which determines a line that minimizes the sum-of-squares vertical distance from the data points that is always perpendicular to the axis of the target variable. Therefore, similar to regression modeling, it is important that the outliers be removed from the analysis since these extreme data points might result in a dramatic shift in the principal component line. Principal components analysis is designed so that the first principal component is perpendicular, orthogonal, and uncorrelated to the second principle component, with the second principle component following the first principal component in explaining the most variability in the data. The number of principal components to select is an arbitrary decision. However, there are various guidelines that are commonly used in selecting the number of principal components. After the best set of principal components is selected, the next step is to, hopefully, determine the characteristic grouping of the linear combination of input variables in each principal component from the principal component scores. This can be achieved by observing the magnitude of the eigenvectors within each principal component. Scatter plots might be constructed in order to observe the variability, outliers, and, possibly, various cluster groupings from the first few principal component scores.

The principal components are comprised of both the eigenvalues and eigenvectors that are computed from the variance–covariance matrix or the correlation matrix. The *eigenvalues* are the diagonal entries in the variance–

covariance matrix or the correlation matrix. The principal components are the linear combination of the input variables with the coefficients equal to the eigenvectors of the corrected variance-covariance matrix or the correlation matrix. The variance-covariance matrix should be used when the input variables are standardized or when the input variables have the same scale of measurement or approximately the same range of values. Otherwise, the correlation matrix is applied.

In principal component analysis, at most times the variability in the linear combination of input variables will be mostly concentrated in the first few principal components, while at the same time a small amount of variability in the data will be concentrated in the rest of the other principal components. Therefore, ignoring the last few principal components will result in a small amount of information that will be lost. If the original input variables are correlated, then only the first few principal components are needed in explaining most of the variability in the data. However, if the original input variables are uncorrelated, then the principal component analysis is useless. The difference between principal components and linear regression is the way in which the models determine the minimum distance between the data points. In traditional least-squares modeling, the best modeling fit is determined by minimizing the distance between the variables in the model and the target response variable. Principal components analysis is basically designed to determine the minimum distance between all the data points in the model. This is illustrated in the following diagram. To better understand the geometric interpretation of principal components, consider the diagram to the left that displays the first principal components line. The line accounts for the largest majority of the variability in the data where the principal component line is not parallel to any of the three variable axes. The elliptical-shape sphere represents the distribution of the data points in a three-dimensional plane and the shadows display the two-dimensional scatter plots, where there is a positive correlation between all pairs of variables in the data.

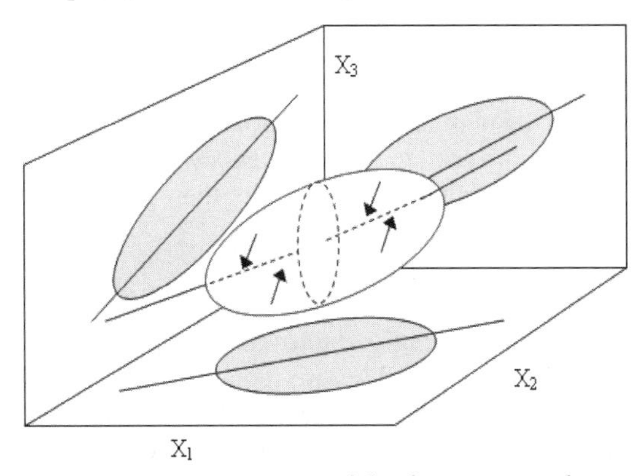

Geometric interpretation of the first principal component explaining most of the variability.

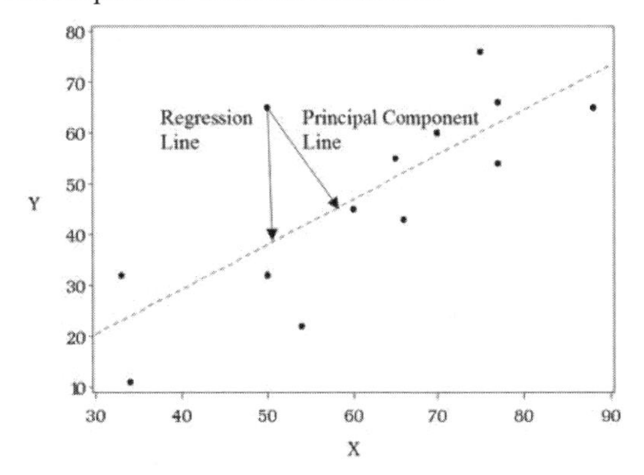

Geometric interpretation of both the least-squares line and the principal components line to minimize.

The purpose of principal component analysis is to reduce the dimensionality and eliminate the collinearity of the variables in the data in order to make the interpretation of the data easier. The principal components are calculated from an uncorrelated linear combination of input variables that is typically calculated from the variance–covariance matrix of the original input variables. However, if the input variables have approximately the same mean and variance, then the principal component scores may also be calculated from the correlation matrix based on the original input variables. The principal components are usually treated as a new set of input variables for subsequent modeling nodes such as the **Memory-Based Reasoning** modeling node. The reason that principal components are needed for this nearest neighbor modeling design is to reduce the dimensionality of the input space in order to accurately estimate the number of neighboring target values across the range of values of the input variables in the model. Principal components might be considered as a preliminary step in cluster analysis, assuming that there are more than two or three input variables in the model. In cluster analysis, the data will lie within the multidimensional plane, whereas applying the principal components between each pair of components in the two-dimensional plane will help visualize the various cluster groupings that would otherwise be extremely difficult to discover. This also holds true in detecting outlying data points in multivariate data. In principal component analysis, it is common practice to create a scatter plot of the scores from the first two principal components to visually observe any cluster groupings that are created. In predictive modeling, principal components might be considered as a variable selection routine in which the first few components explain a large majority of the variability in the data that would then be used as the input variables for the subsequent modeling fit.

One of the most important decisions in principal component analysis is deciding on the number of components to select. Therefore, the node displays various line plots in deciding on the number of principal components for the data mining analysis. The results from the node will display a bar chart of the shared variation and line plots of the log eigenvalues, LEV, scree plots and cumulative variation of the number of principal components. A sharp drop in the shared variation, LEV, or scree plots will indicate the number of principle components to retain in the analysis. The node displays a table listing of the eigenvalues and the associated proportion of explained variation of each principal component that will allow you to select the appropriate number of principle components that sufficiently explains the variability in the data. The same table listing displays the list of the eigenvectors associated with the linear combination of principle components by each input variable in the model. A score file can be created with a listing of the input variables and principle component scores.

There are four basic criteria used in selecting the number of principal components to retain in the analysis:

Scree Plots: A scree plot plots the eigenvalues across the number of principal components that are created and shows a clear elbow in the plot with a sharp drop in the curve that levels off at the elbow.

Proportion of Variance: The proportion of variability explained by each component. It is the proportion of each corresponding eigenvalue over the total number of eigenvalues from the variance–covariance matrix. Since scree plots might not always be clear and decisive in selecting the number of components, the number of components to retain might depend on a predetermined threshold value. Again, a sharp drop will indicate the appropriate number of principle components to select. However, the number of principle components that explains the amount of variability in the data might be determined in advance.

Cumulative Proportion of Variance: The cumulative variability explained by the accumulated eigenvalues over the total number of eigenvalues from the variance –covariance matrix. This is the most commonly used criterion that is selected depending on whether the variance-covariance or the correlation matrix is applied to the analysis.

Eigenvalue > 1: Typically, the sum of the eigenvalues is equal to the number of input variables in the analysis; therefore, a general rule that is based on the eigenvalues is to retain the principal components with their associated eigenvalue greater than one since these components will explain more variability than any one input variable in the model.

Variable Specifications Between Both the Procedures

Both procedures require that the active training data set have at least two input variables. In both principal component analysis and dmneural network training, interval-valued input variables are required to the models. Therefore, the node automatically creates separate dummy variables for each class level of the categorically-valued input variables that are used as interval-valued input variables to the analysis. In other words, categorically-valued input variables with k class levels are replaced by k binary-valued dummy variables using the GLM coding schema. For dmneural training, the interval-valued variables are automatically standardized into z-scores. The big difference between the procedures is that in principal component analysis a target variable to the model is not required in explaining the variability in the data, resulting in unsupervised training. Conversely, the dmneural network training method requires either a binary or an interval-valued target variable in predicting the variability to the target responses, resulting in supervised training. However, if the input data set consists of a nominal or ordinal-valued target variable, or more than two target variables are specified in the active training data set, then principal component analysis is automatically performed.

The Disadvantages to Principal Components

The advantage of principal component analysis is to reduce the variability and the dimensionality in the data. The drawback is that the principal component can be viewed as a mathematical transformation of the set of original input variables. Therefore, interpreting the relationship between the set of principal components is often problematic or impossible. Principal component analysis might be considered if it is assumed that the variables in the model are highly related to each other, which at times might be compounded by the problem of fitting the model to a small number of observations. Therefore, due to a small sample size, this might result in an insufficient degrees of freedom for the error term that will result in an inadequate statistical power of the model. Although principal component analysis always has its advantages in reducing the variability and the dimensionality in the data, the drawback is that the principal component that explains the highest variability in the data might not explain the variability in the target variable in the supervised modeling designs. In addition, removing the smallest principal components might at times result in the removal of very important predictive information from the data. Also, the principal component model requires all nonmissing cases in the input variables in order to calculate the principal component coefficients. Furthermore, it is extremely difficult to interpret the principal components if there exist an enormous number of input variables in the model. Rotating the principal components will aid in the interpretation of the results. However, rotations and adjustments to the principal components are not currently available within the **Princomp/Dmneural** node. Finally, if the interpretation of the components is most important, then you might consider factor analysis as opposed to principal components due to the high collinearity between the input variables in the analysis.

Dmneural Network Training

The basic idea of dmneural network training is to determine the smallest number of the principal components to account for the maximum amount of variability in the target variable by selecting the best subset of input variables in predicting the target values. Dmneural network training is designed to overcome the "*curse of dimensionality*" problem in nonlinear regression modeling, like neural network modeling. In the dmneural training process, as each input variable is entered into the network model, it will result in an additional dimension to the space in which the range of the data resides. Therefore, it is important that input variables that provide very little information be eliminated from the design, which will lead to an increase in the modeling performance and reduction of the dimension of the input variables. The drawback is that the linear relationship between the principal components and the target variable might be uninterruptible. It is also important to mention that the various modeling assumptions must be satisfied in dmneural network training in order to achieve an adequate modeling performance and good generalization.

The following are common problems in neural network modeling designs and the advantages in considering dmneural network training before applying neural network modeling in determining the best set of input variables for the nonlinear model. The following shortcomings are likely to occur when the input variables are highly correlated with each other in the nonlinear model:

- A common problem in many of the optimization techniques applied in nonlinear estimation are inadequate starting values for the parameter estimates. Obtaining inadequate starting values usually leads to finding an inadequate minimum of the optimization function. However, dmneural network training overcomes this deficiency by finding the starting values from a smaller linear combination of input variables in the nonlinear model. The optimization function that you usually want to minimize is easier to find from the grid search procedure that is applied in each iteration.
- Also, the reduction of the number of input variables in the model will dramatically decrease the number of internal calculations called function calls and reduce the amount of computational time for the iterative grid search procedure. Typically, in the iterative grid search procedure several function calls are required at each iteration in order to achieve convergence to the closed form solution, requiring a tremendous amount of computational time. This is especially true with extremely large training data sets that the nonlinear model is trying to fit since the iterative grid search procedure must make several passes through the entire data set for each parameter estimate in order to determine the next-best set of estimates at each iteration. In dmneural network training, a set of grid points are determined by the set of principal components and a multidimensional frequency table that is incorporated to dramatically reduce processing time since segments of the training data set are trained instead of the entire data.
- A serious problem in neural network modeling using the iterative grid search is terminating the procedure too early. In other words, the optimization process has a problem in deciding when the iterative grid search procedure should stop in order to determine the best set of linear combinations of estimates from the input variables in the neural network model. This is overcome by constructing nonlinear models with fewer input variables in the nonlinear model, which corresponds to the reduction in the dimensionality of the error space in finding the optimum minimum in the sum-of-squares error function. The nonlinear estimation problem in common neural network models is seriously under determined, which yields highly rank-deficient Hessian matrices that will result in extremely slow convergence to the nonlinear optimization algorithm.

Dmneural Network Training Algorithm

Initially, the dmneural network training procedure performs principal component analysis of the training data set to determine the best set of principal components from the input variables in the multiple linear regression model. The best set of principal components that best explains the variability in the target variable is determined by the r-square modeling selection criterion, which you want to be as large as possible. The algorithm then obtains a set of grid points from the selected principal components and a multidimensional frequency table from the training data set. The multidimensional frequency table consists of the frequency counts of the selected principal components at a specified number of discrete grid points. An activation function is applied to the linear combination of input variables and eigenvectors, that is, principal components, that best explains the variability in the target variable. The principal components are uncorrelated with each other. The following eight different activation functions are applied independently to the linear combination of principal components, that is, denoted by x, from the training data set at each stage of the dmneural network training algorithm that best explains the target values.

- **Square:** $f(x) = (\beta_0 + \beta_1 \cdot x_i) \cdot x_i$
- **Hyperbolic Tangent:** $f(x) = \beta_0 \cdot \tanh(\beta_1 \cdot x_i)$, where $\tanh(x) = [(e^{\beta x} - e^{-\beta x}) / (e^{\beta x} + e^{-\beta x})]$
- **Arc Tangent:** $f(x) = \beta_0 \cdot [(2/\pi) \tan^{-1}(\beta_1 \cdot x_i)]$
- **Logistic:** $f(x) = \exp(\beta_0 \cdot x_i) / [1 + \exp(\beta_1 \cdot x_i)]$
- **Gauss:** $f(x) = \beta_0 \cdot [\exp(-\beta_1 \cdot x_i^2)]$
- **Sin:** $f(x) = \beta_0 \cdot \beta_1 \cdot [\sin(x_i)]$
- **Cos:** $f(x) = \beta_0 \cdot \beta_1 \cdot [\cos(x_i)]$
- **Exponential:** $f(x) = \beta_0 \cdot \beta_1 \cdot [\exp(x_i)]$

The Activation Function: The activation function is introduced into the model to account for the nonlinearity in the data. The activation function is analogous to neural network modeling and the link function that is applied to transform the output from the model into a desirable range. Dmneural network modeling is essentially nonlinear regression modeling since the nonlinear activation function is introduced into the principal components model. In addition, optimization techniques are applied since the nonlinear activation function that is applied to the model will not result in a closed-form solution to the set of *k* equations of partial derivatives that are set to zero, then solving the *k* principal component estimates. Therefore, the iterative optimization technique is performed in iteratively searching for the minimum of the nonlinear error function. The activation function is a mathematical transformation of the summarized input variables and the associated coefficients that are the eigenvectors from the principal components model to create the fitted values. The activation function is in two parts. The first part applies a combination function that accumulates all the input units into a single value. The combination function is a general-linear combination of each input variable that is multiplied by its associated eigenvectors, that is, the principal components, that are added together into a single value. The second part of the activation function is called the transfer function, which usually applies a nonlinear transformation to these summarized principal components to calculate the fitted values for each case.

The Link Function: A link function is applied to transform the fitted values of the target values in the dmneural network training procedure. The type of link function that is applied depends of the level of measurement of the target variable. For interval-valued target responses, an identity link function, with no transformation, is applied to the fitted values. For binary-valued target responses, a logistic link function is applied that transforms the predicted values into probabilities or proportions, similar to the logit link function that is automatically used in logistic regression modeling.

- **Identity:** $h = f(x)$
- **Logistic:** $h = \exp[f(x)/(1 + \exp(f(x))]$

The Stages of Dmneural Network Training

By default, there are three separate stages that are applied in the dmneural network training procedure. In the first stage, the target variable is applied to determine the best set of principal components in explaining the variability in the target values. After the first stage, the algorithm then uses the residuals from the previous stage playing the role of the target variable. The sum-of-squares error or the misclassification rate is the modeling selection criterion that is applied in determining the best activation function to be used. The reason that the modeling estimation procedure is called additive nonlinear modeling is because an additive stage-wise process is applied. In other words, all the nonlinear models at each stage are added together in creating the final additive nonlinear model for assessment.

The **Princomp/Dmneural** node uses the target values at the first step, then uses the residual values as the target variable for the model in order to retrieve additional information from the residual values in predicting the variability in the target variable. The dmneural model constructs an additive nonlinear model as follows.

$$\hat{y} = \sum_{i=1}^{nstages} g(f(x, \alpha)), \text{ where } f \text{ is the best activation function at the } i^{th} \text{ stage and } g \text{ is the link function.}$$

$$= \alpha + f_1(x, y) + f_2(x, e_1) + f_3(x, e_2), \text{ where } f_i \text{ is the nonlinear model at the } i^{th} \text{ stage for some constant } \alpha.$$

Missing Values

In both principal component and dmneural network modeling, the node will automatically impute missing values in any one of the input variables in the model. For interval-valued input variables, the missing values are calculated from their own mean. For categorical input variables, the missing values are treated as an additional category.

For dmneural network model, when there are missing values in the target variable, then those observations are removed from the model. However, the predicted values of these same observations are still computed. For interval-valued target variables, the predicted values are computed by the average values of the target variable. For categorically-valued target variables, the estimated probabilities for each class level will be automatically set to the prior probabilities. If the prior probabilities are not specified, then the proportion of each class level of the target variable is used in the dmneural network model.

Principal components analysis is discussed in greater detail in Johnson and Wichern (2002), Hastie, Tibshirani, and Friedman (2001), Khattree and Naik (2000), Lattin, Carroll, and Green (2003), Stevens (1982), and Multivariate Statistical Methods: Practical Research Applications Course Notes, SAS Institute (2004).

Data tab

The **Data** tab is designed for you to select the data sets for the analysis. The data sets are usually created from the **Input Data Source** node. You may press the **Select...** button that displays the **Imports Map** window to select an entirely different data set that has been created within the currently opened Enterprise Miner project that can be used as the active training data set to fit the principal component model. The **Properties...** button will allow you to view the file administrative information and a table view of the selected data set. The validation data set might be applied to verify the reported variability in the data and the number of principle components that are selected from the principal component model or determine if the network training estimates are consistent in comparison to the active training data set from the dmneural network model.

Variables tab

Initially, the **Variables** tab will be displayed as you open the **Princomp/Dmneural** node. The **Variables** tab displays a table view of the various properties that are assigned to the listed variables from the metadata sample. The various modeling roles must be specified from either the **Input Data Source** node or the **Data Set Attribute** node. The tab is similar to the **Variables** tabs in the other modeling nodes, therefore, the tab will not be displayed. The tab will give you one last chance in adding or removing input variables from the model.

General tab

The **General** tab is designed for you to select either the dmneural network training or the standard principal components analysis. From the tab, you may either select the **Dmneural** check box to perform dmneural network training or select the **Only do principal components analysis** check box to perform principal components modeling. The default is dmneural network training.

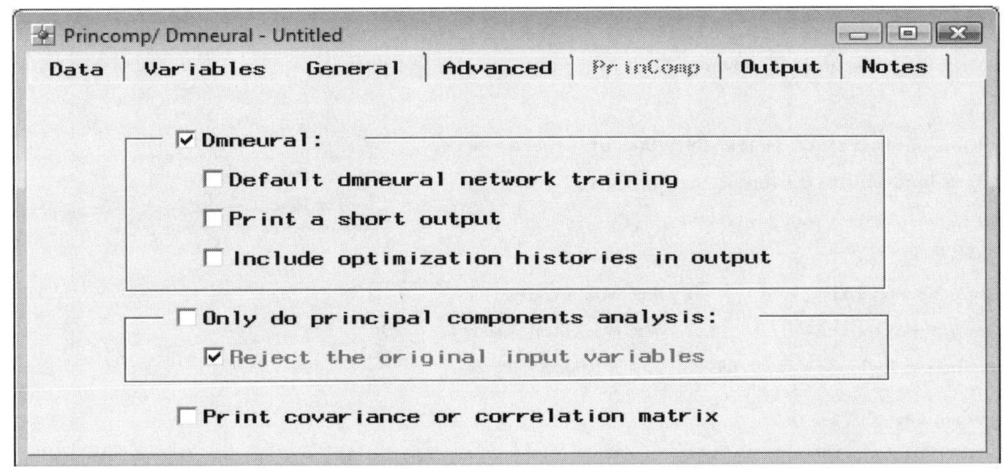

*The **General** tab is used to specify either dmneural network model or principal component analysis.*

- **Dmneural** (default)**:** Selecting the check box will allow you to specify the following three option settings for the dmneural network modeling results. However, clearing the check box will result in the tab automatically selecting the **Only do principal components analysis** option for principal component analysis. In Enterprise Miner, you can not select both types of analysis.
  - **Default dmneural network training:** Clearing the check box will result in **Advanced** tab unavailable for viewing, which will allow you to select the more advanced option settings to the network model. The option settings for dmneural network training will be explained shortly.
  - **Print a short output:** The listed results from dmneural network modeling will print the various goodness-of-fit statistics for each activation function at each stage within the **Results Browser**.
  - **Include optimization histories in output:** The listed results will print the development of the objective function and the number of function calls at each iteration for each activation function at each stage within the **Results Browser** along with all the other procedure output listings.
- **Only do principal components analysis:** Selecting the check box will result in the node performing principal component analysis with the **PrinComp** tab available for you to specify the various configuration option settings for the analysis that will be discussed shortly. Conversely, clearing this option will automatically select the **Dmneural** option for dmneural network training. By default, the principal components are calculated from the eigenvalues and eigenvectors of the corrected variance–covariance matrix. Again, the eigenvalues are the diagonal entries of the corrected variance–covariance matrix. The principal components are the linear combination of the input variables with the coefficients equal to the eigenvectors of the corrected variance–covariance matrix.
  - **Reject the original input variables:** Selecting the check box will result in the node rejecting the input variables from the subsequent analysis. By default, the check box is automatically selected. This will result in the original input variables being removed from the subsequent nodes, which will be automatically set to a variable role of **rejected**. Therefore, these same input variables will be removed from the subsequent modeling fit. However, clearing the check box will instruct the node to include all the input variables and the corresponding principal components that are selected for the subsequent modeling fit.
- **Print covariance or correlation matrix:** Prints the variance–covariance matrix or the correlation matrix that is selected for the analysis.

Advanced tab

Clearing the **Default dmneural network training** check box from the previous **General** tab will allow you to view the various configuration settings for dmneural modeling within the **Advanced** tab. The following options determine the number of principal components to select for the dmneural model. However, it is probably a good idea, especially if you are rather inexperienced in this type of iterative neural network modeling, to rely on the default settings that SAS has carefully selected before making the following adjustments.

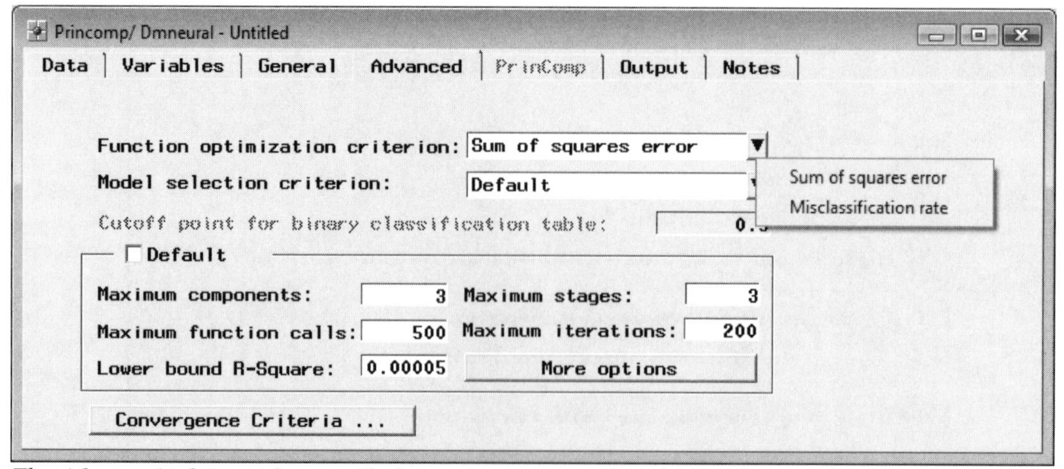

*The **Advanced** tab is used to specify the configuration settings for the dmneural network training process.*

Function optimization criterion: This option will allow you to specify the objective function to be minimized in the iterative grid search procedure. Select the drop-down arrow button from the **Function optimization criterion** display field to select from the following options:

- **Sum of squares error** (default)**:** The sum-of-squares error statistic that you want to minimize that is usually applied to the interval-valued target variable in the model. This is the default.
- **Misclassification error:** The misclassification rate of the observed and predicted target values.

Model selection criterion: This option selects the best model among the eight separate activation functions that are applied at each stage of the dmneural network training procedure.

- **Default:** This is the default option that is based on the specified objective function.
- **Sum of squares error:** Selects the best model based on the smallest sum-of-squares error.
- **Misclassification error:** For interval-valued targets, the best model is selected that produces the largest accuracy rate. The accuracy rate is calculated from the Goodman–Kruskal gamma coefficient that is applied to the observed-predicted frequency table using the percentiles of the target for row and column definitions. The Goodman-Kruskal can have negative values for extremely bad fits.
- **Cutoff point for binary classification table:** This option is only available for binary-valued target variables in the model. The default cutoff probability is .5. For binary-valued target variables, the observation is assigned to the target event, assuming that the fitted probability is greater than the specified cutoff probability. Otherwise, the observation is assigned to the target nonevent.

By default, the **Default** check box is checked. Therefore, clear the **Default** check box and the following options will be made available in order to specify the following option settings for dmneural network training:

- **Maximum components:** Selects the maximum number of principal components that are selected to the model in explaining the variability in the target variable at each stage. The default is 3 principal components, although you may specify the number of components between 2 and 6. However, SAS recommends that better values are between 3 and 5 because of the dramatic increase in processing time for values larger than 5. The option is designed to increase the accuracy to the nonlinear model.
- **Maximum stages:** Specifies the maximum number of stages. The maximum number of stages that can be entered is between 1 and 10. The default is 3. The first stage of the modeling procedure is to predict the target values. Thereafter, from the second-stage and beyond, the residuals of the model replace the target values which the model is trying to predict.
- **Maximum function calls:** The maximum number of function calls at each iteration of the iterative grid search procedure. The default is 500 function calls, which is usually sufficient to reach convergence in the optimization process.
- **Maximum iterations:** The maximum number of iterations in each optimization. The default is 200, which is usually sufficient to reach convergence in the optimization process.
- **Lower bound R-square:** The smallest possible R^2 statistic that determines the set of principal components at each stage of the optimization process. The best set of principal components that best explains the variability in the target variable is determined by the R^2 statistic. The default is .0005.
- **More Options:** Selecting this option button will open the subsequent **More advanced options** window that is illustrated in the following diagram.

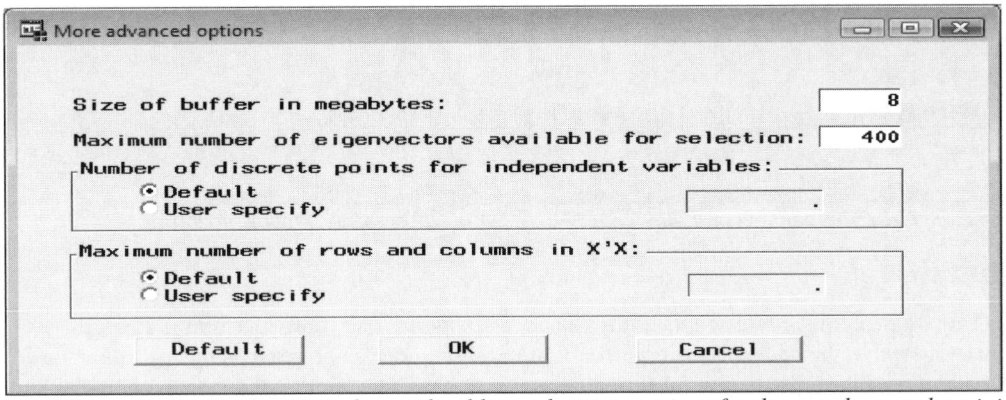

More advanced options

Size of buffer in megabytes: 8

Maximum number of eigenvectors available for selection: 400

Number of discrete points for independent variables:
Default
User specify

Maximum number of rows and columns in X'X:
Default
User specify

Default OK Cancel

*The **More advanced options** window with additional option settings for dmneural network training.*

The following are the various options that may be selected from the **More advanced options** window.

- **Size of buffer in megabyte:** This option will rarely need to be adjusted. However, it will provide you with some performance enhancements for network training and more flexibility in calculating the inverse of the correlation matrix X'X since it takes a lot of memory. The default is 8 megabytes.
- **Maximum number of eigenvectors available for selection:** The maximum number of eigenvectors to select. The default is 400. This option accepts any integer value greater than one.
- **Number of discrete points for independent variables:** Since the dmneural training algorithm initially selects the best set of principal components that best explains the variability in the target variable from the r-square selection criterion, the next step is to determine the set of grid points from the selected number of principal components and a multidimensional frequency table from the training data set. The default value depends on the number of principal components that are selected in the dmneural network training from the **Maximum components** entry field. By default, this value is missing, indicating to you that the maximum number of discrete points is set at run time based on the amount of training data and the complexity of the data. By default, the number of discrete points in SAS is as follows: for 2 or 3 principal components, the default number of discrete points is 17, for 4, 5 and 6 principal components, the default number of discrete points is 15, 13 and 11, respectively.
- **Maximum number of the rows and columns of X'X:** This option is designed to restrict the number of input variables in the X'X matrix, which has nothing to do with the number of observations. For example, suppose you have 10,000 input variables. By default, Enterprise Miner uses the first 3,000 input variables. The default value depends on the size of the random access memory. Select the **User specify** radio button to increase the number of input variables for dmneural training with consideration of the amount of available memory in your computer.

From the **Advanced** tab, select the **Convergence Criteria** button to change the nonlinear convergence criteria value that determines the stopping point in the iterative grid search procedure used in the dmneural training algorithm. Setting the convergence criteria value too high might result in the grid search procedure failing to meet the optimum error. Conversely, setting the convergence criteria value too low will result in an excessive number of iterations of the iterative grid search procedure that might result in an inappropriate fit. The convergence criteria value that is used for dmneural training is the absolute gradient value. The absolute gradient is from the sum-of-squares error function, that is, the first derivative of the sum-of-squares error function with respect to the parameter estimates in the nonlinear model from the current iteration in addition to calculating the maximum of those values. The default gradient convergence criterion is 1E-8 with an absolute gradient convergence of .0005. Convergence is assumed when either convergence criterions is met, that is, the gradient error function falls below the specified convergence criterion value of 0.00000001 or the maximum gradient error function falls below the specified convergence criterion value of .0005. Press the **Default** button to revert back to the displayed convergence criterion values that Enterprise Miner has carefully selected.

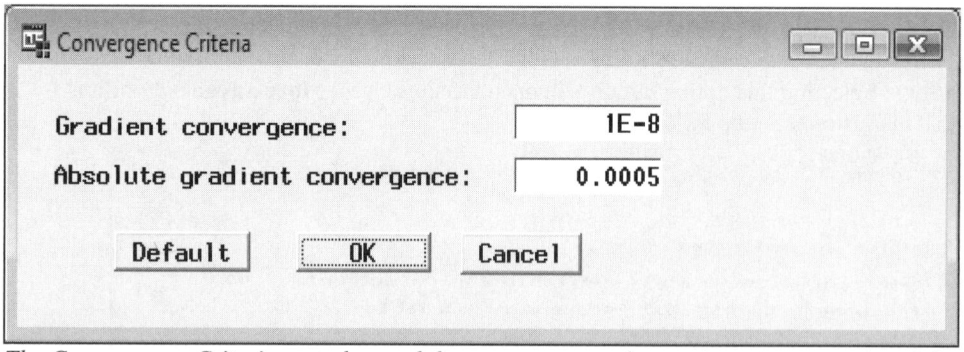

The ***Convergence Criterion*** *window and the convergence values to iterative network training.*

PrinComp tab

Selecting the **Only do principal components analysis** check box will allow you to perform principal components analysis within the node to specify the various configuration option settings for the following principal component results. However, the following adjustments performed within the **Princomp** tab will not affect the principal components selected from the dmneural network training procedure. It is important to realize that at least one principal component is always written to the scored data set and passed along to the subsequent nodes, regardless of the various criteria that are selected.

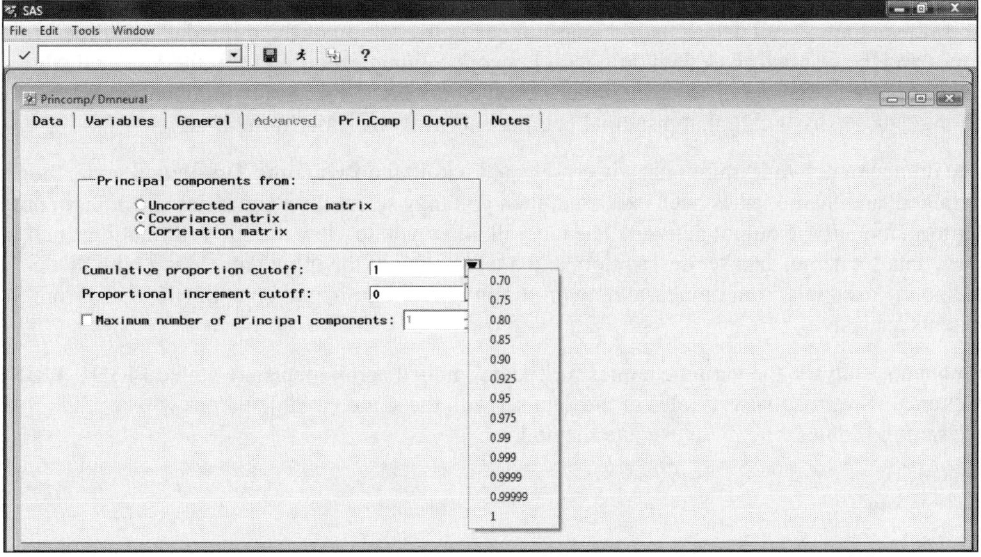

*The **PrinComp** tab and the various cutoff criteria of the principal component model.*

The principal components that are computed are based on the three separate matrices as follows:

- **Uncorrected covariance matrix:** The number of principal components is determined by the linear combination of the input variables in the data that best explains the uncorrected covariance structure. The uncorrected covariance structure is based on the input variables that are left unstandardized.
- **Covariance matrix** (default)**:** This is the default. The number of principal components is determined by the linear combination of the input variables in the data that best explains the variance –covariance structure. It is recommended that the input variables must have the same range of values with approximately the same mean and variance to use the variance–covariance structure.
- **Correlation matrix:** The number of principal components are created by the linear combination of the input variables in the data that best explains the correlation structure.

The total number of principal components depends on the number of input variables and the number of class levels of the categorical input variables. All the scaled principal components are ranked from the highest to the lowest that is defined by their contribution to the total variance. The following cutoff criteria determine the number of principal components that are passed to subsequent nodes and written to the scored data set:

- **Cumulative proportion cutoff:** This option determines the number of principal components based on the cumulative proportion of the total variability in the data that are explained by the principal components. The valid proportions range from .7 to 1. However, the default is 1. In other words, all the principal components with a cumulative proportion of total variability in the data greater than the specified cutoff value are not passed along to the subsequent node.

- **Proportional increment cutoff:** This option determines the cutoff proportion of the explained variability for each principal component. By default, after the cumulative proportional variance has reached 0.9, then the principal components with a proportional increment less than this cutoff value are not passed to subsequent nodes in the process flow. The default value is 0. However, you may specify any proportion between the values of 0 and .1.

- **Maximum number of principal components:** By default, the check box is not selected. However, if you select the check box, then this option will determine the maximum number of principal components that are passed along to the subsequent nodes. The option will allow you to automatically generate the number of principal components written to the output data set that are used as the input variables for the subsequent modeling design such as the nearest neighbor model in the **Memory-Based Reasoning** node. The default is one.

Output tab

The output scored data sets depend on whether dmneural network training or principal components analysis is performed. That is, the output scored data set will consist of each observation of the input data set along with the predicted values and the residuals based on dmneural network training specified from the **General** tab. Otherwise, the output scored data set will contain the principal components along with all other input variables in the active training data set, assuming that principal components analysis is performed.

Assuming that the dmneural network training has been selected within the **Princomp/Dmneural** node, the model has been trained, and the node has been executed, then you may select the **Properties...** button in order to view the file information of the output data set. The tab will allow you to view the file information, such as the data set library, data set name, data set description, and a table view of the input data, along with the predicted values and the residuals from dmneural network training or the principal component scores from principal components analysis.

In principal components analysis, the variable names for the n principal components are called PRIN1, PRIN2, ..., PRINn. However, SAS warns that variables in the data set with the same variable names will result in SAS overwriting these same variables when you execute the node.

Data set details

Information | Table View

☑ Variable labels

	TEAM	FName	LName	POS1	PRIN1	PRIN2	PRIN3	PRIN4	PRIN5	PRIN6	PRIN7
17	BOS	Mark	Bellhorn	2B	361.413685	131.93405558	-191.5756381	-150.1303824	-62.932434	-77.38099099	43.680399809
18	CLE	Ronnie	Belliard	2B	360.73225311	156.9654554	-163.2081827	-218.3784976	-48.10095449	-81.12027651	-8.987046681
19	HOU	Carlos	Beltrán	OF	404.21502446	169.93210385	-172.2162651	-187.2538728	-65.55536743	-59.00279693	-41.14785082
20	LA	Adrián	Béltre	3B	397.91793334	149.67449	-174.4767353	-225.3315755	-30.82421186	-65.05500452	-46.79324657
21	MIL	Gary	Bennett	C	119.73850319	54.537064066	-55.78136452	-74.71041549	-18.3780889	-29.1956929	-2.386911082
22	TOR	Dave	Berg	1B	84.926203694	38.436956801	-42.5334662	-54.88756754	-4.572366627	-16.53223323	1.13155567
23	HOU	Lance	Berkman	OF	389.17662937	133.60136023	-165.9870143	-177.1929228	-77.93034032	-87.82073959	-37.96205327
24	KC	Ángel	Berroa	SS	290.76474646	148.23520364	-134.5129766	-179.806294	-29.73012325	-53.77861566	2.3773566253
25	BAL	Larry	Bigbie	OF	301.70825518	126.53340541	-148.8373167	-158.9317555	-37.55262992	-56.51306313	11.523569441
26	HOU	Craig	Biggio	OF	376.81979721	170.53353426	-168.8810914	-235.7776846	-39.43816022	-71.71891125	-16.70906323
27	CLE	Casey	Blake	3B	378.59645653	147.85975603	-187.1524704	-191.0427479	-48.5584731	-75.81754481	8.8484745392
28	TEX	Hank	Blalock	3B	412.89803628	153.26287722	-203.6099832	-205.0376985	-49.66129831	-82.69571464	3.6961850244
29	MIN	Henry	Blanco	C	177.11935366	76.851424181	-88.31768176	-105.5663744	-17.98994147	-37.92708215	1.0120355376
30	SEA	Willie	Bloomquist	3B	111.73012835	56.93132808	-56.55142564	-59.62601575	-15.26159206	-13.78611281	12.723412121
31	TB	Geoff	Blum	3B	189.11557858	85.110182011	-92.40554997	-115.1738029	-22.43199313	-40.36173543	1.0674280811
32	SF	Barry	Bonds	OF	344.69959022	72.243846398	-113.4143719	-91.06940509	-130.4552368	-99.77851237	-105.0788667
33	SEA	Bret	Boone	2B	361.68523647	149.62679716	-183.028918	-191.6648529	-44.97116237	-68.0420205	13.651296482
34	CHI	Joe	Borchard	OF	116.04706265	47.765815492	-65.82876895	-53.6459804	-16.26661042	-23.62527823	14.780413932
35	LA	Milton	Bradley	OF	325.97828766	134.65292191	-159.9185066	-159.4810476	-55.76963843	-62.55776735	10.943000936
36	MIL	Russell	Branyan	3B	109.32290365	36.464129709	-66.22348102	-40.44221962	-13.3012986	-20.27602491	25.313175336

The scored data set with the various principal component scores from principal components analysis.

Data set details

Information | Table View

☑ Variable labels

	BAD	LOAN	MORTDUE	VALUE	REASON	JOB	YOJ	DEROG	DELINQ	CLAGE	NINQ	CLNO	DEBTINC	Predicted: DEBTINC	Residual: DEBTINC
1	1	1100	25860	39025	HomeImp	Other	10.5	0	0	94.36667	1	9	.	30.064564236	.
2	1	1300	70053	68400	HomeImp	Other	7	0	2	121.8333	0	14	.	32.731557561	.
3	1	1500	13500	16700	HomeImp	Other	4	0	0	149.4667	1	10	.	29.578601862	.
4	1	1500	.	.			.	.	.	.	.	.	.	23.810734475	.
5	0	1700	97800	112000	HomeImp	Office	3	0	0	93.33333	0	14	.	33.876112806	.
6	1	1700	30548	40320	HomeImp	Other	9	0	0	101.466	1	8	37.1136	30.195705952	6.9179076061
7	1	1800	48649	57037	HomeImp	Other	5	3	2	77.1	1	17	.	32.056556459	.
8	1	1800	28502	43034	HomeImp	Other	11	0	0	88.76603	0	8	36.8849	29.233747026	7.6511470672
9	1	2000	32700	46740	HomeImp	Other	3	0	2	216.9333	1	12	.	31.798869083	.
10	1	2000	.	62250	HomeImp	Sales	16	0	0	115.8	0	13	.	33.99177242	.
11	1	2000	22608	.			18	.	.	.	.	.	.	21.838117181	.
12	1	2000	20627	29800	HomeImp	Office	11	0	1	122.5333	1	9	.	30.565601708	.
13	1	2000	45000	55000	HomeImp	Other	3	0	0	86.06667	2	25	.	35.112514224	.
14	0	2000	64536	87400		Mgr	2.5	0	0	147.1333	0	24	.	33.650198304	.
15	1	2100	71000	83850	HomeImp	Other	8	0	1	123	0	16	.	32.791511747	.
16	1	2200	24280	34687	HomeImp	Other	.	0	1	300.8667	0	8	.	28.952405493	.
17	1	2200	90957	102600	HomeImp	Mgr	7	2	6	122.9	1	22	.	34.361475803	.
18	1	2200	23030	.			19	.	.	.	.	.	3.71131	21.786162035	-18.07484974
19	1	2300	28192	40150	HomeImp	Other	4.5	0	0	54.6	1	16	.	32.094948711	.
20	0	2300	102370	120953	HomeImp	Office	2	0	0	90.99253	0	13	31.5885	34.095293963	-2.506790785

The scored data set from the training data set with the fitted values from dmneural network model.

Viewing the Princomp/Dmneural Results

General Layout of the Results Browser within the Princomp/Dmneural Node

- **Model tab**
- **Statistics tab** *(dmneural training only)*
- **Plot tab** *(dmneural training only)*
- **PrinComp tab** *(principal components analysis only)*
- **Log tab**
- **Output tab**
- **Notes tab**

Model tab

The **Model** tab displays the profile information, model creation date, modeling variables for the principal components model, and the various configuration settings that have been previously specified.

General subtab

The **General** subtab displays the file creation date, modification date of the scored data set and target variable to predict in dmneural training, which will not be illustrated.

Variables subtab

The **Variables** subtab displays the standard table listing of the numerous variables involved in either network training or principal components analysis.

Settings subtab

The **Settings** subtab displays the various configuration settings that are specified for either the dmneural network training or the principal component model from the previous node option settings. The following listing shows the various configuration settings from the principal components analysis.

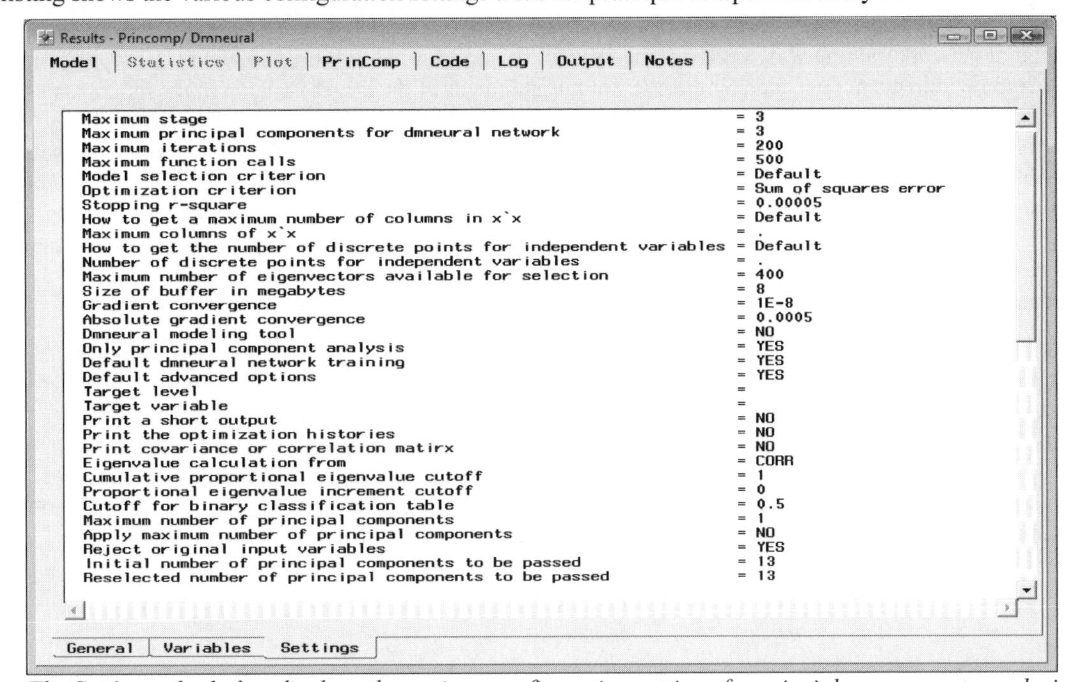

*The **Setting** subtab that displays the various configuration settings for principle components analysis.*

The following results are by fitting the dmneural network model to the home equity data set in predicting the ratio between debt and income in which the various preprocessing routines are performed to the data such as removing outliers from the fit. Conversely, in the subsequent principal components analysis, the 2004 major league baseball data set is applied. Since the various hitting categories have a wide range of values, the principal components were calculated from the correlation matrix.

Statistics tab

The following **Statistics** tab displays the standard goodness-of-fit statistics that are displayed by training the dmneural network model. The **Statistics** tab is only available for viewing when you perform dmneural network training. The various modeling assessment statistics that are displayed in the following table listing are discussed in greater detail from the previous **Regression** node.

Results - Princomp/ Dmneural

Model | Statistics | Plot | PrinComp | Code | Log | Output | Notes

	label	TRAIN	VALID
1	Akaike Information Criterion	10911.639494	.
2	Average Squared Error	49.556382175	48.161134248
3	Average Error Function	49.556382175	48.161134248
4	Degrees of Freedom for Error	2936	.
5	Model Degrees of Freedom	21	.
6	Total Degrees of Freedom	2957	.
7	Error Function	116754.83641	110674.2865
8	Final Prediction Error	50.265295	.
9	Maximum Absolute Error	41.471798618	38.838948452
10	Mean Squared Error	49.910838588	48.161134248
11	Sum of Frequencies	2957	2958
12	Number of Weights	21	.
13	Root Average Squared Error	7.0396294061	6.9398223499
14	Root Mean Squared Error	7.0647603348	6.9398223499
15	Schwarz Bayesian Criterion	11037.470035	.
16	Sum of Squared Error	116754.83641	110674.2865
17	Sum of Case Weights Times Freq	2356	2298

*The **Statistics** tab displays the assessment statistics from dmneural network training for predicting the interval-valued target variable, DEBTINC, from the HMEQ data set.*

Plot tab

Since dmneural network training was specified, then the following **Plot** tab will be displayed. The type of plot that is displayed within the tab depends on the level of measurement of the target variable in the model. The principal component model consists of an interval-valued target variable; therefore, the following predicted plots will be displayed. By default, the scatter plot displays the fitted values against the target values. However, you may view the actual values, predicted values, residuals, and the various input variables in the model. The following plot is used in detecting an inadequate fit to the model with regard to the functional relationships between the modeling terms and the target variable in the model, while at the same time verifying the various modeling assumptions that must be satisfied. The following scatter plot shows little evidence of heteroscedasticity or nonconstant variability in the residual values from the dmneural network model.

The subsequent illustration displays the classification table that is generated when predicting the categorically-valued target variable in the dmneural network training model. The model is based on classifying bad creditors from the home equity data set. The height of the bars is defined by the row percentages of the actual target class levels. Large bars located on the main diagonal will indicate the accuracy of the classification model. From the following frequency bar chart, the actual target class levels are indicated by the F_BAD axis label and the predicted target class levels is indicated by the I_BAD axis label. From the bar chart, the classification model seems to be more accurate in predicting good creditors than bad creditors since the off-diagonal bars display a much higher proportion of the model incorrectly identifying good clients that are actually bad clients that have defaulted on their home loan.

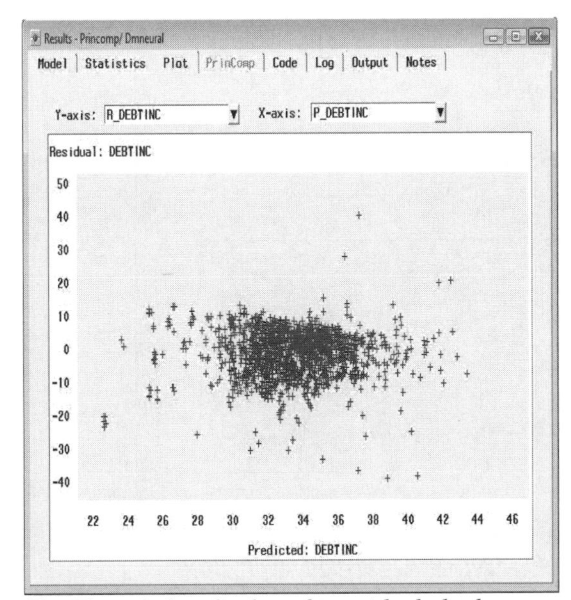

*The **Plot** tab that displays the residual plot by fitting the interval-valued target variable in the dmneural network training model.*

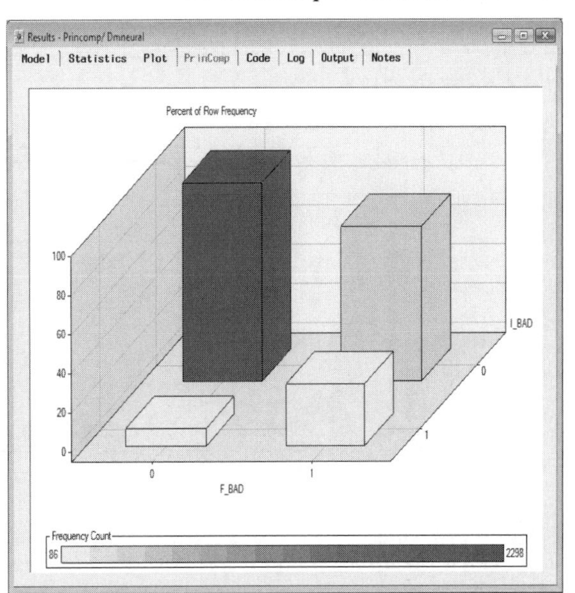

*The **Plot** tab that displays the classification table by fitting the categorically-valued target variable in the dmneural network training model.*

PrinComp Tab

Assuming that principal component analysis is specified from the **General** tab, then the **Result Brower** will display the following **PrinComp** tab. The tab displays the scaled variance, LEV, scree, and cumulative variance plots to graphically view the proportion of variability in the data that is explained by the linear combination of input variables and eigenvectors in the model. The purpose of the following plots is to assist you in determining the number of principal components to select. The following plots are based on the proportion of variability of the input variables in the active training data set across the number of principal components. The principal component cutoff criterion options of **Cumulative proportion cutoff**, **Proportional increment cutoff**, and **Maximum number of principal components** entry fields that are specified from the previous **PrinComp** tab determine the reference line drawn in the following principal component diagrams. The placement of this reference line determines the number of principal components that are passed to the subsequent nodes and written to the output scored data set. Closing the node will result in a window appearing asking you if you would like to write the selected principal components to the output scored data set for further analysis. Generally, a sharp drop in the first occurrence of a leveling-off pattern in the following three plots, that is, the scaled variance, LEV, and scree plots, will indicate the appropriate number principal components to select in explaining the majority of the variability in the data. Conversely, a sharp increase with a leveling-off pattern in the cumulative total variance from the cumulative plot will indicate the appropriate number of principal components to select from that will be written to the output scored data set.

To select the number of principal components to be passed along to the subsequent nodes, select the reference line then hold down the left-click button and drag the reference line over to the desirable number of principal components. Otherwise, select the number of principal components from **The number of principal components to be passed** spin box.

From the following results, the first principal component seems to explain a relative large percentage of nearly 58.2% of the total variability, with the second principal component explaining an additional 13.5% of the total variability in the 2004 major league baseball data. Again, selecting the number of components is a very subjective decision in determining the amount of information that will be lost from the rest of the training data set. For instance, three principal components could be selected since the third principal component will add an additional 10.25% in explaining the variability in the data, and so on. In other words, the number of principal components to select depends on the amount of additional variability that you would like to include, which will explain a significant amount of variability in the data.

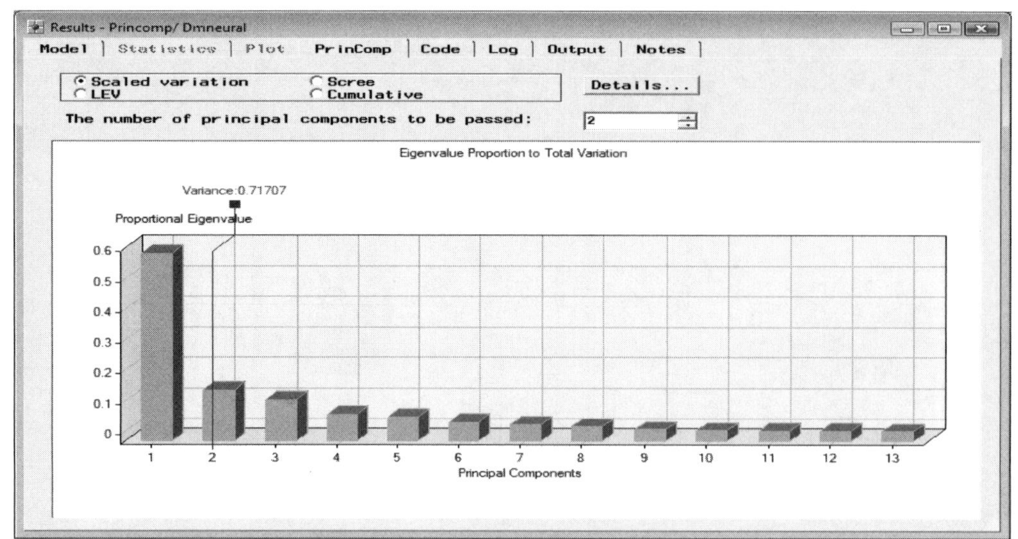

*The **Scaled variation** plot of the proportion of the eigenvalues for each principal component.*

- **Scaled variation:** Plots the proportion of the eigenvalues for each principal component. The number of principal components to select is determined by observing the first occurrence in the bar chart where there is a sharp drop in the size of the bars in comparison to the next adjacent bar or eigenvalue from the correlation matrix. From the plot, this is displayed as the first principal component. However, two principal components should be selected since the second component seems to explain a significant amount of variability. The variability seems to level-off with any additional principal components in the model, which explains the insignificant amount of variability in the data. The proportion of variance explained by each principal component is defined as follows:

$$\frac{\lambda_i}{\sum \lambda_i} = \frac{\lambda_i}{\text{Trace}(\sum)} \text{ for some variance} - \text{covariance matrix } \sum.$$

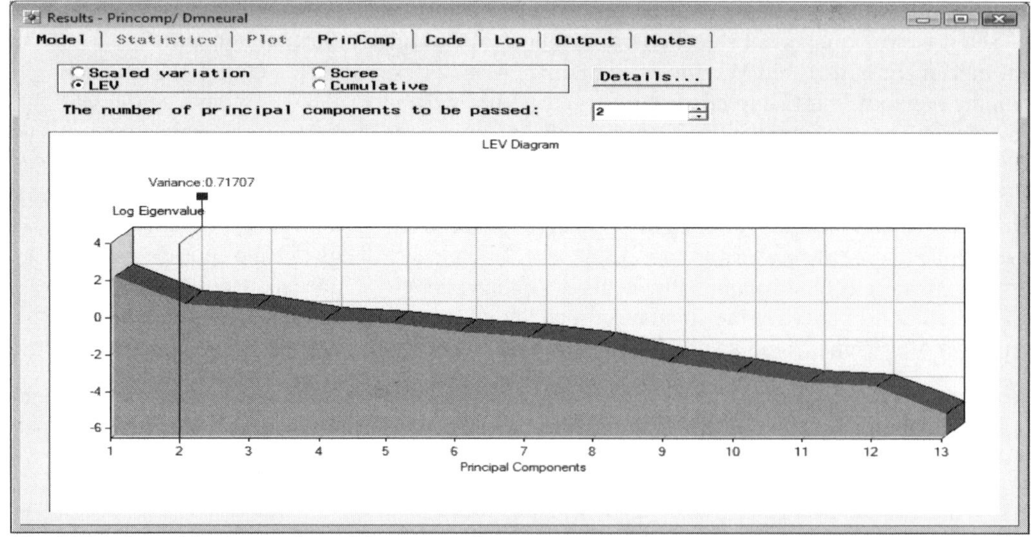

*The **LEV** plot of the logarithm of the eigenvalues against each principal component.*

- **LEV:** Plots the logarithm of the eigenvalues of each principal component. The logarithmic eigenvalue plot is very useful if the eigenvalues generate a wide range of values in the variability of the data. The eigenvalues are calculated from the correlation matrix since there is a wide range of values between the input variables in the model. The number of principal components that should be selected is determined by observing the first occurrence in the plot where the log scores begin to display a leveling-off pattern across each principal component.

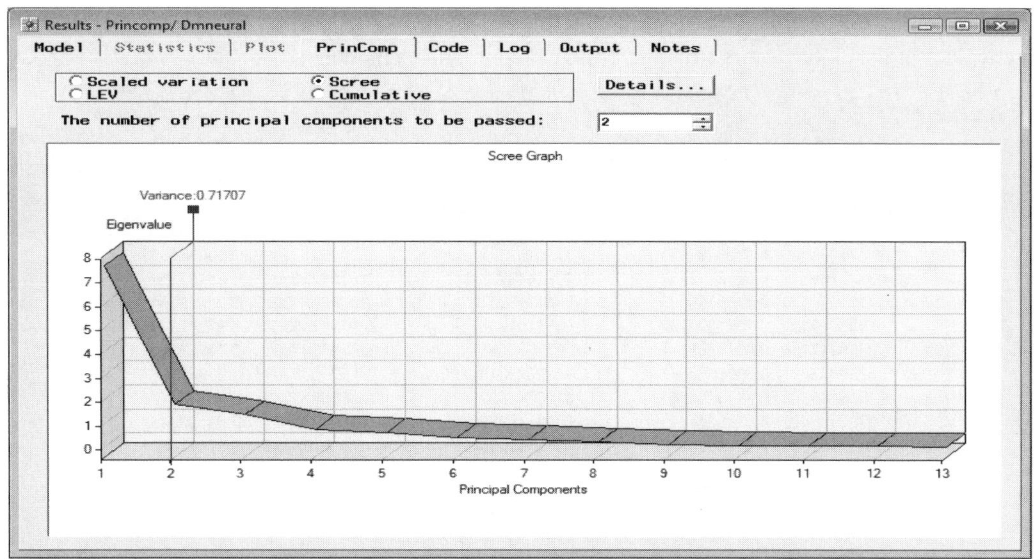

*The **Scree** plot displays the eigenvalues across the number of principal components.*

- **Scree:** Plots the eigenvalues across each principal component. The general rule that is applied in selecting the number of principal components when there is a clear elbow in the curve. This occurs at the second component. In other words, selecting two separate principal components where the elbow is located indicates that adding any additional principal components does not add any significant contribution in explaining the variability in the data.

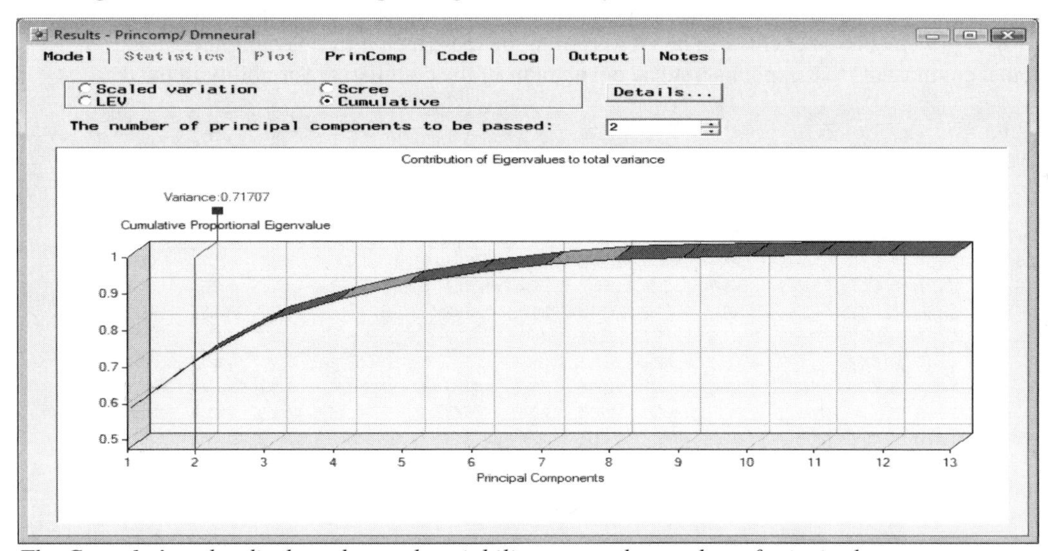

*The **Cumulative** plot displays the total variability across the number of principal components.*

- **Cumulative:** Plots the cumulative proportion of the eigenvalues across the number of principal components. The number of principal components to select is determined by observing a sharp increase in the cumulative proportion with the first occurrence at the point where the plot displays a leveling-off pattern across each adjacent principal component. The cumulative variance explained by each principal component is defined as follows:

$$\text{cumulative variance explained by each principal component} = \frac{\lambda_1 + \ldots + \lambda_k}{\text{Trace}(\Sigma)}$$

Press the **Details...** button that will display the following **Eigenvalues and Eigenvectors** window that lists either the eigenvalues or the eigenvectors from the principal components model.

Eigenvalues and Eigenvectors

(•) Eigenvalues ( ) Eigenvectors

	NAME	EigenValue	Difference	ProportionalEigenvalue	CumProportionalEigenvalue	LogEigenvalue
1	PRIN1	7.5659539145	5.8099168545	0.581996455	0.581996455	2.0236584351
2	PRIN2	1.75603706	0.4229601288	0.1350797738	0.7170762288	0.5630595998
3	PRIN3	1.3330769312	0.6215112439	0.1025443793	0.8196206081	0.2874897524
4	PRIN4	0.7115656874	0.1185166003	0.0547358221	0.8743564302	-0.340287543
5	PRIN5	0.593049087	0.2061273767	0.0456191605	0.9199755908	-0.522478106
6	PRIN6	0.3869217104	0.0974389874	0.0297632085	0.9497387993	-0.949532905
7	PRIN7	0.289482723	0.0986637541	0.0222679018	0.972006701	-1.239659663
8	PRIN8	0.1908189688	0.1139254266	0.0146783822	0.9866850833	-1.656430108
9	PRIN9	0.0768935423	0.0305244813	0.0059148879	0.9925999711	-2.565333382
10	PRIN10	0.0463690609	0.0212590202	0.0035668508	0.996166822	-3.071122832
11	PRIN11	0.0251100408	0.0051178356	0.0019315416	0.9980983636	-3.684487483
12	PRIN12	0.0199922052	0.0152631366	0.0015378619	0.9996362255	-3.912412823
13	PRIN13	0.0047290686	.	0.0003637745	1	-5.354027019

Close

*The **Eigenvalues and Eigenvectors** table listing the eigenvalues of each principal component.*

- **Eigenvalues:** The table listing displays a table view of the eigenvalue, difference between consecutive eigenvalues, proportion of the variance explained by the eigenvalue, cumulative proportion of the variance explained, and the logarithm of each eigenvalue. Each row represents each principal component in the model. A general rule that is typically applied in selecting the number of principal components can be viewed from the table listing by selecting the number of principal components that explains a sufficient amount of the cumulative variability in the data.

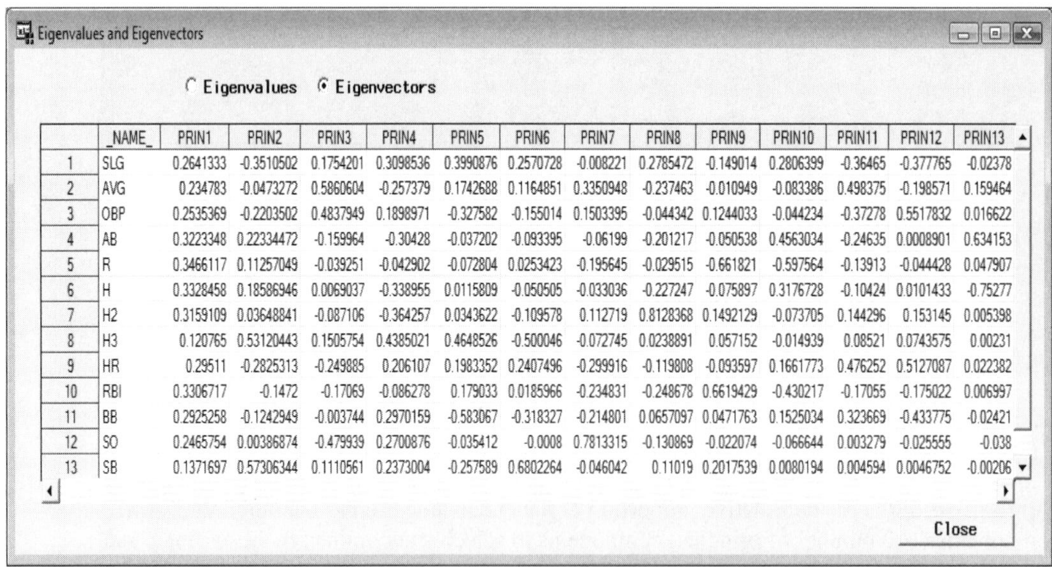

Eigenvalues and Eigenvectors

( ) Eigenvalues (•) Eigenvectors

	NAME	PRIN1	PRIN2	PRIN3	PRIN4	PRIN5	PRIN6	PRIN7	PRIN8	PRIN9	PRIN10	PRIN11	PRIN12	PRIN13
1	SLG	0.2641333	-0.3510502	0.1754201	0.3098536	0.3990876	0.2570728	-0.008221	0.2785472	-0.149014	0.2806399	-0.36465	-0.377765	-0.02378
2	AVG	0.234783	-0.0473272	0.5860604	-0.257379	0.1742688	0.1164851	0.3350948	-0.237463	-0.010949	-0.083386	0.498375	-0.198571	0.159464
3	OBP	0.2535369	-0.2203502	0.4837949	0.1898971	-0.327582	-0.155014	0.1503395	-0.044342	0.1244033	-0.044234	-0.37278	0.5517832	0.016622
4	AB	0.3223348	0.22334472	-0.159964	-0.30428	-0.037202	-0.093395	-0.06199	-0.201217	-0.050538	0.4563034	-0.24635	0.0008901	0.634153
5	R	0.3466117	0.11257049	-0.039251	-0.042902	-0.072804	0.0253423	-0.195645	-0.029515	-0.661821	-0.597564	-0.13913	-0.044428	0.047907
6	H	0.3328458	0.18586946	0.0069037	-0.338955	0.0115809	-0.050505	-0.033036	-0.227247	-0.075897	0.3176728	-0.10424	0.0101433	-0.75277
7	H2	0.3159109	0.03648841	-0.087106	-0.364257	0.0343622	-0.109578	0.112719	0.8128368	0.1492129	-0.073705	0.144296	0.153145	0.005398
8	H3	0.120765	0.53120443	0.1505754	0.4385021	0.4648526	-0.500046	-0.072745	0.0238891	0.057152	-0.014939	0.08521	0.0743575	0.00231
9	HR	0.29511	-0.2825313	-0.249885	0.206107	0.1983352	0.2407496	-0.299916	-0.119808	-0.093597	0.1661773	0.476252	0.5127087	0.022382
10	RBI	0.3306717	-0.1472	-0.17069	-0.086278	0.179033	0.0185966	-0.234831	-0.248678	0.6619429	-0.430217	-0.17055	-0.175022	0.006997
11	BB	0.2925258	-0.1242949	-0.003744	0.2970159	-0.583067	-0.318327	-0.214801	0.0657097	0.0471763	0.1525034	0.323669	-0.433775	-0.02421
12	SO	0.2465754	0.00386874	-0.479939	0.2700876	-0.035412	-0.0008	0.7813315	-0.130869	-0.022074	-0.066644	0.003279	-0.025555	-0.038
13	SB	0.1371697	0.57306344	0.1110561	0.2373004	-0.257589	0.6802264	-0.046042	0.11019	0.2017539	0.0080194	0.004594	0.0046752	-0.00206

Close

*The **Eigenvalues and Eigenvectors** window listing of the eigenvector for each input variable.*

- **Eigenvectors:** The table listing displays a table view of the eigenvectors, that is, the principal components, for each input variable in the analysis. The columns represent each principal component in the model. The eigenvectors listed in each column represent the coefficients in the linear combination of the input variables. Multiplying the listed eigenvectors by each input variable, then summing, will result in the previously listed eigenvalue scores. The listing of the principal component scores will allow you to determine certain characteristics within each principal component that are created from the magnitude of the principal component scores as follows.

Interpretation of the Principal Components

From the previously displayed output listing and the cumulative plot, the first principal component explains nearly 58.2% of the total variability in the data. Since the principal components are uncorrelated, you can add together the proportion of variability in the first two principal components. The first two principal components explain up to 71.7% of the total variability in the input variables. The remaining principal components explain a small amount of the variability in the data.

Since the first principal component explains 58.2% of the total variability in the data, the first principal component seems to be mostly attributed to on base and slugging percentage, number of at bats, runs scored, number of hits, doubles, home runs, runs batted in, and number of walks. Also, the first principal component scores are all positively correlated. The large principal component scores in the first principal component suggest that these baseball hitters are very productive at the plate, with several plate appearances. The second principal component follows the first principal component and accounts for the majority of the remaining variability in the data that seems to be a contrast between the number of at bats, triples, and stolen bases against slugging percentage. The second principal component seems to consist of the established baseball hitters with a lot of speed but not much power, as indicated by a strong positive correlation in the number of at bats, hitting triples, and stealing bases, with a strong negative correlation in slugging percentage. The majority of the variability in the 2004 MLB baseball hitters' data set is summarized by the first two principal components with a reduction of thirteen input variables reduced down to two principal components.

Code tab

The **Code** tab is divided into two separate options. By default, the training code is displayed. The training code will automatically display the data mining PROC DMNEURL procedure. In other words, the DMNEURL procedure will either generate the principal component scores or the dmneural network modeling results and the corresponding scored data sets behind the scenes when you execute the node and train the models. The **Scoring** option button will display the internal SEMMA scoring code that can be copied into any SAS program or applied to a compatible SAS data set in order to calculate new principal component scores or dmneural network estimates.

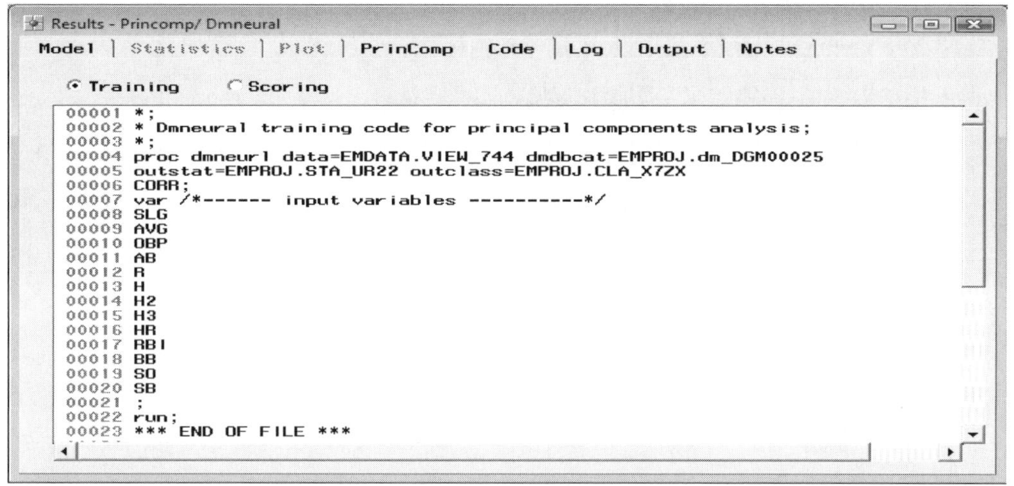

*The **Code** tab displays the SEMMA training code that generates the principal component results.*

The SAS score code from the principal component analysis will not be displayed. The code simply lists the first two principal components that have been previously selected. The principal component scores that can be generated from the subsequent listing will allow you to create scatter plots of the first few principal components that explains a majority of the variability in the data in order to view the variability between each principal component, or view outliers and the various cluster groupings that might be created from the first few principal component scores. Since there are no categorically-valued input variables in the 2004 MLB data set, the scoring code will not display the programming logic in creating separate indicator variables for each class level of the categorically-valued input variables in the analysis. At the beginning, the score code will first impute missing values in any one of the interval-valued input variables in the analysis by its own average values.

The following is the SAS score code from the dmneural network model by selecting the **Scoring** radio option. The scoring code will first display the separate dummy variables that are created for each class level from the nonbinary-valued input variables in the model with the dummy variables that are then standardized. This is followed by imputing missing values from the interval-valued input variables in the model. The interval-valued input variables in the model are then standardized since the input variables have a wide range of values. The code will then display the principal component scores for each input variable in the model at each stage of the iterative model. The code then calculates the fitted values from the squared activation function that is selected at each stage of the model. The predicted values from the additive nonlinear model are calculated by adding the fitted values from the first stage and the residual values in the second and third stage of the model.

Results - Princomp/ Dmneural

Model | Statistics | Plot | PrinComp | Code | Log | Output | Notes

○ Training ◉ Scoring

```
00012 /***************************************************************/
00013 * CREATE STANDARDIZED DUMMY VARIABLES FOR BAD;
00014 /***************************************************************/
00015 _FORMAT=PUT(BAD,BEST12.);
00016 %DMNORMCP(_FORMAT, _NORMFMT);
00017 IF _NORMFMT='0' THEN BAD_1_=0.30032932148039;
00018 /*(1-0.91723259762309)/0.27558881686586 */
00019 ELSE BAD_1_=-3.32826494214934;
00020 /* (0-0.91723259762309)/0.27558881686586  */
00021 IF _NORMFMT='1' THEN BAD_2_=3.32826494214934;
00022 /*(1-0.08276740237691)/0.27558881686586 */
00023 ELSE BAD_2_=-0.30032932148038;
00024 /* (0-0.08276740237691)/0.27558881686586  */
...
00114 /*************************************************************************/
00115 * MISSING VALUE IMPUTATION AND STANDARDIZATION FOR INTERVAL VARIABLES;
00116 /*************************************************************************/
00117 If LOAN=. then _S1_LOAN=0;
00118 else _S1_LOAN=(LOAN-19123.9388794567)/11514.606801211;
00119 If MORTDUE=. then _S2_MORTDUE=0;
00120 else _S2_MORTDUE=(MORTDUE-74603.2812437228)/40818.0784525169;
...
00134 /****************************************/
00135 * Component 1 at Stage 0;
00136 /****************************************/
00137 _SPRIN01=
00138 -0.420285601*BAD_1_+
00139 0.4202856007*BAD_2_+
00140 -0.152927246*REASON_1_+
00141 0.4106357411*REASON_2_+
00142 -0.35884189*REASON_3_+
00143 -0.161943657*JOB_1_+
00144 0.1057312062*JOB_2_+
00145 -0.074102405*JOB_3_+
00146 0.0884206967*JOB_4_+
00147 -0.022936164*JOB_5_+
00148 0.0624006818*JOB_6_+
00149 -0.068656015*JOB_7_+
00150 -0.105620431*NINQ_XA9_1_+
00151 -0.232684043*NINQ_XA9_2_+
00152 0.042361408*NINQ_XA9_3_+
00153 0.3028303*NINQ_XA9_4_+
00154 0.0722775232*_S1_LOAN+
00155 0.0186720344*_S2_MORTDUE+
00156 -0.015711298*_S3_VALUE+
00157 -0.088575003*_S4_YOJ+
00158 -0.113997219*_S5_CLAGE+
00159 0.10488342*_S6_CLNO+
00160 0.1690809609*_S7_INDELINQ+
00161 0.2026048855*_S8_INDEROG;
...
00401 /*****************************************************************/
00402 * Selected activation function at stage 0=SQUARE;
00403 /*****************************************************************/
00404 ;
00405 _YHAT0=0.4270872674
00406 +(0.01437693973772+0.0004981723388*_SPRIN01)*_SPRIN01
00407 +(-0.0108613733431+-0.00153700764569*_SPRIN02)*_SPRIN02
00408 +(0.00882522859131+-0.00154271207078*_SPRIN03)*_SPRIN03
00409 ;
00410 ;
00411 /*****************************************************************/
00412 * Selected activation function at stage 1=ARCTAN;
00413 /*****************************************************************/
00414 ;
00415 _RHAT1=-0.000526002
00416 +-0.0141835887251*ATAN(1.31147421701595*_SPRIN11)
00417 +0.0603263631556*ATAN(-0.10680057945997*_SPRIN12)
00418 +0.0311281647798*ATAN(0.26761700173355*_SPRIN13)
00419 ;
00420 ;
00421 /*****************************************************************/
00422 * Selected activation function at stage 2=SQUARE;
00423 /*****************************************************************/
00424 ;
00425 _RHAT2=-0.001904021
00426 +(0.0054831199943+-0.00073633221385*_SPRIN21)*_SPRIN21
00427 +(-0.00876407496275+0.00125317022432*_SPRIN22)*_SPRIN22
00428 +(0.00599300779161+0.00182477893655*_SPRIN23)*_SPRIN23
00429 ;
00430 ;
00431 _tmpPredict=_YHAT0+_RHAT1+_RHAT2;
00432 P_DEBTINC=_tmpPredict*(78.6543860476344-0.52449921542988)+0.52449921542988;
00433 label P_DEBTINC = 'Predicted: DEBTINC';
00434 R_DEBTINC=DEBTINC-P_DEBTINC;
00435 label R_DEBTINC="Residual: DEBTINC";
00436 DROP _YHAT0 _RHAT1 _RHAT2 _tmpPredict
00437 _S1_LOAN
00438 _S2_MORTDUE
00439 _S3_VALUE
00440 _S4_YOJ
00441 _S5_CLAGE
00442 _S6_CLNO
00443 _S7_INDELINQ
00444 _S8_INDEROG
00445 ;
00446 *--------------------------------------------------------------*;
00447 * Princomp/ Dmneural: DEBTINC: Decisions;
00448 *--------------------------------------------------------------*;
00449 **** NO DECISION CODE;
00450 *** END OF FILE ***
```

The score code that generates the dmneural network estimates to score new data.

Output tab

The **Output** tab displays the procedure output listing based on either dmneural network training or principal components analysis. The principal components analysis results are displayed in the following table. The procedure output listing displays the average values and the standard deviation for each interval-valued input variable and each class level of the categorically-valued input variable that will be followed by the listing of the eigenvalues, the difference between successive eigenvalues, and both the noncumulative and cumulative proportion of total variance of the target responses.

Principal Components Analysis

In our example, there are thirteen interval-valued input variables in the principal components model. The **Princomp/Dmneural** node automatically creates a separate level for each categorically-valued input variable in the model. Also, a separate level is automatically created for all categorical variables with missing values. Therefore, the total number of input variables and the total number of categorical levels in the dmneural network training model determines the total number of principal components. However, at times the number of components might be less than the number of input variables. The reason is that some input variables might have zero variability in their values or some of the input variables in the model might be perfectly correlated to each other. For example, the input variable called on-base percentage and slugging percentage was initially removed from the analysis, or OPS, since it is probably highly correlated with both on-base percentage, or OBP, and slugging percentage, or SLG. In addition, the first eigenvalue will always display the largest value since the linear combination of input variables and the first vector of eigenvectors captures the largest variability in the data. Since the input variables display a wide range of values in their own mean and variance, the correlation matrix was selected to calculate the eigenvalues and corresponding eigenvectors.

The DMNEURL Procedure

Variable	Mean	Std Dev
SLG	0.43441	0.07454
AVG	0.26887	0.03086
OBP	0.33311	0.04083
AB	397.84906	150.98884
R	57.63073	28.90938
H	109.18059	47.67120
H2	22.04313	10.61072
H3	2.24528	2.52805
HR	13.80054	10.40338
RBI	55.19677	28.64350
BB	39.88679	25.51689
SO	71.46361	32.73490
SB	6.47439	9.04112

The descriptive statistics for the input variables in the principle components model.

The DMNEURL Procedure

Eigenvalues of Correlation Matrix

	Eigenvalue	Difference	Proportion	Cumulative
1	7.56595391	5.80991685	0.5820	0.5820
2	1.75603706	0.42296013	0.1351	0.7171
3	1.33307693	0.62151124	0.1025	0.8196
4	0.71156569	0.11851660	0.0547	0.8744
5	0.59304909	0.20612738	0.0456	0.9200
6	0.38692171	0.09743899	0.0298	0.9497
7	0.28948272	0.09866375	0.0223	0.9720
8	0.19081897	0.11392543	0.0147	0.9867
9	0.07689354	0.03052448	0.0059	0.9926
10	0.04636906	0.02125902	0.0036	0.9962
11	0.02511004	0.00511784	0.0019	0.9981
12	0.01999221	0.01526314	0.0015	0.9996
13	0.00472907		0.0004	1.0000

The eigenvalues, the difference in the eigenvalues, the proportion, and the cumulative proportion of eigenvalues from the principle components model.

The following graph displays the first two principal component scores. Since labeling every principal component score will clutter the graph with too many data point labels, the following principal component scores that are labeled to identify the top home run hitters, batting average leaders, and the top base stealers in the major leagues during the 2004 season. The SAS PROC IML code that calculates the principal component scores and the following graph are displayed in my website. From the plot, the top home run hitters that are displayed in circles seem to be mainly clustered in the upper portion of the cluster of points that are closest to the horizontal axis. Conversely, the fastest base runners in the game that are identified by the stars are mainly located at the upper end of the cluster of data points that are closest to the vertical axis. From the principal component plot, Barry Bonds seems to separate himself from all the other hitters in the game, having the second highest batting average along with being the fourth best home run hitter during the 2004 baseball season. This is an indication that the principal component plot can be used to observe outliers in the data.

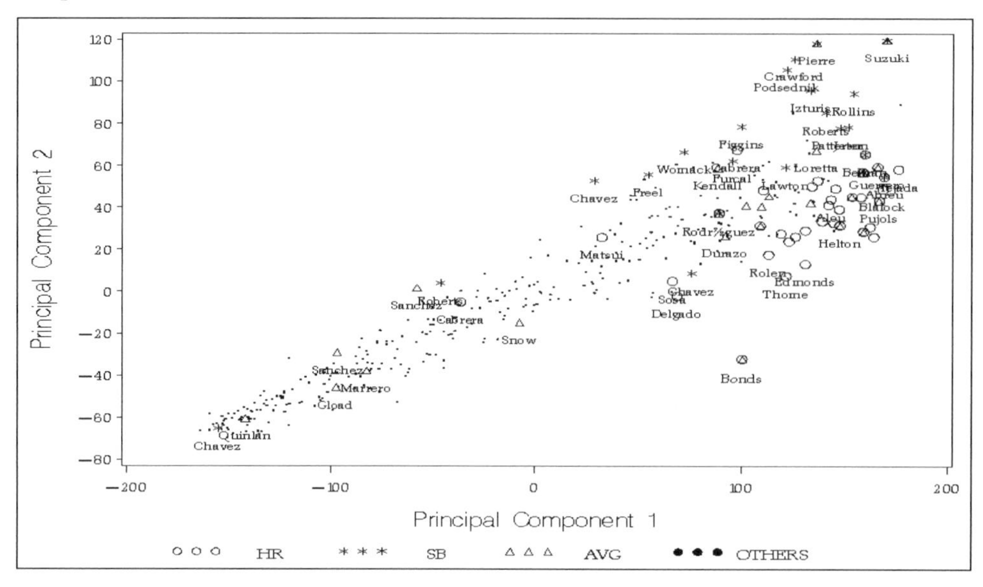

Dmneural Network Training Analysis

Assuming that data mining dmneural network training is requested from the **General** tab, then the following output is displayed from the **Results Browser**. The following listing is based on predicting the debt-to-income ratio, DEBTINC, from the home loan data set by performing a 50–50 split of the input data set. The various preprocessing of the data that was performed from the previous least-squares model in the **Regression** node was applied to the dmneural network model.

The **Output** tab is designed to display the procedure output listing from the underlying PROC DMNEURAL procedure. The following is a partial listing of the **Output** tab from the **Princomp/Dmneural Results Browser** based on the dmneural network training results using the various default settings within the node.

In the first step, principal component analysis is performed in dmneural network training that starts by selecting the best set of predictors or principal components (eigenvalues) in explaining the target values. From the **Advanced** tab, up to three principal components were specified. For the first stage, the second, fourth, and seventh largest eigenvalues were selected that best explained the variability in the target variable that is determined by the r-square modeling selection criterion. The output listing is generated from the PROC DMNEURAL network training procedure that displays the r-square statistic that measures the variability of the target variable explained by each eigenvalue in the model. For example, the second principal component that is the largest eigenvalue in the model captured 2.98% of the variability in the target variable.

The DMNEURL Procedure

Component Selection: SS(y) and R2 (SS_total=441.34596589)

Comp	Eigval	R-Square	F Value	p-Value	SSE
2	5476.058804	0.002986	7.053006	0.0080	440.028123
5	3556.156178	0.000905	2.139626	0.1437	439.628530
1	6237.060377	0.000560	1.324147	0.2500	439.381269

The following procedure output listing displays the various descriptive statistics, the link function that is applied to the target variable, the sum-of-squares selection criterion, maximum number of stages, maximum number of principal components, and the number of grids points in training the additive nonlinear model.

The DMNEURL Procedure

Interval Target	DEBTINC
Number Observations	2957
NOBS w/o Missing Target	2356
Target Range	[0.5244992154, 78.654386048]
Link Function	IDENT
Selection Criterion	SSE
Optimization Criterion	SSE
Estimation Stages	3
Max. Number Components	3
Minimum R2 Value	0.000050
Number Grid Points	17

The following table listing displays the various descriptive statistics for each variable in the predictive model from the training data set. The table listing displays a wide variability in the input variables.

The DMNEURL Procedure

Variable	Mean	Std Dev	Skewness	Kurtosis
DEBTINC	33.47555	7.59988	-0.56679	2.70478
LOAN	23359	11553	2.23514	8.09411
MORTDUE	83950	42543	1.35010	2.57930
VALUE	124821	52473	1.38627	2.21949
YOJ	10.14476	7.52478	0.97913	0.34266
CLAGE	210.68532	80.11704	0.50905	-0.15040
CLNO	25.73345	10.41247	0.82995	1.16282
INDELINQ	0.24745	0.39792	1.52315	0.32019
INDEROG	0.15195	0.32626	2.32444	3.40531

The procedure output listing displays the range of target values that are consolidated into ten separate buckets or groups based on the income-to-debt ratio values from the training data set as a preliminary run through the data. The following listing displays the range of values of the target variable along with a table listing of the number of observations, the cumulative sum and percentages within each group or interval of the transformed categorically-valued target variable.

Percentiles of Target DEBTINC in [0.5244992154 : 78.654386048]

	Nobs	Y Value	Label
1	236	24.016671	0.300680992
2	471	27.573737	0.346208581
3	707	30.082489	0.378318605
4	942	32.657629	0.411278334
5	1178	34.727875	0.437775829
6	1414	36.461211	0.459961134
7	1649	38.139022	0.481435777
8	1885	39.707988	0.501517287
9	2120	41.277127	0.521601011
10	2356	78.654386	1

The table displays the following network training optimization results for each transfer function that is applied at each of the three stages. The following is the optimization table listing that is based on the SQUARE transfer function applied to network training from the first stage. However, the other optimization table listings that apply the other seven transfer functions will not be displayed in order to reduce the amount of listed output. One of the most important result that is displayed in the following table listing is that one of the convergence criteria has been satisfied in the first stage from the squared activation function at the first iteration.

```
------------------- Activation= SQUARE (Stage=0) -------------------

                        The DMNEURL Procedure

                   Levenberg-Marquardt Optimization

                    Scaling Update of More (1978)

                 Parameter Estimates                7

                         Optimization Start

Active Constraints                  0  Objective Function        4.094294962
Max Abs Gradient Element 9.2963085445  Radius                   58.689935569

                         Optimization Results

Iterations                          1  Function Calls                      3
Jacobian Calls                      2  Active Constraints                  0
Objective Function       0.0043075437  Max Abs Gradient Element  1.144856E-13
Lambda                              0  Actual Over Pred Change             1
Radius                   8.3815476969

ABSGCONV convergence criterion satisfied.

          SQUARE: Iter=1 Crit=0.00430754: SSE=123899.446 Acc= 25.7517
```

The following listing displays the minimum optimization function, SSE, listed in descending order from the final iteration for each transfer function applied at the first stage, 0. From the procedure output listing, the squared error function delivered the best goodness-of-fit statistic from the training data set. The Run column displays the number of iterations that were performed since the nonlinear activation function is applied to the model. For interval-valued target variables, the accuracy rate is computed from the Goodman–Kruskal gamma coefficient and the two-way frequency table between the observed and predicted target values. The ordered decile estimates of the actual and predicted target values define the ten separate row and column entries to the two-way frequency table. However, the frequency two-way table will not be displayed to reduce the amount of listed output.

The DMNEURL Procedure

Approximate Goodness-of-Fit Criteria (Stage 0)

Run	Activation	Criterion	SSE	Accuracy
1	SQUARE	0.004308	123899	25.751734
3	ARCTAN	0.004312	124040	24.680383
6	SIN	0.004333	124641	24.727657
2	TANH	0.004335	124676	24.069357
4	LOGIST	0.004336	124716	25.268004
8	EXP	0.004342	124880	24.412919
7	COS	0.004705	135335	34.266285
5	GAUSS	0.004716	135660	-0.520589

The following listing displays the various goodness-of-fit statistics from each stage of the three-stage procedure. That is, the listing displays the activation function, link function, sum-of-squares error, root mean-square error, and modeling assessment statistics, that is, AIC and SBC goodness-of-fit statistics, that you want to minimize. The following summary statistics lists the activation function with the lowest sum-of-squares error at each stage of the iterative network training procedure that is determined by the **Sum of squares error** modeling selection criterion option setting that was selected from the **Advanced** tab. From the procedure output listing, the square activation function resulted in the best fit from the training data set at first and third stage of the dmneural network training procedure.

The DMNEURL Procedure

Summary Table Across Stages

Stage	Activation	Link	SSE	RMSE	Accuracy	AIC	SBC
0	SQUARE	IDENT	123014.3	7.236627	25.163280	9332.769130	9373.122174
1	ARCTAN	IDENT	118997.2	7.128118	30.083983	9268.548542	9349.254630
2	SQUARE	IDENT	116727.0	7.070371	32.454671	9237.167920	9358.227051

The final stage is adding the three separate nonlinear models. The following are the assessment statistics used in measuring the accuracy of the final additive nonlinear model from the partitioned data sets. The validation data set was applied since the parameter estimates are estimated from the training data set. The validation data set was applied since the data set was not used in the development of the predictive model that can be thought of as fitting the model to new data in order to generate a more realistic picture of the performance of the predictive model. From the following procedure output listing, overfitting to the model seems to be of no concern with approximately the same modeling assessment statistics between both partitioned data sets.

Fit Statistics

Statistic	Training	Validation
Akaike Information Criterion	10911.64	.
Average Squared Error	49.56	48.16
Average Error Function	49.56	48.16
Degrees of Freedom for Error	2936.00	.
Model Degrees of Freedom	21.00	.
Total Degrees of Freedom	2957.00	.
Error Function	116754.84	110674.29
Final Prediction Error	50.27	.
Maximum Absolute Error	41.47	38.84
Mean Squared Error	49.91	48.16
Sum of Frequencies	2957.00	2958.00
Number of Weights	21.00	.
Root Average Squared Error	7.04	6.94
Root Mean Squared Error	7.06	6.94
Schwarz Bayesian Criterion	11037.47	.
Sum of Squared Error	116754.84	110674.29
Sum of Case Weights Times Freq	2356.00	2298.00

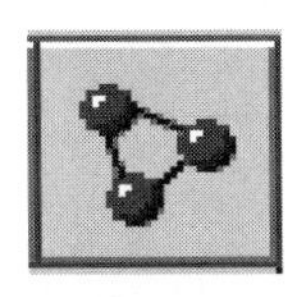

4.6 User Defined Model Node

General Layout of the Enterprise Miner User Defined Model Node

- **Data tab**
- **Variables tab**
- **Predicted Values tab**
- **Output tab**
- **Notes tab**

The purpose of the **User Defined Model** node is that it will allow you to include the fitted values into the process flow from some external modeling design where the fitted values are not generated from the various modeling nodes within Enterprise Miner called *user-defined modeling*. In other words, the node is designed to generate assessment statistics from previously created scored data sets within the process flow. For example, the scored data set can be created from the fitted values that are generated from the numerous modeling procedures that are available in SAS such as the PROC REG, PROC LOGISTIC, or the PROC NLIN modeling procedures, just to name a few. As an alternative, score code can be applied to generate the fitted values from the **SAS Code** node or the scored data set that is created from the **Variable Selection** node. The assessment statistics are calculated within the node by specifying both the target variable and the corresponding fitted values from the scored data set. The **User Defined Model** node will allow you to specify the fitted values from either the predictive or classification model by fitting the interval-valued or categorically-valued target variable to predict. The process of comparing the fitted models from the various modeling nodes and the model that is created from the **User Defined** node is called *integrated assessment.*

One requirement of the node is that the input data set must contain the predicted values. In user-defined modeling, there must be a variable assigned a model role of **predicted** in order to fit the unique target variable to the user-defined model. In predictive modeling, there must be one and only one predicted variable that can be selected as the input variable of the user-defined model. In classification modeling, there must be a predicted variable for each class level of the categorical target variable to predict in the user-defined model. User-defined modeling will not be performed and the **User Defined Model** node will not execute by specifying several predicted variables for each target variable to predict from some previous multivariate regression modeling design. Hence, multivariate regression modeling cannot be performed from the node. For categorically-valued targets, the scored data set will automatically create a separate variable for each target category. Therefore, the user-defined model will contain a separate predicted variable for each target category from the preceding classification model.

One of the most important tabs within the node is the **Predicted Values** tab, which is designed for you to specify the target variable that has been predicted and the corresponding fitted values from some previously fitted model. For categorical valued target variables, the tab will allow you to specify prior probabilities. That is, the node will allow you to adjust the fitted values from some previous modeling fit by the specified prior probabilities for each class level of the target variable. This will enable the node to increase the accuracy of the predicted probabilities that were calculated from some other previous modeling fit, assuming that the correct prior probabilities are specified for the classification model. In addition, incorporating prior probabilities into the model will allow you to determine the accuracy of the user-defined model by comparing the predicted probabilities with the corresponding prior probabilities at each target level. Specifying the correct prior probabilities will generally lead to correct decisions, assuming that the predictive model accurately predicts the variability of the output responses and precisely fits the underlying distribution of the target variable. However, generally, you usually do not know the true nature of the underlying data that the predictive model is trying to fit. Therefore, even specifying the correct prior probabilities for each target level will generally produce incorrect classifications from the predictive model, generating an unsatisfactory fit to the training data set.

Note that another method used in creating the predicted values and the associated target values can be created from a previous modeling node based on the output scored data set that can be specified from the **Output** tab, then connecting the modeling node to the **Data Set Attributes** node in assigning the appropriate model roles to the target and predicted variable in the model. However, there is no benefit in connecting to the **User Defined Model** node to generate the assessment statistics from the corresponding modeling node.

The following are some of the ways of creating a user-defined model from the exported scored data set that is created by some previously connected node within the process flow.

- Perform SAS programming by writing a custom-designed score code within the **SAS Code** node to create the scored data set, then connect the node to the **Data Set Attributes** node in assigning the model roles to the target variable and the predicted variable in the model. Alternatively, the scored data set can be created within the **SAS Code** from the numerous modeling procedures available in SAS such as PROC REG, PROC LOGISTIC, PROC NLIN modeling procedures, and so on.
- Import a scored data set from the **Input Data Source** node with the data role set to RAW from the **Role** display field, then specify the target and predicted variables for the active training data set from the **Variables** tab. For example, one way of creating a scored data set is through the use of the various SAS modeling procedures and the corresponding OUTPUT statement that creates the output data set with the calculated predicted values and the associated target values and input variables.
- The **User Defined Model** node can be used to import the scored data set that is created from the **Variable Selection** node. Simply select the **Score data sets** check box within the **Target Associations** tab from the **Variable Selection** node to create the scored data set. The next step is simply specifying the variable with the fitted values in the scored data set from the **Predicted Values** tab within the **User Defined Model** node. The reason for incorporating the **User Defined Model** node is to allow you to incorporate target profile information into the model, calculate residual values to the existing scored data set, evaluate the accuracy of the modeling fit through the use of the assessment plots that are generated, and store the model in the **Model Manager** from the **User Defined Model** node.

The following is the Enterprise Miner process flow diagram that displays the various ways in which the user-defined model can be created for modeling assessment. Note that the **User Defined Model** node must follow the corresponding node that exported the training, validation, test, or raw data sets that created the prediction or classification model in order to generate the assessment statistics within the **User Defined Model** node.

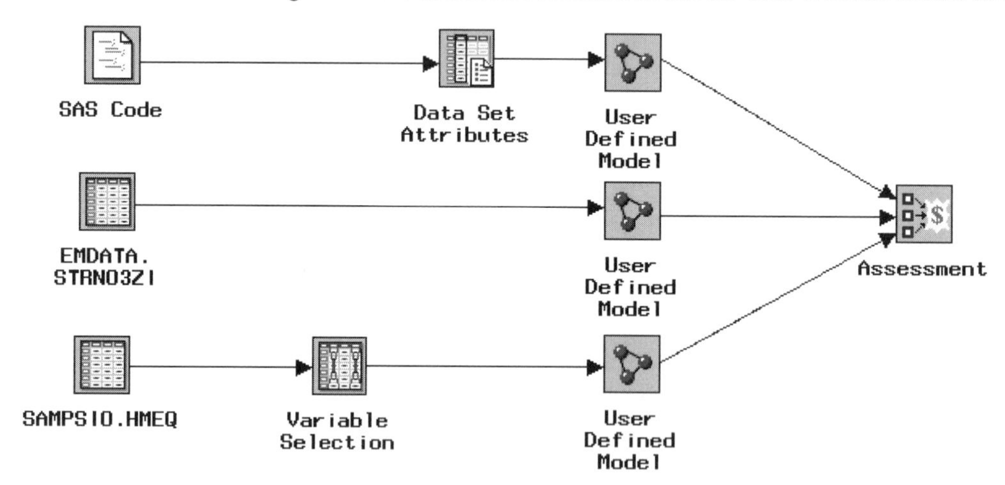

The Enterprise Miner process flow diagram is used to display the various ways to create a user-defined model for modeling assessment.

Data tab

The **Data** tab is similar to the other modeling nodes. The tab will allow you to reassign each one of the selected partitioned data sets. Simply select the **Select...** button that will open the **Imports Map** window to select the corresponding data set connected to the node within the process flow diagram or remove the selected data set from the modeling fit. Selecting the **Properties...** button will allow you to view the file administrative information and a table view of the selected data set to fit. By default, the training data set is automatically used to fit the user-defined model. However, if you have the luxury of a sufficient sample size, then it might not be a bad idea to fit the same model to the validation data set to achieve unbiased estimates by cross-validating the performance of the previous modeling fit.

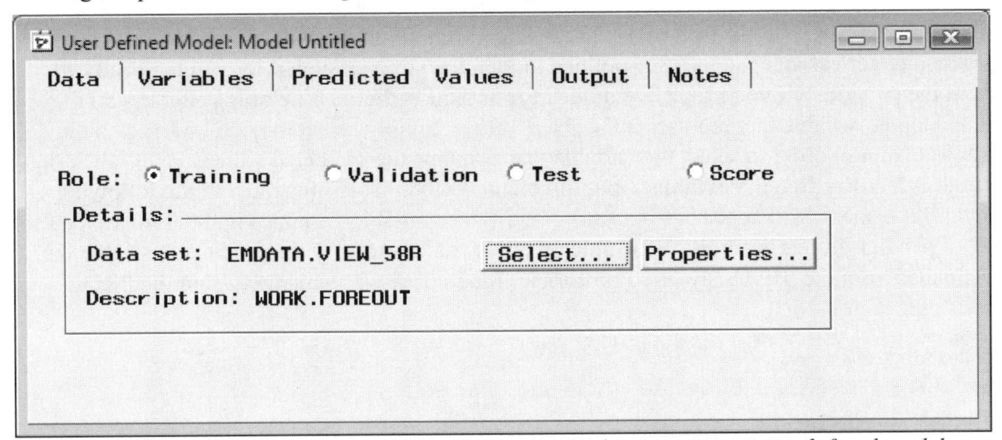

*The **Data** tab to select the active training data set to fit the time series user-defined model.*

Variables tab

The **Variables** tab is designed for you to view the various properties of the listed variables from the scored data set. It is important that there exists a predicted variable in the data set with a model role of **predicted** in order for the user-defined model to perform correctly. Otherwise, the node will not execute. For interval-valued targets, the tab will list the predicted variable from the preceding modeling fit along with all other input variables in the previous model. For categorically-valued targets, the tab will display the predicted variables at each target level from the preceding model. From the **Variables** tab, you may edit, browse, or delete the target profile by fitting the target variable in the user-defined model.

The following illustration displays the variables that are generated from the PROC ARIMA time series procedure by fitting the time series model to the lead production data set that is displayed in the previous **Time Series** node. The difference in successive observations of the target variable was applied to the model in order to achieve stationarity in the time series data over time.

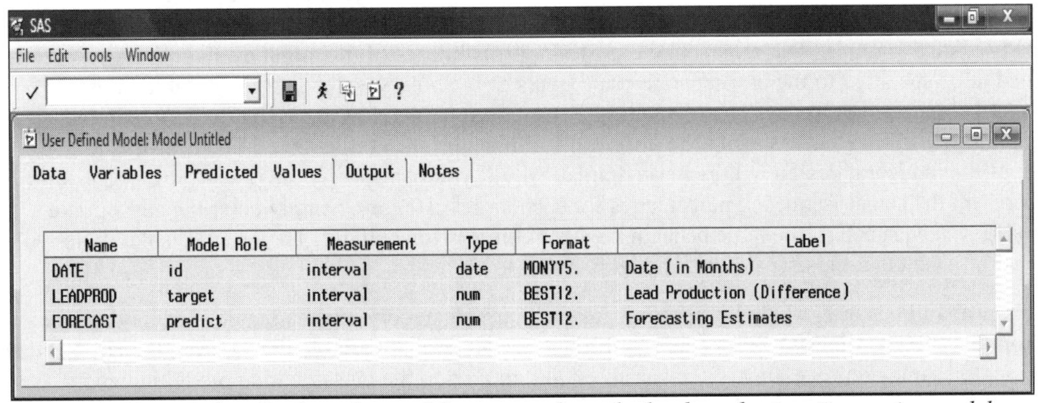

*The **Variables** tab that displays the modeling terms from the lead production time series model.*

Predicted Values tab

Upon opening the **User Defined Model** node, the following **Predicted Values** tab will automatically appear. The **Predicted Values** tab is designed for you to specify the corresponding predicted target variable. The tab will allow you to specify the target variable that has been previously predicted for the user-defined model from the **Target** display field. The predicted variable must be specified within the tab in order for the node to execute and generate the fitted values and assessment statistics. The node has the added option of allowing you to browse the target profile of the previously fitted target variable in the model by simply selecting the **Profile Information...** button. Illustrated in the following two diagrams are the two different layouts of the **Predicted Values** tab by either fitting an interval-valued or a categorically-valued target variable to the user-defined model.

For the interval-valued target variables in the user-defined model, the tab is designed for you to specify the target variable from the previously fitted predictive model. From the **Predicted Variable** column, you may specify the predicted target variable that represents the fitted values. Simply select the cell underneath the **Predicted Variable** column in order to select the variable that contains the predicted values. Right-click the mouse and select the **Select Predicted Variable...** pop-up menu option that is illustrated in the following diagram. A second pop-up menu will appear with a listing of all the available variables in the training data set with a role model of **predict**. In our example, the following tab is used to specify the predicted variable with the forecasting estimates from the previously fitted time series model that was generated from the PROC ARIMA time series procedure.

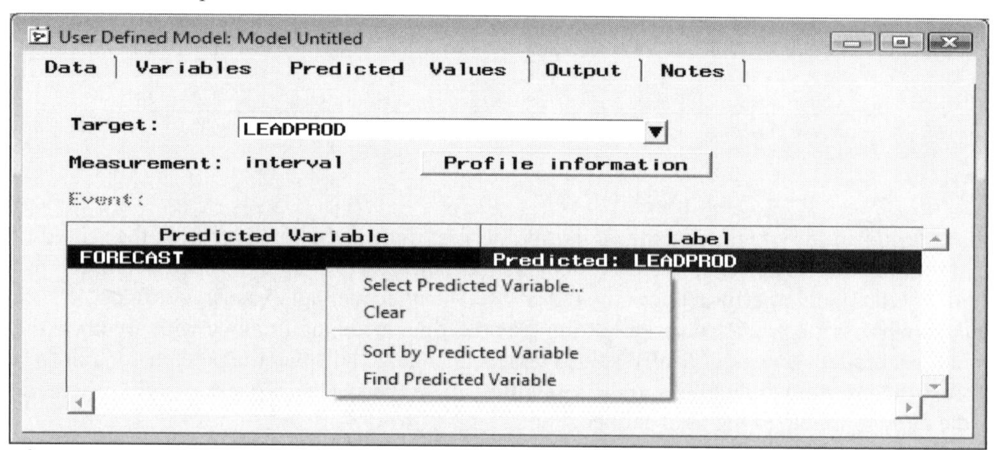

*The **Predicted Values** tab is used to specify the target variable and the predicted values in the model by fitting the interval-valued target variable to the user-defined model.*

For the categorical target variables in the user-defined model, the tab is designed for you to specify the posterior probabilities for each target level from the previously fitted classification model. From the **Predicted Variable** column, simply select the predicted variables for each target level that you want to predict. In our example, the predicted variables that represent the posterior probabilities of the target event and target nonevent need to be specified to the two separate class levels of the binary-valued target variable in the previously fitted classification model. The **Modelling Prior** column that is located next to the **Predicted Variable** column will allow you to specify the prior probabilities for each target level. The classification performance of the modeling design will increase dramatically if you can specify the precise prior probabilities that truly represent the actual frequency percentages for each target category. Simply enter the appropriate prior probabilities in each cell of the table beneath the **Modelling Prior** column. The **Event** display field will indicate the target event that the user-defined model is trying to fit.

The estimated probabilities or estimated proportions of each target category of the user-defined model was generated from the PROC GENMOD procedure by fitting the categorical variable, BAD, from the HMEQ data set. In addition, the same preprocessing of the variables was applied to the classification model in which missing values in the input variables were replaced and the best linear combination of transformed input variables from the GENMOD model were selected.

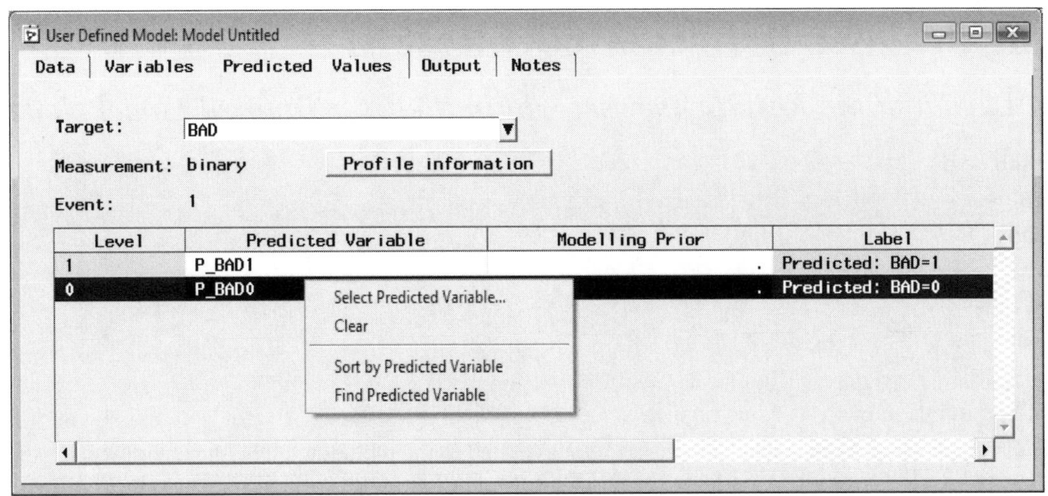

*The **Predicted Values** tab is used to specify the target variable and the predicted values by fitting the categorically-valued target variable to the user-defined model from the PROC GENMOD procedure.*

Output tab

The **Output** tab is similar to the other Enterprise Miner modeling node and will not be shown. The tab is designed for you to create an output scored data set for each partitioned data set that contains the target variable and all other input variables in the data set, along with the following modeling estimates that are created from the user-defined modeling results.

For interval-valued target variables, the output scored data set consists of the following variables:

Variable Name	Description
P_*target variable name*	Predicted values of the target variable.
R_*target variable name*	Residual values that are the difference between the fitted values and the actual target values.

For categorically-valued target variables, the output score data set consists of the following variables:

Variable Name	Description
P_*target variable name <level>*	Predicted probabilities at each class level of the target variable.
R_*target variable name <level>*	Residual values at each class level of the target variable.
F_*target variable name*	Actual target class level.
I_*target variable name*	Classification variable that identifies which class level the target variable is classified from the estimated probabilities and the predetermined cutoff probability.
U_*target variable name*	The unformatted predicted target class level in which the observation is classified.
S_*target variable name*	Standardized input variables by fitting the nonlinear models.
D_*target variable name*	Decision variable.
EP_ or EL_*target variable name*	Expected profit or the expected loss defined by the linear combination of decision entries and posterior probabilities with the reduction in the fixed costs at each decision level. If the target profile is not provided, then the expected profit will be set to the estimated probabilities of the target event.
BP_ or BL_*target variable name*	Best profit or loss from the target profile.
CP_ or CL_*target variable name*	Computed profit or loss from the target profile.

Viewing the User Defined Model Results

General Layout of the Results Browser within the User Defined Model Node

- **Model tab**
- **Plot tab**
- **Code tab**
- **Log tab**
- **Output tab**
- **Notes tab**

Model tab

The **Model** tab displays the name and a short description assigned to the modeling node, the creation date, and the last date on which the node was modified. The **Profile Information** button will allow you to browse the target profile of the categorically-valued target variable of the user-defined model as follows.

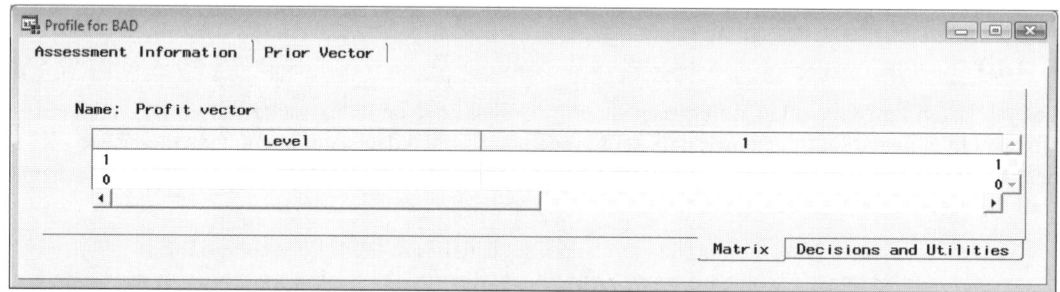

*The **Assessment Information** tab is used to view the decision matrix of the active training data set.*

Plot tab

The **Plot** tab will initially appear once you view the modeling results. For interval-valued target variables, the node displays the two-dimensional scatter plot. The following scatter plot displays the fitted values over time from the PROC ARIMA time series model. The target variable to predict is defined by the difference in the successive target values over time. For categorically-valued target variables, the node will generate a three-dimensional bar chart that displays the classification table in order to view the accuracy of the classification model at each target class level. The classification estimates are generated from the PROC GENMOD procedure. The user-defined model seems to be better at predicting good creditors than bad creditors that is determined by viewing the difference in size of the off-diagonal bars of the chart.

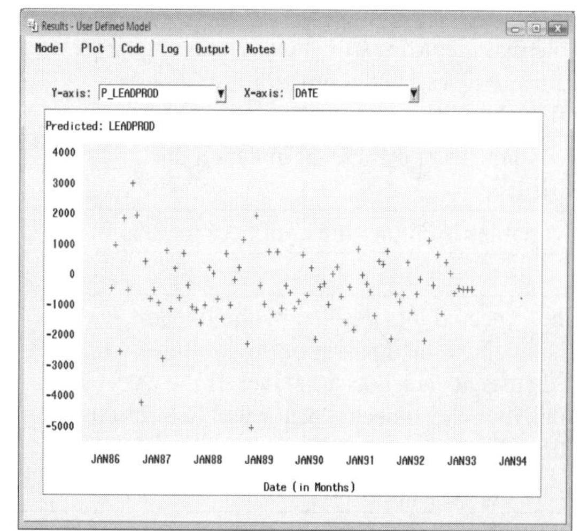

*The **Plot** tab to view the forecasting estimates over time from the user-defined time series model.*

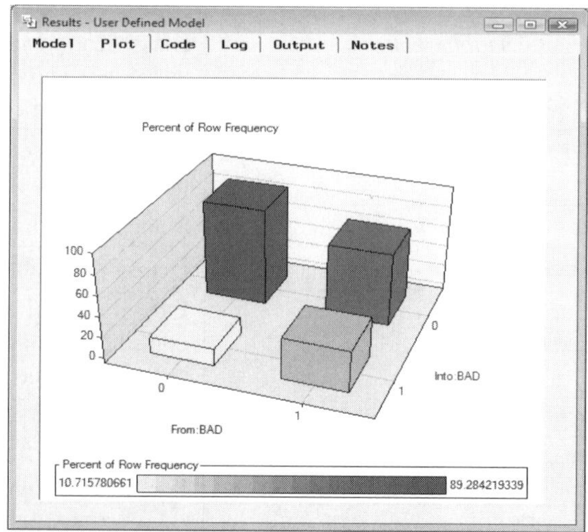

*The **Plot** tab to view the accuracy of the estimated probabilities from the PROC GENMOD procedure.*

Code tab

The **Code** tab displays the scoring code in fitting the interval-valued target variable in the user-defined model. The following scoring code can be used to generate new estimates.

Results - User Defined Model

Model | Plot | Code | Log | Output | Notes

Training ◉ Scoring

```
00001 label P_LEADPROD="Predicted: LEADPROD";
00002 P_LEADPROD = FORECAST;
00003 R_LEADPROD = P_LEADPROD - LEADPROD;
00004 label R_LEADPROD = "Residual:LEADPROD";
00005 *** END OF FILE ***
00006
```

Output tab

The **Output** tab displays the contents of the scored data set. The tab will display the file information of the training data sets used in the user-defined model that is identical to the procedure output listing from the PROC CONTENTS procedure. For categorically-valued target variables, the tab will also display the classification table in order to view the agreement between the actual and predicted class levels. The following is the procedure output listing by fitting the classification model from the PROC GENMOD procedure.

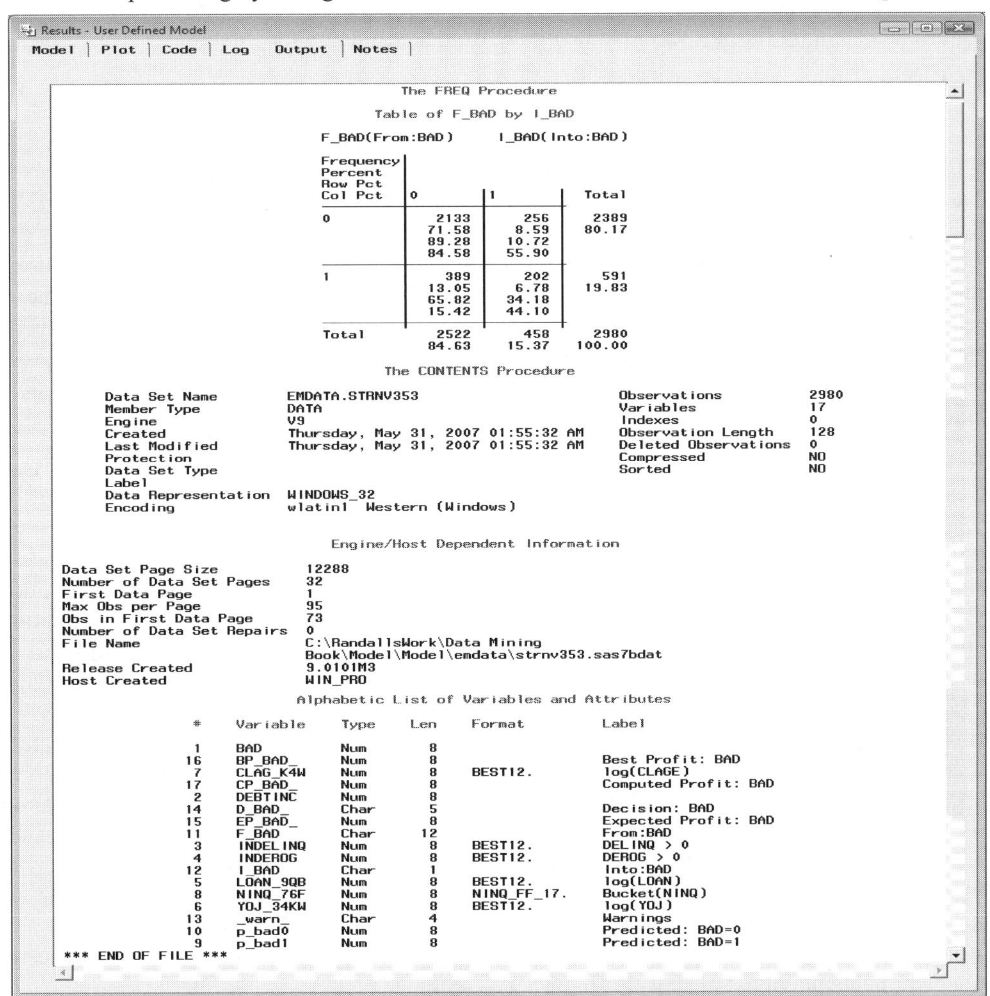

Results - User Defined Model

Model | Plot | Code | Log | Output | Notes

The FREQ Procedure

Table of F_BAD by I_BAD

F_BAD(From:BAD) I_BAD(Into:BAD)

Frequency Percent Row Pct Col Pct	0	1	Total
0	2133 71.58 89.28 84.58	256 8.59 10.72 55.90	2389 80.17
1	389 13.05 65.82 15.42	202 6.78 34.18 44.10	591 19.83
Total	2522 84.63	458 15.37	2980 100.00

The CONTENTS Procedure

Data Set Name	EMDATA.STRNV353	Observations	2980
Member Type	DATA	Variables	17
Engine	V9	Indexes	0
Created	Thursday, May 31, 2007 01:55:32 AM	Observation Length	128
Last Modified	Thursday, May 31, 2007 01:55:32 AM	Deleted Observations	0
Protection		Compressed	NO
Data Set Type		Sorted	NO
Label			
Data Representation	WINDOWS_32		
Encoding	wlatin1 Western (Windows)		

Engine/Host Dependent Information

Data Set Page Size	12288
Number of Data Set Pages	32
First Data Page	1
Max Obs per Page	95
Obs in First Data Page	73
Number of Data Set Repairs	0
File Name	C:\RandallsWork\Data Mining Book\Model\Model\emdata\strnv353.sas7bdat
Release Created	9.0101M3
Host Created	WIN_PRO

Alphabetic List of Variables and Attributes

#	Variable	Type	Len	Format	Label
1	BAD	Num	8		
16	BP_BAD_	Num	8		Best Profit: BAD
7	CLAG_K4W	Num	8	BEST12.	log(CLAGE)
17	CP_BAD_	Num	8		Computed Profit: BAD
2	DEBTINC	Num	8		
14	D_BAD_	Char	5		Decision: BAD
15	EP_BAD_	Num	8		Expected Profit: BAD
11	F_BAD	Char	12		From:BAD
3	INDELINQ	Num	8	BEST12.	DELINQ > 0
4	INDEROG	Num	8	BEST12.	DEROG > 0
12	I_BAD	Char	1		Into:BAD
5	LOAN_9QB	Num	8	BEST12.	log(LOAN)
8	NINQ_76F	Num	8	NINQ_FF_17.	Bucket(NINQ)
6	YOJ_34KW	Num	8	BEST12.	log(YOJ)
13	_warn_	Char	4		Warnings
10	p_bad0	Num	8		Predicted: BAD=0
9	p_bad1	Num	8		Predicted: BAD=1

*** END OF FILE ***

4.7 Ensemble Node

General Layout of the Enterprise Miner Ensemble Node

- **Data tab**
- **Variables tab**
- **Settings tab**
- **Output tab**
- **Notes tab**

The purpose of the **Ensemble** node in Enterprise Miner is to perform *ensemble modeling* by combining the predictive results from several predictive models or performing various resampling techniques. Ensemble modeling is designed to increase the accuracy of the prediction estimates, achieve stability and consistency in the prediction estimates, and improve generalization. In other words, increasing the accuracy of the modeling design by either averaging the predicted values by fitting the continuous target variables or averaging the posterior probabilities by fitting the categorical target variables that you want to predict. The reduction in the error for the model is determined by the overall average of the individual models. The restriction of ensemble modeling is that the model is strictly designed for only one target variable that you want to predict. In addition, it is assumed when training the models that the target variables in each separate model are compatible and have the same level of measurement by fitting each separate model from the same training data set.

Ensembling Methods

There are two separate ways in which the **Ensemble** node creates a new model by averaging the posterior probabilities by fitting a categorically-valued target variable or averaging the predicted values by fitting an interval-valued target variable from several models. The first method is combining the prediction estimates by averaging all the prediction estimates based on separate samples selected from the same data set, which is analogous to bootstrapping. For the bagging and boosting resampling techniques, the **Ensemble** node is used in conjunction with the **Group Processing** node. Bagging and boosting models are created by resampling the active training data set and refitting the model for each sample. The other method in ensemble modeling is by calculating the average estimates based on the predicted values from any number of different predictive models, that is, from neural network modeling to decision tree modeling, by fitting the corresponding models to the same training data set.

In predictive modeling, the model assumes a functional relationship between the input variables and the target variable. In regression modeling, it is assumed that there is a linear relationship between the input variables and the target variable. In decision tree modeling, it is assumed that there is a constant relationship within the range of values of the input variables. In neural network modeling, it is assumed that there is a nonlinear relationship between the input variables and the target variable that depends on the selected architecture and the activation functions that are selected for the neural network design. However, the drawback of combining the prediction estimates from the separate modeling designs is that the ensemble model may produce an entirely different relationship in comparison to the separate models that have been combined. This will restrict you in evaluating the relationship between the variables in the combined model with the main objective of generating more accurate prediction estimates. Ensemble modeling is very efficient and will produce more accurate estimates than the individual models when the assessment statistics, like the squared errors or the classification rates, are much different between the individual models. However, this fact is generally not true with very small training data sets. That is, the accuracy of the ensemble model depends on the sample size of the training data set and, similar to predictive or classification modeling, the predictive and classification performance from the ensemble model increases with an increase in the number of observations. In addition, the ensemble model will perform well assuming that the individual models are selected and combined carefully. Finally, when evaluating the effectiveness of the ensemble model, you should always compare the modeling performance of the ensemble model with the individual models that have been combined.

Advantages of Ensemble Modeling

The key to ensemble modeling is to generally take several measurements by calculating the average posterior probabilities for categorical target values or the predicted values for interval-valued targets from numerous models by fitting a single target variable. The whole idea is that taking several different measurements is usually more accurate than taking a single measurement assuming that the target values have the same level of measurement between the predictive models. Conversely, calculating the weighted average of the predicted values by combining modeling estimates is usually more accurate and more stable than individual models that are inconsistent with one another. However, the performance and the stability of the ensemble model depends on the sample size in comparison to the number of input variables in the training data set. The ensemble model might alleviate instability in the estimates, which is particularly true when fitting complex models to small sampled data sets. Instability in the estimates means that partitioning the same data several times and fitting the subsequent model will result in entirely different estimates. However, one drawback to ensemble modeling is that using a small training data set that is not an appropriate representation of the true population will result in inadequate ensemble modeling estimates. For example in classification designs, calculating inaccurate classification estimates that have been obtained from the various models will lead to ensemble estimates that will also be inaccurate and absolutely useless.

Ensembling might be applied to increase the accuracy of the predictions when the data you want to fit is quite sparse. In other words, you should resample the input data set and create separate predictions, then average these same predictions to the ensemble model. However, it is assumed that the target values that you want to combine have the same level of measurement and the same range of values. In nonlinear predictive modeling, like neural network modeling, there might be an interest in predicting the response from several target variables in the model that you want to predict. However, the neural network model might experience difficulty in network training when the model is trying to find a minimum error of the multiple error functions it is trying to predict, since the error surface of the neural network design can be extremely complex with several local minimums, flat spots, saddle points, and many peaks and valleys. Therefore, an ensemble model might be the best solution to alleviate this optimization phenomenon by consolidating several predictive models into one simple ensemble model. In neural network modeling, ensembling by combining and resampling the data that you want to fit can significantly improve generalization performance. Generalization is a process of choosing the appropriate complexity for the model in generating accurate forecasting predictions based on data that is entirely separate from the data that was used in fitting the predictive model. The ensemble model is designed so that the generalization performance is better in the ensemble model, no matter how much better the prediction estimates are in comparison to the best individual model. However, the ensemble model obviously depends on the accuracy of the other predictive models. The main idea of combining and averaging the predictions across the various models is to reduce the error and increase generalization. The amount of reduction in the error depends on the number of models that are combined.

Transforming Input Variables in the Ensemble Model

If you use the **Transform Variables** node to transform an interval-valued input variable in the ensemble model, you might end up with an incorrect prediction plot, which you may view from the **Model Manager** within the **Ensemble** node. The reason is because the **Transform Variables** node automatically removes the original transformed input variable from the subsequent analysis. When the original input variable is dropped from the analysis, it will result in the values of the transformed variable set to missing in the scored data set, which will result in incorrect predicted values that are calculated in the scored data set within the **Ensemble** node. For example, suppose you want to achieve normality in an input variable by applying a logarithmic transformation to the input variable within the **Transform Variables** node. The problem is that the **Transform Variables** node will automatically drop the input variable from the analysis. Therefore, you must perform the following steps to retain the values of the transformed input variable in the scored data set and calculate the correct predicted values from within the **Ensemble** node.

1. After applying the logarithmic transformation to the input variable in the **Transform Variables** node, retain the original transformed input variable in the data set by changing the variable **Keep** status to Yes. Close the **Transform Variables** node and save your changes.
2. Add the **Data Set Attributes** node to the **Diagram Workspace** and connect the **Transform Variables** node to the **Data Set Attributes** node.
3. Open the **Data Set Attributes** node and select the **Variables** tab. Remove the original input variable that has been transformed from the subsequent model by changing the **New Model Role** from **input** to **rejected**. Close the **Data Set Attributes** node and save your changes.

Bagging and Boosting Resampling Techniques

Bagging and boosting is designed to improve the performance of weak classification designs. Both bagging and boosting were originally designed for decision trees. However, both resampling techniques perform well in neural networks, logistic regression, and *k*-nearest neighbor classification models. Both bagging and boosting techniques tend to generate smoother prediction or classification boundaries as opposed to single models. The performance of both bagging and boosting in the ensemble design are affected by the sample size of the training data set. Bagging is useful for classification designs with a limited amount of available data. Conversely, boosting is designed for an inherently poor classification model given a large amount of data by fitting the categorically-valued target variable. The reason is because the larger the training data, the higher the number of borderline cases. However, weights are applied in order to better classify these borderline cases in the correct target category. In addition, ensemble modeling can be applied to further increase the modeling performance of the design by combining the class probabilities by resampling the training data set. Often, the class probabilities that are combined will result in better classification performance, i.e. lower misclassification rates, in comparison to the class probabilities from a single classification model. However, at times, the combination of models might result predictions that are less accurate than the best model that was combined.

Ensemble Modeling Techniques

The following is the list of ensemble modeling techniques.

Combined Models: This modeling technique combines the predicted values from the interval-valued target variables and the posterior probabilities from categorically-valued target variables that you want to predict. The node will then store the combined model as a single model entry within the **Model Manager** for assessment. The **Ensemble** node can only read one model from any one modeling node. In order to combine the two separate models from the same training data set, you will need to create separate modeling nodes for each corresponding model from the process flow diagram. By default, the **Ensemble** node will combine the modeling estimates, assuming that there is no **Group Processing** node in the process flow. If there is a **Group Processing** node in the process flow, then the type of ensemble model that is created will depend on the option setting in the **Group Processing** node.

Stratified Models: This modeling technique combines the statistical modeling estimates by custom-designed segmentation or partitioning of the training data set that you want to fit. In other words, separate models are created for each level of segmentation or partitioning of the data that you want to fit. The **Ensemble** node can also be used to consolidate the scoring code from different stratified models. Separate scoring code is generated by each level of the categorical variable that is defined from the **Group Processing** node.

Stratified modeling is beneficial when there are exists wide variability between each group, for example, weight gain by sex, by obtaining estimates from each group separately, and then combining the prediction estimates by each group into a single overall estimate. Therefore, the advantage to this modeling technique is if the variability in the target variable is only attributed to the within-group variability as opposed to the between-group variability which might be a large component to the overall variability in the target values. Stratified modeling is also beneficial in controlling for confounding, adjusting for interaction effects, and increase statistical power. The drawback to stratified modeling is the occurrence of an insufficient number of observations within certain groups that will result in low statistical power, unstable estimates, and bad generalization.

Bagging Models: Bagging stands for bootstrap aggregation. This modeling technique is analogous to bootstrapping where separate prediction estimates are created by resampling the data you want to fit. Aggregate bagging is performed, that is, bagging based on bootstrap sampling and aggregation by combining the prediction estimates. Similarly, the ensemble model is based on the predicted values from the continuous target variables and posterior probabilities from the categorical target variables that you want to predict.

Bagging is analogous to bootstrap estimation where a fixed number of independent B samples are created by replacement. The bagged estimates of the predicted values are computed by refitting the statistical model to calculate the fitted values for each bootstrap sample of equal size, then dividing by the number of bootstrap samples as follows:

$\hat{g}_{\text{bag}}(\text{x}) = \frac{1}{B}\sum_{\text{b=1}}^{\text{B}} g(\text{x})$, where $g(\text{x})$ is the bagging estimate from B bootstrap samples.

The bagging estimate from the predictive model would be calculated by the average predicted values from the B bootstrap samples. For classification modeling, the bagged estimates would be the average of the class probabilities from the B bootstrap samples where the predicted target class level are determined by the most often occurring class level from the B bootstrap samples.

The drawback to bagging a model is that the simple structure of the model is lost. In other words, a bagging estimate from a decision tree is no longer a decision tree due to the bagging process. This can restrict your ability to interpret the model and the relationship between the variables, since the main objective to the analysis is to generate more accurate prediction or classification estimates. In addition, bagging will perform poorly when there exists a clear separation in the different class levels that can be separated by simple linear boundaries. The reason is because the bagged estimates that are averaged over the bootstrap samples will result in smooth nonlinear boundaries. However, boosting will perform much better under such circumstances.

From the process flow, bagging models may be performed by specifying the **Unweighted resampling for bagging** option from the **Group Processing** node and connecting the appropriate modeling node to perform bootstrap estimates. Alternatively, the following boosting models may be performed by specifying the **Weighted resampling for boosting** option from the **Group Processing** node.

Boosting Models: In Enterprise Miner, *boosting* fits the categorically-valued target variable where the observations are weighted. That is, observations that performed a poor fit to the model from a previous fit to the data are weighted. The power of the boosting algorithm is that it can dramatically increase the classification performance of even weak classification models. Boosting assigns weights to each one of the observations from the training data set. Initially, all the weighted observations are set to the sample proportion that is based on the sample size of the training data set, meaning that the observations are assigned equal weights. The algorithm then modifies each weighted observation at each successive iteration, then refits the classification model. The weighted observations are modified by increasing the weight estimates for each observation that have been misclassified from the previous fit and the weight estimates are decreased for those observations that have been correctly classified. From the **Group Processing** node, the **General** tab will allow you to specify the number of classification models to fit with the weight estimates adjusting the corresponding estimated probabilities from the previous model. The following formula below is provided for understanding the way in which the weights are calculated in the classification model.

In predictive modeling, this recursive model fitting procedure is very similar to the previous dmneural modeling design in which the model first predicts the target variable, then uses the residual values from the previous fit as the target variable for the iterative model. As an example of the iterative boosting algorithm, suppose you would like to minimize the sum-of-squares errors between an interval-valued target variable and the fitted values from the ensemble model. The first step would be fitting the input data set to the prediction model. The next step is fitting the second model by subtracting the difference between the original target values and the predicted values from the first model that act as the target variable that you want to predict. The third step is taking the difference of the target values and the weighted average of the fitted values of the two previous models acting as the target variable that you want to predict, and so on. It is important to understand that the weights that are used in averaging the individual models and the calculations that are used in creating the target values for the iterative model is determined by the modeling selection criterion that is selected. In the previous example, the calculations that were applied in creating the target values are based on minimizing the sum of squared errors. However, a different method is used when minimizing the sum of absolute errors. Therefore, the process of calculating the target values depends on the modeling selection criterion that is used in the boosting model. And finally, the algorithm increases the weights of the observations that generate large residual values from the previous model and decreases the weights for observations that produce small errors.

For categorical target variables, the classification model is refitted using the weights for each case in minimizing the misclassification error rate. For the ith case, the estimated probabilities are adjusted by the weights that are calculated from the following formula that is provided by Leo Breiman as follows:

$$w(i) = \frac{(1 + m(i)^4)}{\sum_i (1 + m(i)^4)}$$

where $0 \le m(i) \le k$, that is, k is the number of fitted classification models and $m(i)$ is the number of models that misclassified case i in the previous step. The adaptive boosting process is repeated k times in which m models adjust the estimated probabilities, which will result in fewer misclassifications in subsequent fits.

The various ensemble modeling techniques are discussed in greater detail in Hastie, Tibshirani, and Friedman (2001), Hand, Mannila, and Smyth (2001), Giudici (2003), Breiman (1998), and Applying Data Mining Techniques Using Enterprise Miner Course Notes, SAS Institute, (2002).

The following are the various ensemble modeling methods to consider in improving the modeling estimates in the Enterprise Miner process flow.

Combined Modeling Estimation

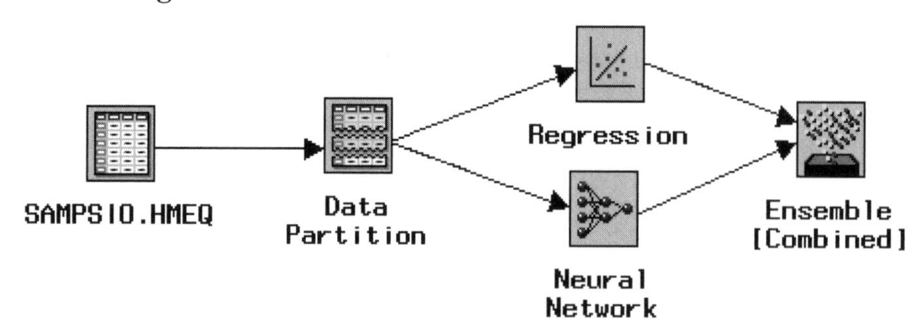

Enterprise Miner process flow diagram of an ensemble model combining the prediction estimates from the neural network and the multiple regression model.

Stratified Modeling Estimation

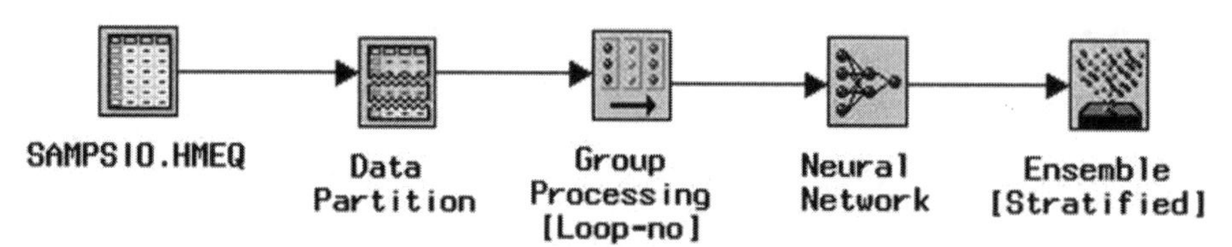

Enterprise Miner process flow diagram of an ensemble model based on the stratification method by specifying the stratification variable from the ***Group Processing*** *node and combining the neural network estimates at each class level of the stratified variable.*

Bagging Modeling Estimation

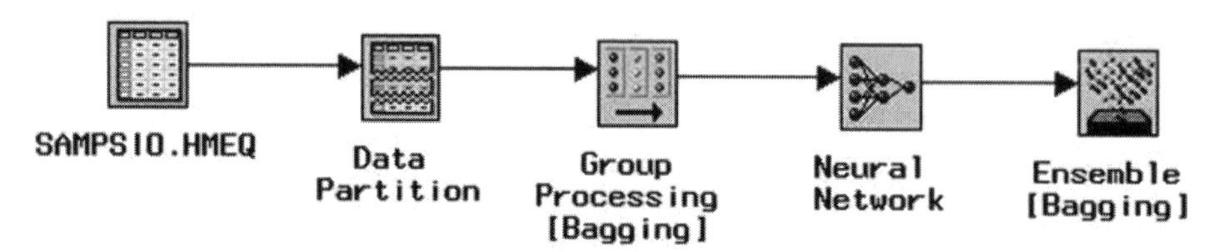

Enterprise Miner process flow diagram of the ensemble model from the bagging method created by specifying the ***Unweighted resampling for bagging*** *and the number of loops from the* ***Group Processing*** *node, with the* ***Ensemble*** *node set to the ensemble mode of bagging or bootstrap estimation.*

Boosting Modeling Estimation

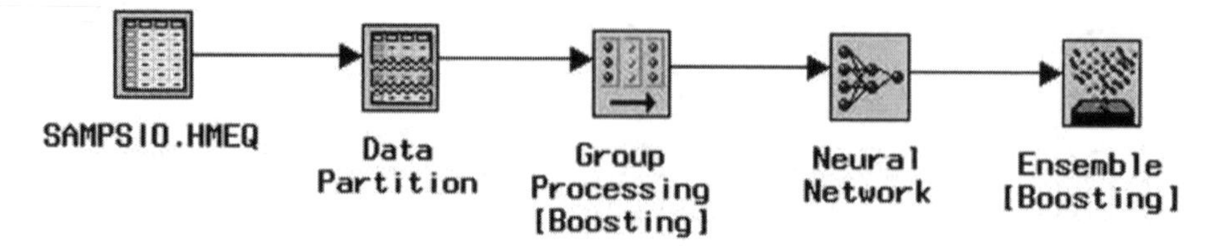

Enterprise Miner process flow diagram of the ensemble model from the boosting method, that is, by specifying the ***weighted resampling for boosting*** *in fitting the binary-valued target variable, that is, BAD, and the number of loops or classification models to fit from the* ***Group Processing*** *node, with the* ***Ensemble*** *node set to the ensemble mode of boosting.*

Data tab

The **Data** tab is designed to reassign the active training data set, view the file administration information or view the table listing of the partitioned data set in the currently opened Enterprise Miner process flow diagram. Ensemble modeling is performed for each one of the partitioned data sets. The purpose of the partitioned data sets is to cross-validate the accuracy of the ensemble model in comparison to the active training data set. The role assignments for the various partitioned data sets can be verified from the tab. By default, the training data set is automatically selected in fitting the ensemble model. Alternatively, the validation and the test data set can be used to provide an unbiased assessment in measuring the accuracy of the modeling fit. At times, a single validation data set might generate incorrect generalization estimates. Therefore, the test data set might be used to obtain unbiased estimates and make a better assessment in fitting the model to new data.

Variables tab

By default, the node is designed to initially display the following **Variables** tab. The difference between the **Variables** tab and the other modeling nodes is that the target variable that the ensemble model is trying to predict is the only variable that is listed within the tab. Since the ensemble model is designed to fit a single target variable, only one target variable may have a variable attribute status of **use**. The tab will allow you to set the variable attribute status of the listed target variable. Changing the variable attribute status of the target variable to **don't use** will prevent ensemble modeling from being performed within the process flow. In addition, the tab will allow you to view the frequency distribution of the target variable from the **View Distribution of <target variable name>** pop-up menu item. The same frequency chart can also be viewed from the **Input Data Source** node.

From the tab, you may edit the target profile from the unique target variable in the ensemble model in order to adjust the posterior probabilities from the specified prior probabilities of the categorically-valued target variable in the predictive model. However, all updates applied within the tab are only reflected within the corresponding modeling node. Simply right-click in any cell of the listed target variable and select the **Edit target profile...** pop-up menu item to define a target profile in the **Ensemble** node. However, you may also browse the target profile by selecting the **Browse target profile...** pop-up menu item if a target profile has already been defined in a previously connected node. Otherwise, delete the target profile by selecting the **Clear target profile** pop-up menu item.

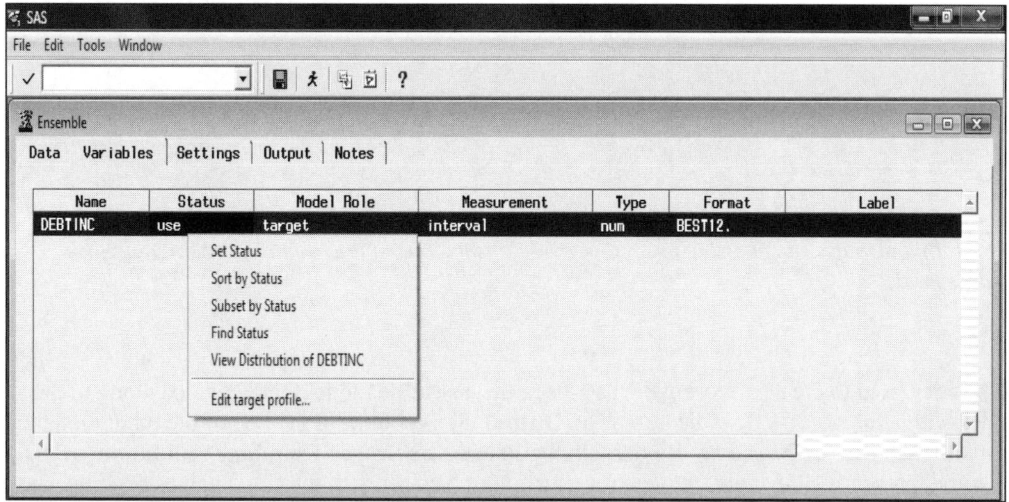

*The **Variables** tab that displays the target variable for the ensemble model that you want to fit.*

Settings tab

The **Settings** tab is designed to specify the type of ensemble method to apply. The default is **Automatic**. The automatic setting determines the type of ensemble model that is applied within the process flow diagram. However, by specifying the appropriate ensemble mode, the ensemble option setting will be displayed in brackets below the **Ensemble** node within the process flow diagram. If there exists a **Group Processing** node in the process flow diagram, then the type of ensemble model that is created depends on the option settings specified from the **Group Processing** node.

- **Ensemble mode:** The following table lists the association between the possible ensemble modes specified within the node with respect to the type of setting specified from the **Group Processing** node that is applied to the modeling design. For example, if you select an ensemble mode that is incompatible with the current configuration of the process flow and then run the process flow diagram, the **Ensemble** node will automatically set the final mode as follows:

Group Process Mode	Initial Ensemble Mode				
	Automatic	**Combined**	**Stratified**	**Bagging**	**Boosting**
Bagging	Bagging	Combined	Bagging	Bagging	Bagging
Boosting	Boosting	Combined	Boosting	Boosting	Boosting
Variable / Inputs	Stratified	Combined	Stratified	Stratified	Stratified
Variable / Target	Combined	Combined	Stratified	Bagging	Boosting
No-loop	Combined	Combined	Stratified	Bagging	Boosting

- **Probability function:** The probability function is used to combine the posterior probabilities or the predicted values of the component models into the ensemble model. For the **Stratified** mode, the function is set to **Identity**. For the other modes, the function is set to **Mean**.

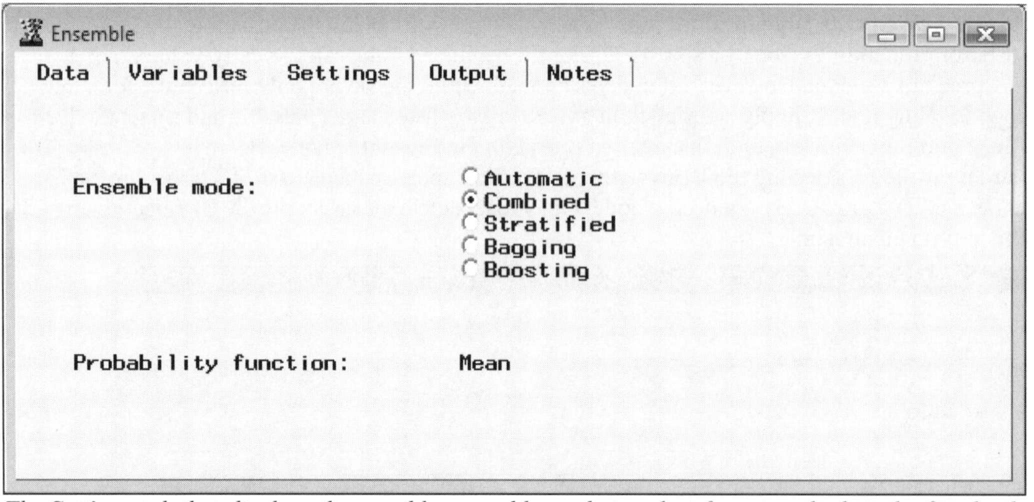

*The **Settings** tab that displays the possible ensemble mode to select from to calculate the fitted values for the ensemble model.*

Output tab

The **Output** tab is designed to create a scoring data set from the ensemble model that is passed along to the subsequent nodes within the process flow diagram. The **Output** tab is similar to the rest of the other modeling nodes. Therefore, the tab will not be shown. By default, the **Process or Score: Training, Validation, and Test** check box is cleared, preventing the node from creating an output data set with the fitted values once the node is executed. Therefore, select the **Process or Score: Training, Validation, and Test** check box to create the output score data set for each partitioned data set. To prevent the node from scoring the score data set, clear the **Process or Score: Training, Validation, and Test** check box. This will also prevent you from viewing the scored output data set. The output produced from the ensemble model will either be the posterior probabilities from the categorical-target variables or the predicted values from the interval-valued target variables that are combined.

Viewing the Ensemble Node Results

General Layout of the Results Browser within the Ensemble Node

- **Model tab**
- **Code tab**
- **Log tab**
- **Output tab**
- **Notes tab**

Model tab

The **Model** tab displays the file information, target variables, configuration settings, and the processing information for the modeling node. The tab consists of the following four subtabs.

- **General subtab**
- **Variables subtab**
- **Settings subtab**
- **Status subtab**

The **General** subtab displays the detailed administration information from the scored data set. The **Variables** subtab displays the target variable that you want to predict. The tab will allow you to view the frequency distribution of the target variable from the metadata sample. In addition, the tab will display the predicted and residual variables by specifying the scored data set from the previous **Output** tab. The **Settings** subtab displays the various option settings specified for the ensemble model settings, such as the type of ensemble mode applied, number of models combined, target variable name, level of measurement, variable type, frequency variable, actual and predicted variable label, and input and output data sets that are illustrated in the following diagram. The **Status** subtab displays various processing information such as the creation dates and processing time in training the model by executing the node.

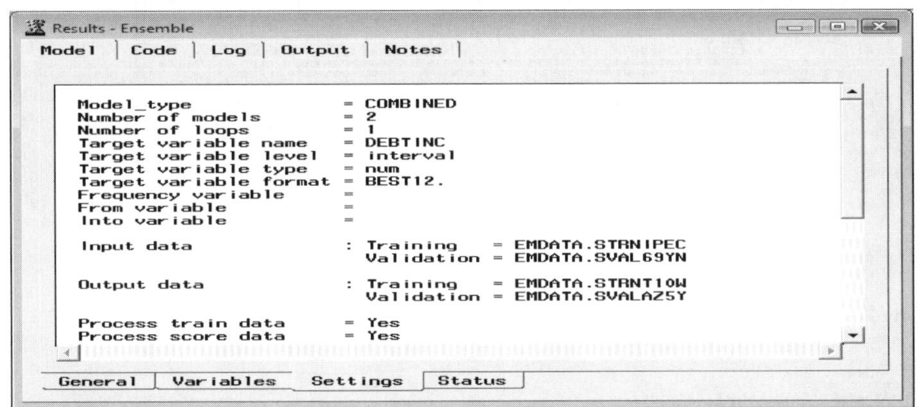

*The **Settings** subtab that displays the configuration settings for the current ensemble model.*

Code tab

The **Code** tab is designed to display the internal SEMMA training code that produced the ensemble prediction estimates and scored data sets by training the ensemble model. The same preprocessing of the data by fitting the least-squares model from the **Regression** node was applied to the both models where outliers were removed from the fitted models and some of the interval-valued input variables were binned into separate categories. In combining models, the scoring code fits each model separately and calculates the prediction estimates by adding each corresponding prediction estimate, then dividing by the number of models combined. The following is a partial listing of the score code that is automatically created by executing the node in calculating the ensemble modeling estimates by combining both the multiple linear regression and the neural network fitted values from the combined ensemble mode. The following scoring code can be used to combine the fitted values from the neural network and least-squares regression models. The code first computes the prediction estimates from the simplest neural network model with two hidden layer units, followed by the prediction estimates from the multiple linear regression model, which then calculates the ensemble prediction estimates by averaging the two separate fitted values from both models.

In stratified modeling, the score code from the **Ensemble** node will combine the scoring code by logically dividing the score code from each combined model by each class level of the categorically-valued stratification variable. This will result in the **Ensemble** node concatenating the scoring code that is consolidated into a single data step by logically dividing the data by simple *if-then* programming code for each group that is created.

Results - Ensemble

Model | Code | Log | Output | Notes

○ Training ◉ Scoring

```
00282 *** If missing inputs, use averages;
00283 if _DM_BAD > 0 then do;
00284    _LP0 =      34.0576025928012;
00285    goto G5044;
00286 end;
00287
00288 *** Compute Linear Predictor;
00289 drop _TEMP;
00290 drop _LP0;
00291 _LP0 = 0;
00292
00293 ***  Effect: BAD ;
00294 _TEMP = 1;
00295 _LP0 = _LP0 + (   -2.26473619821019) * _TEMP * _0_0;
00296
00297 ***  Effect: CLNO ;
00298 _TEMP = CLNO ;
00299 _LP0 = _LP0 + (      0.0786961629595 * _TEMP);
00300
00301 ***  Effect: JOB ;
00302 _TEMP = 1;
00303 _LP0 = _LP0 + (     0.55633207421397) * _TEMP * _2_0;
00304 _LP0 = _LP0 + (     0.72064481357786) * _TEMP * _2_1;
00305 _LP0 = _LP0 + (     0.51301994409476) * _TEMP * _2_2;
00306 _LP0 = _LP0 + (    -1.74804781170533) * _TEMP * _2_3;
00307 _LP0 = _LP0 + (      1.6674257401143) * _TEMP * _2_4;
00308
00309 ***  Effect: LOAN ;
00310 _TEMP = LOAN ;
00311 _LP0 = _LP0 + (     0.00009641409138 * _TEMP);
00312
00313 ***  Effect: MORTDUE ;
00314 _TEMP = MORTDUE ;
00315 _LP0 = _LP0 + (     0.00003840460494 * _TEMP);
00316
00317 ***  Effect: REASON ;
00318 _TEMP = 1;
00319 _LP0 = _LP0 + (   -0.66128701050811) * _TEMP * _1_0;
00320
00321 ***  Effect: VALUE ;
00322 _TEMP = VALUE ;
00323 _LP0 = _LP0 + (   -0.00001558490604 * _TEMP);
00324
00325 ***  Effect: NINQ_XA9 ;
00326 _TEMP = 1;
00327 _LP0 = _LP0 + (   -1.46288907660165) * _TEMP * _3_0;
00328 _LP0 = _LP0 + (     0.05672034033085) * _TEMP * _3_1;
00329 *--- Intercept ---*;
00330 _LP0 = _LP0 + (     31.7713314982493);
00331
00332 G5044:
00333
00334 *** Predicted Value, Error, and Residual;
00335 label P_DEBTINC = 'Predicted: DEBTINC' ;
00336 P_DEBTINC = _LP0;
...
00656 IF _DM_BAD EQ 0 THEN DO;
00657    H11  =        0.4369016976836 * S_CLAGE  +     -2.28682181457172 * S_CLNO
00658         +     -0.63377458995151 * S_INDELINQ  +     -0.55518201819648 *
00659         S_INDEROG  +     -0.44922809693611 * S_LOAN  +     -5.74158620317305 *
00660         S_MORTDUE  +      1.58058100658421 * S_VALUE  +      0.54808888972992 *
00661         S_YOJ ;
00662    H12  =       1.47244034841011 * S_CLAGE  +     -0.43243889067378 * S_CLNO
00663         +     -0.04772412822549 * S_INDELINQ  +     -0.63048127859009 *
00664         S_INDEROG  +      3.11239465269913 * S_LOAN  +      0.12878741953137 *
00665         S_MORTDUE  +     -0.64739601927393 * S_VALUE  +     -0.00147698356098 *
00666         S_YOJ ;
00667    H11  = H11  +      2.59867536889177 * BAD0  +     -1.10324487192957 * JOBMgr
00668         +     -0.32305589679861 * JOBOffice  +     -2.66548981500915 *
00669         JOBOther  +      0.02474436423351 * JOBProfExe
00670         +      4.68351682487483 * JOBSales  +      0.31924689990937 *
00671         REASONDebtCon ;
00672    H12  = H12  +     -2.66760659007132 * BAD0  +     -0.01158489536078 * JOBMgr
00673         +      1.13508322309849 * JOBOffice  +     -1.90217838955762 *
00674         JOBOther  +     -1.50687318058257 * JOBProfExe
00675         +      3.69772304160696 * JOBSales  +     -1.00332107324913 *
00676         REASONDebtCon ;
00677    H11  = H11  +      0.20845362941352 * NINQ_XA90001_low_0_5
00678         +      0.2389386826833 * NINQ_XA90002_0_5_1_5 ;
00679    H12  = H12  +      0.99714814564425 * NINQ_XA90001_low_0_5
00680         +      3.02532547567463 * NINQ_XA90002_0_5_1_5 ;
00681    H11  =     -5.08679447406658 + H11 ;
00682    H12  =      4.66542750856174 + H12 ;
00683    H11  = TANH(H11 );
00684    H12  = TANH(H12 );
00685 END;
00686 ELSE DO;
00687    H11  = .;
00688    H12  = .;
00689 END;
00690 *** *************************;
00691 *** Writing the Node T_INTRVL ;
00692 *** *************************;
00693 IF _DM_BAD EQ 0 THEN DO;
00694    P_DEBTINC  =     -3.02626051634339 * H11  +     2.82896130120174 * H12 ;
00695    P_DEBTINC  =          32.47656391131 + P_DEBTINC ;
00696 END;
00697 ELSE DO;
00698    P_DEBTINC  = .;
00699 END;
00700 IF _DM_BAD EQ 1 THEN DO;
00701    P_DEBTINC  =       34.0576025928012;
00702 END;
00703 ********************************;
00704 *** End Scoring Code for Neural;
00705 ********************************;
00706 *------------------------------------------------------------*;
00707 *  END_CHUNK 1496495910.5:T0X2NNOU                            *;
00708 *------------------------------------------------------------*;
00709 *;
00710 * ENSEMBLE T0YDCB74;
00711 * Probabilities for 2 of 2 imported models;
00712 *;
00713 C_T0YDCB74_P_DEBTINC= C_T0YDCB74_P_DEBTINC + P_DEBTINC;
00714 *------------------------------------------------------------*;
00715 * ENSEMBLE T0YDCB74 : average imported probabilities;
00716 P_DEBTINC= C_T0YDCB74_P_DEBTINC / 2;
00717 C_T0YDCB74_P_DEBTINC= 0.0;
00718 *------------------------------------------------------------*.
```

The score code from the ensemble model that calculates the predictions from the regression and neural network models, then averages the fitted values from both models.

Output tab

The **Output** tab is designed to display the various modeling assessment statistics. However, the procedure output listing is a little different between the various ensemble modeling designs that are selected. By selecting stratified modeling, the procedure output listing will display the predictive model, the target variable to predict, the average squared error between the partitioned data sets by each class level of the stratified variable and the various modeling assessment statistics. By selecting the boosting or bagging techniques, the listing will display the statistical model to fit; the average squared error and the misclassification rate of each classification model that is fit; the classification matrix; the decision matrices, if applicable, that are generated from the final classification model that is fit; and the various modeling assessment statistics. By selecting combining models, the listing will display the numerous models combined, the target variable to predict and the various modeling assessment statistics. This is illustrated in the following procedure output listing.

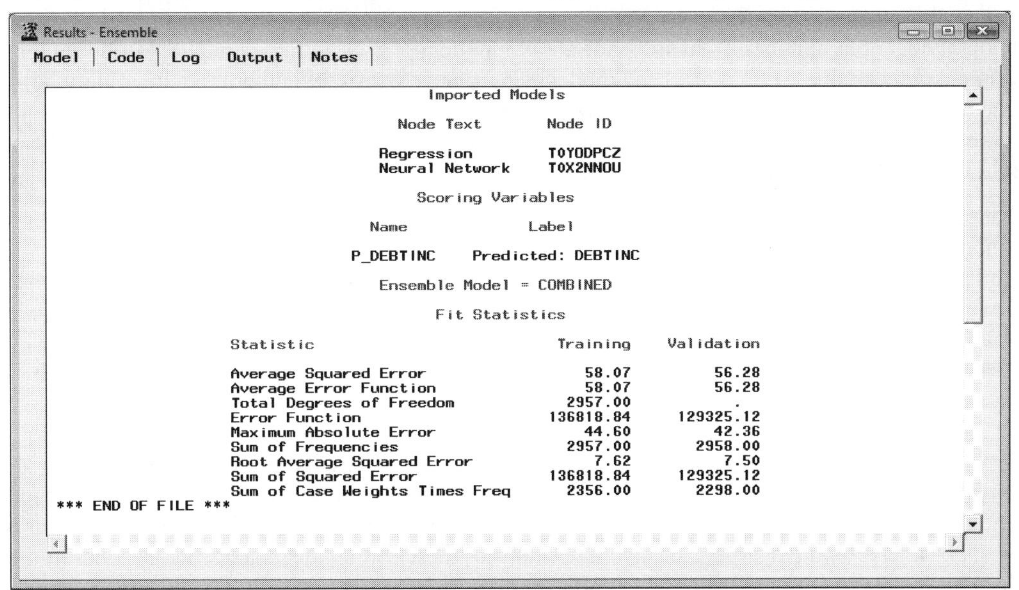

Results - Ensemble

Model | Code | Log | Output | Notes

Imported Models

Node Text	Node ID
Regression	TOYODPCZ
Neural Network	TOX2NNOU

Scoring Variables

Name	Label
P_DEBTINC	Predicted: DEBTINC

Ensemble Model = COMBINED

Fit Statistics

Statistic	Training	Validation
Average Squared Error	58.07	56.28
Average Error Function	58.07	56.28
Total Degrees of Freedom	2957.00	.
Error Function	136818.84	129325.12
Maximum Absolute Error	44.60	42.36
Sum of Frequencies	2957.00	2958.00
Root Average Squared Error	7.62	7.50
Sum of Squared Error	136818.84	129325.12
Sum of Case Weights Times Freq	2356.00	2298.00

*** END OF FILE ***

*The **Output** tab is used to view the assessment statistics by combining the regression and neural network models by fitting the interval-valued target variable, DEBTINC, to the ensemble model.*

Results - Ensemble

Model | Code | Log | Output | Notes

Ensemble Model = COMBINED

Classification Table

From	Into	TRAIN Count	TRAIN Percent	VALID Count	VALID Percent
0	0	2379	79.8322	2359	79.1611
0	1	10	0.3356	23	0.7718
1	0	424	14.2282	421	14.1275
1	1	167	5.6040	177	5.9396

Fit Statistics

Statistic	Training	Validation
Average Profit	0.20	0.20
Average Squared Error	0.10	0.10
Average Error Function	0.33	0.34
Total Degrees of Freedom	2980.00	.
Frequency of Classified Cases	2980.00	2980.00
Error Function	1943.02	2032.75
Maximum Absolute Error	0.97	0.96
Misclassification Rate	0.15	0.15
Sum of Frequencies	2980.00	2980.00
Total Profit	591.00	598.00
Root Average Squared Error	0.32	0.32
Sum of Squared Error	598.41	621.47
Sum of Case Weights Times Freq	5960.00	5960.00
Frequency of Unclassified Cases	0.00	0.00
Number of Wrong Classifications	434.00	444.00

*** END OF FILE ***

*The **Output** tab is used to view the various assessment statistics by combining the classification models by fitting the categorically-valued target variable, BAD, to the ensemble model. The results are based on the various classification models under comparison from the **Assessment** node.*

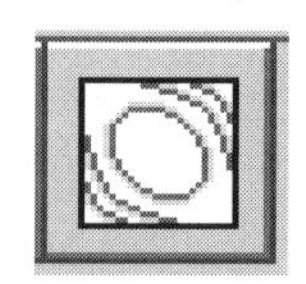

4.8 Memory-Based Reasoning Node

General Layout of the Enterprise Miner Memory-Based Reasoning Node

- **Data tab**
- **Variables tab**
- **Output tab**
- **Notes tab**

The purpose of **Memory-Based Reasoning** node in Enterprise Miner is to perform *k*-nearest neighbor predictive or classification modeling. *Nearest neighbor modeling* is considered nonparametric modeling since there are a small number of assumptions that are made about the functional form that is required in fitting the distribution of the target variable with no inherited model being fitted. Nearest neighbor modeling might be considered when you want to fit a curve to the data, where the functional relationship between the input variables and the target variable is extremely complicated and cannot be determined. Furthermore, this type of modeling might be applied when there are outliers in the data.

The reason that this model is called *k*-nearest neighbor is to emphasize the importance of the smoothing parameter estimate *k* in the nonparametric model. The only estimate that this type of smoothing predictive modeling technique needs is the number of nearest neighbors *k*. In Enterprise Miner, the default is 16 nearest neighbors. However, the number of nearest neighbors *k* is usually determined by trial and error that depends on the distribution of the data and the number of variables in the nonparametric model. The *k*-nearest neighbor estimates are calculated by either the most often occurring target category of the categorically-valued target variable falling within the predetermined region or the average value of data points surrounding the predetermined region by fitting an interval-valued target variable. The *k*-nearest neighbors algorithm is such that each observation consists of the set of input variables and the probe, where the probe has one value for each input variable in the nonparametric model. The first step is computing the squared distance between the observation and the probe *x*. The *k* target values that have the smallest distances to the probe *x* are the *k*-nearest neighbor values. For interval-valued targets, the average values of these same *k*-nearest neighbor values of the target variable are calculated as the prediction of the new observation. For categorically-valued targets, the nearest neighbor method assigns the new observation to the target category that has the most often occurring target class level within the predetermined region of *k*-nearest neighbor values.

MBR modeling performs both predictive and classification modeling. Nearest neighbor predictive modeling can fit either an interval-valued or a categorically-valued target variable. However, it is assumed that there is one and only one target variable in the model that you want to predict. In MBR modeling, the *k*-nearest neighbor estimate acts as a smoothing parameter to the model. In classification modeling, by creating the boundaries between the target groups, higher the *k* values will produce smoother nonlinear decision boundaries and lower *k* values will produce more jagged nonlinear decision boundaries. In predictive modeling, the higher *k* values will produce a smoother prediction line and lower values will produce more a jagged prediction line that will overfit the data. The reason that the nearest neighbor models will have a smoother fit with a larger number of *k* neighbors is because the model will be allowed to average more target values.

Cross-validation techniques are strongly encouraged in determining the correct number of neighbors *k* with the idea of reducing the bias and obtaining unbiased prediction estimates by generating an honest assessment of determining the best *k* value to select from, since the accuracy of the results from the modeling fit depends on the value *k*. In nearest neighbor modeling, you are looking for consistency in the classification or prediction estimates between the separate data sets shown by the different number of neighbors that are specified in order to select the smallest error from the validation data set. The optimal *k* nearest neighbor value that is selected should be large enough so that it will result in the smallest misclassification rate and yet small enough so that there are an adequate number of values within the predetermined region that are close enough to the probe *x*. For noiseless data, a small *k* value will result in the best regression line. For noisy data, a larger *k* value is preferred. The arithmetic mean is calculated by the *k* target values in producing the nearest neighbor estimate. Therefore, as *k* approaches the number of available data points, $n - 1$, then the nearest neighbor estimate approaches the global average that is essentially a straight line. For interval-valued targets, nearest neighbor modeling with the value k set to one is similar to least-squares regression. This will be demonstrated in the following diagram. However, there is no advantage to nearest neighbor modeling in setting the number of neighbors k to one. Generally, smaller values should be selected for the nearest neighbors k, since the dimensionality increases with the number of input variables in the model.

Predictive Modeling: For predictive modeling, the prediction estimates are calculated by the arithmetic mean of the target values within the hypersphere surrounding the probe x that is based on the smallest squared distance between each target value and the probe x. The estimates are computed so that the target values are contained within the specified radius or hypersphere of the k-nearest neighbor values as follows:

$$\hat{y}_i = \frac{\sum_{i=1}^{k} y_i}{k} \quad \text{for some nearest neighbor value } k > 0$$

Classification Modeling: For classification modeling, the fitted probabilities for each target level are determined by the number of occurrences of each target category falling within the hypersphere surrounding the probe x, then dividing the number of occurrences by the total number of data points in the hypersphere.

$$\hat{p}_i = \frac{n_k}{k} \quad \text{where } n_k \text{ is the number of occurrences of the target level within the hypersphere}$$

In other words, the smoothing parameter k determines the size of the hypersphere surrounding the probe x in which the probability estimate is calculated for each target category. Each observation is classified in the target category with the highest estimated probability or the target class level that occurs most often within the given hypersphere of the k-nearest neighbor data points.

In classification modeling, to best understand the way in which k-nearest neighbor modeling works in classifying new observations among a set of given input variables in the model, let us consider the following example of fitting the home equity loan data set, where the goal of the analysis is predicting the target variable BAD, that is, a good or bad creditors, from the pair of input variables, MORTDUE and LOAN. This is illustrated in the following graph that is displayed on the left. The basic approach to classifying a new observation is based on the probe x from a selected number of nearest neighbor values where the goal is to determine whether the new observation should be classified as a good or bad creditor within the surrounding region of predetermined k data points. From the following illustration, if the k-nearest neighbor were set to 1, then the observation would be classified as a bad creditor since it is closest to the probe point x, that is, with probability of $p=1$. Increasing the k-nearest neighbors to 2 will result in an indecisive classification, that is, with probability of $p=.5$. However, increasing the k-nearest neighbors to 3 or 4 will result in the observation being classified as a good creditor, that is, with probability of $p=.667$ or $p=.75$. In regression modeling, the graph that is displayed to the right illustrates the process of predicting radiation exposure over time to show the process of estimating an interval-valued target variable, that is, radiation counts. The vertical line displays the probe x for the input variable time. Therefore, setting the nearest neighbor value to 1 will result in the third data point as the nearest neighbor estimate since the data point is closest to the probe x. By setting the nearest neighbor value to 2 will result in the average of second and third data points, and so on.

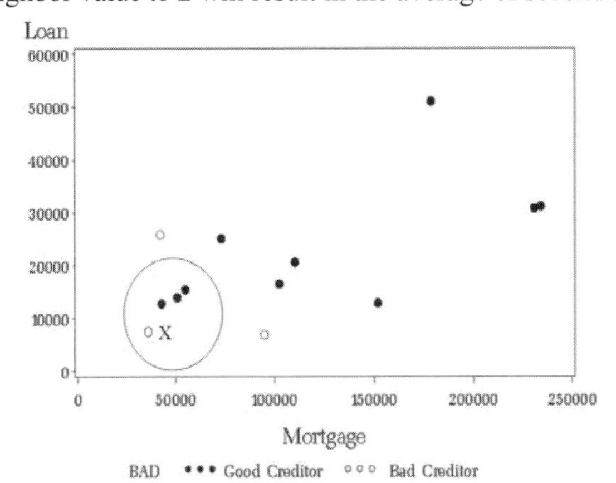

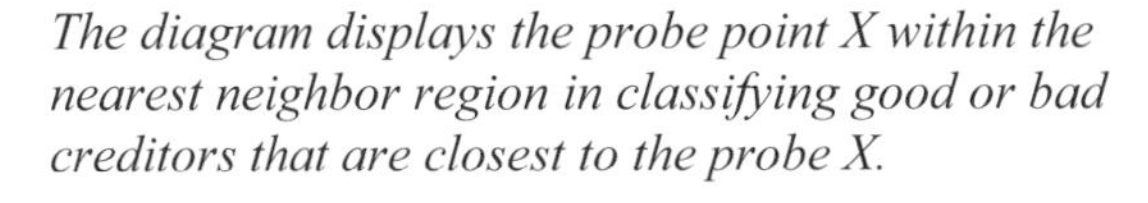
The diagram displays the probe point X within the nearest neighbor region in classifying good or bad creditors that are closest to the probe X.

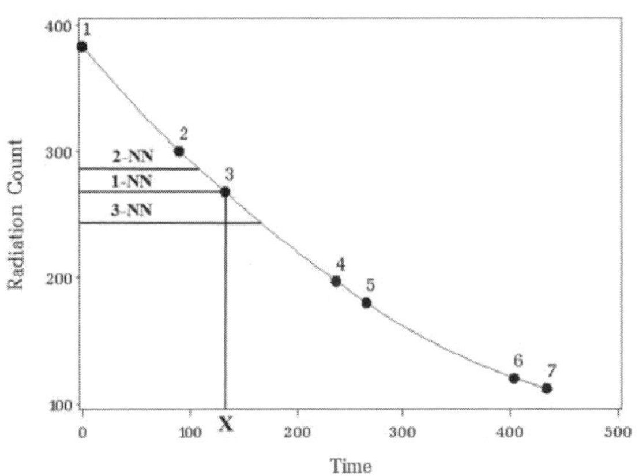

The diagram displays the probe point X in averaging the closest data points to the probe X in calculating the nearest neighbor estimates.

Fundamental Contents to MBR Modeling

In Enterprise Miner, the probe x is defined by the sorted values of the input variables that are created in the SAS data set. For instance, since it is recommended in using the principal component scores with numerous input variables to the analysis, then the probe x is determined by the sorted values of the principal component scores. Therefore, the nearest neighbor modeling estimates are calculated by the average target values or the number of target categories within a predetermined window of k points that lie closest to the current data point to fit in the multidimensional region. In other words, the fitted values are calculated by the average of the k target values that are defined by the sorted values of the input variables in the model. For interval-valued target variables, this type of model fitting is analogous to moving average time series modeling. Moving average time series modeling is basically calculating the average value based on the consecutive observations in the target variable moving over time. Since the squared distance is applied to find the closest k nearest neighbor data points, then it is recommended to remove categorically-valued input variables from the model and replace these same categorically-valued input variables with closely related interval-valued input variables. In classification modeling, it is important that each class level be adequately represented. That is, the training data set should consist of an even number of observations for each target class level. Otherwise, the majority of the nearest neighbor estimates will tend to result in the most frequent target class level. In addition, similar to predictive modeling, the larger the training data set, the better the nearest neighbor estimates.

The reason that this type of technique is called memory-based modeling is that this procedure does not require a predetermined model or a mathematical functional form that you would like to fit. That is, the beauty of k-nearest neighbor predictive modeling is that it does not make any assumptions with regard to the distribution of the data. In this non-parametric modeling technique, the mathematical relationship between the input variables in the model and the target variable that you want to predict is ignored. However, it is important to point out that the various modeling assumptions should be met based on the best smoothing constant that results in the smallest modeling assessment statistic in order to achieve adequate prediction estimates and good generalization.

Input Variables from Principal Components Analysis

The **Memory-Based Reasoning** node performs a singular value decomposition in which the principal components are the input variables of the nearest neighbor nonparametric model. The set of principal components that are used in the model are designed to provide the best linear combination of input variables in explaining the variability in the data. In principal component analysis, the first principal component will have the highest variance in comparison to all possible linear combinations of input variables, and the second principal component will have the next highest variance, with the constraint that the second principal component line is perpendicular to the first principal component line. Therefore, both components are uncorrelated to each other. One of the primary goals of this singular value decomposition technique is to reduce the dimensionality of the data while at the same time preserving the original ranked-order distance between the various input variables in the data along with capturing as much of the variability in the data in order for the nearest neighbor smoothing technique to perform best. Therefore in this nonparametric modeling approach, it is assumed that the interval-valued input variables are standardized to a common numeric scale and are also uncorrelated to each other. The **Princomp/Dmneural** node may be used in creating numeric, orthogonal, and standardized input variables that can then be used as input variables for the following **Memory-Based Reasoning** node. The reason that the input variables must be standardized is that the algorithm depends on the difference between all pairs of Euclidean distances between the separate input variables and the probe x in which the best set of k nearest neighbor values are determined by the range of values in the input variables in the model.

Drawbacks to MBR Modeling

The **Memory-Based Reasoning** node is capable of fitting either a binary, nominal, or interval-valued target variable. However, one drawback to the node is that it is not designed to model an ordinal-valued target variable, but an ordinal-valued target variable can be modeled as an interval-valued target variable.

One of the advantages of this type of nonparametric modeling is that it can fit very complex functions and highly nonlinear classification boundaries. However, a common problem in statistical modeling, like nearest neighbor modeling, is that it is affected by the *curse of dimensionality*. The *curse of dimensionality* problem is such that as an increasing number of input variables are added to the model, it will result in an increase in the dimension of the data, with the complexity in modeling design increasing exponentially with the number of data points needed to calculate the nearest neighbor prediction estimates. The reason is because the increase number of variables will result in an increased dimensionality in the neighborhood of data points and an added complexity of the hypersphere that will tend to make the points further away from one another, resulting in a small number of data points lying within each neighborhood of data points. This will result in a reduction in the performance of the nonparametric model. In classification modeling, the reduction in the modeling performance based on the increased dimensionality of the model will result in a sparse distribution of the data points in the high dimensional space with a fraction of the data points lying closer and closer to the classification boundaries as the dimensionality of the data points grow. This increase in the dimension of the data will result in a smaller number of data points that will be used in determining the closest distance between the best set of k nearest neighbor values and the probe x that are within the predetermined region. Therefore, principal components, that is, those with two principal components that will usually explain a large majority of the variability in the data, are applied to reduce the dimensionally of the data. The main idea is transforming the data set into a two-dimensional plane from the first two principal components that best explains the variability in the data. The drawback is that there is no guarantee that these same principal components might explain the variability in the target values that you want to predict. An alternative method to reducing the number of calculations for the model and increasing the likelihood that the number of observations that will be used in calculating the nearest neighbor estimates in creating the nearest neighbor classification boundary or prediction line is applying cluster analysis in order to condense the data into separate cluster groupings. Assuming that the cluster groupings are well separated from each other, then the cluster means might be used as the data points in fitting the MBR model.

Another disadvantage of this type of modeling design is that it can be computationally intensive due to the number of input variables in the data set in determining the best set of k target values to fit. In memory-based reasoning models, there is no model that you want to fit or underlying function to be estimated. The prediction estimates are driven by the inherited data. In MBR modeling, like all modeling designs, the larger the data is the better results. For categorical target variables, it is important that there be a sufficient number of observations for all possible target levels. However, the drawback is that extremely large data sets with an enormous number of observations will force the **Memory-Based Reasoning** node to search for the best k nearest neighbor values, which can be a very time-consuming process. Therefore, the following searching routines are applied in order to accelerate the searching process and reduce the amount of memory for the estimation process.

The Nearest Neighbor Methods

The two nearest neighborhood methods that are available within the node are the Rd-tree and scan methods. These methods are designed for assisting the nonparametric modeling design in determining the most appropriate neighbors in calculating the prediction estimates.

Scan: Scan is the normal technique that goes through every record in which the k-nearest neighbor estimates are calculated by the squared difference between the k different data points in the input space. The k-nearest neighbor points are determined by the smallest linear squared distance between all possible k different data points. Using the subsequent Kd-tree structure to store the data set is advantageous when the number of input variables in the data set is less than twenty. Beyond that, a simple linear scan through the training data set is almost always more efficient for high-dimensional problems.

Kd-tree (K Dimensionality Tree): The Rd-tree is a technique that was first proposed by J.L. Bentley in 1975. It is an improvement and an enhancement over the Kd-tree algorithm that was invented many years ago. In explaining the Rd-tree method, the Kd-tree method will be first explained since the Rd-tree method is a modification to the Kd-tree method.

The Kd-tree technique is one of the fastest methods for finding the nearest neighbors. The algorithm is designed by recursively partitioning the range of values of the input variables into smaller subsets assuming that the number of input variables to the model is small enough, that is, fewer than ten to twenty input variables. The recursive procedure divides the box called the initial cell or bucket that encloses the range of values of input variables into any number of hyperrectangular cells. The corners of the final cells that are created are the fitted values for the nearest neighbor model. In the first step, the initial cell is split into two halves by the direction of the longest cell edge that is selected as the split direction. The Kd-tree method splits a cell along some dimension, usually the one that has the greatest range of the observations in each cell. The cell is split into two, generally at the median of the selected splitting input variable. Half the observations that have values smaller than the median are grouped into the left cell, and the other half of the observations greater than the median are split into the right cell. The algorithm creates a binary tree by repeatedly partitioning the data by creating several cells or nodes, by dividing and subdividing these same cells and nodes. The binary splitting process continues until the number of observations in a cell is smaller than the specified nearest neighbor value *k*. At each cell, the recursive partitioning procedure determines whether to treat the cell as one group of *k* data points or perform additional partitioning, assuming that the rectangular cell has more than the specified number of *k* data points.

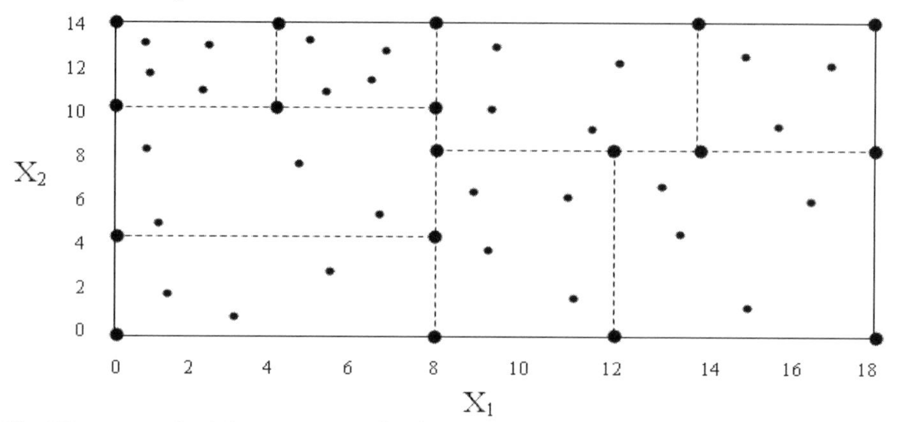

The KD-tree method that partitions the thirty separate data points in two dimensions where several rectangular cells are created from the corner of each cell (the large dots) that are the fitted values for the nearest neighbor model at k = 4.

The following are the steps involved in determining the fitted points in the Kd-tree technique:

1. Select the number of nearest neighbors *k*.
2. The procedure begins by formulating the smallest cell that encompasses the data. In the first step, the first cell is a line segment, assuming that there is one input variable in the model. For two separate input variables, the first cell is a rectangle.
3. Select the direction of the longest cell edge as the split coordinate.
4. Split the cell into two separate cells at the median of the splitting input variable.
5. For each child cell, repeat the process until each cell contains the number of *k* data points, that is, the specified number of neighbors *k*.
6. The corners of the cells that are created determine the fitted values to the model. In other words, the other data points are fitted by interpolating from the data points that are located at the corners of each cell that is formulated.

Rd-tree (Reduced Dimensionality Tree): The Rd-tree method is a modification of the Kd-tree method. Similar to the previous Kd-tree technique, the first step is to determine if the input variables are standardized and orthogonal, with the categorically-valued input variables transformed into an interval-valued scale by creating dummy variables for each class level that will undesirably increase the dimensionality of the model. The partitioning of the data points occurs at the same value, but the difference is the way in which the partitioning of the data points is represented. This change will result in a dramatic reduction in the number of nodes that need to be examined in finding nearest neighbors when the dimensionality increases. Rd-tree usually performs more efficiently than the Kd-tree method, even when the number of input variables is more than 20. The Rd-tree method works even better up to a dimensionality of 100. However, the Rd-tree method performs best with relatively small dimensionality to avoid the boundary effect.

Similar to the previous Kd-tree technique, the binary split will usually occur at the median of the split. However, as opposed to the Kd-tree method that evaluates a split at a specific value, usually at the median, the Rd-tree method evaluates a split in the range of values at each node of the corresponding splitting variable. Each node is based on the range of values of a particular dimension, that is, input variable in the model, to determine the number of nearest neighbors. A recursive binary splitting procedure is performed until each subnode or leaf contains the number of nearest neighbors. The number of neighbors defines the maximum number of data points that are contained in each leaf of the binary tree. In the Rd-tree method, the minimum and maximum values, that is, the range of values, for each node is stored into memory, as opposed to the Kd-tree technique that stores a single splitting value with no range of values. As a new observation is added to the tree, that is, once the observation is assigned to the corresponding node, if necessary, the range of values are updated accordingly, as opposed to the Kd-tree technique in which no values are changed within the node. Unlike the Kd-tree technique, the Rd-tree technique not only stores into memory the data points, but also the range of values in each node along with the dimension, that is, the splitting variable where the split occurred.

The following are the basic steps of the node searching routine in determining the number of neighbors in the Rd-tree technique:

1. **Current Root Node**: At the beginning, the current tree node is the root node.
2. **Node Searching Information from Memory**: At each stage, the algorithm searches all *n* dimensions, that is, each one of the splitting variables, which delivers the greatest variability from the number of nearest neighbors. The split will usually occur at the median. The node searching routine retrieves from memory the range of values in each node that is created to determine which node is nearest to the new observation.
3. **Current Root Node a Leaf Node**: The algorithm determines if the current node is a branch or leaf and evaluates how many data points are contained in the node. The splitting technique will split the current node into two separate nodes or branches so that each branch contains the range of data with the number of nearest neighbors and the dimension in which the split was performed.
4. **Set Current Node**: If the current node is not a leaf node, then the algorithm determines whether to select the current left or right node or select the left or right node to expand based on the dimension that delivers the greatest variability that is denoted by D_i.

   If D_i is greater than the minimum of the right branch, then set the current node to the right branch. If it is not, then determine whether D_i is less than the maximum of the left branch. If it is, then set the current node to the left branch and go to step 6.
5. **Branch to Expand**: When the new observation is between the maximum of the left branch and the minimum of the right branch, then it selects the node with the smallest minimum distance to expand first. In other words, select the left or right branch to expand based on the number of data points on the right branch, N_r, the distance to the minimum value on the right branch, $dist_r$ the number of data points of the left branch, N_l, and distance to the maximum value on the left branch, $dist_l$. When D_i is between the maximum of the left branch and the minimum of the right branch, then the decision rule is to place the new observation in the right branch of the binary tree if $(dist_l / dist_r)/(N_l / N_r) > 1$. Otherwise, place the data point in the left branch. If the data point is place in the right branch, then set the minimum distance of the right branch to D_i. Conversely, if the data point is placed in the left branch, then set the maximum distance of the left branch to D_i. The next step is to determine if the current node is a leaf node.
6. **Add Point to Current Node**: If the node is a leaf node, then the added observation is assigned to the current node to either the left or right node, with the range of values adjusted as follows:
   - **Left branch**: If the current node is a leaf node and the new observation is less than the minimum value of the left branch, then assign the observation to the corresponding node and adjust the range with the assigned value as the updated minimum value to the range of values of the corresponding node.
   - **Right branch**: If the current node is a leaf node and the new observation is greater than the maximum value of the right branch, then assign the observation to the corresponding node and adjust the range with the assigned value as the updated maximum value to the range of values of the corresponding node.
7. **Terminate the Branching Process**: If the current node has less than the number of nearest neighbors, then split the current node, usually at the median, into left and right branches along the dimension with the greatest range. Otherwise, the recursive tree branching process terminates when each node has an adequate number of nearest neighbors.

Weight Adjustments to the Kernal Regression Model

Since that nearest neighbor predictions are determined by the k-nearest neighbor values that have the closest distance to the probe x, the Euclidean squared distance is used in Enterprise Miner. However, improvements to this measurement can be achieved by weighing each nearest neighbor value so that the nearest neighbor values that are closer to the probe x are weighted more heavily. For interval-valued target variables, the node has an additional weight dimension option that is automatically applied to the nonparametric model. The weights are designed to adjust the distance function by favoring some input variables in the model over others. The weighted adjustment is applied to overcome the dimensionality problem.

The general formula of the nearest neighbor nonparametric model is defined so that the target values are weighted by the individual weights to compute the predicted values as follows:

$$\hat{y}_i = \frac{\sum_{i=1}^{k} w_i y_i}{k} \quad \text{where } \hat{y}_i = \text{nearest neighbor estimate at the } i^{\text{th}} \text{ observation}$$

The smoothing parameter estimate k controls the amount of averaging of the target values in calculating the nonparametric estimates. The general formula for the weighted nonparametric regression model is based on each input variable in the model that is weighted by the absolute value of its correlation to the interval-valued target variable with the main idea of eliminating the least relevant input variables from the estimation process. Setting the weight estimate close to zero for the input variables that are not highly correlated with the target variable will contribute very little to the nearest neighbor estimates.

Pictorial Examples

The following scatter plots are displayed to better understand the smoothing behavior or the varying degree of smoothing of the kernal parameter estimate. The diagrams display the different prediction lines and classification boundaries created with different values set to the nearest neighbor kernal parameter estimate. Although there are many theoretical formulas designed to attempt to determine the best choice of the number neighbor estimate k, the best strategy that is recommended is to fit the nonparametric model to many different values and plot the sum-of-squares error or the misclassification rate across the numerous values of k. The following illustration is from the home equity loan data set in predicting the amount of a home loan, LOAN, by the number of years at a job, YOJ, to illustrate the smoothing effect by increasing the value of k in the k-nearest neighbor modeling design in predicting the interval-valued target variable. A random sample of 2% of the data was performed in order to prevent the data points from cluttering the plot so you may then observe the smoothing effect by increasing the nearest neighbor k. Although nearest neighbor modeling is rather robust to outliers in the data, outliers have a profound effect on the estimates, particularly when the smoothing constant k is set at very small values. This is displayed in the line plot to the left.

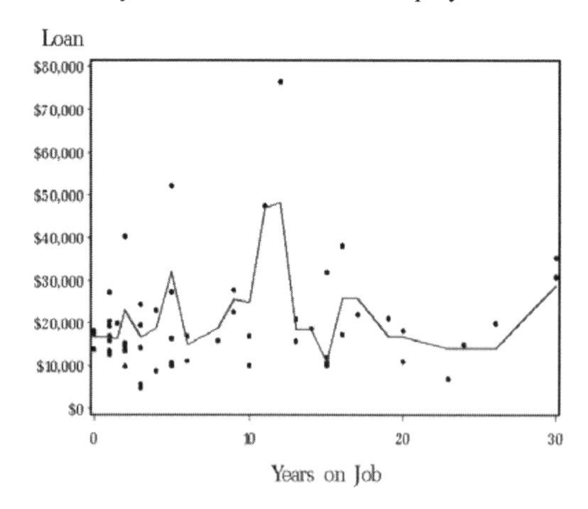

K-nearest neighbor model with k = 3.

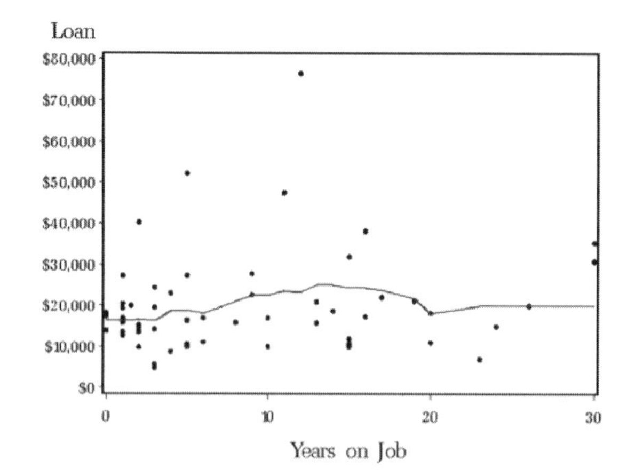

K-nearest neighbor model with k = 16.

MBR modeling can also be applied in classification problems in which the number of nearest neighbors k determines the degree of smoothing of the nonlinear decision boundary. This is illustrated in the following example. The following diagram is based on the iris wildflower data set created by Sir Ronald Fisher in 1936.

The data was used to introduce discriminant and cluster analysis. Notice the three separate cluster groupings that are formed by the different varieties of wildflower that can be identified by the dots, circles, and crosses. Both the petal length and petal width were selected to determine the different type of iris wildflowers since these input variables demonstrated a clear separation between the different varieties of wildflower. Two separate classification boundaries were created to illustrate the different nonparametric classification boundaries that can be created with the number of k-nearest neighbors set at both 2 and 16. Notice the smoothing behavior in the classification boundaries with an increase in the smoothing constant. The same classification boundary that is illustrated to the left is displayed in my first book, *Neural Network Modeling using SAS Enterprise Miner* that compares the classification performance between nearest neighbor modeling and neural network modeling. The various misclassification rates that are displayed from the table listing are derived by fitting several values set to the k-nearest neighbors from an even split applied to the training and validation data sets. An even split was performed for both data sets in order to achieve the same proportionally in the target class levels between each partitioned data set. The best smoothing constant with the lowest misclassification rate in classifying the iris wildflower species seems to be at 2 from the validation data set that is consistent with the training data set.

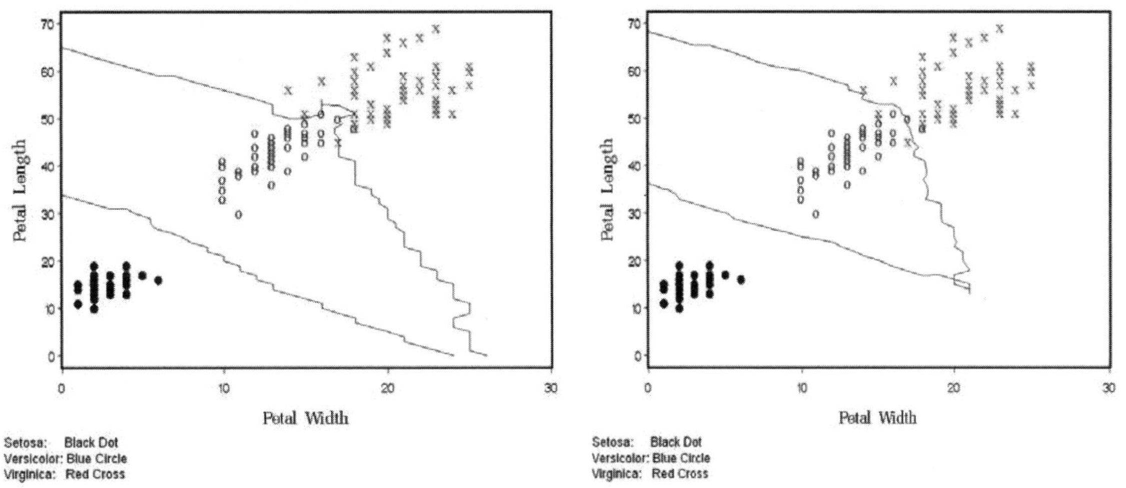

K-nearest neighbor model with k = 2.

K-nearest neighbor model with k = 16.

Missing Values

For interval-valued target variables, when there are missing values in the target variable, then the nearest neighbor model will calculate the fitted values by the average value of the target variable from all nonmissing values of the k target values that are sorted by the values of the input variables in the nonparametric model. Since the fitted values are calculated by the average values of the target variable, therefore missing values in the input variable are not affected by the estimates since the model simply calculates the number of nearest numbers between the missing input values that are recorded in the data. For categorically-valued target variables, missing values in the target variable is treated as an extra class level to predict.

Memory-based reasoning is discussed in greater detail in Hastie, Tibshirani, and Friedman (2001), Berry and Linoff (2004), Hand, Mannila, and Smyth (2001), Bentley (1975), Cox (2005), and Applying Data Mining Techniques Using Enterprise Miner Course Notes, SAS Institute (2002).

Data tab

The **Data** tab is designed for you to select the partitioned data sets and view the file information along with a table view of the active training data set that you want to fit. In the following example, the scored data set from the **Princomp/Dmneural** node was used as the active training data set. The reason is because the first two principal components explained a majority of the variability in the home equity loan data that resulted into a two-dimensional problem for the nonparametric model to fit. The following analysis is performed for each one of the partitioned data sets. Assuming that there is a significantly large sample size, then it is recommended to apply the validation data set to the modeling process. The validation data set can be used to cross-validate the accuracy of the nearest neighbor estimates. Furthermore, the validation data set can be used in determining the most appropriate smoothing constant k that should be selected for the nonparametric modeling design.

Variables tab

The **Variables** tab will automatically appear as you open the **Memory-Based Reasoning** node. The **Variables** tab displays a tabular view of the variables in the input data set. The tab displays the various properties of the listed variables such as the variable name, model role, level of measurement, variable type, format, and label. The tab has the added option of removing certain variables from the model by simply selecting the variable row, then scrolling over to the **Status** column, right-clicking the mouse and selecting the **Set Status** pop-up menu item of **don't use**. Notice that the first two principal components from the **Princomp/Dmneural** node that explained a large majority of the variability in the data are the only two input variables of the nonparametric model that predict the target values, that is, the ratio between debt and income, from the home equity loan data set.

From the **Variables** tab, you may also edit, browse, or delete the target profile of the categorical target variable within the tab. However, updates performed for the target profile will only be reflected within the node. The target profile contains information about the target variable such as the assessment criterion, decision values, prior probabilities, and target event level. The target profile produces optimal decisions from the active decision matrix. For categorically-valued targets, the prior probabilities may be specified for the target class levels in order to adjust the posterior probabilities. In addition, the tab will allow you to view the distribution of each variable in the model since extreme data points will have a profound effect on the final prediction estimates. This phenomenon is particularly true with small smoothing constants.

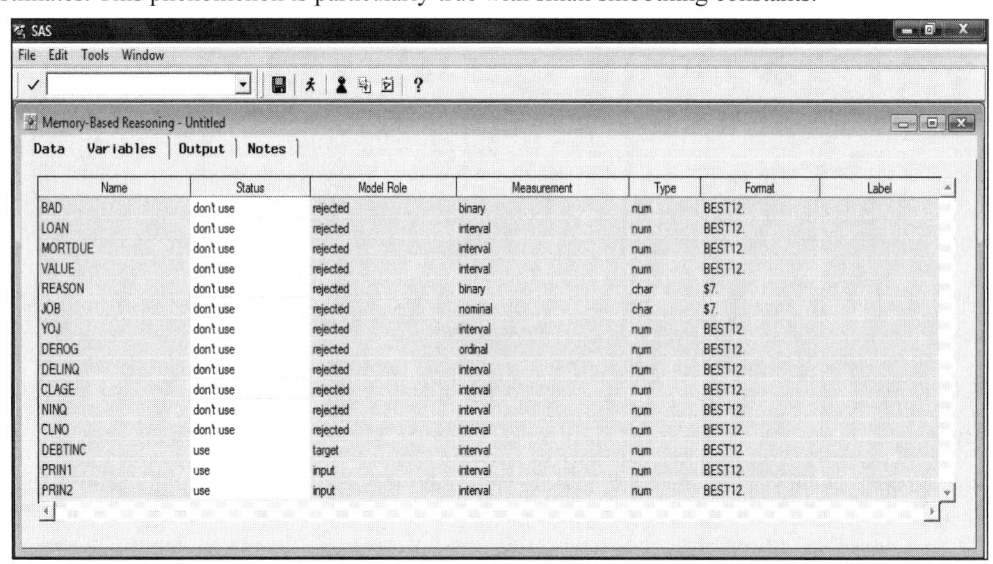

Name	Status	Model Role	Measurement	Type	Format	Label
BAD	don't use	rejected	binary	num	BEST12.	
LOAN	don't use	rejected	interval	num	BEST12.	
MORTDUE	don't use	rejected	interval	num	BEST12.	
VALUE	don't use	rejected	interval	num	BEST12.	
REASON	don't use	rejected	binary	char	$7.	
JOB	don't use	rejected	nominal	char	$7.	
YOJ	don't use	rejected	interval	num	BEST12.	
DEROG	don't use	rejected	ordinal	num	BEST12.	
DELINQ	don't use	rejected	interval	num	BEST12.	
CLAGE	don't use	rejected	interval	num	BEST12.	
NINQ	don't use	rejected	interval	num	BEST12.	
CLNO	don't use	rejected	interval	num	BEST12.	
DEBTINC	use	target	interval	num	BEST12.	
PRIN1	use	input	interval	num	BEST12.	
PRIN2	use	input	interval	num	BEST12.	

*The **Variables** tab is used to view the modeling terms to the k-nearest neighbor model.*

Model Option Setting to the Node

The following **Normalizer: Setting** window will appear by selecting the **Tools** > **Settings** main menu options or selecting the **Settings** toolbar icon from the tools bar menu. By default, the k-nearest neighbor model is automatically set to 16 nearest neighbors. The most important parameter estimate of the model is to determine the smoothing constant k that can be specified from the **K (Number of Neighbors)** entry field. The limitation to the number of neighbors k is that it can be no greater than the number of observations in the data set. However, if you specify the number of neighbors k greater than the number of observations in the data, then Enterprise Miner will automatically reassign the number of neighbors k to the number of observations in the data minus one, which is essentially fitting a straight line between two separate data points. The node is designed to perform two separate k-nearest neighbor techniques. The default method is the Rd-tree procedure. Select any one of the following radio buttons to select the k-nearest neighbor technique that is illustrated in the following diagram.

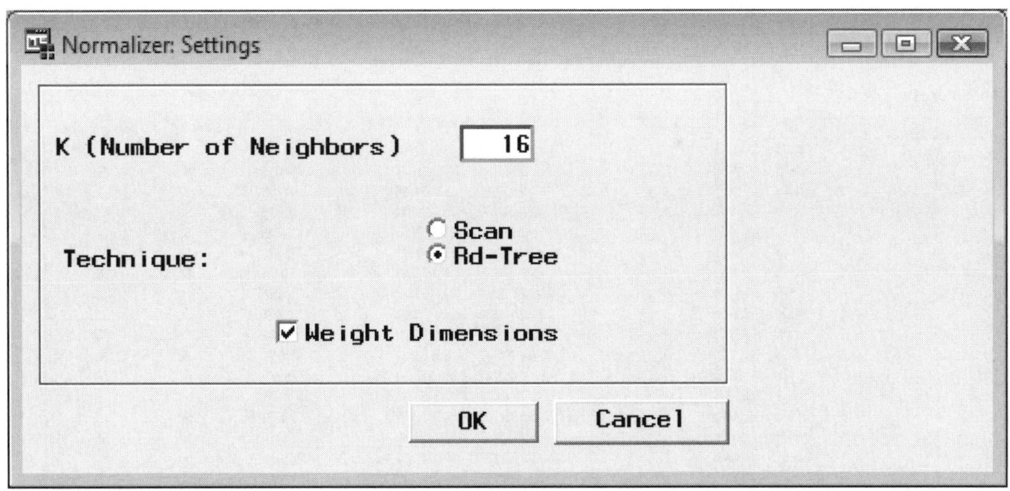

*The **Normalizer: Setting** window is used to specify the various k-nearest neighbor model settings.*

- **K (Number of Neighbors):** Specify the number of nearest neighbors for the model that must be a positive integer number between 1 and 9,999. However, the node will automatically reset the value entered to one less than the number of observations in the training data set. The value determines the degree of smoothing to the design and dictates the way in which the predicted values are computed. For predictive modeling, the number of nearest neighbors *k* determines the number of target values that are combined and averaged to calculate the predicted value. That is, the number of observations that are assigned to each nearest neighbor region. For classification modeling, the number of neighbors *k* determines the degree of smoothing for the decision boundaries between the target groups.

The following are the two separate nearest neighbor techniques that are available within the node:

- **Scan:** When this check box is selected, the node scans through each observation in the training data set when the *k*-nearest neighbors are searched and retrieved. The scan method computes the squared distance for every observation to search for the closest nearest neighbor data points.
- **Rd-tree - (Reduced Dimensionality Tree)** (default)**:** This method is the default nearest neighbor method used. One reason is that this method uses a "smart" search technique as opposed to the **Scan** method. This technique usually performs better than the **Scan** procedure when there are less than twenty variables in the training data set. One reason is because this method examines fewer number of point comparisons in finding the nearest neighbor data points.
- **Weighted Dimension**: This option is designed to weigh each of the input variables when the model fits the interval-valued target variable. If the **Weighted Dimension** check box is selected, then each of the input variables is weighted by the absolute value of its correlation to the target variable. For interval-valued target variables, this option is automatically selected.

Output tab

The **Output** tab will display the output data set that is created for each partitioned data set with the *k*-nearest neighbor prediction estimates along with the other input variables in the partitioned data sets. In order to generate the posterior probabilities from a categorically-valued target variable or the predicted values by fitting an interval-valued target variable for each partitioned data set that is created, you must select the **Process of Score: Training, Validation, and Test** check box from the **Output** tab, which will result in the output scored data set that will be created once you train the nonparametric model by executing the node. By default, the check box is unchecked in order to save disk space since it will instruct the **Memory-Based Reasoning** node to create an output data set when you execute the node. For interval-valued targets, one predicted variable is created for the output data set that is the average target value of the *k*-nearest neighbor data points. For categorically-valued targets, one variable is created for the output data set for each target class level that is the estimated target proportions of the *k*-nearest neighbor data points for each class level of the target variable.

Data set details

Information   Table View

☐ Variable labels

	WARN	I_BAD	U_BAD	P_BAD1	P_BAD0	BAD	PRIN1	PRIN2	D_BAD_	EP_BAD_	BP_BAD_	CP_BAD_
1		0	0	0.375	0.625	1	46695.68971	-2991.49807	1	15.625	50	50
2		1	1	0.5625	0.4375	1	124000.01045	-2475.009044	1	25.9375	50	50
3		0	0	0.125	0.875	0	148267.13595	10253.159556	1	1.875	0	-5
4		0	0	0.25	0.75	1	74853.757875	4173.2801394	1	8.75	50	50
5		1	1	0.75	0.25	1	93442.111515	-42303.83851	1	36.25	50	50
6		0	0	0.375	0.625	1	36254.49973	-1756.472611	1	15.625	50	50
7		0	0	0.1875	0.8125	1	71033.034297	2476.161669	1	5.3125	50	50
8		0	0	0.125	0.875	0	108577.25304	-1403.487027	1	1.875	0	-5
9		0	0	0.4375	0.5625	1	109697.77601	5804.0374393	1	19.0625	50	50
10		0	0	0.4375	0.5625	1	136687.43154	10378.383072	1	19.0625	50	50
11		1	1	0.75	0.25	1	93709.295439	-41999.32331	1	36.25	50	50
12		1	1	0.625	0.375	1	49068.659187	-1992.912312	1	29.375	50	50
13		0	0	0.0625	0.9375	0	158170.49012	8440.3157967	0	-0.625	0	0
14		1	1	0.5625	0.4375	1	59595.326709	1853.1471105	1	25.9375	50	50
15		1	1	0.6875	0.3125	1	49472.395143	-2546.415096	1	32.8125	50	50
16		1	1	0.5625	0.4375	1	58928.759898	-1101.349784	1	25.9375	50	50
17		0	0	0.0625	0.9375	0	152815.44733	7568.5268135	0	-0.625	0	0
18		1	1	0.5625	0.4375	1	25255.258631	-551.5002991	1	25.9375	50	50
19		0	0	0.3125	0.6875	0	39800.089097	-21183.0748	1	12.1875	0	-5
20		0	0	0.0625	0.9375	1	81015.669026	2110.5351514	0	-0.625	50	-10

The scored output data set from the memory-based reasoning model with k=16 by fitting the categorical target variable with two principal components that are the input variables to the model and a target profile with decision entries similar to the following ***Interactive Profit*** *chart.*

For interval-valued target variables, the following are some of the variables that are created in the scored data set along with the existing variables in the active training data set and the two principal components that are the input variables to the *k*-nearest neighbor model.

Variable Name	Description
P_*target variable name*	The predicted values of the target variable.

For categorically-valued target variables, the output score data set consists of the following:

Variable Name	Description
P_*target variable name<level>*	The estimated probabilities at each class level of the target variable.
I_*target variable name*	The classification variable that identifies into which target class level the observation is classified by the largest estimated probabilities or the largest specified prior probabilities.
U_*target variable name*	The unformatted predicted target class level in which the observation is classified.
D_*target variable name*	The decision variable at each decision level.
EP_ *target variable name* or EL_*target variable name*	The expected profit of expected loss defined by the linear combination of decision entries and posterior probabilities with the reduction in the fixed costs at each decision level. If the target profile with the decision entries and prior probabilities are not provided, then the expected profit will be set to the estimated probabilities of the target event.
BP_ *target variable name* or BL_*target variable name*	The best profit or loss values from the target profile.
CP_ *target variable name* or CL_*target variable name*	The computed profit or loss values from the target profile.

Viewing the Memory-Based Reasoning Results

General Layout of the Results Browser within the Memory-Based Reasoning Node

- **Model tab**
- **Code tab**
- **Log tab**
- **Output tab**
- **Notes tab**

The following *k*-nearest neighbor modeling results are based on fitting the interval-valued variable DEBTINC. The **Filter Outliers** node was incorporated into the process flow to remove the extreme values in the interval-valued target variable of DEBTINC, along with removing observations in some of the interval-valued input variables in the model that were well separated from the rest of the other data points.

Model tab

The **Model** tab is designed for you to view the file administrative information for the training data set and view the target profile by fitting the categorically-valued target variable. The tab will allow you to view the various configuration settings that are set to the current nearest neighbor design along with the processing time for training the model in order to calculate the *k*-nearest neighbor estimates. The tab consists of the following four subtabs.

- **General subtab**
- **Variables subtab**
- **Settings subtab**
- **Status subtab**

General subtab

The **General** subtab is designed for you to view the file information of the scored data set. By selecting the **Profile Information...** button, you may then view the target profile of the categorically-valued target variable.

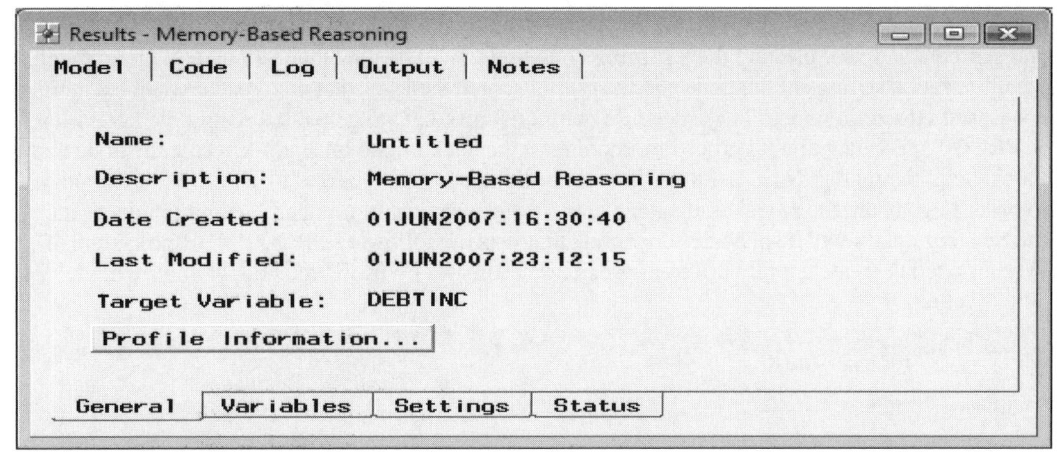

*The **General** subtab displays the file information of the scored data set that is created from the node.*

Variables subtab

The **Variables** tab is basically designed for you to view the variables and the associated model roles, level of measurement, variable type, and format of the variables in the output scored data set. The scored data set will display the two separate principal components that are the input variables to the model, the target variable to predict, and the associated predicted variables along with all other variables in the active training data set. The subtab will allow you to view the frequency distribution of the listed variables within the tab. Simply right-click the mouse and select the **View Distribution of <variable name>** pop-up menu item to display the frequency distribution of the selected variable from the metadata sample.

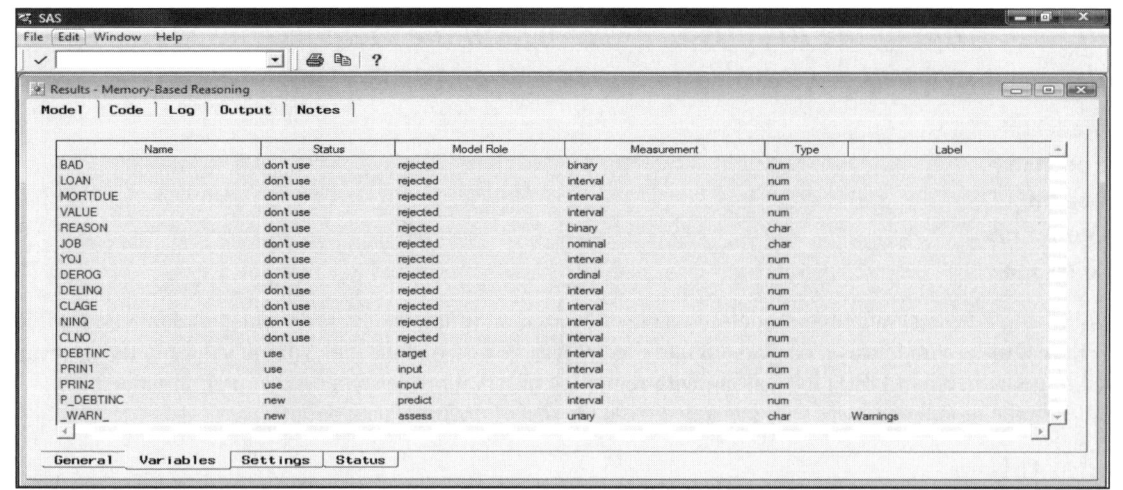

Name	Status	Model Role	Measurement	Type	Label
BAD	don't use	rejected	binary	num	
LOAN	don't use	rejected	interval	num	
MORTDUE	don't use	rejected	interval	num	
VALUE	don't use	rejected	interval	num	
REASON	don't use	rejected	binary	char	
JOB	don't use	rejected	nominal	char	
YOJ	don't use	rejected	interval	num	
DEROG	don't use	rejected	ordinal	num	
DELINQ	don't use	rejected	interval	num	
CLAGE	don't use	rejected	interval	num	
NINQ	don't use	rejected	interval	num	
CLNO	don't use	rejected	interval	num	
DEBTINC	use	target	interval	num	
PRIN1	use	input	interval	num	
PRIN2	use	input	interval	num	
P_DEBTINC	new	predict	interval	num	
WARN	new	assess	unary	char	Warnings

*The **Variables** subtab displays the modeling terms in the scored data set from the k-nearest neighbor model.*

Settings subtab

The **Settings** tab is designed to display the various configuration settings to the *k*-nearest neighbor model that generate the subsequent results. The tab displays the value set to the number of neighbors *k*, nearest neighbor method, weight dimension identifier, fitted data set, data mining catalog, and option settings using in the following PROC PMBR procedure along with the target variable and the principal components that are used as input variables for the nearest neighbor model, Enterprise Miner library, the names that are automatically assigned to the input and output scored data set, and the processing status.

Status subtab

The **Status** subtab is designed to display the various processing times for the training and scored data sets that is similar to the **Status** subtab in many of the other nodes. Therefore, the subtab will not be displayed.

Code tab

The **Code** tab is designed to display the internal SEMMA training code that produced the nearest neighbor estimates and scored data sets. Selecting the **Training** code option will display the following diagram that lists the various configuration settings of the node and the output scored data sets that are created when executing the **Memory-Based Reasoning** node. The procedure output listings that are generated within the node is the result of the PROC PMBR data mining procedure running in the background once you execute the node and train the nearest neighbor model. Note that the following training code can be used to set different smoothing constants in order to compare the model assessment statistics between the training and validation data sets. Selecting the **Scoring** option will display the score code that lists the following PROC PMBR procedure. The score code can be used to create entirely different prediction estimates by fitting the nonparametric model for a new set of input values.

Results - Memory-Based Reasoning

Model | Code | Log | Output | Notes

Training | Scoring

```
00001    * ;
00002    * Memory Based Reasoning Scoring ;
00003    * ;
00004    run;
00005    %let _oldscr=&SYSLAST;
00006    proc pmbr score=_last_ out=work._tmpmbr
00007                 data=EMDATA.STRNSPL0 dmdbcat=EMPROJ.dm_DGM00021 k=16 weighted method=RDTREE
00008                target DEBTINC;
00009                var PRIN1 PRIN2;
00010    data &_oldscr; set work._tmpmbr;
00011 *-------------------------------------------------------------*;
00012 * Memory-Based Reasoning: DEBTINC: Decisions;
00013 *-------------------------------------------------------------*;
00014 **** NO DECISION CODE;
00015 *** END OF FILE ***
```

*The **Code** tab is used to view the SEMMA score code to create different k-nearest neighbor estimates.*

```
Results - Memory-Based Reasoning
Model   Code | Log | Output | Notes |

(•) Training   ( ) Scoring

00001    * ;
00002    * Memory Based Reasoning Training ;
00003    * ;
00004    * ;
00005    * Score training data ;
00006    * ;
00007    proc pmbr score=EMDATA.STRNSPL0 out=work._tmpmbr
00008          data=EMDATA.STRNSPL0 dmdbcat=EMPROJ.dm_DGM00021 k=16 weighted method=RDTREE;
00009          target DEBTINC;
00010          var PRIN1 PRIN2;
00011    data EMPROJ.OUT_N1O1; set work._tmpmbr;
00012 *------------------------------------------------------------*;
00013 * Memory-Based Reasoning: DEBTINC: Decisions;
00014 *------------------------------------------------------------*;
00015 **** NO DECISION CODE;
00016          run;
00017    * ;
00018    * Score validation data ;
00019    * ;
00020    proc pmbr score=EMDATA.SVALOEBD out=work._tmpmbr
00021          data=EMDATA.STRNSPL0 dmdbcat=EMPROJ.dm_DGM00021 k=16 weighted method=RDTREE;
00022          target DEBTINC;
00023          var PRIN1 PRIN2;
00024    data EMPROJ.OUT_C8KV; set work._tmpmbr;
00025 *------------------------------------------------------------*;
00026 * Memory-Based Reasoning: DEBTINC: Decisions;
00027 *------------------------------------------------------------*;
00028 **** NO DECISION CODE;
00029          run;
00030 *** END OF FILE ***
```

*The **Code** tab is used to view the SEMMA training code that produced the k-nearest neighbor results.*

Log tab

The **Log** tab will not be displayed. The tab displays the log listing of the internal SEMMA code that produced the nearest neighbor results. The tab will display the entire SEMMA programming process by initially creating the DMDB data mining data set with its associated data mining catalog and metadata information to check for excessive class levels in the target variable. The PMBR procedure is used to create the scored data sets with the nearest neighbor estimates.

Output tab

The **Output** tab displays the various goodness-of-fit statistics from the nearest neighbor nonparametric model. The listed modeling assessment statistics within the tab are very similar to the other modeling nodes. If the listed assessment statistics are very different between the various partitioned data sets, then it is a strong indication to you that there exists overfitting in the nearest neighbor model that will result in instability and bad generalization. The following table listing indicates that there exists overfitting to the nonparametric model. The various assessment statistics that are displayed for either fitting the interval-valued or categorically-valued target variable are explained in the previous modeling table listings.

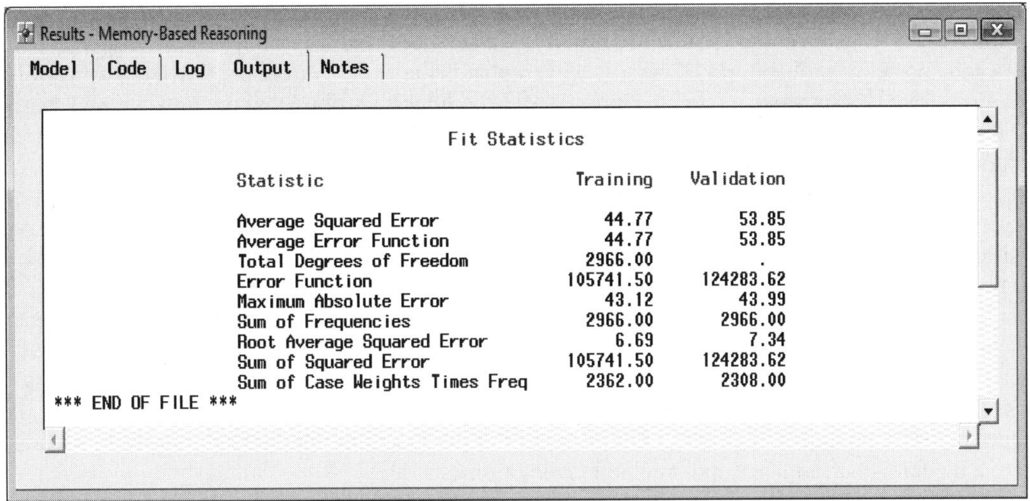

Results - Memory-Based Reasoning

Model | Code | Log | Output | Notes

Fit Statistics

Statistic	Training	Validation
Average Squared Error	44.77	53.85
Average Error Function	44.77	53.85
Total Degrees of Freedom	2966.00	.
Error Function	105741.50	124283.62
Maximum Absolute Error	43.12	43.99
Sum of Frequencies	2966.00	2966.00
Root Average Squared Error	6.69	7.34
Sum of Squared Error	105741.50	124283.62
Sum of Case Weights Times Freq	2362.00	2308.00

*** END OF FILE ***

*The **Output** tab with the modeling assessment statistics from the nearest neighbor model with k=16.*

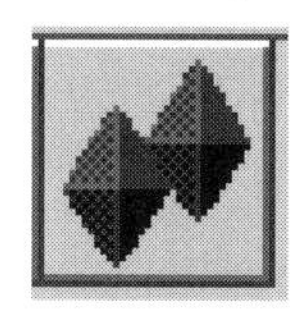

4.9 Two Stage Model Node

General Layout of the Enterprise Miner Two Stage Model Node

- **Data tab**
- **Variables tab**
- **Output tab**
- **Notes tab**

The **Two Stage Model** node in Enterprise Miner is to perform *two-stage modeling* that predicts an interval-valued target variable based on the estimated probabilities of the target event from a separate categorical target variable. In the two-stage modeling design, there are two separate models that are consecutively fitted by combining the probability estimates of the target event from a classification model by fitting the categorically-valued target variable that is followed by a prediction model by fitting the interval-valued target variable to predict. The purpose of two-stage modeling is that at times modeling both the output probability and the target values that you want to predict may generate more accurate estimates. In addition, including the estimated probabilities as one of the input variables in predicting the target values, might at times make it possible to specify a much simpler model. Generally, the target variable in the value model will be associated with the target level in the class model. For example, two-stage modeling might be useful in predicting the total sales of an item based on an identifier of this same item that was purchased, that is, by building a classification model for predicting the probability that the particular item was purchased that is then used in the subsequent prediction model in estimating total sales of this same item. Therefore, it might be advantageous to combine the predictions of both the probability of individuals purchasing the item and the subsequent predicted revenue with the main objective of generating more accurate prediction estimates.

In the first stage, a *class model* is fitted to the categorically-valued target variable. In the second stage, a *value model* is fitted to an interval-valued target variable that generates the final fitted values. Improving the two-stage predictive modeling results requires careful fitting to both the class and the value models. The best results require careful fitting to each modeling component, which is done separately. In order to achieve a more realistic assessment of the accuracy of both models, it is recommended to evaluate the mean square error from the validation data set. The reason is because the validation data set will give a realistic opinion to the performance of the model as opposed to the training data set that is used in fitting the model, which will result in an overly optimistic evaluation to the accuracy of the fit. Furthermore, a test data set might be used in obtaining an unbiased comparison of the accuracy of the estimates since a single validation data set might result in an inaccurate assessment of the modeling results.

The fitted values of the target event to include in the second-stage model are determined by a *transfer function* that either uses the posterior probabilities of the target event that is the default or the predicted classifications of the target event. The type of value model to apply will depend on the distributional relationship between the input variables in the predictive model and the target variable that you want to predict. The **Two Stage Model** node is designed to perform either linear regression modeling, decision tree modeling, and neural network or nonlinear regression modeling. The first-stage model is designed to minimize the misclassification rate based on the posterior probabilities and a predetermined cutoff probability. In Enterprise Miner, the classification model classifies the observation as a target event with the largest estimated probability within each target group from the training data set. The second-stage model is designed to minimize the sum-of-squares error function that is by fitting the interval-valued target variable. In the first stage, it is important that the correct transfer function be applied that follows the range of the predicted values. Conversely, the second stage and fitting an interval-valued target variable requires careful fitting with the added requirement that the various modeling assumptions be satisfied. The various modeling assumptions can be satisfied by simply plotting the residual values across the predicted values. Ignoring these extremely important modeling assumptions leads to poor predictive modeling performance and unreliable prediction estimates.

The **Two Stage Model** node is designed to use all the observations from the training data set in the second stage. However, through the use of the *filter option* you will have the added capability of removing certain observations from the training data set in fitting the subsequent second-stage model. In addition, the node will allow you to optionally adjust the posterior probabilities of the target event from the first-stage model through the use of the *bias adjustment* option in order to calculate the fitted values from the second-stage model. The **Two Stage Model** node also runs a posterior analysis that displays the fitted values from the interval-valued target variable by the actual values and the predictions from the categorically-valued target variable. The score code from the **Two Stage Model** node is a combination of both the class and value models. The fitted values from the interval-valued target variable can then be used in the assessment plots within the **Model Manager** and the **Assessment** node. Obviously, the **Two Stage Model** node requires both a categorically-valued target variable and an interval-valued target variable to perform two-stage modeling.

Selecting the Best Two Stage Model

The common practice in two-stage modeling is using the output from the first stage as input variables in the second stage. The general idea for introducing the predicted values as input variables in the second stage is to reduce the complexity of the model in order to achieve stability and good generalization, while at the same time constructing a predictive model that is easy to interpret. In other words, the output from the first stage might explain the relationship in which several input variables would explain the same variability in the target variable. This will result in a reduction in the overall degrees of freedom in the second-stage model that will result in an increase in power of the statistical test. That is, the predicted values from the first stage would replace many of the input variables used in the second-stage model in explaining the variability in the target values.

The best models are obtained when the individual models are correctly specified. Correct specification requires a careful selection of the input variables, and link function and the correct selection of the distribution of the error values. Proper specification in fitting the range of target values and the distribution of the error values increases the chance of selecting good input variables for the interval-valued target model. The error distribution should account for the relationship between the fitted values and the residuals of the predictive model. The error distribution should have an approximately normal distribution and should be consistent with the range of values that the model is trying to fit. Correctly specifying the distribution of the residual values is extremely important for good modeling performance. Given properly selected input variables, the correct degree of flexibility can be incorporated into the model, which will generate the best predictions. For example, the residual plots might indicate an increasing variability, that is, a fanned-out pattern, across the fitted values where better modeling results might be achieved by specifying such distributions like a lognormal error distribution. This can be achieved by transforming the target values to a logarithm scale. The lognormal distribution is appropriate when the residual values increase in proportion to the square of the fitted values. However, typically the error distribution is assumed to be normally distributed.

By default, the two-stage model fits a decision tree model based on the categorical target variable that you want to predict, followed by fitting the multiple linear regression model that produces the predicted values of the interval-valued target variable. Therefore, in the first stage better probability estimates might be obtained from some other classification modeling design such as logistic regression, neural networks, and so on. Conversely, in the second stage there might be times that there might not even be a linear relationship between the input variables and the target values that you want to predict. Therefore, the default regression model might be unwarranted. In the two-stage model, at times the predicted probabilities from the first stage might under or over-estimate the given probabilities. Therefore, the second stage design might compensate for the inefficiency and generate even better results by paying closer attention to the value model specifications.

The predicted probabilities from the first stage design must fall within a range between zero and one. Therefore, a link function is incorporated in the first stage. The link function is a nonlinear transformation applied to the linear combination of input variables that generates reasonable values of the predicted target values. Specifying the appropriate link function depends on the data that is determined by the distribution of the error values for the target variable. However, once the link transformation is applied, then it is assumed that there is a linear relationship between the estimated probabilities from the target variable and each input variable in the model. It is also assumed that there is no interaction between the input variables in the model. Conversely, the target values in the second stage might be subject to range restrictions, that is, positive values. Therefore, these restrictions to the range of values of the target variable can be achieved by specifying appropriate transformations or link functions for the target variable. Transformations will usually stabilize the variance and achieve normality in the data. An appropriate transformation achieves linearity in the nonlinear regression model, thereby making the linear regression model an appropriate fit. The most common transformation applied to the target variable is the log transformation. Transformations are performed before fitting the model. Again, the idea is to transform the response variable to stabilize the variability, achieve linearity in the model, and, hopefully, normality in the data. The next step is fitting the model in order to calculate the predicted values and confidence intervals for the transformed data set and, finally, applying a retransformation of the predicted values and confidence intervals to its original form. The reason for retransforming the predictive model back to its original form is because the predictive estimates are usually not as interesting or informative as the original predictive estimates.

To summarize, care should be taken in fitting both models. That is, when fitting the value model in the second stage, care should be taken in specifying the correct model. The link function applied in the first stage should match the model output to the range of the target values and the error distribution that is selected should match the variance, skewness, and the range of target values.

Drawbacks to the Two Stage Model Node

No Built-in Modeling Selection Routine: In the **Two Stage Model** node, the same set of input variables are used in both models. Unlike the **Regression** and the **Tree** node, the drawback to the **Two Stage Model** node is that there is no built-in variable selection routine within the node for eliminating irrelevant or redundant input variables for each component model. The advantage in determining the best set of input variables is making the interpretation easier in the modeling terms, based on the linear relationship between the input variables in the predictive model that best explains or predicts the target variable that will result in an increase in the predictive accuracy of the model. Therefore, given that there are several input variables in the model, then it is highly recommended to select some kind of variable selection routine. This is especially true in fitting models to small samples with a small number of degrees of freedoms for the error that will result in a low statistical power of the model. Again, these variable selection routines can be performed within both the **Regression** and **Tree** nodes. Therefore, this limitation of the **Two Stage Model** node can be overcome by building each component model separately through the use of the various modeling nodes and then combine both models in constructing highly accurate two-stage predictive models.

Priors are not Recognized: The node does not recognize the prior probabilities from the target profile created from the **Input Data Source** node. Therefore, the probabilities from the first-stage class model could be biased since the proportion of the target event could be overrepresented in the training data set. In other words, the fitted values from the first stage might calculate probabilities at each class level that are much different than the actual probabilities from the underlying data.

No Built-in Diagnostics: There is no built-in diagnostic statistics that are calculated within the node in order to access the overall fit to the two-stage model. Therefore, the fitted statistics that are passed along to the **Assessment** node could be incorrect.

Modeling Results are Unexplainable: The advantages of the simple modeling designs, such as the regression models and the decision tree models, is that not only do these models generate predictions but these same models have explanations for their predictions. The big advantage to decision tree designs is that the predictions and the various input variables to the model can be explained by observing the development of the recursive if-then splits of the target groups throughout the branches of the tree, from the top of the tree down to the final leaves. In regression modeling, the predictions can be interpreted by observing the parameter estimates that explain the linear relationship between each input variable and the target variable in the model. In logistic regression, the estimated probabilities or proportions can be explained by the associated odds ratio of the input variables. However, in two-stage modeling, even though each modeling component is explainable, the final decision that is achieved by combining both models cannot be explained since the predictions are calculated by a nonlinear combination of the two separate component models combined.

Added Error to the Second-Stage Model: The two-stage modeling design is based on the categorically-valued target variable that is predicted in the first-stage model explaining a large majority of the variability in the interval-valued target variable to the second-stage model, thereby making it possible to specify a much simpler second-stage model. However, the disadvantage to the sequential two-stage modeling approach is that you are introducing error from the first-stage model that is inherited from the posterior probabilities in predicting the target event that plays the role of one of the input variables to the second-stage value model. One option that is available in reducing the added error that is entered in the second stage of the model is through the use of the *training filter* option. This option will allow you to remove all cases in the value model during training that have been incorrectly classified from the class model. The obvious shortcoming of this method is that valuable information might be removed from the subsequent modeling fit.

Steps in the Modeling Design

By default, the two-stage model performs a decision tree design by first fitting the categorical target variable followed by the regression model used in predicting the predicted values. Identical modeling results can be generated in comparison to the **User Defined Model** modeling node by specifying the predicted values from the class model playing the role of the input variable for the user-defined model. To better understand how two-stage modeling calculates its predicted values, we will construct the following process flow diagram by using the **Tree**, **Regression**, and **User-Defined Model** nodes to generate the identical prediction estimates that are calculated from the **Two Stage** modeling node by fitting a predictive model to predict a categorically-valued target variable, then estimating an interval-valued target variable in succession.

Step 1: Select the **Input Data Source** node to read the HMEQ data set to create the training data set. Set the binary-valued variable to all the clients that are defaulting on their home loan, BAD, and the interval-valued variable of ratio-to-income, DEBTINC, as the target variables that you want to predict. Also, select the interval-valued variable DEBTINC to edit the target profile. That is, select the variable DEBTINC, then right-click the mouse and select the **Edit target profile...** pop-up menu item. From the **Assessment Information** tab, select the **Default function** list item, then right-click the mouse and select **Set to use** in order to generate lift charts to compare the classification estimates between both models from the **Assessment** node.

Step 2: Drag and drop the **Two Stage Model** node on to the desktop and connect the **Input Data Source** node to the **Two Stage Model** node to generate the two stage modeling estimates.

Step 3: Drag and drop the **Tree** modeling node on to the desktop and connect the **Input Data Source** node to the **Tree** modeling node.

Step 4: Open the **Tree** node. From the **Score** tab within the **Data** subtab, select the **Process and Score: Training, Validation, and Test** check box to create an output scored data set from the decision tree model.

Step 5: From the **Variables** subtab, clear the **Input variable selection** check box to prevent the **Tree** node from automatically performing the internal variable selection routine. Run the node to create the output scored data set.

Step 6: Drag and drop the **Data Set Attribute** node on to the desktop and connect the **Tree** modeling node to the **Data Set Attribute** node.

Step 7: Open the **Data Set Attribute** node. From the **Data** tab, select the output scored data set as the active data set to export the output scored data set from the decision tree model.

Step 8: From the **Variables** tab, remove the binary-valued variable BAD from the model by assigning the variable a model role of **rejected**. Conversely, set the model role to the predicted value of the target event that you want to predict, P_BAD1, playing the role of the predictor variable to the following multiple linear regression model. That is, set the model role of the variable P_BAD1 from **target** to **input** within the **Variables** tab.

Step 9: Drag and drop the **Regression** node on to the desktop and connect the **Data Set Attribute** node to the **Regression** modeling node.

Step 10: Open the **Regression** node. From the **Score** tab within the **Scored Data Sets** subtab, select the **Process and Score: Training, Validation, and Test** check box to generate an output scored data set from the multiple linear regression model, that is, with the target variable DEBTINC and the input variable P_BAD1, along with all other input variables in the HMEQ data set. Run the node to create the output scored data set from the regression model.

Step 11: Drag and drop the **User Defined Model** node on to the desktop and connect the **Regression** modeling node to the **User Defined Model** node.

Step 12: Open the **User Defined Model** node. From the **Predicted Values** tab, select the **Predicted Variables** cell, and right-click the mouse to select the predicted values from the scored data set, P_ DEBTINC, calculated from the **Regression** node.

Step 13: Drag and drop the **Assessment** node on to the desktop and connect both the **User Defined Model** node and the **Two Stage Model** node to verify that both models generated identical prediction estimates. Run the node to execute the entire process flow diagram.

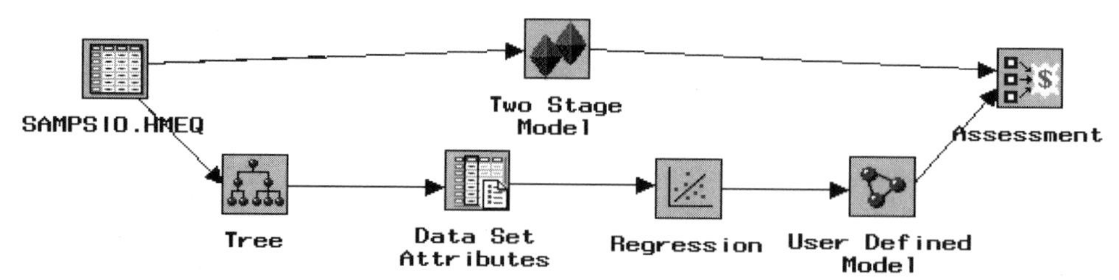

The process flow diagram used in fitting a categorically-valued target variable from the ***Tree*** *node, then an interval-valued target variable from the* ***Regression*** *node in succession to replicate the two stage modeling estimates.*

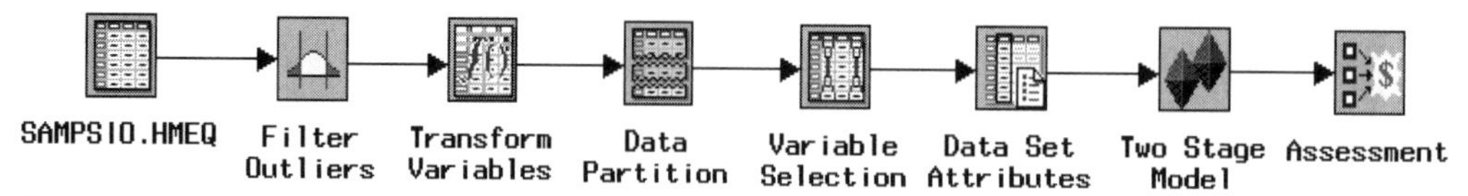

The process flow diagram used to generate the following two-stage modeling results.

Data tab

The **Data** tab is designed to display the file administrative information and description of the partitioned data sets. Select the **Properties...** button to display the file information and a table view of the selected partitioned data set. In other words, if several data sets are created within the process flow, then two-stage modeling is performed for each one the partitioned data sets. If you have the luxury of a large amount of data, then the validation data set can be used to cross-validate the two-stage modeling results.

Variables tab

The **Two Stage Model** node is designed to display the **Variables** tab as you open the node. In two-stage modeling, the node requires that two separate target variables be specified, usually from the **Input Data Source** node. The **Variables** tab is designed to view the various properties of the listed variables such as the variable name, status, model role, measurement level, format, and label of each variable in the active training data set. You may add and drop input variables from the model from the **Status** column. That is, select the corresponding variable row, then scroll over to the **Status** column, and right-click the mouse to specify the variable status of **use** or **don't use**. The input variables with a variable status of **use** are automatically included in both the classification and prediction models. Conversely, variables assigned a variable status of **don't use** are automatically rejected from both models with an associated model role of **rejected**. You may also view the distribution of the selected variable by selecting the **View Distribution of <variable name>** pop-up menu item. The target profile can be edited or browsed within the **Variables** tab. This will allow you to view the assessment criterions, decision values, prior probabilities, and the target event level that the classification model is trying to fit.

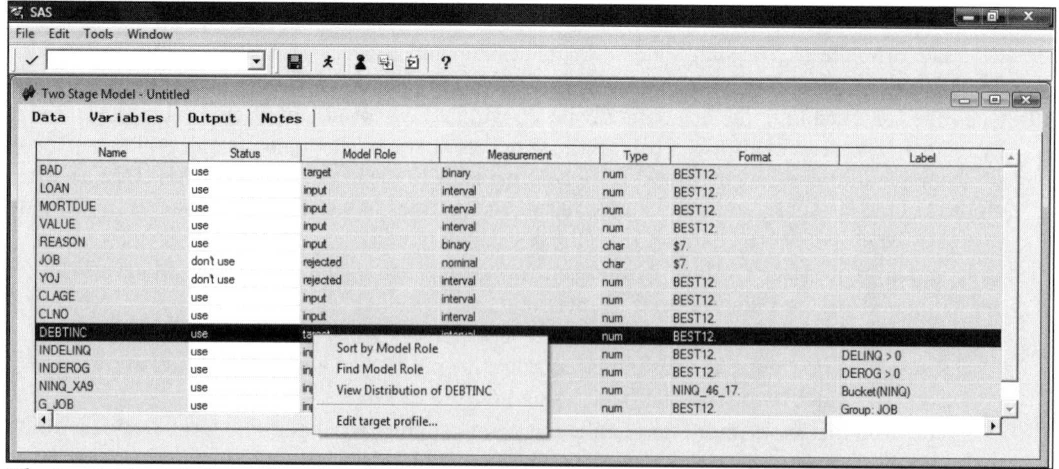

The ***Variables*** *tab is used to view the input variables and the two target variables for the two-stage model.*

Editing or Browsing the Target Profile

Again, you may also edit, browse, or delete the target profile from the **Variables** tab. Simply select any cell from the listed target variable and right-click the mouse, then select the **Edit target profile...** pop-up menu item. If the target profile has been previously defined in a previously connected node, then select the **Browse target profile...** pop-up menu item to browse the existing target profile. The purpose of the target profile is to compute the maximum profit or minimum loss or specify appropriate prior probabilities for the categorical target variable in the model. The target profile contains the decision matrices and the prior probability of the target variable. It is important to point out that since the two-stage model consists of two separate target variables, therefore the target profile is based on the interval-valued target variable from the value model that is displayed in the **Model Manager** for modeling assessment. In other words, the target profile created by fitting the categorically-valued target variable from the class model is not used when performing two-stage modeling.

Output tab

The **Output** tab is designed for you to create an output scored data set based on the results from the two-stage model. Simply select the **Training, Validation, and Test** check box from the **Process or Score** section to score the partitioned data sets that will consist of the corresponding observations along with the predicted values and residuals from the two-stage model. Select the **Properties...** button to view the file information and browse the selected partitioned data set of the output scored data set once the node is compiled. By default, the output scored data sets are exported into the subsequent nodes. One of the purposes of the scored data set is that it will allow you to plot the error values and the fitted values in order to assess the accuracy of the fit.

The following are some of the variables that are written to the output scored data set for each partitioned data set that is created along with the original variables and the frequency variable from the previously connected nodes that depend on the type of model that is selected at each stage.

Variable Name	Description
S_*input variable name*	Standardized input variables by fitting the nonlinear model.
I_*target variable name*	The classification variable that identifies the target class level that the class model has classified. The observation that is used as an input to the value model by specifying the classification transfer function.
U_*target variable name*	The unformatted target class level that the observation is classified.
WARN	Missing value indicator to identify missing values in any one of the input variables in the model in which the interval-valued target variable is imputed by its own average value. For categorically-valued targets, if there are any missing values in any one of the input variables in the model, then the target class levels are imputed by the prior probabilities.
NODE, _LEAF_	The node and leaf identifier for the decision tree model that might be used for further group processing analysis.
H<number>	The hidden unit from the MLP and RBF neural network models.
P_*target variable name <level>*	The posterior probabilities for each class level of the categorically-valued target variable from the class model.
P_*target variable name*	The final fitted values of the interval-valued target variable from the value model.
R_*target variable name*	The residual values from the value model.
EV_*target variable name*	The expected value is determined by the option specified from the **Expected Value Bias** display field. If the **Filter** option is specified, then the value will be set to zero. Conversely, if the setting is set to **None**, then the variable is equivalent to the final fitted values. The expected values are calculated by the adjustments applied to the estimated probabilities from the first-stage model.

Two Stage Model Settings

By selecting the **Tools > Settings** main menu option settings or selecting the **Settings** toolbar icon, the following **Two Stage Model Setting** window will appear in order to specify the various option settings for the two-stage model.

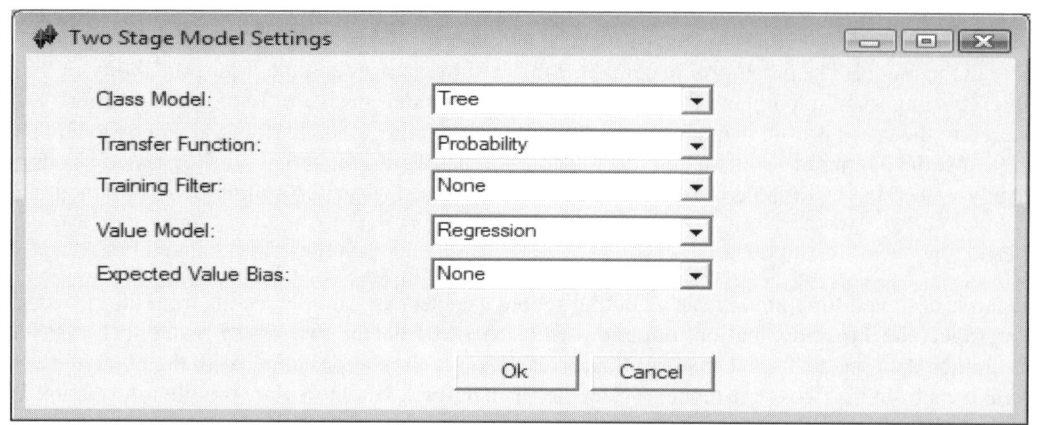

The ***Two Stage Model Settings*** *window is used to specify the option settings for the two-stage model.*

The following is a brief explanation of the various option settings that may be specified within the **Two Stage Model Settings** window.

- **Class Model:** The class model that is applied in the first stage to the categorical target variable that you want to predict that is designed to minimize the misclassification rate.
  - **Tree** (default)**:** Decision tree modeling is used in predicting the categorically-valued target variable, where the target proportions at each leaf are the estimated probabilities to the model. One of the advantages of decision tree modeling is its flexibility and robustness to missing values. By default, the variable selection routine is not performed.
  - **Regression:** Regression modeling is essentially logistic regression modeling with the logit link function that is applied to the output values from the classification model, that is, the expected target values, that will result in the fitted values ranging between the values of zero and one. The default settings from the **Regression** node are applied that depend on the level of measurement of the target variable that you want to predict.
  - **MLP:** Multilayer Perceptron Neural Network Model is the most popular architecture that is used in neural network modeling when it is assumed that there is a nonlinear relationship between the input variables and the target variable you want to predict. The fitted values are usually calculated from an iterative grid search procedure that is performed. The default settings from the **Neural Network** node are applied.
  - **RBF:** Radial Basis Function Neural Network Model is often used in classification modeling that is designed to generate the predicted values within the desired (0, 1) range from the softmax activation function that is applied in the hidden layer. The softmax activation function forces the hidden layer units to sum to one, similar to logistic regression modeling. The softmax basis function that is applied to the NRBF is designed so that each weight estimate from each of the hidden layer units are connected to the output layer, producing expected target values that may be interpreted as an approximation of the posterior probability for each class level of the target variable that sums to one for each input variable.
  - **GLIM:** Generalized Linear Interactive Model Neural Network that has no hidden layers. Both the input variables and the target variable are left unstandardized and the expected target values are left unchanged.
- **Transfer Function:** This option determines the input variables used in the second stage based on the fitted values from the first stage. By default, the **Probability** option is applied that uses the posterior probabilities of the target event from the class model as one of the input variables to fit the subsequent value model. The other choice is the **Classification** method that uses the predicted numeric class level identifiers from the various target groups as the input variable for the second-stage model.

- **Training Filter:** An added option can be selected to specify how you want certain observations excluded during training for the value model in the second stage of the two-stage modeling design.
  - **None** (default)**:** No observations are removed from the analysis in the second stage.
  - **Nonevents:** Observations that are identified as target nonevents are removed during training in the subsequent second-stage model.
  - **Misclassified:** Observations that have been misclassified as target nonevents in the class model are removed from the second-stage model.
  - **Missing Values:** Observations with missing values in the interval-valued target variable are removed from the second-stage model. By default, the missing values of the target variable are estimated by their own estimated proportions at each leaf, since the decision tree model is automatically selected as the first-stage model. In addition, missing values of the target variable and any one of the input variables in the model, then the target values are estimated by their own average value, since the least-squared model is automatically selected as the second-stage model. Otherwise, the target values are estimated from the fitted values generated from the second-stage model.
- **Value Model:** The second-stage model for fitting the interval-valued target variable that is designed to minimize the sum-of-squares error function.
  - **Tree:** Decision tree modeling in which the average values of the target variable at each leaf are the fitted values to the model. By default, the variable selection routine is not performed.
  - **Regression** (default)**:** Regression modeling that assumes that there is a linear relationship between the input variables and the interval-valued target variable in the least-squares model.
  - **MLP:** Multilayer Perceptron Neural Network Model is essentially nonlinear regression modeling. The input variables are automatically standardized in order to generate reasonable initial input-to-hidden layer weight estimates close to zero to assure convergence in the iterative procedure and increase generalization. However, the target variable is left unstandardized.
  - **RBF:** Radial Basis Function Neural Network Model is a two-layer nonlinear design that applies a two-step process in which the input variables are automatically standardized, with a subsequent Gaussian transformation that is applied within the first layer.
  - **GLIM:** Generalized Linear Interactive Model Neural Network. In the second-stage model, the GLIM modeling design is analogous to the previous least-squares regression modeling design in fitting the interval-valued target variable.
- **Expected Value Bias:** This option specifies the method in which the posterior probabilities from the class model and the fitted values from the value model are combined to calculate the final prediction estimates. An additional variable is created in the scored data set that might be interpreted as the final prediction estimate, assuming that the following adjustments are performed on the fitted values in the value model.
  - **None** (default)**:** This is the default.
  - **Multiply:** This option uses the posterior probabilities of the target event to adjust the value predictions from the interval-valued target variable. In other words, the predicted values from the second-stage model are adjusted by the posterior probabilities from the first-stage model. The adjustments applied to the predicted values from the second stage by multiplying the probability estimates of the target event in the first stage are analogous to the estimated probabilities adjusted by the prior probabilities in classification modeling.
  - **Filter:** The predicted values from the observations that have been misclassified as the target nonevent from the class model are set to zero. These same observations are essentially removed from the subsequent modeling fit in the second-stage model. This option will allow you to remove the added error inherited into the second-stage model by removing all observations that have been misclassified from the first-stage model. However, the drawback is that removing these same observations might lead to an undesirable reduction in the accuracy of the subsequent value model.

The following are the various modeling designs that you may select from the **Two Stage Model** node:

- **Tree:** Decision tree modeling where the fitted values are either the target proportions at each leaf by fitting the categorically-valued target variable or the average value at each leaf by fitting the interval-valued target variable. The drawback is that, unlike the **Tree** node, there is no variable selection routine that is performed. There are two separate surrogate rules that are saved with a specified number of observations that are required for a split search.
- **Regression:** Regression modeling where it is assumed that there is a linear relationship between the input variables in the model and the target variable that you want to predict. The default settings are identical to the settings specified from the **Regression** node.
- **MLP:** Multilayer perceptron that is the most popular neural network model to fit.
  - The number of hidden units is a function of the number of parameters in the model as follows:

    h = MIN[24, 1+ INT(df/ 10)], where h = the number of hidden units
  - The hyperbolic tangent transformation is applied within the single hidden layer to the linear combination of input variables in calculating the fitted values.
  - The input variables are standardized in order to assure convergence during training. However, the target variable is left unstandardized.
  - Some randomly generated small numbers determine the initial weight estimates.
  - An iterative grid search procedure is applied in order to minimize the sum-of-squares error function by fitting an interval-valued target variable to the neural network model. Otherwise, the procedure minimizes the multiple Bernoulli likelihood function by fitting a categorically-valued target variable to the nonlinear classification model.
- **RBF:** Radial Basis Function Neural Network Model
  - Uses a two-step process based on two separate layers, where the first step is a nonlinear transformation, usually a Gaussian function, applied in the first layer and the second layer, applying no transformation, that leads to extremely fast convergence. In the first layer, the design initially computes the squared differences between the input values and their centers with a subsequent Gaussian function that is applied. The centers and widths, that is, the mean and variance of the Gaussian function, are usually determined by clustering the input variables into separate nonoverlapping groups to calculate the means and variances within each cluster group. The second layer applies a linear transformation, or no transformation is applied to the linear combination of input variables and associated weight estimates.
  - Uses the radial basis combination function that is applied to the hidden and output layers. The squared Euclidean distance between the input variables and the weight estimates is calculated within the hidden layer.
  - The number of hidden units is a function of the number of parameters in the model as follows:

    h = MAX[2, MIN(8,1+ df/ 10)], where h = the number of hidden units
  - Uses the softmax function that is applied within the hidden layer, forcing the fitted values into rates or proportions.
  - The input variables are standardized in order for all the input variables to span the same range of values since the centers (means), and widths (variances), are initially calculated within each cluster of the input variables in the first layer. However, the target variable is left unstandardized.
- **GLIM:** Generalized Linear Interactive Model
  - In the second stage, the modeling design is essentially least-squares regression modeling.
  - There is no hidden layer. That is, there is no nonlinear transformation applied to the input variables in the model.
  - Both the input variables and target variable are not standardized.
  - Multivariate regression modeling cannot be applied since the node is restricted to one target variable for each model.

Viewing the Two Stage Model Results

General Layout of the Results Browser within the Two Stage Model Node

- **Model tab**
- **Code tab**
- **Log tab**
- **Output tab**
- **Notes tab**

The following two-stage modeling results are based on fitting the categorical variable BAD and the interval-valued variable DEBTINC from the HMEQ data set. The same preprocessing of the data by fitting the least-squares model from the **Regression** node was applied to the two-stage model. In addition, the **Variable Selection** node was incorporated into the process flow diagram to perform the variable selection procedure in order to select the best set of input variables to predict the binary-valued and interval-valued target variables. In other words, the previously displayed process flow diagram was used to generate the following results.

Model tab

The **Model** tab consists of the following four subtabs that display information about the data set, variables, configuration settings that are used in the two-stage model, and processing time in training the model.

- **General subtab**
- **Variables subtab**
- **Settings subtab**
- **Status subtab**

General subtab

The **General** subtab displays the summary information of the model such as the name, description, creation date and modified time of the training data, and the interval-valued target variable name from the second-stage model. The **Profile Information...** button will allow you to open the target profile that will display the target profile information such as the decision matrix, assessment objective, and assessment costs that are specified for the interval-valued target variable in the second-stage model.

Variables subtab

The **Variables** subtab displays a table view of the variables in the output data set created from the two-stage model. However, by creating the scored data set from the previous **Output** tab, the output data set will also contain the residuals, predicted values from both the categorically-valued and interval-valued target variables in the two-stage model, and the expected values that are based on the adjustments applied to the fitted values in the first stage, along with all other variables in the training data set.

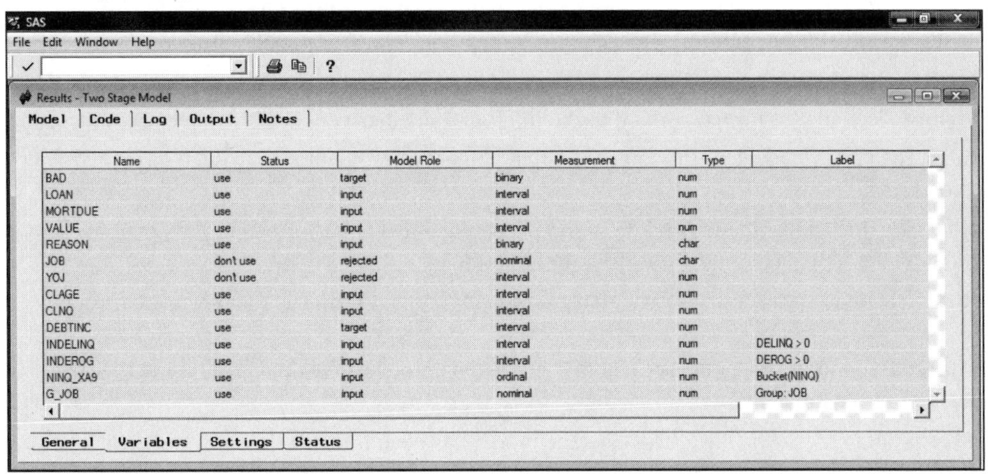

Name	Status	Model Role	Measurement	Type	Label
BAD	use	target	binary	num	
LOAN	use	input	interval	num	
MORTDUE	use	input	interval	num	
VALUE	use	input	interval	num	
REASON	use	input	binary	char	
JOB	dont use	rejected	nominal	char	
YOJ	dont use	rejected	interval	num	
CLAGE	use	input	interval	num	
CLNO	use	input	interval	num	
DEBTINC	use	target	interval	num	
INDELINQ	use	input	interval	num	DELINQ > 0
INDEROG	use	input	interval	num	DEROG > 0
NINQ_XA9	use	input	ordinal	num	Bucket(NINQ)
G_JOB	use	input	nominal	num	Group: JOB

*The **Variables** subtab that displays the variables from the scored data set by executing the node.*

Settings subtab

The **Settings** subtab displays the various configuration settings for the two-stage model as follows:

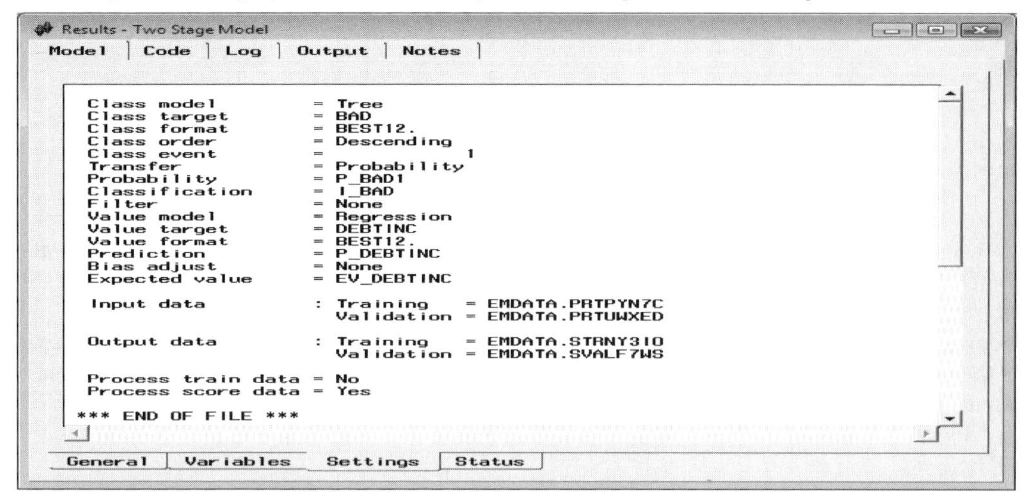

*The **Settings** subtab displaying the option settings for both stages of the two-stage modeling design.*

The following is a brief explanation to the listed output from the **Settings** subtab:

- **Class model:** The type of modeling design applied to the categorically-valued target variable in the first stage. By default, a decision tree model is automatically applied in the first-stage model.
- **Class target:** The categorical target variable name of the class model in the first stage.
- **Class format:** The type of format of the categorically-valued target variable in the first-stage model.
- **Class order:** The ordering level of the categorically-valued target variable in the first-stage model.
- **Class event:** The target event that the class model is trying to predict in the first-stage model.
- **Transfer:** Transfer function that determines the input variables that are used in the second-stage model. By default, the posterior probabilities are applied.
    - **Probability** (default)**:** The variable name to the posterior probabilities to the categorically-valued target variable of the target event that are used as the input variable for the second-stage model.
    - **Classification:** The variable name that identifies the numeric class levels of the classified target variable that is the alternative input variable used in the second-stage model.
- **Filter:** The removal of observations from the input data set that have been misclassified from the class model. By default, no observations are removed from training in the two-stage model.
- **Value model:** The type of modeling design applied to the interval-valued target variable in the second stage. By default, the multiple linear regression model is automatically applied.
- **Value target:** The variable name of the interval-valued target variable in the second-stage model.
- **Value format:** The type of format applied to the interval-valued target variable in the second stage.
- **Prediction:** The variable name of the final predicted values from the interval-valued target variable in the second stage.
- **Bias adjust:** The bias adjustment. By default, no bias adjustment is performed to the posterior probabilities in the first-stage model during training in the two-stage model.
- **Input data:** The input data set that the two-stage model is trying to fit. The default is the training data set.
- **Output data:** The output score data set with the predicted values and residuals, along with all the other variables in the input data set.
- **Process train data:** Indication that the two-stage model has been trained.
- **Process score data:** Indication that the two-stage model has been scored.

Code tab

The **Code** tab displays the internal SEMMA training code used in producing the two-stage modeling results and the corresponding output scored data sets. Press the **Training** or **Scoring** radio button to view the procedure training code or scoring code. The training code will display the corresponding data mining procedures that are used at each separate stage to generate the two-stage modeling results and scored data sets. This is illustrated in the following diagram. The scoring code is a combination of both the class model and the value model. The code will display the prediction equations from the two separate models to calculate entirely different prediction estimates by fitting the model for a new set of input values.

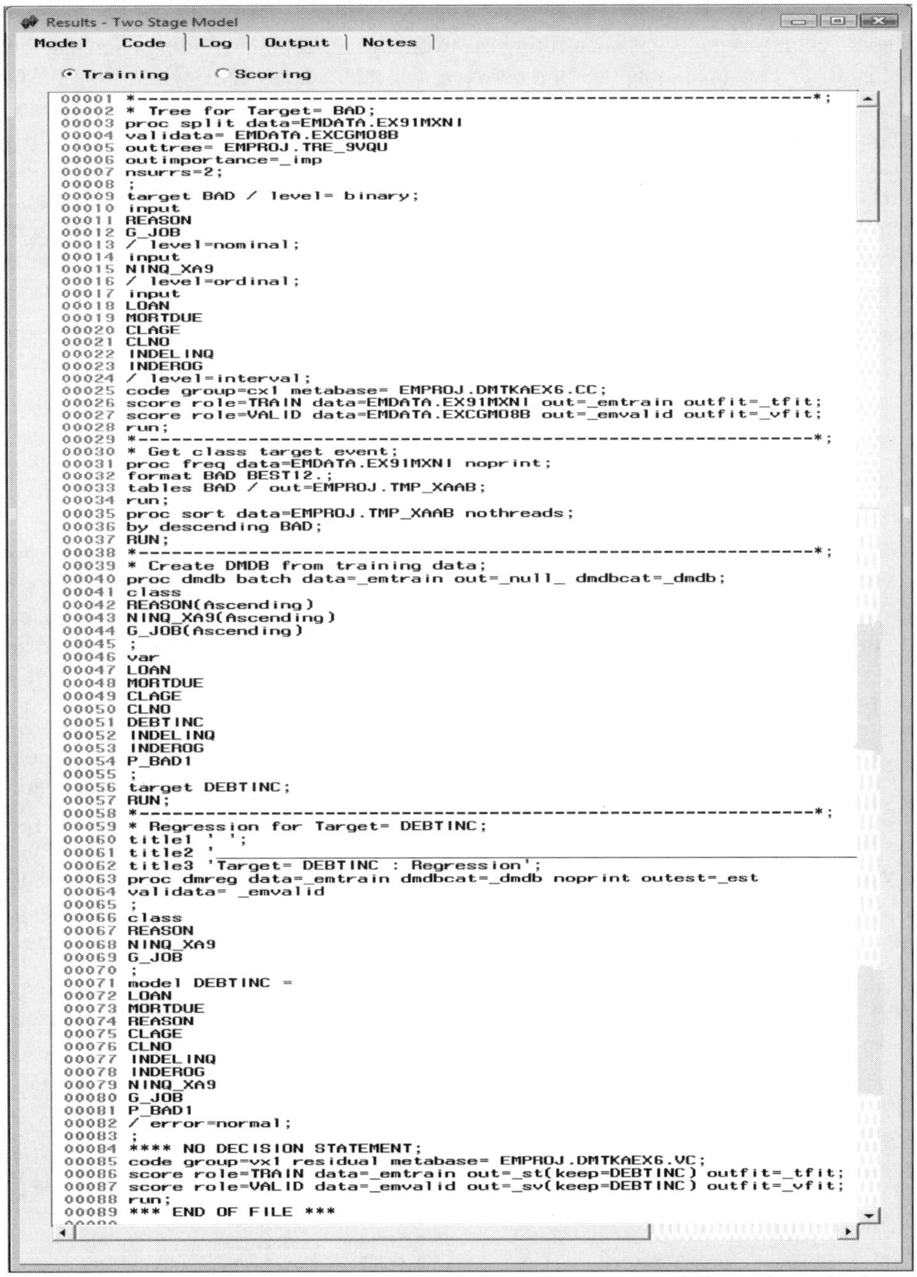

Results - Two Stage Model

Model | Code | Log | Output | Notes

(•) Training ( ) Scoring

```
00001 *------------------------------------------------------------*;
00002 * Tree for Target= BAD;
00003 proc split data=EMDATA.EX91MXNI
00004 validata= EMDATA.EXCGM08B
00005 outtree= EMPROJ.TRE_9VQU
00006 outimportance=_imp
00007 nsurrs=2;
00008 ;
00009 target BAD / level= binary;
00010 input
00011 REASON
00012 G_JOB
00013 / level=nominal;
00014 input
00015 NINQ_XA9
00016 / level=ordinal;
00017 input
00018 LOAN
00019 MORTDUE
00020 CLAGE
00021 CLNO
00022 INDELINQ
00023 INDEROG
00024 / level=interval;
00025 code group=cx1 metabase= EMPROJ.DMTKAEX6.CC;
00026 score role=TRAIN data=EMDATA.EX91MXNI out=_emtrain outfit=_tfit;
00027 score role=VALID data=EMDATA.EXCGM08B out=_emvalid outfit=_vfit;
00028 run;
00029 *------------------------------------------------------------*;
00030 * Get class target event;
00031 proc freq data=EMDATA.EX91MXNI noprint;
00032 format BAD BEST12.;
00033 tables BAD / out=EMPROJ.TMP_XAAB;
00034 run;
00035 proc sort data=EMPROJ.TMP_XAAB nothreads;
00036 by descending BAD;
00037 RUN;
00038 *------------------------------------------------------------*;
00039 * Create DMDB from training data;
00040 proc dmdb batch data=_emtrain out=_null_ dmdbcat=_dmdb;
00041 class
00042 REASON(Ascending)
00043 NINQ_XA9(Ascending)
00044 G_JOB(Ascending)
00045 ;
00046 var
00047 LOAN
00048 MORTDUE
00049 CLAGE
00050 CLNO
00051 DEBTINC
00052 INDELINQ
00053 INDEROG
00054 P_BAD1
00055 ;
00056 target DEBTINC;
00057 RUN;
00058 *------------------------------------------------------------*;
00059 * Regression for Target= DEBTINC;
00060 title1 ' ';
00061 title2 '____________________________________________________________
00062 title3 'Target= DEBTINC : Regression';
00063 proc dmreg data=_emtrain dmdbcat=_dmdb noprint outest=_est
00064 validata= _emvalid
00065 ;
00066 class
00067 REASON
00068 NINQ_XA9
00069 G_JOB
00070 ;
00071 model DEBTINC =
00072 LOAN
00073 MORTDUE
00074 REASON
00075 CLAGE
00076 CLNO
00077 INDELINQ
00078 INDEROG
00079 NINQ_XA9
00080 G_JOB
00081 P_BAD1
00082 / error=normal;
00083 ;
00084 **** NO DECISION STATEMENT;
00085 code group=vx1 residual metabase= EMPROJ.DMTKAEX6.VC;
00086 score role=TRAIN data=_emtrain out=_st(keep=DEBTINC) outfit=_tfit;
00087 score role=VALID data=_emvalid out=_sv(keep=DEBTINC) outfit=_vfit;
00088 run;
00089 *** END OF FILE ***
```

*The **Code** tab that displays the training code from two-stage modeling, where the decision tree model is the class model for predicting bad creditors and the multiple linear regression model is the value model that predicts the interval-valued target variable, that is, debt and income. Notice that the estimated probabilities of the target event from the class model, P_BAD1, is one of the input variables in the second-stage model from the probability transfer function.*

Log tab

The **Log** tab displays the procedure log listing from the two-stage model. The tab will allow you to view the log listing from the two separate modeling procedures that are displayed in the previously listed training code.

Output tab

The **Output** tab will initially display the modeling results from the class model by fitting the categorically-valued target response variable followed by the output results from the value model obtained by fitting the interval-valued target variable.

The following procedure output listing is based on fitting the two-stage model in predicting clients defaulting on a home loan, that is, BAD, then predicting the ratio between debt and income, DEBTINC, from the home equity loan (HMEQ) data set. Therefore, the main objective of the two-stage model is predicting the ratio between debt and income based on all clients who defaulted on their home loan. The following table listing displays the default configuration settings that are automatically applied within the node with the first-stage model fitting the decision tree design and the second-stage model fitting the multiple regression design. In the following assessment listings, the two stage model was fitted to both the training and validation data sets. An equal 50–50 split was performed from the **Data Partition** node, which will result in the same number of observations in both training the model, with the remaining data used in evaluating the accuracy of the fit.

In two-stage modeling, it is important that careful fitting is performed to both stages of the modeling design. Therefore, the variable selection procedure was performed to select the best set of input variables to the two-stage model. However, the limitation of the **Two Stage Model** node is that there are no built-in modeling selection routines. Since it would be astonishing that the same set of input variables would predict both target variables, therefore, the modeling selection procedure was performed in two separate steps. In the first step, a separate process flow diagram was constructed in which the variable selection routine was performed from the **Variable Selection** node to select the set of input variables that best predicted the binary-valued target variable, BAD. From the **Data Set Attributes** node, the input variables LOAN, CLAGE, and the transformed categorical and indicator variables, NINQ (binned), INDEROG, and INDELINQ were forced into the second-stage model. In addition, the **Variable Selection** node was included into the process flow to select the best set of input variables to predict the interval-valued target variable, DEBTINC from the training data set. The linear combination of input variables that were selected from the variable selection procedure were MORTDUE, VALUE, REASON, CLNO, NINQ (binned), and JOB (grouped).

From the first stage, the following table displays the relative importance of the input variables in the decision tree model. Low relative importance values indicate that the input variable poorly divides the target variable into separate groups. Conversely, input variables with large importance statistics are good splitting variables for the decision tree model. In our example, the amount of the loan request, LOAN, has the largest importance value indicating to you that the splitting input variable had the strongest split to the tree in predicting bad creditors from the decision tree configuration. This is followed by the indicator variable of the existence of a delinquent trade line, INDELINQ, that was involved in a single split of the recursive tree branching process, and so on.

Target= BAD : Tree Variable Importance (max=20)

Obs	NAME	LABEL	NUMBER OF RULES IN TREE	RELATIVE IMPORTANCE	VALIDATION IMPORTANCE
1	LOAN		2	1.00000	1.00000
2	INDELINQ	DELINQ > 0	1	0.75139	0.83396
3	CLNO		2	0.58334	0.52380
4	CLAGE		3	0.55579	0.46536
5	MORTDUE		6	0.50119	0.31299
6	INDEROG	DEROG > 0	2	0.47389	0.43331
7	G_JOB	Group: JOB	0	0.31828	0.22377
8	NINQ_XA9	Bucket(NINQ)	1	0.30901	0.32987
9	REASON		1	0.27946	0.27610

The following classification table indicates that the first-stage model has a 78.66% rate of accuracy in predicting clients paying their loan from the training data set as opposed to a 76.95% rate of accuracy from the validation data set. Conversely, there is a 7.01% rate of accuracy in predicting clients defaulting on their loan from the training data set as opposed to a 6.13% rate of accuracy from the validation data set. In addition, both data sets display different assessment statistics, indicating to you that there is overfitting in the class model.

Classification Table

From	Into	TRAIN Count	TRAIN Percent	VALID Count	VALID Percent
0	0	2334	78.6653	2284	76.9542
0	1	60	2.0222	84	2.8302
		.	.	.	.
1	0	365	12.3020	418	14.0836
1	1	208	7.0104	182	6.1321

The following listing displays the various goodness-of-fit statistics from the first-stage model. The class model has a misclassification rate of 17% in predicting clients defaulting on their home loan from the validation data set. These same goodness-of-fit statistics are explained in the **Regression** node. The difference in the misclassification rates between both partitioned data sets indicates overfitting in the first-stage model.

Fit Statistics

Statistic	Training	Validation
Average Squared Error	0.12	0.13
Total Degrees of Freedom	2967.00	.
Divisor for ASE	5934.00	5936.00
Maximum Absolute Error	0.91	1.00
Misclassification Rate	0.14	0.17
Sum of Frequencies	2967.00	2968.00
Root Average Squared Error	0.34	0.36
Sum of Squared Errors	688.74	790.72
Sum of Case Weights Times Freq	5934.00	5936.00

From the second stage, the table listing displays the t-test statistic by decreasing order in absolute values for each input variable in the model in predicting DEBTINC, the ratio between debt-to-income. The value of T-Score is equal to the parameter estimate divided by the corresponding standard error of the estimate of the listed input variable. Large t-test statistics indicate that the input variables are good predictors in explaining the variability in the interval-valued target variable. From the following table listing, all t-test statistics greater than 1.96 will suggest the significance of the input variable in the predictive model at the .05 alpha level. For example, notice that the estimated probabilities in predicting the target event, Predicted: BAD=1, is one of the input variables in the second-stage model which significantly explains the variability in the interval-valued target variable, DEBTINC.

Target= DEBTINC : Regression Effects (max=40)

Obs	Effect	Sign	T-Score
1	Intercept	+	43.7794
2	Bucket(NINQ) 0001:LOW-0.5	-	6.9167
3	MORTDUE	+	6.7727
4	Group: JOB 1	-	4.8622
5	LOAN	+	4.2410
6	Group: JOB 2	+	4.1482
7	CLNO	+	3.8588
8	Predicted: BAD=1	+	3.0671
9	REASON DEBTCON	-	1.6275
10	DELINQ > 0	-	1.3849
11	DEROG > 0	-	1.3478
12	Group: JOB 0	-	1.1506
13	CLAGE	-	0.7293
14	Bucket(NINQ) 0002:0.5-1.5	+	0.0321

The following listing displays the modeling assessment statistics for predicting the ratio between debt and income from the second stage. The smaller the following assessment statistics are, the better the fit. The Schwarz's Bayesian Criterion statistic is considered one of the best goodness-of-fit statistics since it severely punishes models with too many parameters. However, the following table listing indicates that there is overfitting to the second-stage model with large differences in the assessment statistics between both the training and validation data sets.

Fit Statistics

Statistic	Training	Validation
Akaike's Information Criterion	9301.68	.
Average Squared Error	50.46	55.46
Average Error Function	50.46	55.46
Degrees of Freedom for Error	2351.00	.
Model Degrees of Freedom	14.00	.
Total Degrees of Freedom	2365.00	.
Divisor for ASE	2365.00	2307.00
Error Function	119342.10	127938.78
Final Prediction Error	51.06	.
Maximum Absolute Error	40.29	46.72
Mean Square Error	50.76	55.46
Sum of Frequencies	2365.00	2307.00
Number of Estimate Weights	14.00	.
Root Average Sum of Squares	7.10	7.45
Root Final Prediction Error	7.15	.
Root Mean Squared Error	7.12	7.45
Schwarz's Bayesian Criterion	9382.44	.
Sum of Squared Errors	119342.10	127938.78
Sum of Case Weights Times Freq	2365.00	2307.00

The table heading in the following listed output will indicate that the standard five-point summary statistics within each class level are calculated from the validation data set. In addition, the listing displays the target event that the classification model has predicted in the first stage and the bias adjustment performed to the posterior probabilities from the first-stage model. In our example, the target event identifies all clients defaulting on their home loan without any adjustments applied to the estimated probabilities from the first-stage model. The following summary table displays the standard univariate statistics such as the frequency count, mean, standard deviation, and minimum and maximum of the actual, predicted, and expected values by fitting the interval-valued variable DEBTINC (ratio of debt-to-income) within each class level of the binary-valued target variable, BAD. In our example, there were 182 cases that were correctly classified as clients defaulting on a loan and 2284 cases correctly classified as clients paying the loan. Note that the predicted values and expected values are the same since there were no adjustments applied to the posterior probabilities from the first-stage model. The last stages of modeling assessment of the two-stage modeling design would be generating prediction plots from the **Assessment** node to view the accuracy of the prediction estimates and verify the various modeling assumptions that must be satisfied. And, finally, different modeling designs at each stage might be considered in order to achieve more accurate classification and prediction estimates.

Final Prediction Analysis: Validation data
Predicted event: 1
Bias adjustment: None

--------------------------------------- _from=0 _into=0 ---------------------------------------

The MEANS Procedure

Variable	N	Mean	Std Dev	Minimum	Maximum
DEBTINC	2053	33.436	6.791	4.030	45.570
P_DEBTINC	2284	33.903	2.060	27.575	43.614
EV_DEBTINC	2284	33.903	2.060	27.575	43.614

--------------------------------------- _from=0 _into=1 ---------------------------------------

Variable	N	Mean	Std Dev	Minimum	Maximum
DEBTINC	66	30.232	10.728	1.566	42.459
P_DEBTINC	84	35.807	2.511	30.170	45.773
EV_DEBTINC	84	35.807	2.511	30.170	45.773

--------------------------------------- _from=1 _into=0 ---------------------------------------

Variable	N	Mean	Std Dev	Minimum	Maximum
DEBTINC	141	38.604	13.142	0.838	78.654
P_DEBTINC	418	34.003	2.282	28.071	41.162
EV_DEBTINC	418	34.003	2.282	28.071	41.162

--------------------------------------- _from=1 _into=1 ---------------------------------------

Variable	N	Mean	Std Dev	Minimum	Maximum
DEBTINC	47	36.904	11.980	2.594	69.801
P_DEBTINC	182	36.300	2.480	31.038	45.334
EV_DEBTINC	182	36.300	2.480	31.038	45.334

*** END OF FILE ***

Chapter 5

Assess Nodes

Chapter Table of Contents

5.1 Assessment Node

General Layout of the Enterprise Miner Assessment node

- **Models tab**
- **Options tab**
- **Reports tab**
- **Output tab**

The purpose of the **Assessment** node in Enterprise Miner is to evaluate the accuracy of various competing classification or predictive modeling results created from the various modeling nodes. The node will allow you to compare competing models by ignoring such factors as the sample size, modeling terms to each modeling design and the various configuration settings for each modeling node. A common criterion in assessing the accuracy of the predictive models is based on the expected profit or losses that is obtained from the various modeling results. The **Assessment** node and the **Model Manager** will generate the identical charts and modeling assessment listings that are automatically generated once you train the model within the respective modeling node. However, the **Assessment** node has the added flexibility of simultaneously comparing the accuracy of the prediction estimates from several predictive models under assessment. For categorically-valued targets, the node is designed to display several different charts such as lift charts, threshold-based charts and ROC charts that are called *performance charts*. For interval-valued targets, the node will display various diagnostic charts that are basically frequency bar charts for assessing the accuracy of the modeling results and validating the various modeling assumptions, that is, normality, homogeneous variance or constant variability, and independence in error terms that must be satisfied. The **Models** tab will allow you to assess the accuracy of the various listed models under assessment with a variety of listed goodness-of-fit statistics. The performance charts can be created by selecting each model separately or as a group in assessing the accuracy of the posterior probability estimates across the separate classification models. Temporary updates to the target profile may be performed in order to evaluate the modeling performance charts for each model under assessment. In order to evaluate the accuracy of the competing models, the **Assessment** node must be placed after any one of the modeling nodes within the process flow diagram.

The **Assessment** node also creates various performance charts for categorically-valued targets. Performance charts such as lift charts are line plots of the estimated probabilities that are grouped into ten separate intervals that are positioned along the X-axis of the plot. The lift charts display both the estimated cumulative probabilities, and noncumulative probabilities based on the percentage of responses, captured responses, lift value, profit, and return on investments. The lift charts are created by fitting a categorically-valued target variable. The threshold-based charts measure the modeling performance by the various threshold values or cutoff probabilities at each target level and the ROC charts measure the predictive power of the classification model based on the area underneath the curve at each level of the binary-valued target variable. Both the threshold charts and ROC chart measures the classification performance of the model by fitting a binary-valued target variable only.

It is important to understand that the amount of data used from the scored data sets for interactive assessment, that is, modifying the target profile, and the various generated performance charts that are created by each partitioned data set can be controlled by the various options specified from the **Model Manager** within each of the corresponding modeling nodes. The node requires that the model has been trained and the scored data sets be created once you execute the respective modeling nodes in order to generate the goodness-of-fit listings and the various performance charts and plots that are displayed within the node. For categorically-valued targets, the scored data set for interactive assessment will consist of the target values and the posterior probabilities with the input variables removed from the data set. For interval-valued targets, the scored data set for interactive assessment will consist of the target values and the predicted values with the input variables removed from the data set. An alternative objective in the assessment modeling design is maximizing profit or minimizing loss from the profit matrix. The profit matrix is designed to compute the expected revenue, expected profit (the difference between revenue and costs), or the expected loss at each level of the target variable. The expected revenue is computed by the combination of the revenue or cost and the posterior probability at each target level. The **Assessment** node will then determine the accuracy of the modeling estimates by the corresponding cost function. For interactive assessment, you may alter the decision values within the performance charts that will allow you to evaluate the updates that are performed within the node.

In order to avoid overfitting and generating good generalization, it is important that the modeling design also be applied to an entirely separate data set in assessing the accuracy of the classification or predictive modeling estimates. By default, the various modeling assessment listings, diagnostic charts, and performance charts that are displayed are based on fitting the validation data set. However, if there is not enough data available to perform the partitioning, then the training data set is automatically set to the active data set for the analysis. The partitioned data sets can be selected from the **Tools** pull-down menu.

Viewing the results from the **Assessment** node is similar to viewing the results from the modeling node. The advantage of the **Assessment** node is not only to view the table listing of the various assessment statistics from the **Models** tab, but also to give you the ability to select several models in order to visually compare the modeling performance between the selected classification models. From the node, you may simultaneously compare the posterior probability estimates by viewing the various performance charts such as the lift charts, threshold charts, or ROC chart by fitting the categorically-valued target variable. From the **Models** tab, you may also perform temporary updates to the target profile for each model separately by selecting the **View > Target Profile** main menu options.

The sample size used for interactive assessment and the various types of performance charts that are displayed from the **Assessment** node are controlled by the options that are specified within the **Model Manager** window for the respective modeling nodes. Simply select the appropriate modeling node to view the **Model Manager** in order to specify the amount of data to be assessed in the **Assessment** node or the various types of performance charts that you want to display in the **Assessment** node from the allocated data set. By default, the **Assessment** node and the **Model Manager** select a random sample size of 5,000 observations for interactive assessment. It is important to note that in some circumstances the categorically-valued target variable might consist of a certain number of observations that are identified as rare events. Therefore, it is recommended to select a significant amount of data to sample for interactive assessment in order to avoid the predicament of sampling the training data set that does not represent the true distribution of the population of interest with an insufficient amount of data representative of the rare target event. It is important to understand that the entire sample is used to construct the initial lift charts where the random sample of the training data set that is used for interactive assessment can be specified from the **Model Manager** of the corresponding modeling node.

The following process flow diagram is constructed to compare the classification estimates between the logistic regression model, decision tree model, neural network model, and the ensemble model, that is, all three classification models combined, that are then connected to the **Assessment** node to produce the subsequent lift charts that are displayed in evaluating the accuracy of the results. For modeling comparisons, it is essential to compare all the models on the same cases. The reason why the **Replacement** node is a part of the process flow is because the decision tree model will automatically replace missing values during training as opposed to the other two classification models that ignore missing values in any one of the input variables in the data set. The various imputation estimates from the node were applied to the process flow to retain all records in the neural network and logistic regression models in order to compare the various classification models from the same number of observations. From the **Transform Variables** node, the logarithmic transformation and binning was performed to some of the input variables in the logistic model. However, no transformations were applied to the neural network model since there are various transformations that are automatically applied within the nonlinear model that contributes to the inherited flexibility in the modeling design and the decision tree model since the recursive splits in the input variables depend only on the order in which their values are sorted.

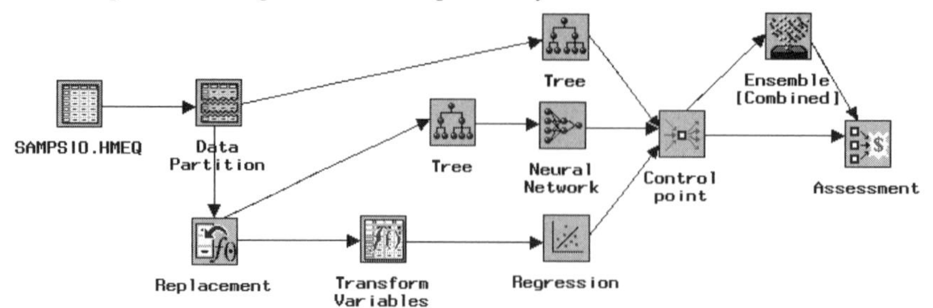

The process flow diagram is used to compare the prediction estimates between the selected models.

Models tab

By default, once you open the **Assessment** node, then the following **Models** tab will be displayed. The **Models** tab is designed for you to view the various goodness-of-fit listings of the various models under assessment. The table listing will allow you to assess the multitude of fitted models under assessment from the listed assessment statistics. In addition, the tab will allow you to select the corresponding models in order to view the various diagnostic or performance charts. In other words, any number of models can be selected to view the various performance charts. Conversely, each model must be selected individually in order to view the various diagnosis plots by fitting an interval-valued target variable to the model. From the **Tools** main menu, you may select from the various diagnostic or performance charts based on the level of measurement of the target variable that you want to predict. Again, it is important to point out that the various performance charts that are created within the **Assessment** node are controlled by the **Model Manager** from the corresponding modeling node. From the following illustration, the decision tree model has the lowest validation misclassification rate that is closely followed by the neural network model.

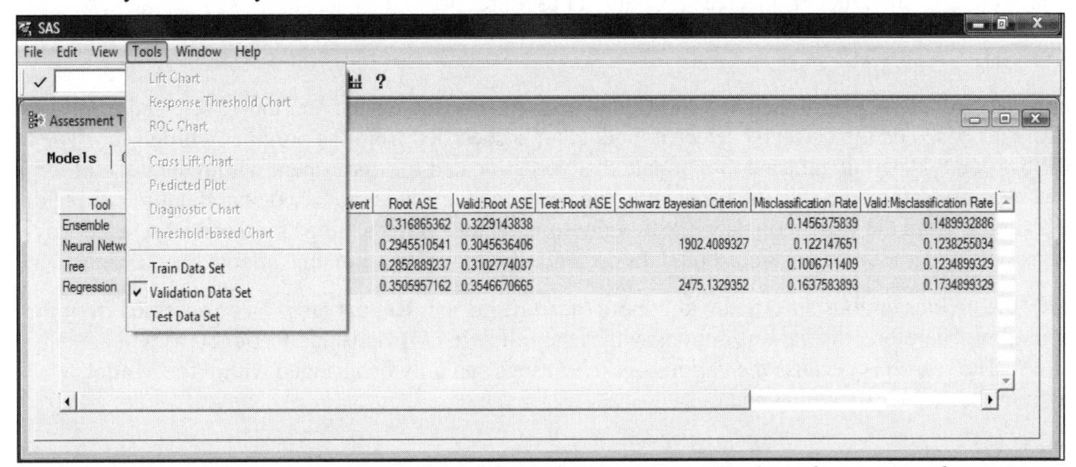

By default, the validation data set is selected from the ***Assessment Tool*** *window to view the various diagnostic and performance charts.*

The **Assessment Tool** window displays the various goodness-of-fit statistics or modeling assessment statistics based on the level of measurement of the target variable. For interval-valued target variables, the table listing displays the square root of the average squared error from the partitioned data sets and the Schwarz Bayesian criterion statistic from the validation data set. For categorically-valued target variables, the tab displays the misclassification rate from the partitioned data sets and the Schwarz Bayesian criterion statistic from the validation data set. By default, the tab displays a partial listing of the various modeling assessment statistics. However, by selecting the **View > Show All Columns** menu option, the node will allow you to view all the available modeling assessment statistics within the **Assessment Tool** window. The various modeling assessment statistics that are listed within the tab from each partitioned data set are displayed in a table view for each model under assessment.

Essential Column Names	Description
Tool	Name of the Enterprise Miner modeling tool or the modeling node used in the process flow diagram.
Name	Model name entered previously in the **Model Manager** window within the modeling node. Otherwise, you may select the name cell and enter the appropriate modeling name.
Description	Description name specified in the **Model Manager** window within the modeling node. Otherwise, you may select the description cell and enter the appropriate modeling description.
Target	Target variable name that the model is trying to predict.
Root ASE	Goodness-of-fit statistic of the square root of the average squared error from the partitioned data sets.

Schwarz Bayesian Criterion	Goodness-of-fit statistic or modeling assessment statistic of the Schwarz Bayesian criterion from the validation data set.
Misclassification Rate	Modeling assessment statistic for categorically-valued targets that calculates the percentage of the estimated probabilities that were incorrectly classified from the validation data set.

Selecting Multiple Models to Assess from the Performance Charts

The advantage of using the **Assessment** node is the ability to compare the accuracy of the modeling performance from several classification models under assessment by fitting the categorically-valued target variable that you want to predict. This can be accomplished by viewing the various performance charts that can be selected within the node. In order to select the various listed models under assessment, simply hold down the Ctrl key to highlight any number of noncontiguous rows to select the appropriate model design within the **Models** tab. By default, the node will automatically display the lift charts, threshold-based charts, and ROC charts by each classification model selected separately or as a group.

Options tab

The **Options** tab will allow you to view the partitioned data sets that have been previously selected from the **Model Manager** based on the currently selected model that is specified from the previous **Models** tab. The tab will also allow you to determine the amount of data that was allocated for assessment within the node in creating the various performance charts. By default, the performance charts are based on a random sample of 5,000 observations from the validation data set. In order to view the option settings for these tabs, select any one of the models under assessment since both tabs are initially grayed-out and unavailable for viewing.

As a review, the various options specified within both the **Options** and **Report** tabs can be assessed from the **Model Manager**. Therefore, the various options within the following **Options** and **Report** tabs will actually be grayed-out. The reason is because the various option settings can only be changed within the **Model Manager** from the various modeling nodes.

Reports tab

The **Reports** tab will allow you to view the type of performance charts that have been previously selected from the **Model Manager**. The tab will indicate the performance charts that can displayed from the **Assessment** node based on the selected model under assessment. By default, all the performance charts are displayed from the **Assessment** node. In order to view the various configuration settings from the tab, simply select any one of the listed models. However, similar to the previous **Options** tab, the option settings to **Reports** tab cannot be changed from the node. To change the option settings to this tab, you must access the **Model Manager** from any one of the modeling nodes. Therefore, the **Reports** tab will allow you to browse the current charts that are available by each selected partitioned data set.

Output tab

The **Output** tab will allow you to select the best model under assessment that can be passed along for subsequent scoring. Simply highlight the row of the corresponding model that you want to pass on to the current process flow diagram. Note that the **Save after model assessment** check box must be selected within the **Model Manager** in order for the following scored data set to be saved after assessment and passed on to the subsequent nodes, such as the **Score** node, within the process flow.

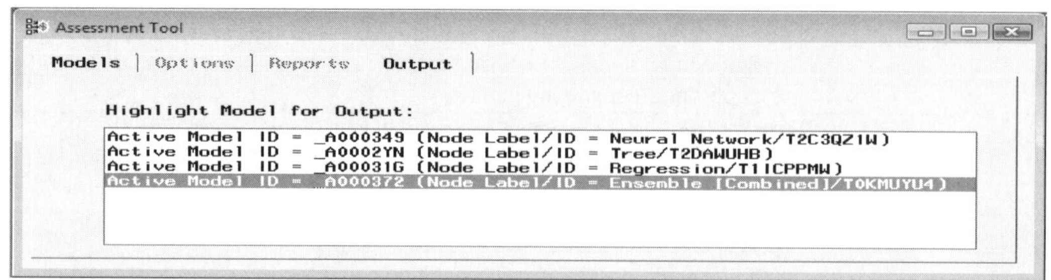

*The **Output** tab is used to select the model to pass along to subsequent nodes within the process flow.*

Note: It is important to make sure that the following assessment plots are not suppressed from the user defined preference options. From the process flow diagram, select the **Options > User preferences...** main menu options that will open the **User Preferences** window. From the **User Preferences** window, make sure that the **Suppress model assessment after training** check box is not selected in order to generate the various lift charts and performance charts. From the **Tools** pull-down menu, the **Assessment** node generates the following charts in assessing the accuracy of the modeling estimates that are based on the level of measurement of the target variable for the predictive or classification model.

From the **Models** tab, the following charts may be viewed by selecting any number of the listed classification models under assessment:

- Lift Charts
  - Cumulative-Non-Cumulative % Response Chart
  - Cumulative-Non-Cumulative % Captured Response Chart
  - Lift Value Charts
  - Profit Charts
  - Return of Investment (ROI) charts
- Receiver Operating Characteristic (ROC) charts
- Response Threshold Charts

From the **Models** tab, the following charts may be viewed by selecting only one of the listed models:

- Cross Lift Charts
- Predicted Plots
- Diagnostic charts
- Threshold-Based Charts
  - Interactive Profit Charts
  - Correct Classification Charts

The following are some of the ways to display the following lift charts within the **Tools Assessment** window:

1. Select the **Line Plot**  icon from the tools bar menu.
2. Select **Tools** > **Lift Chart** from the main menu options.
3. From the **Output** tab, highlight the model row, then right-click the mouse and select the **Tools > Lift Chart** pop-up menu items.

The following are some of the ways to display the following diagnostic charts within the **Tools Assessment** window:

1. Select the **Bar Chart**  icon from the tools bar menu.
2. Select **Tools** > **Diagnostic Chart** from the main menu options.
3. From the **Output** tab, highlight the model row, then right-click the mouse and select the **Tools > Diagnostic Chart** pop-up menu items.

Lift Charts

From the **Assessment Tools** window, select the **Tools > Lift Charts** pull-down menu option to display the various performance charts. The lift charts are designed to display the range of all possible values of the expected profit by sorting the scored data set in descending order of the expected profit. The observations are then grouped into cumulative deciles with the idea of evaluating the classification performance of the modeling design across the various percentile estimates. However if a decision matrix is not specified, then the assessment data set is sorted by the posterior probabilities from the classification model. The observations are then grouped into deciles or intervals at increments of 10%. The lift charts are designed to display both the cumulative and noncumulative values at each decile.

The main idea of the following lift charts is to display the predictive power of the classification model. For binary-valued target variables, you would assume that the model would have absolutely no predictive power at all if the estimated probabilities are uniformly distributed across the various decile groupings. Conversely, the model would display a high predictive power in estimating the target event if the larger estimated probabilities are concentrated at the lower decile estimates.

The **Assessment** node will also generate performance charts of the percent response and the percent captured response that are based on the class levels of the target variable. In order to best interpret both performance plots, it is important to understand the way in which the target event and target nonevent are defined. For binary target variables, the posterior probabilities determine which observation is assigned to the two separate target levels. When the posterior probability is greater than .5, then the observation is assigned to the target event. Otherwise, the observation is classified as the target nonevent. The target event that you want to predict may be verified from the target profile in the **Target** tab under the **Event** display field from the corresponding modeling node. For non-binary targets, each observation is assigned to the target event if the actual profit is greater than the cutoff value. Otherwise, the observation is assigned to the target nonevent, assuming that you have selected **Maximize Profit** option from the **Decision and Utility** window. Conversely, selecting the **Minimize Loss** option from the target profile, then each observation that is less than the cutoff value is assigned to the target event. Otherwise, the observation is assigned to the target nonevent. The cutoff value is computed by calculating the average of the maximum and minimum values of the first decision level from the decision matrix, under the assumption that all the values in the first decision level are nonnegative. Otherwise, the cutoff value is set to zero.

In order to make it easier to understand what the cumulative and noncumulative % captured, response and lift value charts are actually plotting, I will first explain the numerical calculations that are performed behind the scenes in generating the various assessment results for the following lift charts. In this hypothetical example, let us assume that there are a total of 1,000 customers with 200 bad creditors that defaulted on a loan from the validation data set. The following table displays the calculation of both the cumulative and noncumulative percentages of the % captured responses, % responses and the lift values within each decile. Notice that the following lift values are a function of the sample size.

		Cumulative			**Non-Cumulative**		
Decile	**Sample Size**	**% Captured Response**	**% Response**	**Lift Value**	**% Captured Response**	**% Response**	**Lift Value**
10	60	60/200=30.0%	60/100=60.0%	60/(200•10%)=3.00	60/200=30.0%	60/100=60%	60/(200•10%)=3.00
20	30	90/200=45.0%	90/200=45.0%	90/(200•20%)=2.25	30/200=15.0%	30/100=30%	30/(200•10%)=1.50
30	20	110/200=55.0%	110/300=36.6%	110/(200•30%)=1.83	20/200=10.0%	20/100=20%	20/(200•10%)=1.00
40	30	140/200=70.0%	140/400=35.0%	140/(200•40%)=1.75	30/200=15.0%	30/100=30%	30/(200•10%)=1.50
50	20	160/200=80.0%	160/500=32.0%	160/(200•50%)=1.60	20/200=10.0%	20/100=20%	20/(200•10%)=1.00
60	15	175/200=37.5%	175/600=29.2%	175/(200•60%)=1.46	15/200=7.5%	15/100=15%	15/(200•10%)=0.75
70	5	180/200=90.0%	180/700=25.7%	180/(200•70%)=1.29	5/200=2.5%	5/100=5%	5/(200•10%)=0.25
80	5	185/200=92.5%	185/800=23.1%	185/(200•80%)=1.16	5/200=2.5%	5/100=5%	5/(200•10%)=0.25
90	10	195/200=97.5%	195/900=21.7%	195/(200•90%)=1.08	10/200=5.0%	10/100=10%	10/(200•10%)=0.50
100	5	200/200=100%	200/1000=20.0%	200/(200•100%)=1.00	5/200=2.5%	5/100=5%	5/(200•10%)=0.02

Chart	**Description**
% Captured Response	The noncumulative % captured response is the number of responses at each decile divided by the total number of responses. The cumulative % captured response is the cumulative sum of the non-cumulative % captured response.
% Response	The noncumulative % response is the ratio between the number of responses at each decile over the estimated number of responses in the entire population of the validation data set. For example, the estimated total number in the population at the 10th decile is 1000•10% = 100. The cumulative % responses are the % of responses divided by the estimated cumulative sum of the responses in the population at each decile.

Lift Value	The divisor of the lift value is calculated by multiplying the decile estimate by the total number of responses of the target event. For example, the estimated number of total responses at the 10th decile is $200 \cdot 10\% = 20$. Therefore, the lift value is the ratio between the number of responses at each decile over the estimated number of responses at each decile. The cumulative lift value is calculated by the cumulative responses at each decile divided by the estimated number of target events from the cumulative decile estimates.

Cumulative % Response Chart

The following lift charts are based on the home equity loan data set by performing a 50–50 split of the input data set. The various lift charts that are displayed are based on the validation data set. Multiple models were compared to one another in the following performance charts in order to explain the following assessment results. The logistic regression, neural network, and decision tree classification models were compared against each other along with the ensemble model that is the combined estimates of the three separate models. In addition, the following exercise is performed in order to compare the modeling performance of the ensemble model with the individual models that are combined. Lift charts are designed to compare the predictive power between the separate models across the range of the various decile estimates. Lift charts are basically cumulative or noncumulative response charts. From the home equity loan data set, the target variable is called BAD, with the target event that you would like to predict defined as all clients who have defaulted on their home loan as opposed to paying their home loan.

As a review, the main idea of the following lift charts is to display the predictive power of the classification model. From the following lift charts, the classification model with an extremely high predictive power in estimating the target event would display its estimated probabilities concentrated at the highest decile estimates. Conversely, the classification model would display little or no predictive power at all if its estimated probabilities are uniformly distributed at or below the baseline average across the various decile groupings.

For binary-valued target variables, the estimated probabilities from the scored data set are sorted in descending order from the classification model predicting the target event, that is, clients not paying their home loan. The following lift charts are grouped by these posterior probabilities in 10% intervals. The left-most 10% decile responses represent the individuals most likely to default on their home loan with the largest posterior probabilities of the target event that decreases to 100% when every record is identified by the classification model.

By default, the cumulative percentage of responses chart will automatically be displayed once you first open the lift charts from the **Assessment Tools** window. The *percentage of responses chart* is designed to display the cumulative percentage of the observations identified as the target event, that is, clients defaulting on a loan for every observation from the validation data set. The Y-axis represents the cumulative percentage of the observations identified as the target event from the validation data set based on the posterior probabilities across the ordered decile estimates of the top 10%, 20%, 30%, and so on, that is located on the X-axis.

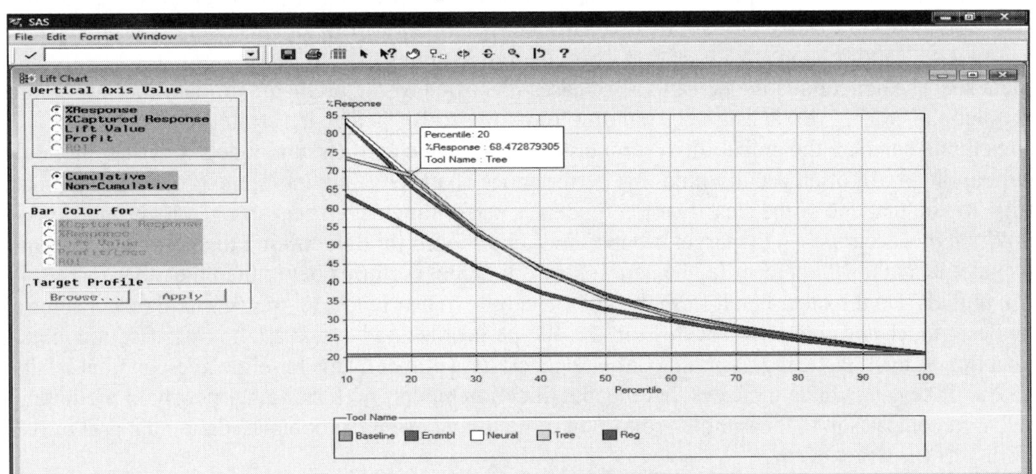

*The **Lift Chart** window that displays the cumulative **%Response** estimates.*

Both the ensemble and neural network models generate comparable estimates across the range of the various decile estimates. Conversely, the decision tree model predicts nearly 75% of the clients defaulting on a loan from the top 10% of the ordered data of the entire sample size from the validation data set. Since the model with the higher response rate is the more desirable model, that is, accurately identifying clients defaulting on a loan, the ensemble and neural network models seems to display better classification performance in comparison to the decision tree and logistic regression models. If you are most interested in the top 10% of the data, with the highest probabilities in identifying bad clients that are most likely to default on their home loan, the neural network model with three separate hidden layered units generates the most accurate estimates that is followed closely by the ensemble model that is based on the three separate classification models that were combined. However, if you are most interested in the top 20% of clients defaulting on their loan, the decision tree model seems to perform slightly better than the other two classification models, and so on. All the classification models generated estimates that are well above the baseline response rate across the ordered decile estimates, indicating to you that the models classify bad creditors better than having no classification model at all. In other words, the *random baseline model* compares the estimated probability of clients defaulting on a loan for a fixed number of customers that were given a loan without the use of a predictive model. By default, the baseline probability is defined by the response rate or the observed sample proportion of the target event from the validation data set. However, the baseline response rate can be changed from the target profile. From the target profile, simply specify the appropriate prior probability of the target event you wish to predict and the baseline probability of the lift charts will change accordingly.

However, displaying the lift charts by selecting each classification model separately will allow you to view the target profile. The **Assessment** node has the added capability of evaluating the change in the performance charts by specifying entirely different prior probabilities or different decision entries for the profit or loss matrix at each class level of the categorically-valued target variable. Simply select the **Target Profile** button in order to access the target profile. To make the appropriate changes in the target profile, select the **Edit** button, which is not shown in this example, to open the **Editing Assessment Profile** window. Close the **Editing Assessment Profile** window and select the **Apply** button to update the lift charts. However, it is important to note that the changes that are made within the target profile are only temporary. Again, the **Editing Assessment Profile** window is designed for interactive assessment by making temporary changes to either the prior probabilities or the decision entries of the decision matrix at each target level within the target profile in order to observe the modeling performance from the corresponding lift charts. The changes that are applied in the **Editing Assessment Profile** window are only temporary and are not saved to the current process flow diagram once the **Lift Chart** window is closed.

Note: From the various assessment plots, simply left-click the mouse along any one of the selected lines and a text box will appear that will display various assessment statistics such as the decile estimate, assessment statistic, and the modeling name. This is illustrated in the previous cumulative % response chart.

Non-Cumulative % Response Chart

The *noncumulative percentage of response chart* is very similar to the previous response chart. However, the difference is that the noncumulative probability estimates being plotted identify clients defaulting on a loan between the separate classification models as they cross the range of the ordered decile estimates from the validation data set. The noncumulative percentage of cases identified by the predicted probabilities are located on the Y-axis and the ordered decile estimates ranging from 10% to 100% that are located on the X-axis. Typically, when you compare the probability estimates between both classification models, overlapping in the estimated probabilities will often occur within this performance chart. Overlapping in the probability estimates will indicating to you that one of the classification models is performing better than another at certain percentiles. When this occurs, then the target groups should be chosen for determining the best classification model. For example, at the 10th percentile, the neural network model performs better than the other three classification models. On the other hand, from the 20th percentile to nearly the 30th percentile the decision tree model is the best model, and so on. However, from the 40th percentile and beyond all the classification models, with the exception to the logistic regression model, are generally well below the baseline average, that is, the remaining 60% of the data, which indicates that the classification models perform rather poorly in predicting the target event in comparison to the sample proportion of the target event from the validation data set or the prior probability of the target event.

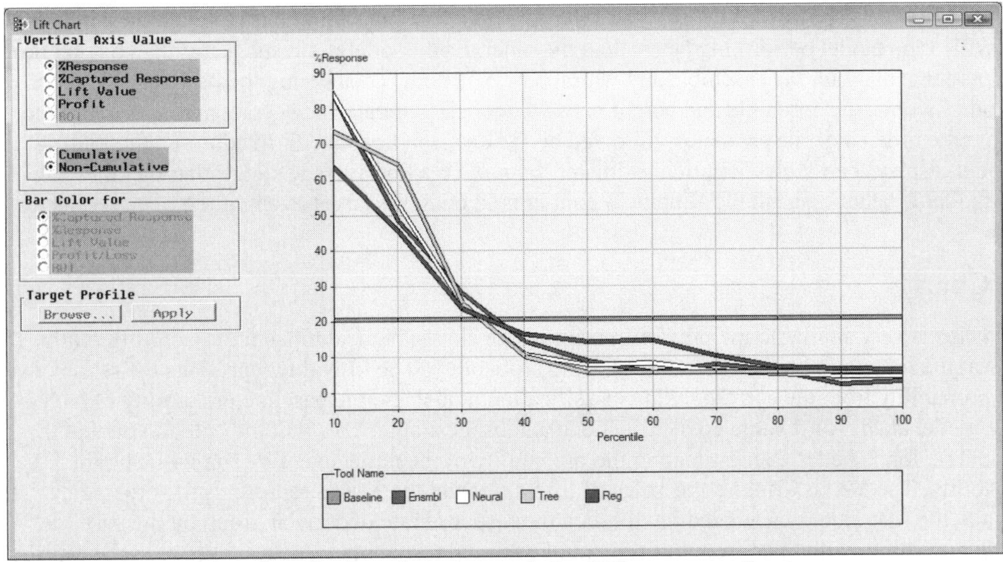

*The **Lift Chart** window that displays the noncumulative **%Response** estimates.*

Cumulative % Captured Response Chart

The *percentage of captured response chart* is based on the total number of observations identified as the target event from validation data set. The percentage of captured responses chart displays the percentage of the observations that are identified as the target level of interest, that is, clients defaulting on a loan, based on the total number of observations that is identified as the target event from the validation data set against the ordered decile estimates. The lines will all meet at the origin, indicating to you that no one has been given a loan and, therefore, no one is predicted to default on their home loan. Conversely, all the lines will meet at the right end of the graph since all the classification models will accurately predict every client defaulting on a loan. The baseline response rate represents the probability that would be identified as defaulting on a loan if you were to take a certain percentage of a random sample of the observations. If no predictive model is used, then you would expect to identify 10% of the target event for 10% of the population, and so on. From the 10th percentile, the decision tree, ensemble, and neural network models will estimate approximately 40% of the responses as opposed to 10% of the sample population without the use of a predictive model with a lift value of approximately 4.0, and so on. Again, the model with the higher captured response rate is the more desirable model which is either the ensemble, neural network, or decision tree models.

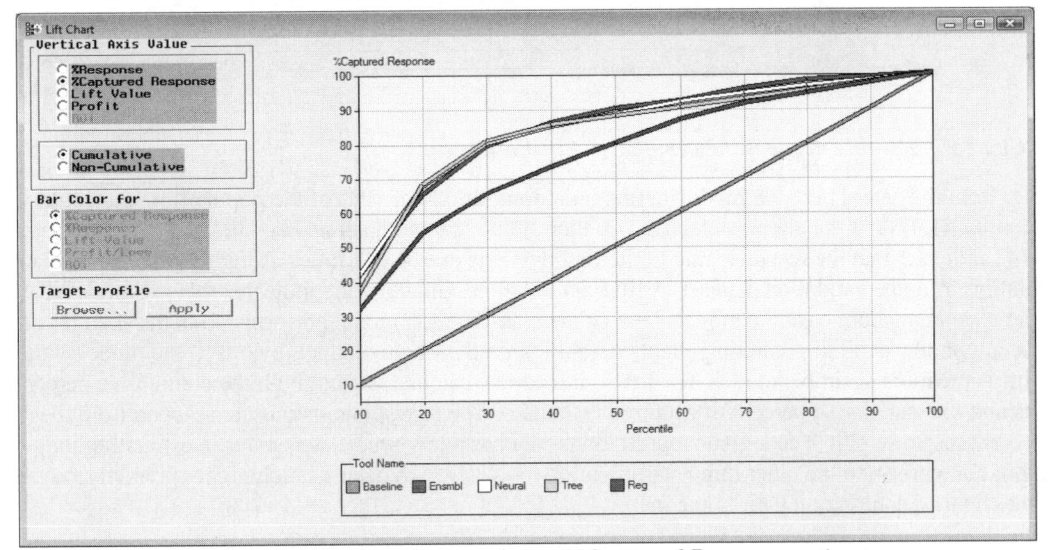

*The **Lift Chart** window that displays the cumulative **%Captured Response** estimates.*

Both the percent of captured responses lift chart and the percentage response lift chart display the same trend, with each classification model performing better than the other. Each model performs better than having no classification model at all, with the ensemble and neural network models displaying the best classification estimates. In other words, the ensemble and neural network models generated very comparable classification estimates across the ordered decile estimates, followed by the decision tree model. In addition, the logistic regression model displays better classification estimates than the baseline average suggesting that the classification model is better than simply flipping a coin in predicting the target event of the binary-valued target variable.

Lift Value Chart

The *lift value chart* is very similar to the previous performance charts, but is defined a little bit differently. The difference is that the lift values are based on the baseline probability. The lift value represents the ratio between the estimated probability from the corresponding classification model and the baseline probability of having any predictive model at all. A lift value greater than one is most desirable. This will indicate to you that the predictive model performs better than estimating the probability of the target event by simply using the baseline probability. The Y-axis displays the range of lift values and the X-axis represents the ordered decile estimates. Again, the lift value is calculated by dividing the response rate of a given group by the baseline average. In other words, the lift value from the percent of captured responses is the ratio between the estimated percentile and the baseline average that is calculated as follows:

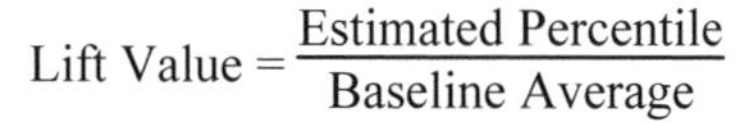

$$\text{Lift Value} = \frac{\text{Estimated Percentile}}{\text{Baseline Average}}$$

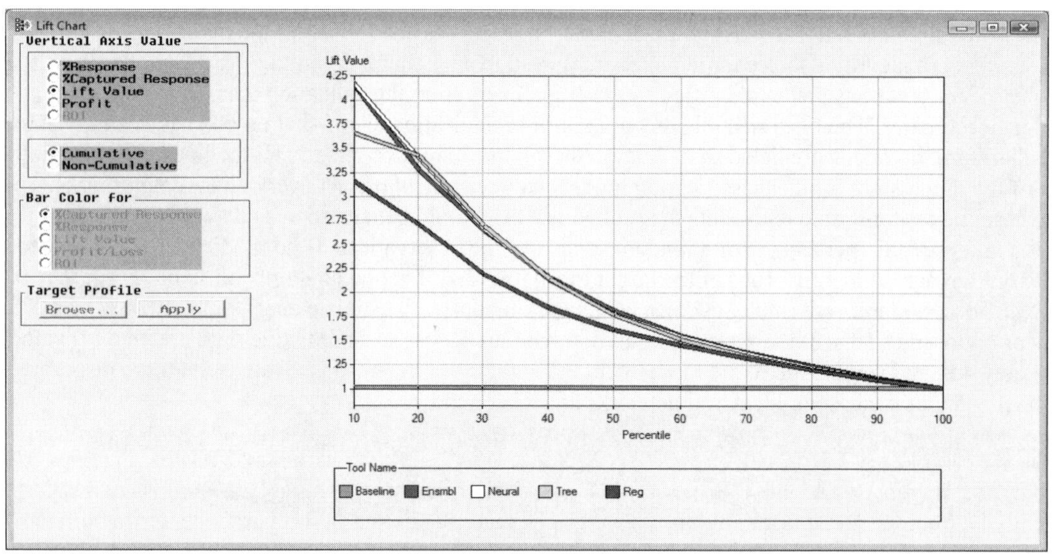

*The **Lift Chart** window that displays the cumulative **Lift Value** estimates.*

For example, from the neural network model, taking a random sample of 10% of the population you would expect to identity 41.47% of the clients defaulting on their home loan with a 41.47%/10% = 4.147 lift. The lift value of 4.147 indicates that the response rate in the first decile is over 4.147 times as high as the response rate in the population from the validation data set. A lift value of approximately one indicates that the model is about the same as the random baseline model. That is, the classification model performs no better than if you were to have any model at all in predicting clients defaulting on their home loan. Obviously, the model with the higher lift is the more desirable model. The lift values can be calculated from both the cumulative percent of responses and the cumulative percent of captured responses. The same conclusions are reached from the previous percent response plot. The logistic regression model has a low predictive power in estimating the target event in comparison to the other three classification models which have a much higher predictive power in estimating clients defaulting on their home loan.

Profit Chart

The *profit chart* displays the cumulative or noncumulative average profit or loss across the ordered percentile estimates. The **Assessment** node will automatically display the appropriate profit or loss chart based on the specified profit or loss matrix from the target profile. That is, there might be times that it might be more important in accurately predicting profit or losses. Therefore, you will be able to determine the correct cutoff points in deciding the amount of profit to be made based on the percentage of people responding or, in our example, the percentage of clients defaulting on a loan. The profit chart uses the information entered in the target profile. The chart is designed to combine the amount of profit or loss entered in the profit or loss matrix along with the prior probabilities from the target profile and the posterior probabilities from the predictive model in generating the expected profit or loss. The following profit plot was generated from the same profit matrix that is specified from the following **Interactive Profit** chart. The table entries will be explained in the subsequent section. In the following profit plot, the ensemble classification model generates the best expected profits of the top 10% of the cases in classifying bad creditors. In other words, if you are only interested in obtaining the highest expected profits from the top 10% of the data, then the ensemble model should be selected. However, if you are interested in the top 15% of the cases, and beyond, in classifying bad creditors, the decision tree model would generate the most desirable profits.

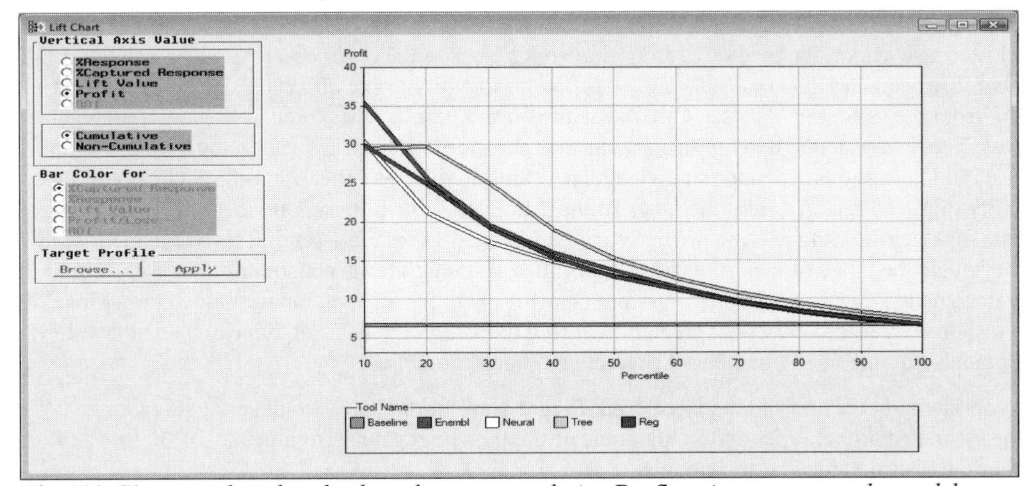

*The **Lift Chart** window that displays the non-cumulative **Profit** estimates among the models.*

ROI Chart

The *ROI chart* displays the return on investment estimates. The chart will display both the cumulative or non-cumulative ROI amounts across the range of the percentile estimates. The chart is based on the ratio between profit and costs. Obviously, a higher rate of return is most desirable. The option in selecting this chart will automatically be dimmed and unavailable for viewing. The reason is because this chart depends on both the amount of revenue and cost from the target profile. From the **Editing Decisions and Utilities** window, you may specify the appropriate cost amounts for the profit matrix. Again, profits are defined to be the difference between the amount of revenue and the fixed cost of revenue.

For example, suppose the total sales of the widget is $2,000, with a total cost of $1,500 to produce the widget within the first and second deciles. Therefore, the actual total profit would be $500, with the cumulative ROI percentage of $100 \cdot [(2000 - 1500) / 1{,}500]\% = 33\%$. From the predictive model, you would expect a 33% return on our investments from the first two ordered decile estimates.

Response Threshold Chart

From the **Assessment Tools** window, select **Tools > Response Threshold Chart** main menu option and the following **Response Threshold Chart** window will appear displaying the response threshold chart for the binary-valued target variable. The *response threshold charts* are designed to display the predictive accuracy of the target event for the classification model across the range of all possible threshold or cutoff values. At each level of the binary-valued target variable, the 3-D response threshold chart displays the rate of accuracy of the given predictive model from the estimated probabilities against the cutoff or threshold probabilities that range between the values of zero and one. Therefore, this plot will allow you to select the correct threshold probability in order to evaluate the classification modeling performance at each target level in comparison to each respective model. The response threshold chart will allow you to determine the modeling consistency across the range of all possible threshold values. The threshold value is simply the cutoff probability set to the binary-valued target event. That is, the observation is assigned to the target event if the posterior or estimated probability is greater than or equal to the threshold probability. Otherwise, the observation is assigned to the target nonevent. From the following threshold charts, the threshold probability is automatically set to .5. This suggests that a given observation has an equal chance of being assigned to either level of the binary-valued target in comparison to the estimated probability. The one big drawback to the following threshold-based classification charts is their dependency on the correct threshold or cutoff value that assigns the given observation their each target group from the estimated probability.

In Enterprise Miner, there exists a color coding schema for distinguishing between the frequency counts across the various cutoff values in the following threshold charts. Red bars will indicate large frequency counts. Conversely, yellow bars will indicate low frequency counts. Therefore, there is this trade-off in achieving a reasonably high rate of accuracy in identifying the target level of interest while at the same time capturing an adequate number of cases within each interval. Large reddish bars at the lower threshold values will indicate that a good portion of the data has been identified to the target level of interest by the classification model with relative ease.

By default in Enterprise Miner, the predictive modeling nodes classify the observation as a target event if the estimated probability is greater than or equal to the baseline probability or the observed sample proportion of the target event from the validation data set. Otherwise, the observation is designated as the target nonevent. Some ideas have been suggested in determining the appropriate cutoff values. Typically, the cutoff probability that is chosen should be based on common knowledge, previous studies, or general intuition. However, the cutoff probability might be selected from the range of the estimated probabilities. One idea for determining the appropriate cutoff value is to run the same predictive model by using a test data set that is independent of the model fit and evaluate the range of the estimated probabilities. Another idea is constructing the ROC curve that is specifically designed to determine the best possible cutoff probability for each binary-valued target level. By default, these response threshold charts are created by fitting the validation data set that can be displayed by selecting one model at a time or any number of predictive models simultaneously.

Select the appropriate target level from the **Predicted Target Level** radio button to observe the rate of accuracy of the estimated probabilities across the range of the threshold values from 0% to 100% for each separate level of the binary-valued target variable.

The process of viewing the frequency values of each bar is similar to viewing the results within the various modeling nodes. That is, simply select anywhere on the bar, then hold down the left-click mouse button and a text box will appear that will list the accuracy rate, threshold value, and frequency count. Otherwise, select the **Select points** black arrow icon from the tools bar icons and select anywhere on the bar to display the corresponding frequency that will be displayed by a small dot appearing along the numeric color bar legend located below the bar chart.

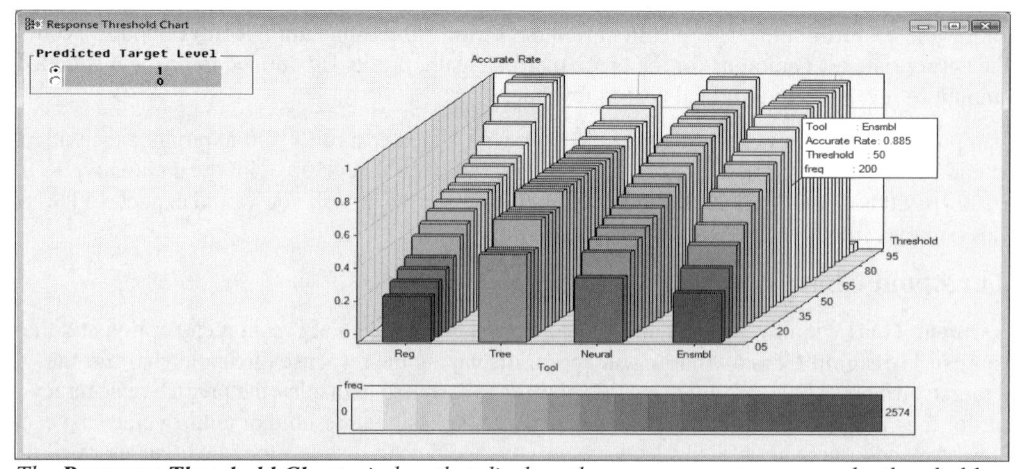

*The **Response Threshold Chart** window that displays the accuracy rates across the threshold probabilities based on the various classification models under comparison.*

The layout of the 3-D response threshold chart is such that the various predictive models under assessment are positioned on the X-axis, the range of all the possible threshold values are positioned on the Y-axis, and the rate of accuracy of the predictive model located on the Z-axis. Obviously, large bars indicate a high rate of accuracy and reddish bars indicate a large frequency count that is most desirable.

From the previous diagram, the ensemble model classified 200 bad creditors at the standard 50% threshold probability with the highest accuracy rate of approximately 88.5% in comparison to the other three classification models, as illustrated in the previous diagram. Conversely, the decision tree model displays approximately the same rate of accuracy across the range of threshold values in predicting bad creditors along with capturing several more cases in comparison to the other three models under assessment. The neural network model displays the same increasing trend of the rate of accuracy in classifying bad creditors, with the logistic regression model identifying a large majority of the observations with a small rate of accuracy in the first few threshold probabilities. Selecting the most appropriate cutoff value for the various threshold-based charts usually requires prior knowledge, general intuition, or previous studies that will result in both capturing a sufficient number of cases while at the same time achieving a high rate of accuracy in identifying the two separate target categories.

ROC Chart

The ROC chart called the *receiver operating characteristic curve* displays the range of all possible values between both the sensitivity and specificity probabilities. The ROC chart requires a binary-valued target variable. In other words, the various probabilities that are displayed within the line plot are calculated by fitting the binary-valued target variable of the classification model. The ROC charts can be displayed by selecting each predictive model separately or as a group.

Classification Matrix			Predicted (test result)		
			Event	Nonevent	
			π_0	π_1	
Observed	Event	π_0	*a*	*b*	δ_0
(disease)	Nonevent	π_1	*c*	*d*	δ_1
			π_0	$1 - \pi_0$	1

Sensitivity	$= \frac{a}{(a+b)}$	Measures the accuracy of predicting the target event, that is, the proportion of the observed target event in predicting the target event, for example, the probability that the symptom is present given that the person has the disease.
Specificity	$= \frac{d}{(c+d)}$	Measures the accuracy of predicting the target nonevent, that is, the proportion of the observed target nonevent in predicting the target nonevent, for example, the probability that the symptom is not present given that the person does not have the disease.
False positive	$= \frac{c}{(c+d)} = 1 - \text{specificity}$	Proportion of the observed nonevent in predicting the target event, for example, the probability that the symptom is present given that the person does not have the disease. (Type II error)
False negative	$= \frac{b}{(a+b)} = 1 - \text{sensitivity}$	Proportion of observed event in predicting the target nonevent, e.g., the probability that the symptom is not present given that the person has the disease. (Type I error)
Prevalence	$= \pi_0$	Proportion of the predicted target event.

The *ROC chart* is a line plot that is designed to measure the predictive accuracy of the classification model. The accuracy of the estimates is calculated from the classification matrix. A classification matrix is a two-way frequency table based on the two levels of the actual and predicted target class levels. The ROC receiver operating characteristic curve displays the sensitivity probability, that is, 1 – type I error rate, in accurately classifying the given observation of the target event correctly across the false positive probabilities, that is, type II error, in classifying the observation incorrectly as the target event by fitting the binary-valued target variable. Since the sensitivity probability is positioned on the Y-axis and the false positive probability is located on the X-axis, therefore, the best possible cutoff point or threshold probability is determined by selecting a point on the curve lying closest to the upper left-hand corner of the curve that represents the best possible situation of 100% sensitivity and 100% specificity, that is, with a false positive probability of zero. In other words, the predictive model accurately predicts both target levels without error. Each point on the ROC curve represents a cutoff probability. The cutoff probability of the ROC chart represents the trade-off between the sensitivity and specificity probabilities. Ideally, the goal would be to have high probabilities for both the sensitivity and specificity. Therefore, this would indicate that the predictive model predicts both target levels extremely well. Furthermore, if you are concerned about reducing the type I error rate that is mistakenly predicting the observed target event, then lower cutoff values should be selected from the ROC chart. The lower cutoff values are located at the upper end of the ROC chart. Conversely, if the aim is reducing the type II error rate, then higher cutoff values should be selected. The higher cutoff values are located at the lower end of the ROC chart. Generally, you are usually interested in reducing the type I error rate. For example, both statistics are often used in determining the relationship between symptoms, that is, test results, and the presence of the disease. In other words, the type I error rate is the probability that the symptom is not present given that the person has the disease.

The curve will usually have a concave upward functional form connecting from the (0, 0) to (1, 1) endpoints. The endpoints of the chart of (0, 0) to (1, 1) represent the cutoff probability points of 1.0 and 0.0 respectively. The curve starts at the point (0, 0) and ends at the point (1, 1). In evaluating the predictive power of the classification model, it is most desirable to have the area underneath the curve as large as possible, indicating to you the superior predictive performance of the classification model. The superiority of the model can be determined by the *ROC area statistic* that is the area between the ROC curve and the imaginary 45° line, that is, the difference between the classification model under assessment in comparison to having no predictive model at all. An absolutely perfect model would result in a 100% sensitivity and 100% specificity, with a corresponding ROC area statistic of one. Conversely, a straight diagonal line indicates that the estimated probability is no better than a random guess or no classification model at all in accurately predicting the binary-valued target level with an associated ROC area statistic of .5.

The **Target Event Level** radio button will allow you to select the separate levels of the target variable. In other words, this will allow you to determine the best possible cutoff probability at each target level by observing the predictive power of the classification model from the area underneath the curve.

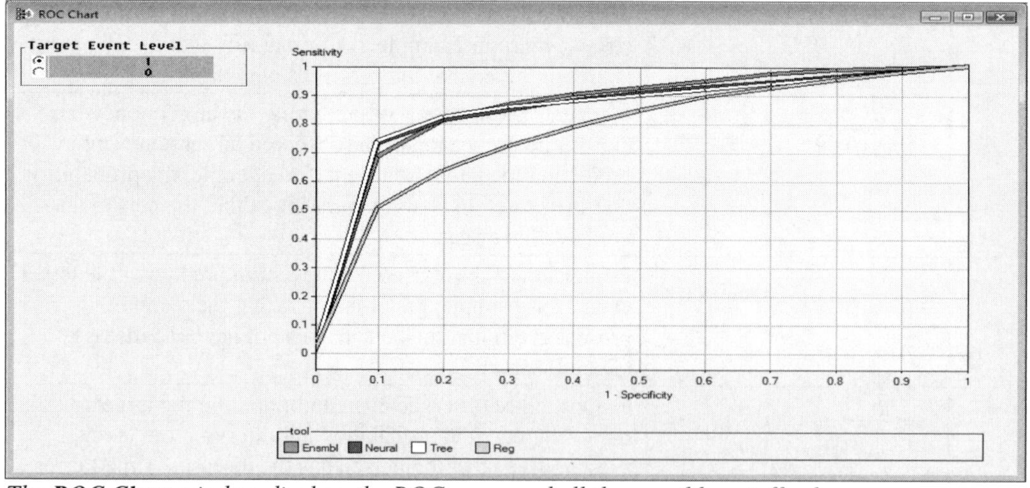

*The **ROC Chart** window displays the ROC curve and all the possible cutoff values.*

From the **Model Manager** window, select **Tools** > **ROC** main menu option and the previous **ROC Chart** window will be displayed. Both the ensemble and neural network models display nearly identical estimates across the various error probabilities. From the ROC chart, the curves from the neural network, decision tree, and ensemble models lies well above the logistic regression model, which indicates a larger ROC area statistic based on the class level of the target event of interest. This suggests that the modeling performance of the three classification models perform better than the logistic regression tree model. In addition, the neural network and decision tree models slightly outperformed the ensemble model at the standard 50% cutoff value. From the upper portion of the ROC chart, if there is more interest in a lower type I error, then the ensemble model should be selected. The best possible cutoff probability from the neural network or decision tree models seems to be a sensitivity of approximately 75% and a specificity of 90% in predicting the target event of interest.

Threshold-based Charts

The *threshold-based charts* display the rate of accuracy in the predictive model based on a given threshold probability. The charts display the rate of accuracy in the modeling design at each level of the binary-valued target variable. Therefore, a threshold value set at .5, which is the default, will result in an observation having an equal chance of being assigned to either target group. As a review, the threshold probability is the arbitrary probability of assigning an observation into one of the two target groups from the estimated probability of the target event from the predictive model. The following threshold-based charts can be selected by choosing one model at a time from the list of models under assessment within the **Tools** tab.

The node is designed to display the threshold-based chart by a 3-D frequency bar chart that represents the classification matrix. The flexibility of the bar chart is due to the ability to change the threshold probability and observe the impact of the classification assignments from the predictive model. Simply select the **Threshold** scroll bar to change the threshold value, which will automatically change the frequency bars within the 3-D bar chart. Therefore, this plot will allow you to select the appropriate threshold probability in measuring the modeling performance of the classification design at each of the two separate class levels of the binary-valued target variable. If the target class levels are perfectly balanced, then an absolutely perfect model would display bars on the main diagonals and the off-diagonals would display no bars at all. From the **Assessment Tools** window, select the **Tools > Threshold-based Chart** main menu option and the following **Threshold-based Charts** window will appear.

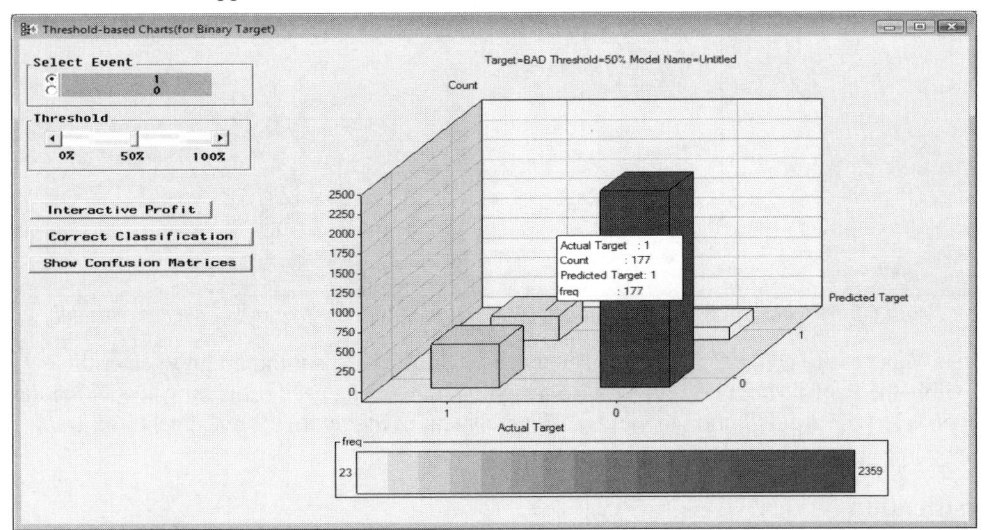

*The **Threshold Based Charts** window displays the classification matrix at each threshold value from the ensemble model.*

Interactive Profit

Select the **Interactive Profit** button to display the following **Interactive Profit Chart** window. The *interactive profit chart* will allow you to enter various decision values from the profit matrix in order to observe the rate of total return across the range of all possible threshold probabilities. For binary-valued target variables, you can interactively change the entries in the profit matrix in order to view the change in the rate of total return across the range of the threshold probabilities. The expected profit is the summation of the product of the decision consequences and the posterior probabilities at each class level of the categorically-valued target variable. Simply select the **Apply** button to update the chart with the entries specified from the profit matrix in order to view the change in the rate of return across all the possible threshold values from the current entries in the profit matrix. By default, an identity profit matrix is applied. Intuitively, the best threshold probability to select from is determined by the highest rate of return. Obviously, a positive rate of total return is most desirable. A rate of total return of zero indicates a breakeven point of the chart.

From the profit matrix, the first row and column indicates that the ensemble model correctly classified a client defaulting on their home loan with an expected profit of $50. The second row and first column indicates a penalty cost in the ensemble model of inaccurately identifying a client defaulting on their loan of $5. The first row and second column represent another penalty cost in the ensemble model of incorrectly classifying a client paying their home loan of $10. The second row and second column indicates that the ensemble model correctly identified clients paying their home loan of $0. Press the **Apply** button in order to observe the impact that the profit matrix has on the average return across the range of threshold probabilities.

In the following interactive profit plot, the line plot indicates that the total rate of return initially increases from $6 to over $7 between the threshold probabilities of 0% to 10%. From the 15% threshold probability, the total rate of return begins to dramatically decrease until the threshold probability of 70% is reached, with a gradual decrease in the total rate of return until the threshold probability of 95%. The standard threshold probability of 50% will result in a profit of nearly 2 from the ensemble model. And, finally, the plot reaches the breakeven point of zero at the threshold probability of appropriately 62.5% with a loss in the total rate of return at the 62.5% threshold probability and beyond from the ensemble model.

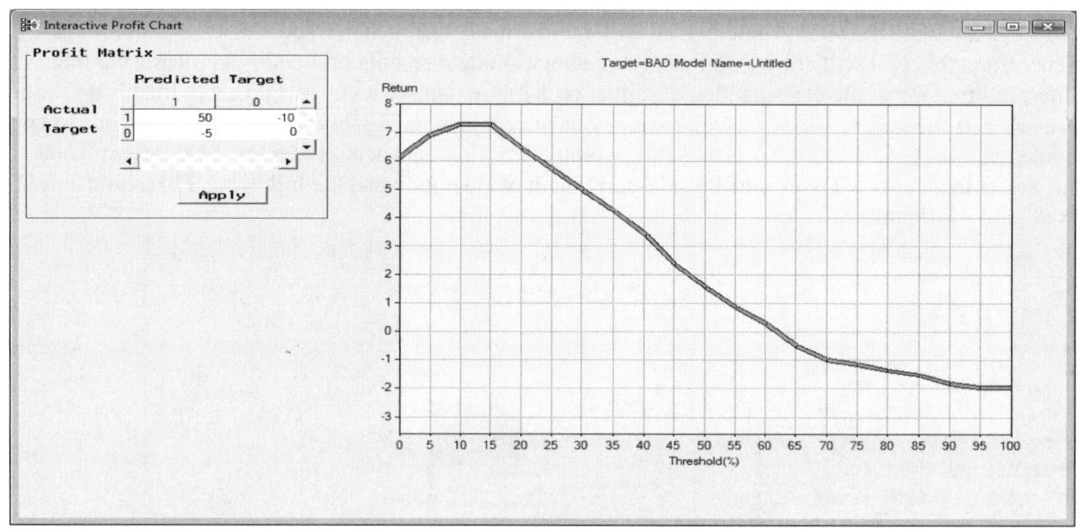

*The **Interactive Profit Chart** window displays the interactive profit estimates from the ensemble model.*

Note: In Enterprise Miner v4, to generate the previous interactive profit chart is it important to enter the decision values within the **Profit Matrix** one at a time when you are inserting each decision value within the **Profit Matrix**, then select the **Apply** button to view the development of the return of investment line graph with each corresponding update of the profit matrix.

Correct Classification

From the **Threshold-based Chart** window, select the **Correct Classification** button that will allow you to view the following correct classification assessment chart. The *correct classification chart* is designed to display the rate of accuracy of the posterior probabilities in predicting each level of the binary-valued target variable in addition to comparing the accuracy by combining both levels across the range of all possible threshold probabilities, that is, combining the two separate target levels by computing the average rate of accuracy between both target levels. The combined target level chart is helpful if the main goal of the analysis is accurately predicting both target levels simultaneously. Again, the following assessment chart will allow you to view the accuracy of the classification performance of the predictive model from the classification matrix across the range of all possible threshold probabilities. It is important to note that the accuracy rate that is display in the following 3-D line plot between both the target event and target nonevent is defined by the row percentages of the actual target class levels from the validation data set. On the other hand, the accuracy rate that is displayed from both target levels is the standard accuracy rate that is calculated from the classification matrix.

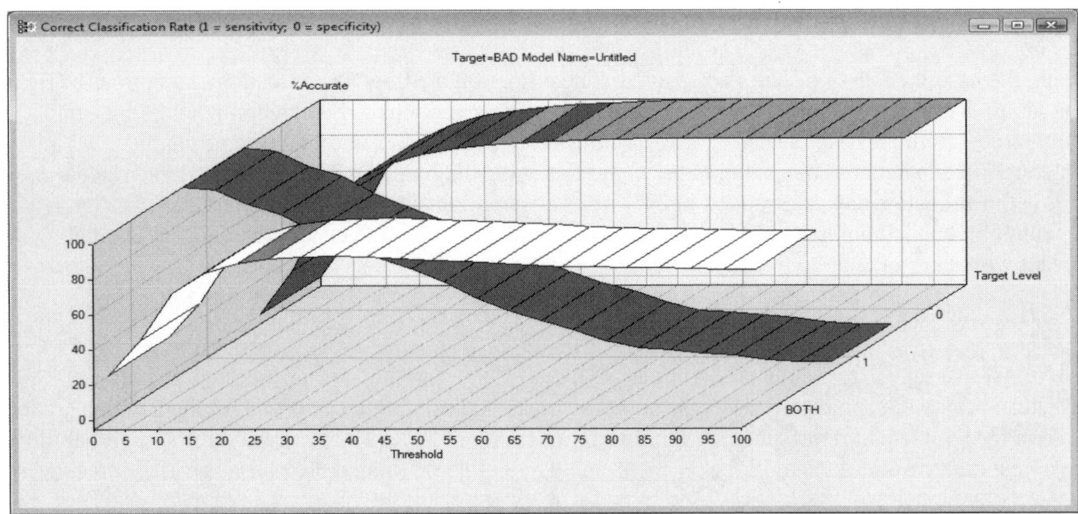

*The **Correct Classification Rate** window displays all possible threshold estimates from the ensemble model.*

From the previous diagram, the reason that the line plot of the target event starts at a 100% rate of accuracy, that is, row percentage at the threshold probability of 0%, is because there are 598 clients that have been identified of defaulting on their loan from the validation data set. Conversely, the opposite holds true for the target nonevent. If the interest in the analysis is to accurately predict the target event, that is, defaulting on a loan, then the 50% threshold probability resulted in a row percentage of 29.60% in predicting the target event. If the interest in the analysis is to accurately predict the target nonevent, then the 50% threshold probability from the ensemble model generated a row percentage 99.03% in predicting clients paying their home loan. If the goal of the analysis is to accurately predict both target levels simultaneously, then the 50% threshold probability resulted in an 85.10% rate of accuracy.

Show Confusion Matrices

Selecting the **Show Confusion Matrices** button from the **Threshold-based Chart** window will display a series of 2x2 classification matrices across the range of the threshold cutoff probabilities. The results will not be displayed. The reason is because the generated results are basically the procedure output listing from the PROC FREQ procedure that displays a series of two-way frequency tables at each value of the threshold probability. The procedure output listing displays several frequency tables from the various threshold cutoff probabilities ranging from 0% to 100% at an increment of 5%. From the series of classification tables, it will allow you to view the row percentages at each level of the binary-valued target variable along with the frequency counts across the various threshold probabilities.

Table View of the Lift Chart Data

From the **Lift Chart** window, press the **View Lift Data** tools bar icon to open a table listing of the lift data. The table listing displays the values used in creating the various performance charts. That table listing displays various assessment statistics such as the expected profit at the various cutoff probabilities, the predicted probabilities, percentiles, lift values, response rate (%), cumulative capture response rate (%), cumulative profit and average profit per observation, and return of investments statistics between each model, the baseline probability, and the exact percentiles. A temporary data set called LIFTDATA is automatically created in SAS that consists of the various lift values. The SAS data set can be opened within the current SAS session in order to perform further analysis. By opening the table listing, it will allow you to save the table listing to an external SAS data set by selecting the **File > Save As** main menu option settings. This will allow you to view the stability in the classification model by comparing the lift values between the training and validation data set from the same classification model by constructing your own performance plots from the two separate SAS data sets. In other words, the stability in the classification model can be determined by viewing small deviations in the lift values between the two separate lines from the training and validation data sets across the ordered decile estimates. Simply construct a process flow with the same partitioning of the source data set, then select the corresponding partitioned data set from the **Model Manager**.

To Display Exact Lift Chart

By default, the baseline 2-D line chart is displayed. Select **Format > Show Exact** main menu options or right-click the mouse and select the **Format > Show Exact** pop-up menu items to display both the baseline and exact percentiles. Both the baseline and exact percentiles can be displayed within the same chart simultaneously or separately. This is illustrated in the following diagram. To best understand the exact probability, the validation data set resulted in 20% of its observations identified by the target event, that is, clients defaulting on their home loan. Therefore from the following plot, the exact probability in predicting the target event without error is up to the 20th percentile estimate.

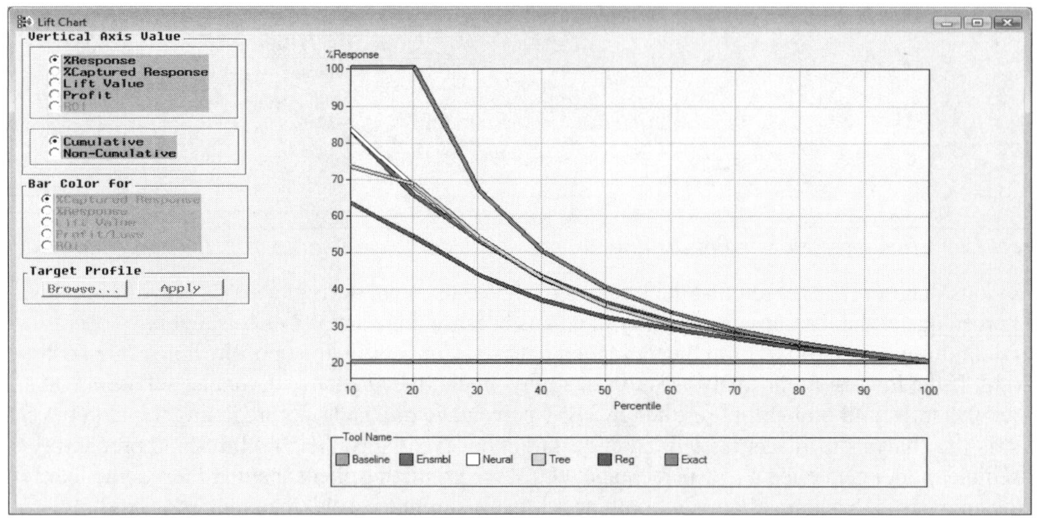

*The **Lift Chart** window that displays the lift chart along with the exact probabilities.*

Diagnostic Charts

The following diagnostic charts can be displayed by fitting an interval-valued target variable or a categorically-valued target variable to the predictive model. The *diagnosis charts* are basically frequency bar charts. For interval-valued targets, the charts are designed to view the distribution of either the target responses or the residuals from the predictive model. For categorically-valued targets, the frequency bar charts display the results from the classification matrix, that is, the two-way frequency table of the actual target class levels against the predicted target class levels. The bar chart will allow you to evaluate the accuracy of the classification model. In other words, if the frequency bar chart of the actual target level across the predicted target levels displays large bars along the main diagonal of the squared grid plot, then it will indicate the accuracy of the predictive model. Conversely, for interval-valued targets, if the frequency bar chart of the residual values against the fitted values displays large bars at zero in the residual values across the fitted values, then it will indicate the accuracy of the predictive model. By default, the tab will display the frequency bar chart of the target values across the fitted values. The accuracy in the predictive model can be determined from the bar chart by observing large bars that are displayed within the same interval between both axes variables. The diagnostic charts can only be viewed by selecting each predictive model separately from the **Model** or **Output** tabs.

There are three different ways to display the diagnostic charts from the **Tools Assessment** window, as follows:

1. Select the **Draw Diagnostic Chart** toolbar icon from the tools bar menu.
2. Select **Tools** > **Diagnostic Chart** from the main menu options.
3. From the **Output** tab, highlight the model row, then right-click the mouse and select **Tools > Diagnostic Chart** from the pop-up menu items.

For an interval-valued target variable:

- **Target by Output** Frequency bar chart of the target variable against the fitted values.
- **Residuals by Output** Frequency bar chart of the residual values against the fitted values.

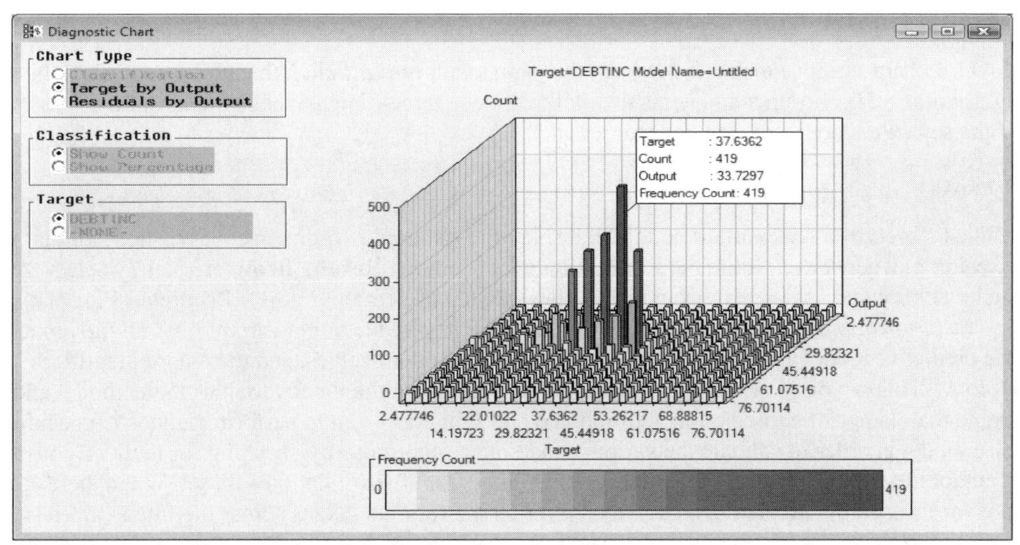

The ***Diagnostic Chart*** *window that displays bar charts between the observed and predicted values from the multiple linear regression model.*

For a categorically-valued target variable:

- Classification bar charts of the frequency counts or percentages of the actual target categories by the predicted target categories, that is, the classification matrix, for visually observing the accuracy in the classification model in predicting each class level of the categorically-valued target variable that is illustrated as follows:

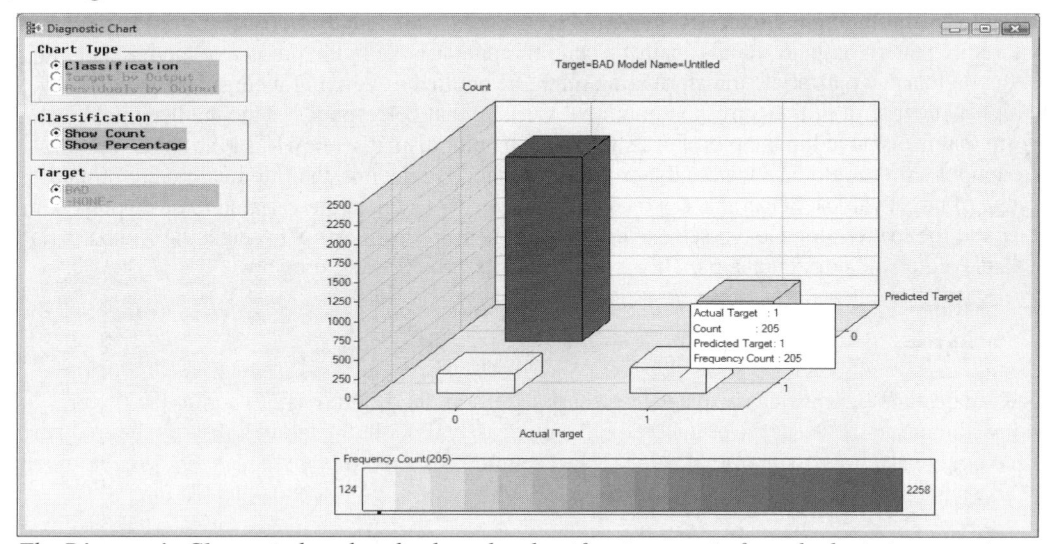

The ***Diagnostic Chart*** *window that displays the classification matrix from the logistic regression model by fitting the categorically-valued target variable.*

Displaying the Actual Values and Frequency Counts

The diagnostic charts are displayed in many different colors, which will allow you to easily distinguish between the range of values of the numerous frequency bars. The high frequency counts are displayed in red, yellow bars represent low counts, and orange bars represent intermediate frequency counts. Select anywhere on the displayed bars by holding down on the left-click button and a text box will automatically appear that is designed for you to observe the target, output axis value, and the frequency of the bar. Click on the **Select Points** black arrow toolbar icon from the toolbar menu, then select anywhere on the bar to view the frequency count of the bar from the numeric color bar legend that is located at the bottom of the chart with a small black dot appearing along the numeric legend.

Viewing 3-D Lift Charts

To view the 3-D lift charts, select **Format > 3D** from the main menu or right-click the mouse and select the corresponding **Format > 3D** pop-up menu items that will display a three-dimensional bar chart of the percentiles by the separate models that are under assessment:

Prediction Plots

For interval-valued targets, the **Assessment** node will display the standard two-dimensional scatter plot that is similar to the scatter plot generated within the **Regression** node from the **Results Browser**. Simply select any one of the listed models under assessment from the **Models** tab, then select the **Tools > Predicted Plot** main menu options. The *prediction plots* or scatter plots will allow you to plot the target variable and all the input variables in the predictive model along with the predicted values, residual values, and the record identifier variable. The plot will allow you to observe the functional form between the input variables in the model and the target variable that you want to predict. In addition, the plot will allow you to perform diagnostic checking of the predictive model in order to validate the various modeling assumptions by checking for outliers, constant variance, and randomness in the residuals across the input values from the validation data set. When there are numerous input variables in the predictive model, then plotting the residual values across the fitted values is advisable in order to satisfy these same modeling assumptions. When the data is collected over time, then the plot will allow you to satisfy the independence assumption by observing any cyclical pattern in the residual values over time. If the residual plot displays a cyclical pattern in the residual values over time, then it may be due to one of two reasons. It might suggest to you that the incorrect functional form has been applied to the predictive model or there exists a high degree of correlation in the data that was collected over time.

For each input variable in the model, the residual plot will allow you to determine the correct functional form to include in the predictive model. For example, if the residual plot displays an identifiable functional form between the residual values and the selected input variable in the model, then this will indicate to you that the appropriate mathematical functional relationship should be added to the predictive model. If the residual plot displays a curvature pattern in the residuals against a certain input variable in the predictive model, then this will suggest to you that an appropriate transformation might be applied or you might consider including an appropriate higher-order term of the corresponding input variable that corresponds to the mathematical functional form that is displayed in the plot. For example, if the plot displays a bowl-like curvature pattern, then a squared input term should be added to the predictive model. By default, the fitted values are plotted across the range of target values. Select the **Y-axis** drop-down arrow to select the variable to be displayed on the vertical axis of the scatter plot. Conversely, select the **X-axis** drop-down arrow to select the X-axis variable that will be displayed on the horizontal axis. This is illustrated in the following diagram.

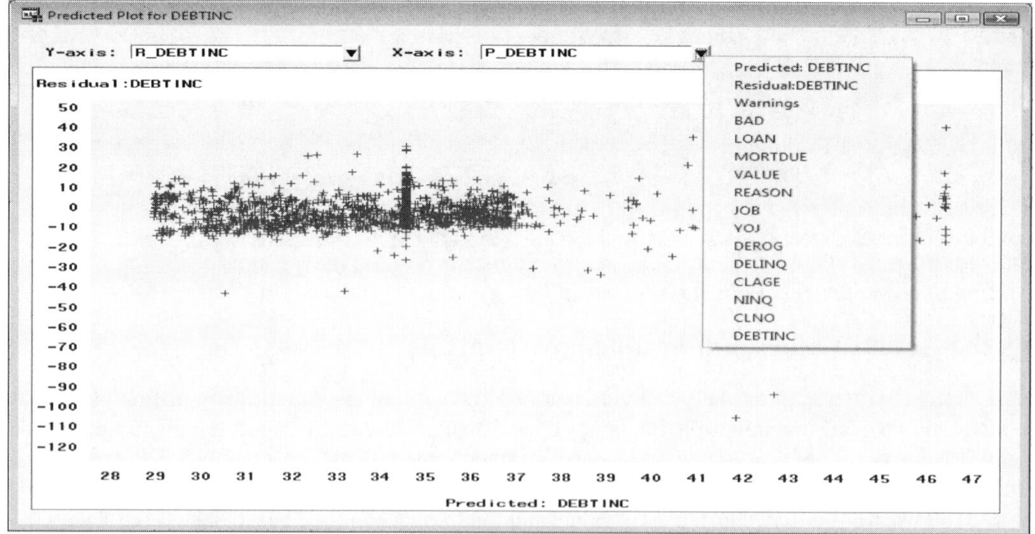

*The **Plot** tab that displays a scatter plot of the variables in the predictive model under assessment from the neural network model by fitting the interval-valued variable, DEBTINC.*

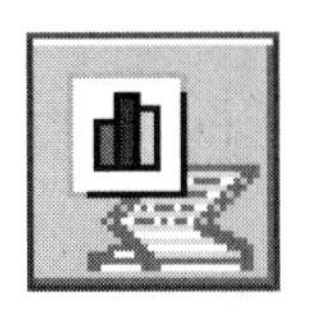

5.2 Reporter Node

General Layout of the Enterprise Miner Reporter node

- **Header tab**
- **Options tab**
- **Registration tab**

The purpose of the **Reporter** node in Enterprise Miner is designed to assemble and consolidate the results from the process flow diagram into a single HTML report that can be viewed by your favorite Web browser. Typically, the **Reporter** node is the last node connected to the process flow diagram. Therefore, running the **Reporter** node will result in the execution of the entire process flow diagram. A picture of the process flow diagram will be displayed at the beginning of the HTML report once you first open the HTML report. The HTML report is organized in various sections, where each section is based on the node that is a part of the process flow diagram connected to the **Reporter** node. Each section of the HTML report will display various hyperlinks. These same hyperlinks will allow you to view both the configuration settings of the node from the various tabs within the corresponding node along with the associated results generated from the node. In general, the HTML layout will display various hyperlinks and graphs. Simply click the various hyperlinks within the HTML file that will allow you to open an associated text file that displays the various configuration settings and the corresponding results generated from the node.

For all the nodes that generate any output, an HTML report is automatically created from every node that is connected to the **Reporter** node. That is, an HTML report is created for all nodes that are placed before the **Reporter** node within the **Diagram Workspace**. However, the restriction is that an HTML report will not be created for all nodes that are placed after the **Reporter** node within the process flow diagram. By executing the **Reporter** node, the flow of execution makes its first pass through the process flow, analyzing the data within each node, followed by a second pass through the process flow to create the HTML report. In the first pass, each node that is analyzed, in succession, will be surrounded in green. In the second pass, each node, in succession, will be surrounded in yellow to create the HTML report.

From the currently opened process flow diagram, select the **Options > User Preference** main menu options, then selecting the **Reports** tab, which will allow you to customize the general layout design of the HTML listing that is created from the **Reporter** node. The tab will give you the capability of specifying the various reporting topics that will be displayed within the HTML report and the order in which the various topics are listed within the HTML report. The **Available report items** listing will display all the possible topics that the **Reporter** node can display. The adjacent **Display in reports** list box will display the order in which the list of selected topics will be displayed in the HTML report.

In Enterprise Miner v4.3, the **Reporter** node has an added option of exporting models into a model repository system. The *model repository* is designed to store and access models from several different Enterprise Miner projects and diagrams in a convenient, centralized location. In order to access this model repository system from the network server, you must either first install the SAS Metadata Sever or the Tomcat Server. Information in setting up the SAS Metadata server can be located through the SAS system documentation.

The following are the various reasons for incorporating the **Reporter** node into the process flow diagram.

- Consolidate all the data mining Enterprise Miner SEMMA process results into a well-organized HTML format to be viewed by your favorite HTML browser. The node summarizes the SEMMA analysis with various diagrams, option settings, data set content listings, procedure code listings, and a wide variety of statistical listings from the various Enterprise Miner nodes that are a part of the process flow diagram.
- Capture the output from the node that can be copied into a separate SAS program for further analysis.
- Export the model and register the metadata information in the model repository.

Selecting the **Reporter** node, right-clicking the mouse, and selecting the **Create Report** pop-up menu item will instruct the node to automatically create an entirely new HTML report without executing the entire process flow diagram.

The general layout of the **Reporter** node consists of the following tabs.

Header tab

The **Header** tab is designed for you to enter a report title, user name, and notes. The information will then be listed in the section heading of the corresponding HTML report when you run the node.

Options tab

By default, a new HTML report is created from the **Reporter** node each time you execute the node. However, the **Options** tab is designed for you to append a new report to the existing HTML file. The tab will also allow you to view the directory where the HTML file is located and the type of format that is used to create the HTML report.

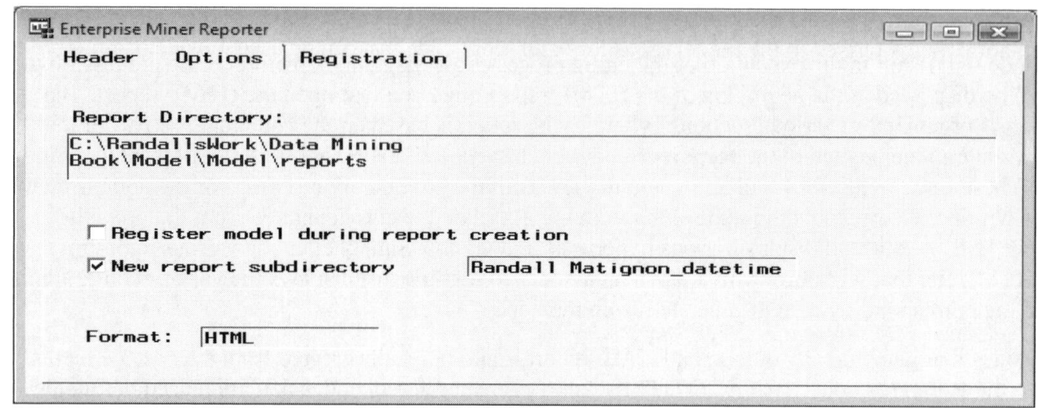

*The **Options** tab is used to either replace or append the existing HTML report of the node results.*

Ways to Create the HTML Report

The following are the various ways to create an HTML report from within the node:

- Create a new HTML report each time you run the **Reporter** node.
- Append the existing HTML results to a designated HTML file.
- Replace the existing HTML report.

Since the **New report subdirectory** check box is automatically selected, a new time-stamp system folder is created each time the **Reporter** node is executed. However, from the **Options** tab, you may clear the **New report subdirectory** check box to prevent Enterpriser Miner from creating a separate folder that will allow you to either replace or append the report to the existing HTML report. Either executing the **Reporter** node or selecting the **Create Report** pop-up menu option will result in the following **Message** window appearing that will allow you to either replace, append, or open the existing HTML report.

*The **Message** window to inform you to either replace, append, or open the existing HTML report.*

Opening the HTML Report

In Enterprise Miner, the following are the various ways to open the HTML report:

- Select the **Reporter** node and right-click the mouse to select the **Open report** pop-up menu item that will open the HTML file from your local Web browser.
- From the **Reporter** node, select **Tools > View Results** main menu options that will open the HTML file.

General Layout of the HTML Report

Each report that is created from the node is conveniently organized and archived in separate entries within the **Reports** tab of the **Project Navigator**. In order to view any of the HTML reports that have been created within the currently opened Enterprise Miner project, simply open the HTML report with the general layout of each report containing the following information.

- **Header Information:** Consists of the report title, user name, creation date of the report, and the corresponding notes of the HTML report.
- **Image of the Work Flow Diagram:** Initially, the report displays an image of the process flow diagram.
- **Sections:** Each section contains various images and hyperlinks based on the numerous configuration option settings from the tabs, along with the corresponding results from each node that is connected.
- **Path Information**: The last section of the HTML report displays the XML code of the metadata information that is passed to the model repository and score code that can be used outside of Enterprise Miner.

File Layout of the Results

The HTML file that is created within the current project is stored in the REPORTS folder which is located within the Enterprise Miner project folder on your machine. Each separate folder contains the various text files and GIF files that comprise the HTML report. A new HTML file is automatically created once you select the **Create report** pop-up menu item from the **Reporter** node. The HTML file contains hyperlinks to the corresponding .TXT text file and the various statistical graphs in the report that are saved in a .GIF file format. XML files are created within each folder that contains XML code of the metadata information that is then passed to the model repository. In addition, CSV (comma separate values) files are created with some of the results that are generated from the nodes that can be placed into Microsoft Excel spreadsheets. A separate folder is created from each HTML report that is created. The naming convention of the folders that are automatically created from the node are assigned a time stamp of the current date and time that the HTML report was created, as illustrated in the following diagram.

The following diagram displays the REPORTS directory that is automatically created once you create an HTML report from the **Reporter** node. In other words, the currently opened process flow must first be executed in order for Enterprise Miner to create the REPORTS folder on your system.

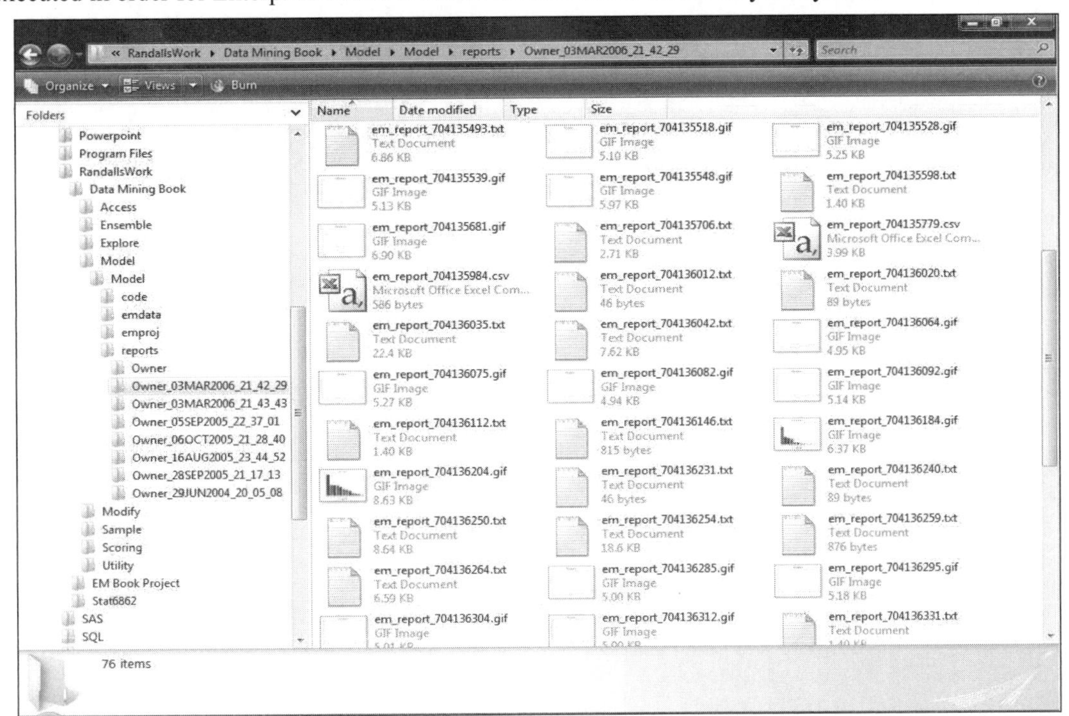

The directory structure of the Enterprise Miner project and the corresponding diagram.

Chapter 6

Scoring Nodes

Chapter Table of Contents

6.1 Score Node

General Layout of the Enterprise Miner Score Node

- **Data tab**
- **Variables tab**
- **Settings tab**
- **Score Code tab**
- **Log tab**
- **Notes tab**

The purpose of the **Score** node in Enterprise Miner is designed to manage and organize SAS score code that is used to generate the predicted values from the trained model. The **Score** node will allow you to either view, edit, save, delete, combine, export, or execute the scoring code program. The training code is the internal SEMMA code that is running behind the scenes when executing the node. The training code is the various data mining procedures that are used in producing the results. *Score code* is basically a SAS program that can be included into any SAS data step. In statistical modeling, score code can be used to score a data set in order to generate the predicted values without the existence of a target variable in the data set. The score code is the internal SEMMA scoring formulas that are generated within the respective nodes that can be used within the SAS session and an associated SAS data step that is outside the Enterprise Miner environment. In predictive modeling, the score code is not only the modeling equations, but also all data manipulation tasks that are performed in preprocessing the data to generate the modeling estimates. From the **Result Browser**, the score code can be viewed from many of the nodes within the **Code** tab by selecting the **Scoring** option button.

The following are examples of some of the score code that are automatically generated in the background once you execute the node within Enterprise Miner, for instance, the various SEMMA scoring formulas that are created from the modeling nodes to generate the prediction equations and the corresponding predicted values from the trained model.; the principal component scores that are the correlation scores from each principal component that is the linear combination of the input variables and eigenvectors in the model that best explains the variability in the data; the cluster score code that calculates the squared distances in assigning the associated clustering assignments to each input variable in the data mining analysis; the interactive grouping score code that uses a series of if-then statements for each input variable in the data set that is optimally grouped into separate categories by the binary-valued target variable from the WOE statistic that is calculated for each group; the transform variables score code that can be used to create new variables from the original variables; the replacement score code that can be used to replace incorrectly recorded values or impute missing values in the data sets; and, finally, the data partition score code that will randomly divide the input data set through the use of a uniform random number along with a counter variable to control the number of observations that are assigned to each partitioned data set.

The **Score** node is designed to manage and organize the scoring code into one data set, similarly to the standard predictive modeling practices. For example, the **Score** node might be used instead of the **Group Processing** node in order to consolidate the separate predictive models for each class level that can be distinguished by a certain group identifier variable. As opposed to the **Ensemble** node, the **Score** code can be used to merge the separate scoring code into one data set in order to distinguish between each stratified model and the associated scoring code through the use of simple *if-then* programming logic. The **Score** node can also be used to update or modify existing scoring code or even create your own custom-designed scoring code. For example, the **SAS Code** node can be used to perform additional processing of the corresponding scoring code.

The node is designed to display either the currently imported or the accumulated group processing runs that are generated from an external file. For example, in predictive modeling the score code can be read into a separate SAS program in order to generate a new set of predicted values from an entirely different set of input values. The node will allow you to observe the internal SEMMA score programming code and the corresponding internal scoring code formulas that are used in producing the prediction estimates for each modeling node that is connected to the **Score** node. The node will also export a scored output DMDB data mining data set that is specified from a previously connected node. The one restriction of the **Score** node is that it must precede the node that produced the score code. The **Score** node will allow you to save the currently listed score code, which is important since the listed score code will be automatically modified once you change the option settings from the previously connected modeling node and recompile the node. The similarity between both the **Score** node and the **Score Converter** node is that the **Score** node is used to score data within SAS as opposed to the **Score Converter** node that creates score code in the C programming or Java programming languages.

Data tab

The **Data** tab is designed for you to view the training data set that has been scored or select the data set for scoring. The **Score** node is designed so that it will only recognize the score code from the training data set. Therefore, the training data set is automatically selected that contains the desired scoring code. The data set will be scored by executing the associated score code within the node. In Enterprise Miner, you can fit several different models to score. From the **Data** tab, you will be able to select which data set that will be used for scoring from the corresponding modeling node that is connected to the **Score** node. For example, the tab will allow you to open the **Imports Map** window that lists each data set that has been created from the previously connected modeling nodes or usually from the **Input Data Set** node, which, in turn, will allow you to select the corresponding data set to score. The tab will also allow you to view the file information and the table view of the selected data set to be scored.

Conversely, select the **Score** radio button in order to score the selected data set. The scored data set is created from the corresponding modeling node or the **Input Data Source** node by selecting the **Score** role assignment. The data set is scored from the following **Score code** tab by selecting the listed score code and compiling the subsequent code that will result in scoring the selected data set. From the modeling nodes, the score data set will consist of the prediction information along with all other variables in the selected data set.

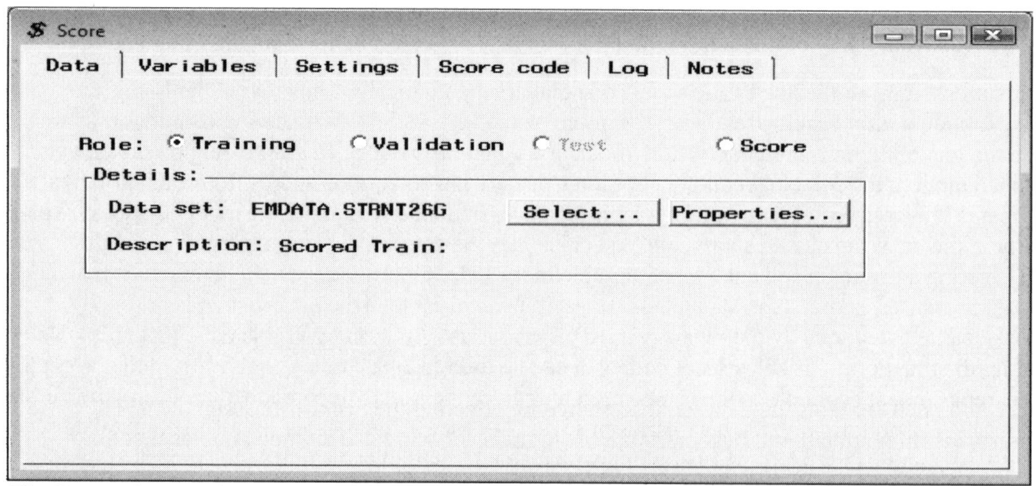

*The **Data** tab is used to select the data set to score by executing the score code within the node.*

Variables tab

The **Variables** tab is designed to display a table view of the variables in the active training data set. The tab is similar to the other **Variables** tab that displays the variable name, model role, level of measurement, format, and variable label. However, the restriction to the tab is that none of the variable settings can be changed.

Settings tab

The **Settings** tab is designed to configure the exportation status of the output scored data set from the node. The subtabs to the **Setting** tab will allow you to specify the type of score file that is created from the **Score** node and control the variables automatically written to the output scored data set. By default, all the variables that are specified from the previous **Variables** tab that are set with a variable status of **Keep** are automatically written to the output scored data set.

- **General subtab**
- **Output Variables subtab**

General subtab

The **General** subtab will initially appear as you open the **Score** node. The **General** subtab is designed for you to control the type of scored data sets that are created from the node. The type of scored data sets that can be specified from the tab are either to create scored data sets created from the training data set, concatenate scored data sets, or merge scored data sets from separate modeling nodes connected to the node.

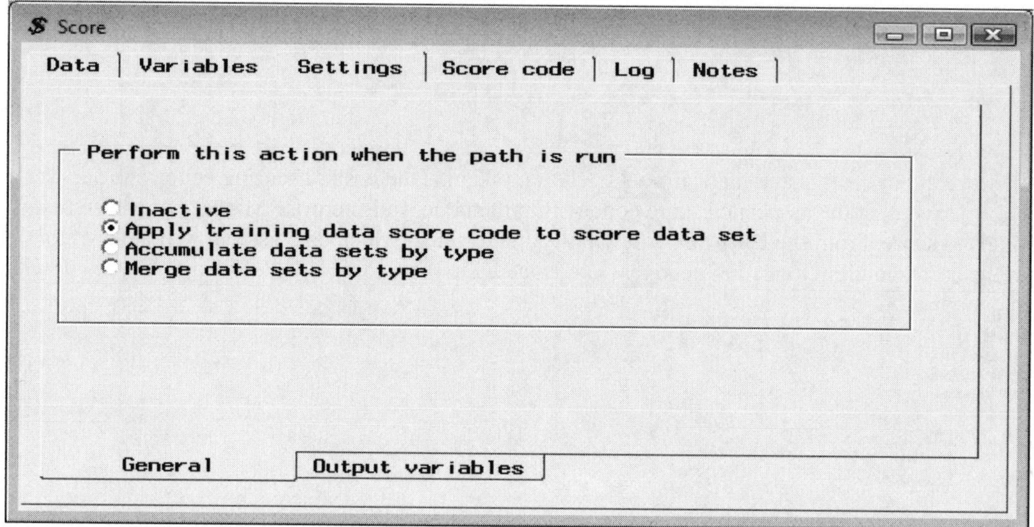

*The **General** subtab is used to specify the way in which the scored data set is created within the node.*

The following are the various radio button options that are available within the tab to configure the status of the scored data set:

- **Inactive** (default)**:** This is the default option that exports the output scored data set by creating an output scored data set for each partitioned data set. The scored output data set is usually created from a previously connected node. The one requirement is that the previously connected node must be executed in order for the corresponding score code to be generated within the node.
- **Apply training data score code to the score data set:** This option adds additional prediction information to the selected score data set such as added missing value indicators, as well as prediction information such as the fitted values that are written to the scored data set that is selected from the previous **Data** tab. However, you must initially select the **Process or Score: Training, Validation, and Test** option from the **Output** tab of the respective modeling node to create the imported scored data sets with the corresponding prediction information or connect the **Input Data Source** node that contains the data set to be scored by selecting the **Score** role assignment. The **Output Variables** subtab will become available in order for you to select the variables that are written to the scored data set.
- **Accumulate data sets by type:** This option copies and exports the data sets that have been imported from the previously connected nodes. For example, the node will accumulate the separate score data sets and corresponding score code by each class level that is created from the **Group Processing** node based on the categorically-valued grouping variable. In other words, the output scored data sets will be concatenated or appended.
- **Merge data sets by type:** The imported data mining data sets that are created from the previously connected node are merged together to create the output scored data set. For example, the output scored data sets from the training data set that are created from two or more modeling nodes that are merged together, assuming that the scored data sets have the same number of observations. Otherwise, an error message will appear to inform you that the number of observations within each merged scored data set are not the same. The merged scored data set will consist of an alphabetical suffix for each variable to distinguish between the separate variables in each data set.

Output Variables subtab

The **Output Variables** tab is designed for you to determine the variables that will be written to the scored data set. In Enterprise Miner, additional variables are automatically included in the output data set by executing the **Score** node. However, the tab will allow you to keep certain variables in the scored data set from the **Variables** tab. The tab is similar to using the KEEP statement in SAS programming by keeping certain variables in the scored data set in order to reduce processing time and save disk space.

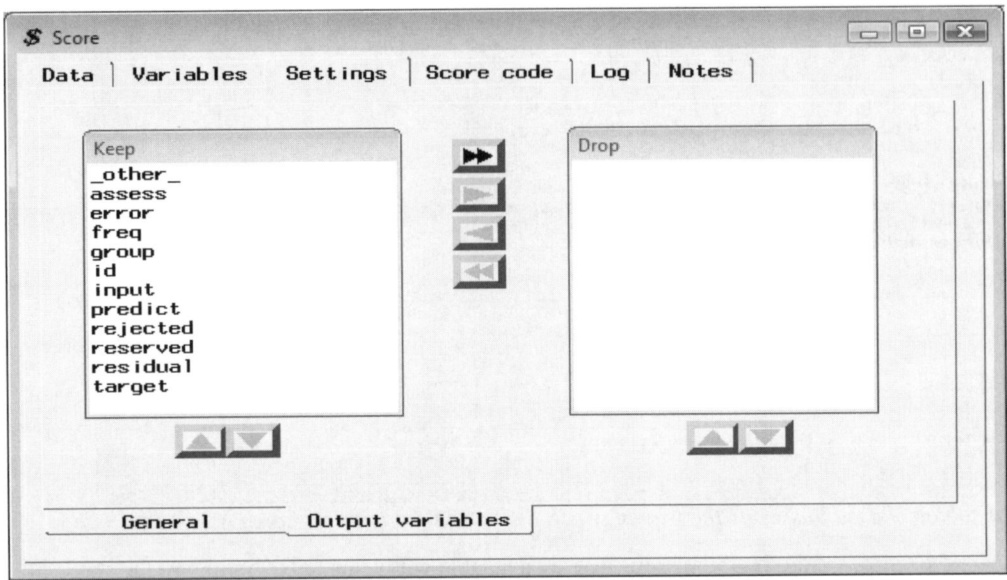

*The **Output variables** subtab is used to specify the variables for the scored output data set.*

Additional variables will be written to the output data sets when you run the **Score** node. You can view the variables written to the output data set from the **Output Variables** subtab. The type of output data set that is created when you run the node depends on the action you set in the **General** subtab. For example, the output data sets can be created from either the partitioned or scored data sets, or the output data set can be created by concatenating or merging the output scored data sets. By default, all of the variables in the **Keep** list box are written to the output data sets. However, you can drop unnecessary variables from the output data set to save disk space.

To drop a variable, select the variable in the **Keep** list box, then click the right arrow control button to move the variables from the **Keep** list box to the **Drop** list box. Similarly, click the left arrow control button to move variables from the **Drop** list box to the **Keep** list box. To select several variables, press the Ctrl key or Shift key or drag your cursor across the adjacent variables. Select the double arrow controls to move all the variables from one list box to the other. The grayed arrow controls that are located beneath the **Keep** and **Drop** list boxes are inactive. These buttons control the way in which the variables appear within each list box.

Score code tab

The **Score code** tab will allow you to manage the score code by either viewing, saving, editing, exporting, or compiling the selected score code. The list box that is positioned to the left of the window will display the listed score code entries from the available score code that is created from corresponding modeling nodes and the associated scored data sets that are connected to the **Score** node. The **Score** window that is displayed to the right of the list box displays the score code associated with the listed entries selected from the list box. By default, the **Current imports** option setting is selected in the list box that is located on the left-hand side of the **Score code** tab, with the most recently compiled score code that will be listed within the **Score** window. The tab will allow you to select the appropriate score code, and compile the code that will instruct the node to create the prediction information for the currently selected data set that is specified from the previous **Data** tab.

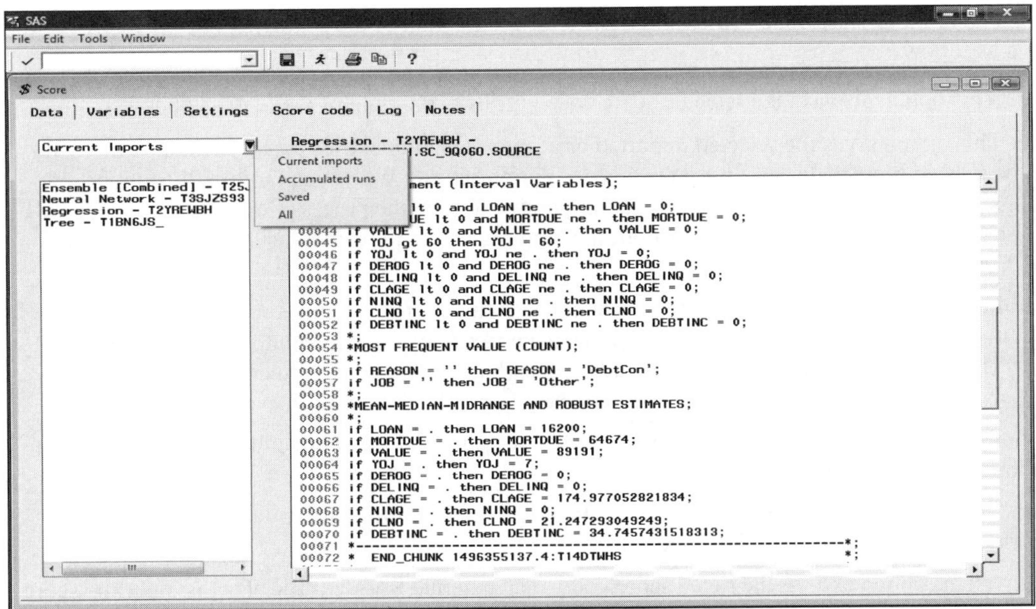

*The **Score code** tab displays the internal SEMMA scored code from the **Regression** modeling node.*

The following options can be specified from the display field that is listed just above the list box that is located to the left of the **Score code** tab. The tab is designed so that the available options specified from the listed entries will be reflected in the **Score** window that is located to the right. Again, the listed score code is basically a SAS program that can be included in any SAS data step.

- **Current imports** (default)**:** This is the default option for the **Score** tab that displays the internal score code within in the **Score** window. The window will display the currently imported score code that is based on the most recent node that has been executed, which is connected to the **Score** node. If there are several modeling nodes connected to the **Score** node, which is illustrated in the previous diagram, then you may select the appropriate modeling node to view the internal SEMMA code that generated the corresponding prediction equations. Select the appropriate predictive model that is displayed in the list box. The corresponding score code will be automatically displayed in the window to the right. The SAS code that is displayed in the **Score** window may be included in an additional SAS data set to produce entirely different prediction estimates from a new set of input values for the predictive or classification model.
- **Accumulated runs:** This option lists scoring code that has been exported by the previously connected modeling nodes from iterative training. If iterative training involves group processing from the **Group Processing** node, then a separate score entry is listed for each group that is created from each previously connected node. This option is the only way to access and view the group processing score code that is generated from the **Group Processing** node.
- **Saved:** This option lists the saved or merged score code entries. Modifying the option settings from the corresponding modeling node and running the node will result in updating the scoring code that is associated with the corresponding model. Therefore, select the score code entry, then right-click the mouse and select the **Save...** pop-up menu item to save the listed score code. A separate **Score code file** dialog window will appear that will allow you to enter a short description of the score code from the **Name** entry field. This will result in the saved score code appearing in the list box with the corresponding entry that you have entered that will be displayed within the listed score code entries. Once the score code is saved, you may then either delete, edit, export, or compile the saved score code, as illustrated in the following diagram.
- **All:** This option lists all the score code entries, that is, all current imports, accumulated runs, and saved score code entries within the list box.

By selecting one or more score code entries and right-clicking the mouse, the following pop-up menu options are available as follows:

- **View:** This option displays the selected score code entries to the right of the **Score** window.
- **Save:** This option saves the selected imported or accumulated run entries to a SAS catalog that will then be a listed score code entry from the **Save...** option setting. When you modify the option settings from the previously connected modeling nodes and run the process flow, the score code will be updated accordingly. Therefore, this option will allow you to prevent the currently listed score code from being modified.
- **Combine:** This option combines multiple accumulated run entries into one single score code entry. From the tab, simply select two or more score code entries to be combined into a single score code entry. The **Score code file** window will then appear, which will allow you to enter an appropriate name for the combined score code that will then be listed in the **Saved** score code entries.
- **Delete:** This option deletes the selected imported or accumulated run score code entry that has been previously saved through the **Save...** option setting.
- **Edit:** This option opens a preview window editing session that will allow you to modify the imported or accumulated run score code entry.
- **Export:** This option exports the saved score code to an external SAS text file. That is, once the score code is saved, you may then select this option to export the saved score code to your operating system. The exported score code can then be used outside of Enterprise Miner in a separate SAS session. In other words, the file can be included into the SAS session through the use of the %INCLUDE statement to calculate new prediction estimates from a new set of input values. In the **Score** node, the data set that you want to score that has been previously selected from the **Data** tab is automatically referenced by the macro variable called &_SCORE and the new data set to be scored with the updated prediction information is automatically referenced by the macro variable &_PREDICT. This is illustrated in the following diagram.
- **Submit code:** This option submits the selected imported or accumulated score code to score the corresponding score data set that is selected from the **Data** tab.

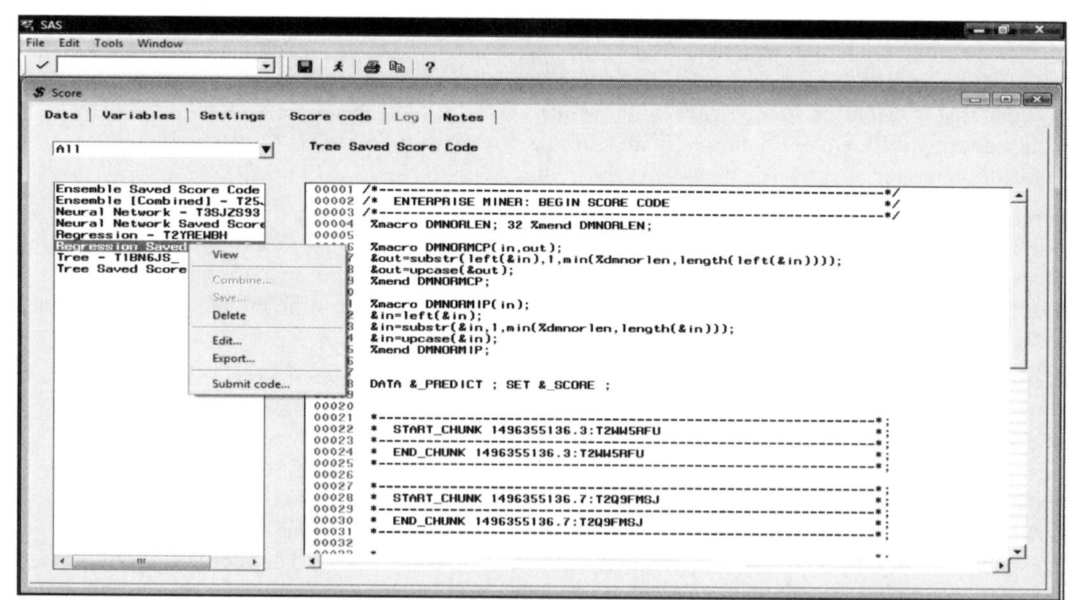

The listed saved scoring code used to specify the source &_SCORE and scored &_PREDICT data sets.

Log tab

The **Log** tab is designed to display the compiled results when observing any compiling errors or warning messages from the compiled score code that has been executed within the node.

Scoring the Data Set within Enterprise Miner

The **Score** node will allow you to score a data set by selecting the best predictive model within the process flow diagram. In the following example, the data set to be scored was created from the **Input Data Source** node. The following steps are performed to create the output scored data set. The process flow diagram used in scoring a data set is illustrated in the following diagram.

Steps in Creating the Scored Data Set

1. Construct a process flow diagram with the appropriate modeling nodes.
2. For each modeling node, select the **Process or Score: Training, Validation, and Test** option from the **Output** tab to create the desired prediction information.
3. Drag and drop the **Score** node on to the process flow diagram.
4. Connect the predictive models to the **Score** node.
5. Drag and drop the **Input Data Source** node on to the process flow diagram to select the data set to be scored. This is illustrated in the following diagram with the **Input Data Source** node that is positioned just above the **Score** node.
6. Read the source data set to create the input data set.
7. From the **Data** tab, change the role of the input data set from RAW to SCORE from the **Role** field, which can be performed within the **Input Data Source** node.
8. Open the **Score** node. From the **Setting** tab, select the **Apply training data score code to score data set** option that will automatically add prediction information, that is, predicted values, residuals, and so on, to the data set that will be scored.
9. From the **Data** tab, open the **Imports Map** window to select the best model to score the training data set that is the temporary home equity home data set.
10. Save the changes, then run the **Score** node.

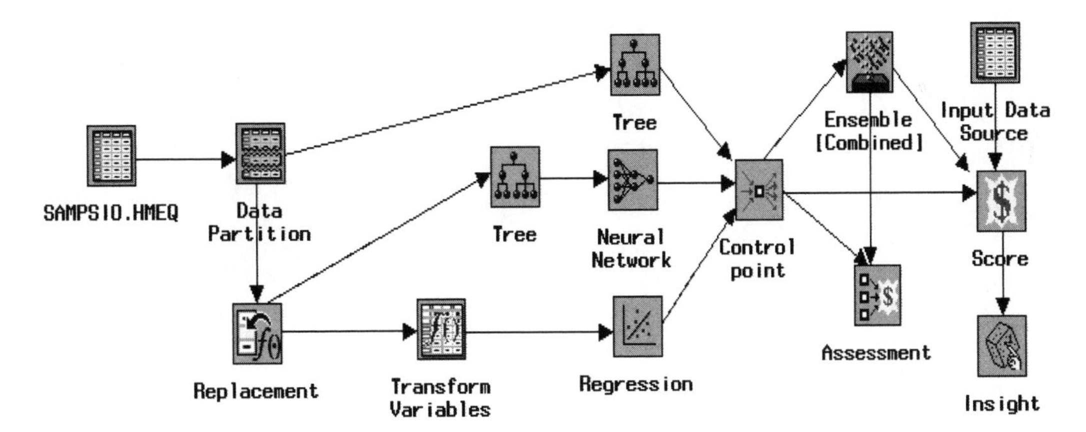

The process flow diagram used in creating a scored data set from the best modeling node.

Chapter 7

Utility Nodes

Chapter Table of Contents

7.1 Group Processing Node

General Layout of the Enterprise Miner Group Processing Node

- **Data tab**
- **General tab**
- **Group Variables tab**
- **Target Variables tab**
- **Weighted Resampling for Bagging tab** *(bagging estimation only)*
- **Unweighted Resampling for Boosting tab** *(boosting estimation only)*
- **Output tab**
- **Notes tab**

The purpose of the **Group Processing** node in Enterprise Miner is designed for you to perform a separate analysis by each class level of some categorical grouping variable, analyze more than one target variable, control the number of times the subsequent nodes will loop in the process flow diagrams that are connected to the node, and perform boosting or bagging by resampling the input data set with replacement.

Group processing is performed within the node, assuming that more than one target variable is in the data set where the node will perform a separate analysis by each target variable in the active training data set. In addition, the node is designed for you to basically perform separate analysis by dividing the active training data set into separate data sets by the class levels of existing categorically-valued grouping variables in the active training data set. For example, you can create several predictive models by the separate class levels of the categorically-valued grouping variable. The variables that are assigned to a model role of **group**, usually from the **Input Data Source** node, will be automatically assigned as the grouping variables for the data mining analysis. In addition, the node is also capable of iteratively processing the incoming data set. By selecting the **Index** processing, then the process of execution of the **Group Processing** node is such that the incoming data set, usually created from the **Input Data Source** node, will iteratively process the same data set in each subsequent pass through the process flow. In other words, the purpose of the **Group Processing** node is to instruct Enterprise Miner to iteratively process the workflow diagram a specified number of times. The node is also capable of performing bagging and boosting resampling techniques. The type of ensemble model that is generated from the **Ensemble** node depends on the resampling methods that are specified from the **Group Processing** node. That is, the bagging or boosting resampling techniques cannot be performed from the **Ensemble** node unless the corresponding resampling techniques are selected within the **Group Processing** node. Finally, the **Group Processing** node can be used to instruct Enterprise Miner to create prediction estimates from each stratified model that can then be combined from the subsequent **Ensemble** node.

Group processing is capable of performing analysis by each separate class level of the categorically-valued grouping variable. Therefore, connecting any one of the modeling nodes to the **Group Processing** node will automatically create separate entries that will be listed in the **Model Manager** for each class level of the target variable. This is illustrated at the end of the section. Furthermore, assuming that multiple target variables are specified from the node, then the **Group Processing** node will create separate model entries between each interval-valued target variable within each class level of the categorical grouping variable.

The general usage of the node is such that one and only one **Group Processing** node can be incorporated in the currently opened process flow. In addition, the node can be connected to any of the nodes in which the **Group Processing** node can accumulate its results, like the modeling nodes or the **Score** node. The following are the various ways to perform group processing within the node.

Group Processing of Each Class Level

The purpose of the **Group Processing** node is to perform the analysis by each class level of a grouping variable. The node can also analyze multiple target variables separately or analyze the same data set repeatedly. In other words, the node is designed to perform predictive modeling by fitting multiple target variables or bootstrap resampling. However, if the multiple target variables are related to each other, for example, weight and height, then SAS recommends that it is better to consolidate the separate target variables into the same predictive model. The **Group Processing** node can be used to train separate models for different target variables or perform predictive modeling for each target group separately. The node can either perform the analysis by grouping variables separately or cross-validation grouping from two separate categorically-valued variables. The node has the option of looping or no looping. Then the grouping process classifies the target variable into a single group. By default, group processing is performed on all the class levels with ten or more observations. The reason is because, in predictive and classification modeling, as a very conservative strategy, assuming that you have a noise-free distribution in the target values, it is recommended that there be at least ten observations for each input variable in the statistical model, and for classification modeling there should be at least ten observations for each class level of the categorically-valued input variable in the statistical model in order to obtain reliable estimates from the statistical model. However, studies have shown that in order to increase the accuracy in the estimates along with obtaining stable estimates with sufficient statistical power in detecting significant effects in the model and better met the various distributional assumptions, there should be at least fifty observations, or even preferably, a hundred observations within each group.

One of the advantages of group processing is that it will allow you to separate the analysis into different groups in order to achieve more accurate results. For example, if you were interested in predicting a person's weight, then it would be to your advantage to divide the data by sex, since men generally weigh more than women or grouping by education levels in predicting personal income, since people with higher education levels tend to make more money. Group processing or stratified modeling is advantageous if there is a large variability with a wide range of values in the target variable between the separate groups along with a strong correlation and functional relationship between the input variables and target variable to predict within each group as opposed to fitting the model to the entire sample. In predictive modeling, group processing is called analysis of covariance. The analysis is a combination of analysis of variance and regression modeling, where the model has both discrete and continuous input variables with an interval-valued target variable that you want to predict. The objective of the analysis is observing the difference in the separate regression lines by creating separate regression models for each grouping input variable that is included in the model to increase the precision in the prediction estimates. One restriction to the analysis is that the slopes of the various regression lines must be the same.

By separating the data into various groups it will allow you to discover trends and patterns in the data between the groups that might not be noticed otherwise. For example, since many of the previous nodes depend on the variables being normally distributed with the squared distance function that is applied in the analysis, therefore, group processing might be considered if the variables might only be normally distributed within each class level of some grouping variable in the data set. Another advantage to using group processing in order to preprocess the data for analysis is that group processing might be performed in order to generate better imputation estimates by filling in missing values across the separate groups that are created.

The hold out method or split sample procedure is most efficient when there is a sufficient amount of data allocated to the validation data set. Conversely, overfitting can occur with a small sample size in the training data set. Overfitting generally means that specifying a model that is too complex will create a model that does fit new data well. Fitting a model to an enormous amount of data eliminates overfitting. However, the data splitting routine is inefficient when the sample size is small. Evaluating the accuracy of the model from a single validation data set can give inaccurate estimates at times. Therefore, the following bootstrap methods are applied. Resampling techniques, such as bootstrapping, should be considered in achieving stable and consistent parameter estimates in predictive models, by building stable predictive models, decision trees or consistent cluster groupings, and so on. These bootstrapping techniques are performed when there is not enough data. The bootstrapping or k-fold cross-validation procedure is performed k different times, each time using a different partitioning of the input data set into a training and validation data set with replacement, that is, placing the selected data point back into the population to be selected again, with the results then averaged by the k number of bootstrap parameter estimates. One of the disadvantages of this allocation procedure is that it requires a large amount of computational time.

Bagging Resampling

Bagging is related to bootstrap sampling. Bootstrap sampling is designed to increase the precision and achieve stability in the estimates. Bagged estimates are calculated by resampling the data set with replacement to calculate the estimates for each bootstrap sample of equal size, then dividing by the number of bootstrap samples. These methods are a remedy to overfitting in model fitting that will lead to improved predictions or classifications if there exists instability in the parameter estimates. When fitting the model to a small subset of the same data, it will result in entirely different estimates. For highly unstable models, such as decision trees, complex neural networks, and nearest neighbor models, bagging will often decrease the modeling error due to averaging the estimates, thereby reducing the variance with the bias left unchanged. Conversely, for simple models, the variance may not decrease since the variability is added and accumulating during the iterative bootstrapping process. One drawback to bagging is that, given a bad prediction model or a bad classification model, bagging can calculate estimates that are even worse. Although Enterprise Miner performs bootstrap resampling, keep in mind that there is no Enterprise Miner node that computes bias corrections, standard errors or confidence intervals based on the bootstrapping estimates.

Boosting Resampling

In Enterprise Miner, boosting adaptively reweights each observation from the training data set to predict the categorically-valued target variable. Initially, all the weighted observations are set to the sample proportion based on the sample size of the training data set. The algorithm then modifies each weighted observation at each successive iteration, then refits the subsequent model by weighting observations that have misclassified the target class level in the previous modeling fit. In other words, the observations are modified by increasing the weight estimates for each observation that has been misclassified from the previous fit. From the **General** tab, the **Number of Loops** entry field will allow you to specify the number of loops and classification models to iteratively fit, in which the weights adjust the estimated probabilities at each successive iteration.

In Enterprise Miner v4.3, boosting is performed from the DMBOOST0, DMBOOST1, DMBOOST2 SAS macros that are located in the !sasroot\SAS 9.1\dmine\sasmacro system folder. The DMBOOST0 is the original macro that gets called to create a data set that corresponds to the training set with an additional _BOOSTWT variable. If no frequency variable is found, then the _ BOOSTWT variable is set to 1. It also creates global macros used by the **Ensemble** node, that is, _SUMFREQ variable that is the sum of the _BOOSTWT variable and the DMBOODIV variable is set to 1. For subsequent iterations, the node reads the scored data set and recalculates the weight variable _BOOSTWT based on the number of misclassifications from the DMBOOST2 macro. Enterprise Miner uses the formula suggested in "Arcing Classifiers" by Leo Breiman in the calculation of the weight variable _BOOSTWT. At the end of each iteration, Enterprise Miner calculates the _RESAMP_ variable as _BOOSTWT / (sum of the _BOOSTWT) using the DMBOOST1 macro. The _RESAMP_ is then used to multiply the estimated probability of each target class level from the classification model in order to generate the estimated probabilities to the boosting model.

Iterative Processing

The node is not designed to run within the node or the **Diagram Workspace**. For example, in order to perform group processing for predictive modeling, you must execute the node from a subsequent modeling node that is connected to the **Group Processing** node. This is illustrated in the following diagram.

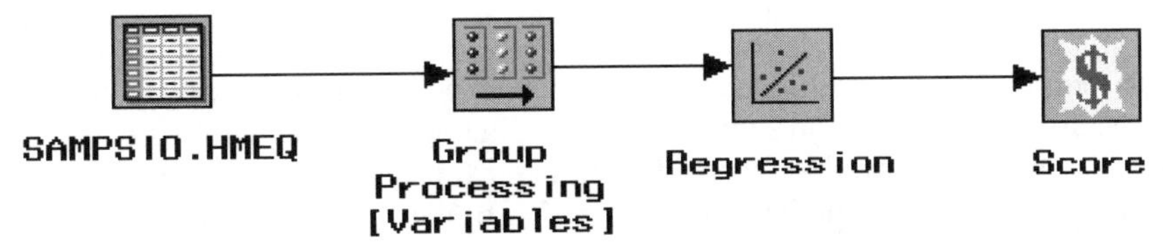

*Group processing of good or bad creditors from the HMEQ data set with two separate models created for assessment from the **Model Manager**, with the **Score** node used to view the score code for each group.*

It is important to point out that the node will not automatically perform looping or iterative processing unless specified otherwise. From the process flow diagram, simply select the **Group Processing** node, then right-click the mouse to select the **Looping-yes** pop-up menu item in order for the node to perform iterative processing by resampling the training data set any number of times.

The following process flow diagram calculates the neural network estimates by performing bootstrap sampling without replacement. The purpose of the **SAS Code** node is that the node can be used to control the type of sampling that is performed for the input data set. In the following example, the node was applied to the process flow to select a random sample of the input data set without replacement for each iterative sample that is created from group processing in calculating bootstrap estimates. The difference between bagging estimates and the corresponding resampling techniques that can be specified from the **Group Processing** node is that these resampling techniques, like bootstrap sampling, are based on sampling with replacement. The following illustration displays the process flow in which the **SAS Code** node is placed next to the **Group Processing** node. The purpose of the **SAS Code** node in the following process flow is to perform a random sample without replacement from the input data set that is iteratively fed through the process flow at each subsequent pass. From the **Group Processing** node, specify the **Index** group process mode within the **Looping Information** subtab. The flow of execution will result in the selection of unique observations from the input data set that is the result of the SAS programming code that is applied within the **SAS Code** node that is then passed on to the **Neural Network** node in fitting the neural network model. From the **Neural Network** node, open the **Imports Map** window within the **Data** tab to select the exported training data set that was created from the **SAS Code** node. The **Ensemble** node was included into the process flow to combine and average all the neural network models.

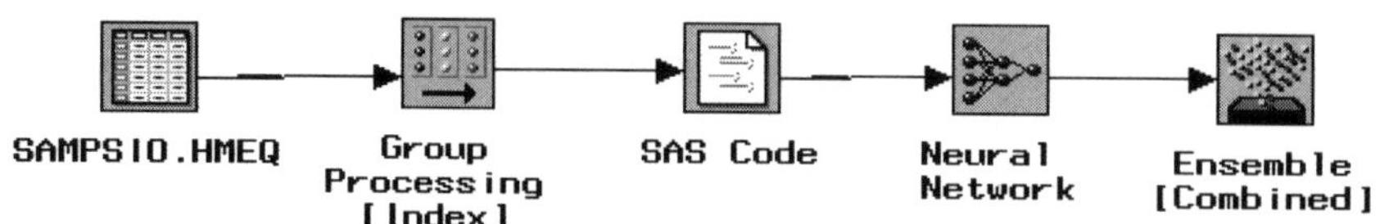

Group processing that performs regression modeling from bootstrap sampling without replacement.

The various resampling techniques are discussed in greater detail in Lunneborg (2000), Efron and Tibshirani (1993), Hastie, Tibshirani, and Friedman (2001), Breiman (1998), and Suess and Trumbo (2007).

Data tab

The **Data** tab is designed to display the name and a short description of the active training data set. The **Test** radio button is dimmed and unavailable since there are no previously created test data sets in the process flow diagram. If there exist partitioned training, validation and test data sets, then the node will perform group processing on each data set. However, the partitioned data sets must be compatible. That is, there must exist identical group variables in each partitioned data set. By default, group processing is performed to each class level with ten or more observations. Therefore, all class levels with less than ten observations will prevent the node from generating a model for the corresponding class level. Note that if the previously created validation, test, or score data sets have no observations in any one of the subgroups, then the **Group Processing** node will not export the data set to the subsequent nodes in the process flow diagram.

Press the **Properties...** button to view both the file administrative information and a table view of the selected active training data set. The **Select...** button will allow you to open the **Imports Map** window in order to select any other training data set that is usually created from the **Input Data Source** node, based on all the nodes that are connected to the **Group Processing** node within the currently opened Enterprise Miner project. The **Data** tab will not be displayed since the tab is similar to the previously displayed windows.

General tab

The **Looping Information** subtab will automatically appear as you open the **Group Processing** node. The **General** tab is designed to specify the configuration settings of the looping process in determining if the modeling results from group processing are saved to the **Model Manager** by controlling the calculation of the assessment statistics from the group processing models. The tab consists of the following two subtabs.

- **Looping Information subtab**
- **Models subtab**

Looping Information subtab

The **Looping Information tab** subtab is designed to display the number of target variables, number of class levels for group processing, and the number of loops. The subtab will allow you to specify the mode of looping, that is, whether the node loops over each class level of the group variables, the number of times group processing is looped through the subsequent nodes that are connected, the number of times the classification model is iteratively fit in the boosting technique, or the number of times the incoming data is resampled in the iterative bootstrapping or bagging technique.

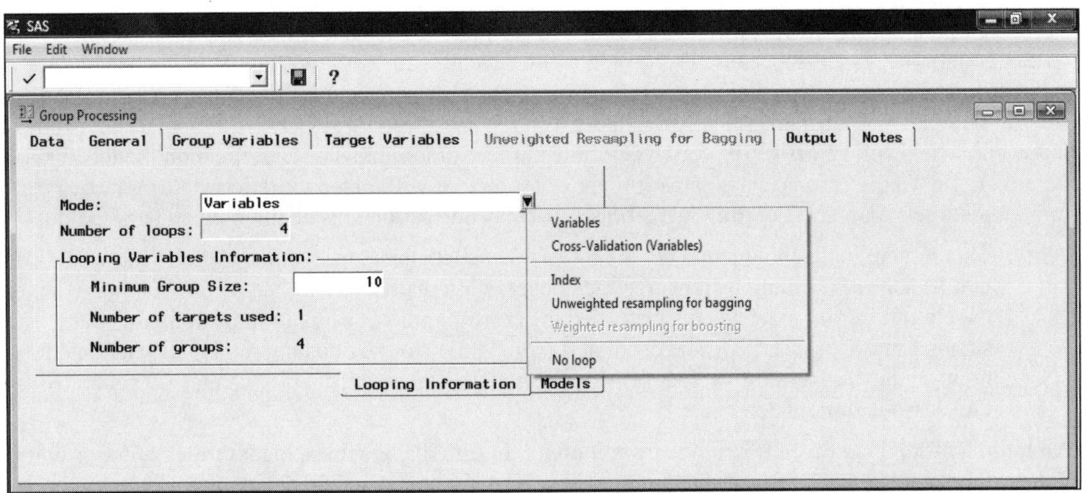

*The **General** tab is used to configure the type of group processing for the analysis within the process flow.*

From the **Mode** option, the following are the various modes of looping that are available within the subtab. Simply select the drop-down arrow from the **Mode** display field in order to select from the following options:

- **Variables** (default)**:** This option performs standard group processing. In other words, the node will automatically loop through each class level of the grouping variables in the active training data set. The **Variables** option is also designed to perform group processing by every possible combination of class levels of the grouping variables, that is, all variables set to a variable role of **group** from the **Input Data Source** node. Selecting this option will result in the availability of the **Looping Variables Information** section. The section is designed for you to specify the minimum number of observations to process within each class level of the grouping variables.
- **Cross-Validation (Variables):** This option is designed to perform group processing based on the complement of all possible combination of class levels of the grouping variables. That is, the node will automatically loop through each class level based on the complement of the grouping variables. For example, the first loop would result in performing analysis based on all records not in Group 1, the second loop would result in performing analysis based on all records not in Group 2, and so on.
- **Index:** This option will allow you to specify the number of times to repeat the group processing procedure. This option is designed to repeatedly process the same data set a specified number of times. Selecting this option will result in the availability of the **Number of loops** entry field in order for you to enter the number of times that the group processing procedure iterates. The default is 10 iterations. For example, this option might be considered in ensemble modeling to control the number of separate predictive models that are iteratively created in which all the predicted values from the entire input data set are combined to create the final predictive model. However, in order to implement the iterative looping routine to the subsequent connected nodes, you must select the **Loop-yes** pop-up menu option item from the process flow diagram. An asterisk will appear next to the menu item that will indicate to you that it is the active looping process. After the process flow is executed, then the **Group Processing** node will automatically revert back to the default setting of no looping, that is, **Loop-no**.
- **Unweighted resampling for bagging:** This option performs bagging bootstrap sampling. Bagging performs random sampling of the active training data set with replacement. In other words, bagging takes repeated random samples of the active training data set in which each observation has an equal chance of being selected from each random sample. Selecting this option will result in the **Unweighted Resampling for Bagging** tab available for viewing in order to specify additional options for the design. The **Number of loops** entry field will become available in order to specify the number of resamples to perform for the active training data set. However, the **Looping Variables Information** section will become inactive since resampling technique does not have anything to do with the class levels of the target variables. That is, the value entered controls the number of times the training data set is sampled. The default is 10 iterations. These *n* samples are then used in calculating the predictive or classification modeling estimates. In order to perform the resampling techniques, you must set the pop-up menu item to **Looping-yes** from the **Diagram Workspace**.

- **Weighted resampling for boosting:** This option is designed to increase the classification performance of the model. The reason is because this option applies weights to the fitted sample by increasing the weights for each observation that was misclassified in the previous sample. The **Weighted Resampling for Boosting** tab will be automatically available for viewing once you select this option. The **Number of loops** entry field will be active for you to enter the number of iterations or classification models to iteratively fit. This option is unavailable if there exists an interval-valued target variable in the active training data set. Also, the **Looping Variables Information** section will be dimmed and grayed-out.

  Note: This resampling technique will create a variable called _resamp_ in the _BOOST_ scored data set that lists the resampling weights for each observation that is determined by the number of times the classification model has misclassified the corresponding observation from the separate samples that are generated. The estimated probabilities to the boosting model are calculated by multiplying the _resamp_ values or weights by the estimated probabilities that are generated from the classification model.

- **No loop:** This option prevents the node from looping. In Enterprise Miner, the **Number of loops** entry field will be automatically grayed-out and unavailable, with the entry field set to one. Alternatively, you may prevent the node from looping by selecting the node from the **Diagram Workspace** and right-clicking the mouse, then selecting the **Loop-no** pop-up menu option.

The following are the various options within the **Looping Variables Information** section that are made available by performing group processing based on the class levels of the categorical-valued grouping variable:

- **Minimum Group Size:** This option will allow you to specify the minimum number of cases that the node will export based on any one class level from the categorically-valued grouping variable. By default, the **Group Processing** node will not process class levels of less than 10 observations. This option will allow you to control the minimum number of cases within each class level that the node will process based on the value entered from the **Minimum Group Size** entry field.
- **Number of targets used:** This option displays the number of target variables used for group processing that are usually specified from either the **Input Data Source** node or the **Data Set Attribute** node.
- **Number of groups:** This option displays the total number of class levels for group processing that is determined by the grouping variables that can be specified from the subsequent **Group Variables** tab.

Model subtab

The **Models** subtab controls the way in which the modeling results are saved to the **Model Manager** based on each loop that is performed by the **Group Processing** node. By default, all the modeling results from group processing are saved to the **Model Manager** with the various assessment statistics that are calculated from the separate group processing models.

Clear the **Save all models to Model Manager** check box to prevent the modeling results from being saved to the **Model Manager**. Conversely, clear the **Assess all models** check box to prevent the assessment statistics from being calculated from the group processing models.

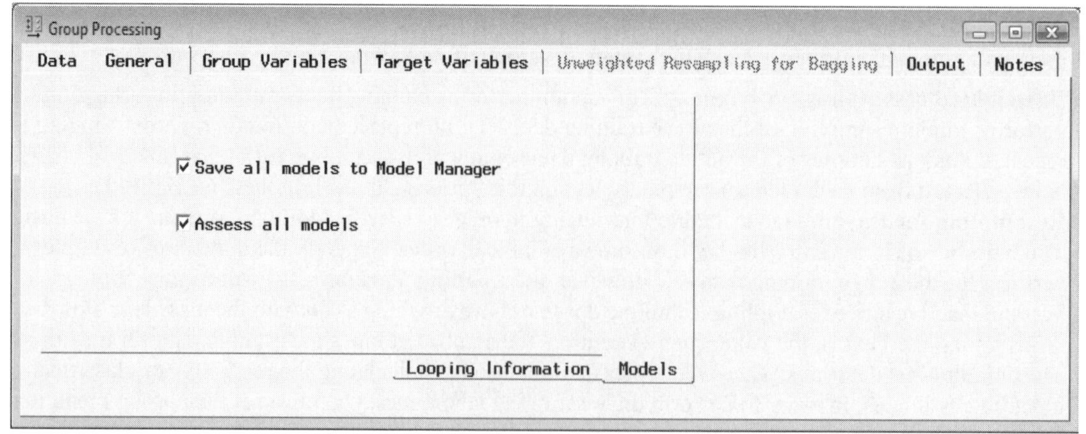

*The **Models** subtab is used to prevent the results from group processing being saved to the **Model Manager**.*

Grouping Variables tab

The **Grouping Variables** tab is designed for you to specify the categorically-valued input variables and the corresponding class levels that are used in group processing by splitting the active training data set into separate groups. The grouping variable must either have a model role of **input** or **group**. By default, all variables assigned a **group** model role are set to a variable status of **use**. Therefore, these variables are automatically assigned as the grouping variables for the data mining analysis. However, all variables assigned a role model of **input** are excluded from group processing with a variable status set to **don't use**. To select the grouping variable for the analysis, select the **Status** cell of the input variable, then right-click the mouse and select the **Set Status** pop-up menu item to set the variable attribute status of the categorically-valued input variable to **use**. Alternatively, you may remove all variables assigned a **group** model role from the analysis by setting the variable status to **don't use**. From the tab, you may view the frequency distribution of all the listed variables. If the variable displays a bimodal distribution, that is, two peaks, then it might suggest that the variable might be a good grouping variable for the training data that indicates a normal distribution between two separate populations. Both the **Grouping Variables** tab and the subsequent **Target Variables** tab are available for viewing assuming that both the first two **Variables** and **Cross-Validation** group processing options are selected from the **General** tab.

Selecting the Group Processing Levels

By default, the node will automatically display the various class levels or discrete values of the selected group variable within the **Selected Values** column. The node automatically performs group processing for all the class levels that are listed in the **Selected Values** column. If the group variable has several class levels, then you might want to perform the analysis on a specific number of class levels. Therefore, from the **Selected Values** column, right-click the mouse and select the **Set Values** pop-up menu item to select the appropriate class levels. The **Group Variable Level Selector** window will appear, which displays a list box of the values that are listed within the **Selected Values** column. Simply press the Ctrl key or drag the mouse over any number of values to select. The analysis will then be subdivided into separate groups that are identified by each value that is selected. Press the **OK** button to return to the **Grouping Variables** tab. The **Group Variable Level Selector** window will not be display.

The **Number of Groups** display field that is located in the upper left-hand corner of the tab will display the total number of class levels that the **Group Processing** node will process. The total number of groups is based on the total number of class levels of the various group variables with a variable status set to **use** and the corresponding class levels that are displayed within the **Selected Values** column. For example, the two separate class levels will be processed from the binary-valued grouping variable, BAD.

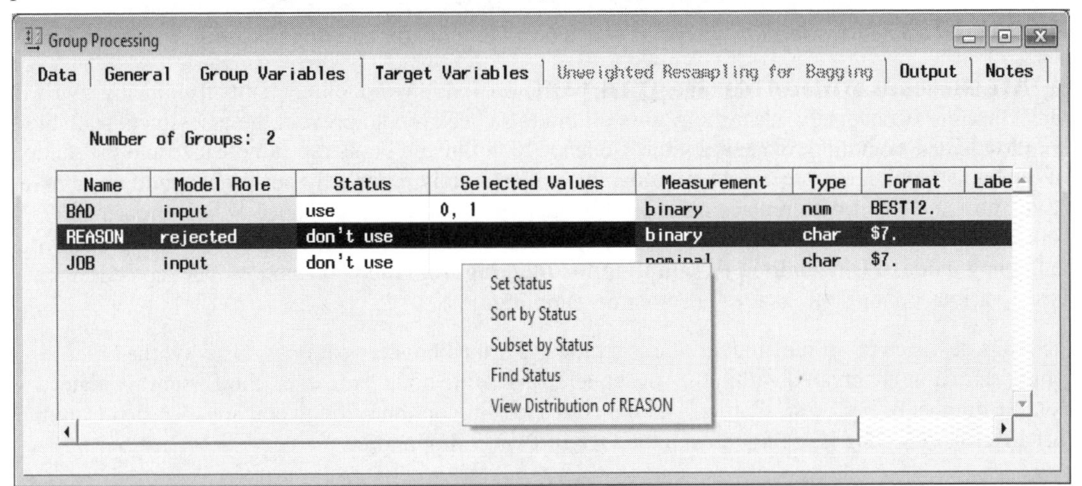

*The **Group Variables** tab is used to specify the group variables to create separate models for each group.*

Target Variables tab

The **Target Variables** tab is designed to display the target variables in the active training data set. By default, every target variable that is previously specified from either the **Input Data Source** node or the **Data Set Attribute** node is automatically included in group processing. The **Group Processing** node is designed to analyze each target variable separately, assuming that the active training data set consists of more than one target variable. In predictive modeling, the node is designed to create separate regression models by each separate level of the categorically-valued grouping variable. However, the tab will allow you to remove certain target variables from the analysis in order to reduce processing time. To remove the target variable from the analysis, select the **Status** cell, then right-click the mouse and select the **Set Status** pop-up menu item to select the **don't use** pop-up menu item. The tab will also allow you to view the frequency distribution of the listed target variables from the metadata sample. Highlight any one of the target variables, then right-click the mouse and select **View Distribution of <target variable>** that will automatically display the standard frequency chart. Again, the **Target Variables** tab will become available for viewing along with the **Grouping Variables** tab, assuming that both the first two **Variables** and **Cross-Validation** group processing options are specified from the **General** tab.

Clear the **Loop over target** check box to prevent group processing from automatically looping over the target variable with a variable status of **use**. However, looping will not be performed unless you specify looping from the pop-up menu item within the **Diagram Workspace**.

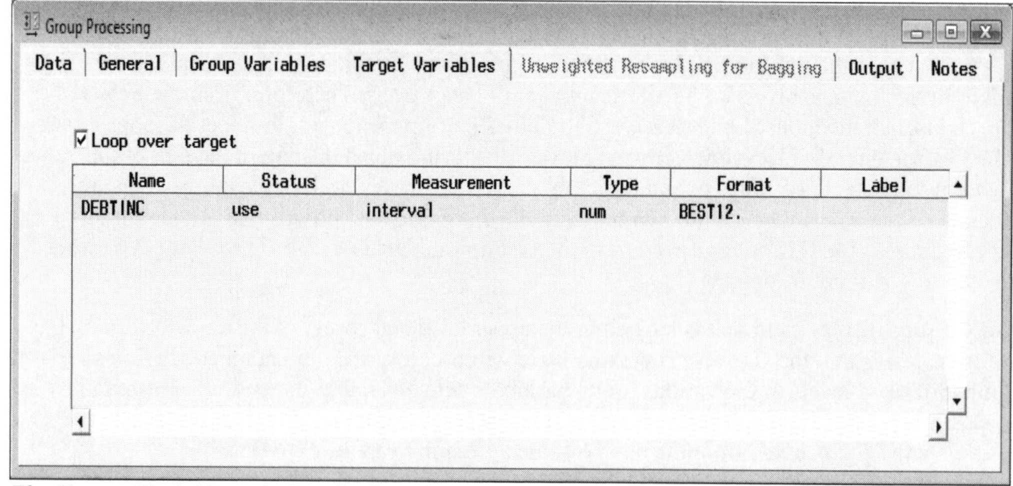

*The **Target Variables** tab that displays the target variables in the active training data set.*

Unweighted Resampling for Bagging tab

The **Unweighted Resampling for Bagging** tab is designed to define the bootstrap sample size and the random sample of the bootstrapping sample. In other words, this option simply repeatedly performs a random sample of the active training data set with replacement. That is, each observation has a chance of being repeatedly drawn from the iterative bootstrap sample. The tab will automatically become available for viewing once you select the looping mode of **Unweighted Resampling for Bagging** from the **General** tab. The tab is designed to specify the various option settings for the bootstrap sampling.

Again, the tab is designed to set the number of observations for the bootstrap samples. Usually, the bootstrapping estimates are created within the subsequent **Ensemble** node. However, the **Ensemble** node depends on the number of bootstrap samples and the corresponding option settings that are specified from the **Unweighted Resampling for Bagging** tab and the **Group Processing** node.

From the **Variables** section, the **Target** display field will be automatically grayed-out since unweighted resampling for bagging does not have anything to do with adjusting the probability estimates of the categorically-valued target variable. However, the **Frequency** display field will display the unique frequency variable in the input data set that cannot be changed. The value of the frequency variable will indicate how many times each record occurs, each observation is assumed to occur n times, where n is the value of the frequency variable for each observation. Assuming that a frequency variable exists, then the way in which the active training data set is sampled is based on the frequency variable to determine the percentage of the observations that are sampled instead of the number of observations that are listed in the data set. The frequency variable that determines the sampling must be assigned a variable role of **freq** from either the **Input Data Source** node or the **Data Attributes** node.

From the **Sample Size** section, the tab will allow you to define the size of each bootstrap sample of equal size from the input data set. Either select the **Percentages** or the **Number** radio button to enter the percentage of observations or the number of observations for each bootstrap sample.

Press the **Generate New Seed** button that determines the random sample. It is important to point out that the **Group Processing** node can generate a different random seed generator in order to draw an entirely different bootstrap sample of the input data set. However, the node will automatically save the seed values that were used to generate your bootstrap samples so you can replicate the same bootstrap samples in an entirely different run of the process flow. Specifying a random seed value of zero will automatically generate a different bootstrap sample each time. The reason is because the computer clock creates a random seed at run time to initialize the seed stream that will result in different bootstrap samples when you rerun the process flow.

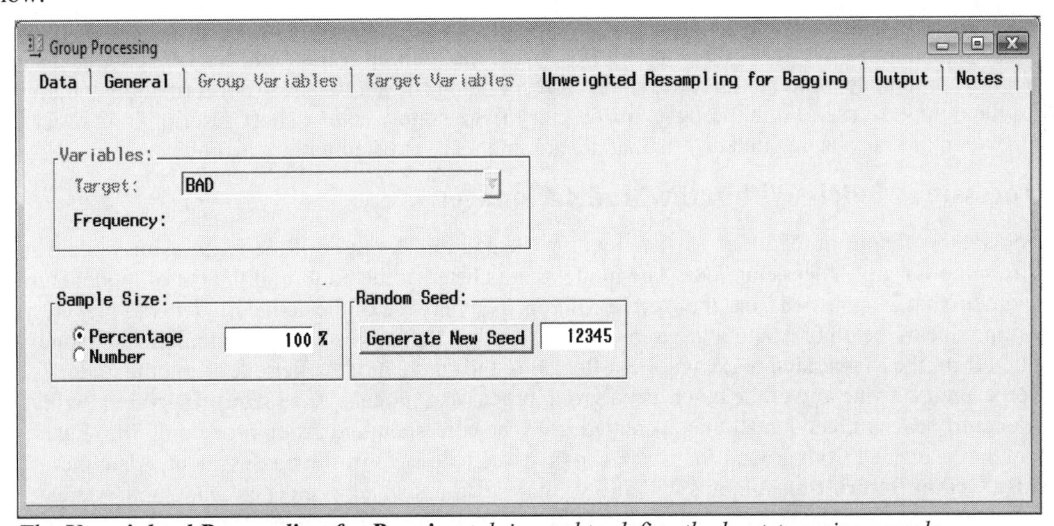

*The **Unweighted Resampling for Bagging** tab is used to define the bootstrapping sample.*

Weighted Resampling for Boosting tab

The **Weighted Resampling for Boosting** tab is designed to configure the boosting options. The following boosting options will allow you to select the categorically-valued target variable to apply the weight estimates to their class probabilities. The weighted resampling is designed to automatically apply weight estimates to the observations in the active training data set. The iterative reweighting algorithm increases the weight estimates for each observation that is misclassified and decreases the weights for each observation that is correctly classified in the previous sample. This tab will automatically become active once you select the **Weighted Resampling for Boosting** looping mode from the **General** tab.

Again, the tab is designed to set the target variable to the boosting model. The boosting model is created within the subsequent **Ensemble** node. However, the **Ensemble** node relies on the various option settings that are specified from the **Weighted Resampling for Boosting** tab and the **Group Processing** node.

From the **Variables** section, the tab will allow you to select the target variable in which its class probabilities will be adjusted. If a frequency variable has been defined in the previous nodes, usually from the **Input Data Source** or the **Data Set Attributes** node, then the frequency variable will be displayed within the **Frequency** display field. The **Samples Size** and **Random Seed** sections will be grayed-out since this option does not have anything to do with sampling the active training data set.

*The **Weighted Resampling for Boosting** tab is used to specify the categorically-valued target variable in which the weight estimates adjust the estimated probabilities at each class level.*

Output tab

The **Output** tab will allow you to view the output data set from the training, validation, test, or score data sets. Simply select the appropriate radio button to view the corresponding file administrative information and the table view of the output data set. From the **Description** entry field, simply enter a short description in order to distinguish between the various output data sets that are automatically created within the node.

Group Processing Models Within the Model Manager

The following are the various models that are listed within the **Model Manager** of the **Regression** node that are created from the **Group Processing** node. The model that is listed at the bottom of the list of model entries is the active model that is identified from the **Active** column. The purpose of the active model is to prevent you from overwriting one of the other existing group processing models that are listed. For example, all changes that are applied from the **Regression** node, will be reflected in the active model after executing the node. However, you cannot activate any of the other listed group processing models. The **Group ID** column will allow you to identify each trained model that is created from the corresponding group processing run. Each trained model that is created from group processing can be passed along for further assessment. Also, the **Target** and **By Group Description** columns will allow you to distinguish between each model that is created within each target group from the **Group Processing** node. To view the various assessment plots, select any one of the group processing models that are listed, then select the corresponding performance charts from the **Tools** main menu options. Both the **Name** and **Description** cells will allow you to either enter a short name or a short description to distinguish between the various group processing models that are created.

In the following example, there were four separate predictive models that were created from the binary-valued grouping variable BAD, that is, defaulting on a home loan, with two separate interval-valued target variables to predict VALUE, (property value) and YOJ (number of years at the present job).

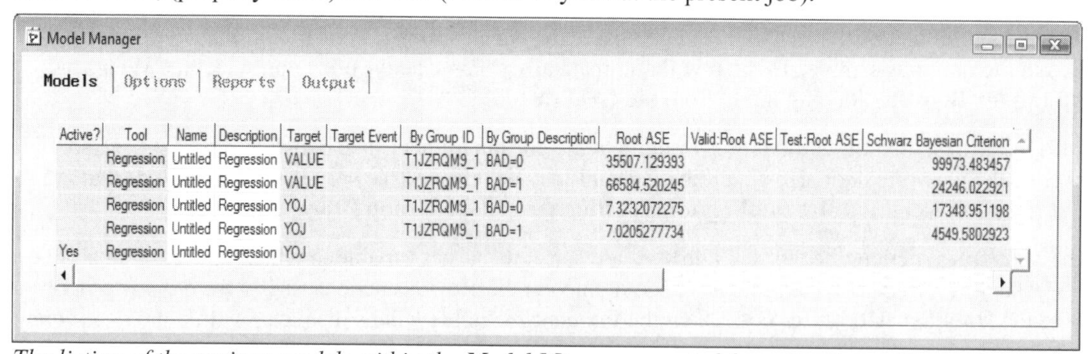

Active?	Tool	Name	Description	Target	Target Event	By Group ID	By Group Description	Root ASE	Valid:Root ASE	Test:Root ASE	Schwarz Bayesian Criterion
	Regression	Untitled	Regression	VALUE		T1JZRQM9_1	BAD=0	35507.129393			99973.483457
	Regression	Untitled	Regression	VALUE		T1JZRQM9_1	BAD=1	66584.520245			24246.022921
	Regression	Untitled	Regression	YOJ		T1JZRQM9_1	BAD=0	7.3232072275			17348.951198
	Regression	Untitled	Regression	YOJ		T1JZRQM9_1	BAD=1	7.0205277734			4549.5802923
Yes	Regression	Untitled	Regression	YOJ							

*The listing of the various models within the **Model Manager** created from the **Group Processing** node.*

7.2 Data Mining Database Node

General Layout of the Enterprise Miner Data Mining Database Node

- **Data tab**
- **Variables tab**
- **Output tab**
- **Notes tab**

The purpose of the **Data Mining Database** node in Enterprise Miner is designed to create the data mining data set that is a snapshot of the original input data set. The node is designed to create a data mining database (DMDB) for batch processing. The role of the DMDB procedure is that it will also compile and compute the metadata information about the input data set from the variable roles. The metadata information is then stored in a metadata catalog. The metadata database and catalog are both designed to optimize data mining batch processing and database storage. The metadata stores both the data set information as well as the statistical information for variables associated with the assigned roles. For categorically-valued variables, the catalog of the metadata information stores important information such as the class level values, class level frequencies, number of missing values for each class level, ordering information, and the target profile information. For interval-valued variables, the metadata will store the range of values of the variables, number of missing values, moments of the variables, and the target profile information that are then used in many of the other Enterprise Miner nodes that require a DMDB data set. As opposed to the previous tabs that list the various descriptive statistics from the metadata sample, the **Data Mining Database** node will allow you to view the distribution of the variables and the corresponding statistics based on all the observations from the source data set.

The DMDB data mining data set is designed to optimize the performance of the **Variable Selection, Regression**, **Neural Network**, and the **Tree** nodes. Therefore, Enterprise Miner automatically creates DMDB data sets for these nodes. Since the previously listed modeling nodes are based on supervised training with a target variable in the database layout, in order for the **Data Mining Database** node to execute there must be a variable in the training data set with a **target** variable role. The DMDB is designed to reduce the number of iterations that the analytical engine needs to make by creating summary statistics for each interval-valued variable and factor-level information for each categorically-valued variable that is stored in the metadata catalog.

The purpose of the **Data Mining Database** node is to create the data mining database for batch processing. The DMDB database is automatically created in nonbatch processing. The node depends on an existing DMDB data set that is usually created from the **Input Data Source** node. The node is designed to automatically create a permanent DMDB data set from the input data set that is generated from the **Input Data Source** node where you must specify the corresponding model roles for the variables in creating a DMDB database or the metadata sample for the Enterprise Miner process flow diagram. For instance, you may want to create a DMDB database with the **Data Mining Database** node from a data set that is created within the **SAS Code** node to gain the visual capacity to view the DMDB database in the process flow diagram.

Managing a Large Number of Class Levels

By default, the maximum number of numeric class levels of the categorically-valued variables in the metadata sample is 128. However, through the use of the DM_MAXLEVEL macro variable, you may specify the maximum number of class levels that the modeling nodes can recognize. In addition, you may specify how many variables can be processed in a single pass of the metadata sample from the DM_METAVARS macro variable. As a review, Enterprise Miner 4.3 creates the DMDB data set from the input data with all the values of the variables stored in memory. Therefore, if you have a large number of numeric class levels for the categorical variables, then the procedure that is used to create the DMDB could easily run out of memory or result in lengthy execution runs. In Enterprise Miner 4.3, the DMDB node uses a new macro variable called DM_MAX_TRAIN_LEVELS to control the number of numeric class levels in the categorical variables of up to 512 class levels. If the DMDB comes across a categorical variable with a tremendous number of class levels, then it compares the number of levels in a variable to the value stored in the macro variable DM_MAX_TRAIN_LEVELS. If the number of class levels of a variable exceeds the threshold value, then it will result in an error and will halt the execution of the process flow. Therefore, you may increase or decrease the maximum number of class levels from the DM_MAX_TRAIN_LEVELS macro variable at any time. In the previous modeling nodes, the PROC DMDB procedure is automatically generated behind-the-scenes in order to check the number of class levels of the target variable that you want to predict in the subsequent model.

Data tab

The tab is designed to display the active input data set that is used in creating the DMDB data mining data set. There are only two roles that can be used to build the DMDB data set: the RAW and TRAIN roles. Select the **Properties...** button to browse the file administrative information and the table view of the active input data set. Press the **Select...** button that will open the **Imports Map** window. The **Imports Map** window will allow you to select an entirely different input data set that will be used to build the DMDB data set.

Variables tab

The **Variables** tab is displayed as you first open the **Data Mining Database** node. The **Variables** tab is designed to view the variables that are a part of the DMDB data set. The tab displays the variable name, variable attribute status, model role, measurement levels, variable type, format, and variable label that is assigned from the metadata sample. In addition, the tab will allow you to view both a frequency table and frequency bar chart based on all observations from the metadata sample. You may remove certain variables from the DMDB data set by selecting the corresponding variable row, then scrolling over to the **Keep** status column and right-clicking the mouse to select the pop-up menu option of **Set Keep** and the subsequent pop-up menu item of **No**. This is illustrated in the following diagram. By default, all the variables are included in the DMDB data mining data set. However, several variables may be changed simultaneously by selecting either the Shift or Ctrl key and selecting the corresponding variable rows.

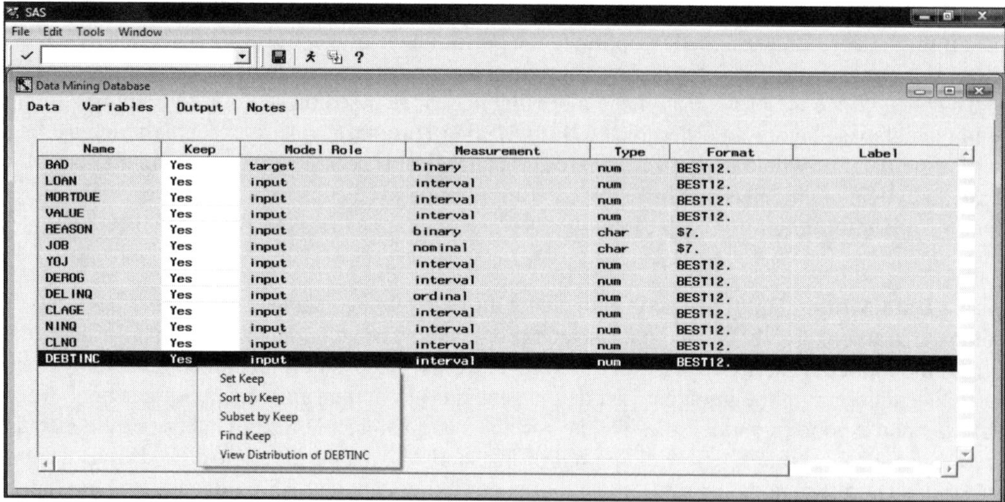

*The **Variables** tab is used to remove variables from the DMDB data mining data set.*

Output tab

By default, the following **Output** tab displays the corresponding library and the name of the DMDB data set that is automatically created within the node. The node must first be executed in order for the following tab to display the corresponding library and name of the DMDB data mining data set that is created. The DMDB data set will be stored in the EMDATA system folder of the currently opened Enterprise Miner project. From the **Description** entry field, you may modify the description of the data mining data set. Select the **Properties...** button to view both the file information and the table view of the DMDB data mining data set.

Notes tab

The **Notes** tab can be used to write important notes about the DMDB data set that is created.

Viewing the Data Mining Database Results

Table View: <DMDB data set name> tab

The tab is designed to display a table view of the DMDB data mining data set that is created by executing the **Data Mining Database** node.

Interval Variables tab

The **Interval Variables** tab is designed to list the various interval-valued variables in the DMDB data mining data set. The tab will also display the various descriptive statistics for each interval-valued variable in the data mining data set. The tab will provide you with a table listing of the various descriptive statistics that are based on all the observations from the input data set.

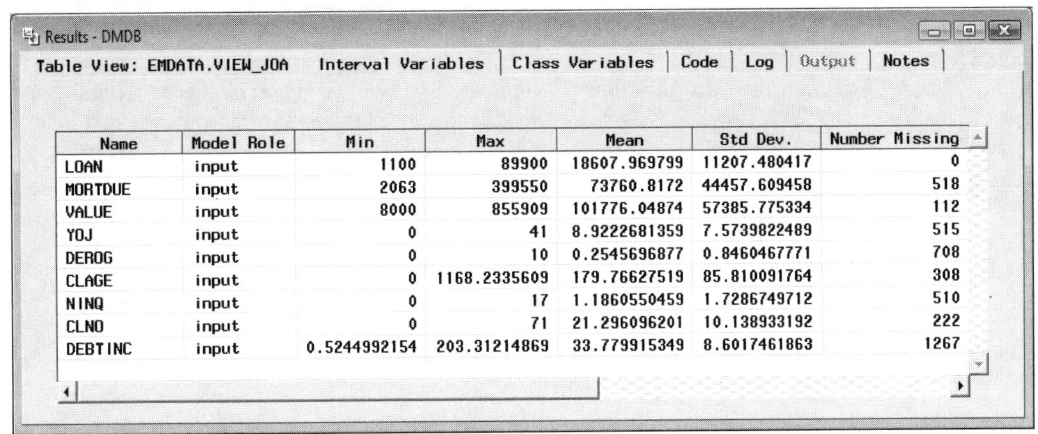

Name	Model Role	Min	Max	Mean	Std Dev.	Number Missing
LOAN	input	1100	89900	18607.969799	11207.480417	0
MORTDUE	input	2063	399550	73760.8172	44457.609458	518
VALUE	input	8000	855909	101776.04874	57385.775334	112
YOJ	input	0	41	8.9222681359	7.5739822489	515
DEROG	input	0	10	0.2545696877	0.8460467771	708
CLAGE	input	0	1168.2335609	179.76627519	85.810091764	308
NINQ	input	0	17	1.1860550459	1.7286749712	510
CLNO	input	0	71	21.296096201	10.138933192	222
DEBTINC	input	0.5244992154	203.31214869	33.779915349	8.6017461863	1267

*The **Interval Variables** tab is used to view the descriptive statistics for the DMDB data set.*

Class Variables tab

The **Class Variables** tab is designed to list the various categorically-valued variables in the DMDB data mining data set. The tab will also display the variable name, model role, number of class levels, measurement level and associated formats for the categorically-valued variables in the data mining data set. The tab will allow you to view both a frequency bar chart and a table listing of frequency counts for each one of the listed categorically-valued variables. Selecting the **View Distribution...** pop-up menu item will allow you to open the **Distribution of <variable name>** window that will display a frequency bar chart from the input data set. As opposed to the **Variables** tabs in the previous nodes, the frequency distribution will be based on all the observations from the input data set.

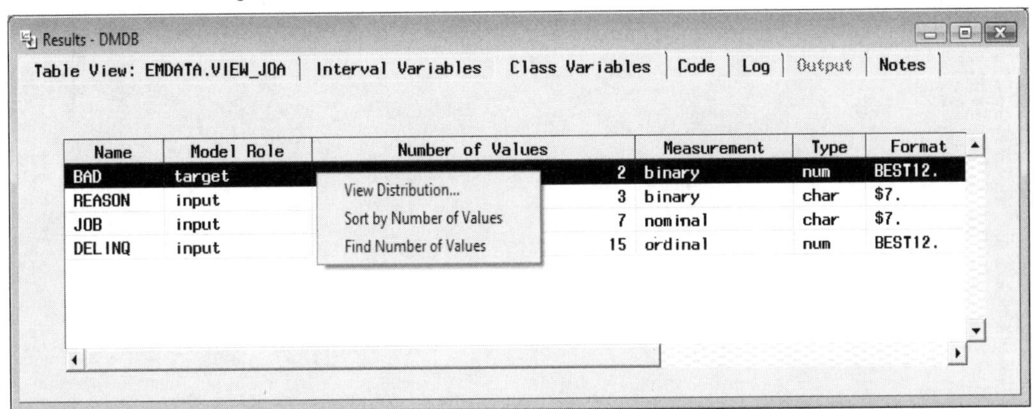

Name	Model Role	Number of Values	Measurement	Type	Format
BAD	target	2	binary	num	BEST12.
REASON	input	3	binary	char	$7.
JOB	input	7	nominal	char	$7.
DELINQ	input	15	ordinal	num	BEST12.

*The **Class Variables** tab is used to view the categorically-valued variables in the DMDB data set.*

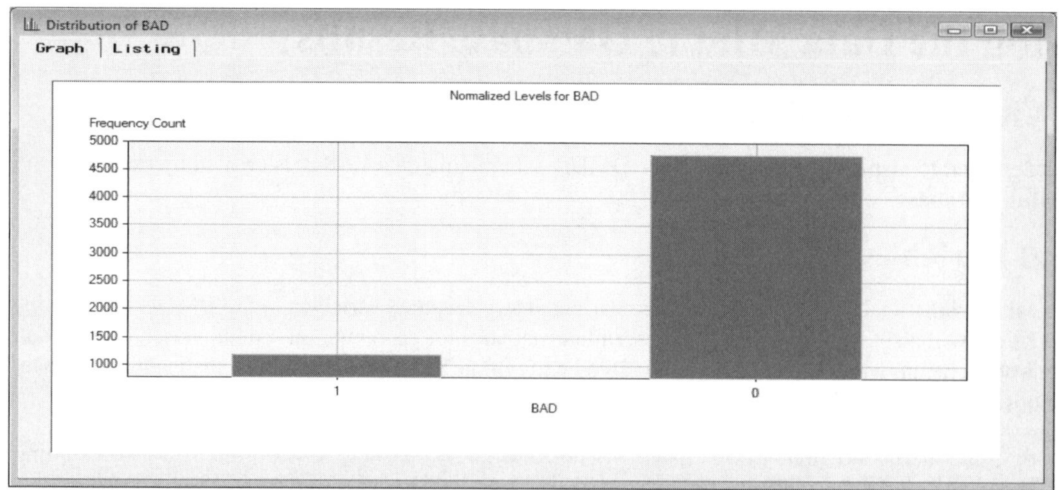

*The **Graph** tab that displays the frequency bar chart to view the distribution of good or bad creditors.*

Distribution of BAD

Graph | Listing

Level	Frequency
1	1189
0	4771

*The **Listing** tab that displays the table listing of the frequency counts of good or bad creditors.*

Code tab

The **Code** tab displays the internal SEMMA training code that created the DMDB data mining data set from the PROC DMDB data mining procedure. Note that the class levels of the categorical target variable are automatically sorted in descending order, and the class levels of all other categorical input variables are sorted in ascending order.

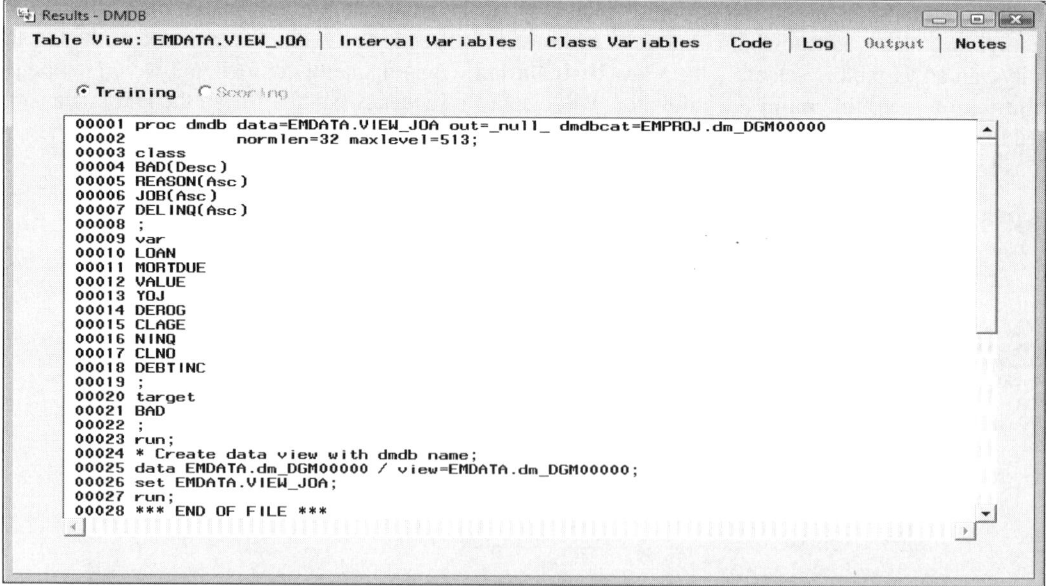

Results - DMDB

Table View: EMDATA.VIEW_JOA | Interval Variables | Class Variables | Code | Log | Output | Notes

Training Scoring

```
00001 proc dmdb data=EMDATA.VIEW_JOA out=_null_ dmdbcat=EMPROJ.dm_DGM00000
00002            normlen=32 maxlevel=513;
00003 class
00004 BAD(Desc)
00005 REASON(Asc)
00006 JOB(Asc)
00007 DELINQ(Asc)
00008 ;
00009 var
00010 LOAN
00011 MORTDUE
00012 VALUE
00013 YOJ
00014 DEROG
00015 CLAGE
00016 NINQ
00017 CLNO
00018 DEBTINC
00019 ;
00020 target
00021 BAD
00022 ;
00023 run;
00024 * Create data view with dmdb name;
00025 data EMDATA.dm_DGM00000 / view=EMDATA.dm_DGM00000;
00026 set EMDATA.VIEW_JOA;
00027 run;
00028 *** END OF FILE ***
```

*The **Code** tab is used to view the internal SEMMA training code that creates the DMDB data set.*

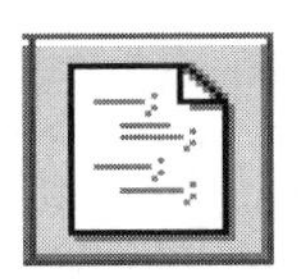

7.3 SAS Code Node

General Layout of the Enterprise Miner SAS Code Node

- **Data tab**
- **Variables tab**
- **Macro tab**
- **Program tab**
- **Export tab**
- **Notes tab**

The purpose of the **SAS Code** node in Enterprise Miner is designed for you to incorporate SAS programming code into the Enterprise Miner process flow diagram. This will give you the ability to access a wide variety of the SAS procedures, generate listings in the SAS system output window, create your own custom-designed scoring code, and manipulate several data sets in the Enterprise Miner process flow diagram by concatenating or merging any number of data sets in the current process flow. The **SAS Code** node is an extremely powerful node in the SEMMA process because of the various SAS programming routines that can be applied within the node are unlimited.

The node is built around a macro facility that is designed for you to reference SAS libraries, data sets, and variables within the **SAS Code** node. Connecting the corresponding node to the **SAS Code** node within the process flow will result in the creation of the macro variables that are used in reference to the corresponding data sets and variables of the connected node. The compiled results, and the various exported data sets that are created from the node, can then be passed along to the subsequent nodes within the currently opened process flow diagram. As opposed to the other nodes, the **SAS Code** node does not require an input data set that is created from the **Input Data Source** node. In other words, there is no need for the **Input Data Source** node to be connected to the **SAS Code** node. Therefore, the **SAS Code** node can be applied at any stage of the process flow. Typically, the **Input Data Source** node is used in the process flow diagram to create the data mining data set by reading the source data set. However, the **SAS Code** node is also designed to create exported output data sets from existing SAS data sets, such as the source data set. This can be done by simply writing SAS programming code to read in the external SAS data set that will create an exported output data set that can be read into the subsequent nodes within the process flow diagram by selecting the appropriate macro variable to reference the corresponding exported data set from the **Exports** tab. In addition, the partitioned training, validation, and test data sets can be processed within the node from the macro reference variables that are automatically created within the node. These newly created exported data sets that are associated with the corresponding partitioned data sets can be passed on to the subsequent nodes within the process flow. For instance, these same data sets that are created within the **SAS Code** node can be assigned variable attributes from the subsequent **Data Attributes** node.

From a wide variety of the Enterprise Miner nodes, the **SAS Code** node will automatically create separate macro variables in reference to the data sets that are automatically scored from the respective nodes or created by selecting the **Process or Score: Training, Validation, and Test** check box within the corresponding modeling node. These same scored data sets are displayed from the **Output** tab within the respective node. That is, a macro variable will be automatically created within the node to reference the score data set that is created from the sampling procedure performed on the input data set from the **Sampling**. Separate macro variables will be automatically created within the node in reference to the partitioned data sets that are created from the **Data Partition** node. A macro variable will be created within the node in reference to the scored data set based on the interactive grouping results, with the assigned class levels and the corresponding WOE statistics for each input variable in the active training data set from the **Interactive Grouping** node. A macro variable will be created within the node in reference to the scored data sets with filtered observations and the observations excluded from the analysis that are created from the **Filter Outliers** node. A macro variable will be created within the node in reference to the scored data set that contains the transformed variables that are created from the **Transform Variables** node. A macro variable will be created within the node in reference to the scored data set with the replacement values and imputed values that are created from the **Replacement** node. A macro variable will be created within the node in reference to the scored data set with the *n*-way association items and the corresponding frequency counts with the various evaluation criterion statistics that are created from the **Association** node. Separate macro variables will be created within the node in reference to the nodes and links data sets that are created from the **Link Analysis**. Separate macro variables will be created within the node in reference to the scored data set with the selected principal components and the transposed data set with the calculated eigenvalues from the **Princomp/Dmneural** node. A macro variable will be created within the node in reference to the scored data set with the various clustering assignments from the **Clustering** and **SOM/Kohonen** nodes. In addition, separate macro variables will be created within the node in reference to the scored data sets with corresponding fitted values from the numerous modeling nodes.

The following are some ideas that might be considered in using the **SAS Code** node in order to further improve the subsequent data mining analysis. As an introduction, the **SAS Code** node can be used within the process flow to create your own custom-designed scoring code that can be connected to the **Data Set Attribute** node in order to assign the appropriate model roles to the predictive model that can then be passed along to the **User Defined Model** node to generate the various assessment statistics. For presentational purposes, maybe you would like to construct your own custom-designed plots and graphs from the scored data sets that are created from the various nodes. Since the **Assessment** node is designed to evaluate the modeling estimates between the same partitioned data set, therefore, the node might be used to generate scatter plots to view the stability in the modeling estimates by viewing small deviations in the fitted values from the training and validation data sets. For group processing analysis, since the **Group Processing** node automatically performs bootstrap sampling with replacement, maybe the **SAS Code** node might be used in the process flow to write your own custom-designed sampling technique that will allow you to control the way in which the input data set is sampled for group processing analysis such as performing bootstrap sampling without replacement instead of with replacement. From the **Link Analysis** node, there might be an interest in determining the accuracy of the clustering assignments that are displayed within the link graph by each class level the categorically-valued target variable in the analysis. This can be accomplished within the **SAS Code** node by specifying the PROC FREQ procedure that will construct the two-way frequency table of the actual and predicted target levels between the partitioned data sets. For cluster analysis, maybe it is important to the analysis to apply some of the other clustering techniques that are not available within the **Clustering** node in order to compare the consistency in the clustering assignments between the separate clustering procedures. In neural network modeling, a standard practice that is often used is retraining or refitting the neural network model and applying the final weight estimates that are calculated from the previous network training run that are then used as the initial weight estimates for the subsequent training run in order to avoid bad local minimums in the error function and achieve convergence in the iterative grid search procedure that will hopefully result in the smallest error from the validation data set. The SAS programming code will be displayed shortly in which the PROC NEURAL procedure is performed. Maybe, it might be important to prune input variables in the neural network model by sequentially removing input variables from the network model one at a time, where its weight estimate is close to zero or below some predetermined threshold value. In two-stage modeling, maybe you would like to use the scored data sets from the various modeling nodes to compare every possible combination from the first-stage and second-stage models in order to select the best model in each of the two separate stages. In nearest neighbor modeling, it might be important to determine the best smoothing constant by fitting the nonparametric model to numerous smoothing constants from the PROC PMBR procedure. In time series modeling, the **Time Series** node might be used to transform the input data set into a time series structure, then applying the various time series procedures, such as the PROC ARIMA procedure, within the **SAS Code** node to generate the time series estimates and connecting the node to the **User Defined Model** node to generate assessment statistics that will allow you to compare several different time series models. For modeling assessment, the **Assessment** node is unable to generate comparison plots from the interval-valued target variable. Therefore, maybe it might be important to the analysis to construct various comparison plots by fitting an interval-valued target variable in order to compare the prediction estimates between the various modeling nodes from the different statistical modeling designs, such as comparing neural network estimates with multiple linear regression estimates, decision tree estimates, nearest neighbor estimates, two-stage modeling estimates, ensemble modeling estimates, or some external modeling procedure in which the **User Defined Model** node can be used to generate the scored data set with the fitted values. In group processing, maybe you would like to create comparison plots in order to view the differences between the prediction estimates that are created from each separate model that is created by each class level of the grouping variable.

Data tab

The **Data** tab will automatically appear once you open the **SAS Code** node. The tab has the same appearance and functionality in comparison to the **Data** tab from the previous Enterprise Miner nodes. The tab displays the file administration information and a table listing of the partitioned data sets or the output scored data set. By default, the node selects the training data set that will allow you to browse both the name and description of the training data set. Simply select the corresponding radio button to designate the other partitioned data sets as the active data set. However, if the node is not connected to the **Input Data Source** node, then the **Data** tab will be grayed-out and unavailable for viewing.

Variables tab

The purpose of the **Variables** tab is to display the variable names, model roles, level of measurements, variable types, formats, and labels from the active training data set. The values displayed within the tab cannot be changed. Similar to the previous **Data** tab, if there is no active data set created from the **Input Data Source** node, then the **Variables** tab will be unavailable for viewing.

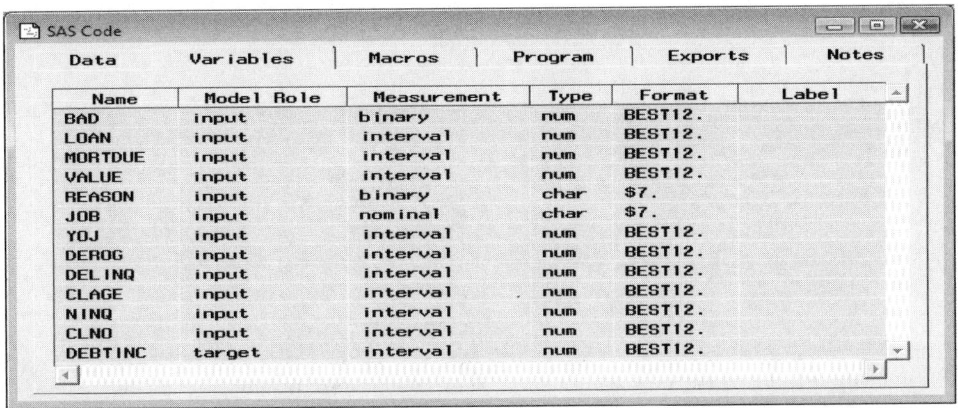

SAS Code

Data | Variables | Macros | Program | Exports | Notes

Name	Model Role	Measurement	Type	Format	Label
BAD	input	binary	num	BEST12.	
LOAN	input	interval	num	BEST12.	
MORTDUE	input	interval	num	BEST12.	
VALUE	input	interval	num	BEST12.	
REASON	input	binary	char	$7.	
JOB	input	nominal	char	$7.	
YOJ	input	interval	num	BEST12.	
DEROG	input	interval	num	BEST12.	
DELINQ	input	interval	num	BEST12.	
CLAGE	input	interval	num	BEST12.	
NINQ	input	interval	num	BEST12.	
CLNO	input	interval	num	BEST12.	
DEBTINC	target	interval	num	BEST12.	

*The **Variables** tab is used to view the variables in the active home equity loan data set.*

Macro tab

The **Macros** tab displays a hierarchical listing of each macro variable that is available in the **SAS Code** node and the currently opened SAS session. The various listed macro variables can be used to write code to either reference the Enterprise Miner data mining libraries, data sets or variables within the **Program** tab. The listed entries in the **Current value** column will indicate the macro variables that are available for processing in the currently opened **SAS Code** node and the currently opened SAS session. From the tab, the macro variables currently available for processing from the node will be identified by non-empty entries listed under the **Current value** column heading. Otherwise, if the corresponding column heading is left blank, then it will indicate to you that the macro variable is undefined and, therefore, cannot be processed in the currently opened SAS session, for example, the &_TRAIN is the SAS macro variable name that is automatically assigned in referencing the training data set when writing code within the **Program** tab. The &_TARGETS macro variable will reference all variables assigned with a role model of **target** and the &_INPUTS macro variable will reference all the input variables in the data set, and so on.

From the **Program** tab, entirely new SAS library references, data sets, and macro variables can be created within the container box. However, SAS warns that it is not advisable to reference the following list of available macro variable names other than the names provided by SAS within the **Macro** tab. Doing so might lead to unpredictable results once you save the currently opened Enterprise Miner diagram. For example, creating entirely different macro variables with same name is not recommended except to reference the automatically created macro names assigned to these same libraries, data sets and corresponding variable roles.

From the **Macro** tab, the macro variables are displayed in three separate folders: the **Libraries**, **Imports**, and the **Exports** folders. The first folder of macro variables available for processing from the **Libraries** folder lists the macro variables to reference the two separate SAS data mining libraries that SAS Enterprise Miner automatically assigns to the currently opened Enterprise Miner project. The **Imports** consists of two separate folders: the **Data sets** and **Variables** folders. The **Data sets** folder displays the available macro variables that reference the various partitioned data sets or the scored data sets that are created from the various modeling nodes or the **Score** node. The **Variables** folder lists the available macro variables that can be processed in identifying the various variable role assignments in the import data set. The last folder of listed macro variables that are available for processing is in reference to the exported data sets that are created from the following **Exports** tab that can be passed on to the subsequent nodes in the process flow. To delete the exported macro variables, simple select the listed macro variable, then right-click the mouse and select the **Delete export** pop-up menu item.

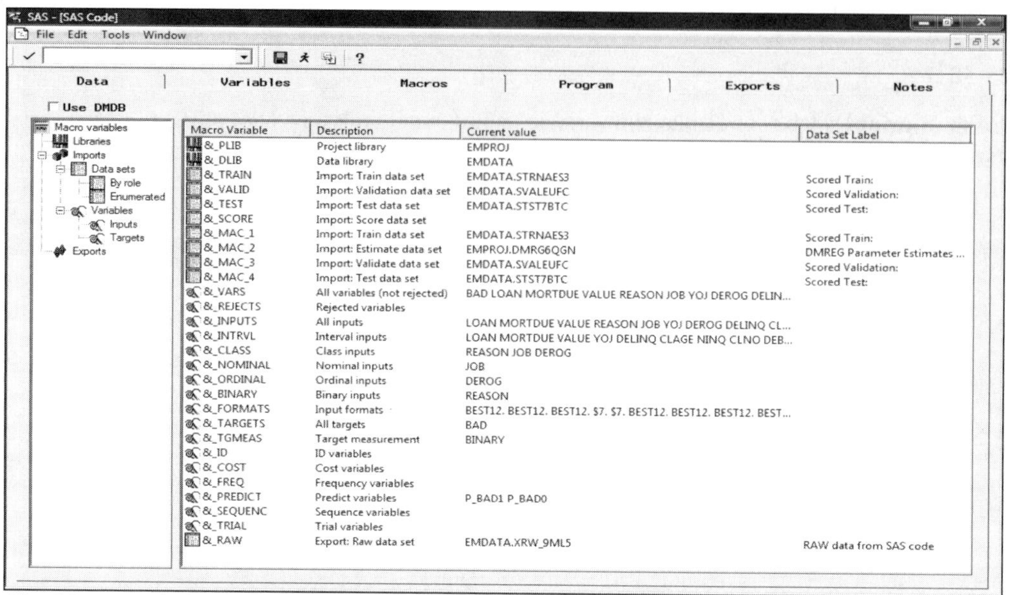

*The **Macros** tab displaying the macro variable names to reference each library, data set, variable, and exported data set that are currently available within the node.*

Check the **Use DMDB** check box if you need to write code that operates on the Data Mining Database (DMDB). A DMDB is a data set that is designed to optimize the performance of the modeling nodes. The DMDB enhances the performance by reducing the number of passes that the analytical engine needs to make through the data. It contains a meta catalog with summary statistics for numeric variables and factor-level information for categorical variables.

Check the **Use DMDB** check box only if you want to perform one of the following tasks:

- Write code that updates the DMDB data set.
- Write code for an Enterprise Miner modeling procedure that requires a DMDB data set and catalog, such as the PROC NEURAL, PROC DMINE, or the PROC SPLIT data mining procedure. The Enterprise Miner modeling nodes automatically create a DMDB-encoded data set and catalog.

Column Name	**Description**
Macro Variable	Macro variable name to use in the program code.
Description	Short description of the listed macro variable.
Current value	The value assigned to the listed macro variable. Therefore if it is blank, then it will indicate to you that the Enterprise Miner macro variable is unavailable for processing in the currently opened SAS session.
Data Set Label	Descriptor label of the listed data set that is based on the label entered in the **Description** entry field within the **Output** tab of the corresponding node. From the tab, if the label column is blank, then it will indicate to you that the macro variable will be unable for processing in reference to the nonexistent data set.

The following is the list of macro variables that are available for processing within the current session. The macro variables are listed in the order in which they are hierarchically displayed within the tab.

The macro variable name and description that references the currently opened Enterprise Miner project libraries:

Macro Data Library Names	Description
&_PUB	Enterprise Miner project library reference name: EMPROJ
&_DLIB	Enterprise Miner data library reference name: EMDATA

The macro variable name and description that references the various partitioned or scored data sets:

Macro Data Set Names	Definition
By Roles	
&_TRAIN	Training data set
&_VALID	Validation data set
&_TEST	Test data set
&_SCORE	Score data set with the predicted values
Enumerated	
&_MAC_1	Scored data set from the training data set.
&_MAC_2	Scored data set from the validation data set.
&_MAC_3	Scored data set from the test data set.

The macro variable name and description that references the various variable roles:

Macro Variable Name	Definition
General Variables	
&_VARS	Macro variable that references every variable in the imported data mining data set.
&_REJECTS	Macro variable that references every rejected or omitted input variables that have been removed from the predictive model.
Input Variables	
&_INPUTS	Macro variable that references every input variable in the data mining analysis.
&_INTRVL	Macro variable that references every interval-valued variable, that is, variables with more than ten different numeric values.
&_CLASS	Macro variable that references every categorically-valued variable in the data mining data set.
&_NOMINAL	Macro variable that references every nominal-valued input variable, that is, input variables with no logical order.
&_ORDINAL	Macro variable that references every ordinal-valued input variable, that is, input variables with a logical order.
&_BINARY	Macro variable that references every binary-valued input variable, that is, input variables with two separate class levels.
&_FORMATS	Macro variable that references the formats assigned to the input variables in the imported data set.

Target Variables	Definition
&_TARGETS	Macro variable that references every target variable to the predictive model.
&_TGMEAS	Macro variable that references the measurement level of the target variable.
Other Variables	
&_ID	Macro variable that references the indicator variable to the data set.
&_COST	Macro variable that references the cost variable that contains decision cost for each observation in the data set.
&_FREQ	Macro variable that references the frequency variable that represents the frequency count for each observation.
&_PREDICT	Macro variable that references the predicted values from the predictive or classification model.
&_SEQUENC	Macro variable that references the sequence variable that is used in both the **Association** and **Time Series** nodes that represents the sequence in time from observation to observation.
&_TRIAL	Macro variable that references the trial variables that contains count data for binomial target variables.

Export Data Sets

The following list of available macro variables refer to the various exported SAS data sets that can be created from the **Exports** tab. These macro variables are used within the **Program** tab that will create the corresponding *exported data set* to be passed on to the subsequent nodes in the process flow. The exported data sets are available for processing within a SAS program by their respective macro variable names. For example, you may want to process the incoming partition data sets with reference to the &_TRAIN, &_VALID, and &_TEST macro variables, therefore, write the subsequent code within the **Program** tab and create the corresponding exported data sets with reference to the &_TRA, &_VAL, and &_TES macro variables that represent the exported training, validation, and test data sets that are then passed on to the subsequent nodes within the process flow. By default, there is no exported SAS data set that is listed in the **Macro** tab since you must first create the reference macro variable of the corresponding exported data set from the **Exports** tab in the **SAS Code** node.

Macro Variable Name	Description
&_TRA	Macro variable that references the exported training data set.
&_VAL	Macro variable that references the exported validation data set.
&_TES	Macro variable that references the exported test data set.
&_SCO	Macro variable that references the exported score data set.
&_PRE	Macro variable that references the exported predict data set.
&_RAW	Macro variable that references the exported raw data set.
&_RES	Macro variable that references the exported results data set.

Program tab

The **Program** tab and the associated *container box* is designed for you to write your own SAS code, import existing SAS code or write your own scoring code. The corresponding SAS code can be submitted within the **SAS Code** node. The SAS code type is determined by the **Event** level setting. The **Event** level settings to select from are either the **Run Event**, **Results Event**, or **Score Event** code types. Select the **Event** drop-down arrow that is positioned just above the container box to set the appropriate SAS code type. By default, the **Run Event** is automatically selected. Select the **Enabled** check box so that the corresponding event level code will be executed when you run the process flow diagram. Conversely, clear the check box in order to prevent the listed code from executing within the process flow diagram.

The following are the various SAS code types to select from the **Events** display field:

- **Run Event** (default): This option will allow you to write SAS code to perform general programming tasks such as data step processing by creating and manipulating SAS data sets in order to access the wide variety of the SAS procedures.
- **Results Event**: This option will allow you to write SAS programming code that can be used to design your own way of listing the results from the **SAS Code** node. For example, the option will allow you to generate PROC REPORT table listings to create a customized report. In other words, simply select this option and write the SAS programming code within the container box that will instruct the node to display the table listing within the SAS system Output window. To execute the **Results Event** code, close and run the node. A message will appear to inform you that the code has completed successfully and ask you if you wish to view the results. Alternatively, right-click the mouse and select the **Results** pop-up menu option from the Enterprise Miner workspace diagram.
- **Score Event**: SAS code that can be written to predict the target variable from a new set of input variable values. An example of the scoring code that can be used within the container box can be viewed from the various modeling nodes or **Score** node. In other words, there might be times that you might want to create your own scoring code that can be passed on to the subsequent **Score** node in creating a scored output data set. For example, adjusting the posterior probabilities for certain class levels from the classification model, imputing missing values of certain input variables in the predictive model to redefine the prediction estimates in the missing input variables, and so on. SAS warns that when you are writing the scoring code, you should not include the DATA, SET, or RUN statement in the body of the scoring code.

Executing the SAS Programming Code

The following are the two separate methods that are available to execute the SAS programming code that is written within the container box from the **Program** tab:

- Click the **Run SAS Code** toolbar icon.
- Select the **Tools > Run SAS Code** from the main menu options.

The following diagram displays the container box that will allow you to write SAS code for a wide variety of uses in your data mining analysis. The following code displays an example of custom-designed scoring code from the HMEQ data set that can be applied to a new set of values from the LOAN and MORTDUE inputs.

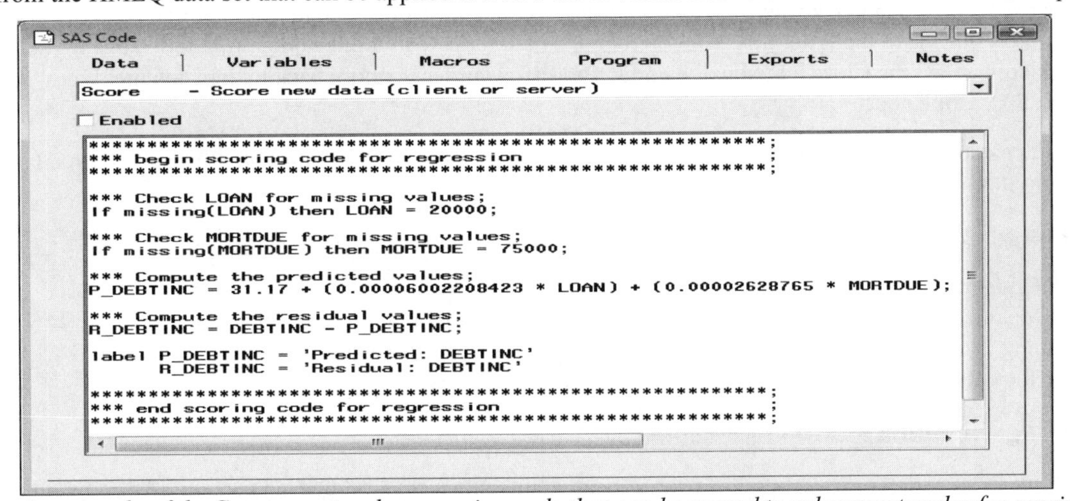

*An example of the **Score event** code or scoring code that can be passed to subsequent nodes for scoring.*

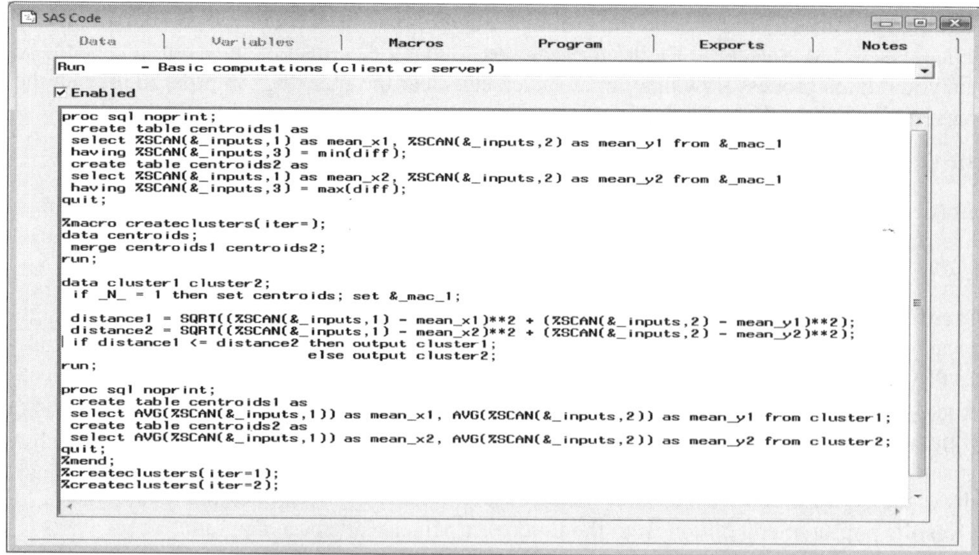

The SAS code that calculates the k-means clustering results by specifying two separate clusters.

The Process in Checking the SAS Programming Results

One of the first things to check after executing the SAS programming code in the **SAS Code** node is to verify that the SAS program ran successfully by viewing the compiled results. This can be done from the **Results Browser**. Select the **Results > Browser** pull-down menu options and the **Results – Code Node** window will appear. The **Results – Code Node** window will allow you to verify that the SAS program executed properly. In addition, the window will allow you to view the listed results by executing SAS code from the **Program** tab. The **Results – Code Node** window will display both the **Log** and **Output** tabs. From the **Log** tab, you may check for errors or warning messages from the compiled SAS program that is similar to the SAS system Log window. Select the **Output** tab to view the compiled results that is similar to the SAS system Output window. To close the **Results – Code Node** window, select the **File > Close** main menu options or select the upper right-hand corner button and the **SAS Code** window will then reappear.

Entering SAS Code in the SAS Program Editor

If you feel uncomfortable in programming in the container box, then the **SAS Code** node is designed to give you the added flexibility of accessing the standard display manager SAS Program Editor window. There are two separate ways in which to write SAS programming code within the **SAS Code** node. By default, the node is designed to enter SAS code within the container box from the **Program** tab. However, select **Edit > SAS Editor** from the main menu options or right-click the mouse within the container box and select the **SAS Editor** pop-up menu item to take advantage of the enhanced editing routines in the SAS Program Editor window that is outside of the Enterprise Miner environment and the **SAS Code** node. From the SAS Program Editor window, select the **Close** button in the upper right-hand corner to return back to the container box within the **SAS Code** node.

From the **Program** tab, the following are the various main menu **File** option settings to either import or export the SAS programming code.

Importing SAS Programming Code

The **SAS Code** node will allow you to import existing SAS programming code that is stored in an external SAS text file or as a SAS catalog entry. Select the **File > Import File** main menu options to import the SAS programming code into the container box of the currently opened **Program** tab. The imported code will overwrite the existing SAS program that is listed in the container box.

Importing the Source Entry

SAS provides you with the capability of saving SAS programs into a SAS catalog. Typically, SAS programs are saved into a text file. However, it might be more convenient to store the SAS program as a category entry.

To import the SAS source entry code into the container box, select **File > Import Source Entry** main menu options that will provide you with a hierarchical listing of the existing SAS libraries, catalogs, and entries that are defined in the currently opened SAS session. That is, this option will allow you to browse the appropriate library that will list the corresponding catalogs in order to select the associated catalog entries and import the source entry in the SAS catalog. Once you select this option, the **Open** window will appear that will list the available SAS libraries and listed catalogs to copy the source entry code into the **Program** tab.

Exporting SAS Programming Code

Select **File > Export File** main menu options to export the currently opened SAS program and save the SAS programming code as an external SAS text file into the selected system folder.

Exporting the Source Entry

Select the **File > Export Source Entry** main menu options to export the source entry into the appropriate SAS catalog from the listed SAS libraries by selecting the desired catalog entry within the SAS catalog that is listed under the corresponding SAS library. This option will display the **Open** window that displays a hierarchical listing of the available SAS libraries and catalogs to export the source code into the selected SAS catalog.

Exports tab

The purpose of the **Exports** tab is to create an exported data set that can be passed along to the subsequent nodes within the process flow. A macro variable is automatically created within the node that is referenced to the corresponding output data set. In addition, the macro variable referenced to the exported data set that is created from the **Exports** tab will then be automatically displayed in the hierarchical listing of the available macro variables from the previous **Macro** tab.

In the following example, the **SAS Code** node was used to create the input data set within the process flow diagram. The macro variable called &_RAW was incorporated in the container box from the **Program** tab to create the input data set within the process flow. The following steps must be applied in order to create the exported data set within the **SAS Code** node. The first step is creating the SAS macro variable from the **Exports** tab. The next step is writing SAS programming code from the **Program** tab to reference the &_RAW macro variable in order to create the input data set within the currently opened process flow diagram.

The **Pass imported data sets to successors** check box that is positioned just below the **Export role** drop-down menu option is designed to prevent the import data sets from being passed to subsequent nodes. By default, the check box is automatically selected, which is actually what you want to do.

To Create the Input Data Set Macro Variable

1. Press the **Add** button and the **Add export data set/view** window will appear.
2. Select **Raw** from the **Export role** pop-up menu items that are illustrated in the following diagram.
3. Press the **OK** button.

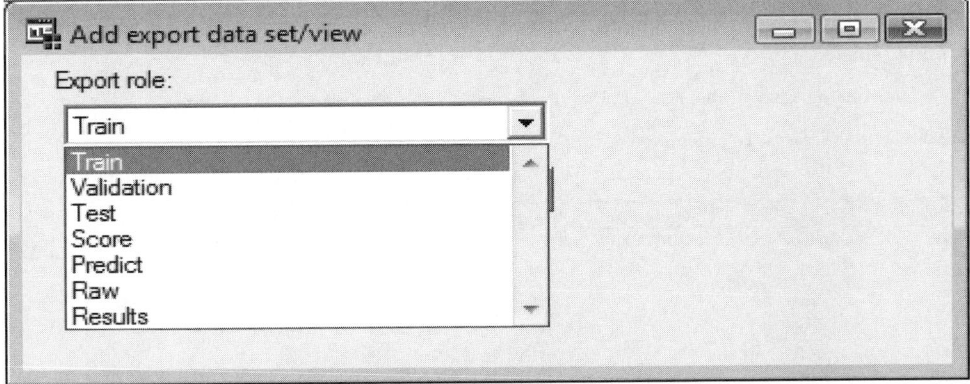

*The **Add export data set/view** window is designed to create the macro variables within the node to be passed on to the subsequent nodes within the process flow diagram.*

Steps in Creating the Data Mining Data Set in the Process Flow Diagram

The following example explains the step-by-step approach to create the input data set for the currently opened process flow diagram.

1. Select the **Exports** tab.

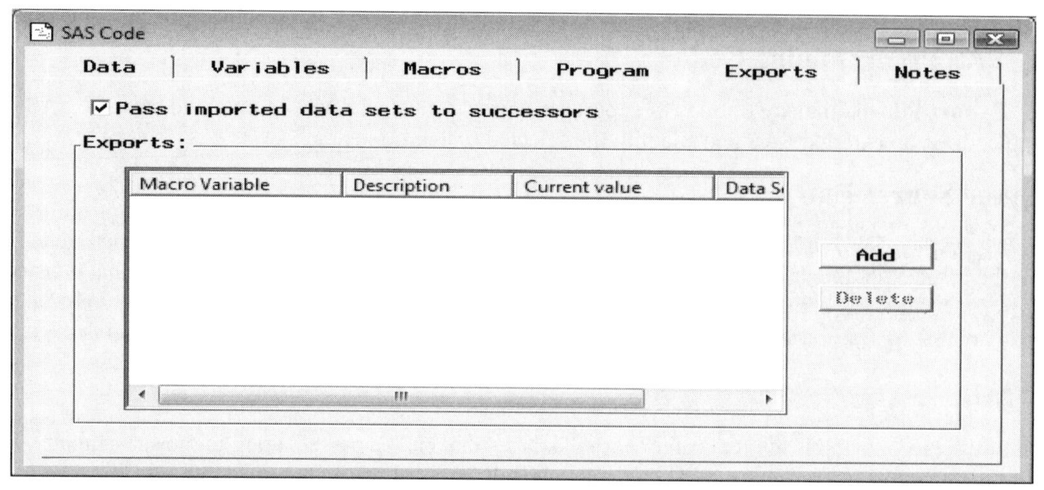

2. Press the **Add** button to create the input data set with the following **Add export data set/view** window appearing.

3. Select **Raw** from the pop-up menu items and the macro variable to reference the input data set will be automatically created in the **Macro** tab with a macro variable name of &_RAW. Similarly, the training data set can be created within the node by selecting the **Train** pop-up menu item, and so on.

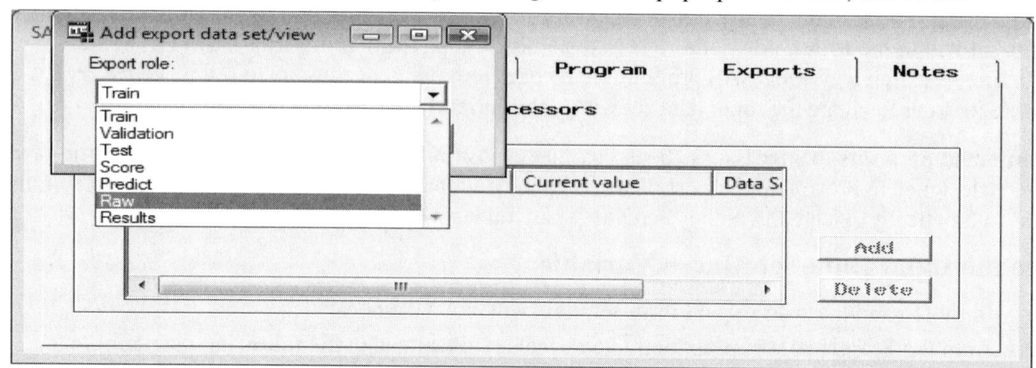

4. Enter the appropriate SAS code in the container box by selecting the default **Run event** code option to read the home equity loan HMEQ data set in order to create the input data set within the **SAS Code** node.

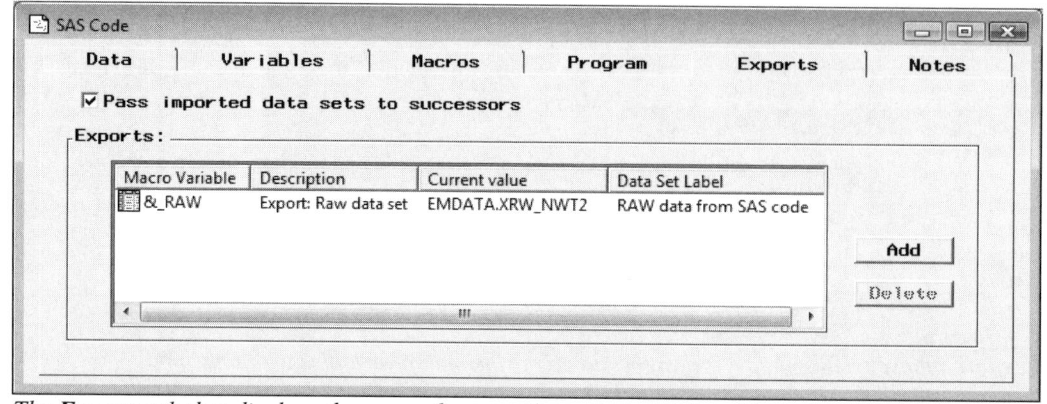

*The **Exports** tab that displays the currently created macro variables of the exported raw data set.*

The following SAS code was written within the **SAS Code** node to calculate the best set of weight estimates from the neural network model that resulted in the smallest validation error. The neural network model is then refitted with these same weight estimates that are set to the initial weight estimates in the subsequent neural network model that produced the final assessment statistics from a single training run as follows:

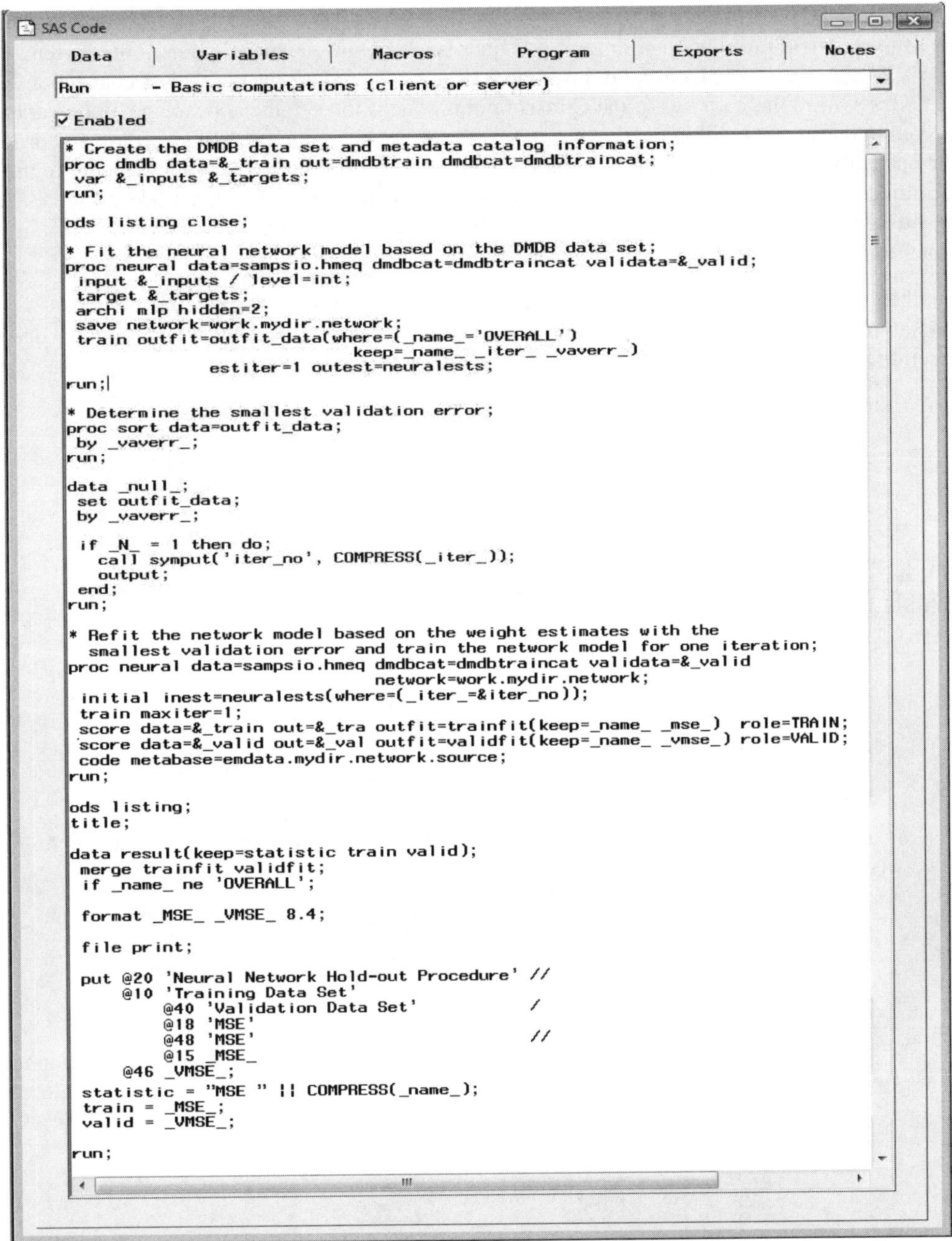


```
* Create the DMDB data set and metadata catalog information;
proc dmdb data=&_train out=dmdbtrain dmdbcat=dmdbtraincat;
 var &_inputs &_targets;
run;

ods listing close;

* Fit the neural network model based on the DMDB data set;
proc neural data=sampsio.hmeq dmdbcat=dmdbtraincat validata=&_valid;
 input &_inputs / level=int;
 target &_targets;
 archi mlp hidden=2;
 save network=work.mydir.network;
 train outfit=outfit_data(where=(_name_='OVERALL')
                          keep=_name_ _iter_ _vaverr_)
              estiter=1 outest=neuralests;
run;

* Determine the smallest validation error;
proc sort data=outfit_data;
 by _vaverr_;
run;

data _null_;
 set outfit_data;
 by _vaverr_;

 if _N_ = 1 then do;
   call symput('iter_no', COMPRESS(_iter_));
   output;
 end;
run;

* Refit the network model based on the weight estimates with the
  smallest validation error and train the network model for one iteration;
proc neural data=sampsio.hmeq dmdbcat=dmdbtraincat validata=&_valid
                                   network=work.mydir.network;
 initial inest=neuralests(where=(_iter_=&iter_no));
 train maxiter=1;
 score data=&_train out=&_tra outfit=trainfit(keep=_name_ _mse_)  role=TRAIN;
 score data=&_valid out=&_val outfit=validfit(keep=_name_ _vmse_) role=VALID;
 code metabase=emdata.mydir.network.source;
run;

ods listing;
title;

data result(keep=statistic train valid);
 merge trainfit validfit;
 if _name_ ne 'OVERALL';

 format _MSE_ _VMSE_ 8.4;

 file print;

 put @20 'Neural Network Hold-out Procedure' //
     @10 'Training Data Set'
         @40 'Validation Data Set'        /
         @18 'MSE'
         @48 'MSE'                        //
         @15 _MSE_
     @46 _VMSE_;
 statistic = "MSE " || COMPRESS(_name_);
 train = _MSE_;
 valid = _VMSE_;

run;
```

The SAS programming code that delivers the best set of weight estimates from the smallest validation error that are then refitted in the subsequent neural network model without iterating.

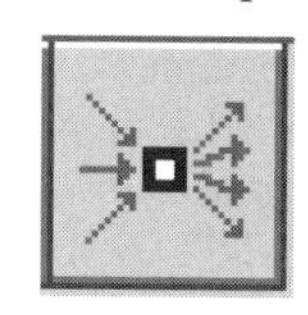

7.4 Control point Node

The purpose of the **Control point** node in Enterprise Miner is to establish a control point in the currently opened process flow diagram. The **Control point** node is designed to reduce the number of connections between the nodes. One of the purposes of the **Control point** node is to keep the appearance of the various nodes connected to one another within the process flow diagram easier to interpret. Simply drag and drop the node into the currently opened process flow diagram. The next step is simply connecting the nodes to the **Control point** node. The **Control point** node will then be connected to the appropriate nodes. Similar to the following **Subdiagram** node, there are no built-in tabs to the node that will allow you to specify various configuration settings within the node.

For example, the **Control point** node was incorporated in the process flow in the previous modeling comparison example in order to limit the number of connections from the three separate modeling nodes to both the **Ensemble** and **Assessment** nodes within the process flow, as illustrated in the following diagram.

Step 1: Construct the entire process flow diagram.

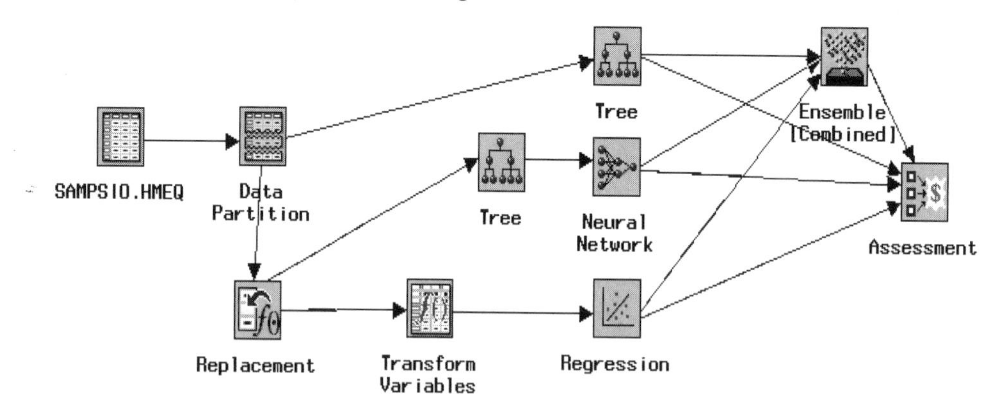

Step 2: Drag and drop the **Control point** node on to the **Diagram Workspace** and place the node between the appropriate connections.

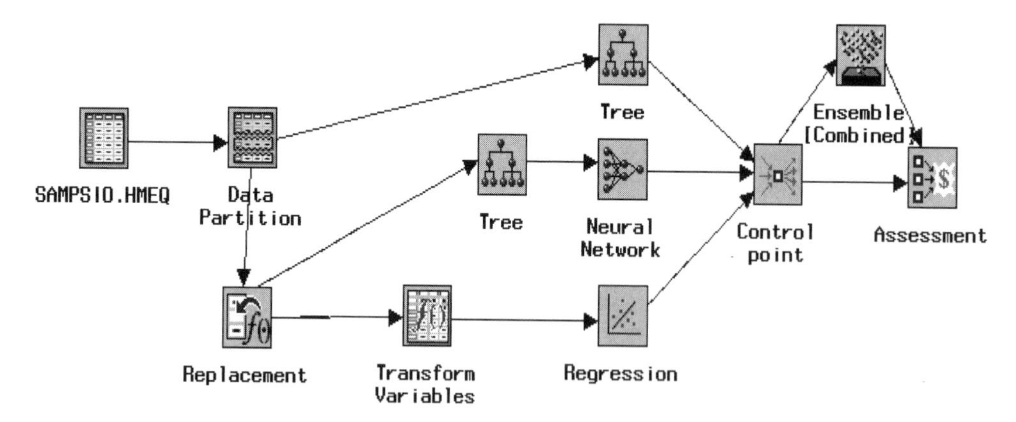

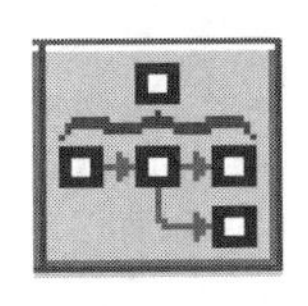

7.5 Subdiagram Node

The purpose of the **Subdiagram** node in Enterprise Miner is to consolidate a portion of the **Diagram Workspace** within a single subdiagram node. There might be times when the diagram workspace might become quite cumbersome, with an enormous number of connected nodes. Therefore, the **Subdiagram** node is designed to condense the display of the numerous connecting nodes into a single **Subdiagram** node icon. The node is designed to give you better control in organizing and simplifying the appearance of the complex process flow diagram. Simply open the single **Subdiagram** node to view the complete structure of the interconnecting process flow diagram.

There are two ways to add a subdiagram to the **Diagram Workspace**. One way is to add a **Subdiagram** node at the beginning of the process flow diagram by adding the various connected nodes within the subdiagram. The **Input Data Source** node would usually be the first node connected to the **Enter** node to begin the process flow subdiagram. Otherwise, once the entire process flow diagram is constructed, you might want to group the collection of nodes in order to make the process flow diagram easier to control.

There are no tabs or resulting output generated from the node. Simply open the **Subdiagram** node, which will then open the **Subdiagram – Utility** window. The window will display the Enterprise Miner workspace with a **Enter** node icon connected to an **Exit** node icon. This is illustrated in the following diagram at Step 3. Creating a subdiagram can be done by simply deleting the connecting lines between these two nodes and then begin creating your personal workflow subdiagram between these two separate nodes.

Copying the Diagram

To copy the entire diagram, simply select the **File > Save diagram as...** main menu options. However, one of the limitations in Enterprise Miner is that it is not designed to select several nodes in order to copy these same nodes to a separate diagram within the currently opened project. Therefore, another advantage of using the **Subdiagram** node is that you may select the node and copy the node along with all other nodes that are a part of the process flow subdiagram to a separate diagram of the currently opened project or any other Enterprise Miner project. This can be performed by simply selecting the various nodes of the currently opened diagram, right-clicking the mouse and selecting the **Create subdiagram** pop-up menu option. The **Subdiagram** node will automatically appear in the diagram workspace. Select the node, right-click the mouse and select the **Export subdiagram...** pop-up menu option. The **Save Diagram As...** window will appear that will allow you to name the copied diagram that will be created in the project folder.

To better understand how the **Subdiagram** node works, consider the following process flow diagram. To view the subdiagram, simply double click the node to view the process flow subdiagram that is displayed in the following illustration. Within each subdiagram, right-click the mouse and select the **Up one level** pop-up menu item to go up one level of the process flow subdiagram. There is no restriction as to the number of subdiagrams that can be created within each subdiagram.

The following is the process flow diagram of the modeling comparison procedure that is displayed in the previous **Assessment** node to compare the modeling estimates from the logistic regression, neural network, decision tree, and ensemble models in predicting clients defaulting on their home loan from the home equity loan data set. The **Data Partition** node is first applied, which is needed to determine overfitting in the models. The **Replacement** node was added to the logistic regression and neural network models to compensate for the surrogate splits that are applied in the decision tree model. The **Transform Variables** node was added to the process flow to transform the variables in the logistic regression model to achieve normality in the data. The modeling selection routine within the **Tree** node was added to the neural network model and the **Variable Selection** node, with the r-square modeling selection routine, was applied to the logistic regression model. The **Regression**, **Neural Network**, and **Tree** nodes are connected to the **Control point** node to reduce the number of connections to the diagram. The **Control point** node connects to the **Ensemble** node to combine the fitted values from the three separate classification modeling designs. In addition, the **Control point** node is connected to the **Assessment** node in order to compare the accuracy of the modeling estimates from the four separate modeling designs.

In Enterprise Miner, there are two separate ways to create a subdiagram within the process flow diagram:

- Drag and drop the **Subdiagram** node on to the **Diagram Workspace**. Double click the **Subdiagram** node, which opens the **Subdiagram** window. The window will allow you to construct your process flow subdiagram.
- Construct the entire process flow diagram. Drag the mouse over the selected nodes and connections that you would like to be a part of the process flow subdiagram.

Method 1

Step 1: Construct the appropriate nodes into the process flow diagram.

Step 2: Drag and drop the **Subdiagram** node on to the process flow diagram. Double-click the node icon or right-click the mouse and select the **Open** pop-up menu item.

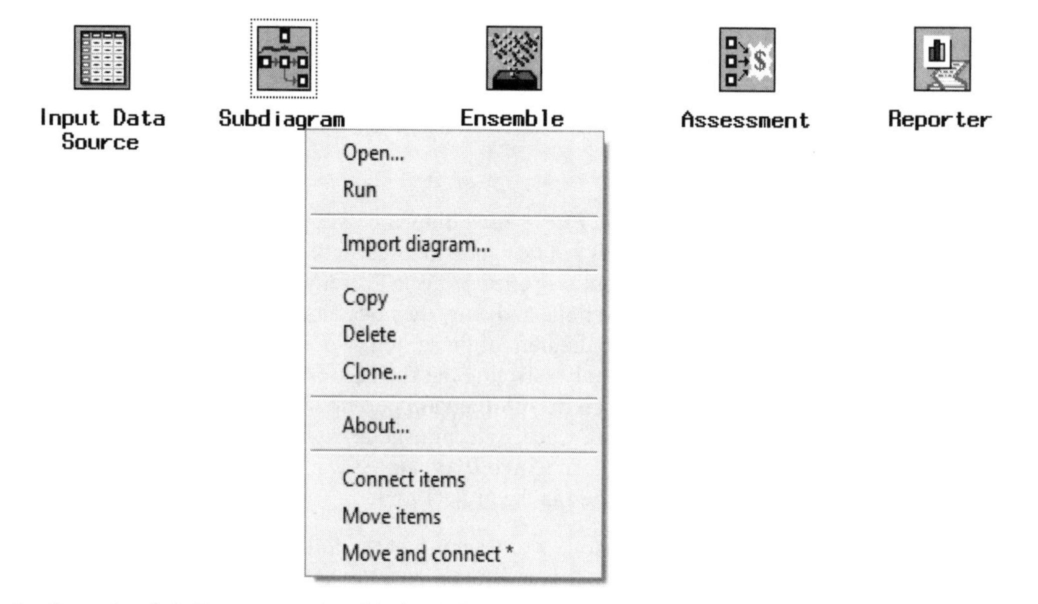

Step 3: Open the **Subdiagram** node with the following **Enter** and **Exit** nodes that will appear. Both nodes will be automatically connected to each other.

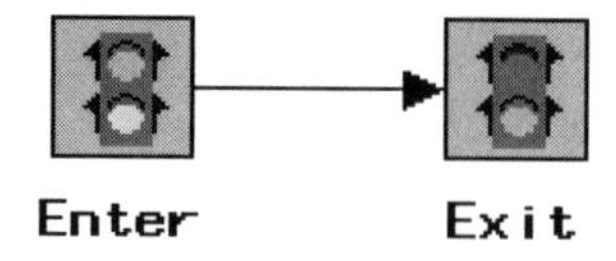

Step 4: Disconnect the lines between the connected **Enter** and **Exit** nodes by selecting the line, then press the **Delete** key or right-click the mouse and select the **Delete** pop-up menu item. The next step is to reposition both nodes in order to give yourself enough room to create the process flow subdiagram as follows:

Step 5: Create the collection of nodes between both the **Enter** and **Exit** nodes. Notice the **Up one level** pop-up menu item that will allow you to navigate between each process flow subdiagram within the Enterprise Miner diagram. Select the **View > Up one level** pop-up menu item to return to the process flow diagram.

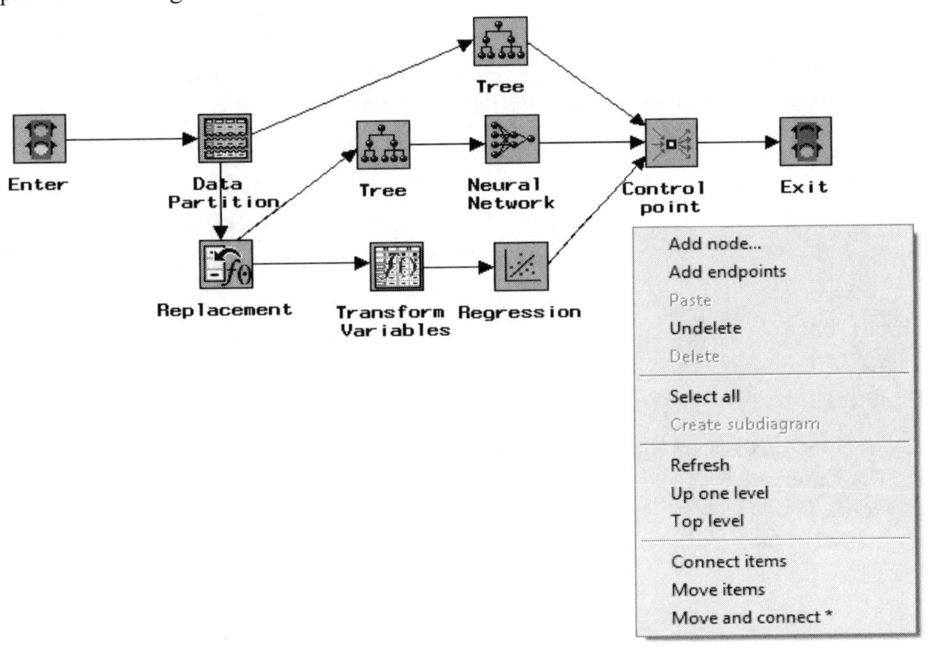

Step 6: Place the nodes to their appropriate positions in the process flow diagram. Select the **Reporter** node to run the entire process flow diagram.

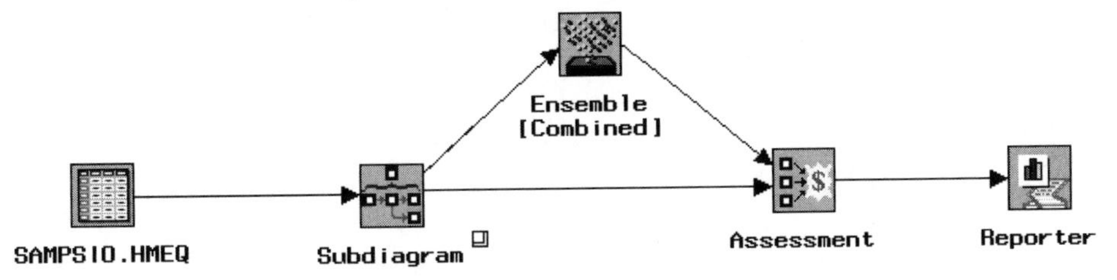

Method 2

Step 1: Construct the entire process flow diagram.

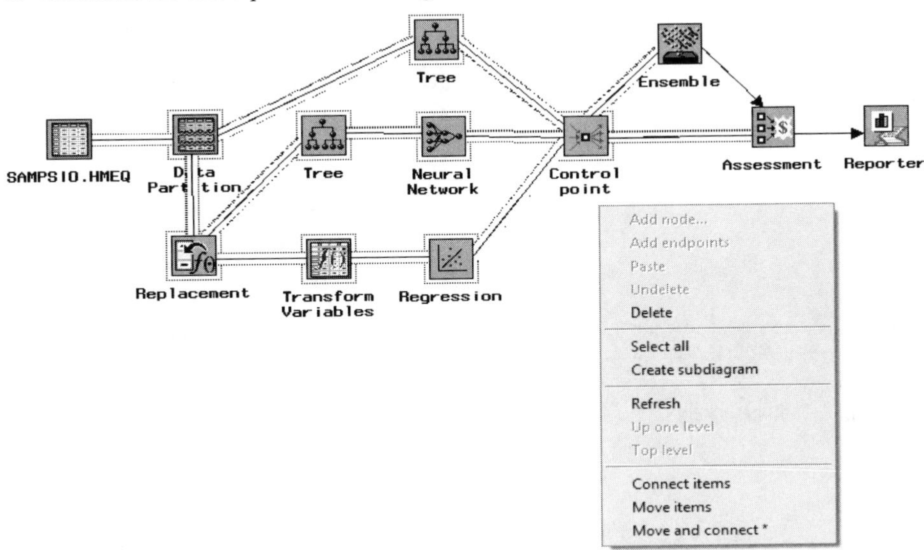

Step 2: Select the Enterprise Miner nodes to create the process flow subdiagram. In this example, the **Data Partition, Replacement, Transform Variable, Regression, Tree,** and **Control point** nodes were selected to create the process flow subdiagram.

Step 3: Right-click the mouse and select the **Create subdiagram** pop-up menu item. The Message window will appear to allow you to condense the currently selected nodes into the subdiagram.

Step 4: Enterprise Miner will automatically create a **Subdiagram** node in the currently opened process flow diagram. By default, the node is not connected to any of the existing nodes within the process flow diagram. Open the subdiagram by double-clicking the **Subdiagram** node.

Step 5: From the subdiagram, right-click the mouse and select the **Add endpoints** pop-up menu option. The **Enter** and **Exit** nodes will appear. Repeat Steps 3–5 from the previous method. In other words, disconnect both nodes and reposition the **Enter** node at the beginning and the **End** node at the end of the subdiagram, as illustrated in the following diagram.

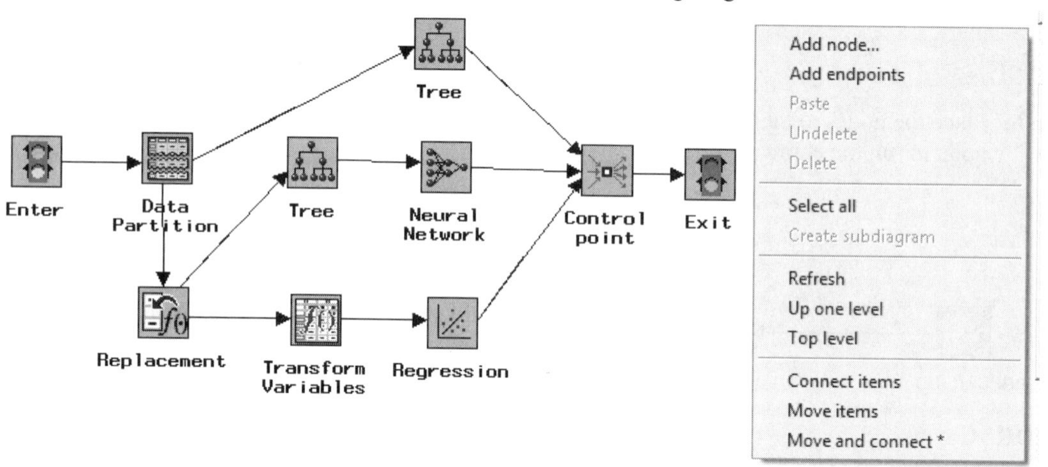

Step 6: From the subdiagram, right-click the mouse and select the **Up one level** pop-up menu option to go back up to the process flow diagram, as illustrated in the previous diagram.

Step 7: From the process flow diagram, connect the **Subdiagram** node to the other existing nodes. From the **Reporter** node, right-click the mouse and select the **Run** pop-up menu option to execute the entire process flow diagram.

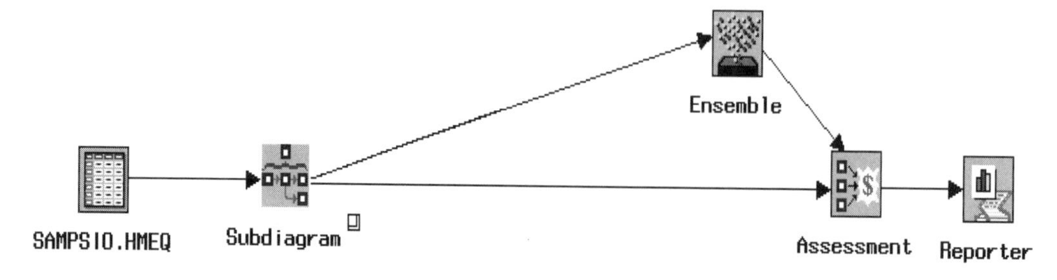

References

Abdi, Hervé. *Neural Networks*, Sage University Paper, 1999.
Agresti, Alan. *Categorical Data Analysis*, 2nd Edition. John Wiley & Sons, Inc., 2002.
Allison, Paul D. *Logistic Regression Using SAS System Theory Application*, SAS Institute, Inc., 1999.
Allison, Paul D. *Missing Data*, Sage University Paper, 2001.
Belle, Gerald van. *Statistical Rules of Thumb*, John Wiley & Sons, 2002.
Bentley, J.L. *Multidimensional Binary Search Trees Used for Associative Searching*, 1975.
Berry, Michael J.A. *Data Mining Techniques for Marketing, Sales and Customer Support*, 1st Ed. John Wiley & Sons, Inc., 1997.
Berry, Michael J.A. *Data Mining Techniques: For Marketing, Sales, and Customer Relationship Management*, 2nd Ed. John Wiley & Sons, Inc., 2004.
Bishop, Christopher M. *Neural Networks for Pattern Recognition*, Oxford University Press, 2002.
Boes, Duane C. *Introduction to the Theory of Statistics,* 3rd Ed. McGraw-Hill, 1974.
Bowerman, Bruce L. *Linear Statistical Models*, 2nd Ed. Duxbury Press, 1990.
Bowerman, Bruce L. *Time Series Forecasting*, 3rd Ed. Duxbury Press, 1993.
Breiman, Leo, *Arcing Classifiers*, The Annals of Statistics, 26(3), 1998.
Brocklebank, John C. *Forecasting Time Series,* 2nd Ed. SAS Institute Inc, 2003.
Burden, Richard L. *Numerical Analysis*, 7th Ed. Prindle, Weber and Schmidt, 2001.
Carpenter, Arthur L. *Quick Results with SAS/Graph Software,* 1st Ed. SAS Institute, 1995.
Carroll, Douglas J. *Analyzing Multivariate Data,* Thomson Learning, Inc., 2003.
Castellan, John N. Jr. *Nonparametrics for the Behavior Sciences,* 2nd Ed. McGraw Hill, Inc., 1988.
Chong, Edwin K.P. *An Introduction to Optimization,* 2nd Ed. John Wiley & Sons, 2001.
Cochran, William G. *Statistical Methods*, 7th Ed. The Iowa University State Press, 1980.
Cochran, William G. *Sampling Techniques*, 3rd Ed. John Wiley & Sons, Inc., 1977.
Cody, Ronald P. *Applied Statistics and the SAS Programming Language*, 4th Ed. SAS Institute, Inc., 1997.
Cox, D. R. *Analysis of Binary Data*, 2nd Ed. Chapman & Hall/CRC, 1992.
Cox, James A. *Nearest Neighbor Data Method and System*, 2005.
Davis, Charles. *Categorical Data Analysis Using SAS System*, SAS Institute, Inc., 2000.
Der, Geoff. *A Handbook of Statistical Analyses Using SAS*, 2nd Ed. SAS Institute, Inc., 2001.
Dickey, David A. *Forecasting Time Series,* 2nd Ed. SAS Institute Inc, 2003.
Dowdy, Shirley. *Statistics for Research*, John Wiley & Sons, Inc., 1983.
Draper, Norman. *Applied Regression Analysis*, 2nd Ed. John Wiley & Sons, 1981.
Edelman, Betty. *Neural Networks*, Sage University Paper, 1999.
Efron, Bradley. *An Introduction to the Bootstrap*, Chapman & Hall, Inc., 1993.
Everitt, Brian. *A Handbook of Statistical Analyses Using SAS*, 2nd Ed. SAS Institute, Inc., 2001.
Faires, Douglas J. *Numerical Analysis*, 7th Ed. Prindle, Weber and Schmidt, 2001.
Fidell, Linda S. *Research Design and Analysis,* Allyn & Bacon, 2001.
Fleiss, Joseph L. *Statistical Methods for Rates and Proportions*, 2nd Ed. John Wiley & Sons, Inc., 1981.
Freund, Rudolf J. *SAS System for Regression*: 3rd Ed. SAS Institute Inc., 2000.
Friedman, Jerome. *The Elements of Statistical Learning*, Springer, 2001.
Friendly, Michael. *SAS System for Statistical Graphics*, 2nd Ed. SAS Institute Inc., 1997.
Glass, Gene V. *Statistical Methods in Education and Psychology,* 3rd Ed. Allyn & Bacon, 1996.
Giudici, Paolo. *Applied Data Mining Statistical Methods for Business and Industry*, John Wiley & Sons, Inc., 2003.
Graybill, Fracklin. A. *Introduction to the Theory of Statistics,* 3rd Ed. McGraw-Hill, 1974.
Green, Paul E. *Analyzing Multivariate Data,* Thomson Learning, Inc., 2003.
Hand, David. *Principles of Data Mining*, The MIT Press, 2001.
Hastie, Trevor. *The Elements of Statistical Learning*, Springer, 2001.
Hill, Thomas. *Statistics: Methods and Applications*, StatSoft, 2006.

Hollander, Myles. *Non-Parametric Statistical Methods*, 2nd Ed. John Wiley & Sons, Inc., 1999.

Hopkins, Kenneth D. *Statistical Methods in Education and Psychology,* 3rd Ed. Allyn & Bacon, 1996.

Johnson, Richard A. *Applied Multivariate Statistical Analysis*, Prentice-Hall, Inc., 2002.

Khattree, Ravindra. *Multivariate Data Reduction and Discrimination with SAS Software*, SAS Institute, Inc., 1999.

Koch, Gary. *Categorical Data Analysis Using SAS System*, SAS Institute, Inc., 2000.

Kutner, Michael H. *Applied Linear Regression Models*, Richard D. Irwin, Inc., 1983.

Lajiness, Michael S. *A Practical Introduction to the Power of Enterprise Miner*, paper 69-26, 2000.

Lattin, James. *Analyzing Multivariate Data,* Thomson Learning, Inc., 2003.

Lewicki, Pawel. *Statistics: Methods and Applications*, StatSoft, 2006.

Linoff, Gordon. *Data Mining Techniques for Marketing, Sales and Customer Support*, 1st Ed. John Wiley & Sons, Inc., 1997.

Linoff, Gordon. *Data Mining Techniques: For Marketing, Sales, and Customer Relationship Management*, 2nd Ed. John Wiley & Sons, Inc., 2004.

Littell, Ramon C*S AS System for Regression*: 3rd Ed. SAS Institute Inc., 2000.

Little, Roderick J.A. *Statistical Analysis of Missing Data*, 2nd Ed. John Wiley & Sons, Inc., 2002.

Lunneborg, Clifford E. *Data Analysis by Resampling: Concepts and Application*, Duxbury Press, 2000.

Lyman, Ott. *Elementary Survey Sampling*, 3rd Ed. Duxbury Press, 1986.

Mannila, Heikki. *Principles of Data Mining*, The MIT Press, 2001.

Matignon, Randall. *Neural Network Modeling using SAS Enterprise Miner,* AuthorHouse, 2005.

McGee, Monnie. *Introduction to Time Series Analysis and Forecasting*, Academic Press, Inc., 2000.

Mendenhall, William. *Elementary Survey Sampling*, 3rd Ed. Duxbury Press, 1986.

Mitchell, Tom M. *Machine Language*, WCB McGraw Hill Series, 1997.

Montgomery, Douglas. *Design and Analysis of Experiments,* 2nd Ed. John Wiley & Sons, 1984.

Mood, Alexander M. *Introduction to the Theory of Statistics,* 3rd Ed. McGraw-Hill, 1974.

Mosteller, Frederick. *Data Analysis and Regression*, Addison-Wesley Publishing Co., Inc., 1977.

Myers, Raymond H. *A First Course in the Theory of Linear Statistical Models*, Duxbury Press, 1991.

Nachtsheim, Christopher. *Applied Linear Statistical Models*, 5th Ed. McGraw Hill, Inc., 2004.

Naik, Dayanand N. *Multivariate Data Reduction and Discrimination with SAS Software*, SAS Institute, Inc., 1999.

Neter, John. *Applied Linear Statistical Models*, 5th Ed. McGraw Hill, Inc., 2004.

Neville, Padraic. *Decision Tree for Predictive Modeling*, SAS Institute, Inc., 1999.

O'Connell, Richard T. *Linear Statistical Models*, 2nd Ed. Duxbury Press, 1990.

O'Connell, Richard T. *Linear Time Series Forecasting*, 3rd Ed. Duxbury Press, 1993.

Potts, William J.E. *Neural Network Modeling Course Notes,* SAS Institute Inc., 2000.

Ripley, Brian D. *Pattern Recognition and Neural Networks*, 6th Ed. Cambridge University Press, 2002.

Sarle, Warren S. *Neural Network FAQ, part 1 of 7: Introduction*, URL: ftp://ftp.sas.com/pub/neural/FAQ.html, 1997

Sarma, Kattamuri S. *Using SAS Enterprise Miner for Forecasting*, Paper 25-26, 2000.

SAS Institute, Inc. *Advanced Predictive Modeling Using SAS Enterprise Miner 5.1 Course Notes*, 2004.

SAS Institute, Inc. *Applied Clustering Techniques Course Notes*, 2003.

SAS Institute, Inc. *Applying Data Mining Techniques Using Enterprise Miner Course Notes*, 2002.

SAS Institute, Inc. *Categorical Data Analysis Using Logistic Regression Course Notes*, 2001.

SAS Institute, Inc. *Data Mining Techniques: Theory and Practice*, 2005.

SAS Institute, Inc. *Data Mining Using Enterprise Miner Software: A Case Study Approach* 1st Ed., 2000.

SAS Institute, Inc. *Decision Tree Modeling Course Notes*, 2001.

SAS Institute, Inc. *Enterprise Miner 4.0 Node: Usage*, 2000.

SAS Institute, Inc. *Enterprise Miner 4.0: User Interface Help*, 2000.

SAS Institute, Inc. *Forecasting Examples for Business and Economics Using the SAS System*, 1996.

SAS Institute, Inc. *Logistic Regression Examples Using the SAS System*, Version 6, 1st Edition. 1995.

SAS Institute, Inc. *Longitudinal Data Analysis with Discrete and Continuous Responses Course Notes*, 2005.

SAS Institute, Inc. *Multivariate Statistical Methods: Practical Research Applications Course Notes*, 2004.

SAS Institute, Inc. *Neural Network Modeling Course Notes,* 2000.

SAS Institute, Inc. *SAS Enterprise Miner 5.13 Help,* 2004.

SAS Institute, Inc. *SAS/ETS Software Applications Guide 1: Time Series Modeling and Forecasting, Financial Reporting and Loan Analysis*, Version 6, 1st Edition. 1992.

SAS Institute, Inc. *SAS User's Guide: Basics* Version 5 Edition. 1985.

SAS Institute, Inc. *SAS User's Guide: Statistics* Version 5 Edition. 1985.

SAS Institute, Inc. *SAS/Insight User's Guide* Version 6, 1st Edition. 1990.

SAS Institute, Inc. *SAS/STAT User's Guide: Statistics* Version 6, 4th Edition. Volume 1 & 2, 1990.

SAS Institute, Inc. *Statistics I: Introduction to ANOVA, Regression, and Logistic Regression Course Notes*, 2004.

SAS Institute, Inc. *Statistics II: Introduction to ANOVA and Regression Course Notes*, 2005.

SAS Institute, Inc. *Two-Stage Modeling Using Enterprise Miner Software Course Notes*, 2003.

Scheaffer, Richard L. *Elementary Survey Sampling*, 3rd Ed. Duxbury Press, 1986.

Searle, S. R. *Linear Models*, John Wiley & Sons, Inc., 1971.

Seber, G. A. F. *Nonlinear Regression*, John Wiley & Sons, Inc., 1989.

Shipp, Charles E. *Quick Results with SAS/Graph Software,* 1st Ed. SAS Institute, 1995.

Shumway, Robert H. *Time Series Analysis and Its Applications,* Springer, 2000.

Siegel, Sidney. *Nonparametrics for the Behavior Sciences,* 2nd Ed. McGraw Hill, Inc., 1988.

Smith, Jeffrey K. *Applied Statistics and the SAS Programming Language*, 4th Ed. SAS Institute, Inc., 1997.

Smith, Harry. *Applied Regression Analysis*, 2nd Ed. John Wiley & Sons, Inc., 1981.

Smyth, Padhraic. *Principles of Data Mining*, The MIT Press, 2001.

Snedecor, George W. *Statistical Methods* 7th Ed. The Iowa University State Press, 1980.

Stevens, James. *Applied Multivariate Statistics for the Social Sciences*, 2nd Ed. 1992.

Stoffer, David S. *Time Series Analysis and Its Applications*, Springer, 2000.

Stokes, Maura. *Categorical Data Analysis Using SAS System*, SAS Institute, Inc., 2000.

Suess, Eric. *Gibbs Sampling and Screening Tests: From Random Numbers to the Gibbs Sampler*, Springer, 2007.

Tabachnick, Barbara G. *Research Design and Analysis,* Allyn & Bacon, 2001.

Thompson, Steven K. *Sampling*, 2nd Ed. Chapman & Hall, Inc., 2002.

Tibshirani, Robert J. *An Introduction to the Bootstrap*, Chapman & Hall, Inc., 1993.

Tibshirani, Robert J. *The Elements of Statistical Learning*, Springer, 2001.

Trumbo, Bruce E. *Learning Statistics with Real Data*, Duxbury, Thomson Learning Publications, 2002.

Trumbo, Bruce E. *Gibbs Sampling and Screening Tests: From Random Numbers to the Gibbs Sampler*, Springer, 2007.

Valentin, Dominique. *Neural Networks*, Sage University Paper, 1999.

Walker, Glenn A. *Common Statistical Methods for Clinical Research* 2nd Ed, SAS Institute, Inc., 2000.

Wasserman, William. *Applied Linear Statistical Models*, 5th Ed. McGraw Hill, Inc., 2004.

Weardon, Stanley. *Statistics for Research*, John Wiley & Sons, Inc., 1983.

Wichern, Dean W. *Applied Multivariate Statistical Analysis*, Prentice-Hall, Inc., 2002.

Wichern, Dean W. *Intermediate Business Statistics: Analysis of Variance, Regression and Time Series*, Holt, Rinechart and Winston, Inc., 1977.

Wild, C. J. *Nonlinear Regression*, John Wiley & Sons, Inc., 1989.

Wolfe, Douglas A. *Non-Parametric Statistical Methods*, 2nd Ed. John Wiley & Sons, Inc., 1999.

Yaffe, Robert. *Introduction to Time Series Analysis and Forecasting*, Academic Press, Inc., 2000.

Zak, Stanislaw H. *An Introduction to Optimization,* 2nd Ed. John Wiley & Sons, 2001.

Index

O

P

Q

R

S